D1084366

Springer Series in Statistics

Springer Series in Statistics

D. F. Andrews and A. M. Herzberg, Data: A Collection of Problems from Many Fields for the Student and Research Worker. xx, 442 pages, 1985.

F. J. Anscombe, Computing in Statistical Science through APL. xvi, 426 pages, 1981.

J. O. Berger, Statistical Decision Theory: Foundations, Concepts, and Methods, 2nd edition. xiv, 425 pages, 1985.

P. Brémaud, Point Processes and Queues: Martingale Dynamics. xviii, 354 pages, 1981.

K. Dzhaparidze, Parameter Estimation and Hypothesis Testing in Spectral Analysis of Stationary Time Series. xii, 300 pages, 1985.

R. H. Farrell, Multivariate Calculation. xvi, 367 pages, 1985.

L. A. Goodman and W. H. Kruskal, Measures of Association for Cross Classifications. x, 146 pages, 1979.

J. A. Hartigan, Bayes Theory. xii, 145 pages, 1983.

H. Heyer, Theory of Statistical Experiments. x, 289 pages, 1982.

H. Kres, Statistical Tables for Multivariate Analysis. xxii, 504 pages, 1983.

H. R. Leadbetter, G. Lindgren and H. Rootzén, Extremes and Related Properties of Random Sequences and Processes. xii, 336 pages, 1983.

R. G. Miller, Jr., Simultaneous Statistical Inference, 2nd edition. xvi, 299 pages, 1981.

F. Mosteller, D. S. Wallace, Applied Bayesian and Classical Inference: The Case of *The Federalist* Papers. xxxv, 301 pages, 1984.

D. Pollard, Convergence of Stochastic Processes. xiv, 215 pages, 1984.

J. W. Pratt and J. D. Gibbons, Concepts of Nonparametric Theory. xvi, 462 pages, 1981.

L. Sachs, Applied Statistics: A Handbook of Techniques. xxviii, 706 pages, 1982.

E. Seneta, Non-Negative Matrices and Markov Chains. xv, 279 pages, 1981.

D. Siegmund, Sequential Analysis: Tests and Confidence Intervals. xii, 272 pages, 1985.

V. Vapnik, Estimation of Dependences based on Empirical Data. xvi, 399 pages, 1982.

K. M. Wolter, Introduction to Variance Estimation. xii, 428 pages, 1985.

James O. Berger

Statistical Decision Theory and Bayesian Analysis

Second Edition

With 23 Illustrations

Springer-Verlag
New York Berlin Heidelberg Tokyo

James O. Berger
Department of Statistics
Purdue University
West Lafayette, IN 47907
U.S.A.

AMS Classification: 60CXX

Library of Congress Cataloging in Publication Data
Berger, James O.
Statistical decision theory and Bayesian analysis.
(Springer series in statistics)
Bibliography: p.
Includes index.
1. Statistical decision. 2. Bayesian
statistical decision theory. I. Title. II. Series.
QA279.4.B46 1985 519.5′42 85-9891

Printed on acid-free paper.

This is the second edition of *Statistical Decision Theory: Foundations, Concepts, and Methods.* © 1980 Springer-Verlag New York Inc.

Typeset by J. W. Arrowsmith Ltd., Bristol, England.
Printed and bound by R. R. Donnelley & Sons, Harrisonburg, Virginia.
Printed in the United States of America.

9 8 7 6 5 4 3 2 (Second Printing, 1988)

ISBN 0-387-96098-8 Springer-Verlag New York Berlin Heidelberg Tokyo
ISBN 3-540-96098-8 Springer-Verlag Berlin Heidelberg New York Tokyo

To Ann, Jill, and Julie

Preface

Statistical decision theory and Bayesian analysis are related at a number of levels. First, they are both needed to solve real decision problems, each embodying a description of one of the key elements of a decision problem. At a deeper level, Bayesian analysis and decision theory provide unified outlooks towards statistics; they give a foundational framework for thinking about statistics and for evaluating proposed statistical methods.

The relationships (both conceptual and mathematical) between Bayesian analysis and statistical decision theory are so strong that it is somewhat unnatural to learn one without the other. Nevertheless, major portions of each have developed separately. On the Bayesian side, there is an extensively developed Bayesian theory of statistical inference (both subjective and objective versions). This theory recognizes the importance of viewing statistical analysis conditionally (i.e., treating observed data as known rather than unknown), even when no loss function is to be incorporated into the analysis. There is also a well-developed (frequentist) decision theory, which avoids formal utilization of prior distributions and seeks to provide a foundation for frequentist statistical theory. Although the central thread of the book will be Bayesian decision theory, both Bayesian inference and non-Bayesian decision theory will be extensively discussed. Indeed, the book is written so as to allow, say, the teaching of a course on either subject separately.

Bayesian analysis and, especially, decision theory also have split personalities with regard to their practical orientation. Both can be discussed at a very practical level, and yet they also contain some of the most difficult and elegant theoretical developments in statistics. The book contains a fair amount of material of each type. There is extensive discussion on how to actually do Bayesian decision theory and Bayesian inference, including how to construct prior distributions and loss functions, as well as how to utilize them. At the other extreme, introductions are given to some of the beautiful theoretical developments in these areas.

The statistical level of the book is formally rather low, in that previous knowledge of Bayesian analysis, decision theory, or advanced statistics is unnecessary. The book will probably be rough going, however, for those without previous exposure to a moderately serious statistics course. For instance, previous exposure to such concepts as sufficiency is desirable. It should also be mentioned that parts of the book are philosophically very challenging; the extreme disagreements that exist among statisticians, concerning the correct approach to statistics, suggest that these fundamental issues are conceptually difficult. Periodic rereading of such material (e.g., Sections 1.6, 4.1, and 4.12), as one proceeds through the book, is recommended.

The mathematical level of the book is, for the most part, at an easy advanced calculus level. Some knowledge of probability is required; at least, say, a knowledge of expectations and conditional probability. From time to time (especially in later chapters) some higher mathematical facts will be employed, but knowledge of advanced mathematics is not required to follow most of the text. Because of the imposed mathematical limitations, some of the stated theorems need, say, additional measurability conditions to be completely precise. Also, less important (but nonignorable) technical conditions for some developments are sometimes omitted, but such developments are called "Results," rather than "Theorems."

The book is primarily concerned with discussing basic issues and principles of Bayesian analysis and decision theory. No systematic attempt is made to present a survey of actual developed methodology, i.e., to present specific developments of these ideas in particular areas of statistics. The examples that are given tend to be rather haphazard, and, unfortunately, do not cover some of the more difficult areas of statistics, such as nonparametrics. Nevertheless, a fair amount of methodology ends up being introduced, one way or another.

This second edition of the book has undergone a title change, with the addition of "Bayesian Analysis." This reflects the major change in the book, namely an extensive upgrading of the Bayesian material, to the point where the book can serve as a text on Bayesian analysis alone. The motivation for this upgrading was the realization that, although I professed to be a "rabid Bayesian" in the first edition (and still am), the first edition was not well suited for a primarily Bayesian course; in particular, it did not highlight the conditional Bayesian perspective properly. In attempting to correct this problem, I fell into the usual revision trap of being unable to resist adding substantial new material on subjects crucial to Bayesian analysis, such as hierarchical Bayes theory, Bayesian calculation, Bayesian communication, and combination of evidence.

For those familiar with the old book, the greatest changes are in Chapters 3 and 4, which were substantially enlarged and almost completely rewritten. Some sections of Chapter 1 were redone (particularly 1.6), and some small subsections were added to Chapter 2. The only significant change to Chapter 5 was the inclusion of an introduction to the now vast field of minimax multivariate estimation (Stein estimation); this has become by far the largest statistical area of development within minimax theory. Only very minor changes were made to Chapter 6, and Chapter 7 was changed only by the addition of a section discussing the issue of optional stopping. A number of changes were made to Chapter 8, in light of recent developments, but no thorough survey was attempted.

In general, no attempt was made to update references in parts of the book that were not rewritten. This, unfortunately, perpetuated a problem with the first edition, namely the lack of references to the early period of decision theory. Many of the decision-theoretic ideas and concepts seem to have become part of the folklore, and I apologize for not making the effort to trace them back to their origins and provide references.

In terms of teaching, the book can be used as a text for a variety of courses. The easiest such use is as a text in a two-semester or three-quarter course on Bayesian analysis and statistical decision theory; one can simply proceed through the book. (Chapters 1 through 4 should take the first semester, and Chapters 5 through 8 the second.) The following are outlines for various possible *single-semester* courses. The first outline is for a master's level course, and has a more applied orientation, while the other outlines also include theoretical material perhaps best suited for Ph.D. students. Of course, quite different arrangements could also be used successfully.

Bayesian Analysis and Decision Theory (Applied)

1 (except 1.4, 1.7, 1.8); 2; 3 (except 3.4, 3.5.5, 3.5.6, 3.5.7); 4 (except 4.4.4, 4.7.4 through 4.7.11, 4.8, 4.11); 7 (except 7.4.2 through 7.4.10, 7.5, 7.6); valuable other material to cover, if there is time, includes 4.7.4, 4.7.5, 4.7.9, 4.7.10, 4.7.11, and 4.11.

Bayesian Analysis and Decision Theory (More Theoretical)

1; 2 (except 2.3, 2.4.3, 2.4.4, 2.4.5); 3 (except 3.4, 3.5.5, 3.5.6, 3.5.7); 4 (except 4.4.4, 4.5.3, 4.6.3, 4.6.4, 4.7.4, 4.7.6, 4.7.7, 4.7.9, 4.7.10, 4.8.3, 4.9, 4.10, 4.11);

(i) With Minimax Option: 5 (except 5.2.3); parts of 8.
(ii) With Invariance Option: 6; parts of 8.
(iii) With Sequential Option: 7 (except 7.4.7 through 7.4.10, 7.5.5, 7.6); parts of 8.

A Mainly Bayesian Course (More Theoretical)

1 (except 1.4, 1.8); 2 (except 2.3); 3 (except 3.5.5 and 3.5.6); 4 (except 4.7.6, 4.7.7); 7 (except 7.4.2 through 7.4.10, 7.5, 7.6); more sequential Bayes could be covered if some of the earlier sections were eliminated.

A Mainly Decision Theory Course (Very Theoretical)

1 (except 1.6); 2 (except 2.3); Sections 3.3, 4.1, 4.2, 4.4, 4.8; 5 (except 5.2.3); 6; 7 (except 7.2, 7.4, 7.7); 8.

I am very grateful to a number of people who contributed, in one way or another, to the book. Useful comments and discussion were received from many sources; particularly helpful were Eric Balder, Mark Berliner, Don Berry, Sudip Bose, Lawrence Brown, Arthur Cohen, Persi Diaconis, Roger Farrell, Leon Gleser, Bruce Hill, Tzou Wu-Jien Joe, T. C. Kao, Jack Kiefer, Sudhakar Kunte, Erich Lehmann, Carl Morris, Herman Rubin, S. Sivaganesan, Bill Studden, Don Wallace, Robert Wolpert, and Arnold Zellner. I am especially grateful to Herman Rubin: he provided most of the material in Subsections 7.4.8 and 7.4.9, and was my "foolishness filter" on much of the rest of the book.

The first edition of the book was typed by Lou Anne Scott, Norma Lucas, Kathy Woods, and Carolyn Knutsen, to all of whom I am very grateful. The highly trying job of typing this revision was undertaken by Norma Lucas, and her skill and cheer throughout the process were deeply appreciated. Finally, I would like to express my appreciation to the John Simon Guggenheim Memorial Foundation, the Alfred P. Sloan Foundation, and the National Science Foundation for support during the writing of the book.

West Lafayette, Indiana
March 1985

JAMES BERGER

Contents

CHAPTER 1

Basic Concepts

1.1. Introduction

Decision theory, as the name implies, is concerned with the problem of making decisions. Statistical decision theory is concerned with the making of decisions in the presence of statistical knowledge which sheds light on some of the uncertainties involved in the decision problem. We will, for the most part, assume that these uncertainties can be considered to be unknown numerical quantities, and will represent them by θ (possibly a vector or matrix).

As an example, consider the situation of a drug company deciding whether or not to market a new pain reliever. Two of the many factors affecting its decision are the proportion of people for which the drug will prove effective (θ_1), and the proportion of the market the drug will capture (θ_2). Both θ_1 and θ_2 will be generally unknown, though typically experiments can be conducted to obtain statistical information about them. This problem is one of decision theory in that the ultimate purpose is to decide whether or not to market the drug, how much to market, what price to charge, etc.

Classical statistics is directed towards the use of sample information (the data arising from the statistical investigation) in making inferences about θ. These classical inferences are, for the most part, made without regard to the use to which they are to be put. In decision theory, on the other hand, an attempt is made to combine the sample information with other relevant aspects of the problem in order to make the best decision.

In addition to the sample information, two other types of information are typically relevant. The first is a knowledge of the possible consequences of the decisions. Often this knowledge can be quantified by determining the loss that would be incurred for each possible decision and for the various possible values of θ. (Statisticians seem to be pessimistic creatures who think in terms of losses. Decision theorists in economics and business talk instead in terms of gains (utility). As our orientation will be mainly statistical, we will use the loss function terminology. Note that a gain is just a negative loss, so there is no real difference between the two approaches.)

The incorporation of a loss function into statistical analysis was first studied extensively by Abraham Wald; see Wald (1950), which also reviews earlier work in decision theory.

In the drug example, the losses involved in deciding whether or not to market the drug will be complicated functions of θ_1, θ_2, and many other factors. A somewhat simpler situation to consider is that of estimating θ_1, for use, say, in an advertising campaign. The loss in underestimating θ_1 arises from making the product appear worse than it really is (adversely affecting sales), while the loss in overestimating θ_1 would be based on the risks of possible penalties for misleading advertising.

The second source of nonsample information that is useful to consider is called prior information. This is information about θ arising from sources other than the statistical investigation. Generally, prior information comes from past experience about similar situations involving similar θ. In the drug example, for instance, there is probably a great deal of information available about θ_1 and θ_2 from different but similar pain relievers.

A compelling example of the possible importance of prior information was given by L. J. Savage (1961). He considered the following three statistical experiments:

1. A lady, who adds milk to her tea, claims to be able to tell whether the tea or the milk was poured into the cup first. In all of ten trials conducted to test this, she correctly determines which was poured first.
2. A music expert claims to be able to distinguish a page of Haydn score from a page of Mozart score. In ten trials conducted to test this, he makes a correct determination each time.
3. A drunken friend says he can predict the outcome of a flip of a fair coin. In ten trials conducted to test this, he is correct each time.

In all three situations, the unknown quantity θ is the probability of the person answering correctly. A classical significance test of the various claims would consider the null hypothesis (H_0) that $\theta = 0.5$ (i.e., the person is guessing). In all three situations this hypothesis would be rejected with a (one-tailed) significance level of 2^{-10}. Thus the above experiments give strong evidence that the various claims are valid.

In situation 2 we would have no reason to doubt this conclusion. (The outcome is quite plausible with respect to our prior beliefs.) In situation 3, however, our prior opinion that this prediction is impossible (barring a belief in extrasensory perception) would tend to cause us to ignore the experimental evidence as being a lucky streak. In situation 1 it is not quite clear what to think, and different people will draw different conclusions according to their prior beliefs of the plausibility of the claim. In these three identical statistical situations, prior information clearly cannot be ignored.

The approach to statistics which formally seeks to utilize prior information is called Bayesian analysis (named after Bayes (1763)). Bayesian analysis and decision theory go rather naturally together, partly because of their common goal of utilizing nonexperimental sources of information, and partly because of some deep theoretical ties; thus, we will emphasize Bayesian decision theory in the book. There exist, however, an extensively developed non-Bayes decision theory and an extensively developed non-decision-theoretic Bayesian viewpoint, both of which we will also cover in reasonable depth.

1.2. Basic Elements

The unknown quantity θ which affects the decision process is commonly called the *state of nature.* In making decisions it is clearly important to consider what the possible states of nature are. The symbol Θ will be used to denote the set of all possible states of nature. Typically, when experiments are performed to obtain information about θ, the experiments are designed so that the observations are distributed according to some probability distribution which has θ as an unknown parameter. In such situations θ will be called the *parameter* and Θ the *parameter space.*

Decisions are more commonly called *actions* in the literature. Particular actions will be denoted by a, while the set of all possible actions under consideration will be denoted $\mathscr{A}$.

As mentioned in the introduction, a key element of decision theory is the loss function. If a particular action a_1 is taken and θ_1 turns out to be the true state of nature, then a loss $L(\theta_1, a_1)$ will be incurred. Thus we will assume a *loss function* $L(\theta, a)$ is defined for all $(\theta, a) \in \Theta \times \mathscr{A}$. For technical convenience, only loss functions satisfying $L(\theta, a) \geq -K > -\infty$ will be considered. This condition is satisfied by all loss functions of interest. Chapter 2 will be concerned with showing why a loss function will typically exist in a decision problem, and with indicating how a loss function can be determined.

When a statistical investigation is performed to obtain information about θ, the outcome (a random variable) will be denoted X. Often X will be a vector, as when $X = (X_1, X_2, \ldots, X_n)$, the X_i being independent observations from a common distribution. (From now on vectors will appear in boldface type; thus $\mathbf{X}$.) A particular realization of X will be denoted x. The set of possible outcomes is the *sample space*, and will be denoted $\mathscr{X}$. (Usually $\mathscr{X}$ will be a subset of R^n, n-dimensional Euclidean space.)

The probability distribution of X will, of course, depend upon the unknown state of nature θ. Let $P_\theta(A)$ or $P_\theta(X \in A)$ denote the probability of the event $A(A \subset \mathscr{X})$, when θ is the true state of nature. For simplicity, X will be assumed to be either a continuous or a discrete random variable, with density $f(x|\theta)$. Thus if X is continuous (i.e., has a density with respect to Lebesgue measure), then

$$P_\theta(A) = \int_A f(x|\theta)dx,$$

while if X is discrete, then

$$P_\theta(A) = \sum_{x \in A} f(x|\theta).$$

Certain common probability densities and their relevant properties are given in Appendix 1.

It will frequently be necessary to consider expectations over random variables. The expectation (over X) of a function $h(x)$, for a given value of θ, is defined to be

$$E_\theta[h(X)] = \begin{cases} \int_{\mathscr{X}} h(x)f(x|\theta)dx & \text{(continuous case)}, \\ \sum_{x \in \mathscr{X}} h(x)f(x|\theta) & \text{(discrete case)}. \end{cases}$$

It would be cumbersome to have to deal separately with these two different expressions for $E_\theta[h(X)]$. Therefore, as a convenience, we will define

$$E_\theta[h(X)] = \int_{\mathscr{X}} h(x)dF^X(x|\theta),$$

where the right-hand side is to be interpreted as in the earlier expression for $E_\theta[h(X)]$. (This integral can, of course, be considered a Riemann–Stieltjes integral, where $F^X(x|\theta)$ is the cumulative distribution function of X. Readers not familiar with such terms can just treat the integral as a notational device.) Note that, in the same way, we can write

$$P_\theta(A) = \int_A dF^X(x|\theta).$$

Frequently, it will be necessary to clarify the random variables over which an expectation or probability is being taken. Superscripts on E or P will serve this role. (A superscript could be the random variable, its density, its distribution function, or its probability measure, whichever is more convenient.) Subscripts on E will denote parameter values at which the expectation is to be taken. When obvious, subscripts or superscripts will be omitted.

The third type of information discussed in the introduction was prior information concerning θ. A useful way of talking about prior information is in terms of a probability distribution on Θ. (Prior information about θ is seldom very precise. Therefore, it is rather natural to state prior beliefs in terms of probabilities of various possible values of θ being true.) The symbol $\pi(\theta)$ will be used to represent a prior density of θ (again for either the continuous or discrete case). Thus if $A \subset \Theta$,

$$P(\theta \in A) = \int_A dF^{\pi}(\theta) = \begin{cases} \int_A \pi(\theta) d\theta & \text{(continuous case)}, \\ \sum_{\theta \in A} \pi(\theta) & \text{(discrete case)}. \end{cases}$$

Chapter 3 discusses the construction of prior probability distributions, and also indicates what is meant by probabilities concerning θ. (After all, in most situations there is nothing "random" about θ. A typical example is when θ is an unknown but fixed physical constant (say the speed of light) which is to be determined. The basic idea is that probability statements concerning θ are then to be interpreted as "personal probabilities" reflecting the degree of personal belief in the likelihood of the given statement.)

Three examples of use of the above terminology follow.

EXAMPLE 1. In the drug example of the introduction, assume it is desired to estimate θ_2. Since θ_2 is a proportion, it is clear that $\Theta = \{\theta_2 : 0 \le \theta_2 \le 1\} = [0, 1]$. Since the goal is to estimate θ_2, the action taken will simply be the choice of a number as an estimate for θ_2. Hence $\mathcal{A} = [0, 1]$. (Usually $\mathcal{A} = \Theta$ for estimation problems.) The company might determine the loss function to be

$$L(\theta_2, a) = \begin{cases} \theta_2 - a & \text{if } \theta_2 - a \ge 0, \\ 2(a - \theta_2) & \text{if } \theta_2 - a \le 0. \end{cases}$$

(The loss is in units of "utility," a concept that will be discussed in Chapter 2.) Note that an overestimate of demand (and hence overproduction of the drug) is considered twice as costly as an underestimate of demand, and that otherwise the loss is linear in the error.

A reasonable experiment which could be performed to obtain sample information about θ_2 would be to conduct a sample survey. For example, assume n people are interviewed, and the number X who would buy the drug is observed. It might be reasonable to assume that X is $\mathcal{B}(n, \theta_2)$ (see Appendix 1), in which case the sample density is

$$f(x|\theta_2) = \binom{n}{x} \theta_2^x (1 - \theta_2)^{n-x}.$$

There could well be considerable prior information about θ_2, arising from previous introductions of new similar drugs into the market. Let's say that, in the past, new drugs tended to capture between $\frac{1}{10}$ and $\frac{1}{5}$ of the market, with all values between $\frac{1}{10}$ and $\frac{1}{5}$ being equally likely. This prior information could be modeled by giving θ_2 a $\mathcal{U}(0.1, 0.2)$ prior density, i.e., letting

$$\pi(\theta_2) = 10 I_{(0.1, 0.2)}(\theta_2).$$

The above development of L, f, and π is quite crude, and usually much more detailed constructions are required to obtain satisfactory results. The techniques for doing this will be developed as we proceed.

EXAMPLE 2. A shipment of transistors is received by a radio company. It is too expensive to check the performance of each transistor separately, so a sampling plan is used to check the shipment as a whole. A random sample of n transistors is chosen from the shipment and tested. Based upon X, the number of defective transistors in the sample, the shipment will be accepted or rejected. Thus there are two possible actions: a_1—accept the shipment, and a_2—reject the shipment. If n is small compared to the shipment size, X can be assumed to have a $\mathcal{B}(n, \theta)$ distribution, where θ is the proportion of defective transistors in the shipment.

The company determines that their loss function is $L(\theta, a_1) = 10\theta$, $L(\theta, a_2) = 1$. (When a_2 is decided (i.e., the lot is rejected), the loss is the constant value 1, which reflects costs due to inconvenience, delay, and testing of a replacement shipment. When a_1 is decided (i.e., the lot is accepted), the loss is deemed proportional to θ, since θ will also reflect the proportion of defective radios produced. The factor 10 indicates the relative costs involved in the two kinds of errors.)

The radio company has in the past received numerous other transistor shipments from the same supplying company. Hence they have a large store of data concerning the value of θ on past shipments. Indeed a statistical investigation of the past data reveals that θ was distributed according to a $\mathcal{B}e(0.05, 1)$ distribution. Hence

$$\pi(\theta) = (0.05)\theta^{-0.95} I_{[0,1]}(\theta).$$

EXAMPLE 3. An investor must decide whether or not to buy rather risky ZZZ bonds. If the investor buys the bonds, they can be redeemed at maturity for a net gain of \$500. There could, however, be a default on the bonds, in which case the original \$1000 investment would be lost. If the investor instead puts his money in a "safe" investment, he will be guaranteed a net gain of \$300 over the same time period. The investor estimates the probability of a default to be 0.1.

Here $\mathcal{A} = \{a_1, a_2\}$, where a_1 stands for buying the bonds and a_2 for not buying. Likewise $\Theta = \{\theta_1, \theta_2\}$, where θ_1 denotes the state of nature "no default occurs" and θ_2 the state "a default occurs." Recalling that a gain is represented by a negative loss, the loss function is given by the following table.

	a_1	a_2
θ_1	-500	-300
θ_2	1000	-300

(When both Θ and $\mathscr{A}$ are finite, the loss function is most easily represented by such a table, and is called a *loss matrix*. Actions are typically placed along the top of the table, and θ values along the side.) The prior information can be written as $\pi(\theta_1)=0.9$ and $\pi(\theta_2)=0.1$.

Note that in this example there is no sample information from an associated statistical experiment. Such a problem is called a *no-data* problem.

It should not be construed from the above examples that every problem will have a well-defined loss function and explicit prior information. In many problems these quantities will be very vague or even nonunique. The most important examples of this are problems of statistical inference. In statistical inference the goal is not to make an immediate decision, but is instead to provide a "summary" of the statistical evidence which a wide variety of future "users" of this evidence can easily incorporate into their own decision-making processes. Thus a physicist measuring the speed of light cannot reasonably be expected to know the losses that users of his result will have.

Because of this point, many statisticians use "statistical inference" as a shield to ward off consideration of losses and prior information. This is a mistake for several reasons. The first is that reports from statistical inferences should (ideally) be constructed so that they can be easily utilized in individual decision making. We will see that a number of classical inferences are failures in this regard.

A second reason for considering losses and prior information in inference is that the investigator may very well possess such information; he will often be very informed about the uses to which his inferences are likely to be put, and may have considerable prior knowledge about the situation. It is then almost imperative that he present such information in his analysis, although care should be taken to clearly separate "subjective" and "objective" information (but see Subsection 1.6.5 and Section 3.7).

The final reason for involvement of losses and prior information in inference is that choice of an inference (beyond mere data summarization) can be viewed as a decision problem, where the action space is the set of all possible inference statements and a loss function reflecting the success in conveying knowledge is used. Such "inference losses" will be discussed in Subsections 2.4.3 and 4.4.4. And, similarly, "inference priors" can be constructed (see Sections 3.3 and 4.3) and used to compelling advantage in inference.

While the above reasons justify specific incorporation of loss functions and prior information into inference, decision theory can be useful even when such incorporation is proscribed. This is because many standard inference criteria can be formally reproduced as decision-theoretic criteria with respect to certain formal loss functions. We will encounter numerous illustrations of this, together with indications of the value of using decision-theoretic machinery to then solve the inference problem.

1.3. Expected Loss, Decision Rules, and Risk

As mentioned in the Introduction, we will be involved with decision making in the presence of uncertainty. Hence the actual incurred loss, $L(\theta, a)$, will never be known with certainty (at the time of decision making). A natural method of proceeding in the face of this uncertainty is to consider the "expected" loss of making a decision, and then choose an "optimal" decision with respect to this expected loss. In this section we consider several standard types of expected loss.

1.3.1. Bayesian Expected Loss

From an intuitive viewpoint, the most natural expected loss to consider is one involving the uncertainty in θ, since θ is all that is unknown at the time of making the decision. We have already mentioned that it is possible to treat θ as a random quantity with a probability distribution, and considering expected loss with respect to this probability distribution is eminently sensible (and will indeed be justified in Chapters 2, 3 and 4).

Definition 1. If $\pi^*(\theta)$ is the believed probability distribution of θ at the time of decision making, the *Bayesian expected loss* of an action a is

$$\rho(\pi^*, a) = E^{\pi^*} L(\theta, a) = \int_{\Theta} L(\theta, a) dF^{\pi^*}(\theta).$$

EXAMPLE 1 (continued). Assume *no* data is obtained, so that the believed distribution of θ_2 is simply $\pi(\theta_2) = 10 I_{(0.1,0.2)}(\theta_2)$. Then

$$\begin{aligned}
\rho(\pi, a) &= \int_0^1 L(\theta_2, a)\pi(\theta_2) d\theta_2 \\
&= \int_0^a 2(a-\theta_2) 10 I_{(0.1,0.2)}(\theta_2) d\theta_2 + \int_a^1 (\theta_2 - a) 10 I_{(0.1,0.2)}(\theta_2) d\theta_2 \\
&= \begin{cases} 0.15 - a & \text{if } a \le 0.1, \\ 15a^2 - 4a + 0.3 & \text{if } 0.1 \le a \le 0.2, \\ 2a - 0.3 & \text{if } a \ge 0.2. \end{cases}
\end{aligned}$$

EXAMPLE 3 (continued). Here

$$\begin{aligned}
\rho(\pi, a_1) &= E^{\pi} L(\theta, a_1) \\
&= L(\theta_1, a_1)\pi(\theta_1) + L(\theta_2, a_1)\pi(\theta_2) \\
&= (-500)(0.9) + (1000)(0.1) = -350,
\end{aligned}$$

$$\begin{aligned}
\rho(\pi, a_2) &= E^{\pi} L(\theta, a_2) \\
&= L(\theta_1, a_2)\pi(\theta_1) + L(\theta_2, a_2)\pi(\theta_2) \\
&= -300.
\end{aligned}$$

We use π^* in Definition 1, rather than π, because π will usually refer to the *initial* prior distribution for θ, while π^* will typically be the final (*posterior*) distribution of θ after seeing the data (see Chapter 4). Note that it is being implicitly assumed here (and throughout the book) that choice of a will not affect the distribution of θ. When the action does have an effect, one can replace $\pi^*(\theta)$ by $\pi_a^*(\theta)$, and still consider expected loss. See Jeffrey (1983) for development.

1.3.2. Frequentist Risk

The non-Bayesian school of decision theory, which will henceforth be called the *frequentist* or *classical* school, adopts a quite different expected loss based on an average over the random X. As a first step in defining this expected loss, it is necessary to define a decision rule (or decision procedure).

Definition 2. A (nonrandomized) *decision rule* $\delta(x)$ is a function from $\mathscr{X}$ into $\mathscr{A}$. (We will always assume that functions introduced are appropriately "measurable.") If $X = x$ is the observed value of the sample information, then $\delta(x)$ is the action that will be taken. (For a no-data problem, a decision rule is simply an action.) Two decision rules, δ_1 and δ_2, are considered equivalent if $P_\theta(\delta_1(X) = \delta_2(X)) = 1$ for all θ.

Example 1 (continued). For the situation of Example 1, $\delta(x) = x/n$ is the standard decision rule for estimating θ_2. (In estimation problems, a decision rule will be called an *estimator.*) This estimator does not make use of the loss function or prior information given in Example 1. It will be seen later how to develop estimators which do so.

Example 2 (continued). The decision rule

$$\delta(x) = \begin{cases} a_1 & \text{if } x/n \le 0.05, \\ a_2 & \text{if } x/n > 0.05, \end{cases}$$

is a standard type of rule for this problem.

The frequentist decision-theorist seeks to evaluate, for each θ, how much he would "expect" to lose if he used $\delta(X)$ repeatedly with varying X in the problem. (See Subsection 1.6.2 for justification of this approach.)

Definition 3. The *risk function* of a decision rule $\delta(x)$ is defined by

$$R(\theta, \delta) = E_\theta^X[L(\theta, \delta(X))] = \int_{\mathscr{X}} L(\theta, \delta(x))\, dF^X(x|\theta).$$

(For a no-data problem, $R(\theta, \delta) \equiv L(\theta, \delta)$.)

To a frequentist, it is desirable to use a decision rule δ which has small $R(\theta, \delta)$. However, whereas the Bayesian expected loss of an action was a single number, the risk is a *function* on Θ, and since θ is unknown we have a problem in saying what "small" means. The following partial ordering of decision rules is a first step in defining a "good" decision rule.

Definition 4. A decision rule δ_1 is *R-better* than a decision rule δ_2 if $R(\theta, \delta_1) \le R(\theta, \delta_2)$ for all $\theta \in \Theta$, with strict inequality for some θ. A rule δ_1 is *R-equivalent* to δ_2 if $R(\theta, \delta_1) = R(\theta, \delta_2)$ for all θ.

Definition 5. A decision rule δ is *admissible* if there exists no R-better decision rule. A decision rule δ is *inadmissible* if there does exist an R-better decision rule.

It is fairly clear that an inadmissible decision rule should not be used, since a decision rule with smaller risk can be found. (One might take exception to this statement if the inadmissible decision rule is simple and easy to use, while the improved rule is very complicated and offers only a slight improvement. Another more philosophical objection to this exclusion of inadmissible rules will be presented in Section 4.8.) Unfortunately, there is usually a large class of admissible decision rules for a particular problem. These rules will have risk functions which cross, i.e., which are better in different places. An example of these ideas is given below.

EXAMPLE 4. Assume X is $\mathcal{N}(\theta, 1)$, and that it is desired to estimate θ under loss $L(\theta, a) = (\theta - a)^2$. (This loss is called *squared-error* loss.) Consider the decision rules $\delta_c(x) = cx$. Clearly

$$\begin{aligned} R(\theta, \delta_c) &= E_\theta^X L(\theta, \delta_c(X)) = E_\theta^X (\theta - cX)^2 \\ &= E_\theta^X (c[\theta - X] + [1 - c]\theta)^2 \\ &= c^2 E_\theta^X [\theta - X]^2 + 2c(1-c)\theta E_\theta^X [\theta - X] + (1-c)^2\theta^2 \\ &= c^2 + (1-c)^2\theta^2. \end{aligned}$$

Since for $c > 1$,

$$R(\theta, \delta_1) = 1 < c^2 + (1-c)^2\theta^2 = R(\theta, \delta_c),$$

δ_1 is R-better than δ_c for $c > 1$. Hence the rules δ_c are inadmissible for $c > 1$. On the other hand, for $0 \le c \le 1$ the rules are noncomparable. For example, the risk functions of the rules δ_1 and $\delta_{1/2}$ are graphed in Figure 1.1. The risk functions clearly cross. Indeed it will be seen later that for $0 \le c \le 1$, δ_c is admissible. Thus the "standard" estimator δ_1 is admissible. So, however, is the rather silly estimator δ_0, which estimates θ to be zero no matter what x is observed. (This indicates that while admissibility may be a desirable property for a decision rule, it gives no assurance that the decision rule is reasonable.)

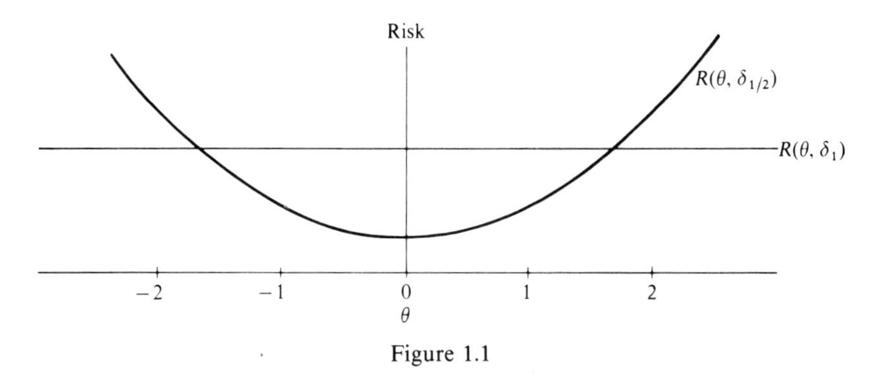

Figure 1.1

EXAMPLE 5. The following is the loss matrix of a particular no-data problem.

	a_1	a_2	a_3
θ_1	1	3	4
θ_2	−1	5	5
θ_3	0	−1	−1

The rule (action) a_2 is R-better than a_3 since $L(\theta_i, a_2) \leq L(\theta_i, a_3)$ for all θ_i, with strict inequality for θ_1. (Recall that, in a no-data problem, the risk is simply the loss.) Hence a_3 is inadmissible. The actions a_1 and a_2 are noncomparable, in that $L(\theta_i, a_1) < L(\theta_i, a_2)$ for θ_1 and θ_2, while the reverse inequality holds for θ_3. Thus a_1 and a_2 are admissible.

In this book we will only consider decision rules with finite risk. More formally, we will assume that the only (nonrandomized) decision rules under consideration are those in the class

$$\mathscr{D} = \{\text{all decision rules } \delta\text{: } R(\theta, \delta) < \infty \text{ for all } \theta \in \Theta\}.$$

(There are actually technical reasons for allowing infinite risk decision rules in certain abstract settings, but we will encounter no such situations in this book; our life will be made somewhat simpler by not having to worry about infinite risks.)

We defer discussion of the differences between using Bayesian expected loss and the risk function until Section 1.6 (and elsewhere in the book). There is, however, one other relevant expected loss to consider, and that is the expected loss which averages over *both* θ and X.

Definition 6. The *Bayes risk* of a decision rule δ, with respect to a prior distribution π on Θ, is defined as

$$r(\pi, \delta) = E^{\pi}[R(\theta, \delta)].$$

EXAMPLE 4 (continued). Suppose that $\pi(\theta)$ is a $\mathcal{N}(0, \tau^2)$ density. Then, for the decision rule δ_c,

$$\begin{aligned} r(\pi, \delta_c) &= E^{\pi}[R(\theta, \delta_c)] = E^{\pi}[c^2 + (1-c)^2\theta^2] \\ &= c^2 + (1-c)^2 E^{\pi}[\theta^2] = c^2 + (1-c)^2\tau^2. \end{aligned}$$

The Bayes risk of a decision rule will be seen to play an important role in virtually any approach to decision theory.

1.4. Randomized Decision Rules

In some decision situations it is necessary to take actions in a random manner. Such situations most commonly arise when an intelligent adversary is involved. As an example, consider the following game called "matching pennies."

EXAMPLE 6 (Matching Pennies). You and your opponent are to simultaneously uncover a penny. If the two coins match (i.e., are both heads or both tails) you win $1 from your opponent. If the coins don't match, your opponent wins $1 from you. The actions which are available to you are a_1—choose heads, or a_2—choose tails. The possible states of nature are θ_1—the opponent's coin is a head, and θ_2—the opponent's coin is a tail. The loss matrix in this game is

	a_1	a_2
θ_1	-1	1
θ_2	1	-1

Both a_1 and a_2 are admissible actions. However, if the game is to be played a number of times, then it would clearly be a very poor idea to decide to use a_1 exclusively or a_2 exclusively. Your opponent would very quickly realize your strategy, and simply choose his action to guarantee victory. Likewise, any patterned choice of a_1 and a_2 could be discerned by an intelligent opponent, who could then develop a winning strategy. The only certain way of preventing ultimate defeat, therefore, is to choose a_1 and a_2 by some random mechanism. A natural way to do this is simply to choose a_1 and a_2 with probabilities p and $1-p$ respectively. The formal definition of such a randomized decision rule follows.

Definition 7. A *randomized decision rule* $\delta^*(x, \cdot)$ is, for each x, a probability distribution on $\mathcal{A}$, with the interpretation that if x is observed, $\delta^*(x, A)$ is the probability that an action in A (a subset of $\mathscr{A}$) will be chosen. In no-data problems a randomized decision rule, also called a *randomized action*, will simply be noted $\delta^*(\cdot)$, and is again a probability distribution on $\mathscr{A}$. Nonrandomized decision rules will be considered a special case of randomized rules, in that they correspond to the randomized rules which, for each x, choose a specific action with probability one. Indeed if $\delta(x)$ is a nonrandomized decision rule, let $\langle\delta\rangle$ denote the equivalent randomized rule given by

$$\langle\delta\rangle(x, A) = I_A(\delta(x)) = \begin{cases} 1 & \text{if } \delta(x) \in A, \\ 0 & \text{if } \delta(x) \notin A. \end{cases}$$

Example 6 (continued). The randomized action discussed in this example is defined by $\delta^*(a_1) = p$ and $\delta^*(a_2) = 1 - p$. A convenient way to express this, using the notation of the preceding definition, is

$$\delta^* = p\langle a_1\rangle + (1-p)\langle a_2\rangle.$$

Example 7. Assume that a $\mathscr{B}(n, \theta)$ random variable X is observed, and that it is desired to test H_0: $\theta = \theta_0$ versus H_1: $\theta = \theta_1$ (where $\theta_0 > \theta_1$). The nonrandomized "most powerful" tests (see Chapter 8) have rejection regions of the form $C = \{x \in \mathscr{X}: x \leq j\} (j = 0, 1, 2, \ldots, n)$. Since $\mathscr{X}$ is discrete and finite, the size of these tests (namely $\alpha = P_{\theta_0}(C)$) can attain only a finite number of values. If other values of α are desired, randomized rules must be used. It suffices to consider the randomized rules given by

$$\delta_j^*(x, a_1) = \begin{cases} 1 & \text{if } x < j, \\ p & \text{if } x = j, \\ 0 & \text{if } x > j, \end{cases}$$

and $\delta_j^*(x, a_0) = 1 - \delta_j^*(x, a_1)$, where a_i denotes accepting H_i. (Thus if $x < j$ is observed, H_0 will be rejected with probability one. If $x > j$ is observed, H_0 will never be rejected (i.e., will always be accepted). If $x = j$ is observed, a randomization will be performed, rejecting H_0 with probability p and accepting with probability $1 - p$. Through proper choice of j and p, a most powerful test of the above form can be found for any given size α.)

The natural way to define the loss function and the risk function of a randomized decision rule is in terms of expected loss. (This will be justified in Chapter 2.)

Definition 8. The loss function $L(\theta, \delta^*(x, \cdot))$ of the randomized rule δ^* is defined to be

$$L(\theta, \delta^*(x, \cdot)) = E^{\delta^*(x, \cdot)}[L(\theta, a)],$$

where the expectation is taken over a (which by an abuse of notation will denote a random variable with distribution $\delta^*(x, \cdot)$). The risk function of

δ^* will then be defined to be

$$R(\theta, \delta^*) = E_\theta^X[L(\theta, \delta^*(X, \cdot))].$$

EXAMPLE 6 (continued). As this is a no-data problem, the risk is just the loss. Clearly

$$\begin{aligned} L(\theta, \delta^*) &= E^{\delta^*}[L(\theta, a)] = \delta^*(a_1)L(\theta, a_1) + \delta^*(a_2)L(\theta, a_2) \\ &= pL(\theta, a_1) + (1-p)L(\theta, a_2) \\ &= \begin{cases} -p + (1-p) = 1 - 2p & \text{if } \theta = \theta_1, \\ p - (1-p) = 2p - 1 & \text{if } \theta = \theta_2. \end{cases} \end{aligned}$$

Note that if $p = \frac{1}{2}$ is chosen, the loss is zero no matter what the opponent does. The randomized rule δ^* with $p = \frac{1}{2}$ thus guarantees an expected loss of zero.

EXAMPLE 7 (continued). Assume that the loss is zero if a correct decision is made and one if an incorrect decision is made. Thus

$$L(\theta_i, a_j) = \begin{cases} 0 & \text{if } i = j, \\ 1 & \text{if } i \neq j. \end{cases}$$

The loss of the randomized rule δ_j^* when $\theta = \theta_0$ is then given by

$$\begin{aligned} L(\theta_0, \delta_j^*(x, \cdot)) &= E^{\delta_j^*(x, \cdot)}[L(\theta_0, a)] \\ &= \delta_j^*(x, a_0)L(\theta_0, a_0) + \delta_j^*(x, a_1)L(\theta_0, a_1) \\ &= \delta_j^*(x, a_1). \end{aligned}$$

Hence

$$\begin{aligned} R(\theta_0, \delta_j^*) &= E^X[L(\theta_0, \delta_j^*(X, \cdot))] = E^X[\delta_j^*(X, a_1)] \\ &= P_{\theta_0}(X < j) + pP_{\theta_0}(X = j). \end{aligned}$$

Similarly, it can be shown that

$$R(\theta_1, \delta_j^*) = P_{\theta_1}(X > j) + (1-p)P_{\theta_1}(X = j).$$

As with nonrandomized rules, we will restrict attention to randomized rules with finite risk.

Definition 9. Let $\mathscr{D}^*$ be the set of all randomized decision rules δ^* for which $R(\theta, \delta^*) < \infty$ for all θ. The concepts introduced in Definitions 4 and 5 will henceforth be considered to apply to all randomized rules in $\mathscr{D}^*$. For example, a decision rule will be said to be admissible if there exists no R-better randomized decision rule in $\mathscr{D}^*$.

Before proceeding, some discussion is in order concerning the use of randomized decision rules. Consider first the following humorous account (given in Williams (1954)) of the use of randomization. The situation is the "Colonel Blotto Problem," which involves a military battle. The exact problem need not concern us, but of interest is the response to Colonel Blotto given by the military aide (knowledgable in statistics) when asked his opinion as to what action to take. He said,

> "... keep your eye on that ant—the one on your map case. When it reaches the grease spot, look at the second hand on your watch. If it points to 6 seconds or less, you should divide your force equally between the threatened points. If it reads between 6 and 30 seconds, give the entire reserve to Dana; if between 30 and 54 seconds, give it to Harry; if between 54 and 60, pick out a new ant."

This example makes randomization look rather silly. For the given situation, however, it might be very reasonable. The military opponent of Colonel Blotto is probably an intelligent adversary, and as in Example 6 it might then be best to randomly choose an action. (This will be discussed further in Chapter 5.) The suggested randomization mechanism is a sensible technique for "on the spot" randomization.

It is relatively rare for an actual decision problem to involve an intelligent adversary. In trying to decide which of two drugs is best, for example, no intelligent opponent is involved. In such situations there seems to be no good reason to randomize, and indeed intuition argues against it. The decision maker should be able to evaluate the relative merits of each possible action, and find the best action. If there is only one best action, there can be nothing gained by randomizing (assuming there is no need to keep an opponent guessing). If there are two or more best actions, one could randomly choose among them if desired, but there seems to be no particular point in doing so. Basically, leaving the final choice of an action up to some chance mechanism just seems ridiculous. This criticism will be seen in Chapter 4 to be valid on a rather fundamental level. Therefore, we will rarely recommend the actual use of a randomized rule.

There are reasons for studying randomized rules other than their usefulness in situations involving intelligent opponents. One such reason has already been indicated in Example 7. If it is desired to use a classical procedure with certain fixed error probabilities (say to fulfill contractual obligations), randomized rules may be necessary. A more important reason for considering randomized rules is that in Chapter 5 they will be seen to be necessary for a proper understanding of minimax theory (a type of decision-theoretic analysis). Nevertheless, randomized rules will rarely be recommended for actual use. Note that the use of randomized rules should not be confused with the use of randomization in experimental design (as in the random assignment of subjects to different treatments), which is a valuable statistical tool (except that use of the randomization probabilities in the final analysis is dubious—see Berger and Wolpert (1984) for discussion and references).

1.5. Decision Principles

In this section we briefly introduce the major methods of actually making a decision or choosing a decision rule.

1.5.1. The Conditional Bayes Decision Principle

The word "conditional" in the title is explained in Section 1.6, and distinguishes Bayesian analysis using Bayesian expected loss from that using Bayes risk (see also Subsection 4.4.1). It is typically very easy to choose an optimal action when one can determine the Bayesian expected loss, $\rho(\pi^*, a)$, for each a. (Recall that π^* is the believed probability distribution for θ at the time of decision making.)

The Conditional Bayes Principle. *Choose an action $a \in \mathscr{A}$ which minimizes $\rho(\pi^*, a)$ (assuming the minimum is attained). Such an action will be called a* Bayes action *and will be denoted a^{π^*}.*

EXAMPLE 1 (continued). In Subsection 1.3.1 it was shown that

$$\rho(\pi, a) = \begin{cases} 0.15 - a & \text{if } a \le 0.1, \\ 15a^2 - 4a + 0.3 & \text{if } 0.1 \le a \le 0.2, \\ 2a - 0.3 & \text{if } a \ge 0.2. \end{cases}$$

Calculus shows that the minimum of this function over a is $\frac{1}{30}$, and is achieved at $a^\pi = \frac{2}{15}$. This, then, would be the estimate for θ_2, the market share of the new drug (assuming no data was available).

EXAMPLE 3 (continued). In Subsection 1.3.1 it was shown that $\rho(\pi, a_1) = -350$ and $\rho(\pi, a_2) = -300$. Clearly a_1 has smaller Bayesian expected loss, and is hence the Bayes action. Thus the risky bonds should be purchased (according to the conditional Bayes principle).

1.5.2. Frequentist Decision Principles

It was remarked in Subsection 1.3.2 that use of risk functions to select a decision rule is difficult, because there are typically many admissible decision rules (i.e., decision rules which can not be dominated in terms of risk). An additional principle must be introduced in order to select a specific rule for use. In classical statistics there are a number of such principles for developing statistical procedures: the maximum likelihood, unbiasedness, minimum variance, and least squares principles to name a few. In decision theory there are also several possible principles that can be used; the three most important being the Bayes risk principle, the minimax principle, and the invariance principle. In this section the basic goal of each of these three principles is stated. In later chapters the methods of implementation of these principles will be discussed.

I. *The Bayes Risk Principle*

In Subsection 1.3.2 it was seen that an alternative method of involving a prior distribution, π, was to look at $r(\pi, \delta) = E^{\pi}R(\theta, \delta)$, the Bayes risk of δ. Since this is a *number*, we can simply seek a decision rule which minimizes it.

The Bayes Risk Principle. *A decision rule δ_1 is preferred to a rule δ_2 if*

$$r(\pi, \delta_1) < r(\pi, \delta_2).$$

A decision rule which minimizes $r(\pi, \delta)$ is optimal; it is called a Bayes rule, *and will be denoted δ^{π}. The quantity $r(\pi) = r(\pi, \delta^{\pi})$ is then called the* Bayes risk for π.

Example 4 (continued). In Subsection 1.3.2 it was calculated that $r(\pi, \delta_c) = c^2 + (1-c)^2\tau^2$, when π is $\mathcal{N}(0, \tau^2)$. Minimizing with respect to c (by differentiating and setting equal to zero) shows that $c_0 = \tau^2/(1+\tau^2)$ is the best value. Thus δ_{c_0} has the smallest Bayes risk among all estimators of the form δ_c. It will be shown in Chapter 4 that δ_{c_0} actually has the smallest Bayes risk among *all* estimators (for the given π). Hence δ_{c_0} is the Bayes rule (or Bayes estimator), and

$$\begin{aligned} r(\pi) = r(\pi, \delta_{c_0}) &= c_0^2 + (1-c_0)^2\tau^2 \\ &= \left(\frac{\tau^2}{1+\tau^2}\right)^2 + \left(\frac{1}{1+\tau^2}\right)^2 \tau^2 = \frac{\tau^2}{1+\tau^2} \end{aligned}$$

is the Bayes risk of π.

Example 3 (continued). Since this is a no-data problem, decision rules are simply actions, and the risk function is simply the loss function. Hence the Bayes risk is simply the Bayesian expected loss, and we solved the problem in Subsection 1.5.1.

The above example points out that, in a no-data problem, the Bayes risk principle will give the same answer as the conditional Bayes decision principle. It will, in fact, be seen in Subsection 4.4.1 that this correspondence always holds. (The π^* used in the conditional Bayes decision principle will usually be a data-modified version of the original prior distribution π used in the Bayes risk principle, but the two approaches will yield the same decision.)

II. *The Minimax Principle*

Complete analysis of problems using the minimax principle generally calls for consideration of randomized decision rules. Thus let $\delta^* \in \mathscr{D}^*$ be a randomized rule, and consider the quantity

$$\sup_{\theta \in \Theta} R(\theta, \delta^*).$$

This represents the worst that can happen if the rule δ^* is used. If it is desired to protect against the worst possible state of nature, one is led to using

The Minimax Principle. *A decision rule δ_1^* is preferred to a rule δ_2^* if*

$$\sup_{\theta} R(\theta, \delta_1^*) < \sup_{\theta} R(\theta, \delta_2^*).$$

Definition 10. A rule δ^{*M} is a *minimax decision rule* if it minimizes $\sup_\theta R(\theta, \delta^*)$ among all randomized rules in $\mathscr{D}^*$, i.e., if

$$\sup_{\theta \in \Theta} R(\theta, \delta^{*M}) = \inf_{\delta^* \in \mathscr{D}^*} \sup_{\theta \in \Theta} R(\theta, \delta^*).$$

The quantity on the right-hand side of the above expression is called the *minimax value* of the problem. (Replacing "inf" by "min" and "sup" by "max" shows the origin of the name "minimax.") For no-data problems, the minimax decision rule will be called simply the *minimax action.*

Sometimes it is of interest to determine the best nonrandomized rule according to the minimax principle. If such a best rule exists, it will be called the *minimax nonrandomized rule* (or *minimax nonrandomized action* in no-data problems).

Example 4 (continued). For the decision rules δ_c,

$$\sup_{\theta} R(\theta, \delta_c) = \sup_{\theta} [c^2 + (1-c)^2\theta^2] = \begin{cases} 1 & \text{if } c = 1, \\ \infty & \text{if } c \neq 1. \end{cases}$$

Hence δ_1 is best among the rules δ_c, according to the minimax principle. Indeed it will be shown in Chapter 5 that δ_1 is a minimax rule and that 1 is the minimax value for the problem. By using the rule $\delta_1(x) = x$, one can thus ensure that the risk is no worse than 1 (and actually equal to 1 for all θ). Note that the minimax rule and the Bayes rule found earlier are different.

Example 3 (continued). Clearly

$$\sup_{\theta} L(\theta, a_1) = \max\{-500, 1000\} = 1000,$$

$$\sup_{\theta} L(\theta, a_2) = \max\{-300, -300\} = -300.$$

Thus a_2 is the minimax nonrandomized action.

Example 6 (continued). The randomized rules can be written

$$\delta_p^* = p\langle a_1\rangle + (1-p)\langle a_2\rangle,$$

which recall means that a_1 is to be selected with probability p, and a_2 is to be chosen with probability $1-p$. The loss (and hence risk) of such a rule was shown to be

$$\begin{aligned} R(\theta, \delta_p^*) &= L(\theta, \delta_p^*) = pL(\theta, a_1) + (1-p)L(\theta, a_2) \\ &= \begin{cases} 1-2p & \text{if } \theta = \theta_1, \\ 2p-1 & \text{if } \theta = \theta_2. \end{cases} \end{aligned}$$

Hence

$$\sup_\theta R(\theta, \delta_p^*) = \max\{1-2p, 2p-1\}.$$

Graphing the functions $1-2p$ and $2p-1$ (for $0 \le p \le 1$) and noting that the maximum is always the higher of the two lines, it becomes clear that the minimum value of $\max\{1-2p, 2p-1\}$ is 0, occurring at $p = \frac{1}{2}$. Thus $\delta_{1/2}^*$ is the minimax action and 0 is the minimax value for the problem.

The above example describes a useful technique for solving two action no-data problems. In such situations, the randomized rules are always of the form δ_p^*, and $R(\theta, \delta_p^*)$ can be graphed as a function of p (for $0 \le p \le 1$). If Θ is finite, one need only graph the lines $R(\theta, \delta_p^*)$ for each θ, and note that the highest line segments of the graph form $\sup_\theta R(\theta, \delta_p^*)$. The minimizing value of p can then be seen directly. Techniques for finding minimax rules in more difficult situations will be presented in Chapter 5.

III. *The Invariance Principle*

The invariance principle basically states that if two problems have identical formal structures (i.e., have the same sample space, parameter space, densities, and loss function), then the same decision rule should be used in each problem. This principle is employed, for a given problem, by considering transformations of the problem (say, changes of scale in the unit of measurement) which result in transformed problems of identical structure. The proscription that the decision rules in the original and transformed problem be the same leads to a restriction to so-called "invariant" decision rules. This class of rules will often be small enough so that a "best invariant" decision rule will exist. Chapter 6 will be devoted to the discussion and application of this principle.

1.6. Foundations

We have defined a variety of expected losses, and decision principles based upon them, without discussing the advantages and disadvantages of each. Such discussion will actually be a recurring feature of the book, but in this section some of the most fundamental issues will be raised. The bulk of the section will be devoted to perhaps the most crucial issue in this discussion (and indeed in statistics), the conditional versus frequentist controversy, but first we will make a few comments concerning the common *misuse* of classical inference procedures to do decision problems. It should be noted that, while easy mathematically, many of the conceptual ideas in this foundations section are *very* difficult. This is a section that should frequently be reread as one proceeds through the book.

1.6.1. Misuse of Classical Inference Procedures

The bulk of statistics that is taught concerns classical inference procedures, and so it is only natural that many people will try to use them to do everything, even to solve clear decision problems. One problem with such use of inference procedures has already been mentioned, namely their failure to involve perhaps important prior and loss information. As another example (cf. Example 1) the loss in underestimation may differ substantially from the loss in overestimation, and any estimate should certaintly take this into account. Or, in hypothesis testing, it is often the case that the loss from an incorrect decision increases as a function of the "distance" of θ from the true hypothesis (cf. Example 1 (continued) in Subsection 4.4.3); this loss cannot be correctly measured by classical error probabilities.

One of the most commonly misused inference procedures is hypothesis testing (or significance testing) of a point null hypothesis. The following example indicates the problem.

Example 8. A sample $X_1, \ldots, X_n$ is to be taken from a $\mathcal{N}(\theta, 1)$ distribution. It is desired to conduct a size $\alpha = 0.05$ test of H_0: $\theta = 0$ versus H_1: $\theta \neq 0$. The usual test is to reject H_0 if $\sqrt{n}|\bar{x}| > 1.96$, where $\bar{x}$ is the sample mean.

Now it is unlikely that the null hypothesis is ever exactly true. Suppose, for instance, that $\theta = 10^{-10}$, which while nonzero is probably a meaningless difference from zero in most practical contexts. If now a very large sample, say $n = 10^{24}$, is taken, then with extremely high probability $\bar{X}$ will be within 10^{-11} of the true mean $\theta = 10^{-10}$. (The standard deviation of $\bar{X}$ is only 10^{-12}.) But, for $\bar{x}$ in this region, it is clear that $10^{12}|\bar{x}| > 1.96$. Hence the classical test is virtually certain to reject H_0, even though the true mean is negligibly different from zero. This same phenomenon exists no matter what size $\alpha > 0$ is chosen and no matter how small the difference, $\varepsilon > 0$, is betweeen zero and the true mean. For a large enough sample size, the classical test will be virtually certain to reject.

The point of the above example is that it is meaningless to state only that a point null hypothesis is rejected by a size α test (or is rejected at significance level α). We *know* from the beginning that the point null hypothesis is almost certainly not exactly true, and that this will always be confirmed by a large enough sample. What we are really interested in determining is whether or not the null hypothesis is approximately true (see Subsection 4.3.3). In Example 8, for instance, we might really be interested in detecting a difference of at least 10^{-3} from zero, in which case a better null hypothesis would be H_0: $|\theta| \leq 10^{-3}$. (There are certain situations in which it is reasonable to formulate the problem as a test of a point null hypothesis, but even then serious questions arise concerning the "final precision" of the classical test. This issue will be discussed in Subsection 4.3.3.)

As another example of this basic problem, consider standard "tests of fit," in which it is desired to see if the data fits the assumed model. (A typical example is a test for normality.) Again it is virtually certain that the model is not exactly correct, so a large enough sample will almost always reject the model. The problem here is considerably harder to correct than in Example 8, because it is much harder to specify what an "approximately correct" model is.

A historically interesting example of this phenomenon (told to me by Herman Rubin) involves Kepler's laws of planetary motion. Of interest is his first law, which states that planetary orbits are ellipses. For the observational accuracy of Kepler's time, this model fit the data well. For todays data, however, (or even for the data just 100 years after Kepler) the null hypothesis that orbits are ellipses would be rejected by a statistical significance test, due to perturbations in the orbits caused by planetary interactions. The elliptical orbit model is, of course, essentially correct, the error caused by perturbations being minor. The concern here is that an essentially correct model can be rejected by too accurate data if statistical significance tests are blindly applied without regard to the actual size of the discrepancies.

The above discussion shows that a "statistically significant" difference between the true parameter (or true model) and the null hypothesis can be an unimportant difference practically. Likewise a difference that is not significant statistically can nevertheless be very important practically. Consider the following example.

EXAMPLE 9. The effectiveness of a drug is measured by $X \sim \mathcal{N}(\theta, 9)$. The null hypothesis is that $\theta \leq 0$. A sample of 9 observations results in $\bar{x} = 1$. This is not significant (for a one-tailed test) at, say, the $\alpha = 0.05$ significance level. It is significant at the $\alpha = 0.16$ significance level, however, which is moderately convincing. If 1 were a practically important difference from zero, we would certainly be very interested in the drug. Indeed if we had to make a decision solely on the basis of the given data, we would probably decide that the drug was effective.

The above problems are, of course, well recognized by classical statisticians (since, at least, Berkson (1938)) who, while using the framework of testing point null hypotheses, do concern themselves with the real import of the results. It seems somewhat nonsensical, however, to deliberately formulate a problem wrong, and then in an adhoc fashion explain the final results in more reasonable terms. Also, there are unfortunately many users of statistics who do not understand the pitfalls of the incorrect classical formulations.

One of the main benefits of decision theory is that it forces one to think about the correct formulation of a problem. A number of decision-theoretic alternatives to classical significance tests will be introduced as we proceed, although no systematic study of such alternatives will be undertaken.

1.6.2. The Frequentist Perspective

On the face of it, it may seem rather peculiar to use a risk (or any other frequentist measure such as confidence, error probabilities, bias, etc.) in the report from an experiment, since they involve averaging the performance of a procedure over all possible data, while it is known *which data* occurred. In this section we will briefly discuss the motivation for using frequentist measures.

Although one can undoubtedly find earlier traces, the first systematic development of frequentist ideas can be found in the early writings of J. Neyman and E. Pearson (cf. Neyman (1967)). The original driving force behind their frequentist development seemed to be the desire to produce measures which did not depend on θ, or any prior knowledge about θ. The method of doing this was to consider a procedure $\delta(x)$ and some criterion function $L(\theta, \delta(x))$ and then find a *number* $\bar{R}$ such that repeated use of δ would yield average long run performance of at least $\bar{R}$.

EXAMPLE 10. For dealing with standard univariate normal theory problems, consider the usual 95% confidence rule for the unknown mean θ,

$$\delta(x) = (\bar{x} - ts, \bar{x} + ts),$$

where $\bar{x}$ and s are the sample mean and standard deviation, respectively, and t is the appropriate percentile from the relevant t distribution. Suppose that we measure the performance of δ by

$$L(\theta, \delta) = 1 - I_{\delta(x)}(\theta) = \begin{cases} 0 & \text{if } \theta \in \delta(x), \\ 1 & \text{if } \theta \notin \delta(x). \end{cases}$$

Note that, if this is treated as a decision-theoretic loss, the risk becomes (including the unknown standard deviation σ as part of the parameter)

$$R((\theta, \sigma), \delta) = E^{X}_{\theta,\sigma} L(\theta, \delta(X)) = P_{\theta,\sigma}(\delta(X) \text{ does not contain } \theta) = 0.05.$$

The idea now is to imagine that we will use δ repeatedly on a series of (independent, say) normal problems with means θ_i, standard deviations σ_i, and data $X^{(i)}$. It is then an easy calculation, using the law of large numbers, to show that (with probability one)

$$\lim_{N\to\infty} \frac{1}{N} \sum_{i=1}^{N} L(\theta_i, \delta(X^{(i)})) = 0.05 \equiv \bar{R}, \tag{1.1}$$

no matter what sequence of (θ_i, σ_i) is encountered.

The above frequentist motivation carries considerable appeal. As statisticians we can "proclaim" that the universal measure of performance that is to be reported with δ is $\bar{R} = 0.05$, since, on the average in repeated use, δ will fail to contain the true mean only 5% of the time. This appealing motivation for the frequentist perspective was formalized as the Confidence Principle by Birnbaum (see Cox and Hinkley (1974) and Birnbaum (1977) for precise formulations). Other relevant works of Neyman on this issue are Neyman (1957 and 1977).

Two important points about the above frequentist justification should be stressed. These are: (i) the motivation is based on repeated use of δ for *different* problems; and (ii) a bound $\bar{R}$ on performance must be found which applies to *any* sequence of parameters from these different problems. The elimination of either of these features considerably weakens the case for the frequentist measure. And, indeed, the risk $R(\theta, \delta)$, that we have so far considered, seems to violate *both* of these conditions; it is defined as the repeated average loss if one were to use δ on a series of data from the *same* problem (since θ is considered fixed), and a report of the function $R(\theta, \delta)$ has not eliminated dependence on θ.

Several justifications for $R(\theta, \delta)$ can still be given in terms of the "primary motivation," however. The first is that risk dominance of δ_1 over δ_2 will usually imply that δ_1 is better than δ_2 in terms of the primary motivation. The second is that $R(\theta, \delta)$ may have an upper bound $\bar{R}$, and, if so, this can typically be shown to provide the needed report for the primary motivation. To see the problem in using just $R(\theta, \delta)$, consider the following example.

EXAMPLE 11. Consider testing the simple null hypothesis H_0: $\theta = \theta_0$ versus the simple alternative hypothesis H_1: $\theta = \theta_1$. If the loss is chosen to be "0–1" loss (see Subsection 2.4.2), the risk function of a test δ turns out to be given by $R(\theta_0, \delta) = \alpha_0 = P_{\theta_0}$ (Type I error) and $R(\theta_1, \delta) = \alpha_1 = P_{\theta_1}$ (Type II error). Suppose now that one always uses the most powerful test of level $\alpha_0 = 0.01$. This would allow one to make the frequentist statement, upon rejecting H_0, "my procedure ensures that only 1% of true null hypotheses will be rejected."

Unfortunately, this says *nothing* about how often one errs when rejecting. For instance, suppose $\alpha_1 = 0.99$ (admittedly terrible Type II error probability, but useful for making the point) and that the null and alternative parameter values occur equally often in repetitive use of the test. (Again, we are imagining repeated use of the $\alpha_0 = 0.01$, $\alpha_1 = 0.99$ most powerful test on a sequence of different simple versus simple testing problems.) Then it can easily be shown that *half* of all rejections of the null will actually be in error. And *this* is the "error" that really measures long run performance of the test (when rejecting). Thus one cannot make useful statements about the actual error rate incurred in repetitive use, without a satisfactory bound on $R(\theta, \delta)$ for all θ.

Other justifications for $R(\theta, \delta)$ can be given involving experimental design and even "Bayesian robustness" (see Subsection 1.6.5 and Section 4.7). It will be important, however, to bear in mind that all these justifications are somewhat secondary in nature, and that assigning inherent meaning to $R(\theta, \delta)$, as an experimental report, is questionable. For more extensive discussion of this issue, see Berger (1984b), which also provides other references.

1.6.3. The Conditional Perspective

The conditional approach to statistics is concerned with reporting data-specific measures of accuracy. The overall performance of a procedure δ is deemed to be of (at most) secondary interest; what is considered to be of primary importance is the performance of $\delta(x)$ for the *actual data* x that is observed in a given experiment. The following simple examples show that there can be a considerable difference between conditional and frequentist measures.

EXAMPLE 12. Suppose that X_1 and X_2 are independent with identical distribution given by

$$P_\theta(X_i = \theta - 1) = P_\theta(X_i = \theta + 1) = \tfrac{1}{2},$$

where $-\infty < \theta < \infty$ is unknown. The procedure (letting $X = (X_1, X_2)$)

$$\delta(X) = \begin{cases} \text{the point } \tfrac{1}{2}(X_1 + X_2) & \text{if } X_1 \neq X_2, \\ \text{the point } X_1 - 1 & \text{if } X_1 = X_2, \end{cases}$$

is easily seen to be a frequentist 75% confidence procedure of smallest size (i.e., $P_\theta(\delta(X) = \theta) = 0.75$ for all θ). However, a conditionalist would reason as follows, depending on the particular x observed: if x has $x_1 \neq x_2$, then we *know* that $\frac{1}{2}(x_1 + x_2) = \theta$ (since one of the observations *must be* $\theta - 1$ and the other *must be* $\theta + 1$), while, if $x_1 = x_2$, the data fails to distinguish in any way between the two possible θ values $x_1 - 1$ and $x_1 + 1$. Hence, conditionally, $\delta(x)$ would be 100% certain to contain θ if $x_1 \neq x_2$, while if $x_1 = x_2$ it would be 50% certain to contain θ.

Careful consideration of this example will make the difference between the conditional and frequentist viewpoints clear. The overall performance of δ, in any type of repeated use, would indeed be 75%, but this arises because half the time the *actual* performance will be 100% and half the time the *actual* performance will be 50%. And, for any given application, one *knows* whether one is in the 100% or 50% case. It clearly would make little sense to conduct an experiment, use $\delta(x)$, and actually report 75% as the measure of accuracy, yet the frequentist viewpoint suggests doing so. Here is another standard example.

EXAMPLE 13. Suppose X is 1, 2, or 3 and θ is 0 or 1, with X having the following probability density in each case:

	x		
	1	2	3
$f(x\|0)$	0.005	0.005	0.99
$f(x\|1)$	0.0051	0.9849	0.01

The classical most powerful test of H_0: $\theta=0$ versus H_1: $\theta=1$, at level $\alpha=0.01$, concludes H_1 when $X=1$ or 2, and this test also has a Type II error probability of 0.01. Hence, a standard frequentist, upon observing $x=1$, would report that the decision is H_1 and that the test had error probabilities of 0.01. This certainly *gives the impression* that one can place a great deal of confidence in the conclusion, but is this the case? Conditional reasoning shows that the answer is sometimes no! When $x=1$ is observed, the likelihood ratio between $\theta=0$ and $\theta=1$ is $(0.005)/(0.0051)$ which is very close to one. To a conditionalist (and to most other statisticians also), a likelihood ratio near one means that the data does very little to distinguish between $\theta=0$ and $\theta=1$. Hence the conditional "confidence" in the decision to conclude H_1, when $x=1$ is observed, would be only about 50%. (Of course $x=1$ is unlikely to occur, but, when it does, should not a sensible answer be given?)

The next example is included for historical reasons, and also because it turns out to be a key example for development of the important Likelihood Principle in the next subsection. This example is a variant of the famous Cox (1958) conditioning example.

EXAMPLE 14. Suppose a substance to be analyzed can be sent either to a laboratory in New York or a laboratory in California. The two labs seem equally good, so a fair coin is flipped to choose between them, with "heads" denoting that the lab in New York will be chosen. The coin is flipped and comes up tails, so the California lab is used. After a while, the experimental results come back and a conclusion and report must be developed. Should this conclusion take into account the fact that the coin could have been heads, and hence that the experiment in New York might have been performed instead? Common sense (and the conditional viewpoint) cries no, that only the experiment *actually performed* is relevant, but frequentist reasoning would call for averaging over all possible data, even the possible New York data.

The above examples were kept simple to illustrate the ideas. Many complex and common statistical situations in which conditioning seems very important can be found in Berger and Wolpert (1984) and the references therein. An example is the use of observed, rather than expected, Fisher information (see Subsection 4.7.8). Examples that will be encountered in this book include hypothesis testing (see Subsection 4.3.3), several decision-theoretic examples, and the very important example of optional stopping, which will be considered in Section 7.7. (The conditional viewpoint leads to the conclusion that many types of optional stopping of an experiment can be ignored, a conclusion that can have a drastic effect on, for instance, the running of clinical trials.)

Savage (1962) used the term *initial precision* to describe frequentist measures, and used the term *final precision* to describe conditional measures. Initially, i.e., before seeing the data, one can only measure how well δ is likely to perform through a frequentist measure, but after seeing the data one can give a more precise final measure of performance. (The necessity for using at least partly frequentist measures in designing experiments is apparent.)

The examples above make abundantly clear the necessity for consideration of conditioning in statistics. The next question, therefore, is—What kind of conditional analysis should be performed? There are a wide variety of candidates, among them Bayesian analysis, fiducial analysis (begun by R. A. Fisher, see Fisher (1935)), various "likelihood methods" (cf. Edwards (1972) and Hinde and Aitkin (1984)), structural inference (begun by D. A. S. Fraser, see Fraser (1968)), pivotal inference (see Barnard (1980)), and even a number of conditional frequentist approaches (see Kiefer (1977a) or Berger (1984b, 1984c)). Discussion of these and other conditional approaches (as well as related conditional ideas such as that of a "relevant subset") can be found in Barnett (1982), Berger and Wolpert (1984), and Berger (1984d), along with many references. In this book we will almost exclusively use the Bayesian approach to conditioning, but a few words should be said about the conditional frequentist approaches because they can provide important avenues for generalizing the book's frequentist decision theory to allow for conditioning (cf. Kiefer (1976, 1977a) and Brown (1978)).

Kiefer (1977a) discussed two types of conditional frequentist approaches, calling them "conditional confidence" and "estimated confidence." The idea behind conditional confidence is to use frequentist measures, but conditioned on subsets of the sample space. Thus, in Example 12, it would be possible to condition on $\{x: x_1 = x_2\}$ and $\{x: x_1 \neq x_2\}$, and then use frequentist reasoning to arrive at the "correct" measures of confidence. And in Example 14, one could condition on the outcome of the coin flip.

The estimated confidence approach does not formally involve conditioning, but instead allows the reported confidence to be data dependent. Thus, in Example 12, one could report a confidence of 100% or 50% as $x_1 \neq x_2$ or $x_1 = x_2$, respectively. The frequentist aspect of this estimated confidence approach is that the average *reported* performance in repeated use will be equal to the *actual* average performance, thus satisfying the primary frequentist motivation. For a rigorous statement of this, and a discussion of the interesting potential that estimated confidence has for the frequentist viewpoint, see Kiefer (1977a) and Berger (1984b and 1984c).

1.6.4. The Likelihood Principle

In attempting to settle the controversies surrounding the choice of a paradigm or methodology for statistical analysis, many statisticians turn to foundational arguments. These arguments generally involve the proposal of axioms or principles that any statistical paradigm should follow, together with a logical deduction from these axioms of a particular paradigm or more general principle that should be followed. The most common such foundational arguments are those that develop axioms of "rational behavior" and prove that any analysis which is "rational" must correspond to some form of Bayesian analysis. (We will have a fair amount to say about these arguments later in the book.) A much simpler, and yet profoundly important, foundational development is that leading to the Likelihood Principle. Indeed the Likelihood Principle, by itself, can go a long way in settling the dispute as to which statistical paradigm is correct. It also says a great deal about how one should condition.

The Likelihood Principle makes explicit the natural conditional idea that *only* the actual observed x should be relevant to conclusions or evidence about θ. The key concept in the Likelihood Principle is that of the likelihood function.

Definition 11. For observed data, x, the function $l(\theta) = f(x|\theta)$, considered as a function of θ, is called the *likelihood function.*

The intuitive reason for the name "likelihood function" is that a θ for which $f(x|\theta)$ is large is more "likely" to be the true θ than a θ for which $f(x|\theta)$ is small, in that x would be a more plausible occurrence if $f(x|\theta)$ were large.

The Likelihood Principle. *In making inferences or decisions about θ after x is observed, all relevant experimental information is contained in the likelihood function for the observed x. Furthermore, two likelihood functions contain the same information about θ if they are proportional to each other (as functions of θ).*

EXAMPLE 15 (Lindley and Phillips (1976)). We are given a coin and are interested in the probability, θ, of having it come up heads when flipped. It is desired to test H_0: $\theta=\frac{1}{2}$ versus H_1: $\theta>\frac{1}{2}$. An experiment is conducted by flipping the coin (independently) in a series of trials, the result of which is the observation of 9 heads and 3 tails.

This is not yet enough information to specify $f(x|\theta)$, since the "series of trials" was not explained. Two possibilities are: (1) the experiment consisted of a predetermined 12 flips, so that $X=[\#$ heads$]$ would be $\mathcal{B}(12, \theta)$; or (2) the experiment consisted of flipping the coin until 3 tails were observed, so that X would be $\mathcal{NB}(3, \theta)$. The likelihood functions in cases (1) and (2), respectively, would be

$$l_1(\theta)=f_1(x|\theta)=\binom{n}{x}\theta^x(1-\theta)^{n-x}=(220)\theta^9(1-\theta)^3$$

and

$$l_2(\theta)=f(x|\theta)=\binom{n+x-1}{x}\theta^x(1-\theta)^n=(55)\theta^9(1-\theta)^3.$$

The Likelihood Principle says that, in either case, $l_i(\theta)$ is all we need to know from the experiment, and, furthermore, that l_1 and l_2 would contain the *same* information about θ since they are proportional as functions of θ. Thus we did not really need to know anything about the "series of trials"; knowing that independent flips gave 9 heads and 3 tails would, by itself, tell us that the likelihood function would be proportional to $\theta^9(1-\theta)^3$.

Classical analyses, in contrast, are quite dependent on knowing $f(x|\theta)$, and not just for the observed x. Consider classical significance testing, for instance. For the binomial model, the significance level of $x=9$ (against $\theta=\frac{1}{2}$) would be

$$\begin{aligned}\alpha_1=P_{1/2}(X\geq 9)&=f_1(9|\tfrac{1}{2})+f_1(10|\tfrac{1}{2})+f_1(11|\tfrac{1}{2})+f_1(12|\tfrac{1}{2})\\&=0.075.\end{aligned}$$

For the negative binomial model, the significance level would be

$$\begin{aligned}\alpha_2=P_{1/2}(X\geq 9)&=f_2(9|\tfrac{1}{2})+f_2(10|\tfrac{1}{2})+\cdots\\&=0.0325.\end{aligned}$$

If significance at the 5% level was desired, the two models would thus lead to quite different conclusions, in contradiction to the Likelihood Principle.

Several important points, illustrated in the above example, should be emphasized. First the correspondence of information from proportional likelihood functions applies *only* when the two likelihood functions are for the *same* parameter. (In the example, θ is the probability of heads for the given coin on a single flip, and is thus defined independently of which experiment is performed. If l_1 had applied to one coin, and l_2 to a different coin, the Likelihood Principle would have had nothing to say.)

A second point is that the Likelihood Principle does *not* say that all information about θ is contained in $l(\theta)$, just that all *experimental* information is. There may well be other information relevant to the statistical analysis, such as prior information or considerations of loss.

The example also reemphasizes the difference between a conditional perspective and a frequentist type of perspective. The significance level calculations involve not just the observed $x=9$, but also the "more extreme" $x \geq 10$. Again it seems somewhat peculiar to involve, in the evaluation, observations that have not occurred. No one has phrased this better than Jeffreys (1961):

> "... a hypothesis which may be true may be rejected because it has not predicted observable results which have not occurred."

Thus, in Example 15, the null hypothesis that $\theta = \frac{1}{2}$ certainly would not predict that X would be larger than 9, and indeed such values *do not occur.* Yet the probabilities of these unpredicted and not occurring observations are included in the classical evidence against the hypothesis.

Here is another interesting example (from Berger and Wolpert (1984)).

EXAMPLE 16. Let $\mathscr{X} = \{1, 2, 3\}$ and $\Theta = \{0, 1\}$, and consider experiments E_1 and E_2 which consist of observing X_1 and X_2, respectively, both having sample space $\mathscr{X}$ and the *same* unknown θ. The probability densities of X_1 and X_2 are (for $\theta = 0$ and $\theta = 1$)

x_1	1	2	3
$f_1(x_1 \mid 0)$	0.90	0.05	0.05
$f_1(x_1 \mid 1)$	0.09	0.055	0.855

x_2	1	2	3
$f_2(x_2 \mid 0)$	0.26	0.73	0.01
$f_2(x_2 \mid 1)$	0.026	0.803	0.171

If, now, $x_1 = 1$ is observed, the Likelihood Principle states that the information about θ should depend on the experiment only through $(f_1(1 \mid 0), f_1(1 \mid 1)) = (0.90, 0.09)$. Furthermore, since this is proportional to $(0.26, 0.026) = (f_2(1 \mid 0), f_2(1 \mid 1))$, the Likelihood Principle states that $x_2 = 1$ would provide the same information about θ as $x_1 = 1$. Another way of stating the Likelihood Principle for testing simple hypotheses, as here, is that the experimental evidence about θ is contained in the likelihood ratio for the observed x. Note that the likelihood ratios for the two experiments are also the same when 2 is observed, and also when 3 is observed. Hence, no matter which experiment is performed, the *same* conclusion about θ should be reached for the given observation.

This example clearly indicates the startling nature of the Likelihood Principle. Experiments E_1 and E_2 are very different from a frequentist perspective. For instance, the test which accepts $\theta = 0$ when the observation is 1 and decides $\theta = 1$ otherwise is a most powerful test with error probabilities (of Type I and Type II, respectively) 0.10 and 0.09 for E_1, and 0.74 and 0.026 for E_2. Thus the classical frequentist would report drastically different information from the two experiments.

The above example emphasizes the important distinction between initial precision and final precision. Experiment E_1 is much more *likely* to provide useful information about θ, as evidenced by the overall better error probabilities (which are measures of initial precision). Once x is at hand, however, this initial precision is no longer relevant, and the Likelihood Principle states that whether x came from E_1 or E_2 is irrelevant. This example also provides a good testing ground for the various conditional methodologies that were mentioned in Subsection 1.6.3. For instance, either of the conditional frequentist approaches has a very hard time in dealing with the example.

So far we have not given any reasons why one *should* believe in the Likelihood Principle. Examples 15 and 16 are suggestive, but could perhaps be viewed as refutations of the Likelihood Principle by die-hard classicists. Before giving the axiomatic justification that exists for the Likelihood Principle, we indulge in one more example in which it would be very hard to argue against the Likelihood Principle.

Example 17 (Pratt (1962)). "An engineer draws a random sample of electron tubes and measures the plate voltages under certain conditions with a very accurate voltmeter, accurate enough so that measurement error is negligible compared with the variability of the tubes. A statistician examines the measurements, which look normally distributed and vary from 75 to 99 volts with a mean of 87 and a standard deviation of 4. He makes the ordinary normal analysis, giving a confidence interval for the true mean. Later he visits the engineer's laboratory, and notices that the voltmeter used reads only as far as 100, so the population appears to be 'censored'. This necessitates a new analysis, if the statistician is orthodox. However, the engineer says he has another meter, equally accurate and reading to 1000 volts, which he would have used if any voltage had been over 100. This is a relief to the orthodox statistician, because it means the population was effectively uncensored after all. But the next day the engineer telephones and says, 'I just discovered my high-range voltmeter was not working the day I did the experiment you analyzed for me.' The statistician ascertains that the engineer would not have held up the experiment until the meter was fixed, and informs him that a new analysis will be required. The engineer is astounded. He says, 'But the experiment turned out just the same as if the high-range meter had been working. I obtained the precise voltages of my sample anyway, so I learned exactly what I would have learned if the high-range meter had been available. Next you'll be asking about my oscilloscope.'"

In this example, two different sample spaces are being discussed. If the high-range voltmeter had been working, the sample space would have effectively been that of a usual normal distribution. Since the high-range voltmeter was broken, however, the sample space was truncated at 100, and the probability distribution of the observations would have a point mass at 100. Classical analyses (such as the obtaining of confidence intervals) would be considerably affected by this difference. The Likelihood Principle, on the other hand, states that this difference should have no effect on the analysis, since values of x which did not occur (here $x \geq 100$) have no bearing on inferences or decisions concerning the true mean. (A formal verification is left for the exercises.)

Rationales for at least some forms of the Likelihood Principle exist in early works of R. A. Fisher (cf. Fisher (1959)) and especially of G. A. Barnard (cf. Barnard (1949)). By far the most persuasive argument for the Likelihood Principle, however, was given in Birnbaum (1962). (It should be mentioned that none of these three pioneers were unequivocal supporters of the Likelihood Principle. See Basu (1975) and Berger and Wolpert (1984) for reasons, and also a more extensive historical discussion and other references. Also, the history of the concept of "likelihood" is reviewed in Edwards (1974).)

The argument of Birnbaum for the Likelihood Principle was a proof of its equivalence with two other almost *universally* accepted natural principles. The first of these natural principles is the sufficiency principle (see Section 1.7) which, for one reason or another, almost everyone accepts. The second natural principle is the (weak) conditionality principle, which is nothing but a formalization of Example 14. (Basu (1975) explicitly named the "weak" version.)

The Weak Conditionality Principle. *Suppose one can perform either of two experiments E_1 or E_2, both pertaining to θ, and that the actual experiment conducted is the* mixed *experiment of first choosing $J = 1$ or 2 with probability $\frac{1}{2}$ each (independent of θ), and then performing experiment E_J. Then the actual information about θ obtained from the overall mixed experiment should depend only on the experiment E_j that is actually performed.*

For a proof that sufficiency together with weak conditionality imply the Likelihood Principle in the case of discrete $\mathscr{X}$, see Birnbaum (1962) or Berger and Wolpert (1984); the latter work also gives a similar development and proof in an extremely general probabilistic setting. The argument poses a serious challenge to all who are unwilling to believe the Likelihood Principle; the only alternatives are to reject the sufficiency principle (which would itself cause havoc in classical statistics) or to reject the weak conditionality principle—yet what could be more obvious?

There have been a number of criticisms of Birnbaum's axiomatic development, including concerns about the existence of the likelihood function(i.e., of $f(x|\theta)$), and even of the existence of "information from an experiment about θ." Also, some of the consequences of the Likelihood Principle are so startling (such as the fact that the Likelihood Principle implies that optional stopping of an experiment should usually be irrelevant to conclusions, see Section 7.7) that many statisticians simply refuse to consider the issue. Basu (1975), Berger and Wolpert (1984), and Berger (1984d) present (and answer) essentially all of the criticisms that have been raised, and also extensively discuss the important consequences of the Likelihood Principle (and the intuitive plausibility of these consequences).

It should be pointed out that the Likelihood Principle does have several inherent limitations. One has already been mentioned, namely that, in designing an experiment, it is obviously crucial to take into account all x that can occur; frequentist measures (though perhaps Bayesian frequentist measures) must then be considered. The situation in sequential analysis is similar, in that, at a given stage, one must decide whether or not to take another observation. This is essentially a design-type problem and, in making such a decision, it may be necessary to know more than the likelihood function for θ from the data observed up until that time. (See Section 7.7 for further discussion.) A third related problem is that of prediction of future observables, in which one wants to predict a future value of X. Again, there may be information in the data beyond that in the likelihood function for θ. Actually, the Likelihood Principle will apply in all these situations if θ is understood to consist of *all* unknowns relevant to the problem, including further random X, and not consist just of unknown model parameters. See Berger and Wolpert (1984) for discussion.

The final, yet most glaring, limitation of the Likelihood Principle is that it does not indicate how the likelihood function is to be used in making decisions or inferences about θ. One proposal has been to simply report the entire likelihood function, and to educate people in its interpretation. This is perhaps reasonable, but is by no means the complete solution. First of all, it is frequently also necessary to consider the prior information and loss, and the interaction of these quantities with the likelihood function. Secondly, it is not at all clear that the likelihood function, by itself, has any particular meaning. It is natural to attempt to interpret the likelihood function as some kind of probability density for θ. The ambiguity arises in the need to then specify the "measure" with respect to which it is a density. There are often many plausible choices for this measure, and the choice can have a considerable effect on the conclusion reached. This problem is basically that of choosing a "noninformative" prior distribution, and will be discussed in Chapter 3.

Of the methods that have been proposed for using the likelihood function to draw conclusions about θ (see Berger and Wolpert (1984) for references), only the Bayesian approach seems generally appropriate. This will be indicated in the next section, and in Chapter 4. (More extensive such arguments can be found in Basu (1975) and Berger and Wolpert (1984).) It will also be argued, however, that a good Bayesian analysis may sometimes require a slight violation of the Likelihood Principle, in attempting to protect against the uncertainties in the specification of the prior distribution. The conclusion that will be reached is that analysis compatible with the Likelihood Principle is an ideal towards which we should strive, but an ideal which is not always completely attainable.

In the remainder of the book, the Likelihood Principle will rarely be used to actually do anything (although conditional Bayes implementation of it will be extensively considered). The purpose in having such a lengthy discussion of the principle was to encourage the "post-experimental" way of thinking. Classical statistics teaches one to think in terms of "pre-experimental" measures of initial precision. The Likelihood Principle states that this is an error; that one should reason only in terms of the actual sample and likelihood function obtained. Approaching a statistical analysis with this viewpoint in mind is a radical departure from traditional statistical reasoning. And note that, while the Likelihood Principle is the "stick" urging adoption of the conditional approach, there is also the "carrot" that the conditional approach often yields great simplification in the statistical analysis: it is usually much easier to work with just the observed likelihood function, rather than having to involve $f(x|\theta)$ for *all* x, as a frequentist must (see also Sections 4.1 and 7.7).

1.6.5. Choosing a Paradigm or Decision Principle

So far we have discussed two broad paradigms, the conditional and the frequentist, and, within each, a number of possible principles or methodologies that could be followed. As these various paradigms and decision principles are discussed throughout the book, considerable effort will be spent in indicating when the methods seem to work and, more importantly, when they do not. The impression that may emerge from the presentation is that statistics is a collection of useful methodologies, and that one should "keep an open mind as to which method to use in a given application." This is indeed the most common attitude among statisticians.

While we endorse this attitude in a certain practical sense (to be made clearer shortly), we do *not* endorse it fundamentally. The basic issue is—How can we *know* that we have a sensible statistical analysis? For example, how can we be certain that a particular frequentist analysis has not run afoul of a conditioning problem? It is important to determine what *fundamentally* constitutes a sound statistical analysis, so that we then have a method of judging the practical soundness and usefulness of the various methodologies.

We have argued that this desired fundamental analysis must be compatible with the Likelihood Principle. Furthermore, we will argue in Chapter 4 that it is conditional Bayesian analysis that is the only fundamentally correct conditional analysis. From a practical viewpoint, however, things are not so clearcut, since the Bayesian approach requires specification of a prior distribution π, for θ, and this can never be done with complete assurance (see Section 4.7). Hence we will modify our position (in Section 4.7) and argue that the fundamentally correct paradigm is the "robust Bayesian" paradigm, which takes into account uncertainty in the prior.

Unfortunately, robust Bayesian analysis turns out to be quite difficult; indeed for many problems it is technically almost impossible. We thus run into the need for what Good (1983) calls "Type II rationality": when time and other realistic constraints in performing a statistical analysis are taken into account, the optimal analysis may be an analysis which is not rigorously justifiable (from, say, the robust Bayesian viewpoint). The employment of any alternative methodology should, however, be justified from this perspective, the justification being that one is in this way most likely to be "close to" the philosophically correct analysis.

With the above reasoning, we will be able to justify a number of uses of frequentist measures such as $R(\theta, \delta)$. Also, recall that partially frequentist reasoning is unavoidable in many statistical domains, such as design of experiments and sequential analysis. A final justification for consideration of $R(\theta, \delta)$ is that, whether we like it or not, the bulk of statistical analyses that will be performed will use prepackaged procedures. Although the primary concern should be to see that such procedures are developed so as to be conditionally sound, the fact that they will see repeated use suggests that verification of acceptable long-run performance would only be prudent. In spite of all these reasons, we would strongly argue that conditional (Bayesian) reasoning should be the primary weapon in a statistician's arsenal.

It should be noted that we did not attempt to justify use of frequentist measures on certain "traditional" grounds such as the desire for "objectivity" or avoidance of use of subjective inputs (such as prior information). Objectivity is clearly very difficult in decision theory, since one cannot avoid subjective choice of a loss function. Even more to the point, strong arguments can be made that one can *never* do truly objective (sensible) statistical analyses; analyses that have the appearance of objectivity virtually always contain hidden, and often quite extreme, subjective assumptions. (For instance, the choice of a model is usually a very sharp subjective input.) Some indications of this will be seen throughout the book, although for more thorough discussions (of this and the other foundational issues), see

Jeffreys (1961), Zellner (1971), Box and Tiao (1973), Good (1983), Jaynes (1983), Berger (1984a), and Berger and Wolpert (1984) (all of which also have other references).

With the exception of Chapters 3 and 4, the book will tend to emphasize methodologies based on $R(\theta, \delta)$. The reason is mainly historical: the bulk of existing statistical decision-theoretic methodology is frequentist in nature. We will often pause, however, to view things from the conditional perspective.

1.7. Sufficient Statistics

The concept of a sufficient statistic (due to Fisher (1920, 1922)) is of great importance in simplifying statistical problems. Intuitively, a sufficient statistic is a function of the data which summarizes all the available sample information concerning θ. For example, if an independent sample $X_1, \ldots, X_n$ for a $\mathcal{N}(\mu, \sigma^2)$ distribution is to be taken, it is well known that $T = (\bar{X}, S^2)$ is a sufficient statistic for $\theta = (\mu, \sigma^2)$. (Here $S^2 = \sum (X_i - \bar{X})^2/(n-1)$.)

It is assumed that the reader is familiar with the concept of sufficiency and with the methods of finding sufficient statistics. We will content ourselves here with a rather brief discussion of sufficiency, including a presentation of the major decision-theoretic result concerning sufficiency. (For an in-depth examination of sufficiency, see Huzurbazar (1976).) The following formal definition of sufficiency uses the concept of a conditional distribution, with which the reader is also assumed familiar.

Definition 12. Let X be a random variable whose distribution depends on the unknown parameter θ, but is otherwise known. A function T of X is said to be a *sufficient statistic* for θ if the conditional distribution of X, given $T(X) = t$, is independent of θ (with probability one).

For understanding the nature of a sufficient statistic and for the development of the decision-theoretic result concerning sufficiency, the concept of a partition of the sample space must be introduced.

Definition 13. If $T(X)$ is a statistic with range $\mathcal{J}$ (i.e., $\mathcal{J} = \{T(x): x \in \mathcal{X}\}$), the *partition of* $\mathcal{X}$ induced by T is the collection of all sets of the form

$$\mathcal{X}_t = \{x \in \mathcal{X}: T(x) = t\}$$

for $t \in \mathcal{J}$.

Note that if $t_1 \neq t_2$, then $\mathcal{X}_{t_1} \cap \mathcal{X}_{t_2} = \varnothing$, and also observe that $\bigcup_{t \in \mathcal{J}} \mathcal{X}_t = \mathcal{X}$. Thus $\mathcal{X}$ is divided up (or partitioned) into the disjoint sets $\mathcal{X}_t$.

Definition 14. A *sufficient partition* of $\mathscr{X}$ is a partition induced by a sufficient statistic T.

Consider now the formal definition of sufficiency given in Definition 12. The conditional distribution of X, given $T(X)=t$, is clearly a distribution giving probability one to the set $\mathscr{X}_t$. Indeed the distribution can usually be represented by a density, to be denoted $f_t(x)$, on $\mathscr{X}_t$. The density does not depend upon θ, since by Definition 12 the conditional distribution is independent of θ. This implies, in particular, that the densities $f_t(x)$ are known, being explicitly calculable from $f(x|\theta)$.

The intuitive reason that a sufficient statistic is said to contain all the sample information concerning θ can be seen from the above considerations. Basically, the random variable X can be thought of as arising first from the random generation of T, followed by the random choice of x from $\mathscr{X}_t$ (t being the observed value of T) according to the density $f_t(x)$. This second stage involves a randomization not involving θ, and so, from a number of intuitive viewpoints, carries no information about θ.

In developing the decision-theoretic result concerning sufficiency, the concept of a conditional expectation will be needed. The conditional expectation of a function $h(x)$, given $T=t$, will be denoted $E^{X|t}[h(X)]$, and, providing the conditional density f_t exists, is given by

$$E^{X|t}[h(X)]=\begin{cases}\displaystyle\int_{\mathscr{X}_t} h(x)f_t(x)dx & \text{continuous case,}\\ \displaystyle\sum_{x\in\mathscr{X}_t} h(x)f_t(x) & \text{discrete case.}\end{cases}$$

We will also need the standard probabilistic result that

$$E^X[h(X)]=E^TE^{X|T}[h(X)].$$

Finally, for any statistic $T(X)$, we will define randomized decision rules $\delta^*(t,\cdot)$, based on T, to be the usual randomized decision rules with $\mathscr{T}$ being considered the sample space. The risk function of such a rule is clearly

$$R(\theta,\delta^*)=E^T[L(\theta,\delta^*(T,\cdot))].$$

Theorem 1. *Assume that T is a sufficient statistic for θ, and let $\delta_0^*(x,\cdot)$ be any randomized rule in $\mathscr{D}^*$. Then (subject to measurability conditions) there exists a randomized rule $\delta_1^*(t,\cdot)$, depending only on $T(x)$, which is R-equivalent to δ_0^*.*

Proof. For $A\subset\mathscr{A}$ and $t\in\mathscr{T}$ define

$$\delta_1^*(t,A)=E^{X|t}[\delta_0^*(X,A)].$$

Thus $\delta_1^*(t,\cdot)$ is formed by averaging δ_0^* over $\mathscr{X}_t$, with respect to the conditional distribution of X given $T=t$. It is easy to check that, for each t, $\delta_1^*(t,\cdot)$

is a probability distribution on $\mathscr{A}$. Assuming it is also appropriately measurable, it follows that δ_1^* is a randomized decision rule based on $T(x)$. Note that the sufficiency of T is needed to ensure that δ_1^* does not depend on θ.

Observe next that

$$L(\theta, \delta_1^*(t, \cdot)) = E^{\delta_1^*(t,\cdot)}[L(\theta, a)] = E^{X|t}E^{\delta_0^*(X,\cdot)}[L(\theta, a)].$$

It follows that

$$\begin{aligned} R(\theta, \delta_1^*) &= E^T[L(\theta, \delta_1^*(T, \cdot))] \\ &= E^T E^{X|T} E^{\delta_0^*(X,\cdot)}[L(\theta, a)] \\ &= E^X E^{\delta_0^*(X,\cdot)}[L(\theta, a)] \\ &= E^X[L(\theta, \delta_0^*(X, \cdot))] = R(\theta, \delta_0^*). \end{aligned}$$ □

The above theorem applies also to a nonrandomized rule δ_0, through the identification of δ_0 and $\langle\delta_0\rangle$ discussed in Section 1.4. Note, however, that even though δ_0 is nonrandomized, the equivalent δ_1^* may be randomized. Indeed it is clear that

$$\begin{aligned} \delta_1^*(t, A) &= E^{X|t}[\langle\delta_0\rangle(X, A)] \\ &= E^{X|t}[I_A(\delta_0(X))] \\ &= P^{X|t}(\delta_0(X) \in A). \end{aligned}$$

When evaluating decision rules through risk functions, Theorem 1 implies that it is only necessary to consider rules based on a sufficient statistic. If a rule is not a function of the sufficient statistic, another rule can be found that is a function of the sufficient statistic and has the same risk function.

It will, in fact, often be the case that a decision rule, which is not solely a function of the sufficient statistic, will be inadmissible. (One such situation is discussed in the next section.) This is an important point because sufficiency is not a universally accepted principle (although it is one of the few points that classical statisticians and Bayesians agree upon). Inadmissibility is a serious criticism, however (even to conditional Bayesians, see Section 4.8), so that violation of sufficiency is hard to justify. (See Berger (1984d) for further discussion and references.)

It should come as no surprise that the Likelihood Principle immediately implies that a sufficient statistic contains all the sample information about θ, because sufficiency was a major component of Birnbaum's *derivation* of the Likelihood Principle. For completeness, however, note that (under mild conditions) the factorization theorem (cf. Lehmann (1959)) for a sufficient statistic will show that the likelihood function can be written as

$$l(\theta) = f(x|\theta) = h(x)g(T(x)|\theta),$$

where h does not depend on θ. Hence the likelihood function is proportional to $g(T(x)|\theta)$, and the Likelihood Principle implies that all decisions and inferences concerning θ can be made through T.

1.8. Convexity

In several places throughout the book, the concepts of convexity and concavity will be used. The needed definitions and properties of convexity are summarized in this section. Some immediate applications concerning sufficiency and randomized rules are also given.

As mentioned earlier, boldface letters explicitly indicate vectors or matrices. Here we will be concerned with vectors $\mathbf{x} \in R^m$, so that $\mathbf{x} = (x_1, x_2, \ldots, x_m)^t$.

Definition 15. A set $\Omega \subset R^m$ is *convex* if for any two points $\mathbf{x}$ and $\mathbf{y}$ in Ω, the point $[\alpha\mathbf{x} + (1-\alpha)\mathbf{y}]$ is in Ω for $0 \le \alpha \le 1$. (Note that $\{[\alpha\mathbf{x} + (1-\alpha)\mathbf{y}]: 0 \le \alpha \le 1\}$ is the line segment joining $\mathbf{x}$ and $\mathbf{y}$. Hence Ω is convex if the line segment between any two points in Ω is a subset of Ω.)

Definition 16. If $\{\mathbf{x}^1, \mathbf{x}^2, \ldots\}$ is a sequence of points in R^m, and $0 \le \alpha_i \le 1$ are numbers such that $\sum_{i=1}^{\infty} \alpha_i = 1$, then $\sum_{i=1}^{\infty} \alpha_i \mathbf{x}^i$ (providing it is finite) is called a *convex combination* of the $\{\mathbf{x}^i\}$. The *convex hull* of a set Ω is the set of all points which are convex combinations of points in Ω. (It is more standard to define the convex hull only in terms of combinations of a finite number of the $\{\mathbf{x}^i\}$, but the definitions turn out to be equivalent for R^m.)

Intuitively, the convex hull of a set is formed by connecting all points of the set by lines, and then filling in the interiors of the surfaces and solids so formed. It is easy to check that the convex hull of a set Ω is itself convex, and (somewhat more difficult) that it is the smallest convex set containing Ω.

Examples of convex sets are ellipses and regular polygons in R^2 (the interiors being considered part of the sets); and solid pyramids, cubes, and balls in R^3. In Figure 1.2, the set Ω_1 is convex, while Ω_2 is not.

If Ω_1 is a finite set of points in R^m, the convex hull of Ω_1 is the polygonal solid formed by joining the points and filling in the interiors. An example in R^2 is shown in Figure 1.3.

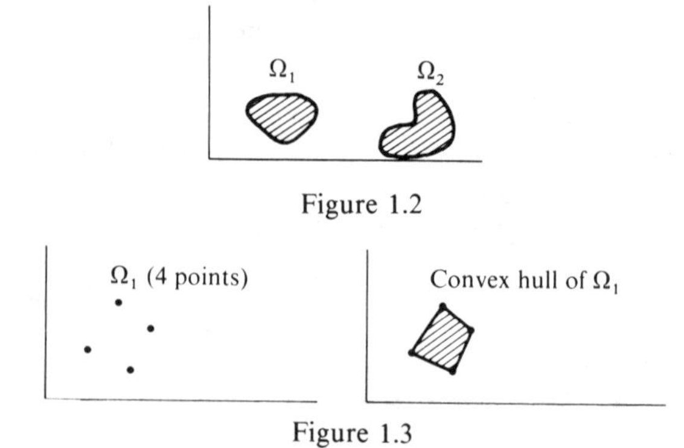

Figure 1.2

Figure 1.3

Definition 17. A real valued function $g(\mathbf{x})$ defined on a convex set Ω is *convex* if

$$g(\alpha\mathbf{x}+(1-\alpha)\mathbf{y})\leq \alpha g(\mathbf{x})+(1-\alpha)g(\mathbf{y})$$

for all $\mathbf{x}\in\Omega$, $\mathbf{y}\in\Omega$, and $0<\alpha<1$. If the inequality is strict for $\mathbf{x}\neq\mathbf{y}$, then g is *strictly convex.* If

$$g(\alpha\mathbf{x}+(1-\alpha)\mathbf{y})\geq \alpha g(\mathbf{x})+(1-\alpha)g(\mathbf{y}),$$

then g is *concave.* If the inequality is strict for $\mathbf{x}\neq\mathbf{y}$, then g is *strictly concave.*

Intuitively, convex functions are bowl shaped, while concave functions are upside-down bowl shaped. The name "convex function" arises from the fact that the set of points lying above the graph of a convex function forms a convex set. Examples of convex functions on R^1 are x^2, $|x|$, and e^x. Indeed x^2 and e^x are strictly convex. Examples of concave functions are $-x^2$ (on R^1) and $\log x$ (on $(0,\infty)$).

Verifying convexity or concavity directly from the definitions above is difficult. The following lemma provides an easy calculational tool for verifying convexity or concavity.

Lemma 1. *Let $g(\mathbf{x})$ be a function defined on an open convex subset Ω of R^m for which all second-order partial derivatives*

$$g^{(i,j)}(\mathbf{x})=\frac{\partial^2}{\partial x_i\,\partial x_j}g(\mathbf{x})$$

exist and are finite. Then g is convex if and only if the matrix G of second-order partial derivatives (i.e., the $(m\times m)$ matrix with elements $g^{(i,j)}(\mathbf{x})$) is nonnegative definite for all $\mathbf{x}\in\Omega$ (i.e., $\mathbf{z}'G\mathbf{z}\geq 0$ for all $\mathbf{z}\in R^m$ and $\mathbf{x}\in\Omega$). Likewise, g is concave if $-G$ is nonnegative definite. If G is positive (negative) definite, then g is strictly convex (strictly concave).

The proof of this lemma is a standard mathematical result and will be omitted. Note that for $m=1$, the lemma says that g is convex (concave) if $g''(x)\geq 0\,(g''(x)\leq 0)$, where g'' is the second derivative of g.

Example 18. For $g(x)=x^2$ and $g(x)=e^x$, it is clear that $g''(x)>0$, so that these functions are strictly convex on R^1. In R^2, the function $g(x_1,x_2)=x_1^2+x_1x_2+x_2^2$ is strictly convex, since its matrix of second-order partial derivatives is

$$\begin{pmatrix}2 & 1\\ 1 & 2\end{pmatrix},$$

which is positive definite.

Several very useful results concerning convex functions can be established from a theorem known as Jensen's inequality. Before giving the theorem, we state a needed lemma. The proofs of the lemma and of Jensen's inequality will be given in Subsection 5.2.5.

Lemma 2. *Let* $\mathbf{X}$ *be an m-variate random vector such that* $E[|\mathbf{X}|]<\infty$ *and* $P(\mathbf{X}\in\Omega)=1$, *where* Ω *is a convex subset of* R^m. *Then* $E[\mathbf{X}]\in\Omega$.

Theorem 2 (Jensen's Inequality). *Let* $g(\mathbf{x})$ *be a convex real-valued function defined on a convex subset* Ω *of* R^m, *and let* $\mathbf{X}$ *be an m-variate random vector for which* $E[|\mathbf{X}|]<\infty$. *Suppose also that* $P(\mathbf{X}\in\Omega)=1$. *Then*

$$g(E[\mathbf{X}])\leq E[g(\mathbf{X})],$$

with strict inequality if g is strictly convex and $\mathbf{X}$ *is not concentrated at a point.* (*Note from Lemma 2 that* $E[\mathbf{X}]\in\Omega$, *so that* $g(E[\mathbf{X}])$ *is defined.*)

EXAMPLE 19. Since $g(x)=x^2$ is strictly convex on R^1, it follows from Jensen's inequality that if X has finite mean and is not concentrated at a point, then

$$(E[X])^2=g(E[X])<E[g(X)]=E[X^2],$$

a well-known probabilistic result.

Our first use of Jensen's inequality will be to show that when the loss function is convex, only nonrandomized decision rules need be considered.

Theorem 3. *Assume that* $\mathscr{A}$ *is a convex subset of* R^m, *and that for each* $\theta\in\Omega$ *the loss function* $L(\theta,\mathbf{a})$ *is a convex function of* $\mathbf{a}$. *Let* δ^* *be a randomized decision rule in* $\mathscr{D}^*$ *for which* $E^{\delta^*(x,\cdot)}[|\mathbf{a}|]<\infty$ *for all* $x\in\mathscr{X}$. *Then* (*subject to measurability conditions*) *the nonrandomized rule*

$$\boldsymbol{\delta}(x)=E^{\delta^*(x,\cdot)}[\mathbf{a}]$$

has $L(\theta,\boldsymbol{\delta}(x))\leq L(\theta,\delta^*(x,\cdot))$ *for all x and* θ.

PROOF. From Lemma 2 it is clear that $\boldsymbol{\delta}(x)\in\mathscr{A}$. Jensen's inequality then gives that

$$L(\theta,\boldsymbol{\delta}(x))=L(\theta,E^{\delta^*(x,\cdot)}[\mathbf{a}])\leq E^{\delta^*(x,\cdot)}[L(\theta,\mathbf{a})]=L(\theta,\delta^*(x,\cdot)).\qquad\square$$

EXAMPLE 20. Let $\Theta=\mathscr{A}=R^1$ and $L(\theta,a)=(\theta-a)^2$. (The decision problem is thus to estimate θ under squared-error loss.) Clearly randomized rules $\delta^*(x,\cdot)$ must have $E^{\delta^*(x,\cdot)}[|a|]<\infty$ in order to have finite loss, so that, by the above theorem, $\delta(x)=E^{\delta^*(x,\cdot)}[a]$ has loss less than or equal to that of $\delta^*(x,\cdot)$ for all x and θ. Indeed since L is strictly convex, it can be shown that $\delta(x)$ has smaller loss than $\delta^*(x,\cdot)$ for all x for which $\delta^*(x,\cdot)$ is nondegenerate.

A second well-known consequence of Jensen's inequality is the Rao–Blackwell theorem.

Theorem 4 (Rao–Blackwell). *Assume that $\mathscr{A}$ is a convex subset of R^m and that $L(\theta, \mathbf{a})$ is a convex function of* $\mathbf{a}$ *for all $\theta \in \Theta$. Suppose also that T is a sufficient statistic for θ, and that $\boldsymbol{\delta}^0(x)$ is a nonrandomized decision rule in $\mathscr{D}$. Then the decision rule, based on $T(x) = t$, defined by*

$$\boldsymbol{\delta}^1(t) = E^{X|t}[\boldsymbol{\delta}^0(X)],$$

is R-equivalent to or R-better than $\boldsymbol{\delta}^0$, provided the expectation exists.

Proof. By the definition of a sufficient statistic, the expectation above does not depend on θ, so that $\boldsymbol{\delta}^1$ is an obtainable decision rule. By Jensen's inequality

$$L(\theta, \boldsymbol{\delta}^1(t)) = L(\theta, E^{X|t}[\boldsymbol{\delta}^0(X)] \le E^{X|t}[L(\theta, \boldsymbol{\delta}^0(X))].$$

Hence

$$\begin{aligned} R(\theta, \boldsymbol{\delta}^1) &= E_\theta^T[L(\theta, \boldsymbol{\delta}^1(T))] \\ &\le E_\theta^T[E^{X|T}\{L(\theta, \boldsymbol{\delta}^0(X))\}] \\ &= E_\theta^X[L(\theta, \boldsymbol{\delta}^0(X))] \\ &= R(\theta, \boldsymbol{\delta}^0). \end{aligned}$$

□

Observe that the Rao–Blackwell theorem could have been obtained directly from Theorems 1 and 3. Theorem 1 and the ensuing discussion show that the rule defined by

$$\delta_1^*(t, A) = E^{X|t}[I_A(\boldsymbol{\delta}^0(X))]$$

is equivalent to $\boldsymbol{\delta}^0$. Letting $\mu_X(A) = I_A(\boldsymbol{\delta}^0(X))$, Theorem 3 then implies that

$$\boldsymbol{\delta}^1(t) = E^{\delta_1^*(t,\cdot)}[\mathbf{a}] = E^{X|t}E^{\mu_X}[\mathbf{a}] = E^{X|t}[\boldsymbol{\delta}^0(X)]$$

is R-equivalent to or R-better than $\boldsymbol{\delta}^0$.

When the loss is convex and there is a sufficient statistic T for θ, Theorems 1 and 3 can be combined to show that (from the viewpoint of risk) only nonrandomized rules based on T need be considered. This results in a great simplification of the problem.

Exercises

Sections 1.3 and 1.5

1. Let X have a $\mathscr{P}(\theta)$ distribution, $\Theta = (0, \infty)$ and $\mathscr{A} = [0, \infty)$. The loss function is $L(\theta, a) = (\theta - a)^2$. Consider decision rules of the form $\delta_c(x) = cx$. Assume $\pi(\theta) = e^{-\theta}$ is the prior density.

(a) Calculate $\rho(\pi, a)$, and find the Bayes action. (Note that this is the optimal Bayes action for the no-data problem in which x is not observed.)
(b) Find $R(\theta, \delta_c)$.
(c) Show that δ_c is inadmissible if $c > 1$.
(d) Find $r(\pi, \delta_c)$.
(e) Find the value of c which minimizes $r(\pi, \delta_c)$.
(f) Find the best rule of the form δ_c in terms of the minimax principle.

2. An insurance company is faced with taking one of the following 3 actions: a_1: increase sales force by 10%; a_2: maintain present sales force; a_3: decrease sales force by 10%. Depending upon whether or not the economy is good (θ_1), mediocre (θ_2), or bad (θ_3), the company would expect to lose the following amounts of money in each case:

		Action Taken		
		a_1	a_2	a_3
State of Economy	θ_1	-10	-5	-3
	θ_2	-5	-5	-2
	θ_3	1	0	-1

(a) Determine if each action is admissible or inadmissible.
(b) The company believes that θ has the probability distribution $\pi(\theta_1) = 0.2$, $\pi(\theta_2) = 0.3$, $\pi(\theta_3) = 0.5$. Order the actions according to their Bayesian expected loss (equivalent to Bayes risk, here), and state the Bayes action.
(c) Order the actions according to the minimax principle and find the minimax nonrandomized action.

3. A company has to decide whether to accept or reject a lot of incoming parts. (Label these actions a_1 and a_2 respectively.) The lots are of three types: θ_1 (very good), θ_2 (acceptable), and θ_3 (bad). The loss $L(\theta_i, a_j)$ incurred in making the decision is given in the following table.

	a_1	a_2
θ_1	0	3
θ_2	1	2
θ_3	3	0

The prior belief is that $\pi(\theta_1) = \pi(\theta_2) = \pi(\theta_3) = \frac{1}{3}$.
(a) What is the Bayes action?
(b) What is the minimax nonrandomized action?

4. A professional baseball team is concerned about attendance for the upcoming year. They must decide whether or not to implement a half-million dollar promotional campaign. If the team is a contender, they feel that \$4 million in

attendance revenues will be earned (regardless of whether or not the promotional campaign is implemented). Letting θ denote the team's proportion of wins, they feel the team will be a contender if $\theta \geq 0.6$. If $\theta < 0.6$, they feel their attendance revenues will be $1+5\theta$ million dollars without the promotional campaign, and $2+\frac{10}{3}\theta$ million dollars with the promotional campaign. It is felt that θ has a $\mathcal{U}(0, 1)$ distribution.
(a) Describe $\mathcal{A}$, Θ, and $L(\theta, a)$.
(b) What is the Bayes action?
(c) What is the minimax nonrandomized action?

5. A farmer has to decide whether or not to plant his crop early. If he plants early and no late frost occurs, he will gain \$5000 in extra harvest. If he plants early and a late frost does occur, he will lose \$2000 as the cost of reseeding. If he doesn't plant early his gain will be \$0. Consulting the weather service, he finds that the chance of a late frost is about 0.6.
(a) Describe $\mathcal{A}$, Θ, the loss matrix, and the prior distribution.
(b) What is the Bayes action?
(c) What is the minimax nonrandomized action?

6. The owner of a ski shop must order skis for the upcoming season. Orders must be placed in quantities of 25 pairs of skis. The cost *per pair* of skis is \$50 if 25 are ordered, \$45 if 50 are ordered, and \$40 if 75 are ordered. The skis will be sold at \$75 per pair. Any skis left over at the end of the year can be sold (for sure) at \$25 a pair. If the owner runs out of skis during the season, he will suffer a loss of "goodwill" among unsatisfied customers. He rates this loss at \$5 per unsatisfied customer. For simplicity, the owner feels that demand for the skis will be 30, 40, 50 or 60 pair of skis, with probabilities 0.2, 0.4, 0.2, and 0.2, respectively.
(a) Describe $\mathcal{A}$, Θ, the loss matrix, and the prior distribution.
(b) Which actions are admissible?
(c) What is the Bayes action?
(d) What is the minimax nonrandomized action?

7. Find the minimax (randomized) action in
(a) Exercise 3.
(b) Exercise 4.
(c) Exercise 5.

8. In Example 2, what would be the Bayes action and what would be the minimax action if no sample information X was obtained?

Section 1.4

9. Assume that $n = 10$ and $\theta_0 = \frac{2}{3}$ in Example 7 (Section 1.4). Find the most powerful randomized test of size $\alpha = 0.05$.

Section 1.6

10. Suppose θ is the percentage change in the yield of a process, under the effect of a very expensive new treatment. To obtain information about θ, we can observe i.i.d. $\mathcal{N}(\theta, 2500)$ observations, $X_1, \ldots, X_n$.

(a) If $n = 25{,}000{,}000$ and $\bar{x} = 0.02$, show that there is statistically significant evidence at the $\alpha = 0.05$ level (one-tailed) that $\theta > 0$. Would it be wise to adopt the new treatment because of the strong evidence that it increases yield?

(b) If $n = 4$ and $\bar{x} = 30$, show that there is *not* statistically significant evidence at the $\alpha = 0.1$ level that $\theta > 0$. Does this imply that it would be wise not to adopt the treatment?

11. Verify that (1.1) holds with probability one for *any* sequence of (θ_i, σ_i) in Example 10.

12. (a) Consider a decision problem and decision rule δ for which

$$E_\theta[(L(\theta, \delta(X)) - R(\theta, \delta))^2] \leq k < \infty \quad \text{for all } \theta.$$

Imagine a sequence of repetitions of the decision problem, where the parameters θ_i can change and the data $X^{(i)}$ (from the density $f(x|\theta_i)$) are independent. Show that, if $R(\theta, \delta) \leq \bar{R}$ for all θ, then

$$\lim_{N\to\infty} \frac{1}{N} \sum_{i=1}^{N} L(\theta_i, \delta(X^{(i)})) \leq \bar{R} \tag{1.2}$$

with probability one for *any* sequence $(\theta_1, \theta_2, \ldots)$.

(b) Construct an example of a decision problem where $\sup_\theta R(\theta, \delta) = \infty$, and yet (1.2) still holds with probability one for any sequence $(\theta_1, \theta_2, \ldots)$.

13. Verify, in Example 11, that the proportion of rejections that are in error (i.e., the proportion of the rejections for which H_0 is true) will tend to $\frac{1}{2}$ with probability one.

14. Let X be $\mathscr{B}(100, \theta)$, and suppose that it is desired to test H_0: $\theta = \frac{1}{3}$ versus H_1: $\theta = \frac{2}{3}$. (Imagine that it is known that θ is one of these two values.) Consider the test which accepts H_0 if $x < 50$, rejects H_0 if $x > 50$, and randomly chooses H_0 or H_1 (with probability $\frac{1}{2}$ each) when $x = 50$.

(a) Calculate the probabilities of Type I and Type II error for this test. (Since these error probabilities are the same, their common value could be considered to be the overall frequentist error probability for the test.)

(b) If $x = 50$ is observed, what is the intuitive (conditional) probability of actually making an error in use of the test.

15. Let X be $\mathscr{U}(\theta, \theta + 1)$, and suppose that it is desired to test H_0: $\theta = 0$ versus H_1: $\theta = 0.9$ (these being the only two values of θ considered possible). Consider the test which rejects H_0 if $x \geq 0.95$, and accepts H_0 otherwise.

(a) Calculate the probabilities of Type I and Type II error for this test.

(b) If $0.9 < x < 1$ is observed, what is the intuitive (conditional) probability of actually making an error in use of the test.

16. Determine the likelihood function of θ, for each x, in:

(a) Exercise 14.

(b) Exercise 15.

(c) Interpret the "message" conveyed by the likelihood function in Exercise 15 for each x.

17. If $X_1, \ldots, X_{20}$ are i.i.d. $\mathcal{N}(\theta, 1)$, and $\bar{x}=3$ is observed, show that, according to the Likelihood Principle, all experimental information about θ is contained in the function $\exp\{-10(\theta-3)^2\}$.

18. Show, by a formal application of the Likelihood Principle, that the condition of the high-range voltmeter should have no effect on the statistical analysis in Example 17.

Sections 1.7 and 1.8

19. Assume a random variable $X \sim \mathcal{B}(n, \theta)$ is observed. It is desired to estimate θ under squared-error loss. Find a nonrandomized decision rule δ which is R-better than the randomized decision rule

$$\delta^* = \tfrac{1}{2}\langle \delta_1 \rangle + \tfrac{1}{2}\langle \delta_2 \rangle,$$

where $\delta_1(x) = x/n$ and $\delta_2(x) = \frac{1}{2}$. Also, calculate $R(\theta, \delta_1)$, $R(\theta, \delta_2)$, and $R(\theta, \delta^*)$.

20. Assume $\mathbf{X} = (X_1, X_2, \ldots, X_n)$ is observed, where the X_i are (independently) $\mathcal{N}(\theta, 1)$. It is desired to estimate θ under a loss $L(\theta, a)$. Let $\tilde{X}$ denote the median of the observations, and note that the mean $\bar{X}$ is sufficient for θ.
    (a) Find the conditional density (on $\mathscr{X}_t$) of $(X_1, X_2, \ldots, X_{n-1})$ given $\bar{X}$.
    (b) Find a randomized rule, based on $\bar{X}$, which is R-equivalent to $\delta(\mathbf{x}) = \tilde{x}$. (The integration need not be performed.)
    (c) If the loss is convex in a, show that $\delta'(\mathbf{x}) = \bar{x}$ is R-better than or R-equivalent to $\delta(\mathbf{x}) = \tilde{x}$. (*Hint*: Show that $E^{\mathbf{X}|\bar{X}=t}[\tilde{X}] = t$. Use a symmetry argument based on Part (a), noting that the $Z_i = X_i - \bar{X}$ $(i = 1, \ldots, n)$ have the same distribution as the $-Z_i$ $(i = 1, \ldots, n)$.)

21. Let $\mathbf{X} = (X_1, X_2, \ldots, X_n)$ be a sample from the $\mathcal{U}(\alpha, \beta)$ distribution. It is desired to estimate the mean $\theta = (\alpha + \beta)/2$ under squared-error loss.
    (a) Show that $T = (\min\{X_i\}, \max\{X_i\})$ is a sufficient statistic for θ. (You may use the factorization theorem.)
    (b) Show that the estimator given by

$$E^{\mathbf{X}|T}[\bar{X}] = \tfrac{1}{2}(\max\{X_i\} + \min\{X_i\})$$

is R-better than or R-equivalent to the sample mean $\bar{X}$.

22. Prove that the following functions are convex:
    (a) e^{cx}, for $-\infty < x < \infty$.
    (b) x^c $(c \geq 1)$, for $0 < x < \infty$.
    (c) $\sum_{i=1}^m c_i x_i$, for $\mathbf{x} = (x_1, \ldots, x_m)^t \in R^m$.

23. Prove that the following functions are concave:
    (a) $\log x$, for $0 < x < \infty$.
    (b) $(1 - e^{-cx})$, for $-\infty < x < \infty$.
    (c) $-\exp\{\sum_{i=1}^m c_i x_i\}$, for $\mathbf{x} = (x_1, \ldots, x_m)^t \in R^m$.

24. Which of the functions in Exercises 22 and 23 are strictly convex or strictly concave.

25. If $y_i \geq 0$, $i = 1, \ldots, m$, and $\alpha_i \geq 0$, $i = 1, \ldots, m$, with $\sum_{i=1}^m \alpha_i = 1$, then prove that $\prod_{i=1}^m y_i^{\alpha_i} \leq \sum_{i=1}^m \alpha_i y_i$. (*Hint*: Let $x_i = \log y_i$ and use Jensen's inequality.)

CHAPTER 2
Utility and Loss

2.1. Introduction

In evaluating the consequences of possible actions, two major problems are encountered. The first is that the values of the consequences may not have any obvious scale of measurement. For example, prestige, customer goodwill, and reputation are important to many businesses, but it is not clear how to evaluate their importance in a concrete way. A typical problem of this nature arises when a relatively exclusive company is considering marketing its "name" product in discount stores. The immediate profit which would accrue from increased sales is relatively easy to estimate, but the longterm effect of a decrease in prestige is much harder to deal with.

Even when there is a clear scale (usually monetary) by which consequences can be evaluated, the scale may not reflect true "value" to the decision maker. As an example, consider the value to you of money. Assume you have the opportunity to do a rather unpleasant task for $100. At your present income level, you might well value the 100 dollars enough to do the task. If, on the other hand, you first received a million dollars, the value to you of an additional $100 would be much less, and you would probably choose not to do the task. In other words, the value of $1,000,100 is probably not the same as the value of $1,000,000 plus the value of $100. As another example, suppose you are offered a choice between receiving a gift of $10,000 or participating (for free) in a gamble wherein you have a 50–50 chance of winning $0 or $25,000. Most of us would probably choose the sure $10,000. If this is the case, then the expected "value" of the gamble is less than $10,000. In the ensuing sections, a method of determining true value will be discussed. This will then be related to the development of the loss function.

2.2. Utility Theory

To work mathematically with ideas of "value," it will be necessary to assign numbers indicating how much something is valued. Such numbers are called *utilities*, and *utility theory* deals with the development of such numbers.

To begin, it is necessary to clearly delineate the possible consequences which are being considered. The set of all consequences of interest will be called the set of *rewards*, and will be denoted by $\mathscr{R}$. Quite frequently $\mathscr{R}$ will be the real line (such as when the consequences can be given in monetary terms), but often the elements of $\mathscr{R}$ will consist of nonnumerical quantities such as mentioned in the introduction.

Often there is uncertainty as to which of the possible consequences will actually occur. Thus the results of actions are frequently probability distributions on $\mathscr{R}$. Let $\mathscr{P}$ denote the set of all such probability distributions. It is usually necessary to work with values and preferences concerning probability distributions in $\mathscr{P}$. This would be easy to do if a real-valued function $U(r)$ could be constructed such that the "value" of a probability distribution $P \in \mathscr{P}$ would be given by the expected utility $E^P[U(r)]$. If such a function exists, it is called a *utility function.*

A precise formulation of the problem begins with the assumption that it is possible to state preferences among elements of $\mathscr{P}$. (If one cannot decide the relative worth of various consequences, there is no hope in trying to construct measures of their value.) The following notation will be used to indicate preferences.

Definition 1. If P_1 and P_2 are in $\mathscr{P}$, then $P_1 < P_2$ means that P_2 is preferred to P_1; $P_1 \approx P_2$ means that P_1 is equivalent to P_2; and $P_1 \preccurlyeq P_2$ means that P_1 is not preferred to P_2.

A reward $r \in \mathscr{R}$ will be identified with the probability distribution in $\mathscr{P}$ which gives probability one to the point r. This probability distribution will be denoted $\langle r \rangle$. (See Section 1.4 for a similar use of this notational device.) Hence the above definition applies also to rewards.

The goal is to find a function $U(r)$ which represents (through expected value) the true preference pattern on $\mathscr{P}$ of the decision maker. In other words, a function U is sought such that if P_1 and P_2 are in $\mathscr{P}$, then P_2 is preferred to P_1 if and only if

$$E^{P_1}[U(r)] < E^{P_2}[U(r)].$$

The function U is then the desired quantification of the decision maker's preference or value pattern, and can be called a utility function.

It is by no means clear that a utility function need exist. We will shortly give a brief discussion of certain conditions which guarantee the existence of a utility function. First, however, a useful method for constructing a utility function (assuming one exists) is given. In the construction, we will be concerned with mixtures of probability distributions of the form $P = \alpha P_1 + (1-\alpha)P_2$, where $0 \le \alpha \le 1$. This, of course, is the probability distribution for which $P(A) = \alpha P_1(A) + (1-\alpha)P_2(A)$, $A \subset \mathcal{R}$. Note, in particular, that $P = \alpha\langle r_1\rangle + (1-\alpha)\langle r_2\rangle$ is the probability distribution giving probability α to r_1 and probability $1-\alpha$ to r_2.

Construction of U

Step 1. To begin the construction of U, choose two rewards, r_1 and r_2, which are not equivalent. Assume they are labeled so that $r_1 < r_2$. Let $U(r_1) = 0$ and $U(r_2) = 1$.

Any choice of r_1 and r_2 is acceptable, but it is best to choose them in a convenient fashion. If there is a worst reward and a best reward, it is often convenient to choose r_1 and r_2 as these. For monetary rewards, choosing $r_1 = 0$ is usually helpful. The choice of r_1 and r_2 really serves only to set the scale for U. However, values of $U(r)$ for other r will be established by comparison with r_1 and r_2. Hence r_1 and r_2 should be chosen to make comparisons as easy as possible.

There are a variety of ways to proceed with the construction of U. One can simply compare each $r \in \mathcal{R}$ with r_1 and r_2, assigning a value, $U(r)$, to r which seems reasonable in comparison with $U(r_1) = 0$ and $U(r_2) = 1$. This is usually a difficult process, however, and it is often useful to consider "betting" situations as an aid to judgement. This is done as follows.

Step 2. For a reward r_3 such that $r_1 < r_3 < r_2$, find the α $(0 < \alpha < 1)$ such that

$$r_3 \approx P = \alpha\langle r_1\rangle + (1-\alpha)\langle r_2\rangle.$$

Define

$$U(r_3) = E^P[U(r)] = \alpha U(r_1) + (1-\alpha)U(r_2) = 1-\alpha.$$

Determining α is the difficult part of the procedure. The idea is that since r_3 is preferred to r_1, while r_2 is preferred to r_3, there should be a "gamble," in which you get r_1 with probability α and r_2 with probability $(1-\alpha)$, which is of the same value as r_3. Another way of stating this is to say that, if you were given reward r_3, you would be willing to "pay" it (but nothing better) in order to play the gamble $\alpha\langle r_1\rangle + (1-\alpha)\langle r_2\rangle$. Such an α will virtually always exist and be unique. It takes practice at introspection to find it, however.

Step 3. For a reward r_3 such that $r_3 < r_1$, find the $\alpha (0 < \alpha < 1)$ such that

$$r_1 \approx P = \alpha\langle r_3\rangle + (1-\alpha)\langle r_2\rangle.$$

Then to have

$$0 = U(r_1) = E^P[U(r)] = \alpha U(r_3) + (1-\alpha)U(r_2) = \alpha U(r_3) + (1-\alpha),$$

we must define

$$U(r_3) = \frac{-(1-\alpha)}{\alpha}.$$

Step 4. For a reward r_3 such that $r_2 < r_3$, find the $\alpha (0 < \alpha < 1)$ such that

$$r_2 \approx P = \alpha\langle r_1\rangle + (1-\alpha)\langle r_3\rangle.$$

Then to have

$$1 = U(r_2) = E^P[U(r)] = \alpha U(r_1) + (1-\alpha)U(r_3) = (1-\alpha)U(r_3),$$

we must define

$$U(r_3) = \frac{1}{1-\alpha}.$$

Step 5. Periodically check the construction process for consistency by comparing new combinations of rewards. For example, assume the utilities of r_3, r_4, and r_5 have been found by the preceding technique, and that $r_3 < r_4 < r_5$. Then find the $\alpha (0 < \alpha < 1)$ such that

$$r_4 \approx P = \alpha(r_3) + (1-\alpha)\langle r_5\rangle.$$

It should then be true that

$$U(r_4) = \alpha U(r_3) + (1-\alpha)U(r_5).$$

If this relationship is not (approximately) satisfied by the previously determined utilities, then an error has been made and the utilities must be altered to attain consistency. This process of comparing and recomparing is often how the best judgements can be made.

Recall that, for U to be a utility function, the expected utilities of the $P \in \mathcal{P}$ must be ordered in the same way as the true preferences concerning the P. For this to be the case (and indeed for the above construction to be possible), the preferences among elements of $\mathcal{P}$ must in some sense be rational. We list here a set of "rationality axioms" which guarantee that the preference pattern is suitable.

Axiom 1. If P_1 and P_2 are in $\mathcal{P}$, then either $P_1 < P_2$, $P_1 \approx P_2$, or $P_2 < P_1$.

Axiom 2. If $P_1 \preccurlyeq P_2$ and $P_2 \preccurlyeq P_3$, then $P_1 \preccurlyeq P_3$.

Axiom 3. If $P_1 < P_2$, then $\alpha P_1 + (1-\alpha)P_3 < \alpha P_2 + (1-\alpha)P_3$ for any $0 < \alpha < 1$ and P_3 in $\mathcal{P}$.

Axiom 4. If $P_1 < P_2 < P_3$, there are numbers $0 < \alpha < 1$ and $0 < \beta < 1$ such that

$$\alpha P_1 + (1-\alpha)P_3 < P_2 \quad \text{and} \quad P_2 < \beta P_1 + (1-\beta)P_3.$$

Axiom 1 simply formalizes the requirement that one must be able to state preferences among the elements of $\mathcal{P}$. Axiom 2 requires a natural transitivity in the preference pattern. Axiom 3 is related to the conditionality principle discussed in Section 1.6, and also seems quite natural. It simply states that if P_2 is preferred to P_1, then, in a choice between two random situations which are identical except that P_2 will occur with probability α in one and P_1 with probability α in the other, the situation involving P_2 will be preferred. Axiom 4 says, in a loose sense, that there is no infinitely desirable or infinitely bad reward (no heaven or hell). If, for example, P_1 was considered to be infinitely bad, then presumably there would be no $\beta > 0$ for which one would risk having P_1 occur (with probability β) in an attempt to achieve P_3 instead of P_2 (i.e., for which $P_2 < \beta P_1 + (1-\beta) P_3$). This axiom might be objected to, on the basis that a "reward" such as death is infinitely bad. If death was really felt to be infinitely bad compared to other consequences, however, one would never risk the additional chance of dying incurred by, say, crossing a street or driving a car.

For a proof that these axioms imply the existence of a utility function, see DeGroot (1970). (Some additional rather technical assumptions are needed to verify that the above construction results in a utility function.) The existence of a utility function is actually assured under generally weaker axioms than those above. (See Fishburn (1981) for a survey, and Rubin (1985) for a very weak axiom system.) Thus the preference pattern of any "rational" person can be described by a utility function. It should be noted, however, that people do not intuitively tend to act in accordance with a utility function (cf. Allais and Hagen (1979)). Thus we are, in essence, *defining* rational behavior for an individual, and suggesting that such behavior is good (see also Shafer (1982b)).

It should be mentioned that utility functions are unique only for a particular scaling, i.e., a particular choice of the initial r_1 and r_2. Indeed if $U(r)$ is a utility function, then $bU(r)+c\,(b>0)$ is also a utility function, since it leads to the same preferences among elements of $\mathcal{P}$. (If $E^{P_1}[U(r)] < E^{P_2}[U(r)]$, then $E^{P_1}[bU(r)+c] < E^{P_2}[bU(r)+c]$ and vice versa.) It is also of interest to note that Axiom 4 requires the utility function to be bounded. To see this, observe that if there exists a sequence of rewards $r_1, r_2, \ldots$ such that $U(r_n) \to \infty$, then a probability distribution $P = p_1\langle r_1 \rangle + p_2 \langle r_2 \rangle + \cdots$ can always be found such that $E^P[U(r)] = \infty$. This will clearly lead to a violation of Axiom 4. (Choose P_3 in the statement of the axiom to be the above P.) A similar argument holds if $U(r_n) \to -\infty$. Although it is very reasonable to have a bounded utility function, a weaker set of axioms can be constructed under which unbounded utility functions are allowed.

EXAMPLE 1. You face the decision of what to do on a Saturday afternoon for recreation. There are two choices available: a_1—see a football game, and a_2—go to a movie. You would really rather see the football game, but there is a 40% chance of rain, which would ruin the afternoon. Let θ_1

denote the state of nature "it rains" and θ_2 denote "no rain." Clearly $\pi(\theta_1)=0.4=1-\pi(\theta_2)$. To carry out a Bayesian analysis as in Chapter 1, it is necessary to determine the loss matrix. Instead, let us consider the "utility matrix" (since loss will just be negative utility).

The four consequences or rewards of interest are $r_1=(a_1, \theta_1)$, $r_2=(a_1, \theta_2)$, $r_3=(a_2, \theta_1)$, and $r_4=(a_2, \theta_2)$. (Here (a_i, θ_j) denotes action a_i being taken and θ_j being the state of nature that occurs.) Clearly the preference ordering is

$$r_1 < r_4 < r_3 < r_2.$$

The reasoning is that r_1 is a ruined afternoon; r_4 is okay, but you will regret somewhat not having gone to the game since it didn't rain; r_3 is an enjoyable afternoon, and you congratulate yourself on making a good decision about the weather; and r_2 is the best afternoon.

To construct U it is natural to use the worst and best rewards, r_1 and r_2, as the initial points. Hence let $U(r_1)=0$ and $U(r_2)=1$. To determine $U(r_4)$, compare r_4 with the gamble $\alpha\langle r_1\rangle+(1-\alpha)\langle r_2\rangle$. After some soul searching, assume it is concluded that r_4 is equivalent to $0.4\langle r_1\rangle+0.6\langle r_2\rangle$. In other words, you would just as soon see the movie and have no rain occur, as you would play the gamble of having a ruined afternoon with probability 0.4 or the best afternoon with probability 0.6. Thus $U(r_4)=(1-\alpha)=0.6$. (It should be noted that there is nothing sacred about thinking in terms of gambles. If you decide r_4 rates as about a 0.6 on a scale from 0 to 1 (r_1 being 0 and r_2 being 1), that is fine. For many people, however, the gambling mechanism proves useful.) Likewise, let us say it is determined that

$$r_3 \approx 0.3\langle r_1\rangle+0.7\langle r_2\rangle.$$

Then $U(r_3)=1-\alpha=0.7$.

It is, at this point, a good idea to check the construction of U by comparing a different combination of rewards, say r_2, r_3, and r_4. Find an α such that

$$r_3 \approx \alpha\langle r_4\rangle+(1-\alpha)\langle r_2\rangle.$$

Let's say $\alpha=0.6$ is felt to be correct. But if the utility function is accurate, we know that

$$0.7=U(r_3)=\alpha U(r_4)+(1-\alpha)U(r_2)=0.6\alpha+(1-\alpha),$$

or $\alpha=0.75$. This does not agree with the value of 0.6 that was obtained from direct comparison. Hence there is an inconsistency. It is thus necessary to go back and re-examine all the comparisons made, until a consistent set of utilities is obtained. Let us assume that the end result of this process is that $U(r_4)=0.6$ and $U(r_3)=0.75$ are decided upon.

Having the utility function, we can proceed directly to find the optimal action in terms of expected utility. (Recall that expected utility is, by construction, the proper way to evaluate uncertain rewards.) Clearly, the expected utility of action a_1 is

$$E^{\pi}[U(r)] = \pi(\theta_1)U(r_1) + \pi(\theta_2)U(r_2)$$
$$= (0.4)(0) + (0.6)(1) = 0.6,$$

while the expected utility of action a_2 is

$$E^{\pi}[U(r)] = \pi(\theta_1)U(r_3) + \pi(\theta_2)U(r_4)$$
$$= (0.4)(0.75) + (0.6)(0.6) = 0.66.$$

Thus the optimal action is to go to the movie.

When the set $\mathcal{R}$ is large, the construction of U becomes much more difficult. The formal construction procedure is time-consuming and can, hence, only be done at a few points of $\mathcal{R}$. Often, however, $U(r)$ at other points can be estimated from the values of U at these few points. A common situation in which this can be done is when $\mathcal{R}$ is an interval of the real line. $U(r)$ will then typically be a smooth function of r which can be graphed from knowledge at a few points. The most important example of this is when the rewards are monetary, so that U is a utility function for money. (This case will be treated separately in the next section.) Other situations in which $\mathcal{R}$ is an interval can arise, however. The following is an example.

EXAMPLE 2. A doctor must control the blood sugar count of a patient. The "reward" for a treatment he administers is the blood sugar count of the patient. The set of all rewards $\mathcal{R}$ can be taken to be the set of all reasonable blood sugar counts, and can be approximated by an interval $(B_{\min}, B_{\max})$. A reasonable utility function for this situation is given in Figure 2.1. B_N is the normal level, and has the highest utility. As the blood sugar count gets higher or lower, the utility drops. $U(r)$ could be reasonably sketched from values at a few points.

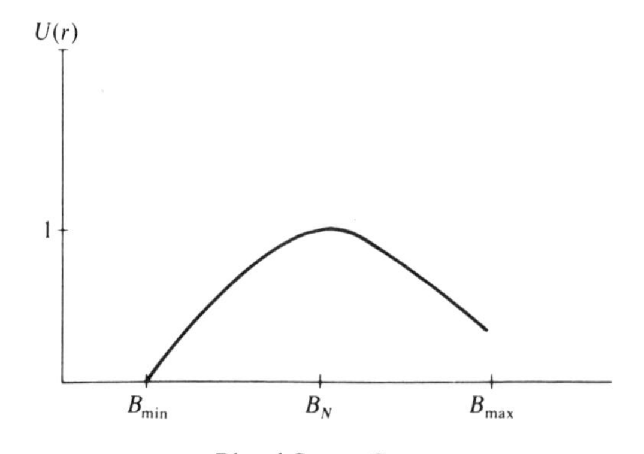

Figure 2.1

To determine a utility function that you know will behave as in Figure 2.1, it is useful to do the construction by segments. In this example, first construct $U(r)$ from $B_{\min}$ to B_N. Next find $U(B_{\max})$ by comparison of $B_{\max}$ to $B_{\min}$ and B_N. Then construct $U(r)$ from B_N to $B_{\max}$, using B_N and $B_{\max}$ as the reference points in the construction. The needed changes in the formulas for $U(r)$ should be clear.

In many problems, the rewards are actually vectors $\mathbf{r}=(r_1, r_2, \ldots, r_m)$. In medical situations, for example, a treatment could have several side effects upon the patient, besides the intended effect upon the illness. The resulting $\mathscr{R}$ would be a rectangle in R^m, and the construction of $U(\mathbf{r})$ on this set would be very difficult. It is often assumed in such situations that

$$U(\mathbf{r}) = \sum_{i=1}^{m} K_i U_i(r_i). \tag{2.1}$$

The problem then simplifies to finding each of the one-dimensional utility functions $U_i(r_i)$, and then determining the *scaling constants* K_i. (These can be found by the same kind of comparison technique that is used in the construction of a utility function.) The simplifying assumption (2.1) need not always be reasonable, of course, especially when there is considerable "interaction" between coordinates of $\mathbf{r}$. (Two side effects of a drug might be acceptable separately, but very dangerous if they occur together.) A more complicated model, such as

$$U(\mathbf{r}) = \sum_{i=1}^{m} K_i U_i(r_i) + \sum_{i=1}^{m} \sum_{j=1}^{m} K_{ij} U_i(r_i) U_j(r_j),$$

might then prove satisfactory. Keeney and Raiffa (1976) deal with this problem in depth.

2.3. The Utility of Money

As indicated in the introduction, the marginal value of money for most people is decreasing. This means, for instance, that the difference in value between $\$(z+100)$ and $\$z$ is decreasing as z increases. An additional \$100 is quite valuable when $z=0$, but of little importance when $z=1{,}000{,}000$. A typical utility function $U(r)$, for positive amounts of money r, is shown in Figure 2.2. Indeed the function typically levels off, i.e., is bounded. (I doubt if anyone would give much value to additional money beyond $\$10^{100}$.)

To get a better feeling for such a utility function, let's consider the construction of a personal utility function for money. In dealing with such a situation, there is a "backwards" method of construction (discussed in Becker, DeGroot, and Marschak (1964)) that is very useful. Begin, as before, by choosing $r_1 < r_2$ and setting $U(r_1)=0$, $U(r_2)=1$. (It is assumed that if

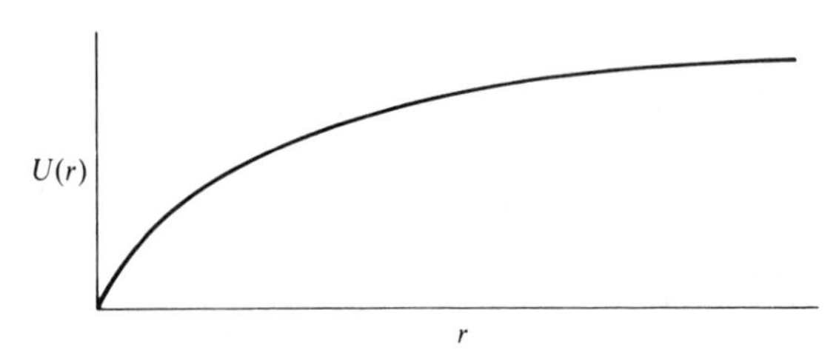

Figure 2.2

$r_1 < r_2$ then $r_1 < r_2$.) The following steps will lead to a graph of $U(r)$ for $r_1 < r < r_2$.

Step 1. Find r_3 so that $U(r_3) = \frac{1}{2}$. Since, for $P = \frac{1}{2}\langle r_1 \rangle + \frac{1}{2}\langle r_2 \rangle$,

$$\tfrac{1}{2} = \tfrac{1}{2}U(r_1) + \tfrac{1}{2}U(r_2) = E^P[U(r)],$$

this is equivalent to finding the reward r_3 which is valued as highly as the gamble $\frac{1}{2}\langle r_1 \rangle + \frac{1}{2}\langle r_2 \rangle$. This type of 50–50 gamble is easy to think of, making the determination of r_3 relatively easy.

Step 2. Find r_4 and r_5 such that $U(r_4) = \frac{1}{4}$ and $U(r_5) = \frac{3}{4}$. Since

$$\tfrac{1}{4} = \tfrac{1}{2}U(r_1) + \tfrac{1}{2}U(r_3) \quad \text{and} \quad \tfrac{3}{4} = \tfrac{1}{2}U(r_3) + \tfrac{1}{2}U(r_2),$$

r_4 is obviously the point equivalent to the gamble $\frac{1}{2}\langle r_1 \rangle + \frac{1}{2}\langle r_3 \rangle$, and r_5 is the point equivalent to $\frac{1}{2}\langle r_3 \rangle + \frac{1}{2}\langle r_2 \rangle$.

Step 3. Since

$$U(r_3) = \tfrac{1}{2} = \tfrac{1}{2}U(r_4) + \tfrac{1}{2}U(r_5),$$

r_3 must be equivalent to the gamble $\frac{1}{2}\langle r_4 \rangle + \frac{1}{2}\langle r_5 \rangle$. This provides a very valuable check on the determinations made so far. If the two are not felt to be equivalent, it is necessary to re-evaluate in Steps 1 and 2.

Step 4. Continue the above process of finding the points with utilities $i/2^n$ and checking for consistency, until a sufficient number of points have been found to enable a graph of $U(r)$ to be made.

The advantages of this second method of finding $U(r)$ are that only simple gambles of the form $\frac{1}{2}\langle r_i \rangle + \frac{1}{2}\langle r_j \rangle$ need be considered, and also that there is an easy check for consistency.

It is informative to construct a typical utility function for money using this approach. Consider a person, Mr. Jones, who earns \$10,000 per year and has \$500 in the bank. Let us construct a reasonable utility function for him on $\mathscr{R} = (-1000, 1000)$. This will be a utility function for the value of *additional* money which Mr. Jones might receive (or lose). A utility function could be constructed for the value Mr. Jones places on his total "fortune." It tends to be somewhat easier, however, to think in terms of the value of changes from the status quo. (Actually, if $U(r)$ is a utility function for total fortune (which happens to be r_0 at the moment), then the utility function for additional money, m, should be

$$U^*(m) = U(m + r_0) - U(r_0).$$

Generally, however, the utility function of a person will change with time, so that, if feasible, it is probably best to develop directly the utility function for additional money in each new situation.)

It is easiest to do the construction in three segments: $(-1000, -500)$, $(-500, 0)$, and $(0, 1000)$. Starting with $(0, 1000)$, set $U(0) = 0$, $U(1000) = 1$. Next, determine r so that $r \approx \frac{1}{2}\langle 0 \rangle + \frac{1}{2}\langle 1000 \rangle$. (Recall that this signifies that the rewards zero and 1000 each have probability $\frac{1}{2}$ of occurring.) Mr. Jones might reasonably decide that a sure reward of $r = 300$ is as good as a 50–50 chance at 0 or 1000. Hence $U(300) = \frac{1}{2}$. Let us say he next decides that $100 \approx \frac{1}{2}\langle 0 \rangle + \frac{1}{2}\langle 300 \rangle$ and $500 \approx \frac{1}{2}\langle 300 \rangle + \frac{1}{2}\langle 1000 \rangle$. Then $U(100) = \frac{1}{4}$ and $U(500) = \frac{3}{4}$. To check for consistency, he determines an r such that $r \approx \frac{1}{2}\langle 100 \rangle + \frac{1}{2}\langle 500 \rangle$. He deems $r = 250$ to be appropriate. Unfortunately this is inconsistent, in that the answer should have been $r = 300$. Through re-examination, Mr. Jones arrives at the consistent choices $U(125) = \frac{1}{4}$, $U(300) = \frac{1}{2}$, $U(550) = \frac{3}{4}$. Using these, he sketches $U(r)$ as indicated in Figure 2.3.

Turning next to the interval $(-500, 0)$, the first step is to determine $U(-500)$. Mr. Jones decides he can best compare -500, 0, and 500. Indeed he feels

$$0 \approx \tfrac{1}{3}\langle -500 \rangle + \tfrac{2}{3}\langle 500 \rangle.$$

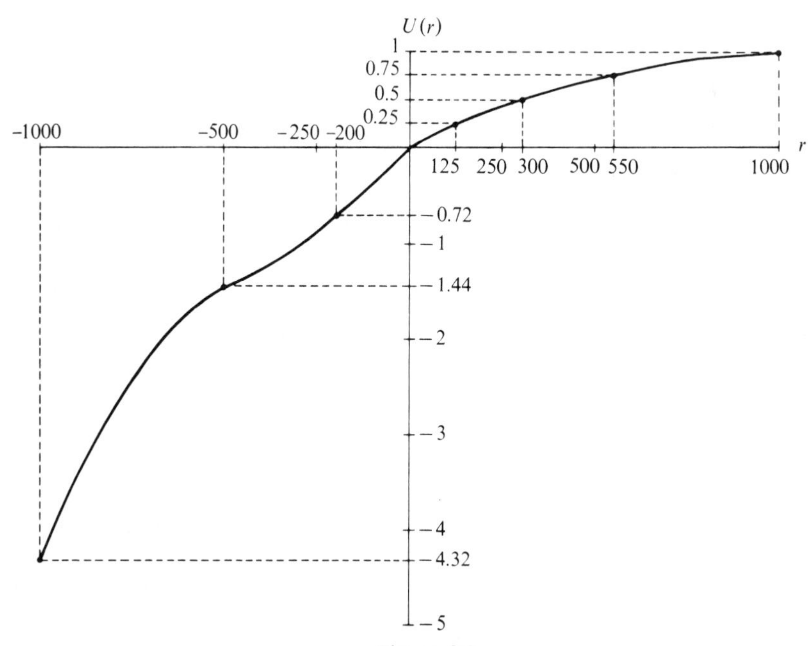

Figure 2.3

Noting from Figure 2.3 that $U(500) \cong 0.72$, it follows that

$$0 = U(0) = \tfrac{1}{3}U(-500) + \tfrac{2}{3}(0.72),$$

or $U(-500) = -1.44$. It is next determined that $-200 \approx \tfrac{1}{2}\langle -500\rangle + \tfrac{1}{2}\langle 0\rangle$, so that $U(-200) = \tfrac{1}{2}U(-500) + \tfrac{1}{2}U(0) = -0.72$. Two more halfway points are determined, and the curve between -500 and 0 sketched.

Finally, the interval $(-1000, -500)$ must be dealt with. In comparing -1000 to -500 and 0, it is determined that $-500 \approx \tfrac{1}{3}\langle -1000\rangle + \tfrac{2}{3}\langle 0\rangle$. (Mr. Jones is very averse to losing more money than he has.) Hence

$$-1.44 = U(-500) = \tfrac{1}{3}U(-1000) + \tfrac{2}{3}U(0),$$

or $U(-1000) = -4.32$. It would be a good idea to check this for consistency by comparison of, say, 0 with -1000 and 1000. Other halfway points are found, and the sketch of $U(r)$ completed.

Note that $U(r)$ is concave, except for the segment from -500 to 0. In this segment, Mr. Jones is apparently willing to engage in slightly unfavorable gambles to possibly avoid a sure loss of money. For example, since $U(0) = 0$, $U(-500) = -1.44$, and $U(-210) \cong -0.75$, it is clear that $\tfrac{1}{2}\langle -500\rangle + \tfrac{1}{2}\langle 0\rangle$ is preferred to -210. ($U(-210) = -0.75 < -0.72 = \tfrac{1}{2}U(-500) + \tfrac{1}{2}U(0)$.) Mr. Jones would rather take a 50–50 chance at losing all or nothing than lose a sure \$210.

Several features of utility functions such as that above are of interest.

(i) $U(r)$ is approximately linear for small values of r. I, personally, would be quite willing to take a 50–50 gamble on winning or losing 25¢. Hence I am sure that my utility function is linear up to 25¢. Professional gamblers would probably feel the same way about amounts up to several hundred or thousand dollars. General Motors probably feels the same way about amounts into the hundreds of thousands or millions of dollars. In other words, General Motors would not be averse to, say, participating in a speculative project where \$500,000 could be lost, as long as the expected outcome was reasonably high. (They would want some expected profit of course.) Thus as a general rule, when r is small compared to, say, the income of the person or business involved, $U(r)$ can be treated as linear.

(ii) $U(r)$ is usually concave (at least for $r > 0$). This is due to the fact that the marginal utility of money is typically decreasing. The marginal utility of money can be thought of as $U'(r) = dU(r)/dr$. If $U'(r)$ is decreasing, then $dU'(r)/dr = U''(r) < 0$, which implies that U is concave.

(iii) $U(r)$ is often quite different for $r \geq 0$ and $r < 0$ (different in shape, not just sign). It is, therefore, generally desirable to construct $U(r)$ separately for $r \geq 0$ and $r < 0$.

(iv) $U(r)$ is typically bounded. (Thus there usually exists a $B < \infty$ such that $|U(r)| \leq B$ for all $r \in \mathscr{R}$.) As an example, few people would feel that additional money beyond $\$10^{100}$ is of much utility, so $U(10^{100})$ is pretty close to an upper bound on U.

This last property of U provides an interesting resolution to the so-called St. Petersburg paradox. This paradox arises from consideration of the following gambling game. A fair coin will be flipped until a tail first appears. The reward will be \$$2^n$, where n is the number of flips it takes until a tail appears. The expected value of the game is

$$\sum_{n=1}^{\infty} 2^n P(n \text{ flips}) = \sum_{n=1}^{\infty} 2^n \cdot \frac{1}{2^n} = \infty.$$

In spite of this, few people are willing to pay very much money to play the game, since the probability of winning much money is small. (For example, the probability of winning 16 or more dollars is only $\frac{1}{8}$.) Utility theory resolves the paradox by showing that, if it costs c dollars to play the game, then the true value of playing is

$$\sum_{n=1}^{\infty} U(2^n - c) \cdot 2^{-n},$$

where U is the utility function for a change in fortune (with, say, $U(0)=0$). One should or should not play the game, depending upon whether this quantity is positive or negative. It will be left as an exercise to show that, for typical utility functions, this quantity is negative when c is large.

A parametric utility theory for money has been suggested. The motivation for such a theory arises from the observation that many real life utility functions can be matched quite well to curves of certain functional forms. The most commonly suggested functional form is

$$U(r) = b \log(cr + 1). \tag{2.2}$$

(Throughout the book, "log" will denote logarithm to the base e.) The constant c is generally chosen to be quite small, so that $U(r) \cong bcr$ for small or moderate $|r|$. Also, $U(r)$ is concave, and, while not bounded, is fairly level. Hence the usual properties of a utility function for money are satisfied by this function. The parametric approach suggests that, rather than going through the difficult procedure of constructing U, one should just choose b and c in (2.2) until $U(r)$ seems reasonable. This can be easily done by obtaining $U(r)$ at two points (other than $r=0$) by the usual method, and then solving for b and c. Keeney and Raiffa (1976) discuss this and other parametric models.

2.4. The Loss Function

2.4.1. Development from Utility Theory

As in Example 1, analysis of a decision problem can be done in terms of a utility function. Each pair (θ, a) (θ the state of nature and a the action) determines some reward r, which has utility $U(r)$. This can, for convenience, be denoted $U(\theta, a)$, and represents the true gain to the statistician when action a is taken and θ turns out to be the true state of nature.

A slight confusion can arise when θ is a statistical parameter. The problem is that then θ and a will seldom directly determine the reward. A typical example is when a business does a marketing survey concerning a new product. From the survey, they hope to obtain an estimate of θ, the proportion of the population that would be interested in buying the product. What they are really interested in, of course is their ultimate profit, which depends upon how many people actually buy the product. This will depend upon a number of extraneous factors, the state of the economy being one. These factors are typically random themselves. Thus the ultimate reward for taking action a when θ is the true state of nature is a random variable, say Z, which has a distribution depending upon θ and a. In such a situation, it is of course proper to define

$$U(\theta, a) = E_{\theta,a}[U(Z)].$$

It would have been logically correct to define θ at the beginning as the overall state of nature, i.e., θ would not only have a component representing the proportion of interested buyers, but would also have components reflecting all the other uncertain factors. In practice, however, it is intuitively easier to separate the components about which statistical information will be obtained, from the components for which only guesses can be made. This is in part because the statistical analysis tends to be more complicated, and in part because the nonstatistical randomness is usually easier to think about when isolated. Thus, in the previous marketing example, it seems best to proceed by (i) determining $U(r)$, where r here stands for profits, (ii) determining the distribution of the random variable Z (which is again profits) for given θ and a, and (iii) calculating $U(\theta, a) = E_{\theta,a}[U(Z)]$. Chapter 3 will present some useful techniques for carrying out step (ii).

In situations such as discussed above, one must be careful to define Z as the *overall* outcome of the random situation. The following example indicates the reason for caution.

EXAMPLE 3. Consider the following game. A coin is to be flipped n times. Each toss is independent of the others, and on any given toss the coin has probability 0.6 of coming up heads. For every toss on which the coin comes up heads, \$1000 is won, while for every tail \$1000 is lost. Let Z_i denote the amount won (1000 or -1000) on the ith flip of the coin. Assume that the utility function for a change, r, in monetary fortune is

$$U(r) = \begin{cases} r^{1/3} & \text{if } r \geq 0, \\ 2r^{1/3} & \text{if } r < 0. \end{cases}$$

For $n = 1$, the outcome of the game is Z_1, and the expected utility of the outcome is

$$E[U(Z_1)] = (0.6)U(1000) + (0.4)U(-1000)$$
$$= (0.6)(10) + (0.4)(-20) = -2.$$

Since this is negative, it would not be desirable to play the game with only one flip.

For $n > 1$, it would be wrong to say that the expected utility is $\sum_{i=1}^{n} E[U(Z_i)] = -2n$. The true outcome is $Z = \sum_{i=1}^{n} Z_i$, and the correct expected utility is $E[U(Z)]$. For large enough n, this will be positive, in which case it would be desirable to play the game. (It will be left as an exercise to determine how large n needs to be.)

The idea in the above example is rather obvious, but it has some disturbing implications. Consider, for instance, a business which is trying to decide whether or not to market a new product, and also whether or not to buy a new computer. If the financial results of given actions in the two decision problems are outcomes Z_1 and Z_2 (possibly random) which will both occur in the same time period of interest, then decisions should be based on overall expected utility, namely $E[U(Z_1 + Z_2)]$. This means that problems for a decision maker cannot be considered in isolation, but must be combined together in one grand mélange. In a sense, this is natural. A business, for instance, must consider the overall financial picture in deciding on specific issues. Nevertheless, this need to combine problems can impose a considerable hardship on the decision maker trying to construct a loss function. (See Shafer (1982b) for discussion.)

The above difficulty does not arise when the utility function is linear. This is because a linear utility function satisfies $U(\sum Z_i) = \sum U(Z_i)$. Hence for linear utility, decision problems can be considered separately.

Once $U(\theta, a)$ has been obtained, the loss function can simply be defined as

$$L(\theta, a) = -U(\theta, a). \tag{2.3}$$

The desire to maximize utility then becomes the desire to minimize loss. By developing L through the utility theory construction, we also have the important fact that expected loss is the proper measure of loss in random situations. This justifies the use of expected loss as a decision criterion when talking about randomized rules, Bayesian expected loss, risks, and Bayes risks.

Note that any linear function of L, such as $bL(\theta, a) + c\,(b > 0)$, could serve as the loss function as well as $L(\theta, a)$. This is because an optimal action or decision rule, obtained from any reasonable decision principle, is the same for all such linearly transformed losses. For example, the Bayes risk principle chooses δ to minimize

$$E^{\pi}[R(\theta, \delta)] = E^{\pi} E_{\theta}^{X} L(\theta, \delta(X)).$$

The δ which minimizes this will clearly also minimize

$$b(E^{\pi}E_{\theta}^{X}[L(\theta, \delta(X))])+c=E^{\pi}E_{\theta}^{X}[bL(\theta, \delta(X))+c],$$

which is the Bayes risk for the linearly transformed loss.

Loss functions are usually bounded from below, i.e.,

$$\inf_{\theta} \inf_{a} L(\theta, a)=B>-\infty.$$

In such situations, the transformed loss function

$$L^*(\theta, a)=L(\theta, a)-B$$

will always be nonnegative. It is convenient to talk in terms of nonnegative losses, so it will often be assumed that the above transformation has been made. Note also that, when derived from a utility function, the loss L^* can be written as

$$L^*(\theta, a)=\sup_{\theta} \sup_{a} U(\theta, a)-U(\theta, a).$$

This actually seems somewhat more sensible as a measure of loss than does the earlier definition in (2.3), in that $L^*(\theta, a)$ measures the true amount "lost" by not having the most favorable possibility occur. The analysis with L^* will not differ from the analysis with L, however. (It could be argued that the true amount "lost" is $\bar{L}(\theta, a)=\sup_a U(\theta, a)-U(\theta, a)$, since one has no control over which θ occurs. This loss is called *regret loss*, and will be discussed in Subsection 5.5.5. Interestingly enough, $\bar{L}$ is equivalent to L and L^* for Bayesian analysis, but can lead to different, and generally more reasonable, results for other decision principles.)

2.4.2. Certain Standard Loss Functions

In making a decision or evaluating a decision rule, the loss function should, ideally, be developed as above. Often, however, analyses of decision rules are carried out for certain "standard" losses. Three of these will be briefly discussed.

I. *Squared-Error Loss*

The loss function $L(\theta, a)=(\theta-a)^2$ is called *squared-error loss*. There are a number of reasons why it it often considered in evaluating decision rules. It was originally used in estimation problems when unbiased estimators of θ were being considered, since $R(\theta, \delta)=E_\theta L(\theta, \delta(X))=E_\theta[\theta-\delta(X)]^2$ would then be the variance of the estimator. A second reason for the popularity of squared-error loss is due to its relationship to classical least squares theory. The similarity between the two makes squared-error loss seem familiar to statisticians. Finally, for most decision analyses, the use of squared-error loss makes the calculations relatively straightforward and simple.

The above justifications for squared-error loss really have very little merit. The question is—does squared-error loss typically reflect the true loss function in a given situation? The initial reaction is, probably, no. As in the discussion of utility theory, one can reason that the loss function should usually be bounded and (at least for large errors) concave. Squared-error loss is neither of these. The convexity of squared-error loss is particularly disturbing. (Large errors are penalized, perhaps, much too severely.)

There are a number of situations, however, in which squared-error loss may be appropriate. For example, in many statistical problems for which a loss symmetric in $(\theta-a)$ is suitable, the exact functional form of the loss is not crucial to the conclusion. Squared-error loss may then be a useful approximation to the true loss. Several problems of this nature will be encountered in later chapters.

Another situation in which squared-error loss can arise is when

$$L(\theta, a)=-U(\theta, a)=-E_{\theta,a}[U(Z)], \tag{2.4}$$

as mentioned earlier. Exercise 14 deals with such a situation, and shows how a loss similar to squared-error loss can occur naturally.

If, more generally, the sample information is quite accurate in a problem with a loss as in (2.4), then $L(\theta, a)$ can frequently be approximated by squared-error loss. For example, assume $Z=h(\theta-a, Y)$, where the distribution of Y does not depend on θ or a. (The reward is thus a function of the accuracy of estimating θ (as measured by $\theta-a$) and some random variable Y which does not depend on θ or a. For example, Y could be a random variable reflecting the future state of the economy.) For convenience, define $g(\theta-a, Y)=U(h(\theta-a, Y))$. Since the sample information about θ is quite accurate, $\theta-a$ will be small. Thus $g(\theta-a, Y)$ can be expanded in a Taylor series about 0, giving

$$g(\theta-a, Y)\cong g(0, Y)+(\theta-a)g'(0, Y)+\tfrac{1}{2}(\theta-a)^2 g''(0, Y).$$

(The derivatives are, of course, with respect to the first argument of g. Higher-order terms will tend to be negligible because they involve higher powers of the small $(\theta-a)$.) Letting

$$K_1=-E^Y[g(0, Y)], \qquad K_2=-E[g'(0, Y)], \quad \text{and} \quad K_3=-\tfrac{1}{2}E[g''(0, Y)],$$

it follows that

$$L(\theta, a)=-E[U(Z)]\cong K_1+K_2(\theta-a)+K_3(\theta-a)^2.$$

Completing squares gives

$$L(\theta, a)\cong K_3\left(\theta-a+\frac{K_2}{2K_3}\right)^2+\left(K_1-\frac{K_2^2}{4K_3}\right).$$

Providing K_3 is a positive constant, this loss is equivalent (for decision making) to the transformed loss

$$L(\theta, a) = \left(\theta - a + \frac{K_2}{2K_3}\right)^2.$$

This would be squared-error loss if it weren't for the constant $K_2/2K_3$. When $K_2 = 0$ (which would occur if $g(0, y)$ was a symmetric function of y, and Y had a symmetric distribution), there is no problem. Otherwise, however, the constant represents the fact that either overestimation or underestimation (depending on the sign of K_2) is desired.

To analyze a decision problem with

$$L(\theta, a) = (\theta - a + c)^2,$$

merely consider the new action space

$$\mathscr{A}^* = \{a - c\colon a \in \mathscr{A}\}.$$

For $a^* \in \mathscr{A}^*$, the loss corresponding to L is $L^*(\theta, a^*) = (\theta - a^*)^2$. The analysis in this transformed problem is thus done with squared-error loss. If δ^* is an optimal decision rule in the transformed problem, then $\delta = \delta^* + c$ will be optimal in the original problem.

A generalization of squared-error loss, which is of interest, is

$$L(\theta, a) = w(\theta)(\theta - a)^2.$$

This loss is called *weighted squared-error loss*, and has the attractive feature of allowing the squared error, $(\theta - a)^2$, to be weighted by a function of θ. This will reflect the fact that a given error in estimation often varies in harm according to what θ happens to be.

The final variant of squared-error loss which will be considered is quadratic loss. If $\boldsymbol{\theta} = (\theta_1, \ldots, \theta_p)^t$ is a vector to be estimated by $\mathbf{a} = (a_1, \ldots, a_p)^t$, and $\mathbf{Q}$ is a $p \times p$ positive definite matrix, then

$$L(\boldsymbol{\theta}, \mathbf{a}) = (\boldsymbol{\theta} - \mathbf{a})^t \mathbf{Q} (\boldsymbol{\theta} - \mathbf{a})$$

is called *quadratic loss*. When $\mathbf{Q}$ is diagonal, this reduces to

$$L(\boldsymbol{\theta}, \mathbf{a}) = \sum_{i=1}^{p} q_i (\theta_i - a_i)^2,$$

and is a natural extension of squared-error loss to the multivariate situation.

II. *Linear Loss*

When the utility function is approximately linear (as is often the case over a reasonable segment of the reward space), the loss function will tend to be linear. Thus of interest is the *linear loss*

$$L(\theta, a) = \begin{cases} K_0(\theta - a) & \text{if } \theta - a \geq 0, \\ K_1(a - \theta) & \text{if } \theta - a < 0. \end{cases}$$

The constants K_0 and K_1 can be chosen to reflect the relative importance of underestimation and overestimation. These constants will usually be different. When they are equal, the loss is equivalent to

$$L(\theta, a) = |\theta - a|,$$

which is called *absolute error loss.* If K_0 and K_1 are functions of θ, the loss will be called *weighted linear loss.* Linear loss (or weighted linear loss) is quite often a useful approximation to the true loss.

III. "0–1" *Loss*

In the two-action decision problem (of which hypothesis testing is an example) it is typically the case that a_0 is "correct" if $\theta \in \Theta_0$, and a_1 is correct if $\theta \in \Theta_1$. (This could correspond to testing H_0: $\theta \in \Theta_0$ versus H_1: $\theta \in \Theta_1$.) The loss

$$L(\theta, a_i) = \begin{cases} 0 & \text{if } \theta \in \Theta_i, \\ 1 & \text{if } \theta \in \Theta_j \end{cases} \qquad (j \neq i),$$

is called "0–1" *loss.* In words, this loss is zero if a correct decision is made, and 1 if an incorrect decision is made. The interest in this loss arises from the fact that, in a testing situation, the risk function of a decision rule (or test) $\delta(x)$ is simply

$$R(\theta, \delta) = E_\theta[L(\theta, \delta(X))] = P_\theta(\delta(X) \text{ is the incorrect decision}).$$

This is either a probability of Type I or Type II error, depending on whether $\theta \in \Theta_0$ or $\theta \in \Theta_1$. And, similarly from the conditional perspective, the Bayesian expected loss is

$$\rho(\pi^*, a_i) = \int L(\theta, a_i) dF^{\pi^*}(\theta) = 1 - P^{\pi^*}(\theta \in \Theta_i),$$

which is one minus the actual (subjective) *probability* that H_i is true (in light of the probability distribution, π^*, for θ).

In practice, "0–1" loss will rarely be a good approximation to the true loss. More realistic losses are

$$L(\theta, a_i) = \begin{cases} 0 & \text{if } \theta \in \Theta_i, \\ k_i & \text{if } \theta \in \Theta_j \end{cases} \qquad (i \neq j),$$

and

$$L(\theta, a_i) = \begin{cases} 0 & \text{if } \theta \in \Theta_i, \\ k_i(\theta) & \text{if } \theta \in \Theta_j \end{cases} \qquad (i \neq j).$$

This last type of loss, with $k_i(\theta)$ being an increasing function of the "distance" of the true θ from Θ_i, is particularly reasonable, in that the harm suffered by an incorrect decision will usually depend on the severity of the mistake. Actually, even when a "correct" decision is made, $L(\theta, a)$ may very well be nonzero, so that the full generality of the loss may be needed. Note, however, that only two functions, $L(\theta, a_0)$ and $L(\theta, a_1)$, must be determined.

For the most part, examples in the book will use the above three losses or variants of them. This is done mainly to make the calculations relatively easy. It should be emphasized that these losses need not necessarily be suitable for a given problem. Indeed, about the only way to decide if they are reasonable is to do a utility analysis.

2.4.3. For Inference Problems

In Section 1.2 we discussed, in general terms, the question of the applicability of decision theory to "statistical inference." We will not pursue the issue of whether or not inference problems really exist (i.e., the argument that *any* conclusion will ultimately be used for something and one should try to anticipate such uses), and will instead concentrate on how decision theory can aid in the process of choosing an inference in a more formal sense.

The most common application of decision theory to inference problems is through the representation of common inference measures as a risk or Bayesian expected loss. For instance, in Subsection 2.4.2 it was shown that frequentist error probabilities in testing correspond to the risk function for "0–1" loss. Another similar example is the confidence set scenario: if C denotes a confidence rule (when x is observed, the set $C(x) \subset \Theta$ will be presented as the confidence set for θ), and one considers the loss function

$$L(\theta, C(x)) = 1 - I_{C(x)}(\theta) = \begin{cases} 1 & \text{if } \theta \notin C(x), \\ 0 & \text{if } \theta \in C(x), \end{cases}$$

then

$$R(\theta, C) = E_\theta[1 - I_{C(X)}(\theta)] = 1 - P_\theta(C(X) \text{ contains } \theta),$$

which is one minus the frequentist coverage probability. Likewise, for a distribution π^* for θ,

$$\rho(\pi^*, C(x)) = E^{\pi^*}[1 - I_{C(x)}(\theta)] = 1 - P^{\pi^*}(\theta \in C(x)),$$

which is one minus the actual (subjective) probability that θ is in the specific set $C(x)$ (for the observed x).

A number of other frequentist and Bayesian inference measures can also be given a formal interpretation as a risk or a Bayesian expected loss. The task of finding a good inference procedure (or conditional inference) then reduces (formally) to a decision theory problem, allowing the extensive machinery of decision theory to be brought to bear. Many good illustrations of this perspective on inference can be found in Lehmann (1959) and Ferguson (1967).

A second, less common, use of decision theory in inference problems is to attempt to formally quantify the purpose of the inference (which is usually to appropriately communicate useful information), and then to find good inferences with respect to this quantification. The formalization is usually carried out by introducing some loss function that measures this "communication of information" goal. The following example is a simple illustration.

EXAMPLE 4. Suppose the inference is to produce a confidence rule C with associated "confidence" $\alpha(x)$ (when x is observed, $\alpha(x)$ will be the reported confidence in the set $C(x)$). For the moment, we will not specify the type of confidence being considered. Note that we allow the confidence to depend on x; this is, of course, fine from the conditional perspective, and can be justified from the frequentist perspective with the idea of "estimated confidence" (discussed briefly in Subsection 1.6.3).

The inference here will consist of the pair $(C(x), \alpha(x))$. To keep matters simple, let us suppose that $C(x)$ is specified, so that all we need to decide upon is a suitable choice of $\alpha(x)$. For instance, in Example 12 of Subsection 1.6.3, the confidence rule (there called $\delta(x)$) was

$$C(x)=\begin{cases}\text{the point } \frac{1}{2}(x_1+x_2) & \text{if } x_1 \neq x_2,\\ \text{the point } x_1-1 & \text{if } x_1=x_2.\end{cases}$$

Two possible "confidences" that were discussed for $C(x)$ were $\alpha_1(x)\equiv 0.75$ (the usual frequentist confidence) and

$$\alpha_2(x)=\begin{cases}1 & \text{if } x_1\neq x_2,\\ 0.5 & \text{if } x_1=x_2\end{cases}$$

(the confidence suggested by almost any form of conditional reasoning).

A decision-theoretic approach to deciding between these two possible inferences would be based on considering some loss function which seems reasonable in light of the purpose of the inference. The purpose of reporting $\alpha(x)$ is (realistically) to give a feeling of the "confidence" that can be placed in $C(x)$, and one way of measuring how well $\alpha(x)$ performs would be to consider a loss function such as

$$L_C(\theta, \alpha(x))=(I_{C(x)}(\theta)-\alpha(x))^2,$$

where again $I_C(\theta)=1$ if $\theta\in C$ and equals zero otherwise. (We index L by C because we are treating C as given; obviously one could also consider choice of C, and develop a more general loss involving this choice.) L_C is a very sensible measure of the performance of C, in that, when θ is in $C(x)$

we would like $\alpha(x)$ to be near 1 (and then $L_C(\theta, \alpha(x))$ would be near zero), while if θ is not in $C(x)$ we would like $\alpha(x)$ to be near zero (in which case L_C would again be small).

One can now treat choice of $\alpha(x)$ as a decision theory problem, employing the conditional Bayes principle or any of the various frequentist principles. The conditional Bayes principle would, in Example 12, yield $\alpha_2(x)$ as the optimal action (if a noninformative prior were used, see Sections 3.3 and 4.3). And, if frequentist risk were the criterion, α_2 would uniformly dominate α_1. Indeed

$$R_C(\theta, \alpha_1) = E_\theta^X L_C(\theta, \alpha_1(X)) = \tfrac{3}{16},$$

while

$$R_C(\theta, \alpha_2) = E_\theta^X L_C(\theta, \alpha_2(X)) = \tfrac{1}{8}.$$

Since α_2 can be given *exactly* the same frequentist justification as α_1, in terms of the long run average "reported confidence" equalling the long run average "actual coverage" (in repeated use), the overall superiority of α_2 (in terms of accurate communication) is clear.

The "inference loss" L_C can be interpreted as the quadratic *scoring rule* of Brier (1950) and deFinetti (1962). Other scoring rules would also serve as useful inference losses in the above example. (A recent reference on scoring rules, from which others can be obtained, is Lindley (1982a).) Further discussion of the situation in this example can be found in Robinson (1979a, b) and Berger (1984b).

Other useful inference losses include information measures such as entropy (see Section 3.4). Recent references on the use of such are Bernardo (1979a) and Good (1983a). Good has long termed losses such as these "quasiutilities."

One caveat should be stated about the use of decision theory in inference problems. To many statisticians (primarily Bayesians) the goal of inference is to produce a distribution $\pi^*(\theta)$ (typically the posterior distribution, see Section 4.2) that describes the uncertainty in θ after all information has been processed. If the goal is to determine such a π^*, then any loss used would have to reflect this, i.e., be of the form $L(\pi^*, a)$ (and not $L(\theta, a)$). Of course, it is still quite possible to use information measures for $L(\pi^*, a)$. A few of the references to this approach are Stein (1965), Bernardo (1979a), Eaton (1982), and Gatsonis (1984).

2.4.4. For Predictive Problems

Predictive problems in statistics are those in which the unknown quantities of interest are future random variables. A typical situation involves the prediction of a random variable Z, having density $g(z|\theta)$ (θ unknown), when there is available data, x, arising from a density $f(x|\theta)$. (Densities are merely considered for convenience.) For instance, x could be the data from a regression study, and it could be desired to predict a future response variable, Z, in the regression setup.

Predictive *inference* raises an interesting set of fundamental questions for classical statistics (cf. Aitchison and Dunsmore (1975) and Hinkley (1979)). Predictive problems pose no fundamental difficulties for decision theory, however (although they may complicate the analysis). To see this, suppose one has a loss $L^*(z, a)$ involving the prediction of Z, and assume that L^* was developed via utility theory. Then, as mentioned at the beginning of Subsection 2.4.1, it would suffice to consider the loss (assuming independence of Z and X, for simplicity)

$$L(\theta, a) = E_\theta^Z L^*(Z, a) = \int L^*(z, a) g(z|\theta) dz,$$

which reduces the decision problem to the standard one involving the unknown θ with data x.

EXAMPLE 5. Suppose Z is $\mathcal{N}(\theta, \sigma^2)$, and that it is desired to estimate Z under the squared-error loss $L^*(z, a) = (z - a)^2$. Then

$$\begin{aligned} L(\theta, a) &= E_\theta^Z (Z - a)^2 = E_\theta^Z (Z - \theta + \theta - a)^2 \\ &= E_\theta^Z (Z - \theta)^2 + E_\theta^Z (\theta - a)^2 = \sigma^2 + (\theta - a)^2. \end{aligned}$$

Working with $L(\theta, a)$ is equivalent to working with squared-error loss for θ. In terms of choosing an action, therefore, one need only consider the problem of estimating θ under squared-error loss and based on X.

Several comments should be made about the reduction of the prediction problem to one involving θ. First, the reduction does not really depend on independence of Z and X, since $g(z|\theta)$ could be replaced by $g(z|\theta, x)$ in the dependent case. (Although this would lead to a loss function dependent on x, such a generalization poses no conceptual problem to decision theory, see also Section 7.8.) Second, this reduction is not dependent on prior information about θ. (Most approaches to predictive inference are dependent on a prior. Of course, one might, in any case, want to involve prior information in the $L(\theta, a)$ reduced problem.) Finally, L^* could itself depend upon θ, and the reduction would still be valid.

What makes the reduction to $L(\theta, a)$ possible, of course, is the assumption that L^* could be treated as a (negative) utility function, ensuring the validity of averaging over Z. One could, alternatively, work directly with (Z, θ) as the unknown quantity, but no easy simplifications then arise. Also, when the *purpose* of the study is to produce a predictive distribution for Z given the data (see Subsection 4.3.4), then the original loss function would be of the form $L^*(\pi^*, a)$ rather than $L^*(Z, a)$ (see Subsection 2.4.3), and no reduction is possible.

The reduction to $L(\theta, a)$ partially explains the tendency of statistical decision-theorists to concentrate on decision problems involving unknown model parameters (such as θ above). Of course, the tendency is to (perhaps inappropriately) work with convenience losses for θ, rather than with losses derived through the reduction from L^* to L.

2.4.5. Vector Valued Loss Functions

It is common in a number of areas of statistics to simultaneously consider two or more criteria. Confidence sets, for instance, are often evaluated by the vector: (size, coverage of θ). Sequential analysis (see Chapter 7) often deals with the vector: (sample size taken, decision loss). Gupta and Panchapakesan (1979) discuss a large number of such vector loss criteria in ranking and selection problems. And even in "pure" decision problems, it may be tempting to keep a vector loss for consequences that are very hard to compare via utility theory (such as "death of a patient" and "cost of a drug").

We will not pursue vector loss functions here for two reasons. First, if one really must take an action, there is often no recourse but to attempt to reconcile the components of the loss function (through utility theory) into a real valued loss function (cf. Keeney and Raifa (1976).) Sometimes, of course, an action might be optimal for *all* components of the loss, in which case there would be no need for such a reduction. Or, possibly, one action could be clearly better for some component of the loss and not much worse for other components, making the choice of an action clear. Indeed, for this reason, it may be best in practice to carry along separate loss components which are hard to compare, performing the difficult utility comparison only if needed. (This will be a theme encountered later in the prior specification also: initial subjective inputs may be rather crude, with refinements made only if necessary.)

From a strictly theoretical viewpoint, it is unnecessary to consider vector valued loss functions because of a simple device discussed and developed in Cohen and Sackrowitz (1984) (see also Lehmann (1959) and Subsection 4.11.2). The device is to consider the *index*, i, of the components of the loss $L=(L_1, \ldots, L_I)$, as a *parameter*, so that one could define a new "loss"

$$L^*((\theta, i), a)=L_i(\theta, a).$$

The loss L^* is real valued, and is a simple one-to-one map of the vector valued loss. Risk dominance with respect to L^* (dominance now for all θ and i) can be easily shown to correspond to (vector) risk dominance for the vector valued loss. Hence many general decision-theoretic results, that were established for real valued losses, can be immediately transferred to vector valued losses. (See Cohen and Sackrowitz (1984) for examples.) Also, Bayesian analysis with L^* corresponds to Bayesian analysis with respect to L, in that (in either case) one must choose the "weights" for the indices, i. (These weights, however, are not the prior probabilities of the various i, but are instead utility theory weights.)

2.5. Criticisms

The use of loss functions is often criticized for (i) being inappropriate for inference problems, (ii) being difficult, and (iii) being nonrobust. The first criticism we have discussed in Section 1.2 and Subsection 2.4.3; decision-theorists can accept the criticism and limit their domain, or decision-theorists can argue any position from the extreme "inference is nonexistent upon close enough examination" to the moderate "better inference can often be done with the aid of decision-theoretic machinery and inference losses." We lean towards the moderate position here.

The second criticism is certainly true: utility analyses and the resulting calculations can be very hard. The amount of time that one spends on a utility analysis will, of course, depend on the importance of the problem, and also on the degree of refinement in utilities that is needed for a conclusion. (If the answer is clear with only very crude utility specifications, there is obviously no reason for a detailed utility construction.) If the losses or utilities do significantly affect the conclusion, however, what choice is there but to perform a suitably refined utility analysis?

The third criticism, that of robustness, raises an issue that will be repeatedly encountered. By *robustness*, we mean the sensitivity of the decision to assumptions (in the analysis) that are uncertain (to paraphrase Herman Rubin (personal communication) and Huber (1981)). Any loss or utility used in an analysis will be uncertain to some degree. (The processes for constructing utilities theoretically require an infinite amount of time and an infinitely accurate mind to obtain a completely accurate specification.) The concern is that some erroneously specified feature of the loss function actually used could cause a bad action to be selected, an action that a less mathematically formal approach would have avoided.

There is no easy answer to the robustness problem. The best solution (if it can be carried out) is to show that reasonable variations in the loss function do not markedly affect the conclusions reached. And, in a sense, this is the *only* possible solution, since if such a robustness study (also called a *sensitivity analysis*) shows that small changes in the loss function do significantly affect the decision, then how can one hope to come to the correct conclusion except by even *more careful* consideration of the loss. We will defer further discussion of robustness with respect to the loss until Subsection 4.7.11.

Much of the criticism of decision theory is aimed at the all-too-frequent use of standard losses, such as those in Subsection 2.4.2, with little or no regard for the actual loss in the problem. (Of course, it is hard to see how such criticism supports completely ignoring the loss.) Also, there is undeniably a need to be concerned with the possible misuse of decision theory, especially since, through a biased choice of prior and loss, a statistician can reach any conclusion he desires! (Just make the other conclusions too costly, or too unlikely to be correct according to the prior information.) In the final analysis, however, we cannot believe that careful consideration of consequences and losses could be a detriment to statistical analysis.

Exercises

1. Determine your utilities for the grade you will receive in this course. (A, B, C, D, and F are the possibilities.) Include at least one check for consistency.

2. Let $\mathscr{R}$ consist of three elements, r_1, r_2, and r_3. Assume that $r_3 < r_2 < r_1$. The utility function is $U(r_3) = 0$, $U(r_2) = u$, and $U(r_1) = 1$, where $0 < u < 1$.
   (a) If $P = (p_1, p_2, p_3)$ and $Q = (q_1, q_2, q_3)$ are two elements of $\mathscr{P}$ (i.e., probability distributions on $\mathscr{R}$), state a numerical condition (in terms of the p_i, q_j, and u) under which $P < Q$.
   (b) Assume $(0.3, 0.3, 0.4) < (0.5, 0, 0.5)$. What can you say about the relationship between $(0.2, 0.5, 0.3)$ and $(0.4, 0.2, 0.4)$? What can you say about u?

3. Sketch your personal utility function for changes in monetary fortune over the interval from \$−1000 to \$1000.

4. Without referring to your utility function as sketched in Exercise 3, decide which gamble you prefer in each of the following pairs:
   (a) $\frac{1}{4}\langle 250\rangle + \frac{3}{4}\langle 0\rangle$ or $\frac{1}{2}\langle 40\rangle + \frac{1}{2}\langle 70\rangle$;
   (b) $\frac{1}{2}\langle 400\rangle + \frac{1}{2}\langle -100\rangle$ or $\frac{2}{3}\langle 150\rangle + \frac{1}{3}\langle 0\rangle$;
   (c) $\frac{1}{2}\langle 1000\rangle + \frac{1}{2}\langle -1000\rangle$ or $\frac{1}{2}\langle 50\rangle + \frac{1}{2}\langle -50\rangle$.
   Now find which gamble would be preferred in parts (a) to (c), as determined by your utility function constructed in Exercise 3. If there is a discrepancy, revise your sketch.

5. Try to find constants b and c so that, for $0 \le r \le 1000$.
   $$U(r) = b \log(cr + 1)$$
   matches the utility function you constructed in Exercise 3.

6. Mr. Rubin has determined that his utility function for a change in fortune on the interval $-100 \le r \le 500$ is
   $$U(r) = (0.62) \log[(0.004)r + 1].$$
   (a) He is offered a choice between \$100 and the chance to participate in a gamble wherein he wins \$0 with probability $\frac{2}{3}$ and \$500 with probability $\frac{1}{3}$. Which should he choose?
   (b) Suppose he is offered instead a chance to *pay* \$100 to participate in the gamble. Should he do it?

7. A person is given a stake of $m > 0$ dollars, which he can allocate between an event A of fixed probability $\alpha (0 < \alpha < 1)$ and its complement A^c. Let $x (0 \le x \le m)$ be the amount he allocates to A, so that $(m - x)$ is the amount allocated to A^c. The person's reward is the amount he has allocated to either A or A^c, whichever actually occurs. Thus he can choose among all gambles of the form $[\alpha\langle x\rangle + (1-\alpha)\langle m - x\rangle]$. Being careful to consider every possible pair of values of α and m, find the optimal allocation of the m dollars when the person's utility function U is defined on the interval $[0, m]$ of monetary gains as follows:
   (a) $U(r) = r^\beta$, where $\beta > 1$.
   (b) $U(r) = r$.
   (c) $U(r) = r^\beta$, where $0 < \beta < 1$.
   (d) $U(r) = \log(r+1)$.

8. Which of the utility functions in Exercise 7 are concave and which are convex?

9. (DeGroot (1970)) Assume that Mr. A and Mr. B have the same utility function for a change, x, in their fortune, given by $U(x) = x^{1/3}$. Suppose now that one of the two men receives, as a gift, a lottery ticket which yields either a reward of r dollars $(r > 0)$ or a reward of 0 dollars, with probability $\frac{1}{2}$ each. Show that there exists a number $b > 0$ having the following property: Regardless of which man receives the lottery ticket, he can sell it to the other man for b dollars and the sale will be advantageous to both men.

10. An investor has \$1000 to invest in speculative stocks. He is considering investing m dollars in stock A and $(1000 - m)$ dollars in stock B. An investment in stock A has a 0.6 chance of doubling in value, and a 0.4 chance of being lost. An investment in stock B has a 0.7 chance of doubling in value, and a 0.3 chance of being lost. The investor's utility function for a change in fortune, x, is $U(x) = \log(0.0007x + 1)$ for $-1000 \le x \le 1000$.
    (a) What is $\mathscr{R}$ (for a fixed m)? (It consists of four elements.)
    (b) What is the optimal value of m in terms of expected utility?
    (*Note*: This perhaps indicates why most investors opt for a diversified portfolio of stocks.)

11. Consider the gambling game described in the St. Petersburg paradox, and assume that the utility function for a change in fortune, x, is given by

$$U(x) = \begin{cases} 100 & \text{if } x > 100, \\ x & \text{if } |x| \le 100, \\ -100 & \text{if } x < -100. \end{cases}$$

    Find the largest amount c that one should be willing to pay to play the game.

12. Consider the gambling game described in the St. Petersburg paradox, and assume that the utility function, $U(x)$, for a change in fortune, x, is bounded, monotonically increasing on R^1, and satisfies $U(0) = 0$. Show that the utility of playing the game is negative for a large enough cost c.

13. For which n in Example 3 is the expected utility positive?

14. An automobile company is about to introduce a new type of car into the market. It must decide how many of these new cars to produce. Let a denote the number of cars decided upon. A market survey will be conducted, with information

obtained pertaining to θ, the proportion of the population which plans on buying a car and would tend to favor the new model. The company has determined that the major outside factor affecting the purchase of automobiles is the state of the economy. Indeed, letting Y denote an appropriate measure of the state of the economy, it is felt that $Z=(1+Y)(10^7)\theta$ cars could be sold. Y is unknown, but is thought to have a $\mathcal{U}(0,1)$ distribution.

Each car produced will be sold at a profit of \$500, unless the supply (a) exceeds the demand (Z). If $a>Z$, the extra $a-Z$ cars can be sold at a loss of \$300 each. The company's utility function for money is linear (say $U(m)=m$) over the range involved. Determine $L(\theta, a)$. (The answer is

$$L(\theta, a)=\begin{cases} -500a & \text{if } \dfrac{a}{10^7}\le\theta, \\[2ex] \dfrac{(4225)(10^7)}{4\theta}\left[\dfrac{8}{13(10^7)}a-\theta\right]^2-\dfrac{(2625)(10^7)\theta}{4} & \text{if } \theta\le\dfrac{a}{10^7}\le 2\theta, \\[2ex] 300a-(1200)(10^7)\theta & \text{if } \dfrac{a}{10^7}\ge 2\theta. \end{cases}$$

Note that in the middle region above, the loss is really a weighted squared-error loss, plus a term involving only θ.)

15. (a) Prove that a decision rule, which is admissible under squared-error loss, is also admissible under a weighted squared-error loss where the weight, $w(\theta)$, is greater than zero for all $\theta\in\Theta$.
    (b) Show that the Bayes action can be different for a weighted squared-error loss, than for squared-error loss.
    (c) Find a weighted squared-error loss for which δ_c in Example 4 of Chapter 1 has a constant risk function. (The weight can depend upon c.)

16. Suppose $\boldsymbol{\theta}=(\theta_1,\ldots,\theta_p)'$ is unknown, and $\boldsymbol{a}=(a_1,\ldots,a_p)'$ has utility $g(\boldsymbol{\theta}-\boldsymbol{a}, Y)$, where all second partial derivatives of g exist and Y is an unknown random variable. For $\boldsymbol{a}$ close to $\boldsymbol{\theta}$, heuristically show that the implied loss for decision making can be considered to be a shifted quadratic loss of the form

$$L(\boldsymbol{\theta}, \boldsymbol{a})\cong(\boldsymbol{\theta}-\boldsymbol{a}+\mathbf{c})'\mathbf{Q}(\boldsymbol{\theta}-\boldsymbol{a}+\mathbf{c}).$$

17. Let θ be the (unknown) total number of cars that will be sold in a given year, and let a be the total number that will be produced. Each car that is sold results in a profit of \$500, while each car that is not sold results in a loss of \$1000. Assuming a linear utility, show that the *regret loss* for the problem (see Subsection 2.4.1) is a linear loss.

18. A chemical in an industrial process must be at least 99% pure, or the process will produce a faulty product. Letting θ denote the percent purity of the chemical, the loss in running the process with $\theta<99$ is \$1000 $(99-\theta)$. (There is zero loss in running with $\theta\ge 99$.) Before use, a test of H_0: $\theta\ge 99$ versus H_1: $\theta<99$ is to be conducted, the outcome of which determines whether the chemical will be used or rejected. If the chemical is rejected, it will be returned to the manufacturer, who will exhaustively test it; if, indeed, $\theta<99$, there is no penalty, but, if $\theta\ge 99$, a penalty of \$1000 will be assessed for incorrectly rejecting good chemical. Give the loss for the decision problem (assuming a linear utility).

19. Suppose we restrict consideration to the class of unbiased estimators for θ (i.e., estimators for which $E_\theta \delta(X) = \theta$ for all θ). Give a decision-theoretic formulation of the "minimum variance unbiased estimator" criterion.

20. In Example 4, verify that $R_C(\theta, \alpha_1) \equiv \frac{3}{16}$ and $R_C(\theta, \alpha_2) \equiv \frac{1}{8}$.

21. In Example 4 it is a fact (see Chapter 4) that, after observing x_1 and x_2, the believed distribution, π^*, of θ to a Bayesian (who initially gave each θ probability density $\pi(\theta) > 0$) will be of the following form: If $x_1 \neq x_2$, then π^* gives probability one to the point $\frac{1}{2}(x_1 + x_2)$; if $x_1 = x_2$, then π^* gives probabilities π_1 and $1 - \pi_1$ to the points $x_1 - 1$ and $x_1 + 1$, respectively, where $\pi_1 = \pi(x_1 - 1)/(\pi(x_1 - 1) + \pi(x_1 + 1))$. Show that, for the loss L_c, the inference α_2 always has smaller Bayesian expected loss than does the inference α_1.

22. Consider the regression setup where

$$Y = \mathbf{b}^t \boldsymbol{\theta} + \varepsilon,$$

$\mathbf{b} = (b_1, \ldots, b_p)^t$ being a vector of regressor variables, $\boldsymbol{\theta} = (\theta_1, \ldots, \theta_p)^t$ being a vector of unknown regression coefficients, and ε being a $\mathcal{N}(0, \sigma^2)$ random error (σ^2 known, for simplicity). Some data, $\mathbf{X}$, is available to estimate $\boldsymbol{\theta}$; let $\boldsymbol{\delta}(\mathbf{x})$ denote the estimator. The goal of the investigation, however, is to predict future (independent) values of Y arising from this model. Indeed, such Y will be predicted (for each corresponding $\mathbf{b}$) by

$$\hat{Y} = \mathbf{b}^t \boldsymbol{\delta}(\mathbf{X}),$$

and the loss in estimating Y by $\hat{Y}$ is squared-error prediction loss, $(Y - \hat{Y})^2$.

(a) Show, for a given $\mathbf{b}$, that choice of $\boldsymbol{\delta}(\mathbf{x})$ in the prediction problem is equivalent to choice of $\boldsymbol{\delta}(\mathbf{x})$ in the problem of estimating $\boldsymbol{\theta}$ under loss

$$L(\boldsymbol{\theta}, \boldsymbol{\delta}) = (\boldsymbol{\theta} - \boldsymbol{\delta})^t \mathbf{b} \mathbf{b}^t (\boldsymbol{\theta} - \boldsymbol{\delta}).$$

(b) Suppose $\mathbf{b} \sim \mathcal{N}_p(\mathbf{0}, \mathbf{Q})$ (independent of future Y and past $\mathbf{X}$). Show that the prediction problem is equivalent to the problem of estimating $\boldsymbol{\theta}$ under loss

$$L(\boldsymbol{\theta}, \boldsymbol{\delta}) = (\boldsymbol{\theta} - \boldsymbol{\delta})^t \mathbf{Q} (\boldsymbol{\theta} - \boldsymbol{\delta}).$$

CHAPTER 3
Prior Information and Subjective Probability

As mentioned in Chapter 1, an important element of many decision problems is the prior information concerning θ. It was stated that a convenient way to quantify such information is in terms of a probability distribution on Θ. In this chapter, methods and problems involved in the construction of such probability distributions will be discussed.

3.1. Subjective Probability

The first point that must be discussed is the meaning of probabilities concerning events (subsets) in Θ. The classical concept of probability involves a long sequence of repetitions of a given situation. For example, saying that a fair coin has probability $\frac{1}{2}$ of coming up heads, when flipped, means that, in a long series of independent flips of the coin, heads will occur about $\frac{1}{2}$ of the time. Unfortunately, this frequency concept won't suffice when dealing with probabilities about θ. For example, consider the problem of trying to determine θ, the proportion of smokers in the United States. What meaning does the statement $P(0.3 < \theta < 0.35) = 0.5$ have? Here θ is simply some number we happen not to know. Clearly it is either in the interval (0.3, 0.35) or it is not. There is nothing random about it. As a second example, let θ denote the unemployment rate for next year. It is somewhat easier here to think of θ as random, since the future is uncertain, but how can $P(3\% < \theta < 4\%)$ be interpreted in terms of a sequence of identical situations? The unemployment situation next year will be a unique, one-time event.

The theory of subjective probability has been created to enable one to talk about probabilities when the frequency viewpoint does not apply. (Some even argue that the frequency concept *never* applies, it being impossible to have an infinite sequence of i.i.d. repetitions of any situation, except in a certain imaginary (subjective) sense.) The main idea of subjective probability is to let the probability of an event reflect the personal belief in the "chance" of the occurrence of the event. For example, you may have a personal feeling as to the chance that θ (in the unemployment example) will be between 3% and 4%, even though no frequency probability can be assigned to the event. There is, of course, nothing terribly surprising about this. It is common to think in terms of personal probabilities all the time; when betting on the outcome of a football game, when evaluating the chance of rain tomorrow, and in many other situations.

The calculation of a frequency probability is theoretically straightforward. One simply determines the relative frequency of the event of interest. A subjective probability, however, is typically determined by introspection. It is worthwhile to briefly discuss techniques for doing this.

The simplest way of determining subjective probabilities is to compare events, determining relative likelihoods. Say, for example, that it is desired to find $P(E)$. Simply compare E with, say, E^c (the complement of E). If E is felt to be twice as likely to occur as E^c, then clearly $P(E)=\frac{2}{3}$ and $P(E^c)=\frac{1}{3}$. This is rather loosely stated, but corresponds, we feel, to the intuitive manner in which people do think about probabilities. As with utility theory, a formal set of axioms can be constructed under which subjective probabilities can be considered to exist and will behave in the fashion of usual probabilities. (See DeGroot (1970) for one such system, and references to others.) Such systems show that the *complexity* of probability theory is needed to quantify uncertainty (i.e., nothing simpler will do), and at the same time indicate that it is not necessary to go beyond the language of probability.

An alternate characterization of subjective probability can be achieved through consideration of "betting" (and the related use of scoring rules, cf. deFinetti (1972) and Lindley (1982a)). In the betting scenario, one determines $P(E)$ by imagining being involved in a gamble wherein z will be *lost* if E occurs and $(1-z)$ will be gained if E^c occurs, where $0 \le z \le 1$. The idea is then to choose z so that the gamble is "fair" (i.e., has overall utility zero), resulting in the equation

$$0 = \text{Expected Utility of the Gamble} = U(-z)P(E) + U(1-z)(1-P(E)).$$

Solving for $P(E)$ yields

$$P(E) = U(1-z)/[U(1-z) - U(-z)].$$

If z is "small," U is probably approximately linear, so that $P(E) \cong 1-z$. One might object that this betting mechanism is circular, since utility functions were *constructed* by considering probabilistic bets, and now we are trying to determine $P(E)$ from knowledge of the utility function. The dilemma can be resolved by noting that, in the construction of the utility function, *any* available probability mechanism can be used (say, a random number table); there is no need to involve probabilities about E.

The betting and scoring rule scenarios are, we feel, very useful in providing additional evidence for the use of subjective probability in quantifying uncertainty. We do not, however, view them as good *operational* devices for *determining* subjective probabilities. They can be of value in ensuring that a probability elicitor is careful or honest (imagining an immediate bet or score will tend to improve concentration and reduce self-delusion), but to us the mechanisms do not seem intuitively accessible. (When involved in betting we have *never* attempted direct intuitive determination of z, but have instead tried to determine z by first calculating $P(E)$—by relative likelihood if it is a subjective problem—and then using the mathematics of probability and utility to process $P(E)$ and determine z.) This type of consideration will be a recurring theme throughout the book. From an operational viewpoint we are very skeptical of grand intuitive leaps (such as one step determination of z above). Instead, we advocate breaking a decision problem into simple components (usually $f(x|\theta)$, L, and π), attempting to separately determine each simple component, and then using the mathematics of decision theory to combine the components and obtain an answer.

We have barely touched the surface of "the meaning of probability." Far deeper discussions, along with references to the very long history of the subject, can be found in Jeffreys (1961), deFinetti (1972), Shafer (1981), Barnett (1982), and Good (1983).

Finally, a few words should be said about the practical difficulties of subjective probability elicitation. The first concern is that, especially with nonstatisticians, care must be taken to ensure "consistency." For instance, it would not be unusual to have a nonstatistician conclude that $P(A)=\frac{1}{3}$, $P(B)=\frac{1}{3}$, and $P(A \text{ or } B \text{ or both})=\frac{3}{4}$, an irrational conclusion to someone trained in probability. A second point (that we will be very concerned with later) is that *everyone* has a great deal of trouble in accurately specifying small probabilities. These and many other difficulties that people experience in probability elicitation are extensively discussed in Kahneman, Slovic, and Tversky (1982), which is highly recommended for its coverage and readability. A scattering of other references to probability elicitation are Winkler (1967a, 1967b, 1972), Savage (1971), Hogarth (1975), and Kadane, Dickey, Winkler, Smith, and Peters (1980).

In the remainder of the chapter we discuss methods of determining a prior density, π, for θ. We will *not* always attempt to discriminate between situations where θ can be considered random in a frequentist sense (such as in the lot inspection example where a sequence of lots with randomly varying proportion of defectives, θ, will be received), and situations where θ can be considered to be random only in a subjective sense. Whether or not the difference is important philosophically, we do not feel it should make any difference operationally (except that the choice of π is often less certain when θ is merely subjectively random, and hence robustness becomes more of a concern).

3.2. Subjective Determination of the Prior Density

Let us first consider the situation in which $\pi(\theta)$ is to be determined wholly subjectively. If Θ is discrete, the problem is simply one of determining the subjective probability of each element of Θ. The techniques in the preceding section should prove adequate for this. When Θ is continuous, the problem of constructing $\pi(\theta)$ is considerably more difficult. Several useful techniques will be discussed.

I. *The Histogram Approach*

When Θ is an interval of the real line, the most obvious approach to use is the histogram approach. Divide Θ into intervals, determine the subjective probability of each interval, and then plot a probability histogram. From this histogram, a smooth density $\pi(\theta)$ can be sketched.

There is no clearcut rule which establishes how many intervals, what size intervals, etc., should be used in the histogram approach. For some problems, only very crude histograms and priors will be needed, while for others, highly detailed versions will be required. The exact needs will be determined by robustness considerations which will be discussed later. Two other difficulties with the histogram approach are that the prior density so obtained is somewhat difficult to work with, and that the prior density has no tails (i.e., gives probability one to a bounded set). The problems caused by this lack of a tail will be indicated later.

II. *The Relative Likelihood Approach*

This approach is also of most use when Θ is a subset of the real line. It consists simply of comparing the intuitive "likelihoods" of various points in Θ, and directly sketching a prior density from these determinations.

Example 1. Assume $\Theta=[0,1]$. It is usually a good idea to begin by determining the relative likelihoods of the "most likely" and "least likely" parameter points. Suppose that the parameter point $\theta=\frac{3}{4}$ is felt to be the most likely, while $\theta=0$ is the least likely. Also, $\frac{3}{4}$ is estimated to be three times as likely to be the true value of θ as is 0. It is deemed sufficient (for an accurate sketch) to determine the relative likelihoods of three other points, $\frac{1}{4}$, $\frac{1}{2}$, and 1. For simplicity, all points are compared with $\theta=0$. It is decided that $\theta=\frac{1}{2}$ and $\theta=1$ are twice as likely as $\theta=0$, while $\theta=\frac{1}{4}$ is 1.5 times as likely as $\theta=0$. Assign the base point $\theta=0$ the prior density value 1.

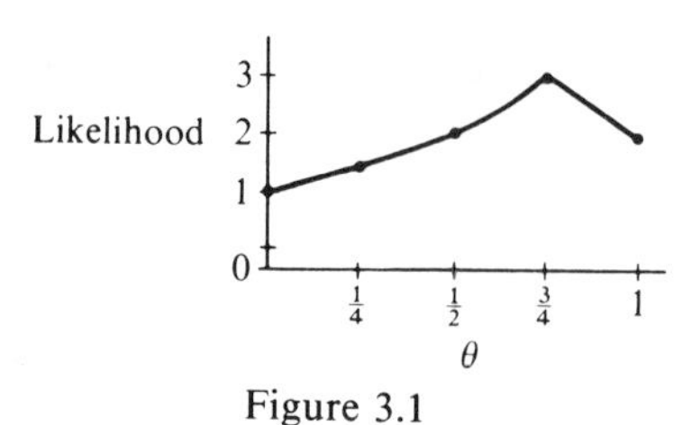

Figure 3.1

Figure 3.1 shows the resulting (unnormalized) prior density. More points could, of course, be included if a more accurate sketch were desired. Note that, as always in such determinations, it is a good idea to check for consistency by comparing other pairs of points. For example, do $\theta = \frac{1}{2}$ and $\theta = 1$ really have the same (subjective) likelihood?

The prior density found in the previous example is not proper, in the sense that it does not integrate to one. A constant c could be found for which $c\pi(\theta)$ is a proper density, but fortunately there is no need to do so. The reason is that the Bayes action is the same whether $\pi(\theta)$ or $c\pi(\theta)$ is used as the prior density. (This is clear since

$$\rho(c\pi, a) = \int_{\Theta} L(\theta, a) c\pi(\theta) d\theta = c\rho(\pi, a),$$

so that any a minimizing $\rho(c\pi, a)$ will also minimize $\rho(\pi, a)$.)

A difficulty is encountered when the relative likelihood approach is used with unbounded Θ. Then, since the relative likelihood determinations can only be done in a finite region, one must decide what to do outside this region. Two particular problems arise. The first is that of determining the shape of the density outside the finite region. (For example, should it be decreasing like θ^{-2} or like $e^{-\theta}$?) Some observations concerning this problem will be presented in the section on Bayesian robustness. The second problem concerns the need to now normalize the density. In particular, it is important that the carefully determined central part of the density have integrated mass correctly proportional to the outside tail of the density. This can be done by simply determining the prior probabilities of the central region and the outer region subjectively, and then making sure that the estimated prior density gives corresponding mass to the two regions.

III. *Matching a Given Functional Form*

This approach is the most used (and misused) approach to determining a prior density. The idea is to simply assume that $\pi(\theta)$ is of a given functional form, and to then choose the density of this given form which most closely matches prior beliefs. In Example 1, for instance, one might assume that

Figure 3.1 shows the resulting (unnormalized) prior density. More points Several ways of choosing such parameters will now be considered.

The easiest way of subjectively determining prior parameters is to calculate them from estimated prior moments. For example, if the prior is assumed to have a $\mathcal{N}(\mu, \sigma^2)$ functional form, one need only decide upon a prior mean and a prior variance to specify the density. Likewise for a $\mathcal{B}e(\alpha, \beta)$ functional form, one can estimate the prior mean, μ, and variance, σ^2, and use the relationships $\mu = \alpha/(\alpha+\beta)$, $\sigma^2 = \alpha\beta/[(\alpha+\beta)^2(\alpha+\beta+1)]$ to determine α and β.

Unfortunately, the estimation of prior moments is often an extremely uncertain undertaking. The difficulty lies in the fact that the tails of a density can have a drastic effect on its moments. For example, if the tail of a density on $(0, \infty)$ behaves like $K\theta^{-2}$, the density has no moments whatsoever (since for any $b > 0$, $\int_b^\infty \theta(K\theta^{-2})d\theta = K\int_b^\infty \theta^{-1}\,d\theta = \infty$). But if K is small, this tail will have almost insignificant probability. Since it is probabilities that can be most reasonably specified subjectively, a tail of small probability cannot realistically be known. Nevertheless its influence on the moments can be great, at least when dealing with unbounded (or very large) parameter spaces. This "simple" method of subjectively determining parameters is thus suspect. (For bounded parameter spaces, such as $\Theta = [0, 1]$, the subjective determination of prior moments is more reasonable, since the tail will then have much less of an effect on the moments.)

A better method of determining prior parameters is to subjectively estimate several fractiles of the prior distribution, and then choose the parameters of the given functional form to obtain a density matching these fractiles as closely as possible. (An α-fractile of a continuous distribution is a point $z(\alpha)$ such that a random variable with this distribution has probability α of being less than or equal to $z(\alpha)$.) Since it is precisely the estimation of probabilities of regions (or equivalently the determination of fractiles) that is easiest to do subjectively, this approach is considerably more attractive than the moment approach. There are available many tables of fractiles of standard densities which facilitate the application of this approach.

Example 2. Assume $\Theta = (-\infty, \infty)$ and that the prior is thought to be from the normal family. It is subjectively determined that the median of the prior is 0, and the quartiles (i.e., $\frac{1}{4}$-fractile and $\frac{3}{4}$-fractile) are -1 and 1. Since, for a normal distribution, the mean and median are equal, it is clear that the desired normal mean is $\mu = 0$. Using tables of normal probabilities, it is also clear that the variance of the normal prior must be $\sigma^2 = 2.19$ (since $P(Z < -1/(2.19)^{1/2}) = \frac{1}{4}$, when Z is $\mathcal{N}(0, 1)$). Hence π will be chosen to be a $\mathcal{N}(0, 2.19)$ density.

If, alternatively, it is assumed that the prior is Cauchy, one finds that the $\mathcal{C}(0, 1)$ density is the appropriate choice. (The median is zero, and it can be checked that $\int_{-\infty}^{-1} (1/\pi[1+\theta^2])d\theta = \frac{1}{4}$.)

From the above example, two interesting observations can be made. First, for a given assumed functional form, only a small number of fractiles need typically be found to determine the specific choice of the prior. Although this makes the analysis quite simple, it is troublesome, since what should be done if other fractiles don't agree with the implied choice? The answer is obvious—if there is unavoidable significant disagreement between the densities of a certain functional form and subjectively chosen fractiles of the prior, that functional form should be discarded as an appropriate model for the prior. Often, this can best be seen by sketching (subjectively) the prior density using methods I or II, and then seeing if a reasonable graphical match can be obtained using the density of the given functional form. (The question unanswered here is what constitutes a "reasonable match" or "significant disagreement." More will be said about this later, but the unfortunate fact is that there is no definitive answer. Basically a "reasonable match" is one in which the differences between the two densities have little effect on the final Bayes rule obtained. Whether or not this is true for a given problem can often be determined only by specific calculation of each of the implied Bayes rules.)

The second observation that can be made from Example 2 is that considerably different functional forms can often be chosen for the prior density. The obvious question is whether or not the choice of functional form is important. It will be seen that the answer is often yes, and certain functional forms will be advocated as generally superior.

One other technique for determining the parameters of a given functional form deserves mention. This technique goes by several names, among them the "technique of equivalent sample size" and the "device of imaginary results." (See Good (1950).) The technique is best understood through an example.

Example 3. Assume a sample $X_1, \ldots, X_n$ from a $\mathcal{N}(\theta, 1)$ distribution is observed. The sample mean, $\bar{X}$, has a $\mathcal{N}(\theta, 1/n)$ distribution. Suppose it is felt that a normal prior density is appropriate. It is quite easy to guess the mean μ of a normal prior, but determination of the prior variance, σ^2, can be difficult. In the next chapter it will be seen that the Bayesian approach results in the combining of the sample information and prior information into what is called the "posterior distribution" for θ. In the situation of this example, the mean of the posterior distribution (and, in a sense, the best "guess" for θ) is

$$\bar{x}\left(\frac{\sigma^2}{\sigma^2+1/n}\right)+\mu\left(\frac{1/n}{\sigma^2+1/n}\right).$$

This suggests that the prior variance, σ^2, plays the same role as $1/n$ in the determination of θ. Hence the idea of equivalent sample size is to determine an n^* such that a sample of that size would make $\bar{x}$ as convincing an estimate of θ as the subjective guess μ. Then $\sigma^2 = 1/n^*$ would be an appropriate prior variance.

This approach, though interesting conceptually, has two major drawbacks. First, it is useful only when certain specific (and often unsuitable) functional forms of the prior are assumed. Second, people who are not extremely well trained statistically do not have good judgement concerning the evidence conveyed by a sample of size n. Indeed people tend to considerably underestimate the amount of information carried by a sample of size n. (See Edwards (1968) and Hogarth (1975) for indications of this.)

IV. *CDF Determination*

Another technique for determining the prior is through subjective construction of the CDF (cumulative distribution function). This can be done by subjectively determining several α-fractiles, $z(\alpha)$, plotting the points $(z(\alpha), \alpha)$, and sketching a smooth curve joining them. This can be a reasonably accurate technique, but will not be considered further since, for the most part, we will be working with prior densities.

Of the four approaches discussed above for subjectively determining a prior distribution, approaches I and II should be the most useful. Personally, the author prefers the relative likelihood approach, but some may have a better feeling for the histogram approach.

Approach III is useful in two situations. The first such situation is when a density of a standard functional form can be found which gives a good match to the prior density obtained by approaches I or II. The use of this standard density can then considerably simplify the ensuing work. The second situation is when only very vague prior information is available. One might just as well then use a standard functional form for the prior density, providing the vague prior information can be incorporated into the standard functional form, *and providing the resulting procedure is robust.* (More will be said of this later.)

The discussion has so far been limited to the univariate situation, in which Θ is a subset of R^1. The problems in trying to determine a multivariate prior density can be considerable. The easiest approach is again the use of a given functional form, since then only a few parameters need be determined subjectively. Also easy is the case in which the coordinates, θ_i, of $\boldsymbol{\theta}$ are thought to be independent. The prior $\pi(\boldsymbol{\theta})$ is then just the product of the univariate prior densities of the θ_i, which can be determined as above. When neither of these simplifying assumptions is thought to be realistic, the best way to proceed is by determining conditional and marginal prior densities. For example, if $\pi(\theta_1, \theta_2)$ is sought, determine by the usual univariate techniques $\pi(\theta_2|\theta_1)$ (the conditional density of θ_2 given θ_1) for various values of θ_1, and also determine $\pi_1(\theta_1)$ (the marginal density of θ_1). Since $\pi(\theta_1, \theta_2) = \pi_1(\theta_1)\pi(\theta_2|\theta_1)$, the joint prior density can thus be approximated.

There has been extensive study of the ability of people (experts and nonexperts) to elicit probability distributions. The studies show that untrained (or unpracticed) elicitors do quite poorly, primarily because of substantial overconfidence concerning their prior knowledge (i.e., the elicited distributions are much too tightly concentrated). Discussion and many references can be found in Alpert and Raiffa (1982) and in Kahneman, Slovic, and Tversky (1982). Besides the obvious solution of attempting to better train people in probability elicitation, one can attempt to partially alleviate the problem through study of "Bayesian robustness" (see Section 4.7).

3.3. Noninformative Priors

3.3.1. Introduction

Because of the compelling reasons to perform a conditional analysis and the attractiveness of using Bayesian machinery to do so (see Chapter 4), there have been attempts to use the Bayesian approach even when no (or minimal) prior information is available. What is needed in such situations is a *noninformative prior*, by which is meant a prior which contains no information about θ (or more crudely which "favors" no possible values of θ over others). For example, in testing between two simple hypotheses, the prior which gives probability $\frac{1}{2}$ to each of the hypotheses is clearly noninformative. The following is a more complex example.

EXAMPLE 4. Suppose the parameter of interest is a normal mean θ, so that the parameter space is $\Theta = (-\infty, \infty)$. If a noninformative prior density is desired, it seems reasonable to give equal weight to all possible values of θ. Unfortunately, if $\pi(\theta) = c > 0$ is chosen, then π has infinite mass (i.e., $\int \pi(\theta)d\theta = \infty$) and is not a proper density. Nevertheless, such π can be successfully worked with. The choice of c is unimportant, so that typically the noninformative prior density for this problem is chosen to be $\pi(\theta) = 1$. This is often called the *uniform density on* R^1, and was introduced and used by Laplace (1812).

As in the above example, it will frequently happen that the natural noninformative prior is an *improper prior*, namely one which has infinite mass. Let us now consider the problem of determining noninformative priors.

The simplest situation to consider is when Θ is a finite set, consisting of say n elements. The obvious noninformative prior is to then give each element of Θ probability $1/n$. One might generalize this (and Example 4) to infinite Θ by giving each $\theta \in \Theta$ equal density, arriving at the uniform noninformative prior $\pi(\theta) \equiv c$. Although this was routinely done by Laplace (1812), it came under severe (though unjustified) criticism because of a lack of invariance under transformation (see Jaynes (1983) for discussion).

Example 4 (continued). Instead of considering θ, suppose the problem had been parameterized in terms of $\eta = \exp\{\theta\}$. This is a one-to-one transformation, and so should have no bearing on the ultimate answer. But, if $\pi(\theta)$ is the density for θ, then the corresponding density for η is (noting that $d\theta/d\eta = d \log \eta/d\eta = \eta^{-1}$ gives the Jacobian)

$$\pi^*(\eta) = \eta^{-1}\pi(\log \eta).$$

Hence, if the noninformative prior for θ is chosen to be constant, we should choose the noninformative prior for η to be proportional to η^{-1} to maintain consistency (and arrive at the same answers in either parameterization). Thus we cannot maintain consistency and choose both the noninformative prior for θ and that for η to be constant.

It could, perhaps, be argued that one usually chooses the most intuitively reasonable parameterization, and that a lack of prior information should correspond to a constant density in this parameterization, but the argument would be hard to defend in general. The lack of invariance of the constant prior has led to a search for noninformative priors which *are* appropriately invariant under transformations. Before discussing the general case in Subsections 3.3.3 and 3.3.4, we present two important examples in Subsection 3.3.2.

3.3.2. Noninformative Priors for Location and Scale Problems

Efforts to derive noninformative priors through consideration of transformations of a problem had its beginnings with Jeffreys (cf. Jeffreys (1961)). It has been extensively used in Hartigan (1964), Jaynes (1968, 1983), Villegas (1977, 1981, 1984), and elsewhere. We present here two illustrations of the idea.

Example 5 (Location Parameters). Suppose that $\mathscr{X}$ and Θ are subsets of R^p, and that the density of $\mathbf{X}$ is of the form $f(\mathbf{x}-\boldsymbol{\theta})$ (i.e., depends only on $(\mathbf{x}-\boldsymbol{\theta})$). The density is then said to be a *location density*, and $\boldsymbol{\theta}$ is called a *location parameter* (or sometimes a *location vector* when $p \geq 2$). The $\mathcal{N}(\theta, \sigma^2)$ (σ^2 fixed), $\mathcal{T}(\alpha, \mu, \sigma^2)$ (α and σ^2 fixed), $\mathscr{C}(\alpha, \beta)$ (β fixed), and $\mathcal{N}_p(\boldsymbol{\theta}, \boldsymbol{\Sigma})$ ($\boldsymbol{\Sigma}$ fixed) densities are all examples of location densities. Also, a sample of independent identically distributed random variables is said to be from a location density if their common density is a location density.

To derive a noninformative prior for this situation, imagine that, instead of observing $\mathbf{X}$, we observe the random variable $\mathbf{Y}=\mathbf{X}+\mathbf{c}$ ($\mathbf{c}\in R^p$). Defining $\boldsymbol{\eta}=\boldsymbol{\theta}+\mathbf{c}$, it is clear that $\mathbf{Y}$ has density $f(\mathbf{y}-\boldsymbol{\eta})$. If now $\mathscr{X}=\Theta=R^p$, then the sample space and parameter space for the $(\mathbf{Y},\boldsymbol{\eta})$ problem are also R^p. The $(\mathbf{X},\boldsymbol{\theta})$ and $(\mathbf{Y},\boldsymbol{\eta})$ problems are thus identical in structure, and it seems reasonable to insist that they have the same noninformative prior. (Another way of thinking of this is to note that observing $\mathbf{Y}$ really amounts to observing $\mathbf{X}$ with a different unit of measurement, one in which the "origin" is $\mathbf{c}$ and not zero. Since the choice of an origin for a unit of measurement is quite arbitrary, the noninformative prior should perhaps be independent of this choice.)

Letting π and π^* denote the noninformative priors in the $(\mathbf{X},\boldsymbol{\theta})$ and $(\mathbf{Y},\boldsymbol{\eta})$ problems respectively, the above argument implies that π and π^* should be equal, i.e., that

$$P^{\pi}(\boldsymbol{\theta}\in A)=P^{\pi^*}(\boldsymbol{\eta}\in A) \tag{3.1}$$

for any set A in R^p. Since $\boldsymbol{\eta}=\boldsymbol{\theta}+\mathbf{c}$, it should also be true (by a simple change of variables) that

$$P^{\pi^*}(\boldsymbol{\eta}\in A)=P^{\pi}(\boldsymbol{\theta}+\mathbf{c}\in A)=P^{\pi}(\boldsymbol{\theta}\in A-\mathbf{c}), \tag{3.2}$$

where $A-\mathbf{c}=\{\mathbf{z}-\mathbf{c}\colon \mathbf{z}\in A\}$. Combining (3.1) and (3.2) shows that π should satisfy

$$P^{\pi}(\boldsymbol{\theta}\in A)=P^{\pi}(\boldsymbol{\theta}\in A-\mathbf{c}). \tag{3.3}$$

Furthermore, this argument applies no matter which $\mathbf{c}\in R^p$ is chosen, so that (3.3) should hold for all $\mathbf{c}\in R^p$. Any π satisfying this relationship is said to be a *location invariant* prior.

Assuming that the prior has a density, we can write (3.3) as

$$\int_A \pi(\boldsymbol{\theta})d\boldsymbol{\theta}=\int_{A-c}\pi(\boldsymbol{\theta})d\boldsymbol{\theta}=\int_A \pi(\boldsymbol{\theta}-\mathbf{c})d\boldsymbol{\theta}.$$

If this is to hold for all sets A, it can be shown that it must be true that

$$\pi(\boldsymbol{\theta})=\pi(\boldsymbol{\theta}-\mathbf{c})$$

for all $\boldsymbol{\theta}$. Setting $\boldsymbol{\theta}=\mathbf{c}$ thus gives

$$\pi(\mathbf{c})=\pi(\mathbf{0}).$$

Recall, however, that this should hold for all $\mathbf{c}\in R^p$. The conclusion is that π must be a constant function. It is convenient to choose the constant to be 1, so the *noninformative prior density for a location parameter* is $\pi(\boldsymbol{\theta})=1$. (This conclusion can be shown to follow from (3.3), even without the assumption that the prior has a density.)

Example 6 (Scale Parameters). A (one-dimensional) *scale density* is a density of the form

$$\sigma^{-1} f\left(\frac{x}{\sigma}\right),$$

where $\sigma > 0$. The parameter σ is called a *scale parameter.* The $\mathcal{N}(0, \sigma^2)$, $\mathcal{T}(\alpha, 0, \sigma^2)$ (α fixed), and $\mathcal{G}(\alpha, \beta)$ (α fixed) densities are all examples of scale densities. Also, a sample of independent identically distributed random variables is said to be from a scale density if their common density is a scale density.

To derive a noninformative prior for this situation, imagine that, instead of observing X, we observe the random variable $Y = cX$ ($c > 0$). Defining $\eta = c\sigma$, an easy calculation shows that the density of Y is $\eta^{-1} f(y/\eta)$. If now $\mathscr{X} = R^1$ or $\mathscr{X} = (0, \infty)$, then the sample and parameter spaces for the (X, σ) problem are the same as those for the (Y, η) problem. The two problems are thus identical in structure, which again indicates that they should have the same noninformative prior. (Here the transformation can be thought of as simply a change in the scale of measurement, from say inches to feet.) Letting π and π^* denote the priors in the (X, σ) and (Y, η) problems, respectively, this means that the equality

$$P^{\pi}(\sigma \in A) = P^{\pi^*}(\eta \in A)$$

should hold for all $A \subset (0, \infty)$. Since $\eta = c\sigma$, it should also be true that

$$P^{\pi^*}(\eta \in A) = P^{\pi}(\sigma \in c^{-1}A),$$

where $c^{-1}A = \{c^{-1}z\colon z \in A\}$. Putting these together, it follows that π should satisfy

$$P^{\pi}(\sigma \in A) = P^{\pi}(\sigma \in c^{-1}A). \tag{3.4}$$

This should hold for all $c > 0$, and any distribution π for which this is true is called *scale invariant.*

The mathematical analysis of (3.4) proceeds as in the preceding example. Write (3.4) (assuming densities) as

$$\int_A \pi(\sigma)\, d\sigma = \int_{c^{-1}A} \pi(\sigma)\, d\sigma = \int_A \pi(c^{-1}\sigma) c^{-1}\, d\sigma,$$

and conclude that, for this to hold for all A, it must be true that

$$\pi(\sigma) = c^{-1}\pi(c^{-1}\sigma)$$

for all σ. Choosing $\sigma = c$, it follows that

$$\pi(c) = c^{-1}\pi(1).$$

Setting $\pi(1) = 1$ for convenience, and noting that the above equality must hold for all $c > 0$, it follows that a reasonable *noninformative prior for a scale parameter* is $\pi(\sigma)=\sigma^{-1}$. Observe that this is also an improper prior, since $\int_0^\infty \sigma^{-1}\,d\sigma=\infty$.

An interesting natural application of this noninformative prior (discussed in Rosenkrantz (1977)) is to the "table entry" problem. This problem arose from the study of positive entries in various "natural" numerical tables, such as tables of positive physical constants, tables of population sizes of cities, etc. The problem is to determine the relative frequencies of the integers 1 through 9 in the *first* significant digit of the table entries. Intuitively, one might expect each digit to occur $\frac{1}{9}$ of the time. Instead, it has been found that i $(i=1,\ldots,9)$ occurs with a relative frequency of about $\log(1+i^{-1})/\log 10$.

Numerous explanations of this phenomenon have been proposed. The explanation of interest to us is that, since the scale of measurement of these positive entries is quite arbitrary, one might expect the distribution of table entries to be scale invariant. This suggests using $\pi(\sigma)=\sigma^{-1}$ to describe the distribution of table entries σ. Since this is not a proper density, it cannot be formally considered the distribution of σ. Nevertheless, one can use it (properly normalized) to represent the actual distribution of σ on any interval (a,b) where $0<a<b<\infty$. For example, consider the interval $(1,10)$. Then the properly normalized version of π is $\pi(\sigma)=\sigma^{-1}/\log 10$. In this region, σ will have first digit i when it lies in the interval $[i,i+1)$. The probability of this is

$$p_i=\int_i^{i+1}[\sigma\log 10]^{-1}\,d\sigma=\frac{\log(i+1)-\log i}{\log 10}=\frac{\log(1+i^{-1})}{\log 10},$$

which is precisely the relative frequency that is observed in reality. One might object to the rather arbitrary choice of the interval $(1,10)$, but the following compelling result can be obtained: if the p_i are calculated for an arbitrary interval (a,b), then as $a\to 0$ or $b\to\infty$ or both, the p_i will converge to the values $\log(1+i^{-1})/\log 10$. This apparent natural occurrence of the noninformative prior may be mere coincidence, but it is intriguing.

The derivations of the noninformative priors in the two previous examples should not be considered completely compelling. There is indeed a logical flaw in the analyses, caused by the fact that the final priors are improper. The difficulty arises in the argument that if two problems have identical structure, they should have the same noninformative prior. The problem here is that, when improper, noninformative priors are not unique. Multiplying an improper prior π by a constant K results in an equivalent prior, in the sense that all decisions and inferences in Bayesian analysis will be identical for the priors π and $K\pi$. Thus there is no reason to insist that π^* and π, in Examples 5 and 6, must be identical. They need only be constant multiples of each other. In Example 5, for instance, this milder restriction will give, in place of (3.1), the relationship

$$P^{\pi}(A)=h(\mathbf{c})P^{\pi^*}(A),$$

where $h(\mathbf{c})$ is some positive function and P is to be interpreted more liberally as a "measure." The analogue of Equation (3.2), namely

$$P^{\pi^*}(A) = P^{\pi}(A - \mathbf{c}),$$

should remain valid, being as it merely specifies that a change of variables should not affect the measure. Combining these equations gives, in place of (3.3), the relationship

$$P^{\pi}(A) = h(\mathbf{c}) P^{\pi}(A - \mathbf{c}).$$

In integral form this becomes

$$\int_A \pi(\boldsymbol{\theta}) d\boldsymbol{\theta} = h(\mathbf{c}) \int_{A-\mathbf{c}} \pi(\boldsymbol{\theta}) d\boldsymbol{\theta} = h(\mathbf{c}) \int_A \pi(\boldsymbol{\theta} - \mathbf{c}) d\boldsymbol{\theta}.$$

For this to hold for all A, it must be true that

$$\pi(\boldsymbol{\theta}) = h(\mathbf{c}) \pi(\boldsymbol{\theta} - \mathbf{c}).$$

Setting $\boldsymbol{\theta} = \mathbf{c}$, it follows that $h(\mathbf{c}) = \pi(\mathbf{c}) / \pi(\mathbf{0})$. The conclusion is that π need only satisfy the functional equation

$$\pi(\boldsymbol{\theta} - \mathbf{c}) = \frac{\pi(\mathbf{0}) \pi(\boldsymbol{\theta})}{\pi(\mathbf{c})}. \tag{3.5}$$

There are many improper priors, besides the uniform, which satisfy this relationship. An example is $\pi(\boldsymbol{\theta}) = \exp\{\boldsymbol{\theta}'\mathbf{z}\}$, where $\mathbf{z}$ is any fixed vector. A prior satisfying (3.5) is called *relatively location invariant.*

The above problem will be encountered in virtually any situation for which improper noninformative priors must be considered. There will be a wide class of "logically possible" noninformative priors. (See Hartigan (1964) and Stein (1965) for general results of this nature.) Selecting from among this class can be difficult. For certain statistical problems, a natural choice does exist, namely the right invariant Haar measure. A general discussion of this concept requires group theory, and will be delayed until Chapter 6. It is a fact, however, that the noninformative priors given in Examples 5 and 6 are the right invariant Haar measures.

3.3.3. Noninformative Priors in General Settings

The type of argument given in Subsection 3.3.2 requires a special "group invariant" structure to the problem (see Chapter 6). For more general problems, various (somewhat *ad hoc*) suggestions have been advanced for determining a noninformative prior. The most widely used method is that of Jeffreys (1961), which is to choose

$$\pi(\theta) = [I(\theta)]^{1/2} \tag{3.6}$$

as a noninformative prior, where $I(\theta)$ is the expected Fisher information;

under commonly satisfied assumptions (cf. Lehmann (1983)) this is given by

$$I(\theta) = -E_\theta\left[\frac{\partial^2 \log f(X|\theta)}{\partial\theta^2}\right].$$

It is easy to calculate that the noninformative priors in Examples 5 and 6 are indeed proportional to (3.6).

If $\boldsymbol{\theta} = (\theta_1, \ldots, \theta_p)^t$ is a vector, Jeffreys (1961) suggests the use of

$$\pi(\theta) = [\det \mathbf{I}(\boldsymbol{\theta})]^{1/2} \tag{3.7}$$

(here "det" stands for determinant), where $\mathbf{I}(\boldsymbol{\theta})$ is the $(p \times p)$ Fisher information matrix; under commonly satisfied assumptions, this is the matrix with (i, j) element

$$I_{ij}(\boldsymbol{\theta}) = -E_{\boldsymbol{\theta}}\left[\frac{\partial^2}{\partial\theta_i\, \partial\theta_j} \log f(X|\boldsymbol{\theta})\right].$$

EXAMPLE 7 (Location-Scale Parameters). A *location-scale* density is a density of the form $\sigma^{-1}f((x-\theta)/\sigma)$, where $\theta \in R^1$ and $\sigma > 0$ are the unknown parameters. The $\mathcal{N}(\theta, \sigma^2)$ and the $\mathcal{T}(\alpha, \theta, \sigma^2)$ (α fixed) densities are the crucial examples of location-scale densities. A sample of independent identically distributed random variables is said to be from a location-scale density if their common density is a location-scale density.

Working with the normal distribution for simplicity, and noting that $\boldsymbol{\theta} = (\theta, \sigma)$ in this problem, the Fisher information matrix is

$$\begin{aligned}
\mathbf{I}(\boldsymbol{\theta}) &= -E_{\boldsymbol{\theta}}\begin{pmatrix} \dfrac{\partial^2}{\partial\theta^2}\left(-\dfrac{(X-\theta)^2}{2\sigma^2}\right) & \dfrac{\partial^2}{\partial\theta\,\partial\sigma}\left(-\dfrac{(X-\theta)^2}{2\sigma^2}\right) \\ \dfrac{\partial^2}{\partial\theta\,\partial\sigma}\left(-\dfrac{(X-\theta)^2}{2\sigma^2}\right) & \dfrac{\partial^2}{\partial\sigma^2}\left(-\dfrac{(X-\theta)^2}{2\sigma^2}\right) \end{pmatrix} \\
&= -E_{\boldsymbol{\theta}}\begin{pmatrix} -1/\sigma^2 & 2(\theta - X)/\sigma^3 \\ 2(\theta - X)/\sigma^3 & -3(X-\theta)^2/\sigma^4 \end{pmatrix} \\
&= \begin{pmatrix} 1/\sigma^2 & 0 \\ 0 & 3/\sigma^2 \end{pmatrix}.
\end{aligned}$$

Hence (3.7) becomes

$$\pi(\boldsymbol{\theta}) = \left(\frac{1}{\sigma^2}\cdot\frac{3}{\sigma^2}\right)^{1/2} \propto \frac{1}{\sigma^2}.$$

(This is improper, so we can again ignore any multiplicative constants.) It is interesting that this prior can also be shown to arise from an invariance-under-transformation argument, as in Subsection 3.3.2.

A noninformative prior for this situation could, alternatively, be derived by assuming *independence* of θ and σ, and multiplying the noninformative priors obtained in Examples 5 and 6. The result is

$$\pi(\theta, \sigma) = \frac{1}{\sigma}; \tag{3.8}$$

this is actually the noninformative prior ultimately recommended by Jeffreys

(1961), and is the right invariant Haar density for the problem (see Chapter 6). It is thus standard to use (3.8) as the noninformative prior for a location-scale problem, and we will do so. (Even Jeffreys felt that (3.7) sometimes gives an inferior noninformative prior.)

One important feature of the Jeffreys noninformative prior is that it is not affected by a restriction on the parameter space. Thus, if it is known in Example 5 that $\theta > 0$, the Jeffreys noninformative prior is still $\pi(\theta) = 1$ (on $(0, \infty)$, of course). This is important, because one of the situations in which noninformative priors prove to be extremely useful is when dealing with restricted parameter spaces (see Chapter 4). In such situations we will, therefore, simply assume that the noninformative prior is that which is inherited from the unrestricted parameter space.

Among the other techniques that have been proposed for determining noninformative priors are those of Novick and Hall (1965), Zellner (1971, 1977), Box and Tiao (1973), Akaike (1978), Bernardo (1979b), and Geisser (1984a). All these approaches seem to work well most of the time. Particularly noteworthy is the approach of Bernardo, in that it seems to work even in examples where the other methods falter. By this we mean that certain examples have been found (see Subsections 4.7.9 and 4.8.2) in which the noninformative priors derived by the other techniques give demonstrably bad answers, while the technique of Bernardo's seems to avoid the pitfalls. Unfortunately, this technique is too involved to consider here.

3.3.4. Discussion

A number of criticisms have been raised concerning the use of noninformative priors. One is that, since methods of deriving noninformative priors depend on the experimental structure (cf. (3.6), (3.7), and Exercise 17), one may violate the Likelihood Principle in using noninformative priors (see also Geisser (1984a)). Another criticism is the "marginalization paradox" of Dawid, Stone, and Zidek (1973) (though see Jaynes (1980) for counter arguments). Other criticisms arise from taking various noninformative priors and attempting to construct examples where they perform poorly (cf. Subsections 4.7.9 and 4.8.2). Perhaps the most embarassing feature of noninformative priors, however, is simply that there are often so many of them.

Example 8. Suppose θ is a binomial parameter. The parameter space is thus $\Theta = [0, 1]$. Four plausible noninformative priors for θ are $\pi_1(\theta) = 1$, $\pi_2(\theta) = \theta^{-1}(1-\theta)^{-1}$, $\pi_3(\theta) \propto [\theta(1-\theta)]^{-1/2}$, and $\pi_4(\theta) \propto \theta^{\theta}(1-\theta)^{(1-\theta)}$. The first is the "natural" uniform density used by Bayes (1763) and Laplace (1812), and supported by Geisser (1984a); the second arises from the approach of Novick and Hall (1965) and the transformation arguments of Jaynes (1968) and Villegas (1977); the third arises from the approaches of Jeffreys (1961), Box and Tiao (1973), Akaike (1978), and Bernardo (1979b);

and the fourth arises from the approach of Zellner (1977). All four possibilities are reasonable; see Geisser (1984a) for discussion. Note that π_1, π_3, and π_4 are proper densities (π_3 and π_4 upon suitable normalization), while π_2 is improper.

There are two common responses to these criticisms of noninformative prior Bayesian analysis. The first response, attempted by some noninformative prior Bayesians, is to argue for the "correctness" of their favorite noninformative prior approach, together with attempts to rebut the "paradoxes" and "counterexamples." The second response is to argue that, operationally, it is rare for the choice of a noninformative prior to markedly affect the answer (unless the noninformative prior blows up in such a way as to cause the answer to blow up, see Section 4.6), so that any reasonable noninformative prior (even Laplace's constant prior) can be used. Indeed, if the choice of noninformative prior does have a pronounced effect on the answer (see Subsection 4.7.9), then one is probably in a situation where it is crucial to involve subjective prior information.

This second response, to which we subscribe, is the view of Bayesians who ultimately believe only in proper prior, subjective Bayesian analysis as a foundation for statistics. Most such Bayesians are very willing to use *ad hoc* approximations to this ideal, however, if they believe that the approximate analysis is likely to yield much the same answer that a complete subjective Bayesian analysis would. (It should be mentioned that some "proper" Bayesians feel that noninformative prior analysis can be fundamentally justified through use of "finitely additive" measures; see Stone (1979), Berger (1984a), and Hill and Lane (1984) for discussion and references.) We would indeed argue that noninformative prior Bayesian analysis is the *single most powerful method of statistical analysis*, in the sense of being the *ad hoc* method most likely to yield a sensible answer for a given investment of effort. And the answers so obtained have the added feature of being, in some sense, the most "objective" statistical answers obtainable (which is attractive to those who feel objectivity is possible). We will see some evidence for these points as we proceed. For more evidence, see Jeffreys (1961), Zellner (1971, 1984b), Bernardo (1979b), Jaynes (1983), and the references therein.

3.4. Maximum Entropy Priors

A situation which lies between those of the two previous subsections often prevails. Frequently partial prior information is available, outside of which it is desired to use a prior that is as noninformative as possible. For example, suppose the prior mean is specified, and among prior distributions with this mean the most noninformative distribution is sought.

A useful method of dealing with this problem is through the concept of entropy (see Jaynes (1968, 1983) and Rosenkrantz (1977)). Entropy is most easily understood for discrete distributions, with which we, therefore, begin.

Definition 1. Assume Θ is discrete, and let π be a probability density on Θ. The *entropy* of π, to be denoted $\mathscr{E}n(\pi)$, is defined as

$$\mathscr{E}n(\pi) = -\sum_{\Theta} \pi(\theta_i)\log \pi(\theta_i).$$

(If $\pi(\theta_i)=0$, the quantity $\pi(\theta_i)\log \pi(\theta_i)$ is defined to be zero.)

Entropy has a direct relationship to information theory, and in a sense measures the amount of uncertainty inherent in the probability distribution. (See Rosenkranz (1977) for an interesting discussion of this.)

Example 9. Assume $\Theta = \{\theta_1, \theta_2, \ldots, \theta_n\}$. If $\pi(\theta_k)=1$, while $\pi(\theta_i)=0$ for $i \neq k$, then clearly the probability distribution describes exactly which parameter point will occur. The "uncertainty" is zero. Correspondingly,

$$\mathscr{E}n(\pi) = -\sum_{i=1}^{n} \pi(\theta_i)\log \pi(\theta_i) = 0.$$

At the other extreme, the "most uncertain" or *maximum entropy* probability distribution is that with $\pi(\theta_i)=1/n$ for all i. For this π,

$$\mathscr{E}n(\pi) = -\sum_{i=1}^{n} \frac{1}{n}\log\left(\frac{1}{n}\right) = \log n.$$

(It can be shown that $\mathscr{E}n(\pi) \leq \log n$ for all proper π.) Note that this maximum entropy distribution is the same as the noninformative prior distribution for a discrete Θ.

Assume now that partial prior information concerning θ is available. It is convenient to consider information in the form of restrictions on the $\pi(\theta_i)$. Indeed assume that

$$E^{\pi}[g_k(\theta)] = \sum_i \pi(\theta_i) g_k(\theta_i) = \mu_k, \qquad k = 1, \ldots, m. \tag{3.9}$$

Two examples should indicate the scope of this type of restriction.

Example 10. Assume $\Theta \subset R^1$, $g_1(\theta)=\theta$, and $g_k(\theta)=(\theta-\mu_1)^k$ for $2 \leq k \leq m$. Then (3.9) corresponds to the specification of the first m central moments, μ_i, of π.

Example 11. Assume $\Theta \subset R^1$, and

$$g_k(\theta) = I_{(-\infty, z_k]}(\theta).$$

Clearly

$$E^{\pi}[g_k(\theta)] = P^{\pi}(\theta \le z_k),$$

so, by (3.9), z_k is the μ_k-fractile of π. Hence this choice of the g_k corresponds to the specification of m fractiles of π.

It seems reasonable to seek the prior distribution which maximizes entropy among all those distributions which satisfy the given set of restrictions (i.e., given prior information). Intuitively this should result in a prior which incorporates the available prior information, but otherwise is as noninformative as possible. The solution to the maximization of $\mathscr{E}n(\pi)$ subject to (3.9) (and, of course, to $\sum \pi(\theta_i) = 1$) is well known, but the derivation is beyond the scope of this book. The solution is given (provided the density defined is proper) by

$$\bar{\pi}(\theta_i) = \frac{\exp\{\sum_{k=1}^{m} \lambda_k g_k(\theta_i)\}}{\sum_i \exp\{\sum_{k=1}^{m} \lambda_k g_k(\theta_i)\}},$$

where the λ_k are constants to be determined from the constraints in (3.9). (The proof of this can be found in many books on the calculus of variations. See, for example, Ewing (1969).)

EXAMPLE 12. Assume $\Theta = \{0, 1, 2, \ldots\}$, and it is thought that $E^{\pi}[\theta] = 5$. This restriction is of the form (3.9) with $g_1(\theta) = \theta$ and $\mu_1 = 5$. The restricted maximum entropy prior is, therefore,

$$\bar{\pi}(\theta) = \frac{e^{\lambda_1 \theta}}{\sum_{\theta=0}^{\infty} e^{\lambda_1 \theta}} = (1 - e^{\lambda_1})(e^{\lambda_1})^{\theta}.$$

This is clearly a $\mathscr{G}e(e^{\lambda_1})$ density, the mean of which (from Appendix 1) is $(1 - e^{\lambda_1})/e^{\lambda_1}$. Setting this equal to $\mu_1 = 5$, and solving, gives $e^{\lambda_1} = \frac{1}{6}$. Hence $\bar{\pi}$ is $\mathscr{G}e(\frac{1}{6})$.

If Θ is continuous, the use of maximum entropy becomes more complicated. The first difficulty is that there is no longer a completely natural definition of entropy. Jaynes (1968) makes a strong case for defining entropy as

$$\mathscr{E}n(\pi) = -E^{\pi}\left[\log \frac{\pi(\theta)}{\pi_0(\theta)}\right] = -\int \pi(\theta) \log\left(\frac{\pi(\theta)}{\pi_0(\theta)}\right) d\theta,$$

where $\pi_0(\theta)$ is the natural "invariant" noninformative prior for the problem. (See Subsection 3.3.) Unfortunately, the difficulties and uncertainties in determining a noninformative prior make this definition somewhat ambiguous. It can still be useful, however.

In the presence of partial prior information of the form

$$E^{\pi}[g_k(\theta)] = \int_{\Theta} g_k(\theta) \pi(\theta) d\theta = \mu_k, \qquad k = 1, \ldots, m, \tag{3.10}$$

the (proper) prior density (satisfying these restrictions) which maximizes $\mathscr{E}n(\pi)$ is given (provided it exists) by

$$\bar{\pi}(\theta)=\frac{\pi_0(\theta)\exp[\sum_{k=1}^{m}\lambda_k g_k(\theta)]}{\int_\Theta \pi_0(\theta)\exp[\sum_{k=1}^{m}\lambda_k g_k(\theta)]d\theta},$$

where the λ_k are constants to be determined from the constraints in (3.10). This is exactly analogous to the discrete case, and can also be derived from arguments involving the calculus of variations.

EXAMPLE 13. Assume $\Theta=R^1$, and that θ is a location parameter. The natural noninformative prior is then $\pi_0(\theta)=1$. It is believed that the true prior mean is μ and the true prior variance is σ^2. These restrictions are of the form (3.10) with $g_1(\theta)=\theta$, $\mu_1=\mu$, $g_2(\theta)=(\theta-\mu)^2$, and $\mu_2=\sigma^2$. The maximum entropy prior, subject to these restrictions, is thus

$$\bar{\pi}(\theta)=\frac{\exp[\lambda_1\theta+\lambda_2(\theta-\mu)^2]}{\int_\Theta \exp[\lambda_1\theta+\lambda_2(\theta-\mu)^2]d\theta},$$

where λ_1 and λ_2 are to be chosen so that (3.10) is satisfied. Clearly

$$\begin{aligned}\lambda_1\theta+\lambda_2(\theta-\mu)^2&=\lambda_2\theta^2+(\lambda_1-2\mu\lambda_2)\theta+\lambda_2\mu^2\\&=\lambda_2\left[\theta-\left(\mu-\frac{\lambda_1}{2\lambda_2}\right)\right]^2+\left[\lambda_1\mu-\frac{\lambda_1^2}{4\lambda_2}\right].\end{aligned}$$

Hence

$$\bar{\pi}(\theta)=\frac{\exp\{\lambda_2[\theta-(\mu-\lambda_1/2\lambda_2)]^2\}}{\int_{-\infty}^{\infty}\exp\{\lambda_2[\theta-(\mu-\lambda_1/2\lambda_2)]^2\}d\theta}.$$

The denominator is a constant, so $\bar{\pi}(\theta)$ is recognizable as a normal density with mean $\mu-\lambda_1/2\lambda_2$ and variance $-1/2\lambda_2$. Choosing $\lambda_1=0$ and $\lambda_2=-1/2\sigma^2$ satisfies (3.10). Thus $\bar{\pi}(\theta)$ is a $\mathcal{N}(\mu,\sigma^2)$ density.

Several difficulties arise in trying to use the maximum entropy approach to determine a prior. The first difficulty has already been mentioned, namely the need (in the continuous case) to use a noninformative prior in the derivation of $\bar{\pi}$. This problem is not too serious, however, as any reasonable noninformative prior will usually give satisfactory results.

A more serious problem is that often $\bar{\pi}$ will not exist. The following example demonstrates this.

EXAMPLE 14. In the situation of Example 13, assume that the only restriction given is $E^{\pi}[\theta]=\mu$. The solution $\bar{\pi}$ must then be of the form

$$\bar{\pi}(\theta)=\frac{\exp\{\lambda_1\theta\}}{\int_{-\infty}^{\infty}\exp\{\lambda_1\theta\}d\theta}.$$

It is clear, however, that $\bar{\pi}$ cannot be a proper density for any λ_1. Hence there is no solution.

The above problem of nonexistence of $\bar{\pi}$ will also occur whenever Θ is unbounded and the specified restrictions are specifications of fractiles of the prior. This is a serious concern, since we have previously argued that subjective knowledge about θ can usually lead to fractile specifications, but *not* to moment specifications. Furthermore, if one does attempt to subjectively specify moments, the resulting maximum entropy prior is likely to be nonrobust. For instance, we will argue in Chapter 4 that the normal $\bar{\pi}$, derived in Example 13, is in many senses nonrobust. The problem here is that specification of a prior variance, σ^2, drastically limits the priors being considered, and in particular eliminates from consideration those with large tails (which will tend to be the robust priors).

Of course, in many physical problems available information *is* in the form of moment restrictions (cf. Jaynes (1983)), in which case maximum entropy is enormously successful. And maximum entropy forms a key element to other broadly applicable techniques for deriving priors: for instance, the technique of Bernardo (1979b) for deriving noninformative priors is based on the maximization of a certain limiting entropy, and Brockett, Charnes, and Paick (1984) consider a modified version of maximum entropy which does allow input of desired prior tails (so that maximum entropy can be done with fractiles). For references to the current explosion of the use of maximum entropy in such fields as spectral analysis, see Jaynes (1983). Other recent references, which also refer to many earlier uses of entropy, are Good (1983) and Zellner and Highfield (1983).

3.5. Using the Marginal Distribution to Determine the Prior

A very important quantity in Bayesian analysis is the marginal distribution of X. After defining and discussing the meaning of the marginal distribution, we will indicate the important role it can play in selection of a prior distribution.

3.5.1. The Marginal Distribution

If X has probability density $f(x|\theta)$, and θ has probability density $\pi(\theta)$, then the *joint* density of X and θ is

$$h(x, \theta) = f(x|\theta)\pi(\theta).$$

We can then define

Definition 2. The *marginal density* of X is

$$m(x|\pi) = \int_\Theta f(x|\theta)dF^\pi(\theta) = \begin{cases} \int_\Theta f(x|\theta)\pi(\theta)d\theta & \text{(continuous case)}, \\ \sum_\Theta f(x|\theta)\pi(\theta) & \text{(discrete case)}. \end{cases} \tag{3.11}$$

Usually we will just write $m(x)$ for $m(x|\pi)$, the dependence on the prior being understood.

EXAMPLE 15. If X (given θ) is $\mathcal{N}(\theta, \sigma_f^2)$ and $\pi(\theta)$ is a $\mathcal{N}(\mu_\pi, \sigma_\pi^2)$ density, then a standard probability calculation shows that $m(x)$ is a $\mathcal{N}(\mu_\pi, \sigma_\pi^2+\sigma_f^2)$ density. (For those who have not seen the calculation, it is given in Example 1 of Section 4.2, in the midst of another calculation.)

The interest in the marginal distribution centers around the fact that, if X has the conditional density $f(x|\theta)$ and θ actually is random with density $\pi(\theta)$, then $m(x)$ is the density according to which X will actually occur. Thus, in Example 15, if X were a test score which was normally distributed about "true ability," θ, and the true abilities in the population varied according to a normal distribution with mean μ_π and variance σ_π^2, then $m(x)$ would be the actual distribution of observed test scores. For this reason $m(x)$ is sometimes called the *predictive distribution* for X, since it describes what one would "predict" that X would be.

Bayesians have long used m to check assumptions. Specifically, if $m(x)$ (for the actual observed data x) turns out to be surprisingly small, then the assumptions (the model f and the prior π) have not "predicted" what actually occurred and are suspect. More discussion of this informal use of m will be given in Subsection 4.7.2.

Here we will be concerned with uses of m in the construction of the prior π. The basic idea is to obtain some information about m, and then use this information (through (3.11)) to assist in the construction of π.

3.5.2. Information About m

There are several possible sources of information about m. The most obvious is subjective knowledge. Indeed, in many problems the unknown θ will have no concrete reality in terms of physical quantities that are easily accessible to intuition, and yet the phenomenon under study may be quite familiar to the investigator, allowing him to partially predict the experimental outcome, X, or (more generally) quantify subjective information about the marginal distribution of X. Many regression situations are of this type; the unknown regression coefficients, θ, are implicitly defined by the model and have no clear intuitive interpretation, but experience may well allow "prediction" of the response variables, X, in the regression. This type of situation is discussed in Kadane, Dickey, Winkler, Smith, and Peters (1980); see also Kadane (1980) and Winkler (1980). Indeed, the original argument of Bayes (1763) was of this type (see Stigler (1982) and Exercise 21).

A second source of possible information about m is the data itself. Suppose, for example, that $\boldsymbol{\theta}$ is a vector, consisting of components $(\theta_1, \ldots, \theta_p)$ that are i.i.d. from the density π_0. (For instance, consider the testing example mentioned in Subsection 3.5.1, in which the θ_i are the "true abilities" of the individuals—here π_0 would be the distribution of such abilities in the entire population.) Suppose also that the data $\mathbf{X}$ consists of independent components $(X_1, \ldots, X_p)$, where each X_i has density $f(x_i|\theta_i)$. (In the testing example, $X_1, \ldots, X_p$ would be the test scores of p random independent subjects.) Then the common marginal distribution of each X_i is

$$m_0(x_i) = \int f(x_i|\theta_i) dF^{\pi_0}(\theta_i), \tag{3.12}$$

and $X_1, \ldots, X_p$ can be considered to be a simple random sample from m_0. Note that this also follows from the direct calculation (assuming continuous densities for convenience)

$$\begin{aligned} m(\mathbf{x}) &= \int f(\mathbf{x}|\boldsymbol{\theta}) \pi(\boldsymbol{\theta}) d\boldsymbol{\theta} \\ &= \int \left[\prod_{i=1}^{p} f(x_i|\theta_i) \right] \left[\prod_{i=1}^{p} \pi_0(\theta_i) \right] d\boldsymbol{\theta} \\ &= \prod_{i=1}^{p} \int f(x_i|\theta_i) \pi_0(\theta_i) d\theta_i = \prod_{i=1}^{p} m_0(x_i). \end{aligned} \tag{3.13}$$

The data $\mathbf{x}$ can thus be used to estimate m_0 (and hence m) or some features thereof.

The above type of situation is typically called an *empirical Bayes* or *compound decision* problem (names due to Robbins (1951, 1955, 1964)). The observations $X_1, \ldots, X_p$ could be *current* data about $\theta_1, \ldots, \theta_p$, with *all* the θ_i being of interest, or the data could be *past* data about previously studied θ_i (such as past data from lot inspections about the proportion of defectives, θ_i, in each lot) from which it is desired to infer π_0 for use on future θ_i. The first case is formally the compound decision problem, while the latter case is the empirical Bayes problem. There is very little operational difference between the two problems, however (at least from the conditional Bayesian perspective), and all such problems have come to be called "empirical Bayes." This and more complicated empirical Bayes situations will be discussed in Section 4.5.

Very rarely it might be the case that actual past values of the θ_i are available. In such situations it is not necessary to work through the marginal distribution; the past data can be used to directly estimate π_0 in any number of standard ways.

3.5.3. Restricted Classes of Priors

To make use of knowledge about m in selecting a prior, it is frequently necessary to restrict consideration to some class of prior distributions. We have already encountered this informally in Section 3.2 when talking about selecting a prior of a given functional form, and in Section 3.4 when talking about choosing a prior (to maximize entropy) from among those satisfying certain constraints. We will make frequent use of classes of priors, and so formalize the concept here. Γ will be used to denote a class of priors. Some common classes follow.

I. *Priors of a Given Functional Form*

For reasons given in Section 3.2, it is often convenient to consider a class of priors of the form

$$\Gamma = \{\pi\colon \pi(\theta) = g(\theta|\lambda),\ \lambda \in \Lambda\}. \tag{3.14}$$

Here g is a prescribed function, so that choice of a prior reduces to the choice of $\lambda \in \Lambda$. The parameter λ (often a vector) is usually called a *hyperparameter* of the prior, particularly in situations where it is considered unknown and to be determined from information about the marginal distribution.

Example 16. Suppose θ is a normal mean. It is felt that the prior distribution π, for θ, can be adequately described by the class of normal distributions, and, in addition, it is certain that the prior mean is positive. Then

$$\Gamma = \{\pi\colon \pi \text{ is } \mathcal{N}(\mu_\pi, \sigma_\pi^2),\ \mu_\pi > 0,\ \sigma_\pi^2 > 0\},$$

so that $\lambda = (\mu_\pi, \sigma_\pi^2)$ is the hyperparameter. (We are not necessarily recommending the use of such a class, see Sections 3.2 and 4.7.)

II. *Priors of a Given Structural Form*

When θ is a vector $\boldsymbol{\theta} = (\theta_1, \ldots, \theta_p)^t$, it will often be the case that some relationship among the θ_i is suspected. For instance, in Subsection 3.5.2 we considered the empirical Bayes situation in which the θ_i were assumed to be i.i.d. This assumption corresponds to the class of priors

$$\Gamma = \left\{ \pi \colon \pi(\boldsymbol{\theta}) = \prod_{i=1}^{p} \pi_0(\theta_i),\ \pi_0 \text{ an arbitrary density} \right\}. \tag{3.15}$$

We will describe such priors and relationships among the θ_i as *structural.* In empirical Bayes analysis, structural relationships are often combined with the given functional form idea to yield classes of priors as in the following example.

EXAMPLE 17. Suppose the X_i are independently $\mathcal{N}(\theta_i, \sigma_f^2)$ (σ_f^2 known) and that the θ_i are likewise felt to be independent with a common $\mathcal{N}(\mu_\pi, \sigma_\pi^2)$ prior distribution (call it π_0), the hyperparameters μ_π and σ_π^2 being completely unknown. Then

$$\Gamma = \left\{ \pi \colon \pi(\boldsymbol{\theta}) = \prod_{i=1}^{p} \pi_0(\theta_i),\ \pi_0 \text{ being } \mathcal{N}(\mu_\pi, \sigma_\pi^2),\ -\infty < \mu_\pi < \infty \text{ and } \sigma_\pi^2 > 0 \right\}. \tag{3.16}$$

(As a concrete illustration, recall the example where X_i is the test score of the ith individual, random about true ability θ_i with known "reliability" σ_f^2, while the true abilities, θ_i, are from an unknown normal population.)

III. *Priors Close to an Elicited Prior*

An attractive idea, particularly for studying robustness with respect to the prior (see Section 4.7), is to elicit a single prior π_0 and, realizing that any prior "close" to π_0 would also be reasonable, choose Γ to consist of all such "close" priors. A rich and calculationally attractive class to work with is the *ε-contamination class*

$$\Gamma = \{\pi \colon \pi(\theta) = (1-\varepsilon)\pi_0(\theta) + \varepsilon q(\theta),\ q \in \mathcal{Q}\}, \tag{3.17}$$

where $0 < \varepsilon < 1$ reflects how "close" we feel that π must be to π_0, and $\mathcal{Q}$ is a class of possible "contaminations."

EXAMPLE 2 (continued). The elicitation process yielded, as one reasonable possibility for π_0, the $\mathcal{N}(0, 2.19)$ distribution. The only features of the prior that were quantitatively considered, however, were the quartiles, and these might even be somewhat uncertain. It is quite possible, therefore, that distributions which have probabilities differing from π_0 by as much as (say) 0.2 would also be plausible priors, so we could choose $\varepsilon = 0.2$. We will defer discussion of the choice of $\mathcal{Q}$ until Subsection 4.7.4.

This class of priors could also be combined with other classes in a variety of ways (cf. Berger and Berliner (1983, 1984)).

3.5.4. The ML-II Approach to Prior Selection

In the discussion following Definition 2, it was pointed out that $m(x|\pi)$ reflects the plausibility of f and π, in light of the data. If we treat f as definitely known, it follows that $m(x|\pi)$ reflects the plausibility of π. Hence when, for the observed data x, it is the case that $m(x|\pi_1) > m(x|\pi_2)$, we can conclude that the data provides more support for π_1 than for π_2. Carrying this one step further, it is reasonable to consider $m(x|\pi)$ as a (possibly subjective) *likelihood function* for π. (If, indeed, π arises as an unknown model for an actually random θ, as in an empirical Bayes setting, then this interpretation of $m(x|\pi)$ as a likelihood function for π would be noncontroversial.) Faced with a "likelihood function" for π, a natural method of choosing π is to use maximum likelihood. In the following we use I. J. Good's nomenclature, see Good (1983) for discussion and his early works on the subject. (Berger and Berliner (1983) also contain a review and additional motivation.)

Definition 3. Suppose Γ is a class of priors under consideration, and that $\hat{\pi} \in \Gamma$ satisfies (for the observed data x)

$$m(x|\hat{\pi}) = \sup_{\pi \in \Gamma} m(x|\pi). \tag{3.18}$$

Then $\hat{\pi}$ will be called the *type* II *maximum likelihood prior*, or ML-II *prior* for short.

The determination of an ML-II prior is quite simple for many classes of priors. For instance, when Γ is the class

$$\Gamma = \{\pi \colon \pi(\theta) = g(\theta|\lambda), \lambda \in \Lambda\},$$

then

$$\sup_{\pi \in \Gamma} m(x|\pi) = \sup_{\lambda \in \Lambda} m(x|g(\theta|\lambda)),$$

so that one simply has to perform a maximization over the hyperparameter λ. (We will call the maximizing hyperparameters the ML-II *hyper-parameters.*)

Example 17 (continued). From the developments in Example 15 and Subsection 3.5.2, it is clear that $m(x|\pi) = \prod_{i=1}^{p} m_0(x_i|\pi_0)$, where m_0 is $\mathcal{N}(\mu_\pi, \sigma_\pi^2 + \sigma_f^2)$. Thus we can write

$$\begin{aligned} m(x|\pi) &= \prod_{i=1}^{p} \frac{1}{[2\pi(\sigma_\pi^2 + \sigma_f^2)]^{1/2}} \exp\left\{-\frac{(x_i - \mu_\pi)^2}{2(\sigma_\pi^2 + \sigma_f^2)}\right\} \\ &= [2\pi(\sigma_\pi^2 + \sigma_f^2)]^{-p/2} \exp\left\{-\frac{\sum_{i=1}^{p}(x_i - \mu_\pi)^2}{2(\sigma_\pi^2 + \sigma_f^2)}\right\} \\ &= [2\pi(\sigma_\pi^2 + \sigma_f^2)]^{-p/2} \exp\left\{\frac{-ps^2}{2(\sigma_\pi^2 + \sigma_f^2)}\right\} \exp\left\{\frac{-p(\bar{x} - \mu_\pi)^2}{2(\sigma_\pi^2 + \sigma_f^2)}\right\}, \end{aligned} \tag{3.19}$$

where $\bar{x}=\sum_{i=1}^{p} x_i/p$ and $s^2=\sum_{i=1}^{p}(x_i-\bar{x})^2/p$. (The last equality in (3.19) is standard and follows from adding and subtracting $\bar{x}$ within the $(x_i-\mu_\pi)^2$.)

We seek to maximize $m(x|\pi)$ over the hyperparameters μ_π and σ_π^2. The maximum over μ_π is clearly attained at $\bar{x}$, regardless of the value of σ_π^2, so that $\hat{\mu}_\pi=\bar{x}$ will always be the ML-II choice of μ_π. Inserting this value into the expression for $m(x|\pi)$, it remains only to maximize

$$\psi(\sigma_\pi^2)=[2\pi(\sigma_\pi^2+\sigma_f^2)]^{-p/2}\exp\left\{\frac{-ps^2}{2(\sigma_\pi^2+\sigma_f^2)}\right\}$$

over σ_π^2. As usual, it is somewhat easier to maximize $\log\psi(\sigma_\pi^2)$. Indeed

$$\frac{d}{d\sigma_\pi^2}\log\psi(\sigma_\pi^2)=\frac{-p/2}{(\sigma_\pi^2+\sigma_f^2)}+\frac{ps^2}{2(\sigma_\pi^2+\sigma_f^2)^2}.$$

This equals zero (and indeed yields the maximum) at $\sigma_\pi^2=s^2-\sigma_f^2$, provided that $s^2\geq\sigma_f^2$. If $s^2<\sigma_f^2$, the derivative is always negative, so that the maximum is achieved at $\sigma_\pi^2=0$. Thus we have that the ML-II estimate of σ_π^2 is

$$(s^2-\sigma_f^2)^+=\max\{0, s^2-\sigma_f^2\}.$$

In conclusion, the ML-II prior, $\hat{\pi}_0$, is

$$\mathcal{N}(\hat{\mu}_\pi, \hat{\sigma}_\pi^2),\quad\text{where}\quad \hat{\mu}_\pi=\bar{x}\quad\text{and}\quad\hat{\sigma}_\pi^2=\max\{0, s^2-\sigma_f^2\}. \tag{3.20}$$

In more complicated situations, an often useful technical tool for finding the ML-II prior is the E–M algorithm of Dempster, Laird, and Rubin (1977). Unfortunately, space precludes discussion of this technique; see, also, Laird (1982), Louis (1982), Leonard (1983b), and Dempster, Selwyn, Patel, and Roth (1984) for development and applications of the algorithm. (There are, of course, many other numerical techniques that can also be used to solve likelihood equations encountered in ML-II analysis; cf. Jennrich and Sampson (1976).)

Surprisingly, when Γ is *infinite* dimensional the calculation of the ML-II prior is again sometimes simple, as the following example shows.

Example 18. For any π in the ε-contamination class

$$\Gamma=\{\pi\colon \pi(\theta)=(1-\varepsilon)\pi_0(\theta)+\varepsilon q(\theta), q\in\mathcal{Q}\},$$

it is clear that

$$\begin{aligned} m(x|\pi)&=\int f(x|\theta)[(1-\varepsilon)\pi_0(\theta)+\varepsilon q(\theta)]d\theta\\ &=(1-\varepsilon)m(x|\pi_0)+\varepsilon m(x|q). \end{aligned}\tag{3.21}$$

Hence, the ML-II prior can be found by maximizing $m(x|q)$ over $q\in\mathcal{Q}$, and using the maximizing $\hat{q}$ in the expression for π.

As a specific example, if $\mathscr{Q}$ is the class of *all* possible distributions (so that we, rather unrealistically, allow π_0 to be contaminated by anything), then

$$m(x|q)=\int f(x|\theta)q(\theta)d\theta$$

will clearly be maximized by choosing q to be concentrated where $f(x|\theta)$ is maximized (as a function of θ). Indeed, if $\hat{\theta}$ maximizes $f(x|\theta)$ (i.e., $\hat{\theta}$ is a maximum likelihood estimate of θ for the given data), then $m(x|q)$ is maximized by having q concentrate all its mass at $\hat{\theta}$. Denoting this unit point mass at $\hat{\theta}$ by $\langle\hat{\theta}\rangle$ (as in Chapter 2), we thus have that the ML-II prior, $\hat{\pi}$, is

$$\hat{\pi}=(1-\varepsilon)\pi_0+\varepsilon\langle\hat{\theta}\rangle. \tag{3.22}$$

(If π_0 is a continuous density, $\hat{\pi}$ is, of course, a mixture of a continuous and a discrete probability distribution.) See Subsection 4.7.9 for further discussion.

A very interesting "infinite dimensional" example of the ML-II approach to prior selection is given in Laird (1978a, 1983) (see also Lindsay (1981, 1983) and Leonard (1984)). These articles consider the structural empirical Bayes situation of (3.15) with the common prior, π_0, assumed to be completely unknown, and calculate the ML-II estimate, $\hat{\pi}_0$, for π_0. This turns out to be calculationally feasible, with the aid of the E–M algorithm, because $\hat{\pi}_0$ can be shown to be a finite discrete distribution. The approach has great potential value in those empirical Bayes settings in which one is quite uncertain as to the form of π_0.

3.5.5. The Moment Approach to Prior Selection

Instead of using maximum likelihood to select $\pi\in\Gamma$, the moment approach could be used. The moment approach applies when Γ is of the "given functional form" type and it is possible to relate prior moments to moments of the marginal distribution, the latter being supposedly either estimated from data or determined subjectively. The following lemma provides one useful relationship between prior and marginal moments.

Lemma 1. *Let $\mu_f(\theta)$ and $\sigma_f^2(\theta)$ denote the conditional mean and variance of X (i.e., the mean and variance with respect to the density $f(x|\theta)$). Let μ_m and σ_m^2 denote the marginal mean and variance of X (with respect to $m(x)$). Assuming these quantities exist, then*

$$\mu_m=E^{\pi}[\mu_f(\theta)],$$
$$\sigma_m^2=E^{\pi}[\sigma_f^2(\theta)]+E^{\pi}[(\mu_f(\theta)-\mu_m)^2].$$

PROOF. For ease in understanding, the proof will only be done when $\pi(\theta)$ is continuous. The proof for the discrete case is similar. Clearly

$$\begin{aligned}\mu_m = E^m[X] &= \int_{\mathscr{X}} xm(x)dx \\ &= \int_{\mathscr{X}} x \int_{\Theta} f(x|\theta)\pi(\theta)d\theta\, dx \\ &= \int_{\Theta} \pi(\theta) \int_{\mathscr{X}} xf(x|\theta)dx\, d\theta \\ &= \int_{\Theta} \pi(\theta)\mu_f(\theta)d\theta = E^\pi[\mu_f(\theta)].\end{aligned}$$

Likewise,

$$\begin{aligned}\sigma_m^2 &= E^m[(X-\mu_m)^2] = \int_{\mathscr{X}} (x-\mu_m)^2 \int_{\Theta} f(x|\theta)\pi(\theta)d\theta\, dx \\ &= \int_{\Theta} \pi(\theta) \int_{\mathscr{X}} (x-\mu_m)^2 f(x|\theta)dx\, d\theta \\ &= E^\pi(E_\theta^f[X-\mu_m]^2) \\ &= E^\pi(E_\theta^f[(X-\mu_f(\theta))+(\mu_f(\theta)-\mu_m)]^2) \\ &= E^\pi(E_\theta^f[(X-\mu_f(\theta))^2+2(X-\mu_f(\theta))(\mu_f(\theta)-\mu_m)+(\mu_f(\theta)-\mu_m)^2]).\end{aligned}$$

Since $E_\theta^f(X-\mu_f(\theta))^2=\sigma_f^2(\theta)$ and

$$E_\theta^f[2(X-\mu_f(\theta))(\mu_f(\theta)-\mu_m)] = 2(E_\theta^f[X]-\mu_f(\theta))(\mu_f(\theta)-\mu_m)=0,$$

it is clear that

$$\sigma_m^2 = E^\pi[\sigma_f^2(\theta)] + E^\pi[(\mu_f(\theta)-\mu_m)^2]. \qquad \square$$

Corollary 1.

(i) *If $\mu_f(\theta)=\theta$, then $\mu_m=\mu_\pi$, where $\mu_\pi=E^\pi[\theta]$ is the prior mean.*
(ii) *If, in addition, $\sigma_f^2(\theta)=\sigma_f^2$ (a constant independent of θ), then $\sigma_m^2=\sigma_f^2+\sigma_\pi^2$, where σ_π^2 is the prior variance.*

PROOF. The result is obvious from Lemma 1. $\square$

The quantities $E^\pi[\mu_f(\theta)]$, $E^\pi[\sigma_f^2(\theta)]$, and $E^\pi[\mu_f(\theta)-\mu_m]^2$ can usually be easily expressed in terms of the hyperparameters of the prior. Since μ_m and σ_m^2 are the marginal mean and variance, they can easily be estimated from data (in the empirical Bayes case), or estimated subjectively (if one has subjective information about m itself). The idea, then, is to equate μ_m and σ_m^2 in the above expressions to these estimates, and solve for the implied prior hyperparameters.

EXAMPLE 19. Suppose $X \sim \mathscr{N}(\theta, 1)$, and that the class, Γ, of all $\mathscr{N}(\mu_\pi, \sigma_\pi^2)$ priors for θ is considered reasonable. Subjective experience yields a "prediction" that X will be about 1, with associated "prediction variance" of 3. Thus we estimate that $\mu_m=1$ and $\sigma_m^2=3$. Using Corollary 1, noting that $\sigma_f^2=1$, we have that $1=\mu_m=\mu_f$ and $3=\sigma_m^2=1+\sigma_\pi^2$. Solving for μ_π and σ_π^2, we conclude that the $\mathcal{N}(1,2)$ prior should be used.

EXAMPLE 17 (continued). We again seek to determine μ_π and σ_π^2. Treating $X_1,\ldots,X_p$ as a sample from m_0, the standard method of moments estimates for μ_{m_0} and $\sigma_{m_0}^2$ are $\bar{x}$ and $s^2=\sum_{i=1}^p(x_i-\bar{x})^2/(p-1)$, respectively. It follows that the moment estimates of μ_π and σ_π^2 are $\hat{\mu}_\pi=\bar{x}$ and $\hat{\sigma}_\pi^2=s^2-\sigma_f^2$. (Note that $\hat{\sigma}_\pi^2$ could be negative, a recurring problem with moment estimates.)

3.5.6. The Distance Approach to Prior Selection

When Γ is not a "given functional form" class and there is considerable information available about m, one might hope to directly estimate m and then use the integral relationship

$$m(x)=\int_\Theta f(x|\theta)dF^\pi(\theta)$$

to determine π. In the empirical Bayes scenario, for instance, if a large amount of data $x_1,\ldots,x_p$ is available, one could directly estimate m using some standard density estimate. One of the simplest such estimates is the discrete distribution which puts mass $1/p$ at each of the observed x_i, i.e.,

$$\hat{m}(x)=\frac{1}{p}[\text{the number of } x_i \text{ equal to } x]. \tag{3.23}$$

The difficulty encountered in using an estimate, $\hat{m}$, is that the equation

$$\hat{m}(x)=\int_\Theta f(x|\theta)dF^\pi(\theta)$$

need have no solution, π. Hence all we can seek is an estimate of π, say $\hat{\pi}$, for which

$$m_{\hat{\pi}}(x)=\int_\Theta f(x|\theta)dF^{\hat{\pi}}(\theta)$$

is close (in some sense) to $\hat{m}(x)$. A reasonable measure of "distance" between two such densities is

$$d(\hat{m},m_{\hat{\pi}})=E^{\hat{m}}\left[\log\frac{\hat{m}(X)}{m_{\hat{\pi}}(X)}\right]=\begin{cases}\displaystyle\int_{\mathscr{X}}\hat{m}(x)\log\left[\frac{\hat{m}(x)}{m_{\hat{\pi}}(x)}\right]dx & \text{(continuous case)},\\ \displaystyle\sum_{\mathscr{X}}\hat{m}(x)\log\left[\frac{\hat{m}(x)}{m_{\hat{\pi}}(x)}\right] & \text{(discrete case)}.\end{cases}$$

This measure of distance is related to the concept of entropy discussed in Section 3.4. Indeed, letting π_0 denote a noninformative prior, it is clear that the entropy of a prior π in the continuous case is simply $-d(\pi, \pi_0)$. For the finite discrete case with n possible parameter points, so that the noninformative prior is $\pi_0(\theta)=1/n$, the entropy of π is $n-d(\pi, \pi_0)$. Thus the entropy of a prior is directly related to how "close" the prior is to a noninformative prior (which recall is supposed to represent a complete lack of knowledge about θ).

Using the above measure of distance, the goal is to seek the $\hat{\pi}$ which minimizes $d(\hat{m}, m_{\hat{\pi}})$. Note that

$$d(\hat{m}, m_{\hat{\pi}})=E^{\hat{m}}\left[\log\frac{\hat{m}(X)}{m_{\hat{\pi}}(X)}\right]=E^{\hat{m}}[\log \hat{m}(X)]-E^{\hat{m}}[\log m_{\hat{\pi}}(X)].$$

Since only the last term of this expression depends on $\hat{\pi}$, it is clear that minimizing $d(\hat{m}, m_{\hat{\pi}})$ (over $\hat{\pi}$) is equivalent to maximizing

$$E^{\hat{m}}[\log m_{\hat{\pi}}(X)]. \tag{3.24}$$

Finding the $\hat{\pi}$ which maximizes (3.24) is a difficult problem. One situation which is relatively easy to deal with is the situation in which Θ is finite, say $\Theta=\{\theta_1, \ldots, \theta_k\}$. Letting $p_i=\hat{\pi}(\theta_i)$, it is clear that, for this situation,

$$m_{\hat{\pi}}(x)=\sum_{i=1}^{k} f(x|\theta_i)p_i.$$

Hence, finding the optimal $\hat{\pi}$ reduces to the problem of maximizing

$$E^{\hat{m}}\left[\log\left(\sum_{i=1}^{k} f(X|\theta_i)p_i\right)\right] \tag{3.25}$$

over all p_i such that $0\le p_i\le 1$ and $\sum_{i=1}^{k} p_i=1$. For the $\hat{m}$ in (3.23), expression (3.25) becomes

$$E^{\hat{m}}\left[\log\left(\sum_{i=1}^{k} f(X|\theta_i)p_i\right)\right]=\sum_{j=1}^{n}\frac{1}{n}\log\left(\sum_{i=1}^{k} f(x_j|\theta_i)p_i\right).$$

The maximization of this last quantity over the p_i is a straightforward linear programming problem.

For continuous θ, little is known about the problem of maximizing (3.24). A very reasonable approach, given in Maritz (1970) (see also Lord and Cressie (1975)), is to only consider $\hat{\pi}$ which are step functions of say N steps (i.e., $\hat{\pi}(\theta)=c_i$ for $a_i\le\theta<a_{i+1}$, where $a_0<a_1<\cdots<a_N$, $c_i\ge 0$, and $\sum_{i=0}^{(N-1)}(a_{i+1}-a_i)\ c_i=1$). One can then seek to maximize (3.24) over the constants a_i and c_i. This can be done on a computer.

3.5.7. Marginal Exchangeability

Through consideration of certain subjective beliefs about the marginal distribution, one can obtain interesting insights into the modelling process itself. So far we have been taking as given the notion of a model $f(x|\theta)$ for the random observation X, but there are deep philosophical issues concerning the origin of models. Take, for instance, the very simple experiment of spinning a coin n times and observing the sequence of heads and tails. It is standard to model this by assuming: (i) that the outcomes of the spins are "independent"; and (ii) that the probability of each outcome being a head is some common value θ. Letting X_i be 1 or 0 as the ith spin results in a head or tail, one then has the model for $\mathbf{X}=(X_1, \ldots, X_n)$ given by

$$f(\mathbf{x}|\theta)=\prod_{i=1}^{n} \theta^{x_i}(1-\theta)^{1-x_i}, \tag{3.26}$$

which upon an application of sufficiency reduces to the usual binomial model.

Consider, however, the two assumptions that went into the above construction. It is difficult, and perhaps impossible, to give precise and operationally realizable definitions of independence and of θ that are not subjective. For instance, the frequentist definition of θ involves some limiting average in a long series of "independent" spins of the coin. Besides the confounding with independence, the definition really involves an imaginary thought experiment, since (in reality) the coin would obviously be changing (minutely) from spin to spin. Likewise, one cannot really define independence without reference to imaginary situations.

Given the impossibility of realizable objective definitions of models, some subjectivists argue that subjective beliefs about actual observables should form the basis of statistics. Thus, in the situation of spinning the coin, the argument goes that attention should be focused on the actual (subjective) distribution of the outcomes $(X_1, \ldots, X_n)$, not on a conditional distribution given imaginary parameters. In other words, the fundamental entity should be a (subjective) probability distribution, m, describing the actual sequence of zeros and ones that would be anticipated. As an example,

$$m((x_1, \ldots, x_n))=2^{-n}$$

would describe a belief that any sequence of zeros and ones was equally likely to occur.

An interesting facet of this subjective approach is that it can provide a rationale for consideration of θ, the model (3.26), and a prior distribution on θ. For instance, a frequently reasonable subjective belief about m is that it should be *exchangeable*, i.e., that

$$m((x_1, \ldots, x_n))=m((x_{i(1)}, x_{i(2)}, \ldots, x_{i(n)})), \tag{3.27}$$

where $(x_{i(1)}, x_{i(2)}, \ldots, x_{i(n)})$ is a permutation of $(x_1, \ldots, x_n)$. (In other words, a distribution is exchangeable if reordering $(X_1, \ldots, X_n)$ does not change the probability distribution.) In the case of spinning the coin, it will usually be the case that the distribution is judged to be exchangeable. (Imagine that someone was to give you the sequence of zeros and ones (heads and tails)—would you attach different probabilities to the possible sequences if you knew the sequence would be given in the order $(X_n, X_{n-1}, \ldots, X_1)$?)

The assumption of exchangeability has a profound consequence. In a fundamental theorem, deFinetti (1937) showed that, if a sequence $(X_1, X_2, \ldots, X_n)$ of 0, 1 valued random variables is judged to be exchangeable for *every n*, then *m must be* of the form

$$m((x_1, \ldots, x_n)) = \int_0^1 \left[\prod_{i=1}^n \theta^{x_i}(1-\theta)^{1-x_i} \right] dF^\pi(\theta), \tag{3.28}$$

where

$$F^\pi(\theta) = \lim_{n\to\infty} P^m\left(\frac{1}{n} \sum_{i=1}^n X_i \le \theta \right).$$

Equating (3.11) with (3.28), it is clear that the assumption of exchangeability for every n is equivalent to the assumption of the model (3.26) for $(X_1, \ldots, X_n)$ and some prior distribution π for θ. This type of relationship between exchangeability for m and the model-prior formulation has been shown to hold in a large number of other situations. Some recent references are Dawid (1982), Diaconis and Freedman (1980, 1984), and Lauritzen (1982).

The implication "(3.27) for all $n \Rightarrow$(3.28)" is philosophically interesting, but its practical utility is somewhat limited, for two reasons. First, it is not clear that a subjective specification of "exchangeability for all n" is really easier than subjective specification of a model such as (3.26). (Note that both require the consideration of an imaginary infinite sequence of experiments.) And even if one can judge exchangeability to hold, it is usually easier to subjectively specify π than to specify m: intuition is simply not very good at combining probabilities (see Kahneman, Slovic, and Tversky (1982)), and m involves a combination of the probability distribution (3.26) and of π. Sometimes, however, considerations of exchangeability will *suggest* model-prior decompositions that are not otherwise apparent. And, since development of new models is one of the most important of all statistical activities, exchangeability (and its generalizations) can prove very useful to statistics. Two recent references in which important new models were generated for practical problems through consideration of exchangeability are Diaconis and Freedman (1981) and Lauritzen (1984). Other indications of the practical importance of exchangeability can be found in Lindley and Novick (1981).

3.6. Hierarchical Priors

An important type of prior distribution is a *hierarchical prior* (so named and long used by I. J. Good, see Good (1983) for discussion and early references), also called a multistage prior (cf. Lindley and Smith (1972)). The idea is that one may have structural *and* subjective prior information at the same time (see Subsection 3.5.2), and it is often convenient to model this in stages. For instance, in the empirical Bayes scenario (discussed in Subsection 3.5.2), structural knowledge that the θ_i were i.i.d. led to the first stage prior description

$$\pi_1(\boldsymbol{\theta}) = \prod_{i=1}^{p} \pi_0(\theta_i).$$

(The subscript 1 on π_1 is to indicate that this is the first stage.) The hierarchical approach would seek to place a "second stage" subjective prior on π_0. Alternatively, following the formalism of Subsection 3.5.3, if the "first stage" of consideration led to a class Γ of priors, then the second stage would be to put a prior distribution on Γ.

The hierarchical approach is most commonly used when the first stage, Γ, consists of priors of a certain functional form (Antoniak (1974), Berry and Christensen (1979), and Kuo (1985) being exceptions). Thus, if

$$\Gamma = \{\pi_1(\boldsymbol{\theta}|\lambda)\colon \pi_1 \text{ is of a given functional form and } \lambda \in \Lambda\}, \tag{3.29}$$

then the second stage would consist of putting a prior distribution, $\pi_2(\lambda)$, on the hyperparameter λ. (Such a second stage prior is sometimes called a hyperprior for this reason.)

Example 17 (continued). The structural assumption of independence of the θ_i, together with the assumption that they have a common normal distribution, led to (where $\lambda \equiv (\mu_\pi, \sigma_\pi^2)$)

$$\Gamma = \left\{ \pi_1(\boldsymbol{\theta}|\lambda)\colon \pi_1(\boldsymbol{\theta}|\lambda) = \prod_{i=1}^{p} \pi_0(\theta_i), \right.$$
$$\left. \pi_0 \text{ being } \mathcal{N}(\mu_\pi, \sigma_\pi^2), -\infty < \mu_\pi < \infty \text{ and } \sigma_\pi^2 > 0 \right\}.$$

A second stage prior, $\pi_2(\lambda)$, could be chosen for the hyperparameters according to subjective beliefs. For instance, in the example where the X_i are test scores measuring the "true abilities" θ_i, one could interpret μ_π and σ_π^2 as the population mean and variance, respectively, of the θ_i. Perhaps it is felt that the "mean true ability," μ_π, is near 100, with a "standard error" of ± 20, while the "variance of true abilities," σ_π^2, is about 200, with a "standard error" of ± 100. A reasonable prior for μ_π would then be the $\mathcal{N}(100, 400)$ distribution, while the $\mathcal{IG}(6, 0.001)$ distribution might be a reasonable prior for σ_π^2. (Note that these distributions have means equal to the "best guesses" for the hyperparameters, and variances equal to the "squared standard errors"—see Appendix 1 for the formulas for means and variances of the various distributions.) If prior independence of μ_π and σ_π^2 is reasonable, the second state prior for λ would thus be

$$\pi_2(\lambda) = \frac{1}{\sqrt{800\pi}} \exp\left\{ -\frac{1}{800} (\mu_\pi - 100)^2 \right\} \frac{1}{\Gamma(6)(0.001)^6 (\sigma_\pi^2)^7} \exp\left\{ -\frac{1000}{\sigma_\pi^2} \right\}. \tag{3.30}$$

It is somewhat difficult to subjectively specify second stage priors, such as $\pi_2(\lambda)$ in the above example, but it will be seen in Chapter 4 that there is "more robustness" in second stage specifications than in single prior specifications (i.e., there is less danger that misspecification of π_2 will lead to a bad answer). Thus, in the above example, we would feel comfortable with the use of moment specifications and priors of a given functional form for λ (whereas we would be wary about such in general, see Section 4.7).

The difficulty of specifying second stage priors has made common the use of noninformative priors at the second stage. This will be discussed further in Section 4.6, where also the many advantages of the hierarchical Bayes approach to empirical Bayes problems will be discussed. Note that there is no theoretical reason for limiting hierarchical priors to just two stages, but more than two are rarely useful in practice. (See also Goel and DeGroot (1981) and Goel (1983).)

As a final comment, note that a hierarchical structure is merely a convenient *representation* for a prior, rather than an entirely new entity; any hierarchical prior can be written as a standard prior. For instance, in the situation of (3.29), the *actual* prior distribution is

$$\pi(\boldsymbol{\theta}) = \int_{\Lambda} \pi_1(\boldsymbol{\theta}|\lambda)\,dF^{\pi_2}(\lambda), \tag{3.31}$$

and any Bayesian analysis will actually be performed with respect to π.

EXAMPLE 17 (continued). Here

$$\begin{aligned}\pi_1(\boldsymbol{\theta}|\lambda) &= \prod_{i=1}^{p} \pi_0(\theta_i)\\ &= \frac{1}{(2\pi\sigma_\pi^2)^{p/2}} \exp\left\{-\frac{1}{2\sigma_\pi^2}\sum_{i=1}^{p}(\theta_i-\mu_\pi)^2\right\}.\end{aligned}$$

Unfortunately, the integration in (3.31) cannot be carried out explicitly with π_2 as in (3.30) (but see Section 4.6 for important aspects of this calculation). For simplicity, therefore, and to give some feeling as to the type of result obtained, let us assume that μ_π is *known* to be equal to 100 exactly, and that σ_π^2 is still described by the $\mathcal{IG}(6, 0.001)$ prior distribution. Then

$$\begin{aligned}\pi(\boldsymbol{\theta}) &= \int_0^\infty \int_{-\infty}^\infty \pi_1(\boldsymbol{\theta}|\lambda)\,dF^{\pi_2}(\lambda)\\ &= \int_0^\infty \frac{1}{(2\pi\sigma_\pi^2)^{p/2}} \exp\left\{-\frac{1}{2\sigma_\pi^2}\sum_{i=1}^{p}(\theta_i-100)^2\right\}\\ &\quad\times \frac{10^{18}}{120(\sigma_\pi^2)^7}\exp\left\{-\frac{1000}{\sigma_\pi^2}\right\} d\sigma_\pi^2\\ &= \frac{10^{18}}{120(2\pi)^{p/2}} \int_0^\infty \frac{1}{(\sigma_\pi^2)^{(7+p/2)}}\end{aligned}$$

$$\times \exp\left\{-\frac{1}{\sigma_\pi^2}\left[1000+\frac{1}{2}\sum_{i=1}^{p}(\theta_i-100)^2\right]\right\} d\sigma_\pi^2 \tag{3.32}$$

$$=\frac{10^{18}}{120(2\pi)^{p/2}}\cdot\frac{\Gamma(6+p/2)}{[1000+\frac{1}{2}\sum_{i=1}^{p}(\theta_i-100)^2]^{(6+p/2)}}$$

$$=\frac{\Gamma([12+p]/2)}{\Gamma(6)(2000\pi)^{p/2}}\left[1+\frac{1}{2000}\sum_{i=1}^{p}(\theta_i-100)^2\right]^{-(12+p)/2},$$

which can be recognized as the p-variate t-distribution with $\alpha = 12$, $\boldsymbol{\mu} = (100, \ldots, 100)'$, and $\boldsymbol{\Sigma} = (500/3)\mathbf{I}$.

Although any hierarchical prior does correspond (through (3.31)) to a standard prior, the conceptual advantages of the hierarchical representation for elicitation purposes should be apparent. (Intuition would be hard pressed to come up with (3.32), or especially the prior based on (3.30), from direct elicitation.) Also, we will see in Section 4.6 that the hierarchical representation has a number of calculational advantages.

3.7. Criticisms

Few statisticians would object to a Bayesian analysis when θ was indeed a random quantity with a known prior distribution, or even when the prior distribution could be estimated with reasonable accuracy (from, say, previous data). The major objections are to use of subjective or "formal" prior distributions, especially when θ is random only in a subjective sense. We discuss here some of these objections.

I. *Objectivity*

To most non-Bayesians, classical statistics is "objective" and hence suitable for the needs of science, while Bayesian statistics is "subjective" and hence (at best) only useful for making personal decisions. Bayesians respond to this in several ways.

The first point to emphasize is that very few statistical analyses are even approximately "objective." This is especially true of decision analyses, as mentioned in Section 2.5, since the choice of a loss function is almost always subjective. (It is indeed rather peculiar that some decision-theorists are ardent anti-Bayesians, precisely because of the subjectivity of the prior, and yet have no qualms about the subjectivity of the loss.) Even in non-decision-theoretic statistical analyses, the only situation in which it is clear that "objectivity" obtains is when there is an overwhelming amount of data, so that virtually *any* analysis would yield the same conclusion. Most statistical investigations are not so fortunate, and then choices of such features as the model (i.e., $f(x|\theta)$) will have a serious bearing on the conclusion. Indeed, in many problems the choice of a model has a *much greater* impact on the answer than the choice of a prior on θ, and is often just as subjective. Thus Box (1980) says

> "In the past, the need for probabilities expressing prior belief has often been thought of, not as a necessity for all scientific inference, but rather as a feature peculiar to Bayesian inference. This seems to come from the curious idea that an outright assumption does not count as a prior belief... I believe that it is impossible logically to distinguish between model assumptions and the prior distribution of the parameters."

More bluntly, Good (1973) says

> "The subjectivist states his judgements, whereas the objectivist sweeps them under the carpet by calling assumptions knowledge, and he basks in the glorious objectivity of science."

The use of a prior distribution does introduce another subjective feature into the analysis, of course, and it might appear that this reduces objectivity (and one cannot deny the usefulness of "appearing to be objective" in many problems). There are two Bayesian responses to this point. The first is that there is simply no choice: to be sensible one *must* take a conditional viewpoint, and only the Bayesian approach to conditioning seems logical. The second, and often overlooked, response is that objectivity can best be sought *through* Bayesian analysis. It is a recurring finding in statistics that, to be sensible, a statistical procedure must correspond (at least approximately) to a Bayesian procedure. (See Section 4.1 and Chapter 8 for some such situations.) But if use of a certain procedure corresponds to assuming that a certain prior, π, describes uncertainty about θ, it would be hard to claim objectivity for the procedure if the prior seemed highly "biased." (See Subsection 5.5.3 for some examples.) This suggests that objectivity must be sought through purposeful choice of "objective" priors; noninformative priors would seem to be the logical choice. The ambiguity in defining a noninformative prior prevents this argument from being completely compelling, but it is hard to argue that one will be more objective "by chance" than by trying to be objective. At the very least, use of noninformative priors should be recognized as being at least as objective as any other statistical technique (although frequentists might insist that frequentist evaluations of the noninformative prior procedures be presented). For more discussion on this issue of objectivity, see, for instance, Jeffreys (1961), Zellner (1971), Box and Tiao (1973), Bernardo (1979b), Good (1983), Jaynes (1983), and Berger and Wolpert (1984).

II. *Misuse of Prior Distributions*

There is no question that use of prior distributions introduces a new possibility for abuse of statistics (although it does eliminate some of the old possibilities for abuse). Consider, for instance, the situation of a person trying to decide whether $\theta \in (0, 1)$ or $\theta \in [1, 5)$. From the very statement of the problem, it is likely that θ is thought to be somewhere around 1, but a person not well versed in Bayesian analysis might choose to use the prior $\pi(\theta) = \frac{1}{5}$ (on $\Theta = (0, 5)$), reasoning that, to be fair, all θ should be given equal weight. The resulting Bayes decision might well be that $\theta \in [1, 5)$, even if the data moderately supports $(0, 1)$, because the prior gives $[1, 5)$ four times the probability of the conclusion that $\theta \in (0, 1)$. The potential for misuse by careless or unscrupulous people is obvious. Perhaps the best reply to this criticism is to point out that the best way to prevent misuse is by proper education concerning prior information. Also, in reporting conclusions from Bayesian analyses, the prior (and data and loss) should be reported separately, in order to allow others to evaluate the reasonableness of the subjective inputs.

III. *Robustness*

The issue of prior *robustness* is a real issue. Do slight changes in the prior distribution cause significant changes in the decision and, if so, what should be done? The following example indicates the problem.

EXAMPLE 2 (continued). The normal and Cauchy densities in Example 2 were both reasonable matches to the given prior information; we would thus hope that they would lead to roughly the same answer. Suppose the problem were to estimate θ under squared-error loss, based on $X \sim \mathcal{N}(\theta, 1)$. For the $\mathcal{N}(0, 2.19)$ prior, the Bayes estimator (see Chapter 4) would be

$$\delta_1(x) = x - \frac{x}{3.19}.$$

For the $\mathscr{C}(0, 1)$ prior, the Bayes estimator *for large* $|x|$ (say $|x| \geq 4$) is approximately

$$\delta_2(x) \cong x - \frac{2x}{1 + x^2}$$

(see Berger and Srinivasan (1978)). For large $|x|$, δ_1 and δ_2 are quite different. For instance, when $x = 4.5$, $\delta_1 = 3.09$ while $\delta_2 \cong 4.09$, a difference of a sample standard deviation. When $x = 10$, $\delta_1 = 6.87$ while $\delta_2 \cong 9.80$, a huge difference. Although an observation of $x = 10$ should alert one to rethink the situation (it is not compatible with the model and prior), the observation $x = 4.5$ is not excessively unusual. In any case, this shows that two "close" priors *can* yield very different answers.

In the above example, note that the difference between the two priors is mainly in the functional form. This causes concern, in that it is precisely the functional form of a prior that is difficult to determine. Further discussion of this issue will be given in Section 4.7. The problem cannot be made to disappear entirely but, through "robust Bayesian" methodology and choice of "robust priors," concern can be reduced.

IV. *Data or Model Dependent Priors*

The idealized Bayesian view is that θ is a quantity about which separate information exists (information that can be quantified by a prior distribution), and that this information is to be combined with that in the data (summarized by $f(x|\theta)$). The approach presumes that the prior, π, does not depend in any way on the data.

There are several reasons why this idealized view is not very realistic. The first is that often the model, $f(x|\theta)$, is only chosen *after* an examination of the data, and indeed the choice of model often *defines* θ. (In regression, for instance, the unknown regression coefficients are effectively defined by the regression model chosen.) With data-snooping going on to set up a model and define θ, how can one then *forget* about the data and determine a prior on θ? As Savage (1962) says, "It takes a lot of self-discipline not to exaggerate the probabilities you would have attached to hypotheses before they were suggested to you."

Even when θ is well defined outside of the experiment, a serious problem arises in multivariate situations: if $\boldsymbol{\theta}$ is a vector with dependent coordinates, the subjective specification of π becomes a very formidable task. Indeed, in practice, it may often be necessary (in multivariate problems) to peek at the data in order to find out where prior elicitation efforts should be concentrated; if certain components of $\boldsymbol{\theta}$ are very well estimated by the data, they can often be ignored (to some extent) in the prior elicitation effort.

Other possible involvements of the data in determination of π were discussed in Section 3.5. And even noninformative prior development is not completely pure, in that virtually all methods for deriving a noninformative prior yield priors dependent on the model $f(x|\theta)$.

In a strictly logical sense, this criticism of (practical) prior dependence on the data cannot be refuted. However, it must be recognized that *all* theories are severely compromised when one gets into actual data analysis. Also, the problem does not seem so bad if a slightly different perspective is adopted. Imagine yourself as the reader of an article reporting a Bayesian analysis. The article should report the model (or class of models) considered, the prior (or class of priors) used, the data (in some suitably summarized form), and the Bayesian conclusions. In evaluating this analysis you will be very concerned with the reasonableness of the model and prior, but if you conclude that they *are* reasonable and that other reasonable choices would likely yield much the same conclusions, then the details of the prior development will not be of much concern. A prior distribution which seemed to excessively mimic the data would naturally appear highly suspicious.

Note, also, the role that robustness studies (of sensitivity of the conclusions to the prior) can play in alleviating concerns about data-dependent priors. See Good (1983) and Berger (1984a) for further discussion.

3.8. The Statistician's Role

For the most part, this book is written as though the statistician is the decision maker in a problem of direct concern to him. Far more frequently, however, the statistician will be acting as a technical consultant to a client who has a statistical decision problem to deal with. A common problem faced by statisticians, in such a situation, is that of eliciting from the client the necessary information to perform the analysis. Besides the classical concerns of modeling and design, the decision theorist and the Bayesian must also worry about obtaining accurate loss functions and priors. This can be difficult if the client has little or no statistical training. For example, the concept of probability is of obvious importance in constructing $\pi(\theta)$, yet many clients may not have a good understanding of it. The statistician will, therefore, often find himself partly in the role of a teacher, explaining certain aspects of utility theory and probability to a client, in an attempt to obtain from him the loss function and the prior. The statistician must be wary in such situations of the tendency to take the easy way out and decide, himself, upon a loss function and a prior. An excellent discussion of this problem, along with an example in which nonstatisticians were guided in constructing their priors, is given in Winkler (1967b).

There are, of course, situations in which only limited prior information and loss information is needed to conduct a decision analysis. The simple no-data Bayesian decision problems solved in Chapter 1 are of this type. Other more complicated examples will be seen later. The information needed from clients in such problems is usually very easy to obtain.

Exercises

Section 3.1

1. Automobiles are classified as economy size (small), midsize (medium), or full size (large). Decide subjectively what proportion of each type of car occurs in your area.

Section 3.2

2. Let θ denote the highest temperature that will occur outdoors tomorrow, near your place of residence. Using the histogram approach, find your subjective prior density for θ.

3. Using the relative likelihood approach, determine your prior density for θ in the situation of Exercise 2.

4. Consider the situation of Exercise 2.
   (a) Determine the $\frac{1}{4}$- and $\frac{1}{2}$-fractiles of your prior density for θ.
   (b) Find the normal density matching these fractiles.
   (c) Find, subjectively, the $\frac{1}{10}$- and $\frac{2}{3}$-fractiles of your prior distribution for θ. (Do not use the normal distribution from (b) to obtain these.) Are these consistent with the normal distribution in (b)? Is the normal density a good fit for your prior density of θ?

5. Repeat Exercise 4(b) and (c), but with "normal distribution" replaced by "Cauchy distribution." *Note*: If $X \sim \mathscr{C}(0, \beta)$, then
$$P(0 < X < s) = \pi^{-1} \tan^{-1}(s/\beta).$$

6. Let θ denote the unemployment rate next year. Determine your subjective probability density for θ. Can it be matched with a $\mathscr{B}e(\alpha, \beta)$ density?

7. For each planet in our solar system, determine (i) the first and third quartiles of your prior distribution for the (average) distance of the planet from the sun (i.e., specify a "central" interval which you think has a 50% chance of containing the distance); (ii) specify a "central" 90% interval for the distance. Next, look up the correct distances, and calculate the percentage of the time that your 50% intervals contained the true distance, and the percentage of the time that your 90% intervals contained the true distance. What can you conclude?

Section 3.3

8. For each of the following densities, state whether the density is a location, scale, or location-scale density, and give the natural noninformative prior for the unknown parameters:
   (a) $\mathscr{U}(\theta - 1, \theta + 1)$,
   (b) $\mathscr{C}(0, \beta)$,
   (c) $\mathscr{T}(\alpha, \mu, \sigma^2)$ (α fixed),
   (d) $\mathscr{P}a(x_0, \alpha)$ (α fixed).

9. By an "invariance under reformulation" argument, show that a reasonable noninformative prior for a Poisson parameter θ is $\pi(\theta) = \theta^{-1}$. (*Hint*: A Poisson random variable X usually arises as the number of occurrences of a (rare) event in a time interval T. The parameter θ is the average number of occurrences in time T. Since the specification of T is rather arbitrary, consider the experiment which would result if the time interval used was cT ($c > 0$). This idea was due to Jaynes (1968).)

10. In the "table entry" problem of Example 6, verify the following statement: As $a \to 0$ or $b \to \infty$ or both, the p_i will converge to the values $[\log(1 + i^{-1})/\log 10]$.

11. In the situation of Example 6, characterize the *relatively scale invariant* priors. (Mimic the reasoning in the discussion of relatively location invariant priors.)

12. Determine the Jeffreys noninformative prior for the unknown parameter in each of the following distributions:
    (a) $\mathscr{P}(\theta)$;

(b) $\mathscr{B}(n, \theta)$ (n given);
(c) $\mathscr{N}\mathscr{B}(m, \theta)$ (m given);
(d) $\mathscr{G}(\alpha, \beta)$ (β given).

13. Determine the Jeffreys noninformative prior for the unknown vector of parameters in each of the following distributions:
(a) $\mathscr{M}(n, \mathbf{p})$ (n given);
(b) $\mathscr{G}(\alpha, \beta)$ (both α and β unknown).

14. Suppose that, for $i = 1, \ldots, p$, $X_i \sim f_i(x_i|\theta_i)$ and $\pi_i(\theta_i)$ is the Jeffreys noninformative prior for θ_i. If the X_i are independent, show that the Jeffreys noninformative prior for $\boldsymbol{\theta} = (\theta_1, \ldots, \theta_p)^t$ is $\pi(\boldsymbol{\theta}) = \prod_{i=1}^p \pi_i(\theta_i)$.

15. In Example 7, verify that $\pi(\theta, \sigma) = \sigma^{-2}$ is also the noninformative prior that would result from an invariance-under-transformation argument which considered the transformed problem defined by $Y = cX + b$, $\eta = c\theta + b$, and $\xi = c\sigma$ ($b \in R^1$ and $c > 0$).

16. Suppose $\mathbf{X} \sim \mathscr{N}_2(\boldsymbol{\theta}, \mathbf{I})$ and that it is known that $\theta_1 > 3\theta_2$. Find a reasonable noninformative prior for $\boldsymbol{\theta}$.

17. Consider the situation of Example 15 in Subsection 1.6.4. *Assume* that use of different priors, with the likelihood $\theta^9(1-\theta)^3$, would result in different answers. Show that use of the Jeffreys noninformative prior can violate the Likelihood Principle (see Exercise 12(b), (c)).

Section 3.4

18. Assume $X \sim \mathscr{N}(\theta, 1)$ is to be observed, but that it is known that $\theta > 0$. It is further believed that θ has a prior distribution with mean μ. Show that the prior density of θ which maximizes entropy, subject to these constraints, is the $\mathscr{E}(\mu)$ density.

19. Assume a scale parameter θ is to be estimated (so that the natural noninformative prior is θ^{-1}). It is believed that $a < \theta < b$, and that the median ($\frac{1}{2}$-fractile) of the prior density is z. Show that the prior density which maximizes entropy, subject to these constraints, is

$$\pi(\theta) = \begin{cases} \theta^{-1}[2\log(z/a)]^{-1} & \text{if } a < \theta < z, \\ \theta^{-1}[2\log(b/z)]^{-1} & \text{if } z < \theta < b. \end{cases}$$

20. Assume θ is from a location density. It is believed that $-K < \theta < K$, and that θ has prior mean $\mu \neq 0$. Show that the prior distribution which maximizes entropy, subject to these constraints, is given by

$$\pi(\theta) = \frac{ze^{z\theta/K}}{2K\sinh(z)} I_{(-K,K)}(\theta),$$

where z is the solution to the equation

$$(K^{-1}\mu z + 1)\tanh(z) - z = 0.$$

(Sinh and tanh stand for hyperbolic sine and hyperbolic tangent, respectively.)

Section 3.5

21. Suppose $X \sim \mathcal{B}(n, \theta)$ and that θ has a $\mathcal{B}e(\alpha, \beta)$ prior distribution.
   (a) Find $m(x)$.
   (b) Show that, if $m(x)$ is constant, then it must be the case that $\alpha = \beta = 1$. (Stigler (1982) reports that this was the motivation for the use of a uniform prior in Bayes (1763).)

22. Suppose X, the failure time of an electronic component, has density (on $(0, \infty)$) $f(x|\theta) = \theta^{-1} \exp\{-x/\theta\}$. The unknown θ has an $\mathcal{IG}(1, 0.01)$ prior distribution. Calculate the (marginal) probability that the component fails before time 200.

23. Suppose that $X_1, \ldots, X_p$ are independent, and that $X_i \sim \mathcal{P}(\theta_i)$, $i = 1, \ldots, p$. If the θ_i are i.i.d. $\mathcal{G}(\alpha, \beta)$, find the marginal density, m, for $\mathbf{X} = (X_1, \ldots, X_p)^t$.

24. Suppose, for $i = 1, \ldots, p$, that $\theta_i = \mu_i + \varepsilon_i$, where the ε_i are i.i.d. $\mathcal{N}(0, \sigma_\pi^2)$, σ_π^2 unknown.
   (a) If $\mu_i = i\xi$, where ξ is in $(0, 1)$ but is otherwise unknown, describe the implied class, Γ, of priors for $\boldsymbol{\theta} = (\theta_1, \ldots, \theta_p)^t$.
   (b) If the $\mu_i = \xi$ for all i, and ξ is known to have a $\mathcal{N}(1, 1)$ distribution (independent of the ε_i), show that the implied class of priors for $\boldsymbol{\theta} = (\theta_1, \ldots, \theta_p)^t$ can be written

$$\Gamma = \{\pi: \pi \text{ is } \mathcal{N}_p(\mathbf{1}, \sigma_\pi^2 \mathbf{I} + (\underline{\mathbf{1}})), \sigma_\pi^2 > 0\},$$

   where $\mathbf{1} = (1, 1, \ldots, 1)^t$ and $(\mathbf{1})$ is the $(p \times p)$ matrix of all ones.

25. Suppose, for $i = 1, \ldots, p$, that $X_i \sim \mathcal{N}(\theta_i, \sigma_f^2)$, and that the X_i are independent.
   (a) Find the ML-II prior in the situation of Exercise 24(a), for any given $\mathbf{x}$.
   (b) Find the ML-II prior in the situation of Exercise 24(b), for any given $\mathbf{x}$.

26. In Exercise 23, suppose that $p = 3$, $x_1 = 3$, $x_2 = 0$, and $x_3 = 5$. Find the ML-II prior.

27. Suppose that $X \sim \mathcal{N}(\theta, 1)$, π_0 is a $\mathcal{N}(0, 2.19)$ prior, and Γ is as in (3.17) with $\varepsilon = 0.2$.
   (a) If $\mathcal{Q}$ is the class of all distributions, find the ML-II prior for any given x.
   (b) If $\mathcal{Q} = \{q: q \text{ is } \mathcal{N}(0, \tau^2), \tau^2 \geq 1\}$, find the ML-II prior for any given x.

28. Let $X_1 \sim \mathcal{N}(\theta_1, 1)$ and $X_2 \sim \mathcal{N}(\theta_2, 1)$ be independent. Suppose θ_1 and θ_2 are i.i.d. from the prior π_0. Find the ML-II prior, over the class of *all* priors,
   (a) when $x_1 = 0$ and $x_2 = 1$;
   (b) when $x_1 = 0$ and $x_2 = 4$.
   (It is a fact, which you may assume to be true, that π_0 gives probability to at most two points, and that π_0 is symmetric about $\frac{1}{2}(x_1 + x_2)$.)

29. Using the moment approach in the situation of Exercise 23, show that estimates of the hyperparameters, α and β, are (when $0 < \bar{x} < s^2$)

$$\hat{\alpha} = \frac{\bar{x}^2}{s^2 - \bar{x}}, \qquad \hat{\beta} = \frac{s^2 - \bar{x}}{\bar{x}},$$

   where $\bar{x}$ and s^2 are the sample mean and variance, respectively.

30. Using the moment approach, find estimates of the hyperparameters in the situation of
    (a) Exercise 25(a).
    (b) Exercise 25(b).

31. In terms of the distance measure $d(f, g)$ given in Subsection 3.5.6, which of the two densities (on R^1),

$$g_1(x) = (6\pi)^{-1/2} e^{-x^2/6} \quad \text{or} \quad g_2(x) = 0.5 e^{-|x|},$$

is closer to $f(x) = (2\pi)^{-1/2} e^{-x^2/2}$.

32. A test of H_0: $\theta = 1$ versus H_1: $\theta = 2$ is to be conducted, where θ is the parameter of a $\mathcal{U}(0, \theta)$ distribution. (Assume θ can only be 1 or 2.) It is desired to estimate the prior probability that $\theta = 1$, i.e., $\pi(1) = 1 - \pi(2)$. There are available n independent past observations $x_1, \ldots, x_n$, each of which had a $\mathcal{U}(0, \theta_i)$ distribution (with the θ_i having been a random sample from the prior distribution).
    (a) Show that, according to the "minimum distance" method, the optimal estimate of $\pi(1)$ is

$$\hat{\pi}(1) = [P^{\hat{m}}(0 < X < 1) - P^{\hat{m}}(1 < X < 2)]^+,$$

    where $\hat{m}$ is the estimated marginal distribution of the X_i.
    (b) If the simple estimate of m in (3.23) is used, show that

$$P^{\hat{m}_0}(1 < X < 2) = \frac{1}{n}[\text{the number of } x_i \text{ between 1 and 2}],$$

$$P^{\hat{m}_0}(0 < X < 1) = \frac{1}{n}[\text{the number of } x_i \text{ between 0 and 1}].$$

33. Suppose that $X_1, X_2, \ldots$ is a series of random variables which assume only the values 0 and 1. Let $m((x_1, \ldots, x_n))$ be the joint distribution of the first n random X_i.
    (a) If $m((x_1, \ldots, x_n)) = 2^{-n}$ for *all* n and $\mathbf{x}^n = (x_1, \ldots, x_n)$, find a representation of the form (3.28).
    (b) Show that there exists an m for which $m((x_1, \ldots, x_n)) = 2^{-n}$ holds for $n = 1$ and 2 (and associated $\mathbf{x}^n$), yet for which there is no representation of the form (3.28).

34. We will observe, for $i = 1, \ldots, p$, independent $X_i \sim \mathcal{N}(\theta_i, 900)$, where θ_i is the unknown mean yield per acre of corn hybrid i. It is felt that the θ_i are similar, to be modelled as being i.i.d. observations from a common population. The common mean of the θ_i is believed to be about 100, the standard error of this guess being estimated to be about 20. Nothing is deemed to be known about the common variance of the θ_i, so it will be given a *constant* (see Section 4.6 for reasons) noninformative prior. Find a reasonable hierarchical model for this prior information.

CHAPTER 4
Bayesian Analysis

4.1. Introduction

This chapter differs from later chapters in scope, because Bayesian analysis is an essentially self-contained paradigm for statistics. (Later chapters will, for the most part, deal with special topics within frequentist decision theory.) In order to provide a satisfactory perspective on Bayesian analysis, we will discuss Bayesian inference along with Bayesian decision theory. Before beginning the study, however, we briefly discuss the seven major arguments that can be given in support of Bayesian analysis. (Later chapters will similarly begin with a discussion of justifications.) Some of these arguments will not be completely understandable initially, but are best placed together for reference purposes.

I. *Important Prior Information May be Available*

This point has already been discussed (cf. Section 1.1), but bears repeating; in a significant fraction of statistical problems, failure to take prior information into account can lead to conclusions ranging from merely inferior to absurd. Of course, most non-Bayesians would agree to the use of reliable and significant prior information, so the impact of this consideration for general adoption of the Bayesian paradigm is unclear. One of the advantages of adopting the Bayesian viewpoint, on the other hand, is that one will be far more likely to recognize *when* significant prior information is available. Also, when significant prior information is available, the Bayesian approach shows how to sensibly utilize it, in contrast with most non-Bayesian approaches. As a simple example, a common situation in statistics is to have a study on several different, but similar, populations, for each of which is needed an estimate of variability. The question is—How should one use the important prior information that the populations are similar? Classical statistics is hard put to answer this. The usual approach, of deciding between separate estimates of variance or a pooled estimate (often based on some significance test) is an extremely crude utilization of the prior information. Bayesian analysis allows much more effective use of such prior information. (See Sections 4.5 and 4.6 for the Bayesian approach to this type of problem.)

II. *Uncertainty Should Be Quantified Probabilistically*

The business of statistics is to provide information or conclusions about uncertain quantities and to convey the extent of the uncertainty in the answer. The language of uncertainty is probability, and only the (conditional) Bayesian approach consistently uses this language to directly address uncertainty.

Consider, for instance, statistical hypothesis testing. The hypotheses are uncertain, and the result of a (conditional) Bayesian analysis will be simply the statement of the believed *probabilities* of the hypotheses (in light of the data and the prior information). In contrast, classical approaches provide "probabilities of Type I or Type II error" or "significance levels (P-values)," all of which are, at best, indirectly related to the *probabilities of the hypotheses* (see Subsection 4.3.3). As another example, we will see that when a Bayesian provides a "confidence set" (to be called a *credible set* in Bayesian language), the reported accuracy will be the believed *probability* that the set actually contains the unknown θ, in contrast to the classical coverage probability (see Subsection 1.6.2).

Of course, direct probability statements about uncertainty essentially *require* Bayesian analysis, and the thrust of classical statistics has been to find alternate ways of indicating accuracy. Indeed, even in classical elementary statistics courses it is common to spend a great deal of effort in pointing out that classical measures are *not* direct probability statements about uncertainty ("a 95% confidence interval is *not* to be interpreted as an interval that has probability 0.95 of containing θ").

There are two issues here, the first philosophically pragmatic and the second pragmatically pragmatic. The philosophically pragmatic issue is—What is the best method of quantifying uncertainty? The literature arguing in favor of direct probabilistic (Bayesian) quantification is vast (cf. Jeffreys (1961), Edwards, Lindman, and Savage (1963), deFinetti (1972, 1974, 1975), Box and Tiao (1973), Lindley (1982a), Good (1983), and Jaynes (1983)). And these contain not *just* philosophical arguments but also very compelling examples. (One such is discussed in Subsection 4.3.3, namely testing a point null hypothesis, where classical error probabilities or significance levels convey a completely misleading impression as to the validity of the null hypothesis.)

The second issue is the very practical issue of how statistical *users* (as opposed to professional statisticians or, at least, those with extensive statistical training) interpret statistical conclusions. Most such users (and probably the overwhelming majority) interpret classical measures in the direct probabilistic sense. (Indeed the only way we have had even moderate success, in teaching elementary statistics students that an error probability is not a probability of a hypothesis, is to teach enough Bayesian analysis to be able to demonstrate the difference with examples.) Among the formal evidence for this misinterpretation of classical measures is an amusing study in Diamond and Forrester (1983). If the majority of users are incapable of interpreting classical measures except, incorrectly, as Bayesian probabilities (whether through our teaching inadequacies or the inherent obscurity of the classical measures), can it be right to provide classical measures?

III. *The Conditional Viewpoint*

In Section 1.6 it was argued that analysis conditional on the observed data, as opposed to frequentist averaging over all potential data, is of crucial importance. There was also a brief mention of strong arguments supporting the Bayesian approach to conditional analysis. Some discussion of these arguments, with examples, will be given later in the chapter, after some needed Bayesian machinery has been developed. (See Berger and Wolpert (1984) for more complete discussion.)

IV. *Coherency and Rationality*

Note that, if one does have a loss function developed via utility theory and a prior distribution for θ, then one should (by the very nature of the utility construction) evaluate an action a by the Bayesian expected loss, and evaluate a decision rule δ by the Bayes risk. This presupposes the existence of a loss and prior, however, and thus may not be a very compelling argument for Bayesian analysis.

As with the axiomatic development of utility theory, however, one can develop various axiomatic bases for statistics itself. These involve the assumption that a preference ordering exists among actions, decision rules, inferences, or statistical procedures (depending on the perceived statistical goal), together with a set of axioms that any "coherent" or "rational" preference ordering should satisfy. Most people find these axioms quite believable (with some exceptions, such as LeCam (1977)), and yet it is invariably found that any rational preference ordering *must* correspond to some type of Bayesian preference ordering. This provides strong support for the Bayesian viewpoint, in that any approach which fails to correspond with a Bayesian analysis must violate some very "common sense" axiom of behavior.

We will not present any of the axiom systems here, partly because there are so many that it is hard to choose among them (Fishburn (1981) reviews over 30 different systems for decision theory alone), and partly because they are not that different in nature from the utility axioms. Some other references to axiom systems are Ramsey (1926) and Savage (1954) (two of the earliest—that in deFinetti (1972, 1974, 1975) was also developed quite early), Ferguson (1967) (a simple and easily understandable case), Rubin (1985) (the most general decision-theoretic system), and Lindley (1982a) and Lane and Sudderth (1983) (and the references therein) which consider "inference" axiom systems. A component of many of the non-decision-theoretic systems is "betting coherency" or the "Dutch book" argument, which will be discussed in Subsection 4.8.3.

There are several subtleties involved in the conclusion of many of these axiomatic developments. One is that the conclusion often requires only that the preference pattern correspond to a Bayesian analysis with respect to what is called a *finitely additive* prior; priors considered in this book are countably additive. Another subtlety is that the developments do not necessarily lead to a separation of the prior from the loss (cf. Savage (1954) and Rubin (1985)). Discussion of either of these issues is beyond the scope of the book.

There is a third subtlety which is very relevant, however, namely that these axiomatic developments do not say that "to be coherent or rational one must do a Bayesian analysis." Instead they say that "to be coherent or rational the analysis *must correspond to* a Bayesian analysis." This is an important logical difference, in that there could be rational methods of choosing a statistical procedure other than the Bayesian method of developing a prior (and loss) and doing a Bayesian analysis. In fact, the method (see Chapter 6 for explanation of terms)—for invariant decision problems with compact parameter space use the best invariant decision rule—can be shown to be "rational," since it corresponds to Bayesian analysis with respect to the prior which is the Haar measure on Θ (see Section 6.6). It is true, however, that no broadly applicable "rational" statistical paradigm, other than the Bayesian paradigm, has been found. And, even if such were found, it would be hard to argue that Bayesian analysis could be ignored: the method, to be rational, must yield an answer corresponding to that for a Bayesian analysis with respect to some prior, and it would be hard to justify the answer if the corresponding prior seemed unreasonable.

Another important aspect of the proper interpretation of the conclusion of the axiomatic developments is that the conclusion does not say that *any* Bayesian analysis is good. A Bayesian analysis may be "rational" in the weak axiomatic sense, yet be terrible in a practical sense if an inappropriate prior distribution is used. Thus Smith (1961) says,

> "Consistency is not necessarily a virtue: one can be consistently obnoxious."

Indeed there is no logical axiomatic guarantee that the best way to be consistent *and* nonobnoxious is through Bayesian analysis. See also Kiefer (1977b) and LeCam (1977) on this matter of interpretation of the rationality developments.

Another common criticism of rationality axioms is that numerous studies (cf. Ellsburg (1961)) have shown that people *do not* act in accordance with these axioms; hence the axioms are supposedly suspect. This criticism misses the point. The purpose of developing statistical methodology is to *improve* the way people act in the face of uncertainty, not to model how they do act. Smith (1984) responds to this behavioral criticism of the axioms by saying,

> "It is rather like arguing against the continued use of formal logic or arithmetic on the grounds that individuals can be shown to perform badly at deduction or long division in suitable experiments."

We have saved for last the most telling criticism of the majority of the axiomatic developments, namely the assumption that a preference ordering on *all* actions (or whatever) even exists. This simply is not going to be the case in what could be termed finite reality, namely the time and calculational constraints of our minds. We delay, however, formal discussion of this until Subsection 4.7.1, since it ties in with the argument that the ideal method of analysis is robust Bayesian analysis.

In spite of the limitations and "weak spots" in the rationality and coherency developments, they provide very powerful evidence that "truth" lies in a Bayesian direction. They also provide devastating weapons in exposing the "irrationality" of many other purported truths in statistics.

V. *Equivalence of Classically Optimal and Bayes Rules*

When a classical optimality principle is proposed, it is natural to reduce consideration to the class of statistical procedures which are acceptable according to this principle. In decision theory, for instance, it is natural to consider only admissible decision rules. As a special case, in simple versus simple hypothesis testing, with the desire for small error probabilities being the optimality principle, it is natural to reduce consideration to the class of "most powerful" tests. In such situations, it has repeatedly been shown that the class of "acceptable" decision rules corresponds to the class of Bayes decision rules (or some subclass or limits thereof). Chapter 8 presents a number of results of this type.

The correspondence between acceptable rules and Bayes rules clearly *suggests* that one should choose from among the acceptable rules through consideration of prior information. Consider, for instance, the situation mentioned above of testing between two simple hypotheses. The class of most powerful tests does essentially correspond to the class of Bayes tests (see Subsection 8.2.4); let us contrast the classical and Bayesian methods of selecting a test from this class. The classical approach is to select a most powerful test by choosing desired error probabilities. Although it is common to select the test having Type I error probability of either $\alpha = 0.05$ or $\alpha = 0.01$, this practice is not particularly endorsed by most statisticians, who instead tend to urge choice of α based on "careful consideration and comparison of the two hypotheses." Unfortunately, this is not much help; the way in which this "careful consideration" can suggest a choice of α is very obscure. Contrast this with the Bayesian (decision-theoretic) approach which says: (i) determine the prior probabilities of each of the two hypotheses; (ii) determine the relative harm in mistakingly concluding each hypothesis; and (iii) use the corresponding Bayes test. This Bayes test will correspond to a most powerful test at some α level, but there is explicit guidance concerning which most powerful test to use. To reiterate, the options here are to either make a *subjective* selection of prior probabilities and decision losses for the hypotheses, or to make a *subjective* selection of the α level; we have no idea how to intuitively do the latter, whereas the Bayesian inputs are intuitively accessible. And this argument can be reversed: the choice of an α level will correspond to certain prior and loss beliefs, and if these beliefs are unreasonable, how can that α level be reasonable? This type of situation is encountered very regularly in classical statistics; even after application of some optimality criterion, there are many possible procedures to use, and it seems hard to argue that an intuitive choice is actually better than attempted quantification of the factors that should be involved in the intuitive choice. Section 5.4 presents an interesting class of such examples.

The above argument, that classical optimality criteria themselves lead to consideration of the class of Bayes rules, is actually very similar to the coherency axiomatics. The big difference is that classical measures still form the basis of the development. Thus, in simple versus simple hypothesis testing, the reported accuracy would still presumably be the relevant error probability, even if α were chosen by Bayesian means (i.e., chosen so as to yield the Bayes test for the available prior information and loss). For this reason, the implications of this argument for Bayesian analysis are somewhat limited in scope. On the other hand, the fact that one seems to end up with Bayes rules, even when starting down a non-Bayesian route, is highly suggestive. Non-Bayesians tend to view this as a mathematical coincidence, but the history of science teaches us that coincidences are usually trying to tell us something.

VI. *Operational Advantages of Bayesian Analysis*

One response, to the above arguments that only Bayesian analysis is completely sensible, is to agree, but state that Bayesian analysis is too hard. After all, one must determine a prior distribution and (for decision theory) a loss function, and we have seen that such determinations are not easy.

Therefore, the argument goes, we should accept nonoptimal, but easier, classical analyses. (Classical decision-theorists clearly have a harder time making this argument, especially because most people find it easier to elicit priors than loss functions.)

This argument is not without some merit. Indeed, in Section 4.7 we will discuss certain situations in which a classical analysis seems to be the best bet because of formidable technical difficulties in conducting a Bayesian analysis (and because the classical analysis can be given a type of Bayesian validation). On the whole, however, we would argue that the situation is just the opposite: for a given investment of effort on a problem, it is Bayesian analysis that is most likely to yield the best answers. Until we see how Bayesian analysis works, of course, it is hard to compare its operational effectiveness with that of classical statistics. And even then the only way to really make a comparison is to go through a large number of situations and actually apply the two paradigms. Although we will see some such comparisons as we proceed, the book's emphasis on ideas (rather than applied methods) prevents a thorough comparison. (In some sense, in any case, one can only be convinced as to the operational advantages of Bayesian analysis by personally trying it on a number of problems, and seeing the clarity that results.) In spite of these provisos it will be helpful to list some of the significant operational advantages of Bayesian analysis:

1. Conditioning on the observed data x introduces great simplifications in the analysis, as mentioned in Section 1.6; one need only work with the observed likelihood function, rather than with averages over all $\mathscr{X}$. The advantages of this include:
   (a) Realistic models can more easily be chosen, since there is less need to have models which allow special frequentist calculations (cf. Rubin (1984)).
   (b) Robustness (of all types) can be dealt with more easily, since (by (a)) model variations cause no essential changes in the needed calculations (cf. Box (1980), Rubin (1984), and Smith (1984)).
   (c) Optional stopping (see Section 7.7) becomes permissible, and as Edwards, Lindman, and Savage (1963) say,

   > "The irrelevance of stopping rules to statistical inference restores a simplicity and freedom to experimental design .... Many experimenters would like to feel free to collect data until they have either conclusively proved their point, conclusively disproved it, or run out of time, money, or patience."

   Note that classical statistics does *not* allow one to stop an experiment when some *unanticipated* conclusive evidence appears.
   (d) Various kinds of censoring of data cause no essential problem for Bayesian analysis, but serious difficulties for classical analysis (cf. Berger and Wolpert (1984)).

2. Bayesian analysis with noninformative priors will be seen to be simple and remarkably successful. If one desires to avoid subjective prior specification, it is rare that one can do better than a Bayesian noninformative prior analysis.
3. Bayesian analysis yields a final distribution (the posterior distribution) for the unknown θ, and from this a large number of questions can be answered simultaneously. For instance, one can not only estimate θ, but can (with little additional effort) obtain accuracy measures for the estimate (or, alternatively, can obtain Bayesian "credible sets" for θ). This is in contrast to classical statistics, for which obtaining estimates for θ and determining the accuracies of these estimates (or confidence sets) are two *very* different problems. (Examples will be given in Sections 4.3 and 4.6.) Another illustration of the ease with which a variety of answers can be obtained from the posterior distribution is multiple hypothesis testing. One can calculate the Bayesian probability of any number of hypotheses (from the posterior distribution), while classical multiple hypothesis testing becomes much more difficult as more hypotheses are involved.
4. Bayesian analysis is an excellent alternative to use of large sample asymptotic statistical procedures. Bayesian procedures are almost always equivalent to the classical large sample procedures when the sample size is very large (see Subsection 4.7.8), and are likely to be more reasonable for moderate and (especially) small sample sizes (where many classical large sample procedures break down). Indeed, unless there have been extensive studies establishing the small and moderate sample size validity of a particular classical large sample procedure, almost any plausible Bayesian analysis would seem preferable (unless calculationally too difficult).

VII. *Objectivity and Scientific Uncertainty*

The most frequent criticism of Bayesian analysis is that different reasonable priors will often yield different answers, a supposedly unappealing lack of objectivity. The issue of objectivity was addressed in Section 3.7, where it was argued that, in attempting to achieve objectivity, there is no better way to go than Bayesian analysis with noninformative priors. We will not repeat these arguments here, but should mention the other side of the coin—when different reasonable priors yield substantially different answers, can it be right to state that there *is* a single answer? Would it not be better to admit that there is scientific uncertainty, with the conclusion depending on prior beliefs?

None of the above seven arguments for Bayesian analysis is completely convincing by itself, although we feel that most of the arguments can be made to be almost compelling, with appropriate "fleshing-out" (cf. the fleshing-out of IV in Berger and Wolpert (1984)). Taken as a whole, the arguments provide strong evidence, indeed, in support of the centrality of the Bayesian viewpoint to statistics. There are, of course, various criticisms of Bayesian analysis, some of which have already been mentioned (here and in Section 3.7), and others which will be encountered in Sections 4.7 through 4.12. These additional criticisms warn the Bayesian not to be too dogmatic. It is crucial to keep sight of the ideal Bayesian goal, but one should be pragmatic about how best to achieve this goal.

4.2. The Posterior Distribution

Bayesian analysis is performed by combining the prior information ($\pi(\theta)$) and the sample information (x) into what is called the posterior distribution of θ given x, from which all decisions and inferences are made. This section discusses the meaning and calculation of this distribution.

4.2.1. Definition and Determination

The *posterior distribution of* θ *given* x (or *posterior* for short) will be denoted $\pi(\theta|x)$, and, as the notation indicates, is defined to be the conditional distribution of θ given the sample observation x. Noting that θ and X have joint (subjective) density

$$h(x, \theta) = \pi(\theta) f(x|\theta),$$

and (as in Subsection 3.5.1) that X has marginal (unconditional) density

$$m(x) = \int_{\Theta} f(x|\theta) dF^{\pi}(\theta),$$

it is clear that (providing $m(x) \neq 0$)

$$\pi(\theta|x) = \frac{h(x, \theta)}{m(x)}.$$

The name "posterior distribution" is indicative of the role of $\pi(\theta|x)$. Just as the prior distribution reflects beliefs about θ *prior* to experimentation, so $\pi(\theta|x)$ reflects the updated beliefs about θ after (*posterior* to) observing the sample x. In other words, the posterior distribution combines the prior beliefs about θ with the information about θ contained in the sample, x, to give a composite picture of the final beliefs about θ. Note that the Likelihood Principle is implicitly assumed in the above statement, in that there is felt to be no sample information about θ other than that contained in $f(x|\theta)$ (for the given x).

In calculating the posterior distribution, it is often helpful to use the concept of sufficiency. Indeed if T is a sufficient statistic for θ with density $g(t|\theta)$, the following result can be established. (The proof is left as an exercise.)

Lemma 1. *Assume $m(t)$ (the marginal density of t) is greater than zero, and that the factorization theorem holds. Then, if $T(x)=t$,*

$$\pi(\theta|x)=\pi(\theta|t)=\frac{\pi(\theta)g(t|\theta)}{m(t)}.$$

The reason for determining $\pi(\theta|x)$ from a sufficient statistic T (if possible) is that $g(t|\theta)$ and $m(t)$ are usually much easier to handle than $f(x|\theta)$ and $m(x)$.

EXAMPLE 1. Assume $X \sim \mathcal{N}(\theta, \sigma^2)$, where θ is unknown but σ^2 is known. Let $\pi(\theta)$ be a $\mathcal{N}(\mu, \tau^2)$ density, where μ and τ^2 are known. Then

$$h(x, \theta)=\pi(\theta)f(x|\theta)=(2\pi\sigma\tau)^{-1}\exp\left\{-\frac{1}{2}\left[\frac{(\theta-\mu)^2}{\tau^2}+\frac{(x-\theta)^2}{\sigma^2}\right]\right\}.$$

To find $m(x)$, note that defining

$$\rho=\tau^{-2}+\sigma^{-2}=\frac{\tau^2+\sigma^2}{\tau^2\sigma^2}$$

and completing squares gives

$$\begin{aligned}\frac{1}{2}\left[\frac{(\theta-\mu)^2}{\tau^2}+\frac{(x-\theta)^2}{\sigma^2}\right]&=\frac{1}{2}\left[\left(\frac{1}{\tau^2}+\frac{1}{\sigma^2}\right)\theta^2-2\left(\frac{\mu}{\tau^2}+\frac{x}{\sigma^2}\right)\theta+\left(\frac{\mu^2}{\tau^2}+\frac{x^2}{\sigma^2}\right)\right]\\&=\frac{1}{2}\rho\left[\theta^2-\frac{2}{\rho}\left(\frac{\mu}{\tau^2}+\frac{x}{\sigma^2}\right)\theta\right]+\frac{1}{2}\left(\frac{\mu^2}{\tau^2}+\frac{x^2}{\sigma^2}\right)\\&=\frac{1}{2}\rho\left[\theta-\frac{1}{\rho}\left(\frac{\mu}{\tau^2}+\frac{x}{\sigma^2}\right)\right]^2-\frac{1}{2\rho}\left(\frac{\mu}{\tau^2}+\frac{x}{\sigma^2}\right)^2\\&\quad+\frac{1}{2}\left(\frac{\mu^2}{\tau^2}+\frac{x^2}{\sigma^2}\right)\\&=\frac{1}{2}\rho\left[\theta-\frac{1}{\rho}\left(\frac{\mu}{\tau^2}+\frac{x}{\sigma^2}\right)\right]^2+\frac{(\mu-x)^2}{2(\sigma^2+\tau^2)}.\end{aligned}$$

Hence

$$h(x, \theta)=(2\pi\sigma\tau)^{-1}\exp\left\{-\frac{1}{2}\rho\left[\theta-\frac{1}{\rho}\left(\frac{\mu}{\tau^2}+\frac{x}{\sigma^2}\right)\right]^2\right\}\exp\left\{-\frac{(\mu-x)^2}{2(\sigma^2+\tau^2)}\right\}$$

and

$$m(x)=\int_{-\infty}^{\infty}h(x, \theta)d\theta=(2\pi\rho)^{-1/2}(\sigma\tau)^{-1}\exp\left\{-\frac{(\mu-x)^2}{2(\sigma^2+\tau^2)}\right\}.$$

It follows that

$$\pi(\theta|x)=\frac{h(x,\theta)}{m(x)}=\left(\frac{\rho}{2\pi}\right)^{1/2}\exp\left\{-\frac{1}{2}\rho\left[\theta-\frac{1}{\rho}\left(\frac{\mu}{\tau^2}+\frac{x}{\sigma^2}\right)\right]^2\right\}.$$

Note, from the above equations, that the marginal distribution of X is $\mathcal{N}(\mu,(\sigma^2+\tau^2))$ and the posterior distribution of θ given x is $\mathcal{N}(\mu(x),\rho^{-1})$, where

$$\mu(x)=\frac{1}{\rho}\left(\frac{\mu}{\tau^2}+\frac{x}{\sigma^2}\right)=\frac{\sigma^2}{\sigma^2+\tau^2}\mu+\frac{\tau^2}{\sigma^2+\tau^2}x=x-\frac{\sigma^2}{\sigma^2+\tau^2}(x-\mu).$$

As a concrete example, consider the situation wherein a child is given an intelligence test. Assume that the test result X is $\mathcal{N}(\theta,100)$, where θ is the true IQ (intelligence) level of the child, as measured by the test. (In other words, if the child were to take a large number of independent similar tests, his average score would be about θ.) Assume also that, in the population as a whole, θ is distributed according to a $\mathcal{N}(100,225)$ distribution. Using the above equations, it follows that, marginally, X is $\mathcal{N}(100,325)$, while the posterior distribution of θ given x is normal with mean

$$\mu(x)=\frac{(100)(100)+x(225)}{(100+225)}=\frac{400+9x}{13}$$

and variance

$$\rho^{-1}=\frac{(100)(225)}{(100+225)}=\frac{900}{13}=69.23.$$

Thus, if a child scores 115 on the test, his true IQ θ has a $\mathcal{N}(110.39,69.23)$ posterior distribution. Note that, as discussed in Subsection 3.5.1, the $\mathcal{N}(100,325)$ marginal distribution of X would be the anticipated distribution of actual test scores in the population.

EXAMPLE 2. Assume a sample $\mathbf{X}=(X_1,\ldots,X_n)$ from a $\mathcal{N}(\theta,\sigma^2)$ distribution is to be taken (σ^2 known), and that θ has a $\mathcal{N}(\mu,\tau^2)$ density. Since $\bar{X}$ is sufficient for θ, it follows from Lemma 1 that $\pi(\theta|x)=\pi(\theta|\bar{x})$. Noting that $\bar{X}\sim\mathcal{N}(\theta,\sigma^2/n)$, it can be concluded from Example 1 that the posterior distribution of θ given $\mathbf{x}=(x_1,\ldots,x_n)$ is $\mathcal{N}(\mu(\mathbf{x}),\rho^{-1})$, where

$$\mu(\mathbf{x})=\frac{\sigma^2/n}{(\tau^2+\sigma^2/n)}\mu+\frac{\tau^2}{(\tau^2+\sigma^2/n)}\bar{x}$$

and $\rho=(n\tau^2+\sigma^2)/\tau^2\sigma^2$.

EXAMPLE 3. A blood test is to be conducted to help indicate whether or not a person has a particular disease. The result of the test is either positive (denoted $x=1$) or negative (denoted $x=0$). Letting θ_1 denote the state of nature "the disease is present" and θ_2 denote the state of nature "no disease is present," assume it is known that $f(1|\theta_1)=0.8$, $f(0|\theta_1)=0.2$, $f(1|\theta_2)=0.3$, and $f(0|\theta_2)=0.7$. According to prior information, $\pi(\theta_1)=0.05$ and $\pi(\theta_2)=0.95$. Then

$$m(1)=f(1|\theta_1)\pi(\theta_1)+f(1|\theta_2)\pi(\theta_2)=0.04+0.285=0.325,$$

$$m(0)=f(0|\theta_1)\pi(\theta_1)+f(0|\theta_2)\pi(\theta_2)=0.01+0.665=0.675,$$

$$\pi(\theta|x=1)=\frac{f(1|\theta)\pi(\theta)}{m(1)}=\begin{cases}\dfrac{0.04}{0.325}=0.123 & \text{if } \theta=\theta_1,\\[2ex] \dfrac{0.285}{0.325}=0.877 & \text{if } \theta=\theta_2,\end{cases}$$

and

$$\pi(\theta|x=0)=\frac{f(0|\theta)\pi(\theta)}{m(0)}=\begin{cases}\dfrac{0.01}{0.675}=0.0148 & \text{if } \theta=\theta_1,\\[2ex] \dfrac{0.665}{0.675}=0.9852 & \text{if } \theta=\theta_2.\end{cases}$$

It is interesting to observe that, even if the blood test is positive, there is still only a 12.3% chance of the disease being present. Note that $m(1)$ and $m(0)$ give the overall proportions of positive and negative tests that can be anticipated. These might be useful for logistic purposes if, say, the positive tests were to be followed up with more elaborate testing; 32.5% of those initially tested would require the more elaborate testing.

In discrete situations, such as Example 3, the formula for $\pi(\theta|x)$ is commonly known as Bayes's theorem, and was discovered by Bayes (1763). The typical phrasing of Bayes's theorem is in terms of disjoint events $A_1, A_2, \ldots, A_n$, whose union has probability one (i.e., one of the A_i is certain to occur). Prior probabilities $P(A_i)$, for the events, are assumed known. An event B occurs, for which $P(B|A_i)$ (the conditional probability of B given A_i) is known for each A_i. Bayes's theorem then states that

$$P(A_i|B)=\frac{P(B|A_i)P(A_i)}{\sum_{j=1}^{n} P(B|A_j)P(A_j)}.$$

These probabilities reflect our revised opinions about the A_i, in light of the knowledge that B has occurred. Replacing A_i by θ_i and B by x, shows the equivalence of this to the formula for the posterior distribution.

Example 4. In airplanes there is a warning light that goes on if the landing gear fails to fully extend. Sometimes the warning light goes on even when the landing gear has extended. Let A_1 denote the event "the landing gear extends" and A_2 denote the event "the landing gear fails to extend." (Note that A_1 and A_2 are disjoint and one of the two must occur.) Let B be the event that the warning light goes on. It is known that the light will go on with probability 0.999 if A_2 occurs (i.e., $P(B|A_2)=0.999$), while $P(B|A_1)=0.005$. Records show that $P(A_1)=0.997$ and $P(A_2)=0.003$. It is desired to determine the probability that the landing gear has extended, even though the warning light has gone on. This is simply $P(A_1|B)$, and from Bayes's theorem is given by

$$P(A_1|B)=\frac{(0.005)(0.997)}{(0.005)(0.997)+(0.999)(0.003)}=0.62.$$

4.2.2. Conjugate Families

In general, $m(x)$ and $\pi(\theta|x)$ are not easily calculable. If, for example, X is $\mathcal{N}(\theta,\sigma^2)$ and θ is $\mathcal{C}(\mu,\beta)$, then $\pi(\theta|x)$ can only be evaluated numerically. A large part of the Bayesian literature is devoted to finding prior distributions for which $\pi(\theta|x)$ can be easily calculated. These are the so called *conjugate priors*, and were developed extensively in Raiffa and Schlaifer (1961).

Definition 1. Let $\mathcal{F}$ denote the class of density functions $f(x|\theta)$ (indexed by θ). A class $\mathcal{P}$ of prior distributions is said to be a *conjugate family* for $\mathcal{F}$ if $\pi(\theta|x)$ is in the class $\mathcal{P}$ for all $f\in\mathcal{F}$ and $\pi\in\mathcal{P}$.

Example 1 shows that the class of normal priors is a conjugate family for the class of normal (sample) densities. (If X has a normal density and θ has a normal prior, then the posterior density of θ given x is also normal.)

For a given class of densities $\mathcal{F}$, a conjugate family can frequently be determined by examining the likelihood functions $l_x(\theta)=f(x|\theta)$, and choosing, as a conjugate family, the class of distributions with the same functional form as these likelihood functions. The resulting priors are frequently called *natural* conjugate priors.

When dealing with conjugate priors, there is generally no need to explicitly calculate $m(x)$. The reason is that, since $\pi(\theta|x)=h(x,\theta)/m(x)$, the factors involving θ in $\pi(\theta|x)$ must be the same as the factors involving θ in $h(x,\theta)$. Hence it is only necessary to look at the factors involving θ in $h(x,\theta)$, and see if these can be recognized as belonging to a particular distribution. If so, $\pi(\theta|x)$ is that distribution. The marginal density $m(x)$ can then be determined, if desired, by dividing $h(x,\theta)$ by $\pi(\theta|x)$. An example of the above ideas follows.

Example 5. Assume $\mathbf{X}=(X_1,\ldots,X_n)$ is a sample from a Poisson distribution. Thus $X_i\sim\mathcal{P}(\theta)$, $i=1,\ldots,n$, and

$$f(\mathbf{x}|\theta)=\prod_{i=1}^{n}\left[\frac{\theta^{x_i}e^{-\theta}}{x_i!}\right]=\frac{\theta^{n\bar{x}}e^{-n\theta}}{\prod_{i=1}^{n}[x_i!]}.$$

Here, $\mathcal{F}$ is the class of all such densities. Observing that the likelihood function for such densities resembles a gamma density, a plausible guess for a conjugate family of prior distributions is the class of gamma distributions. Thus assume $\theta \sim \mathscr{G}(\alpha, \beta)$, and observe that

$$h(\mathbf{x}, \theta) = f(\mathbf{x}|\theta)\pi(\theta) = \frac{e^{-n\theta}\theta^{n\bar{x}}}{\prod_{i=1}^{n}[x_i!]} \cdot \frac{\theta^{\alpha-1}e^{-\theta/\beta}I_{(0,\infty)}(\theta)}{\Gamma(\alpha)\beta^{\alpha}}$$

$$= \frac{e^{-\theta(n+1/\beta)}\theta^{(n\bar{x}+\alpha-1)}I_{(0,\infty)}(\theta)}{\Gamma(\alpha)\beta^{\alpha}\prod_{i=1}^{n}[x_i!]}.$$

The factors involving θ in this last expression are clearly recognizable as belonging to a $\mathscr{G}(n\bar{x}+\alpha, [n+1/\beta]^{-1})$ distribution. This must then be $\pi(\theta|\mathbf{x})$. Since this posterior is a gamma distribution, it follows that the class of gamma distributions is indeed a (natural) conjugate family for $\mathscr{F}$.

In this example, $m(\mathbf{x})$ can be determined by dividing $h(\mathbf{x}, \theta)$ by $\pi(\theta|\mathbf{x})$ and cancelling factors involving θ. The result is

$$m(\mathbf{x}) = \frac{h(\mathbf{x}, \theta)}{\pi(\theta|\mathbf{x})} = \frac{(\Gamma(\alpha)\beta^{\alpha}\prod_{i=1}^{n}[x_i!])^{-1}}{\{\Gamma(\alpha+n\bar{x})[n+1/\beta]^{-(\alpha+n\bar{x})}\}^{-1}}.$$

Besides providing for easy calculation of $\pi(\theta|x)$, conjugate priors have the intuitively appealing feature of allowing one to begin with a certain functional form for the prior and end up with a posterior of the same functional form, but with parameters updated by the sample information. In Example 1, for instance, the prior mean μ gets updated by x to become the posterior mean

$$\mu(x) = \frac{\tau^2}{\sigma^2+\tau^2}x + \frac{\sigma^2}{\sigma^2+\tau^2}\mu.$$

The prior variance τ^2 is combined with the data variance σ^2 to give the posterior variance

$$\rho^{-1} = \left(\frac{1}{\sigma^2} + \frac{1}{\tau^2}\right)^{-1}.$$

This updating of parameters provides an easy way of seeing the effect of prior and sample information. It also makes useful the concept of equivalent sample size discussed in Section 3.2.

These attractive properties of conjugate priors are, however, only of secondary importance compared to the basic question of whether or not a conjugate prior can be chosen which gives a reasonable approximation to the true prior. Many Bayesians say this can be done, arguing for example that, in dealing with a normal mean, the class of $\mathscr{N}(\mu, \tau^2)$ priors is rich enough to include approximations to most reasonable priors. Unfortunately, in Section 4.7 we will encounter reasons to doubt this belief, observing that using a normal prior can sometimes result in unappealing conclusions. Most of the examples and problems in this chapter will make use of conjugate priors, however, due to the resulting ease in calculations. And, at least for initial analyses, conjugate priors such as above can be quite useful in practice.

There do exist conjugate families of priors other than natural conjugate priors. One trivial such class is the class of *all* distributions. (The posterior is certainly in this class.) A more interesting example is the class of finite mixtures of natural conjugate priors. In Example 1, for instance, the class of all priors of the form (for fixed m, say)

$$\pi(\theta)=\sum_{i=1}^{m} w_i\pi_i(\theta), \tag{4.1}$$

where $\sum_{i=1}^{m} w_i=1$ (all $w_i\geq 0$) and the π_i are normally distributed, can be shown to be a conjugate class—the proof will be left for an exercise. The use of such mixtures allows approximations to bimodal and more complicated subjective prior distributions, and yet preserves much of the calculational simplicity of natural conjugate priors. Development and uses of such mixture conjugate classes can be found in Dalal and Hall (1983) and Diaconis and Ylvisaker (1984). Jewell (1983) generalizes natural conjugate priors in a different direction.

4.2.3. Improper Priors

The analysis leading to the posterior distribution can formally be carried out even if $\pi(\theta)$ is an improper prior. For example, if $X\sim\mathcal{N}(\theta,\sigma^2)$ (σ^2 known) and the noninformative prior $\pi(\theta)=1$ is used, then

$$h(x,\theta)=f(x|\theta)\pi(\theta)=f(x|\theta),$$

$$m(x)=\int_{-\infty}^{\infty} f(x|\theta)d\theta=(2\pi)^{-1/2}\sigma^{-1}\int_{-\infty}^{\infty}\exp\left\{\frac{-(x-\theta)^2}{2\sigma^2}\right\}d\theta=1,$$

and

$$\pi(\theta|x)=\frac{h(x,\theta)}{m(x)}=(2\pi)^{-1/2}\sigma^{-1}\exp\left\{\frac{-(\theta-x)^2}{2\sigma^2}\right\}.$$

Hence the posterior distribution of θ given x is $\mathcal{N}(x,\sigma^2)$. Of course, this posterior distribution cannot rigorously be considered to be the conditional distribution of θ given x, but various heuristic arguments can be given to support such an interpretation. For example, taking a suitable sequence of finite priors $\pi_n(\theta)$, which converge to $\pi(\theta)$ as $n\to\infty$, it can be shown that the corresponding posteriors, $\pi_n(\theta|x)$, converge to $\pi(\theta|x)$. Other arguments using finitely additive probability measures can be given to support the informal interpretation of $\pi(\theta|x)$ as the conditional density of θ given x.

4.3. Bayesian Inference

Inference problems concerning θ can easily be dealt with using Bayesian analysis. The idea is that, since the posterior distribution supposedly contains all the available information about θ (both sample and prior information), any inferences concerning θ should consist solely of features of this distribution. Several justifications for this view were mentioned in Sections 1.6 and 4.1.

Statistical inference is not the main subject of this book, so we will only indicate the basic elements of the Bayesian approach to inference. (For more thorough treatments of Bayesian inference, see Jeffreys (1961), Zellner (1971), and Box and Tiao (1973).) Of course, inference can also be treated as a decision-theoretic problem (see Subsection 2.4.3), and in Subsection 4.4.4 we briefly illustrate this possibility.

Statistical inference is often associated with a desire for "objectivity." This issue was discussed in Section 3.7, wherein it was briefly argued that the most reasonable method of "attempting" to be objective is to perform a Bayesian analysis with a noninformative prior. We will, therefore, especially emphasize use of noninformative priors in this section.

Some Bayesians maintain that inference should ideally consist of simply reporting the *entire* posterior distribution $\pi(\theta|x)$ (maybe for a noninformative prior). We do not disagree in principle, since from the posterior one can derive any feature of interest, and indeed a visual inspection of the graph of the posterior will often provide the best insight concerning θ (at least in low dimensions). More standard uses of the posterior are still helpful, however (especially for outside consumption), and will be discussed in this section. Note, once again, that all measures that will be discussed are conditional in nature, since they depend only on the posterior distribution (which involves the experiment only through the observed likelihood function).

4.3.1. Estimation

The simplest inferential use of the posterior distribution is to report a point estimate for θ, with an associated measure of accuracy.

I. *Point Estimates*

To estimate θ, a number of classical techniques can be applied to the posterior distribution. The most common classical technique is maximum likelihood estimation, which chooses, as the estimate of θ, the value $\hat{\theta}$ which maximizes the likelihood function $l(\theta)=f(x|\theta)$. The analogous Bayesian estimate is defined as follows.

Definition 2. The *generalized maximum likelihood* estimate of θ is the largest mode, $\hat{\theta}$, of $\pi(\theta|x)$ (i.e., the value $\hat{\theta}$ which maximizes $\pi(\theta|x)$, considered as a function of θ).

Obviously $\hat{\theta}$ has the interpretation of being the "most likely" value of θ, given the prior and the sample x.

EXAMPLE 1 (continued). When f and π are normal densities, the posterior density was seen to be $\mathcal{N}(\mu(x), \rho^{-1})$. A normal density achieves its maximum value at the mean, so the generalized maximum likelihood estimate of θ in this situation is

$$\hat{\theta} = \mu(x) = \frac{\sigma^2 \mu}{\sigma^2 + \tau^2} + \frac{\tau^2 x}{\sigma^2 + \tau^2}.$$

EXAMPLE 6. Assume

$$f(x|\theta) = e^{-(x-\theta)} I_{(\theta,\infty)}(x),$$

and $\pi(\theta) = [\pi(1+\theta^2)]^{-1}$. Then

$$\pi(\theta|x) = \frac{e^{-(x-\theta)} I_{(\theta,\infty)}(x)}{m(x)(1+\theta^2)\pi}.$$

To find the $\hat{\theta}$ maximizing this quantity, note first that only $\theta \le x$ need be considered. (If $\theta > x$, then $I_{(\theta,\infty)}(x) = 0$ and $\pi(\theta|x) = 0$.) For such θ,

$$\begin{aligned}\frac{d}{d\theta} \pi(\theta|x) &= \frac{e^{-x}}{m(x)\pi} \left[\frac{e^{\theta}}{1+\theta^2} - \frac{2\theta e^{\theta}}{(1+\theta^2)^2} \right] \\ &= \frac{e^{-x}}{m(x)\pi} \frac{e^{\theta}(\theta-1)^2}{(1+\theta^2)^2}.\end{aligned}$$

Since this derivative is always positive, $\pi(\theta|x)$ is increasing for $\theta \le x$. It follows that $\pi(\theta|x)$ is maximized at $\hat{\theta} = x$, which is thus the generalized maximum likelihood estimate of θ.

Other common Bayesian estimates of θ include the mean and the median of $\pi(\theta|x)$. In Example 1 these clearly coincide with $\mu(x)$, the mode. In Example 6, however, the mean and median will differ from the mode, and must be calculated numerically.

The mean and median (and mode) are relatively easy to find when the prior, and hence posterior, are from a conjugate family of distributions. In Example 5, for instance, $\pi(\theta|x)$ is $\mathcal{G}(\alpha + n\bar{x}, [n + 1/\beta]^{-1})$, which has mean $[\alpha + n\bar{x}]/[n + 1/\beta]$. The median can be found using tables of the gamma distribution.

The mean and median of the posterior are frequently better estimates of θ than the mode. It is probably worthwhile to calculate and compare all three in a Bayesian study, especially with regard to their robustness to changes in the prior.

As mentioned at the beginning of this section, Bayesian inference using a noninformative prior is often an easy and reasonable method of analysis. The following example gives a simple demonstration of this in an estimation problem.

EXAMPLE 7. A not uncommon situation is to observe $X \sim \mathcal{N}(\theta, \sigma^2)$ (for simplicity assume σ^2 is known), where θ is a measure of some clearly positive quantity. The classical estimate of θ is x, which is clearly unsuitable when x turns out to be negative. A reasonable way of developing an alternative estimate (assuming no specific prior knowledge is available) is to use the noninformative prior $\pi(\theta) = I_{(0,\infty)}(\theta)$ (since θ is a location parameter). The resulting posterior is

$$\pi(\theta \mid x) = \frac{\exp\{-(\theta - x)^2/2\sigma^2\} I_{(0,\infty)}(\theta)}{\int_0^\infty \exp\{-(\theta - x)^2/2\sigma^2\} d\theta}.$$

Making the change of variables $\eta = (\theta - x)/\sigma$, the mean of the posterior can be seen to be

$$\begin{aligned} E^{\pi(\theta|x)}[\theta] &= \frac{\int_0^\infty \theta \exp\{-(\theta - x)^2/2\sigma^2\} d\theta}{\int_0^\infty \exp\{-(\theta - x)^2/2\sigma^2\} d\theta} \\ &= \frac{\int_{-(x/\sigma)}^\infty (\sigma\eta + x) \exp\{-\eta^2/2\} \sigma \, d\eta}{\int_{-(x/\sigma)}^\infty \exp\{-\eta^2/2\} \sigma \, d\eta} \\ &= x + \frac{(2\pi)^{-1/2} \sigma \int_{-(x/\sigma)}^\infty \eta \exp\{-\eta^2/2\} d\eta}{1 - \Phi(-x/\sigma)} \\ &= x + \frac{(2\pi)^{-1/2} \sigma \exp\{-x^2/2\sigma^2\}}{1 - \Phi(-x/\sigma)}, \end{aligned}$$

where Φ is the standard normal c.d.f. This estimate of θ is quite simple and easy to use.

Example 7 is a simple case of a common type of statistical problem that is quite difficult to handle classically, namely the situation of a restricted parameter space. Restricted parameter spaces can be of many types. The situation in Example 7, where the parameter is known to be positive (or, in higher dimensional settings, where the signs of all coordinates of the parameter are known), is one important case. Among the practical problems which involve such a parameter space are "variance component" problems (cf. Hill (1965, 1977)). Another typical occurrence of restricted parameter spaces is when $\boldsymbol{\theta} = (\theta_1, \ldots, \theta_p)^t$ and the θ_i are known to be ordered in some fashion (cf. Exercise 24). Much more complicated scenarios can also occur, of this same form that θ is known to be in $\Theta' \subset \Theta$.

As in Example 7, a restricted parameter space Θ' can be easily handled using noninformative prior Bayesian analysis. Simply let $\pi(\theta) = \pi_0(\theta) I_{\Theta'}(\theta)$ (recall that $I_{\Theta'}(\theta) = 1$ if $\theta \in \Theta'$ and equals zero, otherwise), where $\pi_0(\theta)$ is an appropriate noninformative prior for unrestricted θ, and proceed. There may be calculational difficulties in determining, say, the posterior mean, but there is no problem conceptually. (See Section 4.9 for discussion of some aspects of Bayesian calculation.)

Sometimes, one may be dealing with situations in which the parameters are not necessarily strictly ordered, but in which they tend to be ordered with "high probability." Such situations are virtually impossible to handle classically, but prior distributions incorporating such "stochastic ordering" can be constructed and used. See Proschan and Singpurwalla (1979, 1980) and Jewell (1979) for development.

II. *Estimation Error*

When presenting a statistical estimate, it is usually necessary to indicate the accuracy of the estimate. The customary Bayesian measure of the accuracy of an estimate is (in one dimension) the posterior variance of the estimate, which is defined as follows.

Definition 3. If θ is a real valued parameter with posterior distribution $\pi(\theta|x)$, and δ is the estimate of θ, then the *posterior variance of* δ is

$$V_\delta^\pi(x) \equiv E^{\pi(\theta|x)}[(\theta-\delta)^2].$$

When δ is the posterior mean

$$\mu^\pi(x) \equiv E^{\pi(\theta|x)}[\theta],$$

then $V^\pi(x) \equiv V_{\mu^\pi}^\pi(x)$ will be called simply the *posterior variance* (and it is indeed the variance of θ for the distribution $\pi(\theta|x)$). The *posterior standard deviation* is $\sqrt{V^\pi(x)}$.

It is customary to use $\sqrt{V_\delta^\pi(x)}$ as the "standard error" of the estimate δ. For calculational purposes, it is often helpful to note that

$$\begin{aligned} V_\delta^\pi(x) &= E^{\pi(\theta|x)}[(\theta-\delta)^2] = E[(\theta-\mu^\pi(x)+\mu^\pi(x)-\delta)^2] \\ &= E[(\theta-\mu^\pi(x))^2] + E[2(\theta-\mu^\pi(x))(\mu^\pi(x)-\delta)] + E[(\mu^\pi(x)-\delta)^2] \\ &= V^\pi(x) + 2(\mu^\pi(x)-\delta)(E[\theta]-\mu^\pi(x)) + (\mu^\pi(x)-\delta)^2 \qquad (4.2) \\ &= V^\pi(x) + (\mu^\pi(x)-\delta)^2. \end{aligned}$$

Observe from (4.2) that the posterior mean, $\mu^\pi(x)$, minimizes $V_\delta^\pi(x)$ (over all δ), and hence is the estimate with smallest standard error. For this reason, it is customary to use $\mu^\pi(x)$ as the estimate for θ and report $\sqrt{V^\pi(x)}$ as the standard error.

Example 1 (continued). It is clear that

$$V^\pi(x) = \rho^{-1} = \frac{\sigma^2\tau^2}{\sigma^2+\tau^2}.$$

Thus, in the example of intelligence testing, the child with $x=115$ would be reported as having as estimated IQ of $\mu^\pi(115)=110.39$, with associated standard error $\sqrt{V^\pi(115)} = \sqrt{69.23} = 8.32$.

The classical estimate of θ for the general normal problem is just $\delta = x$, which (using (4.2)) has

$$\begin{aligned} V_\delta^\pi(x) &= V^\pi(x) + (\mu^\pi(x) - x)^2 \\ &= V^\pi(x) + \left(\frac{\sigma^2 \mu}{\sigma^2 + \tau^2} + \frac{\tau^2 x}{\sigma^2 + \tau^2} - x \right)^2 \\ &= V^\pi(x) + \frac{\sigma^4}{(\sigma^2 + \tau^2)} (\mu - x)^2. \end{aligned} \tag{4.3}$$

Note that, in the IQ example, the classical estimate $\delta = x = 115$ would have standard error (with respect to $\pi(\theta|x)$) of (using (4.2))

$$\sqrt{V_{115}^\pi(115)} = [69.23 + (110.39 - 115)^2]^{1/2} = \sqrt{90.48} = 9.49.$$

Of course, the *classical standard error* of $\delta = x$ is σ, the sample standard deviation. It is interesting to observe from (4.3) that, if $(\mu - x)^2 > (\sigma^2 + \tau^2)$, then $V_\delta^\pi(x) > \sigma^2$, so that a Bayesian who believed in π would feel *dishonest* in reporting the smaller number, σ, as the standard error.

In the above example, a Bayesian would estimate θ by $\mu^\pi(x)$ with standard error $\sqrt{V^\pi(x)} = \rho^{-1/2}$, which is *less than* σ, the classical standard error of the classical estimate $\delta = x$. This is usually (but not always) true for Bayesian estimation: because the Bayesian is using prior information as well as sample information to estimate θ, his estimate will typically have a (Bayesian) standard error that is smaller than the standard error that a classical statistician would report for the classical estimate. (Of course, someone who did not believe that π was a reasonable prior would view the smaller Bayesian standard error as misleading, so we are not claiming this to be an *advantage* of the Bayesian approach.) The major exceptions to this pattern of smaller Bayesian standard error occur when noninformative priors are used, and for certain flat-tailed priors (cf. O'Hagan (1981)).

EXAMPLE 8. In Subsection 4.2.3 it was shown that, if $X \sim \mathcal{N}(\theta, \sigma^2)$ (σ^2 known) and the noninformative prior $\pi(\theta) = 1$ is used, then the posterior distribution of θ given x is $\mathcal{N}(x, \sigma^2)$. Hence the posterior mean is $\mu^\pi(x) = x$, and the posterior variance and standard deviation are σ^2 and σ, respectively.

Example 8 is the first instance of what will be seen to be a common phenomenon: the report (here estimate and standard error thereof) from a noninformative prior Bayesian analysis is often *formally* the same as the usual classical report. The *interpretations* of the two reports differ, but the numbers are formally the same. Many Bayesians maintain (and with considerable justification) that classical statistics has prospered only because, in so many standard situations (such as that of Example 8), the classical numbers reported can be given a sensible noninformative prior Bayesian interpretation (which also coincides with the meaning that nonsophisticates ascribe to the numbers, see Section 4.1). There are many situations in which the two reports do differ, however (and we will encounter several), and the classical report almost invariably suffers in comparison. See Pratt (1965) for further discussion of a number of these issues.

Another point (and this is one of the operational advantages of Bayesian analysis that was alluded to in Section 4.1) is that the calculation of $V^\pi(x)$ is rarely much more difficult than that of $\mu^\pi(x)$. When the calculation is a numerical calculation, there will be essentially no difference in difficulty. And when $\mu^\pi(x)$ can be written in a simple closed form, it is usually also possible to find a reasonably simple closed form for $V^\pi(x)$.

Example 7 (continued). Writing

$$\psi(x)=\frac{(2\pi)^{-1/2}\sigma\exp\{-x^2/(2\sigma^2)\}}{1-\Phi(-x/\sigma)}.$$

it was shown earlier that

$$\mu^\pi(x)=x+\psi(x).$$

To calculate $V^\pi(x)$ easily, note from (4.2) (choosing $\delta=x$) that

$$\begin{aligned}V^\pi(x)&=V_x^\pi(x)-(\mu^\pi(x)-x)^2\\&=V_x^\pi(x)-[\psi(x)]^2,\end{aligned}$$

and that an integration by parts (in the numerator below) gives

$$\begin{aligned}V_x^\pi(x)&=E^{\pi(\theta|x)}[(\theta+x)^2]\\&=\frac{\int_0^\infty(\theta-x)^2\exp\{-(\theta-x)^2/2\sigma^2\}d\theta}{\int_0^\infty\exp\{-(\theta-x)^2/2\sigma^2\}d\theta}\\&=\frac{-\sigma^2x\exp\{-x^2/2\sigma^2\}+\sigma^2\int_0^\infty\exp\{-(\theta-x)^2/2\sigma^2\}d\theta}{\int_0^\infty\exp\{-(\theta-x)^2/2\sigma^2\}d\theta}\\&=-x\psi(x)+\sigma^2.\end{aligned}$$

Hence

$$V^\pi(x)=\sigma^2-[x+\psi(x)]\psi(x). \tag{4.4}$$

The typically straightforward Bayesian calculation of standard error compares favorably (from an operational perspective) with the classical approach. A frequentist must propose an estimator $\delta(x)$, and calculate a "standard error" such as $\sqrt{\bar{V}}$ where, say,

$$\bar{V}=\sup_\theta E_\theta^X[(\delta(X)-\theta)^2]$$

(the maximum "mean squared error" of δ). This calculation can be difficult when no reasonable "unbiased" estimators are available. And sometimes $\bar{V}$ will be very large (even infinite) for all δ (because of the sup over θ), in which case the report of $\sqrt{\bar{V}}$ as the standard error seems highly questionable. Even when $\bar{V}$ appears to be reasonable, its use can be counter-intuitive in some respects, as the following example shows.

Example 7 (continued). A natural frequentist estimator for θ is $\delta(x) = \max\{x, 0\}$, and it can be shown that $\bar{V} = \sigma^2$. Thus one can report $\delta(x)$, together with the frequentist standard error $\sqrt{\bar{V}} = \sigma$. The counter-intuitive feature of this report is that, were it not known that $\theta > 0$, a frequentist would likely use $\delta(x) = x$, which also has $\bar{V} = \sigma^2$. Thus the same standard error would be reported in either case, while (intuitively) it would seem that the knowledge that $\theta > 0$ should result in a smaller reported standard error (at least when x is near zero). Note that the Bayesian analysis, given earlier, does reflect the benefits of this added knowledge. Indeed, $V^\pi(x)$ is increasing in x, with $V^\pi(0) = (2/\pi)\sigma^2$ and $V^\pi(\infty) = \sigma^2$. Thus for small x, the knowledge that $\theta > 0$ is being used to report a substantially smaller standard error than that of the similar analysis in Example 8. It can be shown (using the Cramér–Rao inequality) that *no* estimator will yield frequentist standard error smaller than σ, so the frequentist approach can never take advantage of the additional knowledge. (Conceivably, the estimated frequentist approach mentioned in Subsection 1.6.3 could take advantage of the additional information, but implementation of this approach is very hard.)

III. *Multivariate Estimation*

Bayesian estimation of a vector $\boldsymbol{\theta} = (\theta_1, \theta_2, \ldots, \theta_p)^t$ is also straightforward. The generalized maximum likelihood estimate (the posterior mode) is often a reasonable estimate, although existence and uniqueness difficulties are more likely to be encountered in the multivariate case. The posterior mean

$$\boldsymbol{\mu}^\pi(x) = (\mu_1^\pi(x), \ldots, \mu_p^\pi(x))^t = E^{\pi(\boldsymbol{\theta}|x)}[\boldsymbol{\theta}]$$

is a very attractive Bayesian estimate (providing it can be calculated, see Section 4.9), and its accuracy can be described by the *posterior covariance matrix*

$$\mathbf{V}^\pi(x) = E^{\pi(\boldsymbol{\theta}|x)}[(\boldsymbol{\theta} - \boldsymbol{\mu}^\pi(x))(\boldsymbol{\theta} - \boldsymbol{\mu}^\pi(x))^t]. \tag{4.5}$$

(For instance, the standard error of the estimate $\mu_i^\pi(x)$ of θ_i would be $\sqrt{V_{ii}^\pi(x)}$, where $V_{ii}^\pi(x)$ is the (i, i) element of $\mathbf{V}^\pi(x)$; more sophisticated uses of $\mathbf{V}^\pi$ are discussed in the next subsection.)

The analog of (4.2), for a general estimate $\boldsymbol{\delta}$ of $\boldsymbol{\theta}$, can be shown to be

$$\begin{aligned} \mathbf{V}_{\boldsymbol{\delta}}^\pi(x) &= E^{\pi(\boldsymbol{\theta}|x)}[(\boldsymbol{\theta} - \boldsymbol{\delta})(\boldsymbol{\theta} - \boldsymbol{\delta})^t] \\ &= \mathbf{V}^\pi(x) + (\boldsymbol{\mu}^\pi(x) - \boldsymbol{\delta})(\boldsymbol{\mu}^\pi(x) - \boldsymbol{\delta})^t. \end{aligned} \tag{4.6}$$

Again, it is clear that the posterior mean "minimizes" $\mathbf{V}_{\boldsymbol{\delta}}^\pi(x)$.

Example 9. Suppose $\mathbf{X} \sim \mathcal{N}_p(\boldsymbol{\theta}, \boldsymbol{\Sigma})$ and $\pi(\boldsymbol{\theta})$ is a $\mathcal{N}_p(\boldsymbol{\mu}, \mathbf{A})$ density. (Here $\boldsymbol{\mu}$ is a known p-vector, and $\boldsymbol{\Sigma}$ and $\mathbf{A}$ are known $(p \times p)$ positive definite matrices.) It will be left for the exercises to show that $\pi(\boldsymbol{\theta}|\mathbf{x})$ is a

$\mathcal{N}_p(\boldsymbol{\mu}^\pi(\mathbf{x}), \mathbf{V}^\pi(\mathbf{x}))$ density, where the posterior mean is given by

$$\boldsymbol{\mu}^\pi(\mathbf{x}) = \mathbf{x} - \boldsymbol{\Sigma}(\boldsymbol{\Sigma} + \mathbf{A})^{-1}(\mathbf{x} - \boldsymbol{\mu}) \tag{4.7}$$

and the posterior covariance matrix by

$$\begin{aligned} \mathbf{V}^\pi(\mathbf{x}) &= (\mathbf{A}^{-1} + \boldsymbol{\Sigma}^{-1})^{-1} \\ &= \boldsymbol{\Sigma} - \boldsymbol{\Sigma}(\mathbf{A} + \boldsymbol{\Sigma})^{-1}\boldsymbol{\Sigma}. \end{aligned} \tag{4.8}$$

More sophisticated multivariate Bayesian analyses will be considered in Sections 4.5 and 4.6 and Subsection 4.7.10.

4.3.2. Credible Sets

Another common approach to inference is to present a confidence set for θ. The Bayesian analog of a classical confidence set is called a credible set, and is defined as follows:

Definition 4. A $100(1-\alpha)\%$ *credible set* for θ is a subset C of Θ such that

$$1 - \alpha \leq P(C|x) = \int_C dF^{\pi(\theta|x)}(\theta) = \begin{cases} \int_C \pi(\theta|x)d\theta & \text{(continuous case)}, \\ \sum_{\theta \in C} \pi(\theta|x) & \text{(discrete case)}. \end{cases}$$

Since the posterior distribution is an actual probability distribution on Θ, one can speak meaningfully (though usually subjectively) of the probability that θ is in C. This is in contrast to classical confidence procedures, which can only be interpreted in terms of coverage probability (the probability that the random X will be such the confidence set $C(X)$ contains θ). Discussions of this difference were given in Sections 1.6 and 4.1.

In choosing a credible set for θ, it is usually desirable to try to minimize its size. To do this, one should include in the set only those points with the largest posterior density, i.e., the "most likely" values of θ. (Actually, this minimizes specifically the volume of the credible set. It may be desirable to minimize other types of size, as will be seen shortly.)

Definition 5. The $100(1-\alpha)\%$ HPD *credible set* for θ (HPD stands for highest posterior density), is the subset C of Θ of the form

$$C = \{\theta \in \Theta: \pi(\theta|x) \geq k(\alpha)\},$$

where $k(\alpha)$ is the largest constant such that

$$P(C|x) \geq 1 - \alpha.$$

Example 1 (continued). Since the posterior density of θ given x is $\mathcal{N}(\mu(x), \rho^{-1})$, which is unimodal and symmetric about $\mu(x)$, it is clear that the $100(1-\alpha)\%$ HPD credible set is given by

$$C=\left(\mu(x)+z\left(\frac{\alpha}{2}\right)\rho^{-1/2}, \mu(x)-z\left(\frac{\alpha}{2}\right)\rho^{-1/2}\right),$$

where $z(\alpha)$ is the α-fractile of a $\mathcal{N}(0,1)$ distribution.

In the IQ example, where the child who scores 115 on the intelligence test has a $\mathcal{N}(110.39, 69.23)$ posterior distribution for θ, it follows that a 95% HPD credible set for θ is

$$(110.39+(-1.96)(69.23)^{1/2}, 110.39+(1.96)(69.23)^{1/2})=(94.08, 126.70).$$

Note that since a random test score X is $\mathcal{N}(\theta, 100)$, the classical 95% confidence interval for θ is

$$(115-(1.96)(10), 115+(1.96)(10))=(95.4, 134.6).$$

Example 8 (continued). Since the posterior distribution of θ given x is $\mathcal{N}(x, \sigma^2)$, it follows that the $100(1-\alpha)\%$ HPD credible set for θ is

$$C=\left(x+z\left(\frac{\alpha}{2}\right)\sigma, x-z\left(\frac{\alpha}{2}\right)\sigma\right).$$

This is exactly the same as the classical confidence set for θ, and is another instance of the frequent *formal* similarity of classical and noninformative prior Bayesian answers.

Bayesian credible sets are usually much easier to calculate than their classical counterparts, particularly in situations where simple sufficient statistics do not exist. The following example illustrates this.

Example 10. Assume $\mathbf{X}=(X_1, \ldots, X_n)$ is an i.i.d. sample from a $\mathcal{C}(\theta, 1)$ distribution, and that $\theta>0$. Suppose that a noninformative prior Bayesian analysis is desired. Since θ is a (restricted) location parameter, a reasonable noninformative prior would be $\pi(\theta)=1$ (on $\theta>0$). The posterior density of θ given $\mathbf{x}=(x_1, \ldots, x_n)$ is then given (on $\theta>0$) by

$$\pi(\theta|\mathbf{x})=\frac{\prod_{i=1}^n [1+(\theta-x_i)^2]^{-1}}{\int_0^\infty \prod_{i=1}^n [1+(\theta-x_i)^2]^{-1} d\theta}. \tag{4.9}$$

While this is not an overly attractive posterior to work with, finding a $100(1-\alpha)\%$ HPD credible set on a computer is a relatively simple undertaking. For instance, if $n=5$ with $\mathbf{x}=(4.0, 5.5, 7.5, 4.5, 3.0)$, then the resulting 95% HPD credible set is the interval $(3.10, 6.06)$. In contrast, it is not at all clear how to develop a good classical confidence procedure for this problem. Classical confidence procedures are usually developed with a variety of tricks which do not have general applicability.

The general idea behind numerical calculation of the HPD credible set, in a situation where $\pi(\theta|x)$ is continuous in θ (such as Example 10), is to set up a program along the following lines:

(i) Create a subroutine which, for given k, finds all solutions to the equation $\pi(\theta|x)=k$. The set $C(k)=\{\theta: \pi(\theta|x)\geq k\}$ can typically be fairly easily constructed from these solutions. For instance, if Θ is an infinite interval in R^1 (as in Example 10) and only two solutions, $\theta_1(k)$ and $\theta_2(k)$, are found, then $C(k)=(\theta_1(k), \theta_2(k))$.

(ii) Create a subroutine which calculates

$$P^{\pi(\theta|x)}(C(k))=\int_{C(k)} \pi(\theta|x)d\theta.$$

(iii) Numerically solve the equation

$$P^{\pi(\theta|x)}(C(k))=1-\alpha,$$

calling on the above two subroutines as k varies. (There are, undoubtedly, much more efficient ways to actually write such a program; the main purpose of this general outline has been to simply provide some insight into the process.)

It can happen that an HPD credible set looks unusual. For instance, in Example 10 it could happen that the HPD credible set consists of several disjoint intervals. In such situations, one could abandon the HPD criterion and insist on connected credible sets, but there are very good reasons *not* to do so. Disjoint intervals often occur when there is "clashing" information (maybe the prior says one thing, and the data another), and such clashes are usually important to recognize. Natural conjugate priors typically are unimodal and yield unimodal posteriors, and hence will tend to *mask* these clashes. (This is one of a number of reasons that we will see for being wary of conjugate priors.)

An often useful approximation to an HPD credible set can be achieved through use of the normal approximation to the posterior. It can be shown (see Subsection 4.7.8) that, for large sample sizes, the posterior distribution will be approximately normal. Even for small samples, a normal likelihood function will usually yield a roughly normal posterior, so the approximation can have uses even in small samples. The attractions of using the normal approximation include calculational simplicity and the fact that the ensuing credible sets will have a standard form (for, say, consumption by people who are only comfortable with sets having a familiar shape).

The most natural normal approximation to use, in the univariate case, is to approximate $\pi(\theta|x)$ by a $\mathcal{N}(\mu^{\pi}(x), V^{\pi}(x))$ distribution, where $\mu^{\pi}(x)$ and $V^{\pi}(x)$ are given in Definition 3. (For large samples, $\mu^{\pi}(x)$ and $V^{\pi}(x)$ can be estimated as discussed in Subsection 4.7.8.) The corresponding approximate $100(1-\alpha)\%$ HPD credible set is then

$$C=\left(\mu^{\pi}(x)+z\left(\frac{\alpha}{2}\right)\sqrt{V^{\pi}(x)},\ \mu^{\pi}(x)-z\left(\frac{\alpha}{2}\right)\sqrt{V^{\pi}(x)}\right). \quad (4.10)$$

EXAMPLE 10 (continued). The posterior density in (4.9) is clearly nonnormal and, with only five Cauchy observations, one might imagine that the normal approximation would be somewhat inaccurate. The posterior can be seen to be unimodal, however, and the normal approximation turns out to be excellent out to the 2.5% and 97.5% tails. A numerical calculation gave $\mu^{\pi}(\mathbf{x})=4.55$, $V^{\pi}(\mathbf{x})=0.562$, and actual and approximate percentiles as follows:

Table 4.1. Actual and Approximate Posterior Percentiles.

α	2.5	25	50	75	97.5
αth percentile of $\pi(\theta\|x)$	3.17	4.07	4.52	5.00	6.15
αth percentile of $\mathcal{N}(\mu^{\pi}, V^{\pi})$	3.08	4.05	4.55	5.06	6.02

In the extreme tails the approximation became quite inaccurate, and this indicates an important general limitation of the use of such approximations.

The approximate 95% HDP credible set (using (4.10)) is $C=(3.08, 6.02)$, which is very close to the actual 95% HPD credible set, $(3.10, 6.06)$, calculated earlier. Also, this approximate C has *actual* posterior probability (under (4.9)) of 0.948, which is extremely close to its nominal probability. Similarly, the approximate 90% HPD credible set is $(3.32, 5.78)$, and has *actual* posterior probability of 0.906.

Multivariate Case

For multivariate $\boldsymbol{\theta}$, the definition of an HPD credible set remains the same.

EXAMPLE 9 (continued). The posterior density is $\mathcal{N}_p(\boldsymbol{\mu}^{\pi}(\mathbf{x}), \mathbf{V}^{\pi}(\mathbf{x}))$ which is large when $(\boldsymbol{\theta}-\boldsymbol{\mu}^{\pi}(x))^t\mathbf{V}^{\pi}(\mathbf{x})^{-1}(\boldsymbol{\theta}-\boldsymbol{\mu}^{\pi}(\mathbf{x}))$ is large. Furthermore, this quadratic form has a chi-square distribution with p degrees of freedom, so the $100(1-\alpha)\%$ HPD credible set for $\boldsymbol{\theta}$ is the ellipse

$$C=\{\boldsymbol{\theta}\colon (\boldsymbol{\theta}-\boldsymbol{\mu}^{\pi}(\mathbf{x}))^t\mathbf{V}^{\pi}(\mathbf{x})^{-1}(\boldsymbol{\theta}-\boldsymbol{\mu}^{\pi}(\mathbf{x}))\le\chi^2_p(1-\alpha)\}, \quad (4.11)$$

where $\chi^2_p(1-\alpha)$ is the $(1-\alpha)$-fractile of the chi-square distribution.

In the multivariate case, the use of the normal approximation to the posterior is often especially valuable, because of the increased calculational difficulties. An example of such use is given in Subsection 4.7.10.

Alternatives to the HPD *Credible Set*

Though seemingly natural, the HPD credible set has the unnatural property of not necessarily being invariant under transformations.

EXAMPLE 6 (continued). Slightly modify this example by assuming that the parameter space is $\Theta = (0, \infty)$. Then the posterior becomes

$$\pi(\theta|x) = ce^{\theta}(1+\theta^2)^{-1}I_{(0,x)}(\theta).$$

This was shown to be increasing in θ, so that an HPD credible set will be of the form $(b(\alpha), x)$.

Suppose, now, that, instead of working with θ, we had used $\eta = \exp\{\theta\}$ as the unknown parameter. This is a one-to-one transformation, and should ideally have no effect on the ultimate answer. The posterior density for η, $\pi^*(\eta|x)$, can be found by simply transforming $\pi(\theta|x)$; note that $\theta = \log \eta$ so that $d\theta/d\eta = \eta^{-1}$ is the Jacobian that we must multiply by to obtain the transformed density. Thus

$$\begin{aligned}\pi^*(\eta|x) &= \eta^{-1}ce^{\log \eta}(1+(\log \eta)^2)^{-1}I_{(0,x)}(\log \eta)\\ &= c(1+(\log \eta)^2)^{-1}I_{(1,\exp\{x\})}(\eta).\end{aligned}$$

Note that $\pi^*(\eta|x)$ is decreasing on $(1, \exp\{x\})$, so that the $100(1-\alpha)\%$ HPD credible set for η will be an interval of the form $(1, d(\alpha))$. Transforming back to the θ coordinate system, this interval becomes $(0, \log d(\alpha))$.

The extreme conflict here is apparent; in the original parametrization, the *upper tail* of the posterior is the HPD credible set, while, for a monotonic reparametrization, it is the *lower tail.*

For more examples and discussion of the conflict, see Lehmann (1985). There is no clearcut resolution of the conflict, which is one reason that many Bayesians encourage reporting of the entire posterior, as opposed to just a credible set. Of course, there often is a "preferred" parametrization, in the sense that θ might be in some type of natural unit, and use of an HPD credible set in such a parametrization is not unreasonable.

More formally, one could define a secondary criteria to specify a credible set. One natural such criteria is size. Different measures of size could be considered, however; a quite general formulation of the size of a set C is given by

$$S(C) = \int_C s(\theta)d\theta,$$

where s is a nonnegative function. If $s(\theta) \equiv 1$ is chosen, $S(C)$ is just the usual Lebesgue measure of C ("length" of an interval, "volume" in higher dimensions).

The problem can then be stated as that of finding a $100(1-\alpha)\%$ credible set which has minimum size with respect to S; let us call such a set the

S-optimal $100(1-\alpha)\%$ *credible set.* It is easy to show that, in the continuous case (i.e., when $\pi(\theta|x)$ is a density with respect to Lebesgue measure), the S-optimal $100(1-\alpha)\%$ credible set is given (under mild conditions, see Exercise 35) by

$$C = \{\theta\colon \pi(\theta|x) > ks(\theta)\} \tag{4.12}$$

for some positive k. Note that if $s(\theta) \equiv 1$, then the S-optimal set is simply the HPD credible set, as indicated earlier. This does, therefore, provide a formal justification for the use of the HPD credible set when θ is a "preferred" parametrization, in that we might reasonably want to minimize volume (Lebesgue measure) in this preferred parametrization. An example in which a measure of size other than volume might be desirable is given in Exercise 34.

The utilization of S to specify an optimal confidence set preserves "optimality" under transformation, in that the choice of S will clearly be dependent on the parametrization chosen. Indeed, if S is deemed suitable in the θ parametrization, then the appropriate measure of size in a transformed parametrization is simply that induced by transforming S.

We will not pursue this subject further, because we do not view credible sets as having a clear decision-theoretic role, and therefore are leery of "optimality" approaches to selection of a credible set. We mainly view credible sets as an easily reportable crude summary of the posterior distribution, and, for this purpose, are not unduly troubled by, say, nonuniqueness of HPD credible sets.

4.3.3. Hypothesis Testing

In classical hypothesis testing, a null hypothesis H_0: $\theta \in \Theta_0$ and an alternative hypothesis H_1: $\theta \in \Theta_1$ are specified. A test procedure is evaluated in terms of the probabilities of Type I and Type II error. These probabilities of error represent the chance that a sample is observed for which the test procedure will result in the wrong hypothesis being accepted.

In Bayesian analysis, the task of deciding between H_0 and H_1 is conceptually more straightforward. One merely calculates the posterior probabilities $\alpha_0 = P(\Theta_0|x)$ and $\alpha_1 = P(\Theta_1|x)$ and decides between H_0 and H_1 accordingly. The conceptual advantage is that α_0 and α_1 are the actual (subjective) probabilities of the hypotheses in light of the data and prior opinions. The difficulties in properly interpreting classical error probabilities will be indicated later.

Although posterior probabilities of hypotheses are the primary Bayesian measures in testing problems, the following related concepts are also of interest. Throughout this section we will use π_0 and π_1 to denote the prior probabilities of Θ_0 and Θ_1, respectively.

Definition 6. The ratio α_0/α_1 is called the *posterior odds ratio* of H_0 to H_1, and π_0/π_1 is called the *prior odds ratio.* The quantity

$$B = \frac{\text{posterior odds ratio}}{\text{prior odds ratio}} = \frac{\alpha_0/\alpha_1}{\pi_0/\pi_1} = \frac{\alpha_0\pi_1}{\alpha_1\pi_0}$$

is called the *Bayes factor* in favor of Θ_0.

Many people are more comfortable with "odds" than with probabilities, and indeed it is often convenient to summarize the evidence in terms of posterior odds. (Saying that $\alpha_0/\alpha_1 = 10$ clearly conveys the conclusion that H_0 is 10 times as likely to be true as H_1.) The interest in the Bayes factor is that it can sometimes be interpreted as the "odds for H_0 to H_1 that are *given by the data.*" This is a clearly valid interpretation when the hypotheses are simple, i.e., when $\Theta_0 = \{\theta_0\}$ and $\Theta_1 = \{\theta_1\}$, for then

$$\alpha_0 = \frac{\pi_0 f(x|\theta_0)}{\pi_0 f(x|\theta_0) + \pi_1 f(x|\theta_1)}, \qquad \alpha_1 = \frac{\pi_1 f(x|\theta_1)}{\pi_0 f(x|\theta_0) + \pi_1 f(x|\theta_1)},$$

$$\frac{\alpha_0}{\alpha_1} = \frac{\pi_0 f(x|\theta_0)}{\pi_1 f(x|\theta_1)} \quad \text{and} \quad B = \frac{\alpha_0\pi_1}{\alpha_1\pi_0} = \frac{f(x|\theta_0)}{f(x|\theta_1)}.$$

In other words, B is then just the likelihood ratio of H_0 to H_1, which is commonly viewed (even by many non-Bayesians) as the odds for H_0 to H_1 that are given by the data.

In general, however, B will depend on the prior input. To explore this dependence (and for later developments), it is convenient to write the prior as

$$\pi(\theta) = \begin{cases} \pi_0 g_0(\theta) & \text{if } \theta \in \Theta_0, \\ \pi_1 g_1(\theta) & \text{if } \theta \in \Theta_1, \end{cases} \tag{4.13}$$

so that g_0 and g_1 are (proper) densities which describe how the prior mass is spread out over the two hypotheses. (Recall that π_0 and π_1 are the prior probabilities of Θ_0 and Θ_1.) With this representation, we can write

$$\frac{\alpha_0}{\alpha_1} = \frac{\int_{\Theta_0} dF^{\pi(\theta|x)}(\theta)}{\int_{\Theta_1} dF^{\pi(\theta|x)}(\theta)} = \frac{\int_{\Theta_0} f(x|\theta)\pi_0 dF^{g_0}(\theta)/m(x)}{\int_{\Theta_1} f(x|\theta)\pi_1 dF^{g_1}(\theta)/m(x)}$$
$$= \frac{\pi_0 \int_{\Theta_0} f(x|\theta)\,dF^{g_0}(\theta)}{\pi_1 \int_{\Theta_1} f(x|\theta)\,dF^{g_1}(\theta)}.$$

Hence

$$B = \frac{\int_{\Theta_0} f(x|\theta)\,dF^{g_0}(\theta)}{\int_{\Theta_1} f(x|\theta)\,dF^{g_1}(\theta)},$$

which is the ratio of "weighted" (by g_0 and g_1) likelihoods of Θ_0 to Θ_1. Because of the involvement of g_0 and g_1, this cannot be viewed as a measure of the relative support for the hypotheses provided solely by the data. Sometimes, however, B will be relatively insensitive to reasonable choices of g_0 and g_1, and then such an interpretation is reasonable. The main operational advantage of having such a "stable" Bayes factor is that a scientific report (see Section 4.10) could include this Bayes factor, and any reader could then determine his personal posterior odds by simply multiplying the reported Bayes factor by his personal prior odds.

Example 1 (continued). The child taking the IQ test is to be classified as having below average IQ (less than 100) or above average IQ (greater than 100). Formally, it is thus desired to test H_0: $\theta \le 100$ versus H_1: $\theta > 100$. Recalling that the posterior distribution of θ is $\mathcal{N}(110.39, 69.23)$ a table of normal probabilities yields

$$\alpha_0 = P(\theta \le 100|x) = 0.106, \qquad \alpha_1 = P(\theta > 100|x) = 0.894,$$

and hence the posterior odds ratio is $\alpha_0/\alpha_1 = 8.44$. Also, the prior is $\mathcal{N}(100, 225)$, so that $\pi_0 = P^{\pi}(\theta \le 100) = \frac{1}{2} = \pi_1$ and the prior odds ratio is 1. (Note that a prior odds ratio of 1 indicates that H_0 and H_1 are viewed as equally plausible initially.) The Bayes factor is thus $B = \alpha_0\pi_1/(\alpha_1\pi_0) = 8.44$.

We next explicitly consider one-sided testing, testing of a point null hypothesis, and multiple hypothesis testing, pointing out general features of Bayesian testing and comparing the Bayesian and classical approaches in each case.

I. *One-Sided Testing*

One-sided hypothesis testing occurs when $\Theta \subset R^1$ and Θ_0 is entirely to one side of Θ_1. Example 1 (continued) above is an illustration of this. There are no unusual features of Bayesian testing here. What is of interest, however, is that this is one of the few testing situations in which classical testing, particularly the use of P-values, will sometimes have a Bayesian justification. Consider, for instance, the following example.

Example 8 (continued). When $X \sim \mathcal{N}(\theta, \sigma^2)$ and θ has the noninformative prior $\pi(\theta) = 1$, we saw that $\pi(\theta|x)$ is $\mathcal{N}(x, \sigma^2)$. Consider now the situation of testing H_0: $\theta \le \theta_0$ versus H_1: $\theta > \theta_0$. Then

$$\alpha_0 = P(\theta \le \theta_0|x) = \Phi((\theta_0 - x)/\sigma),$$

where, again, Φ is the standard normal c.d.f.

The classical P-value against H_0 is the probability, when $\theta = \theta_0$, of observing an X "more extreme" than the actual data x. Here the P-value would be

$$P\text{-value} = P(X \ge x) = 1 - \Phi((x - \theta_0)/\sigma).$$

Because of the symmetry of the normal distribution, it follows that α_0 equals the P-value against H_0.

The use of a noninformative prior in this example is, of course, rather disturbing at first glance, since it gives infinite mass to each of the hypotheses (precluding, for instance, consideration of prior odds). It is, however, possible to justify the use of the noninformative prior as an approximation to very vague prior beliefs. Indeed, any proper prior density which is roughly constant over the interval $(\theta_0 - 2\sigma, x + 2\sigma)$ (here we assume $x > \theta_0$), and is not significantly larger outside this interval, would result in an α_0 roughly equal to the P-value.

In a number of other one-sided testing situations, vague prior information will tend to result in posterior probabilities that are similar to P-values (cf. Pratt (1965), DeGroot (1973), Dempster (1973), Dickey (1977), Zellner and Siow (1980), Hill (1982), and Good (1983, 1984)). This is not true for all one-sided testing problems, however. For instance, testing $H_0: \theta = 0$ (or $H_0: 0 \leq \theta \leq \varepsilon$) versus $H_1: \theta > 0$ (or $H_1: \theta > \varepsilon$) is a one-sided testing problem, but P-values and posterior probabilities will tend to differ drastically (in much the same way as in part II of this subsection; see also Berger and Sellke (1984)). Note, also, that classical probabilities of Type I and Type II error do *not* usually have any close correspondence to posterior probabilities of hypotheses; this may partly explain the preference of many classical practitioners for use of P-values instead of probabilities of Type I and Type II errors.

II. *Testing a Point Null Hypothesis*

It is very common in classical statistics (cf. the statistical literature survey in Zellner (1984a)) to conduct a test of $H_0: \theta = \theta_0$ versus $H_1: \theta \neq \theta_0$. Such testing of a point null hypothesis is interesting, partly because the Bayesian approach contains some novel features, but mainly because the Bayesian answers differ *radically* from the classical answers.

Before discussing these issues, several comments should be made about the entire enterprise of testing a point null hypothesis. First of all, as mentioned in Subsection 1.6.1, tests of point null hypotheses are commonly performed in inappropriate situations. It will virtually never be the case that one seriously entertains the possibility that $\theta = \theta_0$ *exactly* (cf. Hodges and Lehmann (1954) and Lehmann (1959)). More reasonable would be the null hypothesis that $\theta \in \Theta_0 = (\theta_0 - b, \theta_0 + b)$, where $b > 0$ is some constant chosen so that all θ in Θ_0 can be considered "indistinguishable" from θ_0. An example in which this might arise would be an attempt to analyze a chemical by observing some feature, θ, of its reaction with a known chemical. If it were desired to test whether or not the unknown chemical is a specific compound, with a reaction strength θ_0 known up to an accuracy of b, then it would be reasonable to test $H_0: \theta \in (\theta_0 - b, \theta_0 + b)$ versus $H_1: \theta \notin (\theta_0 - b, \theta_0 + b)$. A similar example involving forensic science can be found in Lindley (1977) (see also Shafer (1982c)). An example where b might be extremely close to zero is in a test for extra sensory perception, with θ_0 reflecting the hypothesis of *no* extra sensory perception. (The only reason that b would probably be nonzero here is that the experiment designed to test for ESP would not lead to a perfectly well-defined θ_0.) Of course, there are also many decision problems that would lead to a null hypothesis of the above interval form with a *large* b, but such problems will rarely be well approximated by testing a point null.

Given that one should really be testing H_0: $\theta \in (\theta_0 - b, \theta_0 + b)$, we need to know when it is suitable to approximate H_0 by H_0: $\theta = \theta_0$. From the Bayesian perspective, the only sensible answer to this question is—the approximation is reasonable if the posterior probabilities of H_0 are nearly equal in the two situations. A very strong condition under when this would be the case is that the observed likelihood function be approximately constant on $(\theta_0 - b, \theta_0 + b)$. (A more formal statement of this is given in the Exercises.)

EXAMPLE 11. Suppose a sample $X_1, \ldots, X_n$ is observed from a $\mathcal{N}(\theta, \sigma^2)$ distribution, σ^2 known. The observed likelihood function is then proportional to a $\mathcal{N}(\bar{x}, \sigma^2/n)$ density for θ. This will be nearly constant on $(\theta_0 - b, \theta_0 + b)$ when b is small compared to $\sigma/\sqrt{n}$. For instance, in the interesting case where the classical test statistic $z = \sqrt{n}|\bar{x} - \theta_0|/\sigma$ is larger than 1, the likelihood function can be shown to vary by no more than 5% on $(\theta_0 - b, \theta_0 + b)$ if

$$b \le (0.024) z^{-1} \sigma/\sqrt{n}.$$

When $z = 2$, $\sigma = 1$, and $n = 25$, this condition becomes $b \le 0.0024$. Note that the bound on b will depend on $|\bar{x} - \theta_0|$, as well as on $\sigma/\sqrt{n}$.

The point null approximation will be satisfactory in substantially greater generality than the "constant likelihood" situation, but it is usually easier for a Bayesian to deal directly with the interval hypothesis than to check the adequacy of the approximation. Nevertheless, we will develop the Bayesian test of a point null, so as to allow evaluation of the classical test.

To conduct a Bayesian test of the point null hypothesis H_0: $\theta = \theta_0$, one cannot use a continuous prior density, since any such prior will give θ_0 prior (and hence posterior) probability zero. A reasonable approach, therefore, is to give θ_0 a positive *probability* π_0, while giving the $\theta \neq \theta_0$ the *density* $\pi_1 g_1(\theta)$, where $\pi_1 = 1 - \pi_0$ and g_1 is proper. One can think of π_0 as the mass that would have been assigned to the realistic hypothesis H_0: $\theta \in (\theta_0 - b, \theta_0 + b)$, were the point null approximation not being used.

The Bayesian analysis of this situation is quite straightforward, although one must be careful to remember that the prior has discrete and continuous parts. The marginal density of X is

$$m(x) = \int f(x|\theta) dF^{\pi}(\theta) = f(x|\theta_0)\pi_0 + (1 - \pi_0) m_1(x),$$

where

$$m_1(x) = \int_{\{\theta \neq \theta_0\}} f(x|\theta) dF^{g_1}(\theta)$$

is the marginal density of X with respect to g_1. Hence the posterior probability that $\theta = \theta_0$ is

$$\begin{aligned}\pi(\theta_0|x) &= \frac{f(x|\theta_0)\pi_0}{m(x)} \\ &= \frac{f(x|\theta_0)\pi_0}{f(x|\theta_0)\pi_0 + (1-\pi_0)m_1(x)} \\ &= \left[1 + \frac{(1-\pi_0)}{\pi_0} \cdot \frac{m_1(x)}{f(x|\theta_0)}\right]^{-1}.\end{aligned} \tag{4.14}$$

Note that this is α_0, the posterior probability of H_0 in our earlier terminology, and that $\alpha_1 = 1 - \alpha_0$ is hence the posterior probability of H_1. Also, the posterior odds ratio can easily be shown to be (recalling that $\pi_1 = 1 - \pi_0$)

$$\frac{\alpha_0}{\alpha_1} = \frac{\pi(\theta_0|x)}{1 - \pi(\theta_0|x)} = \frac{\pi_0}{\pi_1} \cdot \frac{f(x|\theta_0)}{m_1(x)},$$

so that the Bayes factor for H_0 versus H_1 is

$$B = f(x|\theta_0)/m_1(x). \tag{4.15}$$

EXAMPLE 11 (continued). Reduction to the sufficient statistic $\bar{X}$ yields the effective likelihood function

$$f(\bar{x}|\theta) = \frac{1}{\sqrt{2\pi\sigma^2/n}} \exp\left\{-\frac{n}{2\sigma^2}(\theta - \bar{x})^2\right\}.$$

Suppose that g_1 is a $\mathcal{N}(\mu, \tau^2)$ density on $\theta \neq \theta_0$. (Usually $\mu = \theta_0$ would be an appropriate choice, since θ close to θ_0 would often be deemed more likely *a priori* than θ far from θ_0.) Then, as in Example 1, m_1 is a $\mathcal{N}(\mu, \tau^2 + \sigma^2/n)$ density (the point $\theta = \theta_0$ not mattering in the integration since g_1 is a continuous density). We thus obtain

$$\alpha_0 = \pi(\theta_0|x) = \left[1 + \frac{(1-\pi_0)}{\pi_0} \cdot \frac{\{2\pi(\tau^2 + \sigma^2/n)\}^{-1/2} \exp\{-(\bar{x} - \mu)^2/(2[\tau^2 + \sigma^2/n])\}}{\{2\pi\sigma^2/n\}^{-1/2} \exp\{-(\bar{x} - \theta_0)^2/(2\sigma^2/n)\}}\right]^{-1}. \tag{4.16}$$

In the special case where $\mu = \theta_0$, this reduces to

$$\begin{aligned}\alpha_0 &= \left[1 + \frac{(1-\pi_0)}{\pi_0} \cdot \frac{\exp\{(\bar{x} - \theta_0)^2 n\tau^2/(2\sigma^2[\tau^2 + \sigma^2/n])\}}{\{1 + n\tau^2/\sigma^2\}^{1/2}}\right]^{-1} \\ &= \left[1 + \frac{(1-\pi_0)}{\pi_0} \cdot \frac{\exp\{\frac{1}{2}z^2[1 + \sigma^2/(n\tau^2)]^{-1}\}}{\{1 + n\tau^2/\sigma^2\}^{1/2}}\right]^{-1},\end{aligned} \tag{4.17}$$

where $z=\sqrt{n}|\bar{x}-\theta_0|/\sigma$ is the usual statistic used for testing H_0: $\theta=\theta_0$. For later reference, note that

$$\alpha \geq \left[1+\frac{(1-\pi_0)}{\pi_0}\cdot\frac{\exp\{\frac{1}{2}z^2\}}{\{1+n\tau^2/\sigma^2\}^{1/2}}\right]^{-1}. \tag{4.18}$$

Table 4.2 presents values of α_0 for various z (chosen to correspond to standard classical two-tailed P-values or significance levels for testing a point null) and n, when the prior is specified by $\mu=\theta_0$, $\pi_0=\frac{1}{2}$, and $\tau=\sigma$. The numbers in Table 4.2 are astonishing. For instance, if one observed $z=1.96$, classical theory would allow one to reject H_0 at level $\alpha=0.05$, the smallness of which gives the impression that H_0 is very likely to be false. But the posterior probability of H_0 is quite substantial, ranging from about $\frac{1}{3}$ for small n to nearly 1 for large n. Thus $z=1.96$ actually provides little or no evidence against H_0 (for the specified prior).

The conflict in the above example is of such interest that it deserves further investigation. The Bayesian analysis can, of course, be questioned because of the choice of the prior. The assumption of a normal form for g_1 can be shown to have no real bearing on the issue (unless $|\bar{x}-\theta_0|$ is large), and the choices $\mu=\theta_0$ and $\pi_0=\frac{1}{2}$ are natural. (Indeed, if one were trying to be "objective," these choices of μ and π_0 would seem to be virtually mandatory.) One could certainly question the choice $\tau=\sigma$, however. Jeffreys (1961) argues for such a choice in "objective" testing, but the answer definitely depends crucially on the choice for τ. (Jeffreys actually prefers a Cauchy form for the prior, but again the form matters only in extreme cases. See also Raiffa and Schlaifer (1961), Lindley (1965, 1977), Smith (1965), Zellner (1971, 1984a), Dickey (1971, 1974, 1980), Zellner and Siow (1980), Diamond and Forrester (1983) and Berger and Das Gupta (1985) for similar analyses and generalizations.)

When the answer heavily depends on features of the prior such as τ, a Bayesian has no recourse but to attempt subjective specification of the features. (See Section 4.10 for useful graphical techniques of presenting the Bayesian conclusion as a *function* of such features, allowing easy processing by consumers.) We can, however, expose the enormity of the classical

Table 4.2. Posterior Probability of H_0.

	n						
z (P-value)	1	5	10	20	50	100	1000
1.645 (0.1)	0.42	0.44	0.49	0.56	0.65	0.72	0.89
1.960 (0.05)	0.35	0.33	0.37	0.42	0.52	0.60	0.80
2.576 (0.01)	0.21	0.13	0.14	0.16	0.22	0.27	0.53
3.291 (0.001)	0.086	0.026	0.024	0.026	0.034	0.045	0.124

Bayesian conflict in testing a point null by finding the *minimum* of α_0 over all τ. We will actually go one step further, and find the minimum over *all* g_1. Versions of the following theorem can be found in Edwards, Lindman, and Savage (1963), and Dickey (1973, 1977). (The proof of the theorem is easy and will be left for the Exercises.)

Theorem 1. *For* any *distribution* g_1 *on* $\theta \neq \theta_0$,

$$\alpha_0 = \pi(\theta_0|x) \geq \left[1 + \frac{(1-\pi_0)}{\pi_0} \cdot \frac{r(x)}{f(x|\theta_0)}\right]^{-1}, \tag{4.19}$$

where $r(x) = \sup_{\theta \neq \theta_0} f(x|\theta)$. (*Usually,* $r(x) = f(x|\hat{\theta})$, *where* $\hat{\theta}$ *is a maximum likelihood estimate of* θ.) *The corresponding bound on the Bayes factor for* H_0 *versus* H_1 *is*

$$B = \frac{f(x|\theta_0)}{m_1(x)} \geq \frac{f(x|\theta_0)}{r(x)}. \tag{4.20}$$

Example 11 (continued). With $f(\bar{x}|\theta)$ being a $\mathcal{N}(\bar{x}, \sigma^2/n)$ likelihood, it is clear that the supremum over all $\theta \neq \theta_0$ is $f(\bar{x}|\bar{x})$, so that

$$r(x) = f(\bar{x}|\bar{x}) = \{2\pi\sigma^2/n\}^{-1/2}.$$

Thus

$$\alpha_0 \geq \left[1 + \frac{(1-\pi_0)}{\pi_0} \cdot \frac{\{2\pi\sigma^2/n\}^{-1/2}}{\{2\pi\sigma^2/n\}^{-1/2} \exp\{-(\bar{x}-\theta_0)^2/(2\sigma^2/n)\}}\right]^{-1}$$
$$= \left[1 + \frac{(1-\pi_0)}{\pi_0} \cdot \exp\{\tfrac{1}{2}z^2\}\right]^{-1},$$

and

$$B \geq 1/\exp\{\tfrac{1}{2}z^2\},$$

where $z = \sqrt{n}|\bar{x} - \theta_0|/\sigma$. Table 4.3 gives the values of these bounds for the various z in Table 4.2 (and with $\pi_0 = \frac{1}{2}$ for the bound on α_0). Although the bounds on α_0 are substantially smaller than the particular values given in Table 4.2, they are still substantially larger than the corresponding P-values. Indeed, it can be shown (see Berger and Sellke (1984)) that, for $\pi_0 = \frac{1}{2}$ and $z > 1.68$,

$$\alpha_0 \geq (P\text{-value}) \times (1.25)z. \tag{4.21}$$

Table 4.3. Bounds on α_0 and B.

z	P-value	Bound on α_0	Bound on B
1.645	0.1	0.205	1/3.87
1.960	0.05	0.127	1/6.83
2.576	0.01	0.035	1/27.60
3.291	0.001	0.0044	1/224.83

A pause for recapitulation is in order. It is well understood that a classical error probability or P-value is *not* a posterior probability of a hypothesis. Nevertheless, it is *felt* by the vast majority of users that a P-value of 0.05 means that one can be pretty sure that H_0 is wrong. The evidence, however, is quite to the contrary. Any reasonably fair Bayesian analysis will show that there is at best very weak evidence against H_0 when $z = 1.96$. And even the lower bound on α_0, which was obtained by a Bayesian analysis heavily slanted towards H_1 (the g_1 chosen being that which was most favorable towards H_1), indicates much less real doubt about H_0 than $\alpha = 0.05$ would seem to imply. It thus appears that, for the situation studied, classical error probabilities or P-values are completely misleading descriptions of the evidence against H_0.

For those who are highly suspicious of any Bayesian reasoning, there is still the bound in (4.20) to consider. This bound is clearly the minimum likelihood ratio of H_0 to H_1. Thus the bound on B in Table 4.3 for $z = 1.96$ means that the likelihood ratio of H_0 to H_1 will be at least $1/6.83$; it would seem ridiculous to conclude that the data favors H_1 by more than a factor of 6.83. For more discussion, and also presentation of the conflict in conditional frequentist terms, see Berger and Sellke (1984).

In the face of this overwhelming evidence that classical testing of a point null is misleading, we must seek a better approach. Of course, we basically recommend the subjective Bayesian approach alluded to earlier (especially if implemented as discussed in Section 4.10). It is interesting to ask if there is a more "objective" sensible analysis, however. Noninformative prior Bayesian analyses unfortunately do not work well for testing a point null hypothesis, making impossible an objective Bayesian solution. One can, however, argue for a *standardized* Bayesian solution. Indeed Jeffreys (1961) does so argue for the use of a particular proper prior, one chosen to reflect what he feels will be reasonable uncertainty in many scientific problems. (Zellner and Siow (1980) call these *reference informative priors.*) Although we feel that such a standardized Bayesian approach would definitely be far better than the standardized classical approach, it is much harder to defend than other objective Bayesian analyses.

Another possibility is to present the bounds on α_0 and B, from Theorem 1, as objective evidence. Unfortunately these bounds will typically be much smaller than α_0 and B that are actually obtained through subjective Bayesian calculations; the use of such bounds would thus be limited to showing that there is *no* reason to doubt H_0. In terms of objective appearance, reporting the bound on B is particularly attractive, since it is just a likelihood ratio. (Similar and more extensive developments of likelihood ratios as evidence can be found in Edwards, Lindman, and Savage (1963) and Dempster (1973).)

A quite promising possibility is to find more accurate lower bounds on α_0 and B, by restricting g_1 in some natural fashion. In many applications, for instance, it would be reasonable to restrict consideration to

$$\mathscr{G}=\{g_1: g_1 \text{ is symmetric about } \theta_0 \text{ and nonincreasing in } |\theta-\theta_0|\}. \tag{4.22}$$

Indeed, this could be argued to be a *requirement* of an objective Bayesian analysis. It turns out to be surprisingly easy to work with such $\mathscr{G}$. Indeed, the desired lower bounds on α_0 and B are found in Berger and Sellke (1984) for quite general situations (see also Theorem 5 in Subsection 4.7.9). We content ourselves here with stating the conclusion in the situation of Example 11.

EXAMPLE 11 (continued). It is shown in Berger and Sellke (1984) that, if $z \le 1$, then $\alpha_0 \ge \pi_0$ and $B \ge 1$ for *any* g_1 in (4.22). For $z > 1$, the lower bounds on α_0 and B for this class of priors are

$$\alpha_0 \ge \left\{1+\frac{(1-\pi_0)}{\pi_0}\cdot\frac{[\phi(k+z)+\phi(k-z)]}{2\phi(z)}\right\}^{-1}, \tag{4.23}$$

and

$$B \ge \frac{2\phi(z)}{[\phi(k+z)+\phi(k-z)]}, \tag{4.24}$$

where k is a solution to equation (4.112) in Section 4.7.9, which can be approximated by

$$k = z+\left[2\log\left(\frac{k}{\Phi(k-z)}\right)-1.838\right]^{1/2} \tag{4.25}$$

when $z \ge 1.645$. (Here ϕ and Φ denote the standard normal density and c.d.f., as usual.) Equation (4.25) has the feature of allowing iterative calculation of k: plug in a guess for k on the right-hand side of (4.25), and use the result as a new guess for k to be plugged back in; repeat the process until the guesses for k stabilize, which usually takes only two or three iterations. (A good first guess is $k = z$.)

Table 4.4 gives these lower bounds on α_0 and B for the values of z used in Tables 4.2 and 4.3. The bounds are much larger than the corresponding bounds in Table 4.3, and indeed are close to some of the exact values of α_0 and B obtained from the Jeffreys-type analysis in Table 4.2. Hence these

Table 4.4. $\mathscr{G}$-Bounds on α_0 and B.

z	P-value	$\mathscr{G}$-Bound on α_0	$\mathscr{G}$-Bound on B
1.645	0.10	0.390	1/1.56
1.960	0.05	0.290	1/2.45
2.576	0.01	0.109	1/8.17
3.291	0.001	0.018	1/54.56

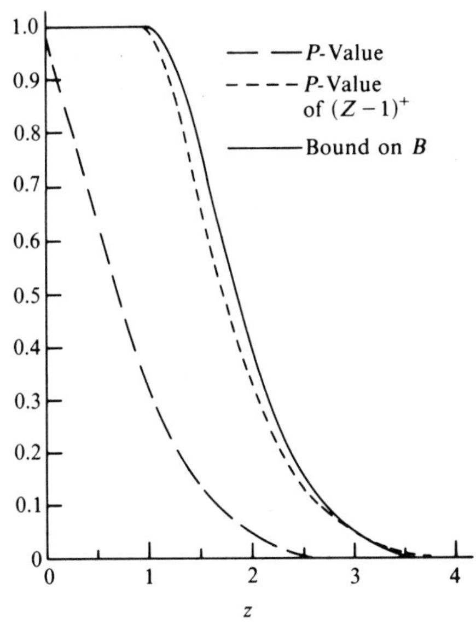

Figure 4.1

lower bounds are indeed somewhat reasonable as "objective" measures of the evidence against H_0, at least for small n.

A rather fascinating "empirical" observation follows from comparing the bound on B in (4.24) with the P-value of $(z-1)^+$ (the positive part of $(z-1)$). These are graphed in Figure 4.1, along with the P-value of z itself. (Note that the bound in (4.23) can be determined from the bound on B.) Clearly, the bound on B is roughly similar to the P-value of $(z-1)^+$. Since the bound on B can be interpreted as a bound on the comparative support of the data for the two hypotheses, the indication is that the common rule-of-thumb that

$$z = \begin{cases} 1 & \text{means only very mild evidence against } H_0, \\ 2 & \text{means significant evidence against } H_0, \\ 3 & \text{means highly significant evidence against } H_0, \\ 4 & \text{means overwhelming evidence against } H_0, \end{cases}$$

should, at the very least, be replaced by the rule-of-thumb

$$z = \begin{cases} 1 & \text{means no evidence against } H_0, \\ 2 & \text{means only very mild evidence against } H_0, \\ 3 & \text{means significant evidence against } H_0, \\ 4 & \text{means highly significant evidence against } H_0. \end{cases}$$

Although we have indicated that the lower bounds on B, or on α_0 when $\pi_0 = \frac{1}{2}$, can be interpreted as *lower bounds* on the comparative support of the data for the two hypotheses, it is questionable if they can be used as actual measures of evidence of comparative support. The reason is that all the bounds are independent of n. For the choice of g_1 leading to (4.18), however, it is clear that $\alpha_0 \to 1$ as $n \to \infty$ with z fixed! (See also Table 4.2.) Thus the bounds on α_0 and B are all likely to be excessively small when n is very large. This phenomenon, that $\alpha_0 \to 1$ as $n \to \infty$ and the P-value is held fixed, actually holds for virtually any fixed prior and point null testing problem. Indeed, if g_1 has a continuous density at θ_0, then, for large (and even moderate) n, an accurate approximation to α_0 when x corresponds to a moderate P-value is

$$\alpha_0 \cong \left[1 + \frac{(1-\pi_0)}{\pi_0} \cdot \frac{g_1(\theta_0)}{f(x|\theta_0)}\right]^{-1}. \tag{4.26}$$

(For an even better approximation, see Berger and Das Gupta (1985).) When n is large and the P-value is not too small, $f(x|\theta_0)$ will be large and α_0 close to one. Thus, for large n, the true "evidence" against H_0 is likely to be much less than that indicated by the lower bounds on α_0 or B. One is then left no recourse but to perform a subjective Bayesian analysis of the problem (but see Section 4.10). It is interesting that there is *no* suitable "noninformative prior" Bayesian analysis here.

This large n phenomenon provides an extreme illustration of the conflict between classical and Bayesian testing of a point null. One could classically reject H_0 with a P-value of 10^{-10}, yet, if n were large enough, the posterior probability of H_0 would be very close to 1. This surprising result has been called "Jeffreys' paradox" and "Lindley's paradox," and has a very long history (cf. Jeffreys (1961), Good (1950, 1965, 1983), Lindley (1957, 1961), and Shafer (1982c)). We will not discuss this "paradox" here because the point null approximation is rarely justifiable for very large n, and because we have already seen enough reasons, when n is small, to reject the use of classical error measures. It is, of course, useful to know that the problem becomes even worse for larger n.

We have concentrated, in this section, on the Example 11 situation of testing a normal mean, so as to provide a reasonably complete view of the issues. Most of the general formulas apply to other point null testing problems, however, and, in one dimension, the numerical results will be very similar. In higher dimensions the lower bounds for *all* g_1 in Theorem 1 can be very small, but lower bounds for $\mathcal{G}$ as in (4.22) will usually be comparable to the one-dimensional bounds. (An alternative to the use of $\mathcal{G}$ as in (4.22) is to replace $f(x|\theta)$ in the analysis by the density of the one-dimensional classical test statistic that would have been used, see Berger and Sellke (1984) for further discussion.) Additional references concerning point null analyses of the type we have considered, but in other situations, include Good (1950, 1958, 1983), Lempers (1971), Leamer (1978), and Smith and Spiegelhalter (1980).

III. *Multiple Hypothesis Testing*

The interesting feature of multiple hypothesis testing is that it is no more difficult, from a Bayesian perspective, than is testing of two hypotheses. One simply calculates the posterior probability of each hypothesis.

EXAMPLE 1 (continued). The child taking the IQ test is to be classified as having below average IQ (less than 90), average IQ (90 to 110), or above average IQ (over 110). Calling these three regions Θ_1, Θ_2, and Θ_3, respectively, and recalling that the posterior is $\mathcal{N}(110.39, 69.23)$, a table of normal probabilities can be used to show that $P(\Theta_1|x=115)=0.007$, $P(\Theta_2|x=115)=0.473$, and $P(\Theta_3|x=115)=0.520$.

4.3.4. Predictive Inference

In Subsection 2.4.4 we discussed the situation of trying to predict a random variable $Z \sim g(z|\theta)$ based on the observation of $X \sim f(x|\theta)$. We will again assume that X and Z are independent, and (for simplicity) that g is a density. (If X and Z are not independent, the necessary change in the following would be to replace $g(z|\theta)$ by $g(z|\theta, x)$.)

The idea of Bayesian predictive inference is that, since $\pi(\theta|x)$ is the believed (posterior) distribution of θ, then $g(z|\theta)\pi(\theta|x)$ is the joint distribution of z and θ given x, and integrating out over θ will give the believed distribution of z given x.

Definition 7. The *predictive density* of Z given x, when the prior for θ is π, is defined by

$$p(z|x) = \int_\Theta g(z|\theta) dF^{\pi(\theta|x)}(\theta).$$

EXAMPLE 12. Consider the linear regression model

$$Z = \theta_1 + \theta_2 Y + \varepsilon, \tag{4.27}$$

where $\varepsilon \sim \mathcal{N}(0, \sigma^2)$ (σ^2 known for simplicity). Available is independent data $((z_1, y_1), \ldots, (z_n, y_n))$ from the regression. A sufficient statistic for $\boldsymbol{\theta} = (\theta_1, \theta_2)$ is the least squares estimator $\mathbf{X} = (X_1, X_2)$, where

$$X_2 = \sum_{i=1}^n (Z_i - \bar{Z})(Y_i - \bar{Y})/SSY, \qquad X_1 = \bar{Z} - X_2\bar{Y},$$

$$SSY = \sum_{i=1}^n (Y_i - \bar{Y})^2, \qquad \bar{Z} = \frac{1}{n}\sum_{i=1}^n Z_i, \quad \text{and} \quad \bar{Y} = \frac{1}{n}\sum_{i=1}^n Y_i.$$

Furthermore, $\mathbf{X} \sim \mathcal{N}_2(\boldsymbol{\theta}, \sigma^2 \boldsymbol{\Sigma})$, where

$$\boldsymbol{\Sigma} = \frac{1}{SSY} \begin{pmatrix} \frac{1}{n} \sum_{i=1}^{n} y_i^2 & -\bar{y} \\ -\bar{y} & 1 \end{pmatrix}.$$

If the noninformative prior $\pi(\boldsymbol{\theta}) = 1$ is used for $\boldsymbol{\theta}$, the posterior distribution, $\pi(\boldsymbol{\theta}|\mathbf{x})$, is easily seen to be $\mathcal{N}_2(\mathbf{x}, \sigma^2 \boldsymbol{\Sigma})$.

Suppose now that one desires to predict a future Z corresponding (through (4.27)) to a given y. Clearly $g(z|\boldsymbol{\theta})$ is then $\mathcal{N}(\theta_1 + \theta_2 y, \sigma^2)$, and the joint density of $(z, \boldsymbol{\theta})$ given $\mathbf{x}$ (and for the given y) is also normal. The desired predictive distribution, $p(z|x)$, is the marginal distribution of Z from this joint normal posterior; this marginal must also be a normal distribution, and calculation gives a mean of $(x_1 + x_2 y)$ and variance $V = \sigma^2(1 + n^{-1} + (\bar{y} - y)^2 / SSY)$.

This predictive distribution can be used for any desired Bayesian inference. For instance, a $100(1-\alpha)\%$ HPD credible set for Z would be

$$\left((x_1 + x_2 y) + z\left(\frac{\alpha}{2}\right)\sqrt{V}, \quad (x_1 + x_2 y) - z\left(\frac{\alpha}{2}\right)\sqrt{V}\right).$$

Note that this happens to be the same set that the classical approach suggests for predicting Z, providing another example of the formal similarity that often exists between answers from classical and noninformative prior Bayesian analyses.

We will not specifically pursue predictive problems in this book, although many of the issues and techniques discussed for posterior distributions will apply equally well to predictive distributions. Among the many good references on predictive inference are Roberts (1965), Geisser (1971, 1980, 1984b), and Aitchison and Dunsmore (1975).

4.4. Bayesian Decision Theory

In this section the influence of the loss function will be considered, as well as that of the prior information. It will be seen that the Bayesian approach to decision theory is conceptually very straightforward.

4.4.1. Posterior Decision Analysis

In Subsections 1.3.1 and 1.5.1 the conditional Bayes decision principle was discussed. One looks at the expected loss of an action for the believed distribution of θ at the time of decision making. We now know that this distribution should be the posterior distribution, $\pi(\theta|x)$. For completeness, we redefine the relevant concepts.

Definition 8. The *posterior expected loss* of an action a, when the posterior distribution is $\pi(\theta|x)$, is

$$\rho(\pi(\theta|x), a) = \int_{\Theta} L(\theta, a) dF^{\pi(\theta|x)}. \tag{4.28}$$

A (*posterior*) *Bayes action*, to be denoted by $\delta^{\pi}(x)$ (for consistency with later definitions), is any action $a \in \mathscr{A}$ which minimizes $\rho(\pi(\theta|x), a)$, or equivalently which minimizes

$$\int_{\Theta} L(\theta, a) f(x|\theta) dF^{\pi}(\theta). \tag{4.29}$$

The simplicity of the Bayesian approach follows from the fact that an optimal action can be found by a simple minimization of (4.28) or (4.29). (The possible advantage of using (4.29) is that $m(x)$ need not be calculated.) Of course, there could be several minimums, and, hence, several Bayes actions.

Another kind of Bayesian analysis was mentioned in Subsections 1.3.2 and 1.5.2, namely choice of a decision rule through minimization of the (frequentist) Bayes risk, $r(\pi, \delta)$. Although minimizing $r(\pi, \delta)$ appears to be a much more difficult problem than minimizing posterior expected loss (choosing a minimizing *function* is usually very hard), the two problems are essentially equivalent.

Result 1. *A Bayes rule δ^{π} (i.e., a rule minimizing $r(\pi, \delta)$) can be found by choosing, for each x such that $m(x) > 0$, an action which minimizes the posterior expected loss* (4.28) (*or equivalently* (4.29)). *The rule can be defined arbitrarily when $m(x) = 0$.*

Since (posterior) Bayes actions need not be unique, δ^{π} need not be unique (and can sometimes even be chosen to be randomized). Also, if $r(\pi, \delta) = \infty$ for *all* δ, then *any* decision rule is a Bayes rule (not just those which correspond to posterior Bayes actions). There are certain technical conditions needed for the validity of Result 1 (cf. Brown and Purves (1973) or Diaconis and Stein (1983)), but the result will hold in all cases of practical interest. Result 1 is an immediate consequence of the following result, which is stated separately for later use (and is only stated for nonrandomized estimators—the generalization to randomized estimators is left for the exercises).

Result 2. *If δ is a nonrandomized estimator, then*

$$r(\pi, \delta) = \int_{\{x:\, m(x) > 0\}} \rho(\pi(\theta|x), \delta(x)) dF^{m}(x). \tag{4.30}$$

Proof. By definition,

$$r(\pi, \delta) = \int_\Theta R(\theta, \delta) dF^\pi(\theta)$$
$$= \int_\Theta \int_{\mathscr{X}} L(\theta, \delta(x)) dF^{X|\theta}(x) dF^\pi(\theta).$$

Since $L(\theta, a) \geq -K > -\infty$ and all measures above are finite, Fubini's theorem can be employed to interchange orders of integration and obtain

$$r(\pi, \delta) = \begin{cases} \int_{\mathscr{X}} \left[\int_\Theta L(\theta, \delta(x)) f(x|\theta) dF^\pi(\theta) \right] dx, \\ \sum_{x \in \mathscr{X}} \left[\int_\Theta L(\theta, \delta(x)) f(x|\theta) dF^\pi(\theta) \right], \end{cases}$$

in the cases of continuous and discrete $\mathscr{X}$, respectively. Finally, noting that, if $m(x)=0$, then $f(x|\theta)=0$ almost everywhere with respect to π, the definitions of $\pi(\theta|x)$ and $\rho(\pi(\theta|x), \delta)$ yield the result. □

Result 1 easily follows from Result 2, since a minimizing δ can be found by minimizing the integrand in (4.30) for each x for which $m(x) > 0$, and this is precisely the earlier definition of a posterior Bayes action. The overall minimization of $r(\pi, \delta)$ has been called the *normal form* of Bayesian analysis, while minimization of $\rho(\pi(\theta|x), a)$ has been called the *extensive form* (cf. Raiffa and Schlaifer (1961)). Operationally, we will virtually always proceed by minimizing $\rho(\pi(\theta|x), a)$, and will for simplicity always call this result the Bayes rule $\delta^\pi(x)$ (to be interpreted as an action or decision rule, as implied by the context). The only case where the normal form of Bayesian analysis need be considered is when a restricted set of decision rules is to be used, so that the overall Bayes rule need not be in the restricted class. (Some examples are discussed in Subsection 4.7.7, see also Haff (1983).)

Note that, from the conditional perspective together with the utility development of the loss, the *correct* way to view the situation is that of minimizing $\rho(\pi(\theta|x), a)$. One should condition on what is known, namely x (incidentally following the Likelihood Principle, since $\pi(\theta|x)$ depends only on the observed likelihood function), and average the utility over what is unknown, namely θ. The desire to minimize $r(\pi, \delta)$ would be deemed rather bizarre from this perspective.

So far we have assumed that π is a proper prior. Even when π is improper, however, it will often make sense to minimize posterior expected loss.

Definition 9. If π is an improper prior, but $\delta^\pi(x)$ is an action which minimizes (4.28) or (4.29) for each x with $m(x) > 0$, then δ^π is called a *generalized Bayes rule.*

We now turn to some standard areas of application of Bayesian decision theory. It should be emphasized that Bayesian decision theory is by no means limited to such standard applications. The methodology of choosing a prior and loss and minimizing the posterior expected loss can be applied in almost any situation.

4.4.2. Estimation

In Bayesian estimation of a real valued parameter θ, the loss which is easiest to deal with is squared-error loss ($L(\theta, a) = (\theta - a)^2$). The posterior expected loss is then

$$\int_\Theta (\theta - a)^2 dF^{\pi(\theta|x)}(\theta).$$

The value of a which minimizes this can be found by expanding the quadratic expression, differentiating with respect to a, and setting equal to zero. The result (assuming all integrals are finite) is

$$\begin{aligned} 0 &= \frac{d}{da}\left[\int_\Theta \theta^2 dF^{\pi(\theta|x)}(\theta) - 2a\int_\Theta \theta dF^{\pi(\theta|x)}(\theta) + a^2 \int_\Theta dF^{\pi(\theta|x)}(\theta)\right] \\ &= -2E^{\pi(\theta|x)}[\theta] + 2a. \end{aligned}$$

Solving for a gives the following result.

Result 3. *If $L(\theta, a) = (\theta - a)^2$, the Bayes rule is*

$$\delta^\pi(x) = E^{\pi(\theta|x)}[\theta],$$

which is the mean of the posterior distribution of θ given x.

EXAMPLE 1 (continued). If f and π are normal, the posterior is $\mathcal{N}(\mu(x), \rho^{-1})$. This has mean $\mu(x)$, so the Bayes rule for squared-error loss is

$$\delta^\pi(x) = \mu(x) = \frac{\sigma^2 \mu}{\sigma^2 + \tau^2} + \frac{\tau^2 x}{\sigma^2 + \tau^2}.$$

For weighted squared-error loss the following result holds. (The proof will be left as an exercise.)

Result 4. *If $L(\theta, a) = w(\theta)(\theta - a)^2$, the Bayes rule is*

$$\begin{aligned} \delta^\pi(x) &= \frac{E^{\pi(\theta|x)}[\theta w(\theta)]}{E^{\pi(\theta|x)}[w(\theta)]} \\ &= \frac{\int \theta w(\theta) f(x|\theta) dF^\pi(\theta)}{\int w(\theta) f(x|\theta) dF^\pi(\theta)}. \end{aligned}$$

From the last expression in Result 4, it is interesting to note that the weight function, $w(\theta)$, plays a role analogous to that of the prior $\pi(\theta)$. This is important to note, in that robustness concerns involving $w(\theta)$ are thus the same as robustness concerns involving the prior.

For the quadratic loss $L(\boldsymbol{\theta}, \mathbf{a})=(\boldsymbol{\theta}-\mathbf{a})^t\mathbf{Q}(\boldsymbol{\theta}-\mathbf{a})$ ($\boldsymbol{\theta}$ and $\mathbf{a}$ are now vectors and $\mathbf{Q}$ is a positive definite matrix), it can be shown that the Bayes estimator is still the posterior mean. (Interestingly, $\mathbf{Q}$ has no effect.)

For absolute error loss, the following result holds. Recall that a median is a $\frac{1}{2}$-fractile of a distribution. (In general, a point $z(\alpha)$ is an α-fractile of the distribution of a random variable X if $P(X\le z(\alpha))\ge\alpha$ and $P(X< z(\alpha))\le\alpha$.)

Result 5. *If $L(\theta, a)=|\theta-a|$, any median of $\pi(\theta|x)$ is a Bayes estimator of θ.*

Proof. Let m denote a median of $\pi(\theta|x)$, and let $a>m$ be another action. Note that

$$L(\theta, m)-L(\theta, a)=\begin{cases} m-a & \text{if } \theta\le m,\\ 2\theta-(m+a) & \text{if } m<\theta<a,\\ a-m & \text{if } \theta\ge a,\end{cases}$$

from which it follows that

$$L(\theta, m)-L(\theta, a)\le(m-a)I_{(-\infty, m)}(\theta)+(a-m)I_{(m,\infty)}(\theta).$$

Since $P(\theta\le m|x)\ge\frac{1}{2}$, so that $P(\theta>m|x)\le\frac{1}{2}$, it can be concluded that

$$\begin{aligned}E^{\pi(\theta|x)}[L(\theta, m)-L(\theta, a)]&\le(m-a)P(\theta\le m|x)+(a-m)P(\theta>m|x)\\ &\le(m-a)\tfrac{1}{2}+(a-m)\tfrac{1}{2}=0,\end{aligned}$$

establishing that m has posterior expected loss at least as small as a. A similar argument holds for $a<m$, completing the proof. □

In the IQ example (Example 1), the posterior was normal with mean $\mu(x)$, which, for a normal distribution, is also the median. Hence the Bayes estimator is the same in this example, whether squared-error or absolute-error loss is used. Indeed, when the posterior is unimodal and symmetric, it can be shown that, for any loss of the form $L(|\theta-a|)$ which is increasing in $|\theta-a|$, the Bayes estimator is the median of the posterior. This partially indicates that, when underestimation and overestimation are of equal concern (merely a verbal statement of the condition that L be a function of $|\theta-a|$), the exact function of $|\theta-a|$ which is used as the loss is not too crucial (i.e., the Bayes rule is robust with respect to this part of the loss). Recall, however, that any weight function, $w(\theta)$, in the loss can have a significant effect.

Perhaps the most useful standard loss is linear loss. The proof of the following result will be left as an exercise.

Result 6. *If*

$$L(\theta, a)=\begin{cases} K_0(\theta-a) & \text{if } \theta-a\ge 0,\\ K_1(a-\theta) & \text{if } \theta-a<0,\end{cases}$$

any $(K_0/(K_0+K_1))$-fractile of $\pi(\theta|x)$ is a Bayes estimate of θ.

EXAMPLE 1 (continued). In estimating the child's IQ, it is deemed to be twice as harmful to underestimate as to overestimate. A linear loss is felt to be appropriate, so the loss in Result 6 is used with $K_0=2$ and $K_1=1$. The $\frac{2}{3}$-fractile of a $\mathcal{N}(0,1)$ distribution is about 0.43, so the $\frac{2}{3}$-fractile of a $\mathcal{N}(110.39, 61.23)$ distribution (which is $\pi(\theta|x)$) is

$$110.39+(0.43)(61.23)^{1/2}=113.97.$$

This is the Bayes estimate of θ.

It should again be emphasized that, while easy to work with, none of the above standard losses need be suitable for a given real problem. If this is the case, it will usually be necessary to calculate and minimize the posterior expected loss numerically. This can be done with a computer or, in many cases, with a programmable hand calculator.

4.4.3. Finite Action Problems and Hypothesis Testing

In estimation there are generally an infinite number of actions to choose from. Many interesting statistical problems involve only a finite number of actions, however. The most important finite action problem is, of course, hypothesis testing.

The finite action Bayesian decision problem is easily solved when considered in extensive form. If $\{a_1, \ldots, a_k\}$ are the available actions and $L(\theta, a_i)$ the corresponding losses, the Bayes action is simply that for which the posterior expected loss $E^{\pi(\theta|x)}[L(\theta, a_i)]$ is the smallest. Several specific finite action problems will now be considered.

In testing H_0: $\theta\in\Theta_0$ versus H_1: $\theta\in\Theta_1$, the actions of interest are a_0 and a_1, where a_i denotes acceptance of H_i. Actually, this is to a degree putting the cart before the horse, since in true decision problems the hypotheses are usually determined by the available actions. In other words, the decision maker is often faced with two possible courses of action, a_0 and a_1. He determines that, if $\theta\in\Theta_0$, then action a_0 is appropriate, while if $\theta\in\Theta_1$, then a_1 is best. While the distinction as to whether the hypotheses or actions come first is important in discussing reasonable formulations of hypothesis testing problems, it makes no difference in the formal analysis of a given problem.

When the loss is "0–1" loss ($L(\theta, a_i)=0$ if $\theta\in\Theta_i$ and $L(\theta, a_i)=1$ if $\theta\in\Theta_j, j\neq i$), then

$$E^{\pi(\theta|x)}[L(\theta, a_1)]=\int L(\theta, a_1)dF^{\pi(\theta|x)}(\theta)=\int_{\Theta_0} dF^{\pi(\theta|x)}(\theta)=P(\Theta_0|x),$$

and

$$E^{\pi(\theta|x)}[L(\theta, a_0)]=P(\Theta_1|x).$$

Hence the Bayes decision is simply the hypothesis with the larger posterior probability.

For the more realistic "0-K_i" loss,

$$L(\theta, a_i) = \begin{cases} 0 & \text{if } \theta \in \Theta_i, \\ K_i & \text{if } \theta \in \Theta_j \ (j \neq i), \end{cases}$$

the posterior expected losses of a_0 and a_1 are $K_0 P(\Theta_1 | x)$ and $K_1 P(\Theta_0 | x)$, respectively. The Bayes decision is again that corresponding to the smallest posterior expected loss.

It is useful to observe the relationship of these Bayesian tests with classical hypothesis tests. In the Bayesian test, the null hypothesis is rejected (i.e., action a_1 is taken) when

$$\frac{K_0}{K_1} > \frac{P(\Theta_0 | x)}{P(\Theta_1 | x)}. \tag{4.31}$$

Usually $\Theta_0 \cup \Theta_1 = \Theta$, in which case

$$P(\Theta_0 | x) = 1 - P(\Theta_1 | x).$$

Inequality (4.31) can then be rewritten

$$\frac{K_0}{K_1} > \frac{1 - P(\Theta_1 | x)}{P(\Theta_1 | x)} = \frac{1}{P(\Theta_1 | x)} - 1,$$

or

$$P(\Theta_1 | x) > \frac{K_1}{K_0 + K_1}.$$

Thus in classical terminology, the rejection region of the Bayesian test is

$$C = \left\{ x \colon P(\Theta_1 | x) > \frac{K_1}{K_0 + K_1} \right\}.$$

Typically, C is of exactly the same form as the rejection region of a classical (say likelihood ratio) test. An example of this follows.

Example 1 (continued). Assume f and π are normal and that it is desired to test $H_0\colon \theta \geq \theta_0$ versus $H_1\colon \theta < \theta_0$ under "0-K_i" loss. Noting that $\pi(\theta | x)$ is a $\mathcal{N}(\mu(x), \rho^{-1})$ density, the Bayes test rejects H_0 if

$$\frac{K_1}{K_0 + K_1} < P(\Theta_1 | x) = \left(\frac{\rho}{2\pi}\right)^{1/2} \int_{-\infty}^{\theta_0} \exp\left\{\frac{-\rho(\theta - \mu(x))^2}{2}\right\} d\theta$$

$$= (2\pi)^{-1/2} \int_{-\infty}^{\rho^{1/2}(\theta_0 - \mu(x))} \exp\left\{\frac{-\eta^2}{2}\right\} d\eta$$

(making the change of variables $\eta = \rho^{1/2}(\theta - \mu(x))$). Letting $z(\alpha)$ denote the α-fractile of a $\mathcal{N}(0, 1)$ distribution, it follows that the Bayes test rejects

H_0 if

$$\rho^{1/2}(\theta_0 - \mu(x)) > z\left(\frac{K_1}{K_0 + K_1}\right).$$

Recalling that

$$\mu(x) = \frac{\tau^2}{\sigma^2 + \tau^2} x + \frac{\sigma^2}{\sigma^2 + \tau^2} \mu$$

and rearranging terms, gives the equivalent condition

$$x < \theta_0 + \frac{\sigma^2}{\tau^2}(\theta_0 - \mu) - \sigma^2 \rho^{1/2} z\left(\frac{K_1}{K_0 + K_1}\right).$$

The classical uniformly most powerful size α tests are of the same form, rejecting H_0 when

$$x < \theta_0 + \sigma z(\alpha).$$

In classical testing, the "critical value" of the rejection region is determined by α, while in the Bayesian test it is determined by the loss and prior information.

In situations such as that above, the Bayesian method can be thought of as providing a rational way of choosing the size of the test. Classical statistics provides no such guidelines, with the result being that certain "standard" sizes (0.1, 0.05, 0.01) have come to be most frequently used. Such an *ad hoc* choice is clearly suspect in a true decision problem. Indeed even classical statisticians will tend to say that, in a true decision problem, the size should be chosen according to "subjective" factors. This is, of course, precisely what the Bayesian approach does.

There are many decision problems with more than two possible actions. For instance, a frequently faced situation in hypothesis testing is the existence of an indifference region. The idea here is that, besides the actions a_0 and a_1, which will be taken if $\theta \in \Theta_0$ or $\theta \in \Theta_1$, a third action a_2 representing indifference will be taken if $\theta \in \Theta_2$. For example, assume it is desired to test which of two drugs is the most effective. Letting θ_1 and θ_2 denote the probabilities of cures using the two drugs, a reasonable way to formulate the problem is as a test of the three hypotheses H_0: $\theta_1 - \theta_2 < -\varepsilon$, H_1: $\theta_1 - \theta_2 > \varepsilon$, and H_2: $|\theta_1 - \theta_2| \le \varepsilon$, where $\varepsilon > 0$ is chosen so that when $|\theta_1 - \theta_2| \le \varepsilon$ the two drugs are considered equivalent.

Even in classical hypothesis testing there are usually three actions taken: a_0—accept H_0, a_1—accept H_1, and a_2—conclude there is not significant evidence for accepting either H_0 or H_1. The choice among these actions is classically made through an informal choice of desired error probabilities. Attacking the problem from a Bayesian decision-theoretic viewpoint (including the specification of $L(\theta, a_2)$) seems more appealing.

Another type of common finite action problem is the classification problem, in which it is desired to classify an observation as belonging to one of several possible categories. An example folllows.

EXAMPLE 1 (continued). For the IQ example in which it is desired to classify the child as a_1—below average ($\theta<90$), a_2—average ($90\le\theta\le110$), or a_3—above average ($\theta>110$), the following losses are deemed appropriate:

$$L(\theta, a_1)=\begin{cases}0 & \text{if } \theta<90,\\ \theta-90 & \text{if } 90\le\theta\le110,\\ 2(\theta-90) & \text{if } \theta>110,\end{cases}$$

$$L(\theta, a_2)=\begin{cases}90-\theta & \text{if } \theta<90,\\ 0 & \text{if } 90\le\theta\le110,\\ \theta-110 & \text{if } \theta>110,\end{cases}$$

$$L(\theta, a_3)=\begin{cases}2(110-\theta) & \text{if } \theta<90,\\ 110-\theta & \text{if } 90\le\theta\le110,\\ 0 & \text{if } \theta>110.\end{cases}$$

(These could arise, for example, if children are put into one of three reading groups (slow, average, and fast) depending on their IQ classification.) Since $\pi(\theta|x)$ is $\mathcal{N}(110.39, 69.23)$, the posterior expected losses are

$$\begin{aligned}E^{\pi(\theta|x)}[L(\theta, a_1)]&=\int_{90}^{110}(\theta-90)\pi(\theta|x)d\theta+\int_{110}^{\infty}2(\theta-90)\pi(\theta|x)d\theta\\ &=6.49+27.83=34.32,\end{aligned}$$

$$\begin{aligned}E^{\pi(\theta|x)}[L(\theta, a_2)]&=\int_{-\infty}^{90}(90-\theta)\pi(\theta|x)d\theta+\int_{110}^{\infty}(\theta-110)\pi(\theta|x)d\theta\\ &=0.02+3.53=3.55,\end{aligned}$$

$$\begin{aligned}E^{\pi(\theta|x)}[L(\theta, a_3)]&=\int_{-\infty}^{90}2(110-\theta)\pi(\theta|x)d\theta+\int_{90}^{110}(110-\theta)\pi(\theta|x)d\theta\\ &=0.32+2.95=3.27.\end{aligned}$$

(The preceding integrals are calculated by first transforming to a $\mathcal{N}(0. 1)$ density, and then using normal probability tables and the fact that

$$\int_a^b \theta e^{-\theta^2/2}\,d\theta=-e^{-\theta^2/2}\Big|_a^b=e^{-a^2/2}-e^{-b^2/2}.)$$

Thus a_3 is the Bayes decision.

4.4.4. With Inference Losses

In Subsection 2.4.3 we discussed two of the ways in which decision theory can be useful for inference problems. First, many common inference measures can be given a formal decision-theoretic representation. For instance, the choice of "0–1" loss in testing leads to standard Bayesian testing measures; the posterior expected loss of rejecting a hypothesis is then the posterior probability of the hypothesis. As another example, if

$$L(\theta, C(x)) = 1 - I_{C(x)}(\theta),$$

where $C(x) \subset \Theta$, then the posterior expected loss of $C(x)$ would be the posterior probability that θ is not in $C(x)$. Such formal relationships allow the use of decision-theoretic machinery in solving inference problems.

The second use of decision theory in inference, also discussed in Subsection 2.4.3, was the use of loss functions to represent the actual success of an inference in communicating information. Thus, in Example 4 of Subsection 2.4.3, it was suggested that, since one reports $\alpha(x)$ as a measure of the "confidence" with which it is felt that θ is in $C(x)$, it would be reasonable to measure the accuracy of the report by

$$L_C(\theta, \alpha(x)) = (I_{C(x)}(\theta) - \alpha(x))^2.$$

This loss could be used, decision-theoretically, to suggest a choice of the report $\alpha(x)$. For instance, if $C(x)$ is the credible set to be used in Bayesian inference, it would be reasonable to choose $\alpha(x)$ to minimize the posterior expected loss for L_C. As in the development of Result 3, it is easy to show that the Bayes choice of $\alpha(x)$ is then

$$\begin{aligned}\alpha^{\pi}(x) &= E^{\pi(\theta|x)}[I_{C(x)}(\theta)] \\ &= P^{\pi(\theta|x)}(\theta \in C(x)).\end{aligned}$$

Thus we have been led to the "obvious," that the best (to a Bayesian) report, for the confidence to be placed in $C(x)$, is the posterior probability of the set. More sophisticated (and interesting) uses of inference losses can be found in the references in Subsection 2.4.3.

4.5. Empirical Bayes Analysis

4.5.1. Introduction

An empirical Bayes problem, as discussed in Subsection 3.5.2, is one in which known relationships among the coordinates of the parameter vector $\boldsymbol{\theta} = (\theta_1, \ldots, \theta_p)^t$ allow use of the data to estimate some features of the prior distribution. Such problems occur with moderate frequency in statistics, generally in one of two related situations. The most obvious such situation is when the θ_i arise from some common population, so what we can imagine creating a probabilistic model for the population and can interpret this model as the prior distribution. The simplest version of this situation is when the θ_i are i.i.d. from (the prior) π_0, which is the (perhaps partially unknown) model for the population. Examples include:

(i) the θ_i are the proportions of defectives in a series of lots of parts from the same supplier;
(ii) the θ_i are the mean bushels of corn per acre for a random selection of farms from a given county;
(iii) the θ_i are the mean worker accident rates from a sample of similar companies in a given industry;
(iv) the θ_i are the bone loss rates of a random sample of individuals in an osteoporosis high risk group.

One might care to question the i.i.d. assumption in any of these examples, and devise more elaborate models, but the θ_i can be expected to be significantly related and this relationship can be used to greatly augment the information provided by the data.

A situation, in which an empirical Bayes analysis is particularly desirable, is when one is debating between a higher and a lower dimensional model for the θ_i. For instance, in many statistical problems one has to estimate a set of variances from, say, related treatments, and a decision must be made concerning whether or not to use a pooled variance estimate. In other words, one is deciding between assumption of the high-dimensional model that the variances are distinct, and the low-dimensional model that the variances are equal. (See Hui and Berger (1983) for a discussion of this particular problem.) Empirical Bayes analysis is ideally suited for such situations.

These issues will be discussed further as we proceed; indeed, the emphasis will be on constructing appropriate models for the θ_i. When viewed from this "modelling" perspective, empirical Bayes problems can be considered to be problems in classical statistics (indeed, classical *random effects* models are of this type), and as such have a very long history (cf. Whitney (1918), Kelly (1927), Lush (1937), Good (1953)). Such models also now go by many other names, such as mixed models, random coefficient regression, principle components estimation, credibility theory, factor analysis regression, and filtering theory. (See Berger (1986) for references.) Extensive development of empirical Bayes methodology (and the name) began with Robbins (1951, 1955, 1964).

Empirical Bayes methods can be categorized in two different ways. One division is between *parametric empirical Bayes* (PEB) and *nonparametric empirical Bayes* (NPEB), names due to Morris (1983a). In the former, one assumes that the prior distribution of $\boldsymbol{\theta}$ is in some parametric class with unknown hyperparameters, while, in the latter, one typically assumes only that the θ_i are i.i.d. Examples of PEB analysis will be given in Subsections 4.5.2 and 4.5.3. Subsection 4.5.4 deals with NPEB.

A different categorization of empirical Bayes analysis can be given according to its operational focus. The most natural focus is to use the data to estimate the prior distribution or the posterior distribution. Once this is done, the analysis can proceed in a typical Bayesian fashion. (The analyses in Subsections 4.5.2 and 4.5.3 are of this type.) A second possible operational focus is to represent the Bayes rule in terms of the unknown prior, and then use the data to estimate the Bayes rule directly (as opposed to using the estimated prior or posterior to construct the Bayes rule). This possibility will be discussed in Subsection 4.5.4, it usually being associated with NPEB.

We will be able to do no more than give a few examples of empirical Bayes methodology. Subsections 4.5.2 and 4.5.3 present a fairly thorough analysis of the PEB approach to estimating a normal mean vector, and Subsection 4.5.4 presents a NPEB example with Poisson means. Many other examples (and references) can be found in Aitchison and Martz (1969), Copas (1969), Lord (1969), Maritz (1970), Efron and Morris (1973, 1976b), Lord and Cressie (1975), Bishop, Fienberg, and Holland (1975), Rao (1977), Haff (1980), Hoadley (1981), Cressie (1982), Martz and Waller (1982), Leonard (1983), Hui and Berger (1983), Morris (1983a, b), Robbins (1983), Strenio, Weisberg, and Bryk (1983), Albert (1984a), Berliner (1984b), Dempster, Selwyn, Patel, and Roth (1984), Jewell and Schnieper (1984), Rubin (1984), Hudson (1985), and Berger (1986). The Morris (1983a), Rubin (1984), and Berger (1986) works contain other references, including references to many actual applications of empirical Bayes methodology.

There is also a large *theoretical* literature on empirical Bayes analysis. This literature explores the question of *asymptotic optimality* of empirical Bayes procedures: an empirical Bayes procedure is asymptotically optimal if, as the dimension $p\to\infty$, the procedure is as good (in some sense) as the Bayes procedure were the prior actually known. There are very interesting theoretical issues here, but space (and our applied orientation) preclude discussion. Introductions and references to these issues can be found in Robbins (1951, 1955, 1964), Copas (1969), Maritz (1970), O'Bryan (1979), Stijnen (1980), Gilliland, Boyer, and Tsao (1982), and Susarla (1982).

4.5.2. PEB for Normal Means—The Exchangeable Case

A very basic empirical Bayes situation, discussed in Example 17 in Section 3.5, is that in which $X_i \sim \mathcal{N}(\theta_i, \sigma_f^2)$ (independently, $i=1,\ldots,p$) and the θ_i are considered to be exchangeable, modelled by supposing that the θ_i are independently $\mathcal{N}(\mu_\pi, \sigma_\pi^2)$, the hyperparameters μ_π and σ_π^2 being unknown. We assume, until the end of the subsection, that σ_f^2 is known.

In Subsection 3.5.4 we estimated the unknown hyperparameters, μ_π and σ_π^2, by maximum likelihood, the results being

$$\hat{\mu}_\pi = \bar{x} \quad \text{and} \quad \hat{\sigma}_\pi^2 = \max\left\{0, \frac{1}{p}s^2 - \sigma_f^2\right\}, \quad \text{where} \quad s^2 = \sum_{i=1}^{p}(x_i - \bar{x})^2. \tag{4.32}$$

(The definition of s^2 here differs from that in Subsection 3.5.4, for later convenience.) If, instead, the moment method of Subsection 3.5.5 were used to estimate the hyperparameters, one would have $\hat{\mu}_\pi = \bar{x}$ and $\hat{\sigma}_\pi^2 = \sum_{i=1}^p (x_i - \bar{x})^2/(p-1) - \sigma_f^2$. The difference in these estimates of σ_π^2 will be minor unless p is small. (There is actually some advantage in dividing the sum of squares by $p-1$, but the moment estimator has the disadvantage of possibly being negative.)

Formally, one could then pretend that the θ_i are (independently) $\mathcal{N}(\hat{\mu}_\pi, \hat{\sigma}_\pi^2)$, and proceed with a Bayesian analysis. This indeed works well when p is large. For small or moderate p, however, such an analysis leaves something to be desired, since it ignores the fact that μ_π and σ_π^2 were estimated; the errors undoubtedly introduced in the hyperparameter estimation will not be reflected in any of the conclusions. This is indeed a general problem with the empirical Bayes approach, and will lead us to recommend the hierarchical Bayes approach (see Section 4.6) when p is small or moderate. For the situation of this subsection, however, Morris (1983a) (see also Morris (1983b)) has developed empirical Bayes approximations to the hierarchical Bayes answers which do take into account the uncertainty in $\hat{\mu}_\pi$ and $\hat{\sigma}_\pi^2$. These approximations can best be described, in empirical Bayes terms, as providing an estimated *posterior* distribution (rather than an estimated prior).

To describe these approximations, recall that the posterior distribution of θ_i, for given μ_π and σ_π^2, is $\mathcal{N}(\mu_i(x_i), V)$, where, letting $B = \sigma^2/(\sigma^2 + \sigma_\pi^2)$,

$$\mu_i(x_i) = x_i - B(x_i - \mu_\pi)$$

and

$$V = \frac{\sigma_\pi^2 \sigma_f^2}{(\sigma_f^2 + \sigma_\pi^2)} = \sigma_f^2(1-B).$$

The estimates Morris (1983a) suggests for $\mu_i(x_i)$ and V are (when $p \geq 4$)

$$\mu_i^{\text{EB}}(\mathbf{x}) = x_i - \hat{B}(x_i - \bar{x}), \tag{4.33}$$

and

$$V_i^{\text{EB}}(\mathbf{x}) = \sigma_f^2\left(1 - \frac{(p-1)}{p}\hat{B}\right) + \frac{2}{(p-3)}\hat{B}^2(x_i - \bar{x})^2, \tag{4.34}$$

where the estimate of B is

$$\hat{B} = \left(\frac{p-3}{p-1}\right)\frac{\sigma_f^2}{(\sigma_f^2 + \tilde{\sigma}_\pi^2)}, \qquad \tilde{\sigma}_\pi^2 = \max\left\{0, \frac{s^2}{(p-1)} - \sigma_f^2\right\}. \tag{4.35}$$

In way of explanation, $\tilde{\sigma}_\pi^2$ is just a slight modification of $\hat{\sigma}_\pi^2$ in (4.32); the factor $(p-3)/(p-1)$ in (4.35) has to do with adjusting for the error in the estimation of σ_π^2, and $(p-1)/p$ in (4.34) and the last term in (4.34) have to do with the error in estimating μ_π. The terms in (4.33) and (4.34) can also be described from the intuitive viewpoint that the empirical Bayes formulation provides a "compromise" between the model where the θ_i are completely unrelated, and that where all the θ_i are assumed to be equal. We delay discussion of this point until Subsection 4.5.3, however.

The resultant $\mathcal{N}(\mu_i^{EB}(\mathbf{x}), V_i^{EB}(\mathbf{x}))$ (estimated) posterior for θ_i can be used in the standard Bayesian ways. The natural estimates for the θ_i are, of course, just $\mu_i^{EB}(\mathbf{x})$. The (estimated) $100(1-\alpha)\%$ HPD credible set for θ_i is

$$C_i^{EB}(\mathbf{x}) = \left(\mu_i^{EB}(\mathbf{x}) + z\left(\frac{\alpha}{2}\right)\sqrt{V_i^{EB}(\mathbf{x})},\ \mu_i^{EB}(\mathbf{x}) - z\left(\frac{\alpha}{2}\right)\sqrt{V_i^{EB}(\mathbf{x})}\right). \quad (4.36)$$

A $100(1-\alpha)\%$ HPD credible set for $\boldsymbol{\theta}$ can be formed by noting that the estimated posterior for $\boldsymbol{\theta}$ is $\mathcal{N}_p(\boldsymbol{\mu}^{EB}(\mathbf{x}), \mathbf{V}^{EB}(\mathbf{x}))$, where $\boldsymbol{\mu}^{EB}(\mathbf{x}) = (\mu_1^{EB}(\mathbf{x}), \ldots, \mu_p^{EB}(\mathbf{x}))^t$ and $\mathbf{V}^{EB}(\mathbf{x})$ is the diagonal matrix with (i, i) element $V_i^{EB}(\mathbf{x})$. Employing (4.11) from Subsection 4.3.2 yields the ellipsoid

$$C^{EB}(\mathbf{x}) = \left\{\boldsymbol{\theta}: \sum_{i=1}^{p} (\theta_i - \mu_i^{EB}(\mathbf{x}))^2 / V_i^{EB}(\mathbf{x}) \leq x_p^2(1-\alpha)\right\}. \quad (4.37)$$

(There is a better approximation than $\mathbf{V}^{EB}$ to the posterior covariance matrix—see Section 4.6.)

EXAMPLE 13. Consider the example of the child who scores $x_7 = 115$ on a $\mathcal{N}(\theta_7, 100)$ IQ test. Also available are intelligence test scores of the child for six previous years. These six scores, 105, 127, 115, 130, 110, and 135, are assumed to be observations, $X_1, \ldots, X_6$, from independent $\mathcal{N}(\theta_i, 100)$ distributions. It may be tempting to assume that all θ_i are equal to the child's "true" IQ, but this is probably unwise, since IQs can be expected to vary from year to year (either because of changes in the individual or changes in the tests). On the other hand, it is very reasonable to assume that the true yearly IQs, θ_i, are from a common prior distribution. (A little care should be taken here. If the past observations seem to be nonrandom, say are increasing, this may indicate a trend in the θ_i. The assumption that the θ_i are a sample from a common distribution is then inappropriate. See Subsection 4.5.3 for an example.)

It can be calculated, for the above data, that $\bar{x} = 121$ and $s^2 = 762$. Since $\sigma_f^2 = 100$ in this example, (4.35), (4.33), and (4.34) become (for estimating θ_7)

$$\tilde{\sigma}_\pi^2 = 27.000, \qquad \hat{B} = (\tfrac{4}{6})(\tfrac{100}{127}) = 0.525,$$

$$\mu_7^{EB}(\mathbf{x}) = 115 - (0.525)(115 - 121) = 118.150, \quad \text{and}$$

$$V_7^{EB}(\mathbf{x}) = 100[1 - (\tfrac{6}{7})(0.525)] + (\tfrac{2}{4})(0.525)^2(115 - 121)^2 = 59.96.$$

The (estimated) 95% HPD credible set for θ_7 is

$$C_7^{EB}(\mathbf{x}) = (118.15 \pm (1.96)\sqrt{59.96}) = (102.97,\ 133.33).$$

Note that the classical (or noninformative prior Bayesian) 95% confidence set for θ_7, based on X_7 alone, is

$$C(x_7) = (115 \pm (1.96)\sqrt{100}) = (95.40,\ 134.60).$$

EXAMPLE 14. Suppose it is desired to estimate the IQs of 20 children who are in a special class for the gifted. It would clearly be inappropriate to presume that the θ_i come from the overall population prior, but it might be reasonable to assume that the θ_i of the 20 children come from some common normal prior. If the 20 children have $\bar{x}=135$ and $s^2=3800$, then calculation gives $\tilde{\sigma}_\pi^2=100$, $\hat{B}=0.447$,

$$\mu_i^{\text{EB}}(\mathbf{x})=x_i-(0.447)(x_i-135)=(0.553)x_i+60.395$$

and

$$V_i^{\text{EB}}(\mathbf{x})=57.500+(0.024)(x_i-135)^2.$$

An (estimated) $100(1-\alpha)\%$ HPD credible set for $\boldsymbol{\theta}$ would be given by (4.37).

Note that, in Example 13, we were interested in inference about θ_7 with available information about previous θ_i, while in Example 14 the entire vector $\boldsymbol{\theta}$ was of interest. As mentioned in Subsection 3.5.2, the first type of problem is sometimes called an *empirical Bayes* problem, while the latter is often called a *compound decision* problem. Since we will make no operational distinction between the two types of problems, we will call both empirical Bayes.

So far we have assumed that σ_f^2 is known. Usually σ_f^2 will be unknown, but there will be available samples $(X_i^1,\ldots,X_i^n)$ from $\mathcal{N}(\theta_i,\sigma^2)$ distributions, from which the variance can be estimated. Indeed one can reduce such data (by sufficiency) to consideration of $X_i=\bar{X}_i\sim\mathcal{N}(\theta_i,\sigma_f^2)$, where $\sigma_f^2=\sigma^2/n$, and estimate σ_f^2 by

$$\hat{\sigma}_f^2=\frac{1}{n(n-1)p}\sum_{i=1}^{n}\sum_{j=1}^{n}(x_i^j-\bar{x}_i)^2.$$

Unless both n and p are small, $\hat{\sigma}_f^2$ will be an accurate estimate of σ_f^2, and the empirical Bayes procedures will suffer little if σ_f^2 is replaced by $\hat{\sigma}_f^2$. If n and p are small, it is not quite clear what to do, but a plausible *ad hoc* fix would be to use $\hat{\sigma}_f^2$ in place of σ_f^2, and to replace $z(\alpha/2)$ in confidence intervals by $t_\nu(\alpha/2)$, the corresponding fractile of the t-distribution with ν degrees of freedom, ν being the degrees of freedom of $\hat{\sigma}_f^2(\nu=(n-1)p$ in the above scenario).

It should be mentioned that much of the empirical Bayes literature on normal means has concentrated on frequentist evaluations of the empirical Bayes procedures. It is perhaps surprising that these procedures (or related procedures) have frequentist risks (for sum of squares-error loss) and coverage probability comparable to or better than those of the corresponding classical procedures (unless p is very small). Some recent references include Rao (1977), Morris (1983a, b), Casella and Hwang (1983), and Reinsel (1983, 1984). (Morris considers a frequentist Bayes risk criterion, for priors in a given class, see Subsection 4.7.5 for discussion.)

4.5.3. PEB for Normal Means—The General Case

It will be somewhat rare in practice for all the symmetry assumptions of the previous subsection to apply. For instance, it will often be the case that the X_i are independently $\mathcal{N}(\theta_i, \sigma_i^2)$, $i=1,\ldots,p$, but with unequal σ_i^2. Such a situation will occur, for example, when (as discussed at the end of Subsection 4.5.2) the X_i are sample means from normal populations with means θ_i, but the sample sizes from each of the p populations differ (or the populations themselves have different variances). We will again assume that the σ_i^2 are known, but in practice it will typically be necessary to replace them by estimates (as discussed at the end of Subsection 4.5.2.)

A second type of generalization that is frequently needed is to replace the assumption that the θ_i are i.i.d., with the assumption that the θ_i arise from some statistical model. We will consider, here, the situation in which the θ_i arise from a regression model

$$\theta_i = \mathbf{y}_i^t \boldsymbol{\beta} + \varepsilon_i, \tag{4.38}$$

where $\boldsymbol{\beta} = (\beta_1, \ldots, \beta_l)^t$ is a vector of unknown regression coefficients ($l < p-2$), $\mathbf{y}_i = (y_{i1}, \ldots, y_{il})^t$ is a known set of regressors for each i, and the ε_i are independently $\mathcal{N}(0, \sigma_\pi^2)$.

Example 15. Consider a modification of the IQ example, in which the 7 yearly IQ scores (the x_i) of a child are (earliest to latest), 105, 110, 127, 115, 130, 125, and 137, and that it is desired to estimate the child's current IQ (corresponding to the test score 137). Suppose it is deemed plausible for the IQs to be linearly increasing with time, so that a reasonable model for θ_i (the IQ at time i) is

$$\theta_i = \beta_1 + \beta_2 i + \varepsilon_i,$$

β_1 and β_2 being unknown and the ε_i being, say, independent $\mathcal{N}(0, \sigma_\pi^2)$ random variables. This clearly corresponds to (4.38) with $l=2$ and $\mathbf{y}_i = (1, i)^t$. In addition, suppose that the test scores were $\mathcal{N}(\theta_i, 100)$ for the first 4 years, but that more accurate $\mathcal{N}(\theta_i, 50)$ tests were used for the last 3 years. Thus $\sigma_1^2 = \sigma_2^2 = \sigma_3^2 = \sigma_4^2 = 100$ and $\sigma_5^2 = \sigma_6^2 = \sigma_7^2 = 50$.

The model (4.38) amounts to assuming that the θ_i are $\mathcal{N}(\mathbf{y}_i^t\boldsymbol{\beta}, \sigma_\pi^2)$, and is thus a natural extension of the i.i.d. model for the θ_i. The simplest empirical Bayes analysis entails estimating the hyperparameters, $\boldsymbol{\beta}$ and σ_π^2, and using the estimates in the above prior. Again, this estimation would be based on the marginal density of $\mathbf{X} = (X_1, \ldots, X_p)^t$ which, since the X_i have independent $\mathcal{N}(\mathbf{y}_i^t\boldsymbol{\beta}, \sigma_i^2 + \sigma_\pi^2)$ marginal distributions, is given by

$$m(\mathbf{x}) = \left(\prod_{i=1}^{p} [2\pi(\sigma_i^2 + \sigma_\pi^2)]^{-1/2} \right) \exp\left\{ -\frac{1}{2} \sum_{i=1}^{p} (x_i - \mathbf{y}_i^t\boldsymbol{\beta})^2 / (\sigma_i^2 + \sigma_\pi^2) \right\}. \tag{4.39}$$

Following the ML-II approach, we can estimate $\boldsymbol{\beta}$ and σ_π^2 by differentiating $m(\mathbf{x})$ with respect to the β_i and σ_π^2, and setting the equations equal to zero. Letting $\hat{\boldsymbol{\beta}}$ and $\hat{\sigma}_\pi^2$ denote these ML-II estimates, the equations obtained can be written

$$\hat{\boldsymbol{\beta}} = (\mathbf{y}^t\mathbf{V}^{-1}\mathbf{y})^{-1}(\mathbf{y}^t\mathbf{V}^{-1}\mathbf{x}), \tag{4.40}$$

where $\mathbf{y}$ is the $(p\times l)$ matrix with rows $\mathbf{y}_i^t$ and $\mathbf{V}$ is the $(p\times p)$ diagonal matrix with diagonal elements $V_{ii} = \sigma_i^2 + \hat{\sigma}_\pi^2$, and

$$\hat{\sigma}_\pi^2 = \frac{\sum_{i=1}^p \{[(x_i - \mathbf{y}_i^t\hat{\boldsymbol{\beta}})^2 - \sigma_i^2]/[\sigma_i^2 + \hat{\sigma}_\pi^2]^2\}}{\sum_{i=1}^p (\sigma_i^2 + \hat{\sigma}_\pi^2)^{-2}}. \tag{4.41}$$

(Note that (4.40) also defines the least squares estimate of $\boldsymbol{\beta}$ when σ_π^2 is known.) Unfortunately, (4.41) does not provide a closed form expression for $\hat{\sigma}_\pi^2$; (4.40) and (4.41) do, however, provide an easy iterative scheme for calculating $\hat{\boldsymbol{\beta}}$ and $\hat{\sigma}_\pi^2$. Start out with a guess for σ_π^2, and use this guess in (4.40) to calculate an approximate $\hat{\boldsymbol{\beta}}$. Then plug the guess and the approximate $\hat{\boldsymbol{\beta}}$ into the right-hand side of (4.41), obtaining a new estimate for $\hat{\sigma}_\pi^2$. Repeat this procedure with the updated estimates until the numbers stabilize (which usually takes only a few iterations). If the convergence is to a negative value of $\hat{\sigma}_\pi^2$, the ML-II estimate of σ_π^2 is probably zero, and $\hat{\beta}$ is then given by (4.40) with $\hat{\sigma}_\pi^2 = 0$.

Empirical Bayes analyses based on pretending that the θ_i have $\mathcal{N}(\mathbf{y}_i^t\hat{\boldsymbol{\beta}}, \hat{\sigma}_\pi^2)$ priors will work well if $p-l$ is large (and the modelling assumptions are valid). For smaller p, however, the estimation of $\boldsymbol{\beta}$ and σ_π^2 can again introduce substantial errors that must be taken into account. Morris (1983a) accordingly develops approximations to the *posterior* means and variances of the θ_i which do take these additional errors into account. The approximations are given by (compare with (4.33) through (4.35))

$$\mu_i^{\mathrm{EB}}(\mathbf{x}) = x_i - \hat{B}_i(x_i - \mathbf{y}_i^t\hat{\boldsymbol{\beta}}) \tag{4.42}$$

and

$$V_i^{\mathrm{EB}}(\mathbf{x}) = \sigma_i^2\left[1 - \frac{(p-\hat{l}_i)}{p}\hat{B}_i\right] + \frac{2}{(p-l-2)}\hat{B}_i^2\left(\frac{\bar{\sigma}^2 + \tilde{\sigma}_\pi^2}{\sigma_i^2 + \tilde{\sigma}_\pi^2}\right)(x_i - \mathbf{y}_i^t\hat{\boldsymbol{\beta}})^2, \tag{4.43}$$

where $\hat{\boldsymbol{\beta}}$ and $\tilde{\sigma}_\pi^2$ are defined by (4.40) and

$$\tilde{\sigma}_\pi^2 = \frac{\sum_{i=1}^p \{[(p/(p-l))(x_i - \mathbf{y}_i^t\hat{\boldsymbol{\beta}})^2 - \sigma_i^2]/[\sigma_i^2 + \tilde{\sigma}_\pi^2]^2\}}{\sum_{i=1}^n (\sigma_i^2 + \tilde{\sigma}_\pi^2)^{-2}} \tag{4.44}$$

(again replacing $\tilde{\sigma}_\pi^2$ by zero, as the estimate and in (4.40), if the answer from the iteration is negative),

$$\hat{B}_i = \frac{(p-l-2)}{(p-l)}\cdot\frac{\sigma_i^2}{\sigma_i^2 + \tilde{\sigma}_\pi^2},$$

$$\hat{l}_i = p[\mathbf{y}(\mathbf{y}^t\mathbf{V}^{-1}\mathbf{y})^{-1}\mathbf{y}^t]_{ii}/(\sigma_i^2 + \tilde{\sigma}_\pi^2)$$

and

$$\bar{\sigma}^2 = \frac{\sum_{i=1}^{p} \sigma_i^2/(\sigma_i^2 + \tilde{\sigma}_\pi^2)}{\sum_{i=1}^{p} 1/(\sigma_i^2 + \tilde{\sigma}_\pi^2)}.$$

(Actually, (4.44) is only one of the possible ways Morris (1983a) suggests for defining $\tilde{\sigma}_\pi^2$.) One then assumes that θ_i has a $\mathcal{N}(\mu_i^{\text{EB}}(\mathbf{x}), V_i^{\text{EB}}(\mathbf{x}))$ *posterior* distribution, and proceeds with the analysis.

EXAMPLE 15 (continued). In this example, note that

$$\mathbf{y}^t = \begin{pmatrix} 1 & 1 & 1 & 1 & 1 & 1 & 1 \\ 1 & 2 & 3 & 4 & 5 & 6 & 7 \end{pmatrix},$$

$V_{ii} = 100 + \tilde{\sigma}_\pi^2$ for $i = 1, 2, 3, 4$, $V_{ii} = 50 + \tilde{\sigma}_\pi^2$ for $i = 5, 6, 7$, and $\mathbf{x} = (105, 110, 127, 115, 130, 125, 137)^t$. Equations (4.40) and (4.44) become

$$\hat{\boldsymbol{\beta}} = \left[\frac{20}{V_{11}^2} + \frac{170}{V_{11}V_{55}} + \frac{6}{V_{55}^2}\right]^{-1} \begin{pmatrix} \left[\frac{2050}{V_{11}^2} + \frac{17452}{V_{11}V_{55}} + \frac{658}{V_{55}^2}\right] \\ \left[\frac{94}{V_{11}^2} + \frac{788}{V_{11}V_{55}} + \frac{21}{V_{55}^2}\right] \end{pmatrix}$$

and

$$\begin{aligned} \tilde{\sigma}_\pi^2 = \left(\frac{4}{V_{11}^2} + \frac{3}{V_{55}^2}\right)^{-1} \bigg\{ & \frac{1}{V_{11}^2} [\tfrac{7}{5}([105 - (1,1)\hat{\boldsymbol{\beta}}]^2 \\ & + [110 - (1,2)\hat{\boldsymbol{\beta}}]^2 + [127 - (1,3)\hat{\boldsymbol{\beta}}]^2 \\ & + [115 - (1,4)\hat{\boldsymbol{\beta}}]^2) - 400] + \frac{1}{V_{55}^2} [\tfrac{7}{5}([130 - (1,5)\hat{\boldsymbol{\beta}}]^2 \\ & + [125 - (1,6)\hat{\boldsymbol{\beta}}]^2 + [137 - (1,7)\hat{\boldsymbol{\beta}}]^2) - 150] \bigg\}. \end{aligned}$$

Starting the iterative scheme for determining $\hat{\boldsymbol{\beta}}$ and $\tilde{\sigma}_\pi^2$ with a guess of $\tilde{\sigma}_\pi^2 = 100$, one immediately obtains negative values for subsequent $\tilde{\sigma}_\pi^2$. Thus one should use $\tilde{\sigma}_\pi^2 = 0$, which results in $\hat{\boldsymbol{\beta}} = (103.09, 4.57)^t$. Calculation of (4.42) and (4.43) then yields the values of $\mu_i^{\text{EB}}(\mathbf{x})$ and $V_i^{\text{EB}}(\mathbf{x})$ given in Table 4.5.

Also listed in Table 4.5, in addition to the x_i themselves and their associated variances σ_i^2, are regression estimates $\mu_i^{\text{R}}(\mathbf{x})$ and their variances $V_i^{\text{R}}(\mathbf{x})$. The regression estimates are those that would result from the assumption that the regression model, $\theta_i = \mathbf{y}_i^t\boldsymbol{\beta}$, held *exactly*, so that the X_i would be $\mathcal{N}(\mathbf{y}_i^t\boldsymbol{\beta}, \sigma_i^2)$. Note that this assumption is equivalent to the assumption that $\sigma_\pi^2 = 0$; the regression estimates are thus also the estimates that would result from the cruder empirical Bayes analysis which, after finding $\hat{\sigma}_\pi^2 = 0$ to be the ML-II hyperparameter estimate, used the $\mathcal{N}_p(\mathbf{y}\hat{\boldsymbol{\beta}}, \hat{\sigma}_\pi^2\mathbf{I})$ prior. These estimates are simply $\mu_i^{\text{R}} = \mathbf{y}_i^t\hat{\boldsymbol{\beta}}$. The natural variances of these estimates (based on the uncertainty in $\hat{\boldsymbol{\beta}}$) are, from either classical regression theory

Table 4.5. Estimates and Variances for Example 15.

x_i	105	110	127	115	130	125	137
μ_i^{EB}	106.6	111.3	120.9	118.8	127.6	128.3	135.8
μ_i^{R}	107.7	112.2	116.8	121.4	125.9	130.5	135.1
σ_i^2	100	100	100	100	50	50	50
V_i^{EB}	67.4	57.4	67.5	53.4	31.8	39.2	36.2
V_i^{R}	43.8	27.6	16.7	10.9	10.4	15.1	25.0

or noninformative prior Bayesian analysis,

$$\begin{aligned} V_i^{\mathrm{R}} &= \mathbf{y}_i^t(\mathbf{y}^t\mathbf{\Sigma}^{-1}\mathbf{y})^{-1}\mathbf{y}_i \\ &= (0.0384)^{-1}[(2.5 \text{-} -(0.92)i+(0.1)i^2], \end{aligned}$$

where $\mathbf{\Sigma}$ is the diagonal matrix with diagonal elements σ_i^2.

The x_i are the standard estimates of θ_i for the general model where $\boldsymbol{\theta}$ is completely unrestricted, while the μ_i^{R} are the estimates from the lower dimensional regression model. It is clear that the empirical Bayes estimates are a compromise between these two extremes. This feature is one of the great advantages of empirical Bayes (or next section's hierarchical Bayes) analysis; it allows one to postulate several possible models, and estimate the θ_i by a weighted average of the different model estimates, where the weights are dependent on how well the data supports that model. The variances V_i^{EB} are particularly interesting in this regard; for instance, $V_3^{\mathrm{EB}} = 67.5$ is weighted quite heavily towards $\sigma_3^2 = 100$, rather than $V_3^R = 16.7$, largely because $x_3 = 127$ is quite far from where it should be under the regression model.

Of course, the μ_i^{EB} asnd V_i^{EB} can be used to construct credible sets for the θ_i. For instance, an (estimated) 95% HPD credible set for θ_7 (the current IQ) would be

$$C_7^{\mathrm{EB}}(\mathbf{x}) = (135.8 \pm (1.96)\sqrt{36.2}) = (124.0, 147.6).$$

Several further comments about the empirical Bayes analysis in Example 15 are in order. First, as always with modelling, one can feel much more comfortable with a model proposed on intuitive or theoretical grounds, as opposed to ones created by "data snooping." Of course, often the data will suggest a model which, upon reflection, is reasonable on other grounds. Without such outside justification, however, empirical Bayes modelling can be as suspect as all data-dependent modelling. (Formally, if the model has a small posterior probability of being approximately correct, which may be the case if its prior probability was small, then answers obtained through its use are suspect.)

A second point raised by Example 15 is the interpretation of having $\tilde{\sigma}_\pi^2 = 0$. It is not the case that we really feel that the prior variance is near zero. Indeed, Hill (1965, 1977) shows that such situations may imply a *lack* of information about σ_π^2, i.e., the likelihood function for σ_π^2 may be quite flat; and, in fact, the likelihood in Example 15 (see Exercise 75) decreases slowly from a maximum at $\sigma_\pi^2 = 0$ to about 10% of the maximum at $\sigma_\pi^2 = 100$. (Moderately flat likelihood functions for certain parameters are not uncommon in statistics, and are difficult to handle by any method other than Bayesian analysis; see Meinhold and Singpurwalla (1983) for a more extreme example.) Note that formula (4.43) does recognize the problem, however, and "corrects" the posterior variance to take the uncertainty in σ_π^2 into account.

The final comment about this example relates to another important property of empirical Bayes estimation in the unequal variance case, a property that could be called *order reversal.* Imagine that one has p different treatments to compare, it being desired to determine which treatments have large means, θ_i. Available are sample means X_i, for each θ_i, but the sample variances, σ_i^2, differ. Were the σ_i^2 equal, it would be natural to simply select those θ_i with largest X_i. This is not necessarily a good idea when the σ_i^2 are unequal, however. To see this, imagine that X_1 through X_{p-2} are between -5 and 5, $X_{p-1} = 10$, $X_p = 10.1$, $\sigma_i^2 = 1$ for $1 \le i \le p-1$, and $\sigma_p^2 = 100$. One can then be very confident that θ_{p-1} is large, but not very confident that θ_p is large; if one had to select a single treatment, the $(p-1)$st would clearly seem optimal.

Such problems are very suited to empirical Bayes analysis. One might model the θ_i as being i.i.d. $\mathcal{N}(\mu_\pi, \sigma_\pi^2)$ (or use a more complicated regression structure if appropriate), and obtain the estimates $\mu_i^{\text{EB}}(\mathbf{x})$ for the θ_i. It can easily happen that $x_i > x_j$ but $\mu_i^{\text{EB}}(\mathbf{x}) < \mu_j^{\text{EB}}(\mathbf{x})$, obtaining a reversal of the naive ranking of θ_i and θ_j. In Example 15 (continued), $x_3 = 127 > 125 = x_6$, but $\mu_3^{\text{EB}}(\mathbf{x}) = 120.9 < 128.3 = \mu_6^{\text{EB}}(\mathbf{x})$. (The magnitude of the reversal is here due to the increasing regression model for the θ_i, but reversals can occur even if it is only assumed that the θ_i are i.i.d. Note, however, that order reversals cannot occur in the symmetric situation of Subsection 4.5.2.)

Before moving on, it should be reemphasized that we have only given an introduction to a particular empirical Bayes analysis. The references in Subsection 4.5.1 provide a wide range of applications of the empirical Bayes method to different statistical problems. And even within the domain of estimating normal means, $(\theta_1, \ldots, \theta_p)$, quite different analyses could have been performed, including:

(i) Incorporating a loss function, i.e., using the $\mathcal{N}(\mu_i^{\text{EB}}(\mathbf{x}), \mathrm{V}_i^{\text{EB}}(\mathbf{x}))$ posteriors in a true decision problem. (Although for some decision problems even this reduction may not be desirable, cf. Louis (1984).)

(ii) Modelling the θ_i via alternate structural assumptions on the prior. For instance, one could let the prior variances differ according to some model, or let the θ_i be stochastically dependent in some fashion. Even more intricate structural assumptions become possible when the θ_i are vectors. A few of the many references to such generalizations are Efron and Morris (1976b), Rao (1977), Haff (1978), and Reinsel (1983, 1984).

(iii) Basing the analysis on normal errors for the ε_i in (4.38) is calculationally convenient, but may not be optimal. In particular, assuming that the ε_i are i.i.d. from some flatter-tailed distribution, like a t-distribution, can lead to differing (and in some ways more appealing) conclusions.

(iv) As discussed in Section 3.5, the classes of priors typically employed in empirical Bayes analyses are rather inadequate and could be refined or augmented. One possible refinement would be to eliminate from consideration priors that are unrealistic (say, priors with means or variances outside reasonable bounds). A valuable augmentation would be to include "mixture" priors, allowing simultaneous consideration of various structural assumptions on the prior. (See Section 4.6 and Berger and Berliner (1984) for more discussion.)

4.5.4. Nonparametric Empirical Bayes Analysis

Up until now, we have focused on particular parametric classes of priors, using the data to assist in estimation of the prior hyperparameters or various functions thereof. The nonparametric empirical Bayes approach supposes that a large amount of data is available to estimate the prior (or some function thereof), and hence places no (or minimal) restrictions on the form of the prior. One can try for direct estimation of the prior, as discussed in Subsections 3.5.4 and 3.5.6, and, if implementable, this is probably the best approach (cf. Laird (1978, 1983), Lindsay (1981, 1983), and Leonard (1984)).

A *mathematically* appealing alternative (introduced in Robbins (1955)), is to seek a representation of the desired Bayes rule in terms of the marginal distribution, $m(\mathbf{x})$, of $\mathbf{x}$, and then use the data to estimate m, rather than π.

EXAMPLE 16. Suppose $X_1, \ldots, X_p$ are independent $\mathcal{P}(\theta_i)$, $i=1, \ldots, p$, random variables, and that the θ_i are i.i.d. from a common prior π_0. Then $X_1, \ldots, X_p$ can (unconditionally) be considered to be a sample from the marginal distribution

$$m(x_i) = \int f(x_i|\theta)\pi_0(\theta)d\theta,$$

where $f(x_i|\theta)$ is the $\mathcal{P}(\theta)$ density. This m can be estimated in any number of ways; one reasonable possibility is the empirical density estimate

$$\hat{m}(j) = (\text{the number of } x_i \text{ equal to } j)/p. \tag{4.45}$$

Suppose now that it is desired to estimate θ_p using the posterior mean. Observe that the posterior mean can be written

$$\begin{aligned}\delta^{\pi_0}(x_p) &= E^{\pi_0(\theta_p|x_p)}[\theta_p] = \int \theta_p \pi_0(\theta_p|x_p)d\theta_p \\ &= \int \theta_p f(x_p|\theta_p)\pi_0(\theta_p)d\theta_p/m(x_p) \\ &= \int \theta_p^{(x_p+1)}\exp\{-\theta_p\}(x_p!)^{-1}\pi_0(\theta_p)d\theta_p/m(x_p) \\ &= (x_p+1)\int f(x_p+1|\theta_p)\pi_0(\theta_p)d\theta_p/m(x_p) \\ &= (x_p+1)m(x_p+1)/m(x_p).\end{aligned}$$

Replacing m by the estimate in (4.45) results in the estimated Bayes rule

$$\delta^{\text{EB}}(\mathbf{x}) = \frac{(x_p+1)(\#\text{ of } x_i \text{ equal to } x_p+1)}{(\#\text{ of } x_i \text{ equal to } x_p)}.$$

The general approach, of which Example 16 is an illustration, can be stated as:

(i) find a representation (for the Bayes rule) of the form $\delta^{\pi}(x) = \psi(x, \varphi(m))$, where ψ and φ are known functionals;
(ii) estimate $\varphi(m)$ by $\widehat{\varphi(m)}$; and
(iii) use $\delta^{\text{EB}}(x) = \psi(x, \widehat{\varphi(m)})$.

This program is often calculationally easier than that of estimating π (as in Subsections 3.5.4 and 3.5.6) and then using $\delta^{\hat{\pi}}(x)$. Which approach is statistically more satisfactory, however?

The first consideration is that it may not be possible to find a representation of the form $\delta^{\pi}(x) = \psi(x, \varphi(m))$. For instance, such a representation for the posterior mean exists mainly when $f(x|\theta)$ is in the exponential family. Perhaps more seriously, one needs to find a *different* representation for each desired Bayesian output. For instance, if one also wanted the posterior variance or wanted to conduct a Bayesian test, a new analysis would have to be done. The approach of finding $\hat{\pi}$, as in Subsections 3.5.4 and 3.5.6 (or $\hat{\pi}(\theta|x)$ as in Subsections 4.5.2 and 4.5.3), allows for a wide variety of Bayesian outputs.

A second difficulty with use of $\delta^{\text{EB}}(x) = \psi(x, \widehat{\varphi(m)})$ is that it will tend to be very "jumpy." In Example 16, for instance, if none of $x_1, \ldots, x_{p-1}$ equal 10, but two of them equal 11, then $\delta^{\text{EB}}(\mathbf{x}) = 0$ when $x_p = 9$ and $\delta^{\text{EB}}(\mathbf{x}) = 22$ when $x_p = 10$. Such erratic behavior is clearly unsatisfactory, and would necessitate some "smoothing" or "monotonization." This latter idea will be discussed in Section 8.4. The technique is difficult enough, however, that the entire approach becomes calculationally questionable, when compared with the approach of estimating π directly.

The issue also arises of comparison of parametric and nonparametric empirical Bayes methods. From a practical perspective, the parametric approach will usually be better. This is simply because it is relatively rare to encounter large enough p to hope to be able to estimate the finer features of π, and, even when one does encounter such a p, the calculational problems are usually significant. It may well turn out, of course, that certain particular problems can best be handled by NPEB methods. As a general tool, however, PEB analysis will prove more useful.

4.6. Hierarchical Bayes Analysis

4.6.1. Introduction

There is a full Bayesian approach to the type of problem discussed in Section 4.5, an approach that compares very favorably with empirical Bayes analysis (see Subsection 4.6.4 for such comparison). The approach is based on the use of hierarchical priors, which were discussed in Section 3.6, a rereading of which might be in order. (Since most applications are to multiparameter problems, we will switch to vector notation in this section).

Attention will be restricted to two stage priors. The first stage, $\pi_1(\boldsymbol{\theta}|\boldsymbol{\lambda})$, where $\boldsymbol{\lambda}$ is a hyperparameter in Λ, can be thought of as the unknown prior in the empirical Bayes scenario (e.g., $\theta_1, \ldots, \theta_p$ might be i.i.d. $\mathcal{N}(\mu_\pi, \sigma_\pi^2)$.) Instead of estimating $\boldsymbol{\lambda}$, as in empirical Bayes analysis, however, $\boldsymbol{\lambda}$ will be given a second stage prior distribution $\pi_2(\boldsymbol{\lambda})$. This could be a proper prior but is often chosen to be a suitable noninformative prior. It is frequently calculationally helpful to write $\boldsymbol{\lambda} = (\boldsymbol{\lambda}^1, \boldsymbol{\lambda}^2)$, and represent π_2 as

$$\pi_2(\boldsymbol{\lambda}) = \pi_{2,1}(\boldsymbol{\lambda}^1|\boldsymbol{\lambda}^2)\pi_{2,2}(\boldsymbol{\lambda}^2). \tag{4.46}$$

The reason is that some of the necessary calculations (integrations over $\boldsymbol{\lambda}^1$) can often be carried out explicitly, while the remaining calculations (integrations over $\boldsymbol{\lambda}^2$) may require numerical integration.

Of interest will be the posterior distribution of $\boldsymbol{\theta}$, or some features thereof. The key calculational tool is the following result, expressing the posterior distribution in terms of the posterior distributions at the various stages of the hierarchical structure. (The proof is left as an exercise.)

Result 7. *Supposing all densities below exist and are nonzero,*

$$\pi(\boldsymbol{\theta}|\mathbf{x}) = \int_\Lambda \pi_1(\boldsymbol{\theta}|\mathbf{x}, \boldsymbol{\lambda})\pi_{2,1}(\boldsymbol{\lambda}^1|\mathbf{x}, \boldsymbol{\lambda}^2)\pi_{2,2}(\boldsymbol{\lambda}^2|\mathbf{x})d\boldsymbol{\lambda}. \tag{4.47}$$

Here

$$\pi_1(\boldsymbol{\theta}|\mathbf{x}, \boldsymbol{\lambda}) = \frac{f(\mathbf{x}|\boldsymbol{\theta})\pi_1(\boldsymbol{\theta}|\boldsymbol{\lambda})}{m_1(\mathbf{x}|\boldsymbol{\lambda})},$$

$$\text{where} \quad m_1(\mathbf{x}|\boldsymbol{\lambda}) = \int f(\mathbf{x}|\boldsymbol{\theta})\pi_1(\boldsymbol{\theta}|\boldsymbol{\lambda})d\boldsymbol{\theta}, \tag{4.48}$$

$$\pi_{2,1}(\boldsymbol{\lambda}^1|\mathbf{x}, \boldsymbol{\lambda}^2) = \frac{m_1(\mathbf{x}|\boldsymbol{\lambda})\pi_{2,1}(\boldsymbol{\lambda}^1|\boldsymbol{\lambda}^2)}{m_2(\mathbf{x}|\boldsymbol{\lambda}^2)},$$

$$\text{where} \quad m_2(\mathbf{x}|\boldsymbol{\lambda}^2) = \int m_1(\mathbf{x}|\boldsymbol{\lambda})\pi_{2,1}(\boldsymbol{\lambda}^1|\boldsymbol{\lambda}^2)d\boldsymbol{\lambda}^1, \tag{4.49}$$

$$\pi_{2,2}(\boldsymbol{\lambda}^2|\mathbf{x}) = \frac{m_2(\mathbf{x}|\boldsymbol{\lambda}^2)\pi_{2,2}(\boldsymbol{\lambda}^2)}{m(\mathbf{x})},$$

$$\text{where} \quad m(\mathbf{x}) = \int m_2(\mathbf{x}|\boldsymbol{\lambda}^2)\pi_{2,2}(\boldsymbol{\lambda}^2)d\boldsymbol{\lambda}^2. \tag{4.50}$$

Example 17. In Example 13, we had available seven independent IQ test scores $X_i \sim \mathcal{N}(\theta_i, 100)$, and assumed the θ_i were independently from a common $\mathcal{N}(\mu_\pi, \sigma_\pi^2)$ distribution. Thus $\mathbf{X} = (X_1, \ldots, X_7)^t \sim \mathcal{N}_7(\boldsymbol{\theta}, 100\mathbf{I})$ (defining $f(\mathbf{x}|\boldsymbol{\theta})$) and $\boldsymbol{\theta} = (\theta_1, \ldots, \theta_7)^t \sim \mathcal{N}_7((\mu_\pi, \ldots, \mu_\pi)^t, \sigma_\pi^2\mathbf{I})$ (defining $\pi_1(\boldsymbol{\theta}|\boldsymbol{\lambda})$, $\boldsymbol{\lambda} = (\mu_\pi, \sigma_\pi^2)$). Rather than estimating $\boldsymbol{\lambda}$ as in empirical Bayes analysis, we can put a second stage hyperprior on $\boldsymbol{\lambda}$. It is natural to give μ_π a $\mathcal{N}(100, 225)$ prior distribution, this being the overall population distribution of IQs; denote this density by $\pi_{2,1}(\mu_\pi)$. Our knowledge about σ_π^2 might be very vague, so that an (improper) constant density $\pi_{2,2}(\sigma_\pi^2) = 1$ would seem appropriate (see Subsection 4.6.2 for the reason for avoiding the standard noninformative prior). Thus the second stage hyperprior on $\boldsymbol{\lambda}$ would be (assuming independence of μ_π and σ_π^2), $\pi_2(\boldsymbol{\lambda}) = \pi_{2,1}(\mu_\pi)\pi_{2,2}(\sigma_\pi^2)$. From previous calculations, we know that

$$\pi_1(\boldsymbol{\theta}|\mathbf{x}, \boldsymbol{\lambda}) \text{ is } \mathcal{N}_7\left(\left[\mathbf{x} - \frac{100}{(100+\sigma_\pi^2)}\{\mathbf{x} - (\mu_\pi, \ldots, \mu_\pi)^t\}\right], \frac{100\sigma_\pi^2}{(100+\sigma_\pi^2)}\mathbf{I}\right)$$

and

$$m_1(\mathbf{x}|\boldsymbol{\lambda}) \text{ is } \mathcal{N}_7((\mu_\pi, \ldots, \mu_\pi)^t, (100+\sigma_\pi^2)\mathbf{I}).$$

Since m_1 is normal and $\pi_{2,1}(\lambda^1) = \pi_{2,1}(\mu_\pi)$ is $\mathcal{N}(100, 225)$, it can also be shown (see Subsection 4.6.2) that

$$\pi_{2,1}(\mu_\pi|\mathbf{x}, \sigma_\pi^2) \text{ is } \mathcal{N}\left(\left[\bar{x} - \frac{(100+\sigma_\pi^2)}{(1675+\sigma_\pi^2)}(\bar{x} - 100)\right], \frac{(100+\sigma_\pi^2)(225)}{(1675+\sigma_\pi^2)}\right)$$

and

$$m_2(\mathbf{x}|\sigma_\pi^2) \text{ is } \mathcal{N}_7((100, \ldots, 100)^t, (100+\sigma_\pi^2)\mathbf{I} + 225(\underline{\mathbf{1}})),$$

where $(\underline{\mathbf{1}})$ is the matrix of all ones. Hence, in determination of $\pi(\boldsymbol{\theta}|\mathbf{x})$ via

Result 7, all densities within the integral in (4.47) are known (up to the constant $m(\mathbf{x})$).

Suppose now that one is interested in the posterior expectation of some function $\psi(\boldsymbol{\theta})$, and imagine that the expectations

$$\psi_1(\mathbf{x}, \boldsymbol{\lambda}) = E^{\pi_1(\boldsymbol{\theta}|\mathbf{x},\boldsymbol{\lambda})}[\psi(\boldsymbol{\theta})] \tag{4.51}$$

and

$$\psi_2(\mathbf{x}, \boldsymbol{\lambda}^2) = E^{\pi_{2,1}(\boldsymbol{\lambda}^1|\mathbf{x},\boldsymbol{\lambda}^2)}[\psi_1(\mathbf{x}, \boldsymbol{\lambda})] \tag{4.52}$$

are known. Then, from Result 7, it is clear that (under the mild conditions of Fubini's theorem)

$$\begin{aligned} E^{\pi(\boldsymbol{\theta}|\mathbf{x})}[\psi(\boldsymbol{\theta})] &= \int_\Theta \int_\Lambda \psi(\boldsymbol{\theta})\pi_1(\boldsymbol{\theta}|\mathbf{x}, \boldsymbol{\lambda})\pi_{2,1}(\boldsymbol{\lambda}^1|\mathbf{x}, \boldsymbol{\lambda}^2)\pi_{2,2}(\boldsymbol{\lambda}^2|\mathbf{x})d\boldsymbol{\lambda}\, d\boldsymbol{\theta} \\ &= \int_\Lambda \left[\int_\Theta \psi(\boldsymbol{\theta})\pi_1(\boldsymbol{\theta}|\mathbf{x}, \boldsymbol{\lambda})d\boldsymbol{\theta}\right]\pi_{2,1}(\boldsymbol{\lambda}^1|\mathbf{x}, \boldsymbol{\lambda}^2)d\boldsymbol{\lambda}^1\pi_{2,2}(\boldsymbol{\lambda}^2|\mathbf{x})d\boldsymbol{\lambda}^2 \\ &= \int_\Lambda \psi_1(\mathbf{x}, \boldsymbol{\lambda})\pi_{2,1}(\boldsymbol{\lambda}^1|\mathbf{x}, \boldsymbol{\lambda}^2)d\boldsymbol{\lambda}^1\pi_{2,2}(\boldsymbol{\lambda}^2|\mathbf{x})d\boldsymbol{\lambda}^2 \\ &= \int_{\{\text{all } \boldsymbol{\lambda}^2\}} \psi_2(\mathbf{x}, \boldsymbol{\lambda}^2)\pi_{2,2}(\boldsymbol{\lambda}^2|\mathbf{x})d\boldsymbol{\lambda}^2. \end{aligned} \tag{4.53}$$

This scenario includes many important examples, such as Example 17, since the known form of $\pi_1(\boldsymbol{\theta}|\mathbf{x}, \boldsymbol{\lambda})$ and $\pi_{2,1}(\boldsymbol{\lambda}^1|\mathbf{x}, \boldsymbol{\lambda}^2)$ will often make calculation of ψ_1 and ψ_2 easy. Note that, in hierarchical Bayes analysis, one frequently tries to *choose* π_1 and π_2 to be of functional forms that make ψ_1 and ψ_2 easily calculable. (The goal is to avoid numerical calculation of high-dimensional integrals over $\boldsymbol{\theta}$.) Hierarchical priors are reasonably robust (see Subsection 4.7.9), so such "convenient" choices need not be greatly feared.

We will apply (4.53) to $\psi(\boldsymbol{\theta}) = \boldsymbol{\theta}$, since this will yield the posterior mean, $\boldsymbol{\mu}^\pi(\mathbf{x})$, and to $\psi(\boldsymbol{\theta}) = (\boldsymbol{\theta} - \boldsymbol{\mu}^\pi(\mathbf{x}))(\boldsymbol{\theta} - \boldsymbol{\mu}^\pi(\mathbf{x}))^t$, since this will yield the posterior covariance matrix, $\mathbf{V}^\pi(\mathbf{x})$. We will indeed proceed in the examples by simply calculating $\boldsymbol{\mu}^\pi(\mathbf{x})$ and $\mathbf{V}^\pi(\mathbf{x})$, and using the $\mathcal{N}_p(\boldsymbol{\mu}^\pi(\mathbf{x}), \mathbf{V}^\pi(\mathbf{x}))$ approximation to the true posterior of $\boldsymbol{\theta}$ for reaching conclusions. Since, again, only normal mean problems will be considered, we feel reasonably comfortable with this approximation (see the discussion in Subsection 4.3.2). More accurate approximations involving higher moments have been considered for some related situations in Van Der Merwe, Groenewald, Nell, and Van Der Merwe (1981).

Subsection 4.6.2 deals with the exchangeable normal mean situation of Subsection 4.5.2, while Subsection 4.6.3 gives an analysis of the general normal mean case, as in Subsection 4.5.3. Subsection 4.6.4 compares the empirical Bayes and hierarchical Bayes approaches. Note that we are again considering only two specific hierarchical Bayes situations. There is a huge literature on hierarchical Bayes analysis for a very wide range of other problems. A few references, in which many others can be found, are, Good (1950, 1965, 1980, 1983), Tiao and Tan (1965, 1966), Box and Tiao (1968, 1973), Lindley (1971b), Zellner (1971), Jackson, Novick, and Thayer (1971), Lindley and Smith (1972), Novick, Jackson, Thayer, and Cole (1972), Smith (1973a, b), Antoniak (1974), Dickey (1974), Bishop, Fienberg, and Holland (1975), Leonard (1976), Hill (1977), Berry and Christensen (1979), Chen (1979), Goel and DeGroot (1981), Deely and Lindley (1981), Woodroofe (1982), DuMouchel and Harris (1983), Kuo (1983, 1985), Press (1982, 1984), Press and Shigemasu (1984), Reilly and Sedransk (1984), and Berger (1986), which contains a general survey. Among the references which contain the material in the following sections (in some fashion) are Lindley (1971b), Lindley and Smith (1972), Smith (1973a, b), DuMouchel and Harris (1983), Morris (1983a), and Berger (1986).

4.6.2. For Normal Means—The Exchangeable Case

As in Subsection 4.5.2, suppose that $\mathbf{X}=(X_1,\ldots,X_p)'$, the X_i being independently $\mathcal{N}(\theta_i,\sigma_f^2)$ (σ_f^2 known), while the θ_i are thought to be exchangeable, modelled as i.i.d. $\mathcal{N}(\mu_\pi,\sigma_\pi^2)$. Suppose the hyperparameter μ_π is given a $\mathcal{N}(\beta_0, A)$ distribution (β_0 and A being given constants), and σ_π^2 is assumed to be independent of μ_π. In the notation of Subsection 4.6.1, we thus have that

$$f(\mathbf{x}|\boldsymbol{\theta}) \text{ is } \mathcal{N}_p(\boldsymbol{\theta},\sigma_f^2\mathbf{I}), \qquad \text{where } \boldsymbol{\theta}=(\theta_1,\ldots,\theta_p)';$$

$$\pi_1(\boldsymbol{\theta}|\boldsymbol{\lambda}) \text{ is } \mathcal{N}_p(\mu_\pi\mathbf{1},\sigma_\pi^2\mathbf{I}), \quad \text{where } \boldsymbol{\lambda}=(\mu_\pi,\sigma_\pi^2) \text{ and } \mathbf{1}=(1,\ldots,1)';$$

$$\pi_2(\boldsymbol{\lambda})=\pi_{2,1}(\mu_\pi)\cdot\pi_{2,2}(\sigma_\pi^2), \text{ where } \pi_{2,1}(\mu_\pi) \text{ is } \mathcal{N}(\beta_0,A).$$

Note that Example 17 is of this form with $p=7$, $\sigma_f^2=100$, $\beta_0=100$, $A=225$, and $\pi_{2,2}(\sigma_\pi^2)=1$.

Theorem 2. *For the hierarchical prior π, defined by π_1 and π_2 above, the posterior mean, $\boldsymbol{\mu}^\pi(\mathbf{x})$, and covariance matrix, $\mathbf{V}^\pi(\mathbf{x})$, are*

$$\boldsymbol{\mu}^\pi(\mathbf{x})=E^{\pi_{2,2}(\sigma_\pi^2|\mathbf{x})}[\boldsymbol{\mu}^*(\mathbf{x},\sigma_\pi^2)] \tag{4.54}$$

and

$$\mathbf{V}^\pi(\mathbf{x})=E^{\pi_{2,2}(\sigma_\pi^2|\mathbf{x})}\bigg[\sigma_f^2\mathbf{I}-\frac{\sigma_f^4}{(\sigma_f^2+\sigma_\pi^2)}\mathbf{I}+\frac{\sigma_f^4A}{(\sigma_f^2+\sigma_\pi^2)(pA+\sigma_f^2+\sigma_\pi^2)}(\underline{1})$$
$$+(\boldsymbol{\mu}^*(\mathbf{x},\sigma_\pi^2)-\boldsymbol{\mu}^\pi(\mathbf{x}))(\boldsymbol{\mu}^*(\mathbf{x},\sigma_\pi^2)-\boldsymbol{\mu}^\pi(\mathbf{x}))'\bigg], \tag{4.55}$$

where $\mathbf{1}=(1,\ldots,1)'$, $(\underline{1})$ is the matrix of all ones,

$$\boldsymbol{\mu}^*(\mathbf{x},\sigma_\pi^2)=\mathbf{x}-\frac{\sigma_f^2}{(\sigma_f^2+\sigma_\pi^2)}(\mathbf{x}-\bar{x}\mathbf{1})-\frac{\sigma_f^2}{(pA+\sigma_f^2+\sigma_\pi^2)}(\bar{x}-\beta_0)\mathbf{1},$$

and

$$\pi_{2,2}(\sigma_\pi^2|\mathbf{x}) = K\frac{\exp\left\{-\frac{1}{2}\left[\frac{s^2}{(\sigma_f^2+\sigma_\pi^2)}+\frac{p(\bar{x}-\beta_0)^2}{(pA+\sigma_f^2+\sigma_\pi^2)}\right]\right\}}{(\sigma_f^2+\sigma_\pi^2)^{(p-1)/2}(pA+\sigma_f^2+\sigma_\pi^2)^{1/2}A^{-1/2}}\pi_{2,2}(\sigma_\pi^2). \tag{4.56}$$

In (4.56), $s^2=\sum_{i=1}^p(x_i-\bar{x})^2$ *and* K *is the appropriate normalizing constant, found by integrating over* σ_π^2 *and setting equal to* 1 (*assuming the integral is finite, which is guaranteed if* $\pi_{2,2}(\sigma_\pi^2)$ *is bounded and* $p\geq 3$).

Proof. To calculate $\boldsymbol{\mu}^\pi(\mathbf{x})$, we will use (4.53) with $\psi(\boldsymbol{\theta})=\boldsymbol{\theta}$. Note first that (see subsection 4.3.1)

$$\pi_1(\boldsymbol{\theta}|\mathbf{x},\boldsymbol{\lambda}) \text{ is } \mathcal{N}_p(\boldsymbol{\mu}^1(\mathbf{x},\boldsymbol{\lambda}),\mathbf{V}^1(\sigma_\pi^2)), \tag{4.57}$$

where

$$\boldsymbol{\mu}^1(\mathbf{x},\boldsymbol{\lambda})=\mathbf{x}-\frac{\sigma_f^2}{(\sigma_f^2+\sigma_\pi^2)}(\mathbf{x}-\mu_\pi\mathbf{1}), \qquad \mathbf{V}^1(\sigma_\pi^2)=\sigma_f^2\mathbf{I}-\frac{\sigma_f^4}{(\sigma_f^2+\sigma_\pi^2)}\mathbf{I}.$$

Hence (see (4.51))

$$\psi_1(\mathbf{x},\boldsymbol{\lambda})=E^{\pi_1(\boldsymbol{\theta}|\mathbf{x},\boldsymbol{\lambda})}[\boldsymbol{\theta}]=\boldsymbol{\mu}^1(\mathbf{x},\boldsymbol{\lambda}).$$

To proceed to the second stage, note that $m_1(\mathbf{x}|\boldsymbol{\lambda})$ is $\mathcal{N}_p(\mu_\pi\mathbf{1},(\sigma_f^2+\sigma_\pi^2)\mathbf{I})$. It is convenient to use the familiar fact that $\Sigma(x_i-\mu_\pi)^2=p(\bar{x}-\mu_\pi)^2+s^2$, so that, as a function of $\boldsymbol{\lambda}$,

$$m_1(\mathbf{x}|\boldsymbol{\lambda})\propto m_{1,1}(\bar{x}|\boldsymbol{\lambda})\cdot m_{1,2}(s^2|\sigma_\pi^2), \tag{4.58}$$

where

$$m_{1,1}(\bar{x}|\boldsymbol{\lambda})=(\sigma_f^2+\sigma_\pi^2)^{-1/2}\exp\left\{-\frac{p(\bar{x}-\mu_\pi)^2}{2(\sigma_f^2+\sigma_\pi^2)}\right\}$$

and

$$m_{1,2}(s^2|\sigma_\pi^2)=(\sigma_f^2+\sigma_\pi^2)^{-(p-1)/2}\exp\left\{-\frac{s^2}{2(\sigma_f^2+\sigma_\pi^2)}\right\}.$$

Since $m_{1,2}$ does not depend on μ_π and $m_{1,1}$ is proportional to a $\mathcal{N}(\mu_\pi,p^{-1}(\sigma_f^2+\sigma_\pi^2))$ density for $\bar{X}$, and since $\pi_{2,1}(\mu_\pi)$ is $\mathcal{N}(\beta_0,A)$, it follows that

$$\pi_{2,1}(\mu_\pi|\mathbf{x},\sigma_\pi^2) \text{ is } \mathcal{N}(\mu^2(\mathbf{x},\sigma_\pi^2),V^2(\sigma_\pi^2)), \tag{4.59}$$

where

$$\mu^2(\mathbf{x},\sigma_\pi^2)=\bar{x}-\frac{(\sigma_f^2+\sigma_\pi^2)}{(\sigma_f^2+\sigma_\pi^2+pA)}(\bar{x}-\beta_0), \qquad V^2(\sigma_\pi^2)=\frac{A(\sigma_f^2+\sigma_\pi^2)}{(\sigma_f^2+\sigma_\pi^2+pA)}.$$

Thus (see (4.52) and (4.59))

$$\begin{aligned}\psi_2(\mathbf{x},\sigma_\pi^2)&=E^{\pi_{2,1}(\mu_\pi|\mathbf{x},\sigma_\pi^2)}[\boldsymbol{\mu}^1(\mathbf{x},\boldsymbol{\lambda})]\\&=E^{\pi_{2,1}(\mu_\pi|\mathbf{x},\sigma_\pi^2)}\left[\mathbf{x}-\frac{\sigma_f^2}{(\sigma_f^2+\sigma_\pi^2)}(\mathbf{x}-\mu_\pi\mathbf{1})\right]\\&=\mathbf{x}-\frac{\sigma_f^2}{(\sigma_f^2+\sigma_\pi^2)}(\mathbf{x}-\mu^2(\mathbf{x},\sigma_\pi^2)\mathbf{1})\\&=\boldsymbol{\mu}^*(\mathbf{x},\sigma_\pi^2).\end{aligned}$$

Combining this with (4.53) and (4.54), we have verified the formula for $\boldsymbol{\mu}^{\pi}(\mathbf{x})$.

To calculate $\mathbf{V}^{\pi}(\mathbf{x})$, let $\psi(\boldsymbol{\theta})=(\boldsymbol{\theta}-\boldsymbol{\mu}^{\pi}(\mathbf{x}))(\boldsymbol{\theta}-\boldsymbol{\mu}^{\pi}(\mathbf{x}))^t$, and note from (4.51) and the usual trick of adding and subtracting the mean that

$$\begin{aligned}\psi_1(\mathbf{x},\boldsymbol{\lambda})&=E^{\pi_1(\boldsymbol{\theta}|\mathbf{x},\boldsymbol{\lambda})}[(\boldsymbol{\theta}-\boldsymbol{\mu}^1+\boldsymbol{\mu}^1-\boldsymbol{\mu}^{\pi})(\boldsymbol{\theta}-\boldsymbol{\mu}^1+\boldsymbol{\mu}^1-\boldsymbol{\mu}^{\pi})^t]\\&=E^{\pi_1(\boldsymbol{\theta}|\mathbf{x},\boldsymbol{\lambda})}[(\boldsymbol{\theta}-\boldsymbol{\mu}^1)(\boldsymbol{\theta}-\boldsymbol{\mu}^1)^t]+(\boldsymbol{\mu}^1-\boldsymbol{\mu}^{\pi})(\boldsymbol{\mu}^1-\boldsymbol{\mu}^{\pi})^t\\&=\mathbf{V}^1(\sigma_\pi^2)+(\boldsymbol{\mu}^1(\mathbf{x},\boldsymbol{\lambda})-\boldsymbol{\mu}^{\pi}(\mathbf{x}))(\boldsymbol{\mu}^1(\mathbf{x},\boldsymbol{\lambda})-\boldsymbol{\mu}^{\pi}(\mathbf{x}))^t.\end{aligned}$$

To calculate (4.52) based on (4.59), we use the same trick to obtain

$$\begin{aligned}\psi_2(\mathbf{x},\sigma_\pi^2)&=E^{\pi_{2,1}(\mu_\pi|\mathbf{x},\sigma_\pi^2)}[\psi_1(\mathbf{x},\boldsymbol{\lambda})]\\&=\mathbf{V}^1(\sigma_\pi^2)+E^{\pi_{2,1}}[(\boldsymbol{\mu}^1-\boldsymbol{\mu}^*+\boldsymbol{\mu}^*-\boldsymbol{\mu}^{\pi})(\boldsymbol{\mu}^1-\boldsymbol{\mu}^*+\boldsymbol{\mu}^*-\boldsymbol{\mu}^{\pi})^t]\\&=\mathbf{V}^1(\sigma_\pi^2)+E^{\pi_{2,1}}[(\boldsymbol{\mu}^1-\boldsymbol{\mu}^*)(\boldsymbol{\mu}^1-\boldsymbol{\mu}^*)^t]+(\boldsymbol{\mu}^*-\boldsymbol{\mu}^{\pi})(\boldsymbol{\mu}^*-\boldsymbol{\mu}^{\pi})^t\\&=\mathbf{V}^1(\sigma_\pi^2)+E^{\pi_{2,1}}\left[\frac{\sigma_f^4}{(\sigma_f^2+\sigma_\pi^2)^2}(\mu_\pi-\mu^2(\mathbf{x},\sigma_\pi^2))^2\mathbf{1}\mathbf{1}^t\right]\\&\quad+(\boldsymbol{\mu}^*-\boldsymbol{\mu}^{\pi})(\boldsymbol{\mu}^*-\boldsymbol{\mu}^{\pi})^t\\&=\mathbf{V}^1(\sigma_\pi^2)+\frac{\sigma_f^4}{(\sigma_f^2+\sigma_\pi^2)^2}V^2(\sigma_\pi^2)(\underline{1})+(\boldsymbol{\mu}^*-\boldsymbol{\mu}^{\pi})(\boldsymbol{\mu}^*-\boldsymbol{\mu}^{\pi})^t,\end{aligned}$$

which upon simplification yields the argument in (4.55) and establishes (through (4.53)) the validity of (4.55).

It remains only to verify the formula for $\pi_{2,2}(\sigma_\pi^2|\mathbf{x})$. It is clear from (4.56) that we need only calculate $m_2(\mathbf{x}|\sigma_\pi^2)$ up to proportionality constants. Hence we can use (4.58) to obtain

$$\begin{aligned}m_2(\mathbf{x}|\sigma_\pi^2)&\propto\int m_{1,1}(\bar{x}|(\mu_\pi,\sigma_\pi^2))m_{1,2}(s^2|\sigma_\pi^2)\pi_{2,1}(\mu_\pi)d\mu_\pi\\&=m_{1,2}(s^2|\sigma_\pi^2)\int m_{1,1}(\bar{x}|(\mu_\pi,\sigma_\pi^2))\pi_{2,1}(\mu_\pi)d\mu_\pi\\&\propto m_{1,2}(s^2|\sigma_\pi^2)(pA+\sigma^2+\sigma_\pi^2)^{-1/2}\exp\left\{-\frac{p(\bar{x}-\beta_0)^2}{(pA+\sigma_f^2+\sigma_\pi^2)}\right\},\end{aligned}$$

the last step following from the fact that a normal $m_{1,1}$ and normal $\pi_{2,1}$ result in a normal marginal. This yields (4.56) directly (the $A^{-1/2}$ being an irrelevant constant included for consistency with a later expression). □

A number of comments about this theorem are in order. First, the expressions for $\boldsymbol{\mu}^{\pi}(\mathbf{x})$ and $\mathbf{V}^{\pi}(\mathbf{x})$ are really quite easy to calculate, only numerical integration over σ_π^2 being needed. Second, the expressions are not written quite as concisely as possible, but there are several advantages to their present format. One advantage is that the expressions involve no matrix inversions, in contrast with some of the more elegent representations. Another advantage is that the various terms are rather fun to contemplate, and allow easy intuitive understanding of the effect of the prior distribution.

Thus the second term in $\boldsymbol{\mu}^\pi$ (or $\boldsymbol{\mu}^*$) is roughly the adjustment to $\mathbf{x}$ caused by the assumption that the θ_i are i.i.d. (resulting in shrinkage of the x_i towards $\bar{x}$, the estimate if all θ_i were *known* to be equal), while the third term arises from the prior belief that $\mu_\pi \sim \mathcal{N}(\beta_0, A)$. And, similarly, the terms of $\mathbf{V}^\pi(\mathbf{x})$ can be roughly identified as arising from various sources of error in the estimation of θ_i. It is, indeed, often of value to explicitly calculate these terms separately, to see the effects of the various stages of the prior.

A related point is that our proof of the theorem was rather plodding, in comparison, say, with the elegant approach of Lindley and Smith (1972). Our reasons for the plodding approach were:

(i) it yields the "most convenient" representation for $\boldsymbol{\mu}^\pi$ and $\mathbf{V}^\pi$ directly, without having to go through trying matrix manipulations; and
(ii) it is the approach most likely to be useful in other hierarchical Bayes analyses.

To emphasize this last point, by reiterating what was said in Subsection 4.6.1, hierarchical Bayes analyses usually require a step by step approach, as in (4.51) to (4.52) to (4.53), the initial steps depending on recognizing and using functional forms for π_i which allow easy explicit calculation (and thus avoiding the need for high-dimensional numerical integration).

Example 17 (continued). This is exactly the situation of Theorem 2, with $p=7$, $\sigma_f^2=100$, $\bar{x}=121$, $s^2=762$, $\beta_0=100$, $A=225$, and $\pi_{2,2}(\sigma_\pi^2)\equiv 1$. With $x_7=115$, numerical calculation in (4.54) and (4.55) gives

$$\begin{aligned}\mu_7^\pi(\mathbf{x}) &= 115 - E\left[\frac{100}{(100+\sigma_\pi^2)}\right](115-121) - E\left[\frac{100}{(1675+\sigma_\pi^2)}\right](121-100)\\ &= 115-(0.8088)(-6)-(0.0588)(21)=118.61,\end{aligned}$$

and

$$\begin{aligned}V_{77}^\pi(\mathbf{x}) &= 100 - E\left[\frac{(100)^2}{(100+\sigma_\pi^2)}\right] + E\left[\frac{(100)^2(225)}{(100+\sigma_\pi^2)(1675+\sigma_\pi^2)}\right]\\ &\quad + E[(\mu_7^*(\mathbf{x}, \sigma_\pi^2)-118.61)^2]\\ &= 100-80.88+10.71+0.17=30.00.\end{aligned}$$

These are the posterior mean and variance of θ_7; thus the estimate for the child's IQ would be 118.61, and an (approximate) 95% credible set would be

$$(118.61 \pm (1.96)\sqrt{30.00}) = (107.87, 129.35).$$

A comment is in order concerning the choice $\pi_{2,2}(\sigma_\pi^2)\equiv 1$ in the above example. First, there is no known functional form for $\pi_{2,2}$ which allows simple analytic evaluation of the expectations in (4.54) and (4.55). (We would have used such a functional form if it existed.) Numerical evaluation, however, is feasible for almost any proper prior, and if prior information did exist about σ_π^2 (cf. Example 17 (continued) in Section 3.6) such should be used. Knowledge about hyperparameters is often quite vague, however, so that π_2 is frequently chosen to be at least partially noninformative. In the above example, it would thus have been tempting to use the standard noninformative prior, $\pi_{2,2}(\sigma_\pi^2)=1/\sigma_\pi^2$, for σ_π^2, but care must be taken. Indeed, if $\pi_{2,2}(\sigma_\pi^2)=1/\sigma_\pi^2$ is used in (4.56), it is clear that the right-hand side does not define a proper posterior because of nonintegrability as $\sigma_\pi^2\to 0$. For this reason, we chose to use $\pi_{2,2}(\sigma_\pi^2)=1$ as the noninformative prior. Because of such potential problems in hierarchical Bayes analysis, it is indeed often best to simply choose constant noninformative priors on hyperparameters. (Perhaps Laplace was right, in a practical sense, to simply pretend that unknown parameters had constant priors.)

The prior $\pi_{2,1}(\mu_\pi)$ is, itself, often chosen to be noninformative, i.e., $\pi_{2,1}(\mu_\pi)\equiv 1$. The results for this choice of $\pi_{2,1}$ can be found from Theorem 2 by letting $A\to\infty$ in all expressions. (As $A\to\infty$, the $\mathcal{N}(\beta_0, A)$ prior becomes effectively constant for our purposes.) These results are summarized in the following corollary.

Corollary 1. *If $\pi_2((\mu_\pi, \sigma_\pi^2))\equiv 1$ and $p\geq 4$ (so that $\pi_{2,2}(\sigma_\pi^2|\mathbf{x})$ below is proper), then, letting $B(\sigma_\pi^2)=\sigma_f^2/(\sigma_f^2+\sigma_\pi^2)$,*

$$\boldsymbol{\mu}^\pi(\mathbf{x})=\mathbf{x}-[EB(\sigma_\pi^2)](\mathbf{x}-\bar{x}\mathbf{1}), \tag{4.60}$$

and

$$\mathbf{V}^\pi(\mathbf{x})=\sigma_f^2\mathbf{I}-\sigma_f^2[EB]\left(\mathbf{I}-\frac{1}{p}(\underline{1})\right)+([EB^2]-[EB]^2)(\mathbf{x}-\bar{x}\mathbf{1})(\mathbf{x}-\bar{x}\mathbf{1})^t, \tag{4.61}$$

where E stands for expectation with respect to

$$\pi_{2,2}(\sigma_\pi^2|\mathbf{x})=K(\sigma_f^2+\sigma_\pi^2)^{-(p-1)/2}\exp\left\{-\frac{s^2}{2(\sigma_f^2+\sigma_\pi^2)}\right\}. \tag{4.62}$$

Explicit expressions for $[EB]$ and $([EB^2]-[EB]^2)$ are

$$EB=\frac{(p-3)\sigma_f^2}{s^2}\left[1-H_{p-3}\left(\frac{s^2}{2\sigma_f^2}\right)\right], \tag{4.63}$$

and

$$(E[B^2]-[EB]^2)=\frac{2(p-3)\sigma_f^4}{s^4}\left[1+\left\{\frac{s^2}{2\sigma_f^2}([EB]-1)-1\right\}H_{p-3}\left(\frac{s^2}{2\sigma_f^2}\right)\right], \tag{4.64}$$

where H_{p-3} is a function defined and discussed in Appendix 2.

Proof. The verification of (4.60) and (4.61) follows easily from Theorem 2, and is left as an exercise.

To establish (4.63) observe that, by definition (where K in (4.62) is the inverse of the denominator below),

$$EB = \frac{\sigma_f^2 \int_0^\infty (\sigma_f^2+\sigma_\pi^2)^{-(p+1)/2} \exp\left\{-\frac{s^2}{2(\sigma_f^2+\sigma_\pi^2)}\right\} d\sigma_\pi^2}{\int_0^\infty (\sigma_f^2+\sigma_\pi^2)^{-(p-1)/2} \exp\left\{-\frac{s^2}{2(\sigma_f^2+\sigma_\pi^2)}\right\} d\sigma_\pi^2}.$$

Making the change of variables $y = s^2/[2(\sigma_f^2+\sigma_\pi^2)]$ and defining $v = s^2/(2\sigma_f^2)$, yields

$$EB = \frac{\int_0^v y^{(p-3)/2} e^{-y}\, dy}{v \int_0^v y^{(p-5)/2} e^{-y}\, dy}. \tag{4.65}$$

Integration by parts in the numerator yields

$$EB = \frac{y^{(p-3)/2}(-e^{-y})\big|_0^v + \int_0^v \frac{(p-3)}{2} y^{(p-5)/2} e^{-y}\, dy}{v \int_0^v y^{(p-5)/2} e^{-y}\, dy}$$
$$= \frac{(p-3)}{2v}\left(1 - \frac{2v^{(p-3)/2} e^{-v}}{(p-3)\int_0^v y^{(p-5)/2} e^{-y}\, dy}\right),$$

and (4.63) follows from the representations in Appendix 2.

The verification of (4.64) will be left as an exercise. □

It can be seen from the representations in Appendix 2 that H_{p-3} is essentially an exponentially decreasing function of s^2, so that the first terms of (4.63) and (4.64) dominate when s^2 is moderately large. It is interesting to reanalyze Example 14 (continued) from Subsection 4.5.2 using Corollary 1, and to compare the answers so obtained with the answers from the empirical Bayes analysis.

EXAMPLE 14 (continued). Recall that $p = 20$, $\sigma_f^2 = 100$, $\bar{x} = 135$, and $s^2 = 3800$. Using any of the formulas in Appendix 2, one obtains $H_{p-3}(s^2/(2\sigma_f^2)) = 0.00349$. Thus

$$[EB] = \frac{17}{2(19)}[1 - 0.00349] = 0.4458,$$

and

$$E[B^2] - [EB]^2 = \frac{2(17)(100)^2}{(3800)^2}\left[1 + \left\{\frac{3800}{200}(0.4458-1) - 1\right\}(0.00349)\right] = 0.0226.$$

Hence, from Corollary 1 we have that

$$\boldsymbol{\mu}^\pi(\mathbf{x}) = \mathbf{x} - (0.4458)(\mathbf{x} - 135\mathbf{1})$$

and

$$\mathbf{V}^\pi(\mathbf{x}) = 100\mathbf{I} - (44.58)(\mathbf{I} - \tfrac{1}{20}(\underline{\mathbf{1}})) + (0.0226)(\mathbf{x} - 135\mathbf{1})(\mathbf{x} - 135\mathbf{1})^t$$
$$= (55.42)\mathbf{I} + (2.23)(\underline{\mathbf{1}}) + (0.0226)(\mathbf{x} - 135\mathbf{1})(\mathbf{x} - 135\mathbf{1})^t.$$

The posterior variance of each θ_i is thus

$$V_{ii}^\pi(\mathbf{x}) = 57.65 + (0.0226)(x_i - 135)^2,$$

from which individual credible sets could be formed. Observe that these answers are essentially equal to the Morris empirical Bayes answers obtained in Subsection 4.5.2. It is also possible to obtain, as in Subsection 4.3.2, an approximate $100(1-\alpha)\%$ credible set for the entire vector $\boldsymbol{\theta} = (\theta_1, \ldots, \theta_p)^t$, namely the ellipsoid

$$C = \{\boldsymbol{\theta}: (\boldsymbol{\theta} - \boldsymbol{\mu}^\pi(\mathbf{x}))^t \mathbf{V}^\pi(\mathbf{x})^{-1}(\boldsymbol{\theta} - \boldsymbol{\mu}^\pi(\mathbf{x})) \le \chi^2_{20}(1-\alpha)\}.$$

The center of this ellipsoid is, of course, $\boldsymbol{\mu}^\pi(\mathbf{x})$, and the directions of the axes from this center are specified by the eigenvectors of $\mathbf{V}^\pi(\mathbf{x})$, which are $\mathbf{1}$, $(\mathbf{x} - \bar{x}\mathbf{1})$, and the set of orthonormal vectors orthogonal to these two vectors. It is interesting to note that the *lengths* of the axes of C are $2[\varphi(\mathbf{w})\chi^2_{20}(1-\alpha)]^{1/2}$, where $\varphi(\mathbf{w})$ is the eigenvalue of the eigenvector $\mathbf{w}$. For $\mathbf{w} = \mathbf{1}$, φ is specified by

$$\mathbf{V}^\pi(\mathbf{x})\mathbf{1} = (55.42)\mathbf{1} + (2.23)(\underline{\mathbf{1}})\mathbf{1} + 0$$
$$= [55.42 + (2.23)(20)]\mathbf{1} \equiv \varphi(\mathbf{1})\mathbf{1},$$

so that $\varphi(\mathbf{1}) = [55.42 + (2.23)(20)] = 100.02$. Similarly, $\varphi(\mathbf{x} - 135\mathbf{1}) = 55.42 + (0.0226)s^2 = 141.68$, and the remaining 18 eigenvectors have $\varphi(\mathbf{w}) = 55.42$. The square roots of these eigevalues are 10.00, 11.90, and 7.44, showing that C is substantially wider in the directions $\mathbf{1}$ and $(\mathbf{x} - \bar{x}\mathbf{1})$.

As a final comment, the situation of unknown σ_f^2 causes no conceptual difficulty in the analysis. Often, as indicated in Subsection 4.5.2, an accurate estimate, $\hat{\sigma}_f^2$, of σ_f^2 will be available, and can just be inserted in all expressions. If, however, the available estimate is not highly accurate, an additional integration may be necessary. Indeed, if the available estimate, $\hat{\sigma}_f^2$, has density $g(\hat{\sigma}_f^2|\sigma_f^2)$ and is independent of $\mathbf{X}$ (see Subsection 4.5.2 for the reason why this will usually be the case), and if $\pi^*(\sigma_f^2)$ is the assumed prior density of σ_f^2 (taking $\pi^*(\sigma_f^2) = 1$ or $\pi^*(\sigma_f^2) = 1/\sigma_f^2$ will both usually be satisfactory in the noninformative case), then the results in Theorem 2 hold with the relatively minor change of replacing $\pi_{2,2}(\sigma_\pi^2|\mathbf{x})$ by

$$\pi^*(\sigma_\pi^2, \sigma_f^2|\mathbf{x}, \hat{\sigma}_f^2) = \frac{K^* \exp\{\ \}\pi_{2,2}(\sigma_\pi^2) g(\hat{\sigma}_f^2|\sigma_f^2)\pi^*(\sigma_f^2)}{(\sigma_f^2 + \sigma_\pi^2)^{(p-1)/2}(pA + \sigma^2 + \sigma_\pi^2)^{1/2} A^{-1/2}},$$

where $\exp\{\ \}$ is the exponential term in (4.56) and K^* is the appropriate normalizing constant. The calculations now involve two-dimensional numerical integration, of course. (The cruder alternative of using confidence intervals based on $t_\nu(\alpha/2)$ fractiles could also be employed, see the discussion at the end of Subsection 4.5.2.)

4.6.3. For Normal Means—The General Case

In this subsection, the more general situation of Subsection 4.5.3 will be considered. Thus we assume that the X_i are independently $\mathcal{N}(\theta_i, \sigma_i^2)$, σ_i^2 known, and that the θ_i are independently $\mathcal{N}(\mathbf{y}_i^t\boldsymbol{\beta}, \sigma_\pi^2)$, the $\mathbf{y}_i$ being known vectors of regressors; we will vary from Subsection 4.5.3, however, by giving $\boldsymbol{\lambda} = (\boldsymbol{\lambda}^1, \lambda^2) \equiv (\boldsymbol{\beta}, \sigma_\pi^2)$ a second stage prior distribution. Converting to the notation of Subsection 4.6.1, we have that $f(\mathbf{x}|\boldsymbol{\theta})$ is $\mathcal{N}_p(\boldsymbol{\theta}, \boldsymbol{\Sigma})$, where $\boldsymbol{\Sigma}$ is the diagonal matrix with diagonal elements σ_i^2, and $\pi_1(\boldsymbol{\theta}|\boldsymbol{\lambda})$ is $\mathcal{N}_p(\mathbf{y}\boldsymbol{\beta}, \sigma_\pi^2\mathbf{I})$, where $\mathbf{y}$ is the $(p \times l)$ matrix with rows equal to $\mathbf{y}_i^t$. (Without loss of generality, $\mathbf{y}$ will be assumed to be of rank l; if not, the dimension of $\boldsymbol{\beta}$ can be reduced.) The second stage prior density, π_2, will be assumed to be of the form $\pi_2((\boldsymbol{\beta}, \sigma_\pi^2)) = \pi_{2,1}(\boldsymbol{\beta}) \cdot \pi_{2,2}(\sigma_\pi^2)$, where $\pi_{2,1}(\boldsymbol{\beta})$ is $\mathcal{N}_l(\boldsymbol{\beta}^0, \mathbf{A})$ ($\boldsymbol{\beta}^0$ a known vector and $\mathbf{A}$ a known positive definite $(l \times l)$ matrix), and $\pi_{2,2}(\sigma_\pi^2)$ is left arbitrary. Example 15 in Subsection 4.5.3 is an illustration of this scenario, although no second stage prior naturally suggests itself. (We will thus later analyze Example 15 with a noninformative π_2.)

The seemingly more general situation where $\boldsymbol{\Sigma}$ above is nondiagonal and the covariance matrix in π_1 is of the form $\sigma_\pi^2\mathbf{B}$ ($\mathbf{B}$ known) can be reduced to the $\boldsymbol{\Sigma}$ diagonal, $\mathbf{B} = \mathbf{I}$ case by a simultaneous diagonalization of the two matrices. Also, the case of unknown σ_i^2 can be dealt with by simply inserting estimates $\hat{\sigma}_i^2$ into all following expressions, providing the estimates are accurate; otherwise additional integrations over the σ_i^2 may be needed, as discussed at the end of Subsection 4.6.2 for the exchangeable case. (Some type of approximation might, however, be needed if p is large and the σ_i^2 differ, high-dimensional integration being difficult. One possibly reasonable approximation, when calculating μ_i^π and V_{ii}^π for θ_i, is to replace the σ_j^2 by the $\hat{\sigma}_j^2$ for all $j \neq i$, but carry out the integration over σ_i^2. Another possibility, as discussed at the end of Subsection 4.5.2, is simply to use $t_{\nu_i}(\alpha/2)$ fractiles in confidence intervals for the θ_i, where ν_i is the degrees of freedom of the estimate $\hat{\sigma}_i^2$.)

The analog of Theorem 2 for this general setting is

Theorem 3. *For the hierarchical prior defined by π_1 and π_2 above, the posterior mean, $\boldsymbol{\mu}^\pi(\mathbf{x})$, and covariance matrix, $\mathbf{V}^\pi(\mathbf{x})$, are*

$$\boldsymbol{\mu}^\pi(\mathbf{x}) = E^{\pi_{2,2}(\sigma_\pi^2|\mathbf{x})}[\boldsymbol{\mu}^*(\mathbf{x}, \sigma_\pi^2)] \tag{4.66}$$

and

$$\mathbf{V}^\pi(\mathbf{x}) = E^{\pi_{2,2}(\sigma_\pi^2|\mathbf{x})}[\boldsymbol{\Sigma} - \boldsymbol{\Sigma}\mathbf{W}\boldsymbol{\Sigma} + \boldsymbol{\Sigma}\mathbf{W}\mathbf{y}(\mathbf{y}^t\mathbf{W}\mathbf{y} + \mathbf{A}^{-1})^{-1}\mathbf{y}^t\mathbf{W}\boldsymbol{\Sigma} + (\boldsymbol{\mu}^* - \boldsymbol{\mu}^\pi(\mathbf{x}))(\boldsymbol{\mu}^* - \boldsymbol{\mu}^\pi(\mathbf{x}))^t], \tag{4.67}$$

where

$$W = (\mathbf{\Sigma} + \sigma_\pi^2 \mathbf{I})^{-1},$$

$$\boldsymbol{\mu}^* = \boldsymbol{\mu}^*(\mathbf{x}, \sigma_\pi^2) = \mathbf{x} - \mathbf{\Sigma}\mathbf{W}(\mathbf{x} - \mathbf{y}\hat{\boldsymbol{\beta}}) - \mathbf{\Sigma}\mathbf{W}\mathbf{y}(\mathbf{y}'\mathbf{W}\mathbf{y} + \mathbf{A}^{-1})^{-1}\mathbf{A}^{-1}(\hat{\boldsymbol{\beta}} - \boldsymbol{\beta}^0), \quad (4.68)$$

$$\hat{\boldsymbol{\beta}} = (\mathbf{y}'\mathbf{W}\mathbf{y})^{-1}\mathbf{y}'\mathbf{W}\mathbf{x}, \quad (4.69)$$

$$\pi_{2,2}(\sigma_\pi^2 | \mathbf{x}) = K \frac{\exp\{-\frac{1}{2}[\|\mathbf{x} - \mathbf{y}\hat{\boldsymbol{\beta}}\|_*^2 + \|\hat{\boldsymbol{\beta}} - \boldsymbol{\beta}^0\|_{**}^2]\}}{[\det \mathbf{W}]^{-1/2}[\det(\mathbf{y}'\mathbf{W}\mathbf{y} + \mathbf{A}^{-1})]^{1/2}} \cdot \pi_{2,2}(\sigma_\pi^2), \quad (4.70)$$

$$\|\mathbf{x} - \mathbf{y}\hat{\boldsymbol{\beta}}\|_*^2 = (\mathbf{x} - \mathbf{y}\hat{\boldsymbol{\beta}})'\mathbf{W}(\mathbf{x} - \mathbf{y}\hat{\boldsymbol{\beta}}),$$

$$\|\hat{\boldsymbol{\beta}} - \boldsymbol{\beta}^0\|_{**}^2 = (\hat{\boldsymbol{\beta}} - \boldsymbol{\beta}^0)'([\mathbf{y}'\mathbf{W}\mathbf{y}]^{-1} + \mathbf{A})^{-1}(\hat{\boldsymbol{\beta}} - \boldsymbol{\beta}^0),$$

and K is the appropriate normalizing constant for $\pi_{2,2}$. (Again, we are implicitly assuming that (4.70) *defines a proper density, i.e., does not have infinite mass; this is guaranteed if $\pi_{2,2}(\sigma_\pi^2)$ is bounded and $p \geq 3$.)*

Proof. The proof exactly parallels that of Theorem 2. Equation (4.57) is still true, with

$$\boldsymbol{\mu}^1(\mathbf{x}, \boldsymbol{\lambda}) = \mathbf{x} - \mathbf{\Sigma}\mathbf{W}(\mathbf{x} - \mathbf{y}\boldsymbol{\beta}), \qquad \mathbf{V}^1(\sigma_\pi^2) = \mathbf{\Sigma} - \mathbf{\Sigma}\mathbf{W}\mathbf{\Sigma}.$$

The analog of (4.58) follows from noting that $m_1(\mathbf{x}|\boldsymbol{\lambda})$ is $\mathcal{N}_p(\mathbf{y}\boldsymbol{\beta}, \mathbf{W}^{-1})$, and that the quadratic form in this density can be written (by adding and subtracting $\mathbf{y}\hat{\boldsymbol{\beta}}$ appropriately) as

$$(\mathbf{x} - \mathbf{y}\boldsymbol{\beta})'\mathbf{W}(\mathbf{x} - \mathbf{y}\boldsymbol{\beta}) = \|\mathbf{x} - \mathbf{y}\hat{\boldsymbol{\beta}}\|_*^2 + (\hat{\boldsymbol{\beta}} - \boldsymbol{\beta})'(\mathbf{y}'\mathbf{W}\mathbf{y})(\hat{\boldsymbol{\beta}} - \boldsymbol{\beta}).$$

(This is just the usual partition in regression analysis.) Thus we have that

$$m_1(\mathbf{x}|\boldsymbol{\lambda}) \propto m_{1,1}(\hat{\boldsymbol{\beta}}|\boldsymbol{\beta}, \sigma_\pi^2) \cdot m_{1,2}(\|\mathbf{x} - \mathbf{y}\hat{\boldsymbol{\beta}}\|_*^2 | \sigma_\pi^2),$$

where

$$m_{1,1}(\hat{\boldsymbol{\beta}}|\boldsymbol{\beta}, \sigma_\pi^2) = [\det(\mathbf{y}'\mathbf{W}\mathbf{y})]^{1/2} \exp\{-\tfrac{1}{2}(\hat{\boldsymbol{\beta}} - \boldsymbol{\beta})'(\mathbf{y}'\mathbf{W}\mathbf{y})(\hat{\boldsymbol{\beta}} - \boldsymbol{\beta})\}$$

and

$$m_{1,2}(\|\mathbf{x} - \mathbf{y}\hat{\boldsymbol{\beta}}\|_*^2 | \sigma_\pi^2) = \frac{\exp\{-\frac{1}{2}\|\mathbf{x} - \mathbf{y}\hat{\boldsymbol{\beta}}\|_*^2\}}{[\det \mathbf{W}]^{-1/2}[\det(\mathbf{y}'\mathbf{W}\mathbf{y})]^{1/2}}.$$

Since $\pi_{2,1}(\boldsymbol{\beta})$ is $\mathcal{N}_l(\boldsymbol{\beta}^0, \mathbf{A})$, the analog of (4.59) is

$$\pi_{2,1}(\boldsymbol{\beta}|\mathbf{x}, \sigma_\pi^2) \text{ is } \mathcal{N}_l(\boldsymbol{\mu}^2(\mathbf{x}, \sigma_\pi^2), \mathbf{V}^2(\sigma_\pi^2)),$$

where

$$\begin{aligned}\boldsymbol{\mu}^2(\mathbf{x}, \sigma_\pi^2) &= \hat{\boldsymbol{\beta}} - (\mathbf{y}'\mathbf{W}\mathbf{y})^{-1}[(\mathbf{y}'\mathbf{W}\mathbf{y})^{-1} + \mathbf{A}]^{-1}(\hat{\boldsymbol{\beta}} - \boldsymbol{\beta}^0) \\ &= \hat{\boldsymbol{\beta}} - [\mathbf{A}^{-1} + (\mathbf{y}'\mathbf{W}\mathbf{y})]^{-1}\mathbf{A}^{-1}(\hat{\boldsymbol{\beta}} - \boldsymbol{\beta}^0), \\ \mathbf{V}^2(\sigma_\pi^2) &= [\mathbf{y}'\mathbf{W}\mathbf{y} + \mathbf{A}^{-1}]^{-1}.\end{aligned}$$

The remainder of the proof is exactly like the proof of Theorem 2, with obvious modifications. (In the calculation of $m_2(\mathbf{x}|\sigma_\pi^2)$, one uses the fact that

$$[\det(\mathbf{y}'\mathbf{W}\mathbf{y})]^{1/2}[\det((\mathbf{y}'\mathbf{W}\mathbf{y})^{-1} + \mathbf{A})]^{1/2} = [\det(\mathbf{y}'\mathbf{W}\mathbf{y} + \mathbf{A}^{-1})]^{1/2}[\det \mathbf{A}]^{1/2};$$

and $[\det \mathbf{A}]^{1/2}$ is absorbed into K in (4.70).) □

Again, one can exhibit $\boldsymbol{\mu}^\pi$ and $\mathbf{V}^\pi$ more elegantly, but (4.66) and (4.67) have the intuitive interpretive advantage discussed in Subsection 4.6.2, and are also the easiest forms to calculate, involving only inversions of the $(l \times l)$ matrices $(\mathbf{y}'\mathbf{W}\mathbf{y})$, $\mathbf{A}$, and $[\mathbf{y}'\mathbf{W}\mathbf{y}+\mathbf{A}^{-1}]$ (note that $\mathbf{W}$ is diagonal). Since l is usually quite small, these inversions (in the midst of the numerical integrations) should pose no problem.

If prior information about $\boldsymbol{\beta}$ and σ_π^2 is very vague, it is appealing to use the noninformative prior $\pi_2((\boldsymbol{\beta}, \sigma_\pi^2)) \equiv 1$. The results for this choice of π_2 can be found by setting $\mathbf{A}^{-1} = \mathbf{0}$ in Theorem 3, and are summarized in the following corollary.

Corollary 2. *If $\pi_2((\boldsymbol{\beta}, \sigma_\pi^2)) \equiv 1$ and $p \geq l+3$ (so that $\pi_{2,2}(\sigma_\pi^2|\mathbf{x})$ below is proper), then*

$$\boldsymbol{\mu}^\pi(\mathbf{x}) = E^{\pi_{2,2}(\sigma_\pi^2|\mathbf{x})}[\boldsymbol{\mu}^*(\mathbf{x}, \sigma_\pi^2)] \tag{4.71}$$

and

$$\begin{aligned}\mathbf{V}^\pi(\mathbf{x}) = E^{\pi_{2,2}(\sigma_\pi^2|\mathbf{x})}[&\boldsymbol{\Sigma} - \boldsymbol{\Sigma}\mathbf{W}\boldsymbol{\Sigma} + \boldsymbol{\Sigma}\mathbf{W}\mathbf{y}(\mathbf{y}'\mathbf{W}\mathbf{y})^{-1}\mathbf{y}'\mathbf{W}\boldsymbol{\Sigma} \\ &+ (\boldsymbol{\mu}^* - \boldsymbol{\mu}^\pi(\mathbf{x}))(\boldsymbol{\mu}^* - \boldsymbol{\mu}^\pi(\mathbf{x}))'],\end{aligned} \tag{4.72}$$

where

$$\mathbf{W} = (\boldsymbol{\Sigma} + \sigma_\pi^2 \mathbf{I})^{-1},$$
$$\boldsymbol{\mu}^* = \boldsymbol{\mu}^*(\mathbf{x}, \sigma_\pi^2) = \mathbf{x} - \boldsymbol{\Sigma}\mathbf{W}(\mathbf{x} - \mathbf{y}\hat{\boldsymbol{\beta}}),\ \hat{\boldsymbol{\beta}} = (\mathbf{y}'\mathbf{W}\mathbf{y})^{-1}\mathbf{y}'\mathbf{W}\mathbf{x},$$

and

$$\pi_{2,2}(\sigma_\pi^2|\mathbf{x}) = K \frac{\exp\{-\frac{1}{2}(\mathbf{x} - \mathbf{y}\hat{\boldsymbol{\beta}})'\mathbf{W}(\mathbf{x} - \mathbf{y}\hat{\beta})\}}{[\prod_{i=1}^p (\sigma_i^2 + \sigma_\pi^2)]^{1/2}[\det(\mathbf{y}'\mathbf{W}\mathbf{y})]^{1/2}}.$$

Example 15 (continued). Here $p = 7$, $l = 2$, $\mathbf{x} = (105, 110, 127, 115, 130, 125, 137)'$, $\sigma_i^2 = 100$ for $i = 1, 2, 3, 4$, $\sigma_i^2 = 50$ for $i = 5, 6, 7$, and

$$\mathbf{y}' = \begin{pmatrix} 1 & 1 & 1 & 1 & 1 & 1 & 1 \\ 1 & 2 & 3 & 4 & 5 & 6 & 7 \end{pmatrix}.$$

Numerical integration, using the formulas in Corollary 2, yields

$$\boldsymbol{\mu}^\pi(\mathbf{x}) = (106.4, 111.2, 121.3, 118.5, 128.2, 127.4, 136.1)',$$

$$\mathbf{V}^\pi(\mathbf{x}) = \begin{pmatrix} 69.6 & 19.9 & 12.1 & 9.7 & 1.6 & -0.8 & -6.5 \\ 19.9 & 60.1 & 10.2 & 8.9 & 2.5 & 1.0 & -3.2 \\ 12.1 & 10.2 & 60.9 & 2.7 & 6.9 & -1.7 & 1.8 \\ 9.7 & 8.9 & 2.7 & 53.6 & 2.9 & 6.7 & 2.9 \\ 1.6 & 2.5 & 6.9 & 2.9 & 33.6 & 3.3 & 6.1 \\ -0.8 & 1.0 & -1.7 & 6.7 & 3.3 & 36.8 & 7.3 \\ -6.5 & -3.2 & 1.8 & 2.9 & 6.1 & 7.3 & 39.1 \end{pmatrix}. \tag{4.73}$$

To get a feeling for the contribution of each term of (4.72) to $\mathbf{V}^\pi(\mathbf{x})$, the breakdown of $V_{11}^{\pi}(\mathbf{x})$ was

$$V_{11}^{\pi}(\mathbf{x}) = 100 - 55.9 + 25.0 + 0.5 = 69.6.$$

An (estimated) 95% credible set for θ_7 would be

$$C = (136.1 \pm (1.96)\sqrt{39.1}) = (123.8, 148.4).$$

The comments in Subsection 4.5.3 about interpretation, modelling, and generalizations of empirical Bayes theory apply with natural modifications to the hierarchical Bayes theory discussed here.

4.6.4. Comparison with Empirical Bayes Analysis

Empirical Bayes and hierarchical Bayes analyses are naturally very related. Before making some general observations about the similarities and differences between the two approaches, it is helpful to compare the answers obtained in Example 15.

Example 15 (continued). Summarized, for convenience, in Table 4.6 are the estimates of $\theta_1, \ldots, \theta_7$ corresponding to the data $\mathbf{x} = (105, 110, 127, 115, 130, 125, 137)^t$. The table lists the estimates for the empirical Bayes approach of Subsection 4.5.3 (μ_i^{EB}), the hierarchical Bayes approach of the last subsection (μ_i^{π}), and the regression estimates (μ_i^{R}) defined in Example 15 in Subsection 4.5.3. The corresponding variances of the estimates are also given in the table.

Recall that the raw data estimates, x_i, and the regression estimates, μ_i^{R}, can be considered two extremes; the x_i would be the natural estimates if *no* relationships among the θ_i were suspected, while the μ_i^{R} would be the estimates under the specific lower dimensional regression model. The empirical Bayes and hierarchical Bayes estimates are, of course, averages of these two extremes; note that the empirical Bayes estimates are slightly closer to the regression estimates. Similarly, the raw data variances, σ_i^2, and

Table 4.6. Estimates for Example 15.

x_i	105	110	127	115	130	125	137
μ_i^{EB}	106.6	111.3	120.9	118.8	127.6	128.3	135.8
μ_i^{π}	106.4	111.2	121.3	118.5	128.2	127.4	136.1
μ_i^{R}	107.7	112.2	116.8	121.4	125.9	130.5	135.1
σ_i^2	100	100	100	100	50	50	50
V_i^{EB}	67.4	57.4	67.5	53.4	31.8	39.2	36.2
V_{ii}^{π}	69.6	60.1	60.9	53.6	33.6	36.8	39.1
V_i^{R}	43.8	27.6	16.7	10.9	10.4	15.1	25.0

the variances, V_i^R, from the regression estimates can be considered to be two extremes; the additional structure of the regression model yields much smaller variances (only valid if the model is actually correct, of course). Again, the empirical Bayes and hierarchical Bayes variances are an average of these extremes, and are reasonably similar. Note that the hierarchical Bayes variances tend to be smaller than the empirical Bayes variances for middle i, but larger otherwise; they thus mimic the pattern of the regression variances more closely than do the empirical Bayes variances.

While the Morris empirical Bayes and the hierarchical Bayes answers are very similar in the situations we have considered, hierarchical Bayes and "naive" empirical Bayes analyses can differ substantially. Indeed, in the discussion of Example 15 in Subsection 4.5.3, it was pointed out that the μ_i^R were the empirical Bayes estimates that would result from the ML-II approach of estimating the hyperparameters by maximum likelihood, and then using the prior with these estimates in a standard Bayesian fashion. We criticized this because of a failure to consider hyperparameter estimation error. Empirical Bayes theory does not by itself indicate how to incorporate the hyperparameter estimation error in the analysis. Hierarchical Bayesian analysis incorporates such errors automatically, and is, hence, generally the more reasonable of the approaches. Sophisticated empirical Bayes procedures, such as that of Morris in Subsections 4.5.2 and 4.5.3, are usually developed by trying to approximate the hierarchical Bayes answers (compare (4.33) and (4.34) with (4.54) and (4.55), or (4.42) and (4.43) with (4.66) and (4.67)).

Another advantage of the hierarchical approach is that, with only slight additional difficulty, one can incorporate actual subjective prior information at the second stage. There is often available both structural prior information (leading to the first stage prior structure) and subjective prior information about the location of $\boldsymbol{\theta}$; the hierarchical Bayes approach allows the use of both types of information (see Theorems 2 and 3). This can be especially valuable for smaller p.

A third advantage of the hierarchical approach is that it easily produces a greater wealth of information. For instance, the posterior covariances presented in (4.73) are substantial, and knowledge of them would be important for a variety of statistical analyses. These covariances are easily calculable in the hierarchical Bayes analysis, but would require work to derive in a sophisticated empirical Bayes fashion.

From a calculational perspective the comparison is something of a toss-up. Standard empirical Bayes theory requires solution of likelihood equations such as (4.40) and (4.41), while the hierarchical Bayes approach requires numerical integration. Solution of likelihood equations is probably somewhat easier, particularly when the needed numerical integration is higher dimensional, but numerical issues are never clearcut (e.g., one has to worry about uniqueness of solutions to the likelihood equation).

In conclusion, it appears that the hierarchical Bayes approach is the superior methodology for general application. When p is large, of course, there will be little difference between the two approaches, and whichever is more convenient can then be employed. Also, for some (non-Bayesian) audiences a sophisticated empirical Bayes formulation, such as in Subsections 4.5.2 and 4.5.3, might be more suitable for the sake of appearances. Indeed Morris (1983a, b) argues that the empirical Bayes analyses in these sections actually have a type of frequentist justification (see also Subsection 4.7.5), making such analyses more attractive to non-Bayesians. Further discussion of a number of these issues can be found in Deely and Lindley (1981), Morris (1983a), and Berger (1986).

4.7. Bayesian Robustness

4.7.1. Introduction

This section is devoted to studying the robustness (or sensitivity) of Bayesian analysis to possible misspecification of the prior distribution. (Subsection 4.7.11 mentions related issues of model and loss robustness.) Bayesian robustness has been an important element of the philosophies of a number of Bayesians (cf. Good (1950, 1962, 1965, 1983), Dempster (1968, 1975, 1976), Hill (1965, 1975, 1980a, b), and Rubin (1971, 1977)), and has recently received considerable study. A detailed review of the subject and its literature can be found in Berger (1984a).

It is helpful to begin with a reexamination of Example 2 (continued) from Section 3.7.

Example 18. We observe $X \sim \mathcal{N}(\theta, 1)$, and subjectively specify a prior median of 0 and prior quartiles of ± 1. Either the $\mathcal{C}(0, 1)(\pi_C)$ or $\mathcal{N}(0, 2.19)(\pi_N)$ densities are thought to be reasonable matches to prior beliefs. Does it matter whether we use π_C or π_N?

This question can be answered in a number of ways. The most natural way is to simply see if it actually makes a difference. For instance, suppose θ is to be estimated under squared-error loss, so that the posterior means, δ^C and δ^N (for π_C and π_N, respectively), will be used. Table 4.7 presents values of these estimates for various x. For small x (i.e., $x \leq 2$), it appears that $\delta^C(x)$ and $\delta^N(x)$ are quite close, indicating some degree of robustness with respect to choice of the prior. For moderate or large x, however, there can be a substantial difference between δ^C and δ^N, indicating that the answer is then *not* robust to reasonable variation in the prior. Note the dependence of robustness on the actual x which occurs.

Table 4.7. Posterior Means.

x	0	1	2	4.5	10
$\delta^C(x)$	0	0.52	1.27	4.09	9.80
$\delta^N(x)$	0	0.69	1.37	3.09	6.87

The above type of robustness is called *posterior robustness*, since it considers the robustness of the actual action taken in the posterior Bayesian analysis. Subsections 4.7.3 and 4.7.4 deal with this (most natural) type of Bayesian robustness. A different approach to robustness evaluation is through consideration of Bayes risk criteria. For instance, we could imagine using an *estimator* δ in Example 18, and look at

$$\max\{r(\pi_C, \delta), r(\pi_N, \delta)\}.$$

Such an evaluation might be of interest for a number of reasons (see Subsections 4.7.5 through 4.7.7). In Example 18, it can be shown that

$$\max\{r(\pi_C, \delta^C), r(\pi_N, \delta^C)\} = r(\pi_N, \delta^C) = 0.746,$$

while

$$\max\{r(\pi_C, \delta^N), r(\pi_N, \delta^N)\} = r(\pi_C, \delta^N) = \infty.$$

(This last fact follows easily from the calculation that

$$R(\theta, \delta^N) = (2.19/3.19)^2 + (3.19)^{-2}\theta^2,$$

and that $\int \theta^2 \pi_C(\theta) d\theta = \infty$.) Thus, when looking at Bayes risks in Example 18, δ^C seems quite satisfactory, even when π_N is the actual prior, while δ^N is terrible when π_C is the actual prior. Since we are uncertain as to whether π_C or π_N best describes our prior beliefs, use of δ^C seems indicated.

This example also illustrates the most commonly used technique for investigating robustness; simply try different reasonable priors and see what happens. This is often called *sensitivity analysis*.

More formal approaches to robustness will also be discussed. One is the (theoretical) identification of inherently robust (and inherently nonrobust) situations, together with the related development of inherently robust priors. The idea here is to remove the burden of performing an extensive robustness study by utilizing procedures which are theoretically known to be robust. Subsections 4.7.8 through 4.7.10 consider some such developments.

A second formal approach to robustness is through consideration of classes of priors. The idea here is to acknowledge our prior uncertainty by specifying a class, Γ, of possible prior distributions, and then investigating the robustness of a proposed action (or decision rule) as π varies over Γ. Several possible classes of priors were discussed in Subsection 3.5.3. The most commonly used class is the class of priors of a given functional form, with hyperparameters restricted in some fashion.

EXAMPLE 18 (continued). One might seek to acknowledge uncertainty in the $\mathcal{N}(0, 2.19)$ prior by considering

$$\Gamma = \{\pi: \pi \text{ is } \mathcal{N}(\mu, \tau^2), -0.1 < \mu < 0.1 \text{ and } 2.0 < \tau^2 < 2.4\}. \qquad (4.74)$$

The appeal of a class such as that in (4.74) is that it can be easy to work with (cf. Hartigan (1969) and Goldstein (1980)). The disadvantage is that it typically will fail to include many priors which are plausible possibilities. Thus, in Example 18, it may well be the case that the posterior mean is robust as π varies over a class such as that in (4.74), but changes markedly for π_C. Indeed, one of the chief sources of nonrobustness in estimation will be seen to be the degree of flatness of the prior tail; classes such as (4.74) do not admit much variation in the prior tail, and hence may provide a false illusion that robustness obtains.

For the above reason, we prefer using the ε-contamination class of priors (see Subsection 3.5.3)

$$\Gamma = \{\pi: \pi = (1-\varepsilon)\pi_0 + \varepsilon q, q \in \mathcal{Q}\}. \qquad (4.75)$$

EXAMPLE 18 (continued). Suppose we elicit the $\mathcal{N}(0, 2.19)$ prior (this will be π_0 in (4.75)), but feel we could be as much as 10% off (in terms of implied probabilities of sets); this would suggest the choice $\varepsilon = 0.1$ in (4.75). The set, $\mathcal{Q}$, of allowed contaminations could be chosen in any number of ways. The choice $\mathcal{Q} = \{\text{all distributions}\}$ certainly makes Γ large enough to include a reasonable range of priors (including ones with flat tails), but such a $\mathcal{Q}$ is typically unnecessarily large and a smaller set may be desirable (see Subsections 4.7.4 and 4.7.9).

Classes of prior distributions other than the given functional form and ε-contamination classes are sometimes used in robustness studies. One important example is DeRobertis and Hartigan (1981). See Berger (1984a) for other references.

Before proceeding, it is worthwhile to pause for a little philosophizing. There are Bayesians who argue, in varying degrees, that formal prior robustness investigations are not needed. The most extreme of these views is that the prior one elicits is the only prior that needs to be considered, being the best available summary of prior beliefs. The inadequacy of this argument is indicated by Example 18; the only prior features elicited were the median, quartiles, and the prescription that the prior is roughly symmetric (about the median) and unimodal. The decision as to whether to use a $\mathcal{N}(0, 2.19)$ or $\mathcal{C}(0, 1)$ distribution (or any t-distribution with the correct median and quartiles) calls for a fine distinction that most elicitors find almost impossible to make. And if the conclusion depends heavily on this fairly arbitrary choice of the functional form, how can one feel comfortable with the conclusion?

One reason that many Bayesians resist deviation from single-prior Bayesian analysis is that the coherency arguments discussed in Section 4.1

seem to imply that coherent behavior corresponds to single-prior Bayesian analysis. We indicated in Section 4.1 that such conclusions need cautious interpretation, however. Even more to the point, at a very deep level it seems clear that such coherency developments are faulty. The difficulty lies in the almost universally made assumptions that (i) any two objects can be compared, and (ii) comparisons are transitive. Unfortunately, infinitely precise comparison of objects is clearly a practical impossibility. Thus Good (1980), in discussing axiom systems of probability elicitation, says,

> "For it would be only a joke if you were to say that the probability of rain tomorrow (however sharply defined) is 0.3057876289."

One might attempt to escape from the impossiblity of needing arbitrarily accurate comparisons in the axiom systems by deciding that objects which are "close" are *equivalent.* (The axioms do allow two objects to be judged equivalent.) This, however runs afoul of the transitivity axiom. For instance, any two objects judged to be within 10^{-10} of each other could be called equivalent, but, by transitivity, we could then set up a chain of equivalence between *all* actions.

Defenders of the single-prior axioms systems often respond to the above argument by emphasizing that the axioms and their consequences are an ideal towards which we should strive, and that our efforts should be directed towards increasing our elicitation capabilities. Thus Lindley, in the discussion of Berger (1984a), states (see also Lindley, Tversky, and Brown (1979)),

> "My own view ... is that we should learn to measure probabilities. Physicists, presented with Newtonian mechanics for the first time, did not dismiss it because they could not measure accelerations; they learnt to do so. Surveyors do not deplore Euclidean geometry because they cannot measure distances without error: they use techniques like least-squares. And they discover that angles are easier to 'elicit' than distances ... "

This argument of Lindley's is certainly persuasive, but the analogies can be questioned. If surveyors could measure distances only to within an accuracy of 10%, should they report to bridge builders only their actual measurement of the span of a river, or should they report the interval of possible values? And can we even expect routine statistical users (or experts, for that matter) to improve their elicitation abilities so as to be always accurate to within 10%? (See Kahneman, Slovic, and Tversky (1982) for discussion of the difficulties of accurate probability elicitation.)

To a large extent, this entire argument is unresolvable, boiling down to the practical question of whether one's limited time and resources can best be spent on more refined elicitation of a single prior, or on carrying out a formal (or informal) robustness study with a cruder prior. Actually, we will argue that practical considerations usually suggest a combination of the two. One may start out with a (perhaps) crude prior and check, via a robustness study, if the suggested conclusion is clearcut; if not, further refinement of the prior should be attempted. The process (ideally) iterates until a clearcut robust conclusion is reached.

We have actually been somewhat unfair to the coherency approach, because there do exist axiom systems which admit that not all objects can be compared, i.e., that only a *partial preference ordering* might be available. The typical conclusion from such a system is again that coherency corresponds with Bayesian analysis, but now with respect to a *class* of priors (and class of utilities) resulting from the incomplete initial ordering. Thus we indeed do have fundamental axiomatic justification for being a "robust Bayesian." See Smith (1961), Good (1962), Fine (1973), Giron and Rios (1980), Gärdenfors and Sahlin (1982), Wolfenson and Fine (1982), and Good (1983) for such axiomatic developments; other references can be found in Berger (1984a).

The remainder of the section consists of brief introductions to a variety of robust Bayesian techniques. Some of the techniques address the problem of determining whether or not robustness is present, and some address the issue of what to do when robustness is lacking or cannot be properly investigated (presumably because of technical or time limitations). Among the latter category of techniques will be included some that are (at least partly) frequentist in nature. We do not necessarily feel "incoherent" in using such, since the "correct" coherency arguments only lead to classes of priors, and, if a conclusion does not emerge from a study of this class, we are left in limbo. At the same time, we are highly sympathetic to the claim of many Bayesians that one will tend to do better by spending additional effort in refinement of the prior, than by trying to involve frequentist measures.

In light of the above comments, we should mention one technique that will *not* be discussed in this section, namely the technique of putting a second stage prior on the class, Γ, of possible priors. This would be the natural Bayesian way of dealing with Γ. It is also probably a very good *ad hoc* method, whenever the calculations can be carried out. The word *ad hoc* is used because, when Γ is the end result of the elicitation effort, no more true subjective information exists to form a second stage prior. A noninformative second stage prior would likely work quite well, however, assuming the analysis is tractable. Unfortunately, for realistically large Γ (such as the ε-contamination class), hierarchical Bayes calculations become technically overwhelming; and there is no clear answer as to whether it is better to do a hierarchical Bayes analysis with a limited Γ, or use an even more *ad hoc* method with a larger Γ.

4.7.2. The Role of the Marginal Distribution

A natural way of attempting to deal with uncertainty in the prior is to seek to eliminate from consideration priors which seem to be ruled out by the data. (The same can, of course, be said for model uncertainty.) In Section

Table 4.8. Values of m.

x	0	4.5	6.0	10
$m(x\|\pi_N)$	0.22	0.0093	0.00079	3.5×10^{-8}
$m(x\|\pi_C)$	0.21	0.018	0.0094	0.0032

3.5 it was pointed out that the marginal density, $m(x|\pi)$, is invaluable in this regard, since it can (for given x) be interpreted as the likelihood of π (in some subjective sense, perhaps).

EXAMPLE 18 (continued). Suppose that only π_N and π_C are under consideration. Note that $m(x|\pi_N)$ is a $\mathcal{N}(0, 3.19)$ density, while

$$m(x|\pi_C)=\int_{-\infty}^{\infty}\frac{1}{\sqrt{2\pi}}\exp\{-\tfrac{1}{2}(x-\theta)^2\}\cdot\frac{1}{\pi(1+\theta^2)}\,d\theta$$

(which must be calculated by numerical integration). Table 4.8 gives values of $m(x|\pi_N)$ and $m(x|\pi_C)$ for various x. For small x, π_N and π_C are equally well supported by the data; at $x=4.5$, π_C is "twice as likely" as π_N; at $x=6.0$, π_C is over ten times as likely as π_N; and at $x=10$ there is essentially no doubt that π_N must be wrong. The indication is that, for $x\geq 6$, the data supports π_C much more strongly then π_N, and suggests that we use π_C.

This rather informal use of m can be given a type of Bayesian interpretation. Indeed if, in Example 18, we were to assign π_N and π_C probability $\frac{1}{2}$ each of being correct (defining a second stage prior, in effect), yielding an overall prior of $\pi(\theta)=\frac{1}{2}\pi_N(\theta)+\frac{1}{2}\pi_C(\theta)$, then a calculation (see Subsection 4.7.4) gives the posterior as

$$\pi(\theta|x)=\lambda(x)\pi_N(\theta|x)+[1-\lambda(x)]\pi_C(\theta|x), \tag{4.76}$$

where

$$\lambda(x)=\left[1+\frac{m(x|\pi_C)}{m(x|\pi_N)}\right]^{-1}. \tag{4.77}$$

When $\lambda(x)$ is small (as will be the case if $x\geq 6$ in Example 18), the overall posterior will essentially be equal to that from π_C.

This Bayesian motivation, for comparing priors via m, also explains why it is not correct to completely avoid prior elicitation and simply choose a prior to make $m(x|\pi)$ large; the subjective probabilities of the priors being true are also relevant. Thus, if π_N and π_C had differing subjective probabilities of being true (say, p_N and $p_C=1-p_N$, respectively), then (4.76) would hold with

$$\lambda(x)=\left[1+\frac{p_C m(x|\pi_C)}{p_N m(x|\pi_N)}\right]^{-1}.$$

Thus intuitive comparison (and possible choice or elimination) of priors, based on use of $m(x|\pi)$, should be restricted to priors having approximately equal subjective believability.

One can formalize the above use of $m(x|\pi)$ by stating that one should choose the ML-II prior in Γ, i.e., the prior $\hat{\pi}$ which maximizes $m(x|\pi)$ over all $\pi \in \Gamma$. (See Subsection 3.5.4 for discussion and examples.) We will see indications that this approach works well if Γ contains only reasonable priors, but can give bad results if Γ is too large (i.e., contains unreasonable priors).

Another possible use of m is to alert one that a robustness problem may exist. Suppose a particular prior π_0 is elicited for use, but that it is found that $m(x|\pi_0)$ is surprisingly small (for the observed data, of course). There may well then be a robustness problem (although it could be a problem with the model, and not with π_0, see Subsection 4.7.11). The difficulty here is in determining what is "surprisingly small". Some Bayesians (cf. Box (1980)) suggest use of significance tests based on the density $m(x|\pi_0)$. As discussed in Sections 1.6 and 4.3.3, however, the use of significance tests is questionable on several counts. An alternative, somewhat more in accord with the conditional viewpoint, is to look at the *relative likelihoods*

$$m^*(x|\pi_0) = \frac{m(x|\pi_0)}{\sup_x m(x|\pi_0)} \quad \text{or} \quad m^{**}(x|\pi_0) = \frac{m(x|\pi_0)}{E^m[m(X|\pi_0)]}.$$

These measures seek to provide a base of comparison for the size of $m(x|\pi_0)$, not by averaging $m(x|\pi_0)$ over "extreme" x which did not occur (as a significance test must), but by directly seeing if the observed x is unusual compared to the "most likely" x or the "average" x, respectively.

EXAMPLE 18 (continued). Note that both $m(x|\pi_N)$ and $m(x|\pi_C)$ are maximized at $x=0$, so that (from Table 4.8) $\sup_x m(x|\pi_N)=0.22$ and $\sup_x m(x|\pi_C)=0.21$. Also, calculation gives that $E^m[m(X|\pi_N)]=0.16$ and $E^m[m(X|\pi_C)]=0.12$. From Table 4.8, we can then calculate the values of m^* and m^{**} for π_N and π_C. The results are given in Table 4.9. We are not *comparing* π_N and π_C here, but are imagining what we would think if each prior was being considered in isolation. The observation $x=4.5$ is not too implausible under π_C, but is questionable under π_N. If $x=6$ were observed,

Table 4.9. Relative Likelihoods.

x	0	4.5	6	10
$m^*(x\|\pi_N)$	1	0.042	0.0036	1.6×10^{-7}
$m^{**}(x\|\pi_N)$	1.4	0.059	0.0051	2.3×10^{-7}
$m^*(x\|\pi_C)$	1	0.086	0.045	0.015
$m^{**}(x\|\pi_C)$	1.8	0.15	0.079	0.026

π_C would not be completely ruled out, while π_N probably would be. At $x=10$, neither π_C nor (especially) π_N seems plausible.

The use of m^* or m^{**} to indicate the plausibility of π_0 is still not strictly justifiable from a Bayesian perspective. The basic fact is that a hypothesis should not be ruled out just because something unusual has happened; Bayesian reasoning teaches us to reject a hypothesis only if an alternative explanation for the unusual occurrence has been found. The danger in rejecting an hypothesis solely on the basis of unusual x was illustrated in Subsection 4.3.3, and is clarified by the following simple example.

EXAMPLE 19. Suppose an urn contains 1 red, 50 green, and θ blue balls (θ unknown). We conduct an experiment by drawing 3 balls with replacement; denote the data by $\mathbf{x}=(x_R, x_G, x_B)$, where x_R is the number of red balls, x_G the number of green balls, and x_B the number of blue balls among the 3 drawn. Then $f(\mathbf{x}|\theta)$ is the multinomial distribution specified by

$$f(\mathbf{x}|\theta)=\frac{6}{x_R!\,x_G!\,x_B!}\left[\frac{1}{51+\theta}\right]^{x_R}\left[\frac{50}{51+\theta}\right]^{x_G}\left[\frac{\theta}{51+\theta}\right]^{x_B},$$

and, for any prior π,

$$m(\mathbf{x}|\pi)=\frac{6(50)^{x_G}}{x_R!\,x_G!\,x_B!}\sum_\theta\frac{\theta^{x_B}}{(51+\theta)^3}\pi(\theta).$$

Suppose now that $\mathbf{x}=(3,0,0)$ is observed (i.e., all balls in the sample are red). Then

$$m((3,0,0)|\pi)=\sum_\theta(51+\theta)^{-3}\pi(\theta).$$

But, since

$$\sup_{\mathbf{x}} m(\mathbf{x}|\pi)\geq m((0,3,0)|\pi)=(50)^3\sum_\theta(51+\theta)^{-3}\pi(\theta),$$

it follows that

$$\begin{aligned}m^*((3,0,0)|\pi)&=\frac{m((3,0,0)|\pi)}{\sup_{\mathbf{x}} m(\mathbf{x}|\pi)}\\&\leq\frac{m((3,0,0)|\pi)}{m((0,3,0)|\pi)}\\&=(50)^{-3}.\end{aligned}$$

Since this is very small for *any* prior, it is clearly untenable to reject a proposed prior simply because of a small value of m^*. (The model, of course, comes under substantial suspicion with data such as this.)

Although Example 19 indicates that caution must be used in interpreting m^* and m^{**}, they will often give a closer correspondence to a true Bayesian answer than will tail area probabilities of significance tests. For instance, if $\bar{X} \sim \mathcal{N}(\theta, \sigma^2/n)(\sigma^2$ known) and θ is $\mathcal{N}(\mu, \tau^2)$ (so that $m(\bar{x}|\pi)$ is $\mathcal{N}(\mu, \sigma^2/n+\tau^2)$), then

$$\begin{aligned} m^*(\bar{x}|\pi) &= \frac{m(\bar{x}|\pi)}{\sup_{\bar{x}} m(\bar{x}|\pi)} \\ &= \frac{m(\bar{x}|\pi)}{m(\mu|\pi)} \\ &= \exp\{-\tfrac{1}{2}z^2\}, \end{aligned} \tag{4.78}$$

where $z = (\bar{x}-\mu)/[\sigma^2/n+\tau^2]^{1/2}$, and

$$\begin{aligned} m^{**}(\bar{x}|\pi) &= \frac{m(\bar{x}|\pi)}{E^m[m(\bar{X}|\pi)]} \\ &= \frac{m(\bar{x}|\pi)}{[4\pi(\sigma^2/n+\tau^2)]^{-1/2}} \\ &= \sqrt{2}\exp\{-\tfrac{1}{2}z^2\}. \end{aligned} \tag{4.79}$$

A classical significance test that $m(\bar{x}|\pi)$ was the true distribution of $\bar{X}$ would be based on the tail area

$$\alpha = P^m(|\bar{X}-\mu| > |\bar{x}-\mu|) = P(|Z| > |z|),$$

where Z is $\mathcal{N}(0, 1)$. But, as shown in Subsection 4.3.3, the expressions (4.78) and (4.79) correspond more closely to Bayesian posterior probabilities that π is true (or, more precisely, Bayes factors for π) than does α. (The correspondence with posterior probabilities need not be close in higher dimensions; m^* can be particularly bad, see Exercise 84.)

When all is said and done, however, nothing is a completely satisfactory replacement for honest Bayesian analysis. Thus, while m^* or m^{**} might provide useful indications of a possible robustness problem, they should not lead one to reject use of a prior unless another plausible prior has a substantially larger value of $m(x|\pi)$.

There is a substantial literature on the above, or related, uses of m in robustness. A few references (from which others can be obtained) are Jeffreys (1961), Box and Tiao (1973), Dempster (1975), Geisser and Eddy (1979), Box (1980), Good (1983), Berger and Berliner (1983), and Rubin (1984).

4.7.3. Posterior Robustness: Basic Concepts

Suppose a Bayesian is considering choice of an action, a, and is concerned about its robustness as π varies over a class Γ of priors. If $L(\theta, a)$ is the loss function and $\rho(\pi(\theta|x), a)$ the posterior expected loss of a, it is natural to evaluate the robustness of a by considering

$$\left(\inf_{\pi\in\Gamma} \rho(\pi(\theta|x), a), \sup_{\pi\in\Gamma} \rho(\pi(\theta|x), a)\right); \tag{4.80}$$

this gives the range of possible posterior expected losses. Through the device of using inference losses, robustness in inference can also be considered using this framework.

EXAMPLE 20. Suppose the action, a, is to choose a credible set $C\subset\Theta$. Defining $L(\theta, C)=1-I_C(\theta)$, so that

$$\rho(\pi(\theta|x), C)=1-P^{\pi(\theta|x)}(\theta\in C)=P^{\pi(\theta|x)}(\theta\notin C),$$

the interval in (4.80) determines the range of posterior probabilities of the complement of C.

As a specific example, suppose $X\sim\mathcal{N}(\theta, 1)$,

$$\Gamma=\{\pi\colon \pi \text{ is } \mathcal{N}(\mu, \tau^2), 1\le\mu\le 2, 3\le\tau^2\le 4\},$$

and $x=0$ is observed. Suppose the credible set $C=(-1, 2)$ is to be reported, and it is desired to determine its minimum and maximum probabilities of containing θ as π ranges over Γ. Note that, for a $\mathcal{N}(\mu, \tau^2)$ prior, the posterior is normal with mean and variance

$$\mu^{\pi}(x)=\frac{1}{(1+\tau^2)}\mu+\frac{\tau^2}{(1+\tau^2)}x=\frac{1}{(1+\tau^2)}\mu, \quad\text{and}\quad V^{\pi}=\frac{\tau^2}{(1+\tau^2)}.$$

Thus the range of *posteriors* corresponding to Γ is

$$\Gamma^*(x)=\left\{\pi(\theta|0)\colon \pi(\theta|0) \text{ is } \mathcal{N}\left(\frac{\mu}{(1+\tau^2)}, \frac{\tau^2}{(1+\tau^2)}\right), 1\le\mu\le 2, 3\le\tau^2\le 4\right\}.$$

We seek to determine the minimum and maximum of $P^{\pi(\theta|0)}((-1, 2))$ over this set of posteriors. The maximum is achieved by choosing $\mu=2$ and $\tau^2=3$, the corresponding $\mathcal{N}(0.5, 0.75)$ posterior clearly being the one in Γ^* that is most concentrated in the interval $(-1, 2)$. (Its mean is at the center of the interval and its variance is as small as possible.) Calculation gives that the probability of $(-1, 2)$ for this prior is 0.916. The minimum probability of $(-1, 2)$ is harder to determine. Noting that $0.2\le\mu/(1+\tau^2)\le 0.5$, for the given range of μ and τ^2, and observing that we want to choose $\pi(\theta|0)$ to put as much mass outside $(-1, 2)$ as possible (to minimize the probability), it becomes clear that $\mu=1$ should be chosen (since 0.5 is the middle of the interval, and the mean of $\pi(\theta|0)$ should be chosen as far from this center as possible). We are left with the problem of minimizing $P^{\pi(\theta|0)}((-1, 2))$ over τ^2 when $\mu=1$. Differentiating with respect to τ^2 leads, after a laborious calculation, to the conclusion that $P^{\pi(\theta|0)}((-1, 2))$ is decreasing in τ^2 over the indicated range. Hence, $\mu=1$ and $\tau^2=4$ will yield the minimum value, which can be calculated to be 0.888. Thus

$$0.888\le P^{\pi(\theta|0)}(C)\le 0.916, \quad\text{or}\quad 0.084\le\rho(\pi(\theta|0), C)\le 0.112.$$

The narrowness of this range would leave one quite comfortable with stating, say, that $C=(-1, 2)$ is a robust 90% credible set.

When Γ is chosen to be a class of given functional form, as in Example 20, the minimization and maximization in (4.80) will simply be minimization and maximization over the prior hyperparameters (within their given ranges, of course). This will usually be calculationally straightforward (although numerical minimization and maximization might well be needed). Larger and more realistic classes of priors can also be dealt with, however, as indicated by De Robertis and Hartigan (1981) and in the next subsection.

Various formal definitions related to posterior robustness can be given (see Berger (1984a)). Here we present only two examples.

Definition 10. The Γ-*posterior expected loss of* a_0 is

$$\rho_\Gamma(a)=\sup_{\pi\in\Gamma}\rho(\pi(\theta|x), a).$$

EXAMPLE 20 (continued). We calculated that $\rho_\Gamma(C)=0.112$. This corresponds to saying that C is *at least* an 88.8% credible set.

A different concept of robustness arises from asking that the selected action be close to the optimal Bayes action for every prior in Γ.

Definition 11. An action a_0 is ε-*posterior robust* with respect to Γ if, for all $\pi\in\Gamma$,

$$|\rho(\pi(\theta|x), a_0)-\inf_a\rho(\pi(\theta|x), a)|\leq\varepsilon. \tag{4.81}$$

EXAMPLE 21. Suppose it is desired to estimate θ under $L(\theta, a)=(\theta-a)^2$. From (4.2) we have that

$$\rho(\pi(\theta|x), a_0)=V^\pi(x)+(\mu^\pi(x)-a_0)^2,$$

where μ^π and V^π are the posterior mean and variance. Also, we know that the minimum of $\rho(\pi(\theta|x), a)$ is achieved at $a=\mu^\pi(x)$, which has posterior expected loss $V^\pi(x)$. Hence

$$|\rho(\pi(\theta|x), a_0)-\inf_a\rho(\pi(\theta|x), a)|=(\mu^\pi(x)-a_0)^2.$$

Thus a_0 is ε-posterior robust if it is within $\pm\sqrt{\varepsilon}$ of all the posterior means corresponding to priors in Γ. In Example 20, the posterior means range over the interval [0.2 to 0.5], so $a_0=0.35$ would be $(0.15)^2=(0.0225)$-posterior robust.

It is tempting to suggest choosing an action a_0 which minimizes the supremum (over all $\pi\in\Gamma$) of the left-hand side of (4.81). In Example 21, this would mean choosing a_0 to minimize $\sup_{\pi\in\Gamma}(\mu^\pi(x)-a_0)^2$, which, if $\mu^\pi(x)$ ranges over an interval, can easily be shown to be achieved by choosing a_0 to be the midpoint of the interval. This is not necessarily a good idea, however, as can be seen from Example 18. We partially indicated that the posterior mean for the Cauchy prior is a substantially better estimate than the posterior mean for the normal prior, when x is moderate or large. Hence it would not do to merely choose an action midway between the two posterior means. This is also emphasized by a real Bayesian analysis of the problem: if (4.76) defines the actual overall posterior, then the actual posterior mean will be

$$\begin{aligned} E^{\pi(\theta|x)}[\theta] &= \int \theta[\lambda(x)\pi_N(\theta|x)+(1-\lambda(x))\pi_C(\theta|x)]d\theta \\ &= \lambda(x)\delta^N(x)+(1-\lambda(x))\delta^C(x), \end{aligned}$$

which will be strongly weighted towards $\delta^C(x)$ when x is moderate or large.

There is a large literature on posterior robustness, most of which is surveyed in Berger (1984a). This literature includes Good (1965, 1983a, b), Hill (1965, 1975, 1980a, b), Fishburn (1965), Fishburn, Murphy, and Isaacs (1968), Dickey (1976), Leamer (1978, 1982), Rukhin (1978, 1984a), De Robertis and Hartigan (1981), Polasek (1982, 1984, 1985), Kadane and Chuang (1984), and Chuang (1984). The works on "stable estimation," mentioned in Subsection 4.7.8, and some of the material on testing a point null hypothesis, in Subsection 4.3.3, can also be interpreted in this light.

4.7.4. Posterior Robustness: ε-Contamination Class

The ε-contamination class of priors, defined by

$$\Gamma=\{\pi: \pi=(1-\varepsilon)\pi_0+\varepsilon q, q\in\mathcal{Q}\},$$

is particularly attractive to work with when investigating posterior robustness. We have previously discussed the attractive richness and flexibility of the class (through appropriate choice of $\mathcal{Q}$, one can ensure that Γ contains all plausible priors and no implausible ones), and it also has substantial calculational advantages. The following lemma presents the basic equations needed to work with this Γ. The proof is straightforward, and is left as an exercise.

Lemma 2. *Suppose $\pi=(1-\varepsilon)\pi_0+\varepsilon q$, that the posterior densities $\pi_0(\theta|x)$ and $q(\theta|x)$ exist, and that $m(x|\pi)>0$. Then*

$$\pi(\theta|x)=\lambda(x)\pi_0(\theta|x)+[1-\lambda(x)]q(\theta|x), \tag{4.82}$$

where

$$\begin{aligned} \lambda(x) &= \frac{(1-\varepsilon)m(x|\pi_0)}{m(x|\pi)} \\ &= \left[1+\frac{\varepsilon m(x|q)}{(1-\varepsilon)m(x|\pi_0)}\right]^{-1}. \end{aligned} \tag{4.83}$$

Furthermore, in a decision problem,

$$\begin{aligned}\rho(\pi(\theta|x), a) &= E^{\pi(\theta|x)}[L(\theta, a)] \\ &= \lambda(x)\rho(\pi_0(\theta|x), a)+[1-\lambda(x)]\rho(q(\theta|x), a). \quad (4.84)\end{aligned}$$

EXAMPLE 22. To find the posterior mean, $\mu^{\pi}(x)$, for $\pi \in \Gamma$, set $L(\theta, a) \equiv \theta$ in (4.84), yielding

$$\begin{aligned}\mu^{\pi}(x) &= \lambda(x)E^{\pi_0(\theta|x)}[\theta]+[1-\lambda(x)]E^{q(\theta|x)}[\theta] \\ &= \lambda(x)\mu^{\pi_0}(x)+[1-\lambda(x)]\mu^{q}(x).\end{aligned} \quad (4.85)$$

To find the posterior variance, $V^{\pi}(x)$, set $L(\theta, a) \equiv (\theta-\mu^{\pi}(x))^2$ in (4.84), yielding

$$\begin{aligned}V^{\pi}(x) &= \lambda(x)E^{\pi_0(\theta|x)}[(\theta-\mu^{\pi})^2]+[1-\lambda(x)]E^{q(\theta|x)}[(\theta-\mu^{\pi})^2] \\ &= \lambda(x)[V^{\pi_0}+(\mu^{\pi_0}-\mu^{\pi})^2]+[1-\lambda(x)][V^{q}+(\mu^{q}-\mu^{\pi})^2]\end{aligned}$$

(the last step following, as usual, by adding and subtracting the posterior means, under π_0 and q, in the respective expectations). Replacing μ^{π} by the expression in (4.85) and simplifying yields

$$\begin{aligned}V^{\pi}(x) &= \lambda(x)V^{\pi_0}(x)+[1-\lambda(x)]V^{q}(x) \\ &\quad +\lambda(x)[1-\lambda(x)][\mu^{\pi_0}(x)-\mu^{q}(x)]^2. \quad (4.86)\end{aligned}$$

The comparative ease of use of the ε-contamination Γ is due in part to the fact that $\pi_0(\theta|x)$ and $m(x|\pi_0)$ will be known constants (recall that π_0 is the elicited prior), so that minimizations and maximizations of $\rho(\pi(\theta|x), a)$ over $\pi \in \Gamma$ depend only on the variation in $m(x|q)$ and $\rho(q(\theta|x), a)$ as q varies over $\mathcal{Q}$. The following theorem provides useful results for the important case where L is an indicator function and $\mathcal{Q}$ is the class of all distributions. The theorem was given in Huber (1973) (see also Berger and Berliner (1983)).

Theorem 4. *Suppose* $\mathcal{Q} = \{\text{all distributions}\}$ *and* $L(\theta, a) = I_C(\theta)$, *so that*

$$\rho(\pi(\theta|x), a) = P^{\pi(\theta|x)}(\theta \in C).$$

Then

$$\inf_{\pi \in \Gamma} P^{\pi(\theta|x)}(\theta \in C) = P_0\left[1+\frac{\varepsilon \sup_{\theta \notin C} f(x|\theta)}{(1-\varepsilon)m(x|\pi_0)}\right]^{-1}, \quad (4.87)$$

and

$$\sup_{\pi \in \Gamma} P^{\pi(\theta|x)}(\theta \in C) = 1-(1-P_0)\left[1+\frac{\varepsilon \sup_{\theta \in C} f(x|\theta)}{(1-\varepsilon)m(x|\pi_0)}\right]^{-1}, \quad (4.88)$$

where $P_0 = P^{\pi_0(\theta|x)}(\theta \in C)$.

Proof. Let $\bar{C}$ denote the complement of C. Also, for any $q \in \mathcal{Q}$, let

$$z_q(A) = \int_A f(x|\theta) q(d\theta).$$

Clearly

$$P^{\pi(\theta|x)}(\theta \in C) = \frac{(1-\varepsilon)m(x|\pi_0)P_0 + \varepsilon z_q(C)}{(1-\varepsilon)m(x|\pi_0) + \varepsilon z_q(C) + \varepsilon z_q(\bar{C})}. \tag{4.89}$$

Consider the function

$$h(z) = \frac{(K_1 + z)}{(K_2 + z + g(z))}.$$

It is straightforward to check that h is increasing in $z \geq 0$ when $K_2 \geq K_1 \geq 0$ and g is a positive, decreasing function of z. Setting $K_1 = (1-\varepsilon)m(x|\pi_0)P_0$ and $K_2 = (1-\varepsilon)m(x|\pi_0)$, it follows that (4.89) can be decreased by taking any mass that q assigns to C and giving it to $\bar{C}$; thus (4.89) is minimized when $Z_q(C) = 0$. Furthermore,

$$\begin{aligned}\inf_{\pi \in \Gamma} P^{\pi(\theta|x)}(\theta \in C) &= \inf_{\{q: z_q(C)=0\}} \frac{(1-\varepsilon)m(x|\pi_0)P_0}{[(1-\varepsilon)m(x|\pi_0) + \varepsilon z_q(\bar{C})]} \\ &= \frac{(1-\varepsilon)m(x|\pi_0)P_0}{(1-\varepsilon)m(x|\pi_0) + \varepsilon \sup_{\{q: z_q(C)=0\}} z_q(\bar{C})}.\end{aligned}$$

But

$$\sup_{\{q: z_q(C)=0\}} z_q(\bar{C}) = \sup_{\theta \in \bar{C}} f(x|\theta),$$

and (4.87) follows. Formula (4.88) follows from applying (4.87) to $\bar{C}$. □

The major applications of Theorem 4 are to robustness of credible sets (where C is the credible set), and to hypothesis testing (where C defines a hypothesis). We present examples of each application.

Example 23. Suppose that $X \sim \mathcal{N}(\theta, \sigma^2)$ and that π_0 is $\mathcal{N}(\mu, \tau^2)$. We know that the $100(1-\alpha)\%$ HPD credible set for θ under π_0 is

$$C = \left(\mu^{\pi}(x) + z\left(\frac{\alpha}{2}\right)\sqrt{V^{\pi}(x)},\ \mu^{\pi}(x) - z\left(\frac{\alpha}{2}\right)\sqrt{V^{\pi}(x)}\right), \tag{4.90}$$

where

$$\mu^{\pi}(x) = x - \left[\frac{\sigma^2}{(\sigma^2+\tau^2)}\right](x - \mu_0), \qquad V^{\pi}(x) = \frac{\sigma^2\tau^2}{(\sigma^2+\tau^2)}.$$

To investigate the robustness of C over the ε-contamination class of priors, with $\mathcal{Q} = \{\text{all distributions}\}$, we use Theorem 4.

Note first that, if $x \in C$, then

$$\sup_{\theta \in C} f(x|\theta) = f(x|x) = (2\pi\sigma^2)^{-1/2}, \tag{4.91}$$

while

$$\sup_{\theta \notin C} f(x|\theta) = f(x|\text{endpoint of } C \text{ closest to } x)$$

$$= (2\pi\sigma^2)^{-1/2} \exp\left\{-\frac{1}{2\sigma^2}(\text{endpoint} - x)^2\right\} \tag{4.92}$$

$$= (2\pi\sigma^2)^{-1/2} \exp\left\{-\frac{1}{2\sigma^2}\left(|\mu^\pi(x) - x| + z\left(\frac{\alpha}{2}\right)\sqrt{V^\pi}\right)^2\right\}.$$

Since $P_0 = 1 - \alpha$ and

$$m(x|\pi_0) = [2\pi(\sigma^2 + \tau^2)]^{-1/2} \exp\left\{-\frac{\frac{1}{2}(x-\mu)^2}{(\sigma^2+\tau^2)}\right\},$$

we have determined all the quantities in (4.87) and (4.88). If $x \notin C$, then the right-hand sides of (4.91) and (4.92) should be interchanged.

As a concrete example, suppose that $\sigma^2 = 1$, $\tau^2 = 2$, $\mu = 0$, and $\varepsilon = 0.1$. First suppose that $x = 1$ is observed. Then $\mu^\pi(1) = \frac{2}{3}$, $V^\pi = \frac{2}{3}$, and the 95% HPD credible interval is

$$C = (\tfrac{2}{3} \pm (1.96)(\tfrac{2}{3})^{1/2}) = (-0.93, 2.27),$$

so that $x = 1 \in C$. Hence (4.91) and (4.92) give

$$\sup_{\theta \in C} f(1|\theta) = (2\pi)^{-1/2} = 0.40, \quad \text{and}$$

$$\sup_{\theta \notin C} f(1|\theta) = (2\pi)^{-1/2} \exp\{-\tfrac{1}{2}(|\tfrac{2}{3} - 1| - (1.96)(\tfrac{2}{3})^{1/2})^2\} = 0.18.$$

Also, $P_0 = 1 - \alpha = 0.95$ and $m(1|\pi_0) = 0.19$, so that Theorem 4 yields

$$\inf_{\pi \in \Gamma} P^{\pi(\theta|1)}(\theta \in C) = (0.95)\left[1 + \frac{(0.1)(0.18)}{(0.9)(0.19)}\right]^{-1} = 0.86,$$

$$\sup_{\pi \in \Gamma} P^{\pi(\theta|1)}(\theta \in C) = 1 - (1 - 0.95)\left[1 + \frac{(0.1)(0.40)}{(0.9)(0.19)}\right]^{-1} = 0.96.$$

Thus our "confidence" in C actually ranges between 0.86 and 0.96 as π ranges over Γ.

Suppose, instead, that $x = 3$ is observed. Then $\mu^\pi(3) = 2$ and $C = (0.40, 3.60)$. Again $x = 3 \in C$, so we use (4.91) and (4.92) directly. Calculation gives that

$$\inf_{\pi\in\Gamma} P^{\pi(\theta|3)}(\theta\in C)=(0.95)\left[1+\frac{(0.1)(0.33)}{(0.9)(0.051)}\right]^{-1}=0.55, \quad \text{and}$$

$$\sup_{\pi\in\Gamma} P^{\pi(\theta|3)}(\theta\in C)=1-(1-0.95)\left[1+\frac{(0.1)(0.40)}{(0.9)(0.051)}\right]^{-1}=0.97.$$

EXAMPLE 24. Again suppose that $X\sim\mathcal{N}(\theta,\sigma^2)$ and $\theta\sim\mathcal{N}(\mu,\tau^2)$, but now we desire to test H_0: $\theta\le\theta_0$ versus H_1: $\theta>\theta_0$. Defining $C=(-\infty,\theta_0)$, we have that the posterior probability of H_0 is simply that of C. The formulas in Theorem 4 and Example 23 thus apply directly to this problem.

As a concrete example, suppose $\sigma^2=1$, $\mu=0$, $\tau^2=2$, $\theta_0=0$, and $x=2$ is observed. Then the posterior probability of C under π_0 is

$$P_0=P^{\pi_0(\theta|2)}(C)=\Phi([0-\mu^{\pi}(2)]/\sqrt{V^{\pi}})=\Phi(-1.63)=0.052$$

(where Φ denotes the standard normal c.d.f.). Note that $x=2\notin C$, so that the formulas (4.91) and (4.92) must be applied with the right-hand sides interchanged. Thus

$$\sup_{\theta\in C} f(2\,|\,\theta)=(2\pi)^{-1/2}\exp\{-\tfrac{1}{2}(0-2)^2\}=0.054, \quad \text{and}$$

$$\sup_{\theta\notin C} f(2\,|\,\theta)=(2\pi)^{-1/2}=0.40.$$

Also $m(2\,|\,\pi_0)=0.12$, so that Theorem 4 yields

$$\inf_{\pi\in\Gamma} P^{\pi(\theta|2)}(\theta\in C)=(0.052)\left[1+\frac{(0.1)(0.40)}{(0.9)(0.12)}\right]^{-1}=0.038, \quad \text{and}$$

$$\sup_{\pi\in\Gamma} P^{\pi(\theta|2)}(\theta\in C)=1-(1-0.052)\left[1+\frac{(0.1)(0.054)}{(0.9)(0.12)}\right]^{-1}=0.097.$$

An important feature of posterior robustness, that can be seen from the previous examples, is that it is generally highly dependent on the x observed. Thus, for $x=1$ in Example 23, the nominal 95% credible set C maintains reasonably high probability for all $\pi\in\Gamma$. For $x=3$, however, the probability of the nominal 95% credible set can drop to as low as 0.55 for $\pi\in\Gamma$. The importance of recognizing this feature is that, since robustness is not easy to determine, it is often necessary to wait until x is at hand before a robustness check becomes feasible.

What is to be done when posterior robustness is found to be lacking as π varies over Γ? The first consideration should be to determine if the lack of posterior robustness is due to Γ containing unreasonable prior distributions. For the ε-contamination class with $\mathcal{Q}=\{\text{all distributions}\}$, a lack of robustness may very well be due to this cause. For instance, in Example 23 when $x=3$, the minimum probability is attained at the prior in Γ which is a 90% mixture of π_0 and 10% of a point mass at 3.6 (the endpoint of

C). Now, while it is certainly plausible to imagine that 10% of the prior mass could be misspecified in π_0, it is unnatural to suppose that concentrating it all at a single point in the tail of π_0 would be sensible *a priori.* Far more reasonable, would be to only allow this mass to be spread out in a moderately smooth fashion. An extremely attractive such Γ, when π_0 is unimodal, is the ε-contamination class with $\mathcal{Q}$ restricted so as to yield only *unimodal* π. (We may often feel *very* confident that π is unimodal, in addition to thinking that it is reasonably close to π_0.) It is somewhat remarkable that such a Γ can be successfully worked with (see Subsection 4.7.9, Berger and Berliner (1983), and Berliner (1984a) for indications of this), but the results are a bit too involved to consider here.

Unfortunately, it often will be the case that $\mathcal{Q}=\{\text{all distributions}\}$ is too large to allow verification of robustness in confidence set problems and estimation problems. The situation in hypothesis testing problems is somewhat more favorable, since "extreme x," which cause this $\mathcal{Q}$ to have a particularly unreasonable effect, yield clearcut decisions anyway (see also Subsection 4.7.8). The application of $\mathcal{Q}=\{\text{all distributions}\}$ to testing a point null hypothesis will, in particular, often be satisfactory.

EXAMPLE 25. It is desired to test H_0: $\theta=\theta_0$ versus H_1: $\theta\neq\theta_0$, based on observation of $X\sim f(x|\theta)$. The prior π_0 is specified by a positive probability (not just density) $\pi_0(\theta_0)$ of H_0 (we simply used π_0 to denote this quantity in Subsection 4.3.3), and a density $g_0(\theta)$, for $\theta\neq\theta_0$, which describes the spread of the probability $(1-\pi_0(\theta_0))$ given to H_1. Then (4.87) and (4.88) yield, for the posterior probability of H_0 (i.e., when $C=\{\theta_0\}$),

$$\inf_{\pi\in\Gamma} P^{\pi(\theta|x)}(\theta_0)=P_0\left[1+\frac{\varepsilon f(x|\hat{\theta})}{(1-\varepsilon)m(x|\pi_0)}\right]^{-1}$$
$$=P_0\left[1+\frac{\varepsilon}{(1-\varepsilon)\pi_0(\theta_0)}\cdot P_0\cdot\frac{f(x|\hat{\theta})}{f(x|\theta_0)}\right]^{-1}$$

and

$$\sup_{\pi\in\Gamma} P^{\pi(\theta|x)}(\theta_0)=1-(1-P_0)\left[1+\frac{\varepsilon f(x|\theta_0)}{(1-\varepsilon)m(x|\pi_0)}\right]^{-1}$$
$$=1-(1-P_0)\left[1+\frac{\varepsilon}{(1-\varepsilon)\pi_0(\theta_0)}\cdot P_0\right]^{-1},$$

where $P_0=P^{\pi_0(\theta|x)}(\theta_0)=\pi_0(\theta_0)f(x|\theta_0)/m(x|\pi_0)$, and $\hat{\theta}$ is an m.l.e. for θ.

In the situation where $X=\bar{X}\sim\mathcal{N}(\theta,\sigma^2/n)$ and π_0 is $\mathcal{N}(\mu,\tau^2)$, P_0 is given by (4.16) (with $\pi_0=\pi_0(\theta_0)$) and $f(x|\hat{\theta})/f(x|\theta_0)=\exp\{\frac{1}{2}z^2\}$, where $z=\sqrt{n}(\bar{x}-\theta_0)/\sigma$. When $\varepsilon=0.1$, $n=10$, $\pi_0(\theta_0)=\frac{1}{2}$, $\tau=\sigma$, $\mu=\theta_0$, and $z=2.576$, so that $P_0=0.14$ (Table 4.2), the inf is 0.08 and the sup is 0.17.

A relatively simple alternative to using the "too large" Γ in Theorem 4 is to use the ε-contamination class, with $\mathcal{Q}$ chosen to be a *parametric* class of distributions. This may be a "too small" class, but, if $\mathcal{Q}$ is sensibly chosen, the results should give a good indication of actual robustness.

EXAMPLE 26. Again suppose that $X \sim \mathcal{N}(\theta, \sigma^2)$ and that π_0 is $\mathcal{N}(\mu, \tau^2)$. Consider the ε-contamination class, Γ, with

$$\mathcal{Q} = \{q_k\colon q_k \text{ is } \mathcal{U}(\mu - k, \mu + k), k > 0\}.$$

Any $\pi \in \Gamma$ will be a pretty sensible prior, and the class does contain priors with *effective* tails substantially larger than π_0. (The part of the tail of π that is "effective" is the part near the likelihood function.) The class is admittedly rather small, however, and can not incontrovertibly establish posterior robustness.

To find the range of posterior probabilities of an interval $C = (c_1, c_2)$ as π ranges over Γ (or k varies over $(0, \infty)$), we need to consider

$$P^{\pi(\theta|x)}(\theta \in C) = \lambda_k(x) P_0 + (1 - \lambda_k(x)) Q_k, \tag{4.93}$$

where

$$P_0 = P^{\pi_0(\theta|x)}(\theta \in C),$$

$$\lambda_k(x) = \left[1 + \frac{\varepsilon}{(1-\varepsilon)} \cdot \frac{m(x|q_k)}{m(x|\pi_0)}\right]^{-1},$$

$$m(x|q_k) = \int_{\mu-k}^{\mu+k} f(x|\theta) \cdot \frac{1}{2k}\, d\theta = \frac{1}{2k}\left[\Phi\left(\frac{\mu+k-x}{\sigma}\right) - \Phi\left(\frac{\mu-k-x}{\sigma}\right)\right],$$

$$Q_k(x) = P^{q_k(\theta|x)}(\theta \in C)$$

$$= \frac{1}{m(x|q_k)} \int_{c^*}^{c^{**}} f(x|\theta) \cdot \frac{1}{2k}\, d\theta$$

$$= \frac{1}{2km(x|q_k)}\left[\Phi\left(\frac{c^{**}-x}{\sigma}\right) - \Phi\left(\frac{c^*-x}{\sigma}\right)\right]^+,$$

$$c^* = \max\{c_1, -k\}, \qquad c^{**} = \min\{c_2, k\},$$

"+" stands for the positive part, and Φ denotes the standard normal c.d.f. The expression in (4.93) can easily be minimized and maximized numerically.

As a concrete example, consider $\varepsilon = 0.1$, $\sigma^2 = 1$, $\tau^2 = 2$, $\mu = 0$, $x = 1$, and $C = (-0.93, 2.27)$, as in Example 23. Then numerical calculation gives

$$\inf_{\pi \in \Gamma} P^{\pi(\theta|x)}(\theta \in C) = 0.945 \quad \text{(achieved at } k = 3.4\text{)},$$

$$\sup_{\pi \in \Gamma} P^{\pi(\theta|x)}(\theta \in C) = 0.956 \quad \text{(achieved at } k = 0.93\text{)}.$$

Thus we have excellent posterior robustness. For the case $x = 3$, with $C = (0.40, 3.60)$, the inf is 0.913 (achieved at $k = 5.2$) and the sup is 0.958 (achieved at $k = 3.6$). Recalling that the corresponding inf from Example

23 was 0.55, it seems clear that the apparent nonrobustness in that Example (for $x=3$) was due to the unreasonable Γ. It would thus seem quite safe to call C at least a 90% credible set.

Although we have restricted ourselves to providing ranges for posterior probabilities, it should be clear that similar developments are possible for other features, such as posterior means, variances, and expected losses (see Berliner (1984a)).

4.7.5. Bayes Risk Robustness and Use of Frequentist Measures

Although attaining posterior robustness is the natural Bayesian goal in statistical analyses, there are some situations in which it is useful to consider frequentist measures in robustness investigations. Most such considerations center around the Bayes risk $r(\pi, \delta)$.

Definition 12. The Γ-*Bayes risk* of a procedure δ is

$$r_\Gamma(\delta) = \sup_{\pi \in \Gamma} r(\pi, \delta).$$

The most common uses of Bayes risk are in situations where $R(\theta, \delta) \leq C$ for all θ, so that $r(\pi, \delta) = E^\pi[R(\theta, \delta)] \leq C$ for *all* π, and $r_\Gamma(\delta) \leq C$ for any Γ.

Example 27. Suppose $\bar{X} \sim \mathcal{N}(\theta, 1/n)$, and that it is desired to estimate θ under squared-error loss. The estimator $\delta(\bar{x}) = \bar{x}$ has risk $R(\theta, \delta) \equiv 1/n$, so that $r(\pi, \delta) = 1/n$ for *all* π. Hence $r_\Gamma(\delta) = 1/n$ for any Γ.

Example 28. Suppose $X_1, \ldots, X_{15}$ are i.i.d. from a completely unknown continuous density f on R^1. (Here we identify θ with the unknown f, so Θ is the set of all continuous densities on R^1.) Of interest is the median of f, to be denoted by η, for which a credible set is desired. Converting to decision-theoretic language, a procedure $\delta(\mathbf{x})$ (where $\mathbf{x} = (x_1, \ldots, x_{15})$) is simply a set in R^1 for each $\mathbf{x}$, and the "loss" can be considered to be

$$L(\theta, \delta(\mathbf{x})) = 1 - I_{\delta(\mathbf{x})}(\eta),$$

so that $\rho(\pi(\theta|\mathbf{x}), \delta(\mathbf{x}))$ is the posterior probability that η is not in $\delta(\mathbf{x})$, and $R(\theta, \delta)$ is one minus the coverage probability of δ (for η).

Consider now the procedure $\delta(\mathbf{x}) = [x_{(4)}, x_{(12)}]$, where $x_{(1)} < x_{(2)} < \cdots < x_{(15)}$ are the increasingly ordered values of the x_i. (Because f is a continuous density, we need not worry about equality of any of the x_i.) Observe that

$$\begin{aligned} R(\theta, \delta) &= E_\theta^{\mathbf{X}} L(\theta, \delta(\mathbf{X})) \\ &= 1 - P_\theta([X_{(4)}, X_{(12)}] \text{ contains } \eta) \\ &= P_\theta(X_{(4)} > \eta) + P_\theta(X_{(12)} < \eta). \end{aligned}$$

The event $\{X_{(4)} > \eta\}$ is precisely the event that 12, 13, 14, or 15 of the X_i are greater than η. But, letting Y denote the number of X_i which are greater than η, it is clear that Y is a $\mathscr{B}(15, \frac{1}{2})$ random variable (since $P(X_i > \eta) = \frac{1}{2}$ and the X_i are independent). Thus, from a binomial table we obtain

$$P_\theta(X_{(4)} > \eta) = P(12 \le Y \le 15) = 0.0176.$$

By symmetry, it is clear that $P_\theta(X_{(12)} < \eta) = 0.0176$, so that $R(\theta, \delta) = 0.0352$, $r(\pi, \delta) = 0.0352$ for *all* π, and $r_\Gamma(\delta) = 0.0352$ for any Γ. (Note that δ is a 96.48% confidence procedure.)

The main interest in $r_\Gamma(\delta)$ (from a conditional Bayesian perspective) stems from the following relationship, obtained as Result 2 in Subsection 4.4.1:

$$r(\pi, \delta) = E^m[\rho(\pi(\theta|X), \delta(X))]. \tag{4.94}$$

To understand the value of this relationship, suppose that $L(\theta, a) \ge 0$ (so that $\rho(\pi(\theta|x), \delta(x)) \ge 0$) and that $r_\Gamma(\delta)$ is small. Then, since (4.94) and Definition 12 imply that

$$E^m[\rho(\pi(\theta|X), \delta(X))] \le r_\Gamma(\delta) \tag{4.95}$$

for all $\pi \in \Gamma$, it must be the case that $\rho(\pi(\theta|X), \delta(X))$ is small *with high probability* (with respect to m). This knowledge would be of greatest use in situations where it is impossible to perform a posterior robustness investigation (involving $\rho(\pi(\theta|x), \delta(x))$) for the actual observed x. Small $r_\Gamma(\delta)$ would nevertheless make it likely that $\rho(\pi(\theta|x), \delta(x))$ actually is small for all $\pi \in \Gamma$. There is no guarantee of this, of course (the observed x could be "unlucky"), but small $r_\Gamma(\delta)$ would certainly be comforting.

Example 28 provides a good illustration of this. Because of the complexity of Θ, it is enormously difficult to do *any* Bayesian analysis whatsoever, much less a conclusive posterior robustness study. And yet $r_\Gamma(\delta) = 0.0352$ is so small that it seems highly likely that $\rho(\pi(\theta|x), \delta(x))$ is small for our "true" prior, and that we can hence place substantial credibility in $\delta(x)$.

It is important not to read too much into the above argument. We are *not* assigning primary meaning to the measures $r(\pi, \delta)$ or $r_\Gamma(\delta)$; posterior expected loss is still the acknowledged criteria. If, however, we cannot directly determine the posterior expected loss, or check posterior robustness, use of $r_\Gamma(\delta)$ as above may be helpful. Of course, the more actual posterior Bayesian calculations done, the better. Since small $r_\Gamma(\delta)$ does not guarantee that $\rho(\pi(\theta|x), \delta(x))$ is small for the observed x, it would, at a minimum, be a good idea to calculate $\rho(\pi(\theta|x), \delta(x))$ for *some* reasonable π, in an attempt to detect a conditioning problem. Even this may be too difficult in situations such as Example 28, however.

There are other justifications that can be given for looking at $r_\Gamma(\delta)$ as a secondary measure. For instance, it seems unlikely that it will ever be possible to train most nonstatisticians to do careful robustness studies in their statistical analyses. Such users will have to be provided with methods of analyses that are inherently robust. Consider the situation of Example 18, for instance. We may reasonably ask the user to specify his prior median and quartiles, and then fit a prior to these specifications. Assuming a bell-shaped form is deemed appropriate, it seems clear that *automatic* use of π_C is preferable to *automatic* use of π_N. The point here is that, if we are talking about repeatedly using a procedure, then its frequentist properties become of interest (though still, perhaps, of only secondary interest). Thus, in Example 18, the fact that $r_\Gamma(\delta^C)=0.746$, while $r_\Gamma(\delta^N)=\infty$ (for $\Gamma=\{\pi_C, \pi_N\}$), indicates that δ^C will perform much better in repeated use than will δ^N. And recall that it takes a fairly sophisticated posterior analysis (involving consideration of $m(x|\pi)$) to recognize the superiority of δ^C.

To many statisticians, of course, frequentist justification is important in a primary sense. Some such views, together with extensive discussion of many of the issues raised here, can be found in the Dicussion of Berger (1984a). Another reference (with Discussion) is Morris (1983a) (see also Morris (1983b)). Indeed, Morris recommends adopting $r_\Gamma(\delta)$ as the *primary* accuracy measure; in his (empirical Bayes) setting, Γ is the (usually conjugate) class of priors known up to hyperparameters, δ is an empirical Bayes estimator or confidence procedure, and the losses considered are, say, sum of squares error or "0–1," respectively. Thus, for confidence procedures,

$$r_\Gamma(\delta)=1-\inf_{\pi\in\Gamma} E^\pi[P_\theta(\delta(X) \text{ contains } \theta)].$$

If this is small, we know that $\delta(X)$ will tend to contain θ in repeated use (with θ varying according to π). Morris calls this "empirical Bayes confidence." Among the other excellent references concerning Bayesian justification of use of frequentist measures, are Pratt (1965), Good (1976, 1983), Rubin (1984), and Berger (1984e). Other references to use of frequentist risk concepts in Bayesian robustness can be found in Martz and Waller (1982) and Berger (1984a). In Subsections 4.7.6 and 4.7.7 some specific frequentist approaches to Bayesian robustness will be considered.

4.7.6. Gamma-Minimax Approach

A theoretically popular frequentist Bayes approach to robustness is to choose δ to minimize $r_\Gamma(\delta)$. It may be the case that the minimum is attained by a randomized rule (see Section 1.4), so we will use δ^* to denote a decision rule in the following definitions (and $\mathscr{D}^*$ to denote the set of all finite risk randomized decision rules). The basic concepts in this subsection were developed in Robbins (1951, 1964) and Good (1952). Other general discussions can be found in Menges (1966), Blum and Rosenblatt (1967), Kudo (1967), and Berliner (1984b).

Γ-minimax Principle. *A decision rule* δ_1^* *is preferred to a rule* δ_2^* *if*

$$r_\Gamma(\delta_1^*) < r_\Gamma(\delta_2^*).$$

Definition 13. The Γ-*minimax value* of the problem is defined to be

$$r_\Gamma = \inf_{\delta^* \in \mathscr{D}^*} r_\Gamma(\delta^*) = \inf_{\delta^* \in \mathscr{D}^*} \sup_{\pi \in \Gamma} r(\pi, \delta^*).$$

A rule δ^* is said to be Γ-*minimax* if $r_\Gamma(\delta^*) = r_\Gamma$.

These ideas are very related to the minimax principle discussed in Section 1.5. Indeed, if Γ is chosen to be the set of *all* priors, it can be shown that

$$\sup_{\pi \in \Gamma} r(\pi, \delta^*) = \sup_{\theta \in \Theta} R(\theta, \delta^*). \tag{4.96}$$

(The proof is left as an exercise.) In this situation, Γ-minimaxity clearly corresponds exactly with usual minimaxity. At the other extreme, if Γ is chosen to contain only one prior distribution, π_0, then the Γ-minimax principle corresponds to the Bayes risk principle. Reasonable choices of Γ fall somewhere between these two extremes.

EXAMPLE 29. Assume $X \sim \mathcal{N}(\theta, 1)$, and that it is desired to estimate θ under squared-error loss. Since the loss is convex, it follows from Theorem 3 of Section 1.8 that only nonrandomized rules need be considered. Assume it is felt that the prior has mean μ and variance τ^2, but that otherwise nothing is known. It is then reasonable to let

$$\Gamma = \{\pi \colon E^\pi(\theta) = \mu \text{ and } E^\pi(\theta - \mu)^2 = \tau^2\}.$$

Consider the estimator

$$\delta^N(x) = \frac{\tau^2}{1+\tau^2} x + \frac{1}{1+\tau^2} \mu.$$

Clearly

$$\begin{aligned} R(\theta, \delta^N) &= E_\theta \left(\frac{\tau^2}{1+\tau^2} X + \frac{1}{1+\tau^2} \mu - \theta \right)^2 \\ &= E_\theta \left(\frac{\tau^2}{1+\tau^2} [X - \theta] + \frac{1}{1+\tau^2} [\mu - \theta] \right)^2 \\ &= \frac{\tau^4}{(1+\tau^2)^2} E_\theta [X - \theta]^2 + \frac{1}{(1+\tau^2)^2} [\mu - \theta]^2 \\ &= \frac{\tau^4}{(1+\tau^2)^2} + \frac{[\mu - \theta]^2}{(1+\tau^2)^2}. \end{aligned}$$

Hence, for any $\pi \in \Gamma$,

$$r(\pi, \delta^N) = E^\pi[R(\theta, \delta^N)] = \frac{\tau^4}{(1+\tau^2)^2} + \frac{\tau^2}{(1+\tau^2)^2} = \frac{\tau^2}{1+\tau^2}. \tag{4.97}$$

Note, however, that δ^N is the Bayes rule with respect to π_N, the $\mathcal{N}(\mu, \tau^2)$ prior. Hence for *any* rule δ,

$$r(\pi_N, \delta) \geq r(\pi_N, \delta^N) = \frac{\tau^2}{1+\tau^2}.$$

This implies that

$$\sup_{\pi \in \Gamma} r(\pi, \delta) \geq r(\pi_N, \delta) \geq \frac{\tau^2}{1+\tau^2}.$$

Combining this with (4.97), it can be concluded that, for all rules δ,

$$\sup_{\pi \in \Gamma} r(\pi, \delta) \geq \frac{\tau^2}{1+\tau^2} = \sup_{\pi \in \Gamma} r(\pi, \delta^N).$$

Thus δ^N is Γ-minimax. (This result is a special case of a theorem in Jackson, Donovan, Zimmer, and Deely (1970).)

The above example has the unfortunate feature of dealing with a class, Γ, of priors that involves moment specifications. While prior information may sometimes be available in the form of moments (particularly when dealing with physical systems), it was pointed out in Chapter 3 that specification of fractiles and shape features of the prior is all that is feasible for most problems. The bulk of the literature deals either with Γ as in Example 29, or with Γ consisting of priors of a given functional form. This literature is too large to review here. See Berger (1984a) for references.

Recently, there have been several Γ-minimax analyses dealing with more realistic class of priors. One approach, taken in Rummel (1983) and Lehn and Rummel (1985), deals with estimation problems with Γ being, more or less, the set of priors with specified probabilities of certain sets. Efficient numerical methods are found, in these papers, for calculating the Γ-minimax rule. (Note that no general such methods are known; indeed it is often very difficult to come up with *any* method of calculating a Γ-minimax rule—see also Chapter 5.) Results in selection problems with such classes of priors can be found in R. Berger (1979) and Gupta and Hsiao (1981). There has also been progress in dealing with the ε-contamination class of priors, because then

$$r_\Gamma(\delta^*) = (1-\varepsilon) r(\pi_0, \delta^*) + \varepsilon \sup_{q \in \mathcal{Q}} r(q, \delta^*).$$

If, in fact, $\mathcal{Q} = \{\text{all distributions}\}$ (or merely a sufficiently large class of distributions, see Exercise 98), then (4.96) implies that

$$r_\Gamma(\delta^*) = (1-\varepsilon) r(\pi_0, \delta^*) + \varepsilon \sup_{\theta} R(\theta, \delta^*). \tag{4.98}$$

This convenient form often allows (at least approximate) minimization of $r_\Gamma(\delta^*)$ over all δ^*. An example is given in the next subsection. Some recent references include Marazzi (1985), Berger (1982b, c), Bickel (1984), and Berger and Berliner (1984).

One criticism of the Γ-minimax principle is that the quantity $r_\Gamma(\delta^*)$ is unduly responsive to unfavorable π. To be more precise, if there is a prior π_0 for which $r(\pi_0)$ (the Bayes risk of π_0) is exceptionally large, a Γ-minimax rule will tend to be Bayes (or nearly Bayes) with respect to π_0. (This will certainly happen, for example, if

$$r_\Gamma(\delta^*) = \sup_{\pi \in \Gamma} r(\pi, \delta^*) = r(\pi_0, \delta^*)$$

for all rules δ^*.) This is unreasonable in that, if π_0 is the true prior, one can *at best* suffer the risk $r(\pi_0)$. It is more sensible when looking at $r(\pi_0, \delta^*)$, therefore, to see how much worse it is than $r(\pi_0)$. In other words, a more natural quantity to consider is $r(\pi_0, \delta^*) - r(\pi_0)$. (An extensive discussion of the similar idea of regret loss is given in Chapter 5.)

Definition 14. Define the Γ-*minimax regret risk* of a decision rule δ^* to be

$$r_\Gamma^*(\delta^*) = \sup_{\pi \in \Gamma} [r(\pi, \delta^*) - r(\pi)].$$

Γ-minimax Regret Principle. *A decision rule δ_1^* is preferred to a rule δ_2^* if $r_\Gamma^*(\delta_1^*) < r_\Gamma^*(\delta_2^*)$.*

Definition 15. The Γ-*minimax regret value* of the problem is defined to be

$$r_\Gamma^* = \inf_{\delta^* \in \mathscr{D}^*} r_\Gamma^*(\delta^*) = \inf_{\delta^* \in \mathscr{D}^*} \sup_{\pi \in \Gamma} [r(\pi, \delta^*) - r(\pi)].$$

A rule δ^* is said to be a Γ-*minimax regret rule* if $r_\Gamma^*(\delta^*) = r_\Gamma^*$.

Examples of application of this principle can be found in Albert (1983, 1984).

When a Γ-minimax or Γ-minimax regret analysis can be carried out, the answer is usually quite attractive. The Γ-minimax rule is usually a Bayes rule that is compatible with the specified prior information, and it is usually robust (in some sense) over Γ. To a Bayesian, of course, Γ-minimaxity would be only an uncertain substitute for posterior robustness.

4.7.7. Uses of the Risk Function

The formal calculation of $r_\Gamma(\delta)$ might be difficult, partly because of the need to worry about specification of Γ. Sometimes, therefore, it might just be easier to look at the risk function, $R(\theta, \delta)$, and process it intuitively.

EXAMPLE 18 (continued). The risk functions of the two Bayes rules δ^N and δ^C are plotted in Figure 4.2. It is immediately clear that $r_\Gamma(\delta^C)$ is never much more than 1, while $r_\Gamma(\delta^N)$ is likely to be very bad. Thus, without any formal analysis, use of δ^C seems indicated.

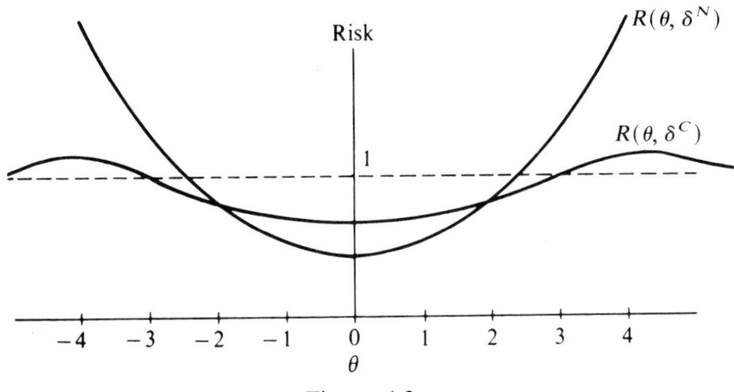

Figure 4.2

A more formal use of the risk function is the *restricted risk Bayes* approach of Hodges and Lehmann (1952). The idea is to restrict consideration to decision rules, δ^*, from the class

$$\mathscr{D}_C^* = \{\delta^*: R(\theta, \delta^*) \le C \text{ for all } \theta \in \Theta\}, \tag{4.99}$$

and then choose a $\tilde{\delta}^* \in \mathscr{D}_C^*$ which has small Bayes risk with respect to an elicited prior π_0. The rationale is that any $\delta^* \in \mathscr{D}_C^*$ will have $r_\Gamma(\delta^*) \le C$, ensuring a degree of Bayes risk robustness, and the selected $\tilde{\delta}$ will hopefully also have respectable Bayesian performance for priors near π_0.

EXAMPLE 30 (Efron and Morris (1971)). Suppose that $X \sim \mathcal{N}(\theta, \sigma^2)$ (σ^2 known), that π_0 is $\mathcal{N}(\mu, \tau^2)$, and that it is desired to estimate θ under squared-error loss. Consider the *limited translation* estimator

$$\delta^M(x) = \begin{cases} x - [\sigma^2/(\sigma^2+\tau^2)](x-\mu) & \text{if } (x-\mu)^2 < M(\sigma^2+\tau^2), \\ x - (\text{sgn}[x-\mu])[M\sigma^4/(\sigma^2+\tau^2)]^{1/2} & \text{otherwise.} \end{cases}$$

Note that this estimator is the Bayes rule with respect to π_0 when x is near μ, but is merely x shifted (towards μ) by a constant when x is far from μ. It can be shown (see Efron and Morris (1971)) that

$$R(\theta, \delta^M) = E_\theta(\theta - \delta^M(X))^2 \le \sigma^2 + \frac{\sigma^4 M}{(\sigma^2+\tau^2)}. \tag{4.100}$$

Note that the usual estimator, $\delta^0(x) = x$, has $R(\theta, \delta^0) \equiv \sigma^2$, so that M controls the excess risk (over δ^0) that would be allowable. (These bounds are directly translatable into $r_\Gamma(\delta)$, and hence into bounds on Bayes risk robustness.) Efron and Morris (1971) also show that

$$r(\pi_0, \delta^M) = \frac{\sigma^2\tau^2}{(\sigma^2+\tau^2)}\left(1 + \frac{\sigma^2}{\tau^2} H(M)\right), \tag{4.101}$$

where $H(M) = 2[(M+1)(1-\Phi(\sqrt{M})) - \sqrt{M}\phi(\sqrt{M})]$, ϕ and Φ denoting the standard normal density and c.d.f., as usual. Some values of $H(M)$ are given in Table 4.10.

Table 4.10. $H(M)$ For Limited Translation Rules.

M	0	0.02	0.10	0.2	0.4	0.6	1.0	1.4	4.0	∞
$H(M)$	1	0.80	0.58	0.46	0.32	0.24	0.16	0.10	0.0115	0

Note that, as M increases, so does the bound on $R(\theta, \delta^M)$ (and hence the bound on $r_\Gamma(\delta^M)$), while $r(\pi_0, \delta^M)$ decreases. The limiting cases, $M=0$ and $M=\infty$, correspond to use of δ^0 and the Bayes rule, δ^{π_0}, respectively; the former could be called "most robust" (r_Γ is smallest) while the latter exhibits the best performance for π_0.

As a specific example, if $\sigma^2=1$, $\tau^2=2$, and $M=0.6$, then $R(\theta, \delta^{0.6}) \le 1.2$ while $r(\pi_0, \delta^{0.6}) = (\frac{2}{3})(1.12)$. Thus $\delta^{0.6}$ is only 12% from optimal with respect to π_0, and only 20% worse than optimal in terms of the bound on $R(\theta, \delta)$ or $r_\Gamma(\delta)$. (It can be shown—see Chapter 5—that 1 is the smallest obtainable upper bound on $R(\theta, \delta)$.) Thus $\delta^{0.6}$ should be good for the elicited π_0, and is a good bet to also be robust.

EXAMPLE 31. Suppose $p \ge 3$ and $\mathbf{X} = (X_1, \ldots, X_p)^t \sim \mathcal{N}_p(\boldsymbol{\theta}, \sigma^2\mathbf{I})$, and that it is desired to estimate $\boldsymbol{\theta} = (\theta_1, \ldots, \theta_p)^t$ under sum of squares-error loss. Suppose π_0 is $\mathcal{N}_p(\boldsymbol{\mu}, \tau^2\mathbf{I})$, $\boldsymbol{\mu}$ an arbitrary vector. Consider the estimator

$$\boldsymbol{\delta}^S(\mathbf{x}) = \mathbf{x} - \min\left\{\frac{\sigma^2}{\sigma^2+\tau^2}, \frac{2(p-2)\sigma^2}{|\mathbf{x}-\boldsymbol{\mu}|^2}\right\}(\mathbf{x}-\boldsymbol{\mu}). \tag{4.102}$$

It can be shown (see Subsection 5.4.3) that $R(\boldsymbol{\theta}, \boldsymbol{\delta}^S) < p\sigma^2$ (so that $r_\Gamma(\boldsymbol{\delta}^S) \le p\sigma^2$), while $\sup_{\boldsymbol{\theta}} R(\boldsymbol{\theta}, \boldsymbol{\delta}) \ge p\sigma^2$ for *all* $\boldsymbol{\delta}$. Hence $\boldsymbol{\delta}^S$ is optimally Bayes risk robust (at least for $\Gamma = \{\text{all priors}\}$). Also, $\boldsymbol{\delta}^S$ is better than the usual estimator, $\boldsymbol{\delta}^0(\mathbf{x}) = \mathbf{x}$, which has $R(\boldsymbol{\theta}, \boldsymbol{\delta}^0) \equiv p\sigma^2$. The Bayes risk, $r(\pi_0, \boldsymbol{\delta}^S)$, is calculated in Berger (1982b), and is shown to be of the form

$$r(\pi_0, \boldsymbol{\delta}^S) = \frac{p\sigma^2\tau^2}{(\sigma^2+\tau^2)}\left(1 + \frac{\sigma^2}{\tau^2}K(p)\right), \tag{4.103}$$

where $K(p)$ is given in Table 4.11 for various p.

A moment's reflection reveals the astonishing nature of these numbers. The optimal Bayes risk for π_0 is $p\sigma^2\tau^2/(\sigma^2+\tau^2)$, and, since σ^2/τ^2 is usually less than one, $\boldsymbol{\delta}^S$ will only be slightly worse than optimal for π_0. (If $p=5$,

Table 4.11. $K(p)$ For $\boldsymbol{\delta}^S$.

p	3	4	5	6	7	8	9	10	20
$K(p)$	0.296	0.135	0.073	0.043	0.027	0.017	0.012	0.008	0.0004

for instance, the percentage increase in Bayes risk is only $[(7.3)\times(\sigma^2/\tau^2)]\%$.) But we also saw that $\boldsymbol{\delta}^S$ was optimally Bayes risk robust (for $\Gamma=\{\text{all priors}\}$). To be nearly optimal with respect to π_0 and optimally robust against all other priors (in the r_Γ sense) is astonishing. Further discussion of this phenomenon, and generalization to arbitrary multivariate normal situations, can be found in Berger (1982b, 1982c, 1984a), Chen (1983), Berger and Berliner (1984), and Bickel (1984); the development in Subsection 5.4.3 is also related.

It should be reemphasized that, even when a procedure appears excellent from a Bayes risk (and Bayes risk robustness) viewpoint, it still should be checked for conditional soundness. The estimators in Examples 30 and 31 do pass such a check, being the (conditional) Bayes estimates with respect to π_0 when x is near the prior mean, and corresponding to Bayes estimates for flatter tailed versions of π_0 when x is far from the prior mean (cf. Strawderman (1971), Brown (1979), and Berger (1980a)).

Hodges and Lehmann (1952) actually proposed the following formal decision principle:

Restricted Risk Bayes Principle. *Choose a* $\delta^*\in\mathscr{D}_C^*$ *which minimizes* $r(\pi_0,\delta^*)$ *over all* $\delta^*\in\mathscr{D}_C^*$.

Unfortunately, it is generally *very* difficult to determine such a δ^*; furthermore, this "optimal" δ^* is usually extremely messy and difficult to work with. Marazzi (1985) approximately calculates the optimal δ^* for the situation of Example 30. Interestingly, the much simpler δ^M seems to have performance very close to that of the optimal δ^* (see Efron and Morris (1971) and Marazzi (1985)).

Sometimes it is possible to find a functional, $\mathscr{R}$, on $\mathscr{D}_C^*$ such that, for all reasonable δ^*,

$$R(\theta,\delta^*)=E_\theta[\mathscr{R}(\delta^*)(X)]. \tag{4.104}$$

(See Subsection 5.4.2 for illustrations.) Rather than working with $\mathscr{D}_C^*$, one might then want to consider

$$\tilde{\mathscr{D}}_C^*=\{\delta^*: \mathscr{R}(\delta^*)(x)\le C \text{ for all } x\}.$$

Any $\delta^*\in\tilde{\mathscr{D}}_C^*$ clearly has $R(\theta,\delta^*)\le C$ (i.e., $\tilde{\mathscr{D}}_C^*\subset\mathscr{D}_C^*$). The reason for considering this reduced class of decision rules is that the Restricted Risk Bayes Principle may be much easier to apply when $\mathscr{D}_C^*$ is replaced by $\tilde{\mathscr{D}}_C^*$. Indeed, the estimators in Examples 30 and 31 can be obtained in this way. See Berger (1982c) and Chen (1983) for development.

The restricted risk Bayes approach seems less desirable (from the Bayesian perspective) than the Γ-minimax approach, in that it effectively demands Bayes risk robustness over all priors ($R(\theta,\delta)\le C\Rightarrow r(\pi,\delta)\le C$ for *all* π), while the Γ-minimax approach only demands robustness over the "sensible" priors in Γ. For instance, if one wants to do well with respect to π_0, and yet be robust to reasonable departures from π_0, the Γ-minimax approach, with Γ being the ε-contamination class, would seem ideal (from a Bayes risk perspective, of course). It is interesting that the two problems often turn out to be *equivalent*, in the sense that a monotonic function g can be found for which, if δ^{ε} is the solution of the Γ-minimax problem (for appropriate $\mathcal{Q}$ in the ε-contamination class), then δ^{ε} is also the solution to the restricted risk Bayes problem for $C = g(\varepsilon)$. Examples 30 and 31 are situations where this equivalence holds (cf. Marazzi (1985) and Bickel (1984)). Thus δ^M and $\boldsymbol{\delta}^S$ can also be considered to be approximate solutions to the Γ-minimax problem. (Actually, $\boldsymbol{\delta}^S$ is the approximate solution to the Γ-minimax problem with $\varepsilon = 1$.)

EXAMPLE 30 (continued). The δ^M are approximate solutions to the Γ-minimax problem, with $\Gamma = \{\pi: \pi = (1-\varepsilon)\pi_0 + \varepsilon q, q \in \mathcal{Q}\}$, providing $\mathcal{Q}$ is large enough so that, for any δ,

$$\sup_{q\in\mathcal{Q}} r(q, \delta) = \sup_{\theta} R(\theta, \delta) \tag{4.105}$$

(see Exercise 98). Thus (4.100) and (4.101) yield

$$\begin{aligned}\sup_{\pi\in\Gamma} r(\pi, \delta^M) &= (1-\varepsilon)r(\pi_0, \delta^M) + \varepsilon \sup_{\theta} R(\theta, \delta^M)\\ &= (1-\varepsilon)\left[\frac{\sigma^2\tau^2}{(\sigma^2+\tau^2)}\left(1+\frac{\sigma^2}{\tau^2}H(M)\right)\right] + \varepsilon\left[\sigma^2 + \frac{\sigma^4}{(\sigma^2+\tau^2)}M\right]\\ &= \frac{\sigma^2\tau^2}{(\sigma^2+\tau^2)}\left(1+\frac{\sigma^2}{\tau^2}[(1-\varepsilon)H(M)+\varepsilon(M+1)]\right).\end{aligned}$$

The optimal M can be found by differentiating with respect to M and setting equal to zero. The resulting equation can be most conveniently written

$$\sqrt{M} = \phi(\sqrt{M})\Big/\left[\frac{\varepsilon}{2(1-\varepsilon)} + 1 - \Phi(\sqrt{M})\right],$$

which can be iteratively solved for $\sqrt{M}$. Table 4.12 presents various values of these quantities (with $\psi(M) = [(1-\varepsilon)H(M) + \varepsilon(M+1)]$). Note that $(\psi(M)\sigma^2/\tau^2)$ reflects the maximum percentage that $r(\pi, \delta^M)$ could be worse than $r(\pi_0, \delta^{\pi_0}) = \sigma^2\tau^2/(\sigma^2+\tau^2)$ for *any* $\pi \in \Gamma$. Thus if $\sigma^2 = 1$, $\tau^2 = 2$, and $\varepsilon = 0.1$, the maximum harm could only be $(0.33/2) = 16\frac{1}{2}\%$. Of course, this

Table 4.12. Γ-Minimax Values.

ε	0.01	0.05	0.1	0.2	0.3	0.4	0.5	0.7	0.9	1.0
M	3.80	1.96	1.30	0.74	0.46	0.30	0.19	0.063	0.006	0
$\psi(M)$	0.06	0.20	0.33	0.51	0.65	0.75	0.83	0.94	0.99	1.0

maximum harm increases as ε increases (there then being greater uncertainty in π_0).

When $\mathcal{Q}$ is *not* large enough for (4.105) to hold, the restricted risk Bayes approach can be a poor approximation to the Γ-minimax approach. Such situations typically arise when Γ consists of priors of known *structural* form, such as priors which are i.i.d. in the coordinates of $\boldsymbol{\theta}$.

4.7.8. Some Robust and Nonrobust Situations

It is reasonably clear that, if the likelihood function for a parameter is quite flat, then the conclusion will be highly dependent on the prior input. Thus one important set of distinctions will be between situations with flat and sharp likelihood functions. In multivariate problems, it will often happen that the likelihood is quite concentrated in some directions and flat in others (especially when multicolinearity is present). It can then be important to identify the "problem" directions so as to concentrate prior elicitation efforts where needed (see also Subsection 4.7.11). Another class of problems, in which flat likelihood functions for some parameters are often encountered, is variance component problems (cf. Hill (1965, 1977, 1980)).

A second consideration concerning prior robustness is whether the "body" or "tail" of the prior provides the operational influence on the conclusion. If the likelihood function is concentrated in the body of the prior, there is less concern about prior robustness, because the body of a prior will usually be comparatively easy to specify. If, however, the likelihood function is concentrated in the tail of the prior, there are real problems; the form of the tail is very hard to specify, and different choices can drastically change the conclusion, as we have seen. In this subsection, we briefly run through several situations involving these considerations.

Stable Estimation

Common sense would argue that, in the presence of overwhelmingly accurate sample information, the prior will have little or no effect. There will then automatically be considerable robustness with respect to the prior. A more formal statement is that if the likelihood function, $l(\theta)=f(x|\theta)$, is very sharply concentrated in a small region Ω, over which $\pi(\theta)$ is essentially constant (which should be the case if Ω is small enough), then

$$\pi(\theta|x)\cong\frac{l(\theta)}{\int_\Omega l(\theta)d\theta}\quad\text{for }\theta\in\Omega. \tag{4.106}$$

Hence the prior will have little impact on the conclusion, and we have almost complete prior robustness. This situation is called *stable estimation*, and is often emphasized in the Bayesian literature (cf. Edwards, Lindman, and Savage (1963)).

In situations of stable estimation, it will also typically be the case that $\pi(\theta|x)$ is approximately a normal distribution.

Result 8. *Suppose that* $X_1, X_2, \ldots, X_n$ *are* i.i.d. *from the density* $f_0(x_i|\boldsymbol{\theta})$, $\boldsymbol{\theta} = (\theta_1, \ldots, \theta_p)^t$ *being an unknown vector of parameters.* (*We will write* $\mathbf{x} = (x_1, \ldots, x_n)^t$ *and* $f(\mathbf{x}|\boldsymbol{\theta}) = \prod_{i=1}^n f_0(x_i|\boldsymbol{\theta})$, *as usual.*) *Suppose* $\pi(\boldsymbol{\theta})$ *is a prior density, and that* $\pi(\boldsymbol{\theta})$ *and* $f(\mathbf{x}|\boldsymbol{\theta})$ *are positive and twice differentiable near* $\hat{\boldsymbol{\theta}}$, *the* (*assumed to exist*) *maximum likelihood estimate of* $\boldsymbol{\theta}$. *Then, for large n and under commonly satisfied assumptions, the posterior density,*

$$\pi_n(\boldsymbol{\theta}|\mathbf{x}) = \frac{f(\mathbf{x}|\boldsymbol{\theta})\pi(\boldsymbol{\theta})}{m(\mathbf{x})}, \tag{4.107}$$

can be approximated in the following four ways:

(i) π_n *is approximately* $\mathcal{N}_p(\boldsymbol{\mu}^\pi(\mathbf{x}), \mathbf{V}^\pi(\mathbf{x}))$, *where* $\boldsymbol{\mu}^\pi$ *and* $\mathbf{V}^\pi$ *are the posterior mean and covariance matrix*;

(ii) π_n *is approximately* $\mathcal{N}_p(\hat{\boldsymbol{\theta}}^\pi, [\mathbf{I}^\pi(\mathbf{x})]^{-1})$, *where* $\hat{\boldsymbol{\theta}}^\pi$ *is the generalized maximum likelihood estimate for* $\boldsymbol{\theta}$ (*i.e., the* m.l.e. *for* $l^*(\boldsymbol{\theta}) = f(\mathbf{x}|\boldsymbol{\theta})\pi(\boldsymbol{\theta})$), *and* $\mathbf{I}^\pi(\mathbf{x})$ *is the* ($p \times p$) *matrix having* (i, j) *element*

$$I_{i,j}^\pi(\mathbf{x}) = -\left[\frac{\partial^2}{\partial\theta_i\,\partial\theta_j}\log(f(\mathbf{x}|\boldsymbol{\theta})\pi(\boldsymbol{\theta}))\right]_{\boldsymbol{\theta}=\hat{\boldsymbol{\theta}}^\pi};$$

(iii) π_n *is approximately* $\mathcal{N}_p(\hat{\boldsymbol{\theta}}, [\hat{\mathbf{I}}(\mathbf{x})]^{-1})$, *where* $\hat{\mathbf{I}}(\mathbf{x})$ *is the* observed (*or* conditional) Fisher information matrix, *having* (i, j) *element*

$$\begin{aligned}\hat{I}_{ij}(\mathbf{x}) &= -\left[\frac{\partial^2}{\partial\theta_i\,\partial\theta_j}\log f(\mathbf{x}|\boldsymbol{\theta})\right]_{\boldsymbol{\theta}=\hat{\boldsymbol{\theta}}} \\ &= -\sum_{i=1}^n \left[\frac{\partial^2}{\partial\theta_i\,\partial\theta_j}\log f(x_i|\boldsymbol{\theta})\right]_{\boldsymbol{\theta}=\hat{\boldsymbol{\theta}}};\end{aligned} \tag{4.108}$$

(iv) π_n *is approximately* $\mathcal{N}_p(\hat{\boldsymbol{\theta}}, [\mathbf{I}(\hat{\boldsymbol{\theta}})]^{-1})$, *where* $\mathbf{I}(\boldsymbol{\theta})$ *is the* expected Fisher information *matrix with* (i, j) *element*

$$I_{ij}(\boldsymbol{\theta}) = -nE_{\boldsymbol{\theta}}^{X_1}\left[\frac{\partial^2}{\partial\theta_i\,\partial\theta_j}\log f(X_1|\boldsymbol{\theta})\right]. \tag{4.109}$$

PROOF. We give only a heuristic proof (since we have not stated precise conditions), and for simplicity consider only part (iii) and $p = 1$. A Taylors series expansion of $[\log f(\mathbf{x}|\theta)]$ about $\hat{\theta}$ gives, for θ close to $\hat{\theta}$ (so that $\pi(\theta)$ is approximately constant),

$$\begin{aligned}\pi_n(\theta|\mathbf{x}) &= \frac{\exp\{\log f(\mathbf{x}|\theta)\}\pi(\theta)}{\int \exp\{\log f(\mathbf{x}|\theta)\}\pi(\theta)d\theta} \\ &\cong \frac{\exp\{\log f(\mathbf{x}|\hat{\theta}) - \frac{1}{2}(\theta-\hat{\theta})^2\hat{I}(\mathbf{x})\}\pi(\hat{\theta})}{\int \exp\{\log f(\mathbf{x}|\hat{\theta}) - \frac{1}{2}(\theta-\hat{\theta})^2\hat{I}(\mathbf{x})\}\pi(\hat{\theta})d\theta} \\ &= \frac{\exp\{-\frac{1}{2}(\theta-\hat{\theta})^2\hat{I}(\mathbf{x})\}}{[2\pi(\hat{I}(\mathbf{x}))^{-1}]^{1/2}}.\end{aligned}$$

It can also be shown that only θ near $\hat{\theta}$ contribute significantly to the posterior. This essentially yields part (iii) of the result. The other parts follow from arguments that

$$\mu^{\pi}(\mathbf{x}) \cong \hat{\theta}^{\pi} \cong \hat{\theta} \quad \text{and} \quad V^{\pi}(\mathbf{x}) \cong [I^{\pi}(\mathbf{x})]^{-1} \cong [\hat{I}(\mathbf{x})]^{-1} \cong [I(\hat{\theta})]^{-1}. \qquad \square$$

Results of the above type are Bayesian versions of the central limit theorem. Explicit results, with conditions and other references, can be found in Laplace (1812), Jeffreys (1961), LeCam (1956), Johnson (1967, 1970), Walker (1969), DeGroot (1970), Heyde and Johnstone (1979), Ghosh, Sinha, and Joshi (1982), and Hartigan (1983). Note that it will generally be the case that the approximations in Result 8 are each less accurate than the approximation in the preceding part. Note also that the approximations from i.i.d. observations, and π could be a noninformative prior, yielding "objective" approximate posteriors.

EXAMPLE 10 (continued). Since $X_1, \ldots, X_5$ are i.i.d. $\mathscr{C}(\theta, 1)$ and $\pi(\theta) \equiv 1$,

$$I^{\pi}(\mathbf{x}) = \hat{I}(\mathbf{x}) = -\sum_{i=1}^{n} \left[\frac{d^2}{d\theta^2} \log\{1 + (\theta - x_i)^2\}^{-1} \right]_{\theta = \hat{\theta}}$$

$$= 2 \sum_{i=1}^{n} \frac{[1 - (\hat{\theta} - x_i)^2]}{[1 + (\hat{\theta} - x_i)^2]^2}.$$

For the data $\mathbf{x} = (4.0, 5.5, 7.5, 4.5, 3.0)$, calculation gives $\hat{\theta} = 4.45$ and $I^{\pi}(\mathbf{x}) = \hat{I}(\mathbf{x}) = 2.66$. Thus the approximation, via parts (ii) and (iii) of Result 8, is that π_n is $\mathscr{N}(4.45, 0.38)$. (The very accurate part (i) approximation was $\mathscr{N}(4.55, 0.56)$.)

For the part (iv) approximation, note that

$$I(\theta) = -5E_{\theta}^{X_1} \left[\frac{d^2}{d\theta^2} \log\{1 + (X_1 - \theta)^2\}^{-1} \right]$$

$$= 10E_{\theta}^{X_1} \left[\frac{\{1 - (X_1 - \theta)^2\}}{\{1 + (X_1 - \theta)^2\}^2} \right]$$

$$= 10E_{\theta}^{X_1} \left[\frac{2}{\{1 + (X_1 - \theta)^2\}^2} - \frac{1}{\{1 + (X_1 - \theta)^2\}} \right]$$

$$= \frac{10}{\pi} \int_{-\infty}^{\infty} \left[\frac{2}{(1 + x^2)^3} - \frac{1}{(1 + x^2)^2} \right] dx$$

$$= \frac{10}{\pi} \left[\frac{6\pi}{8} - \frac{\pi}{2} \right] = 2.5.$$

Hence this would yield a $\mathscr{N}(4.45, 0.40)$ approximate posterior. Note that the approximations from parts (ii), (iii), and (iv) are essentially the same here, but are not exceptionally accurate because of the small n. Note the remarkable success of the part (i) approximation, in comparison.

As with all posterior robustness considerations, verification that stable estimation obtains will often be data dependent. Consider the following simple illustration.

EXAMPLE 32. Suppose $X_1, \ldots, X_n$ are i.i.d. $\mathcal{N}(\theta, 1)$ and π is $\mathcal{N}(0, 1)$. Then $\pi_n(\theta \mid \mathbf{x}=(x_1, \ldots, x_n))$ is $\mathcal{N}(n\bar{x}/(1+n), 1/(1+n))$. The asymptotic approximations to the posterior, from parts (iii) and (iv) of Result 8, are both $\mathcal{N}(\bar{x}, 1/n)$. Whether or not this is a good approximation to the actual posterior, however, depends on the magnitude of $|\bar{x}|/(1+n)$ (the difference between the actual and approximate posterior means). Of course, for large n it will be rare to observe an $\bar{X}$ for which $|\bar{X}|/(1+n)$ is not small (since $m(\bar{x})$ is $\mathcal{N}(0, (n+1)/n)$). Indeed, one can generally show that "with high probability with respect to the marginal distribution of $\mathbf{X}$," the approximations to the posterior in parts (iii) and (iv) of Result 8 will be quite accurate for large n.

The above examples emphasize the advantage of using the $\mathcal{N}_p(\boldsymbol{\mu}^\pi(\mathbf{x}), \mathbf{V}^\pi(\mathbf{x}))$ approximation to the posterior. It will be essentially equivalent to the other asymptotic approximations for very large samples, and will usually be a much better approximation for small samples. Of course, it can be more difficult to calculate.

There is also a large literature on "asymptotic Bayes risk" approximations. See Berger (1984a) for references.

Estimation

In estimation problems, posterior robustness will typically exist when the likelihood function is concentrated in the central portion of the prior. If the likelihood function is concentrated in the tail of the prior, posterior robustness is usually not obtainable. In this latter situation, Bayes risk robustness is typically achieved by flat-tailed priors. Example 18 illustrates these comments.

Testing and Associated Decision Problems

If one is trying to decide between several hypotheses (or associated actions), there is usually substantial posterior robustness. If the likelihood function is concentrated near the center of the prior, the posterior probabilities of the hypotheses (and the optimal posterior action) will usually be similar for all $\pi \in \Gamma$. Of course, these posterior probabilities may be highly dependent on the choice of this "center." When the likelihood function is concentrated in the tail of the prior, one is usually in the situation of essentially knowing which hypothesis or action is correct (the center of the prior typically being near the boundary between the hypotheses). The actual posterior probabilities may then vary substantially as π ranges over Γ, but they will all imply essentially the same conclusion.

Example 18 (continued). Recall that $X \sim \mathcal{N}(\theta, 1)$ and $\Gamma = \{\pi_C, \pi_N\}$, where π_C is $\mathcal{C}(0, 1)$ and π_N is $\mathcal{N}(0, 2.19)$. Suppose it is desired to test H_0: $\theta \le 0$ versus H_1: $\theta > 0$. If $x = 1$ is observed (the likelihood function is then mainly concentrated in the center of the prior), a normal theory calculation gives (recalling that $\pi_N(\theta|x)$ is $\mathcal{N}((0.687)x, 0.687)$)

$$P^{\pi_N(\theta|1)}(\theta \le 0) = \Phi\left(\frac{(0-0.687)}{\sqrt{0.687}}\right) = 0.204.$$

Numerical integration yields

$$P^{\pi_C(\theta|1)}(\theta \le 0) = \frac{\int_{-\infty}^{0}[1/(1+\theta^2)]\exp\{-\frac{1}{2}(\theta-1)^2\}d\theta}{\int_{-\infty}^{\infty}[1/(1+\theta^2)]\exp\{-\frac{1}{2}(\theta-1)^2\}d\theta} = 0.238.$$

These are reasonably close.

For $x = 6$, on the other hand, where the likelihood function is concentrated in the tail of the prior, calculation gives

$$P^{\pi_N(\theta|6)}(\theta \le 0) = 3.30 \times 10^{-7} \quad \text{and} \quad P^{\pi_C(\theta|6)}(\theta \le 0) = 3.21 \times 10^{-8}.$$

While the evidence is clearly overwhelmingly against H_0, the actual posterior probabilities differ by a factor of 10.

One testing situation in which robustness can be a problem is testing a point null hypothesis with a moderately large sample size, n. When n is small, it is reasonable to approximate a realistically fuzzy point null by a sharp point null (as discussed in Subsection 4.3.3), and specification of the prior probability of H_0 and the prior density under H_1 is reasonably easy. When n is very large, θ will be determined with great accuracy, and there will rarely be a serious problem in deciding what to do. For moderately large n, however, the point null approximation may be inadequate, while the accuracy with which θ is determined may not be great enough to allow one to ignore the prior; the difficulty then is that it will usually be very hard to specify a prior density on the realistic fuzzy H_0. This example is interesting because, when n is small, one will tend to have robustness because the needed prior inputs (probability of H_0 and density on H_1) are relatively easy to specify; when n is very large one will tend to have "stable estimation" robustness; but when n is moderately large there may be a robustness problem (difficult to specify features of the prior being relevant). See Rubin (1971) and Berger (1984a) for further discussion.

Design of Experiments

Optimal Bayesian designs are usually robust with respect to small changes in the prior, such as changes in the tail. This is because design is a pre-experimental undertaking, necessitating use of overall average performance measures (such as frequentist Bayes risk), and these overall averages will tend to be dominated by the contributions from the central portion of the prior. Of course, after taking the data and being faced with the need to make posterior conclusions, robustness may have to be completely reevaluated.

In design with a known sampling model, there may be real technical advantages in thinking of the Bayes risk as $R(\theta, \delta)$ averaged over the prior, rather than as $\rho(\pi(\theta|X), \delta(X))$ averaged over the marginal distribution of X (which is more instinctively appealing to a Bayesian). This is because any (pre-experimental) robustness analysis will involve the uncertain π only at the stage of averaging $R(\theta, \delta)$; the other approach involves the uncertain π both in $\rho(\pi(\theta|X), \delta(X))$ and in the marginal m.

4.7.9. Robust Priors

Because of the difficulty of formal robustness analysis (i.e., of finding the range of a criterion function as π ranges over Γ), it is of interest to develop priors which are inherently robust in some sense. We discuss general methods of doing so here, and give some specific results in Subsection 4.7.10.

Robustness of Natural Conjugate Priors

We have already seen considerable evidence that natural conjugate priors are not automatically robust: they have tails that are typically of the same form as the likelihood function, and will hence remain influential when the likelihood function is concentrated in the (prior) tail. This was seen to sometimes give intuitively unappealing results and bad Bayes risk robustness. We also argued that, since $m(x|\pi)$ will often be exceptionally small in such situations, posterior Bayesian reasoning supports the case against natural conjugate priors. The literature discussing this issue includes Anscombe (1963b), Tiao and Zellner (1964), Dawid (1973), Hill (1974), Dickey (1974), Meeden and Isaacson (1977), Rubin (1977), Berger (1984a), and Berger and Das Gupta (1985).

Of course, if the likelihood function is concentrated in the central portion of the prior, use of a natural conjugate form will usually be reasonably robust. If one can identify that such a robust situation (or any other robust situation discussed in Subsection 4.7.8) exists, then the calculational simplicity of natural conjugate priors makes their use attractive.

There is a large literature which presumes to avoid robustness problems by considering only a restricted class of decision rules, selection from which requires only a few features of the prior distribution (and hence is supposedly robust).

EXAMPLE 33. Suppose $X \sim \mathcal{N}(\theta, \sigma^2)$ (σ^2 known) and that it is desired to estimate θ under squared-error loss. Available is knowledge of the prior mean μ and variance τ^2, and attention is to be restricted to linear estimators $\delta(x) = ax + b\mu$. The Bayes risk of δ is

$$\begin{aligned} r(\pi, \delta) &= E^{\pi} E_{\theta}^{X}(aX + b\mu - \theta)^2 \\ &= E^{\pi} E_{\theta}^{X}(a(X-\theta) + (a-1)(\theta-\mu) + (b+a-1)\mu)^2 \\ &= E^{\pi}(a^2\sigma^2 + [(a-1)(\theta-\mu) + (b+a-1)\mu]^2) \\ &= a^2\sigma^2 + (a-1)^2\tau^2 + (b+a-1)^2\mu^2. \end{aligned}$$

Minimizing this over a and b yields $b = 1 - a$ and $a = \tau^2/(\sigma^2 + \tau^2)$.

The appeal of the approach in the above example is that one apparently avoids assuming a (say) conjugate form for the prior, and can hence claim robustness with respect to this form. There are two problems with this "claim," however. The first is that the assumption that a linear estimator is to be used is *equivalent* (for the situation of Example 33) to assuming a natural conjugate form for the prior. Indeed, this equivalence holds quite generally in exponential families (cf. Diaconis and Ylvisaker (1979)). The second problem is that prior moments, μ and τ^2, are often very difficult to specify (see Chapter 3), and the "linear estimator" approach requires knowledge of the prior moments. In conclusion, therefore, the use of "best linear estimators" is no more robust than is use of natural conjugate priors.

Flat-Tailed Priors

Substantial evidence has been presented to the effect that priors with tails that are flatter than those of the likelihood function tend to be fairly robust. It is thus desirable to develop fairly broad classes of flat-tailed priors for use in "standard" Bayesian analyses. Examples of such developments will be seen in Subsection 4.7.10.

Noninformative Priors

In a sense, use of noninformative priors can claim to be robust, since no subjective prior beliefs are assumed. One can, however, ask if: (i) the statistical conclusion depends on the choice of the noninformative prior; and (ii) the noninformative prior used is a good reflection of actual prior beliefs.

It will generally be the case that the conclusion is not highly dependent on the choice of the noninformative prior. Let us reconsider Example 8 from Section 3.3, for instance.

EXAMPLE 34. Here $X \sim \mathcal{B}(n, \theta)$, and it is desired to estimate θ under squared-error loss. For the noninformative priors $\pi_1(\theta) = 1$, $\pi_2(\theta) = \theta^{-1}(1-\theta)^{-1}$, and $\pi_3(\theta) = \theta^{-1/2}(1-\theta)^{-1/2}$, the posterior means are

$$\mu^{\pi_1}(x) = \frac{\int_0^1 \theta \cdot \theta^x (1-\theta)^{n-x}\, d\theta}{\int_0^1 \theta^x (1-\theta)^{n-x}\, d\theta} = \frac{(x+1)}{(n+2)},$$

$$\mu^{\pi_2}(x) = \frac{\int_0^1 \theta \cdot \theta^x (1-\theta)^{n-x} \theta^{-1}(1-\theta)^{-1}\, d\theta}{\int_0^1 \theta^x (1-\theta)^{n-x} \theta^{-1}(1-\theta)^{-1}\, d\theta} = \frac{x}{n},$$

$$\mu^{\pi_3}(x) = \frac{\int_0^1 \theta \cdot \theta^x (1-\theta)^{n-x} [\theta(1-\theta)]^{-1/2}\, d\theta}{\int_0^1 \theta^x (1-\theta)^{n-x} [\theta(1-\theta)]^{-1/2}\, d\theta} = \frac{x+0.5}{n+1}.$$

These are very close, unless n is small. Of course, for small n they can differ substantially. For instance, when $n = 1$ and $x = 0$, the estimates are $\frac{1}{3}$, 0, and $\frac{1}{4}$, respectively. (Actually, the posterior for π_2 does not exist when x is 0 or n; μ^{π_2} is, however, the generalized Bayes rule under squared-error loss.)

When the choice of noninformative prior does matter, as for small n in the above example, one is usually in the situation wherein subjective beliefs are important and must be considered. This is not really a lack of real robustness, therefore.

The second question concerning noninformative priors is more difficult to answer. "Standard" noninformative priors can, indeed, sometimes be inappropriate representations of vague prior beliefs. Consider the following example, due to Charles Stein.

EXAMPLE 35. Suppose that $\mathbf{X} \sim \mathcal{N}_p(\boldsymbol{\theta}, \mathbf{I})$, and that it is desired to estimate $\eta = \sum_{i=1}^p \theta_i^2$. The natural noninformative prior for $\boldsymbol{\theta}$ is $\pi(\boldsymbol{\theta}) = 1$, for which the posterior expected value of η is (noting that $\pi(\boldsymbol{\theta}|\mathbf{x})$ is $\mathcal{N}_p(\mathbf{x}, \mathbf{I})$)

$$\begin{aligned} \delta^\pi(\mathbf{x}) = E^{\pi(\boldsymbol{\theta}|\mathbf{x})}[\eta] &= \sum_{i=1}^p E^{\pi(\boldsymbol{\theta}|\mathbf{x})}[\theta_i^2] \\ &= \sum_{i=1}^p E^{\pi(\boldsymbol{\theta}|\mathbf{x})}[((\theta_i - x_i) + x_i)^2] \\ &= \sum_{i=1}^p (1 + x_i^2) = p + \sum_{i=1}^p x_i^2. \end{aligned}$$

This seems somewhat counter-intuitive. For instance, as $p \to \infty$ and under mild conditions, the law of large numbers shows that

$$\frac{1}{p} \sum_{i=1}^p x_i^2 \to \frac{1}{p} \sum_{i=1}^p \theta_i^2 + 1,$$

so that

$$\frac{1}{p} \delta^\pi(\mathbf{x}) = 1 + \frac{1}{p} \sum_{i=1}^p x_i^2 \to 1 + \frac{1}{p} \sum_{i=1}^p \theta_i^2 + 1 = 2 + \frac{\eta}{p};$$

this seems to indicate that $\delta^\pi(\mathbf{x})$ substantially *overestimates* η. (Another indication of this problem is that $\delta^\pi(\mathbf{x})$ is seriously inadmissible under squared-error loss, see Section 4.8.)

The reason that $\pi(\boldsymbol{\theta})=1$ seems to give bad results here can be seen by transforming $\pi(\boldsymbol{\theta})$ to a density π^* on η. The result is that $\pi^*(\eta)\propto\eta^{(p-2)/2}$. Thus, by using $\pi(\boldsymbol{\theta})\equiv 1$, one is implicitly assuming that very large η are more likely than small η. In particular, since the likelihood function is concentrated in the area where η is near $\sum_{i=1}^p x_i^2$, it is the part of $\pi^*(\eta)$ near $\sum_{i=1}^p x_i^2$ that is operationally relevant, and the prior bias towards large η effects a pronounced shift upwards in the estimate.

In the above example, we are not arguing that anything is clearly right or wrong (although we would argue that it is wrong to *always* use $\pi(\boldsymbol{\theta})=1$). The point being made is that standard choices of noninformative priors may not always be accurate reflections of true vague beliefs. It *could* be the case that $\pi(\boldsymbol{\theta})=1$ is reasonable in a given realization of Example 35 (i.e., one could imagine situations where one was essentially equally uncertain as to all $\boldsymbol{\theta}$ near the observed $\mathbf{x}$), but very often it may be more natural to assume, say, that η has a uniform prior. In this connection, the noninformative prior analyses of Bernardo (1979b) are very interesting; the analyses differ according to the parameter ($\boldsymbol{\theta}$ or η, say) of interest, and seem to avoid the kind of problem encountered in Example 35. See also Rabena (1983).

In conclusion, noninformative prior Bayesian analyses are not automatically sensible. It is certainly desirable to determine if the choice of noninformative prior has a pronounced effect, and if the prior does approximate prior beliefs (especially in high dimensions). Overall, however, we judge Bayesian noninformative prior analysis to generally be very safe and robust.

Hierarchical Priors

Hierarchical priors are usually robust because they are usually flat-tailed priors. In Example 17 in Section 3.6, for instance, it was shown that the hierarchical prior could be written as a single stage t-prior. This has flatter tails than the normal likelihood function. Indeed, the most natural method of constructing a hierarchical prior is to mix (at the second stage) over hyperparameters of (first stage) natural conjugate priors; this mixing will generally result in tails flatter than natural conjugate prior tails, and hence flatter than the tails of the likelihood function.

The hierarchical analysis may be sensitive to the functional form chosen for the first stage prior, in the sense that a wrong choice could destroy the benefits of hierarchical modelling. Thus, in the examples of Section 4.6, if the θ_i were really generated according to independent Cauchy distributions and p was large, it can be shown that the estimates would tend to collapse back to $\mathbf{X}$, the standard estimate. Efforts at robustifying such estimates against long-tailed first stage priors are discussed, for a particular situation, in the next subsection. One could, of course, simply attempt a hierarchical or empirical Bayes analysis with long-tailed first stage priors, but the calculations can be formidable. (See Gaver (1985) for an example.)

The choice of a form for the second stage (or higher stage) prior seems to have relatively little effect. If using a noninformative second stage prior, recall the warning from Section 4.6 to the effect that choosing a constant noninformative prior may be prudent.

Maximum Entropy Priors

We will merely review the relevant comments from Section 3.4. Maximum entropy priors derived from moment considerations often are natural conjugate priors, and, therefore, may raise concerns about robustness. It should be mentioned that if one *knows* or has accurate *estimates* of the moments, there is less concern than if the moments were simply chosen subjectively. This is because there is then some evidence that the prior is not extremely long tailed.

For maximum entropy priors developed from fractile specifications, the robustness concerns again center about tail behavior. Maximum entropy seeks to assign constant tails to the prior, which may be reasonable for bounded Θ, but is impossible for unbounded Θ. Further development is needed to allow for specification of prior tails in maximum entropy theory (cf. Brockett, Charnes, and Paick (1984)).

ML-II *Priors*

In Section 3.5 we discussed the possibility of selecting $\pi \in \Gamma$ according to the ML-II principle. To what extent is the prior so selected robust? The answer depends on the use of the prior and on the nature of Γ.

We restrict attention here to discussion of the ε-contamination class

$$\Gamma = \{\pi\colon \pi(\theta) = (1-\varepsilon)\pi_0(\theta) + \varepsilon q(\theta),\ q \in \mathcal{Q}\}.$$

For this class, recall that the ML-II prior, $\hat{\pi}$, is simply $\hat{\pi} = (1-\varepsilon)\pi_0 + \varepsilon\hat{q}$, where $\hat{q}$ maximizes $m(x|q)$. The first point to emphasize is that, when $\mathcal{Q} = \{\text{all priors}\}$, $\hat{\pi}$ may very well be *nonrobust.*

EXAMPLE 36. Recall from Example 18 (in Subsection 3.5.4) that, if $\mathcal{Q} = \{\text{all priors}\}$, then $\hat{\pi} = (1-\varepsilon)\pi_0 + \varepsilon\langle\hat{\theta}\rangle$, where $\langle\hat{\theta}\rangle$ is the distribution giving a unit point mass to the m.l.e. $\hat{\theta}$. To clearly see the nonrobustness of $\hat{\pi}$, note from (4.82) and (4.83) that the resulting posterior is

$$\hat{\pi}(\theta|x) = \lambda(x)\pi_0(\theta|x) + [1-\lambda(x)]\langle\hat{\theta}\rangle,$$

where

$$\lambda(x) = \left[1 + \frac{\varepsilon f(x|\hat{\theta})}{(1-\varepsilon)m(x|\pi_0)}\right]^{-1}.$$

Now, as x gets "extreme," it will typically be the case that $f(x|\hat{\theta})/m(x|\pi_0) \to \infty$ (indeed $f(x|\hat{\theta})$ will often stay bounded away from zero, while $m(x|\pi_0) \to 0$). Then $\lambda(x) \to 0$ and $\hat{\pi}(\theta|x) \to \langle\hat{\theta}\rangle$. Thus, for extreme x where robustness is often a problem, $\hat{\pi}$ will be essentially a point mass at the m.l.e. $\hat{\theta}$. This is ridiculous, implying that we become *certain* that $\theta = \hat{\theta}$.

We encountered a similar instance of the nonrobustness of the ML-II prior in Example 15 of Subsection 4.5.4. That ML-II prior also had (estimated) prior variance of zero.

The ML-II approach can fail, as above, when Γ contains unreasonable priors, one of which is selected as $\hat{\pi}$. In Example 36 it will typically be unreasonable to allow a point mass at $\hat{\theta}$, and in Example 15 it will typically be unreasonable to suppose that σ_π^2 can equal zero. The obvious solution is to restrict Γ to contain only reasonable priors.

One fairly reasonable class, when π_0 is unimodal with mode at θ_0, is the ε-contamination class with

$$\mathcal{Q} = \{\text{densities of the form } q(|\theta - \theta_0|),\ q \text{ nonincreasing}\}. \quad (4.110)$$

Thus, only symmetric contaminations of π_0 are allowed, and only those which preserve unimodality of π_0. The point is that the most reasonable contaminations are those that preserve the essential nature of π_0, and one might be quite confident that θ_0 is the most likely value of θ and that values of θ more distant from θ_0 are less likely. (The symmetry assumption on q is not necessarily natural, and is made here for mathematical convenience. For an analysis with the symmetry constraint removed, see Berger and Berliner (1983).)

Theorem 5. *If* $\Theta = (-\infty, \infty)$, $\Gamma = \{\pi\colon \pi = (1-\varepsilon)\pi_0 + \varepsilon q,\ q \in \mathcal{Q}\}$, *and* $\mathcal{Q}$ *is given by* (4.110), *then an* ML-II *prior is* $\hat{\pi} = (1-\varepsilon)\pi_0 + \varepsilon\hat{q}$, *where* $\hat{q}$ *is* $\mathcal{U}(\theta_0 - k, \theta_0 + k)$, *k being chosen to maximize* (*slightly abusing notation*)

$$m(x|k) = \int_{\theta_0 - k}^{\theta_0 + k} \frac{1}{2k} f(x|\theta)\, d\theta \quad (4.111)$$

(*assuming the maximum is attained*).

PROOF (Berger and Sellke (1984)). Any density in $\mathcal{Q}$ can be written as

$$q(|\theta - \theta_0|) = \int_0^\infty \frac{1}{2k} I_{[0,k)}(|\theta - \theta_0|)\, dF(k),$$

where F is some distribution function on $[0, \infty)$. Thus

$$\begin{aligned} m(x|q) &= \int_{-\infty}^{\infty} f(x|\theta)q(|\theta-\theta_0|)d\theta \\ &= \int_{-\infty}^{\infty}\int_{0}^{\infty} f(x|\theta)\frac{1}{2k}I_{[0,k)}(|\theta-\theta_0|)dF(k)d\theta \\ &= \int_{0}^{\infty}\left[\int_{-\infty}^{\infty} f(x|\theta)\frac{1}{2k}I_{[0,k)}(|\theta-\theta_0|)d\theta\right]dF(k) \\ &= \int_{0}^{\infty} m(x|k)dF(k). \end{aligned}$$

This is clearly maximized by choosing F to be a point mass at a k maximizing $m(x|k)$; the corresponding q is indeed the indicated $\hat{q}$. □

EXAMPLE 37. If $X \sim \mathcal{N}(\theta, \sigma^2)$ (σ^2 known), we seek to maximize

$$\begin{aligned} m(x|k) &= \frac{1}{2k}\int_{\theta_0-k}^{\theta_0+k} \frac{1}{(2\pi\sigma^2)^{1/2}}\exp\left\{-\frac{1}{2\sigma^2}(x-\theta)^2\right\}d\theta \\ &= \frac{1}{2k}\left[\Phi\left(\frac{\theta_0+k-x}{\sigma}\right) - \Phi\left(\frac{\theta_0-k-x}{\sigma}\right)\right]. \end{aligned}$$

It can be shown that the maximum is unique and can be found by differentiating with respect to k and setting equal to zero (unless $|x-\theta_0|/\sigma \le 1$, in which case $k=0$ yields the maximum). Since

$$\begin{aligned} \frac{d}{dk}m(x|k) = &-\frac{1}{2k^2}\left[\Phi\left(\frac{\theta_0+k-x}{\sigma}\right) - \Phi\left(\frac{\theta_0-k-x}{\sigma}\right)\right] \\ &+\frac{1}{2k\sigma}\left[\phi\left(\frac{\theta_0-k-x}{\sigma}\right) - \phi\left(\frac{\theta_0-k-x}{\sigma}\right)\right], \end{aligned}$$

the relevant equation is, setting $z = |x-\theta_0|/\sigma$ and $k^* = k/\sigma$,

$$[\Phi(k^*-z) - \Phi(-(k^*+z))] = k^*[\phi(k^*-z) - \phi(-(k^*+z))].$$

A rewriting of this equation, which is useful for iterative calculation, is

$$k^* = z + \left[-2\log\left(\sqrt{2\pi}\left\{\frac{1}{k^*}[\Phi(k^*-z) - \Phi(-(k^*+z))] - \phi(-(k^*+z))\right\}\right)\right]^{1/2}. \tag{4.112}$$

Recall that, for iterative calculation, we plug in a guess for k^* on the right, carry out the calculation to obtain a revised k^*, and iterate until the value stabilizes. (This formula is actually the exact version of (4.25) in Example 11 of Subsection 4.3.3 (k there is k^* here).) Figure 4.3 is a graph of k^* for z between 0 and 4.

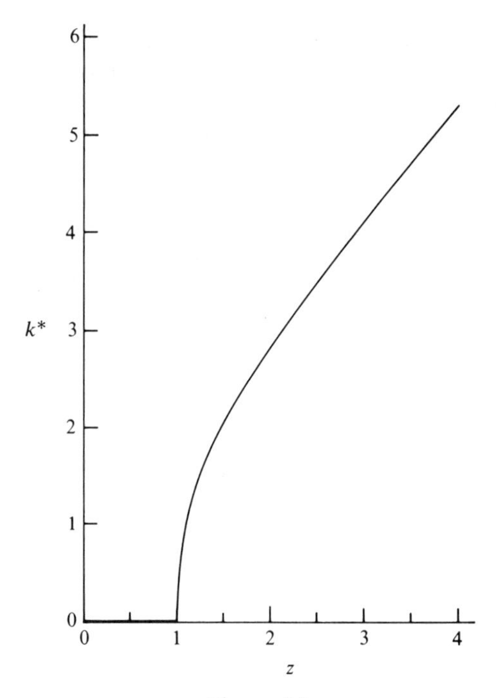

Figure 4.3

As a specific example, suppose $\sigma^2 = 1$, π_0 is $\mathcal{N}(0, 2.19)$ (so $\theta_0 = 0$), $\varepsilon = 0.1$, and $x = 1.96$. Then (4.112) (or Figure 4.3) yields $k^* = 2.78$. Thus

$$\hat{\pi} = (0.9)[2\pi(2.19)]^{-1/2} \exp\left\{\frac{-\theta^2}{2(2.19)}\right\} + (0.1)\frac{1}{2(2.78)} I_{(-2.78,2.78)}(\theta).$$

The posterior is given by (4.82), where (using (4.83))

$$\begin{aligned}\lambda(1.96) &= \left[1 + \frac{(0.1)}{(0.9)} \times \frac{m(1.96|2.78)}{m(1.96|\pi_0)}\right]^{-1} \\ &= \left[1 + \frac{(0.1)}{(0.9)} \times \frac{(0.143)}{(0.122)}\right]^{-1} = 0.885.\end{aligned}$$

Calculations with this posterior are straightforward.

The ML-II prior, arising from Theorem 5, will be robust in various senses, because $\hat{\pi}$ will effectively have a long tail. (The $\mathcal{U}(\theta_0 - k, \theta_0 + k)$ contamination will typically pass through the bulk of the likelihood function, yielding an effective tail for $\hat{\pi}$ of at least $\varepsilon/2k$.) Indeed, as $|x - \theta_0| \to \infty$, it will typically happen that $k \to \infty$ in such a way that $\lambda(x) \to 0$ and (see (4.82)) $\hat{\pi}(\theta|x)$ becomes essentially the posterior for the constant noninformative prior.

From a purely Bayesian perspective, the use of the ML-II prior with a sensible Γ seems likely to be quite robust. At the one extreme, when the likelihood function is in the tails of the priors in Γ, the ML-II approach selects a sensible prior having large $m(x|\pi)$, and this seems likely to be close to a complete (say, hierarchical) Bayesian analysis; such an analysis would naturally give most of the weight to priors having large $m(x|\pi)$ (assuming that all $\pi \in \Gamma$ have roughly equal initial probability, an implicit definition of a "sensible" Γ). At the other extreme, when the likelihood function is in the central portion of the priors in Γ (again, all central portions will be roughly the same if Γ is "sensible"), the Bayesian conclusions will be roughly the same for all $\pi \in \Gamma$; the ML-II prior is then rather trivially robust.

The above discussion explains why even a strict Bayesian need not strongly object to the data-dependent choice of π; the ultimate answer can be considered to be an approximation to a fully Bayesian analysis. Recall, also, why we are *not* performing a fully Bayesian analysis: methods of putting a, say, "uniform" second stage prior on Γ and carrying out a full Bayesian analysis are not known. Because of the Bayesian justification and the use of realistically large Γ, such ML-II priors have great promise in robust Bayesian analysis.

4.7.10. Robust Priors for Normal Means

It behooves Bayesians to provide simple "standardized" Bayesian procedures with built in robustness. While it is certainly desirable to perform specific robustness studies or use complicated robust priors, as outlined in Subsections 4.7.4 through 4.7.9, such will require a fairly high level of sophistication. For routine use by nonexperts, Bayesian procedures are needed which require minimal prior inputs and are, in some sense, inherently robust.

As an illustration of such a development, we consider in this subsection the important case in which $\mathbf{X} \sim \mathcal{N}_p(\boldsymbol{\theta}, \boldsymbol{\Sigma})$, and in which inferences or decisions concerning the unknown $\boldsymbol{\theta} = (\theta_1, \ldots, \theta_p)^t$ are to be made. Such a setup is, of course, the canonical form for many common statistical problems, such as:

(i) $\mathbf{X}^{(1)}, \ldots, \mathbf{X}^{(n)}$ are an i.i.d. sample from a common multivariate normal distribution, in which case $\mathbf{X}$ will be the vector of sample means;

(ii) $\mathbf{X}$ is the least squares estimator,

$$\mathbf{X} = (\mathbf{D}^t\mathbf{D})^{-1}\mathbf{D}^t\mathbf{Y},$$

arising from a regression model,

$$\mathbf{Y} = (Y_1, \ldots, Y_n)^t = \mathbf{D}\boldsymbol{\theta} + \boldsymbol{\varepsilon},$$

where $\mathbf{Y}$ is the vector of observations, $\boldsymbol{\theta}$ is the unknown vector of regression coefficients, $\mathbf{D}$ is a known $(n \times p)$ design matrix of rank $p < n$, and $\boldsymbol{\varepsilon} = (\varepsilon_1, \ldots, \varepsilon_p)^t$ is a $\mathcal{N}_n(0, \sigma^2 \mathbf{I})$ random error; in this case $\boldsymbol{\Sigma} = \sigma^2(\mathbf{D}^t\mathbf{D})^{-1}$.

We will initially assume that $\boldsymbol{\Sigma}$ is known. Later, the case where $\boldsymbol{\Sigma} = \sigma^2 \boldsymbol{\Sigma}_0$, $\boldsymbol{\Sigma}_0$ known but σ^2 unknown (as in the typical regression scenario), will be considered.

The goal here is to develop robust Bayesian procedures based on very simple prior inputs. Perhaps the simplest prior inputs are a "guess," $\boldsymbol{\mu}$, for $\boldsymbol{\theta}$, together with a matrix, $\mathbf{A}$, reflecting the accuracies of these guesses. Because of familiarity with natural conjugate prior analysis, we will call $\boldsymbol{\mu}$ the prior mean and $\mathbf{A}$ the prior covariance matrix. More realistically, however, one should think of $\boldsymbol{\mu}$ and $\mathbf{A}$ as, respectively, the prior median, and a matrix of scale factors derived through consideration of prior fractiles. As an illustration, if the "guess" for θ_1 is 5, and the (subjective) quartiles for θ_1 are 4 and 6, matching a normal prior to the information would yield a prior mean of $\mu_1 = 5$ and prior variance of $A_{11} = 2.19$. We will assume that such a conversion to means and variances has taken place (although we will *not* assume normality of the prior). Another way of thinking of $\boldsymbol{\mu}$ and $\mathbf{A}$ is to imagine the specification of an ellipse,

$$\{\boldsymbol{\theta}: (\boldsymbol{\theta} - \boldsymbol{\mu})^t \mathbf{A}^{-1} (\boldsymbol{\theta} - \boldsymbol{\mu}) \leq p - 0.6\},$$

which is believed to have a 50% chance of containing $\boldsymbol{\theta}$ (such an ellipse having probability of roughly 0.5 under a $\mathcal{N}_p(\boldsymbol{\mu}, \mathbf{A})$ prior).

In this section we are *not* envisaging an empirical or hierarchical Bayes situation, where the θ_i are thought to arise in some fashion from a common distribution; we are merely trying to robustify the subjective conjugate prior analysis given in Example 9 of Subsection 4.3.1. The type of situation to imagine is the regression situation, with the coordinates of $\boldsymbol{\theta}$ being quite different in nature.

The inputs, $\boldsymbol{\mu}$ and $\mathbf{A}$, concerning the location of $\boldsymbol{\theta}$, will be utilized through a symmetric (about $\boldsymbol{\mu}$) flat-tailed prior, the tail chosen to yield robustness. (If symmetry about $\boldsymbol{\mu}$ is not deemed to be a reasonable assumption, the ensuing analysis is, of course, suspect.) The particular robust prior that will be used is actually a hierarchical prior; by such a choice we can obtain robustness while keeping the calculations relatively simple. This prior, to be denoted by $\pi_p(\theta)$, can be defined in two stages as follows:

(i) Given λ, $\boldsymbol{\theta}$ is $\mathcal{N}_p(\boldsymbol{\mu}, \mathbf{B}(\lambda))$, where

$$\mathbf{B}(\lambda) = \rho\lambda^{-1}(\boldsymbol{\Sigma} + \mathbf{A}) - \boldsymbol{\Sigma}, \quad \text{and} \quad \rho = \frac{(p+1)}{(p+3)};$$

(ii) λ has density $\pi_2(\lambda) = \frac{1}{2}\lambda^{-1/2}$ on $(0, 1)$.

This prior is similar to one used in Berger (1980b) (see also Strawderman

(1971)). The rather odd looking choice of $\mathbf{B}(\lambda)$ substantially simplifies the calculations, and will be argued to yield very reasonable robust Bayesian procedures whenever

$$\operatorname{ch}_{\max}(\mathbf{A}^{-1}\mathbf{\Sigma}) \le \tfrac{1}{2}(p+1), \tag{4.113}$$

this being the condition needed for π_p to be a proper prior (here $\operatorname{ch}_{\max}$ denotes maximum characteristic root). The following lemma presents the key calculational results concerning the posterior distribution $\pi_p(\boldsymbol{\theta}|\mathbf{x})$.

Lemma 3. *The posterior mean and covariance matrix for* $\pi_p(\boldsymbol{\theta}|\mathbf{x})$ *are given, respectively, by*

$$\boldsymbol{\mu}^{\pi_p}(\mathbf{x}) = \mathbf{x} - h_p(\|\mathbf{x}\|^2)\mathbf{\Sigma}(\mathbf{\Sigma}+\mathbf{A})^{-1}(\mathbf{x}-\boldsymbol{\mu}) \tag{4.114}$$

and

$$\mathbf{V}^{\pi_p}(\mathbf{x}) = \mathbf{\Sigma} - h_p(\|\mathbf{x}\|^2)\mathbf{\Sigma}(\mathbf{\Sigma}+\mathbf{A})^{-1}\mathbf{\Sigma} + g_p(\|\mathbf{x}\|^2) \times \mathbf{\Sigma}(\mathbf{\Sigma}+\mathbf{A})^{-1}(\mathbf{x}-\boldsymbol{\mu})(\mathbf{x}-\boldsymbol{\mu})^t(\mathbf{\Sigma}+\mathbf{A})^{-1}\mathbf{\Sigma}, \tag{4.115}$$

where $\|\mathbf{x}\|^2 = (\mathbf{x}-\boldsymbol{\mu})^t(\mathbf{\Sigma}+\mathbf{A})^{-1}(\mathbf{x}-\boldsymbol{\mu})$,

$$h_p(\|\mathbf{x}\|^2) = \frac{(p+1)}{\|\mathbf{x}\|^2}\left[1 - H_{p+1}\left(\frac{\|\mathbf{x}\|^2}{2\rho}\right)\right],$$

$$g_p(\|\mathbf{x}\|^2) = \frac{2(p+1)}{\|\mathbf{x}\|^4}\left[1 + \left\{\frac{\|\mathbf{x}\|^2}{2\rho}(\rho h_p(\|\mathbf{x}\|^2)-1)-1\right\} H_{p+1}\left(\frac{\|\mathbf{x}\|^2}{2\rho}\right)\right],$$

$\rho = (p+1)/(p+3)$, *and* H_{p+1} *is defined in Appendix* 2. *Also, the marginal density of* $\mathbf{X}$ *is given by*

$$m(\mathbf{x}|\pi_p) = \left[(p+1)(2\pi\rho)^{p/2}\{\det(\mathbf{\Sigma}+\mathbf{A})\}^{1/2} \times \exp\left\{\frac{\|\mathbf{x}\|^2}{2\rho}\right\} H_{p+1}\left(\frac{\|\mathbf{x}\|^2}{2\rho}\right)\right]^{-1}. \tag{4.116}$$

PROOF. The proof is essentially identical to that of Corollary 1 in Subsection 4.6.2. The key to the simplification is that the first stage posterior is normal with mean $[\mathbf{x} - \mathbf{\Sigma}(\mathbf{\Sigma}+\mathbf{B}(\lambda))^{-1}(\mathbf{x}-\boldsymbol{\mu})]$ and covariance matrix $[\mathbf{\Sigma} - \mathbf{\Sigma}(\mathbf{\Sigma}+\mathbf{B}(\lambda))^{-1}\mathbf{\Sigma}]$, and that

$$\mathbf{\Sigma} + \mathbf{B}(\lambda) = \mathbf{\Sigma} + \rho\lambda^{-1}(\mathbf{\Sigma}+\mathbf{A}) - \mathbf{\Sigma} = \rho\lambda^{-1}(\mathbf{\Sigma}+\mathbf{A}).$$

Hence, λ factors out as a scale factor in the first stage posterior (and also in the first stage marginal), so that the second stage calculation is straightforward. The details are omitted. □

As $\|\mathbf{x}\|^2 \to 0$, it can be shown that $h_p(\|\mathbf{x}\|^2) \to 1$ (indeed $h_p(\|\mathbf{x}\|^2) \cong \min\{1, (p+1)/\|\mathbf{x}\|^2\}$ is a reasonably good approximation to h_p) and $g_p(\|\mathbf{x}\|^2) \to 4/(p^2+6p+5)$. Hence $\boldsymbol{\mu}^{\pi_p}(\mathbf{x}) \to \mathbf{x} - \mathbf{\Sigma}(\mathbf{\Sigma}+\mathbf{A})^{-1}(\mathbf{x}-\boldsymbol{\mu})$ and $\mathbf{V}^{\pi_p}(\mathbf{x}) \to \mathbf{\Sigma} - \mathbf{\Sigma}(\mathbf{\Sigma}+\mathbf{A})^{-1}\mathbf{\Sigma}$, the posterior mean and covariance matrix for the conjugate $\mathcal{N}_p(\boldsymbol{\mu}, \mathbf{A})$ prior.

At the other extreme, since H_{p+1} is an exponentially decreasing function as $\|\mathbf{x}\|^2 \to \infty$, h_p and g_p are approximately $(p+1)/\|\mathbf{x}\|^2$ and $2(p+1)/\|\mathbf{x}\|^4$, respectively, for large $\|\mathbf{x}\|^2$. Thus $\boldsymbol{\mu}^{\pi_p}$ and $\mathbf{V}^{\pi_p}$ act in what we have called a robust fashion, being typical Bayes rules when $\|\mathbf{x}\|^2$ is as expected, but collapsing back to $\mathbf{X}$ and $\boldsymbol{\Sigma}$ when $\|\mathbf{x}\|^2$ is unexpectedly large.

The impropriety of π_p, when $\text{ch}_{\max}(\mathbf{A}^{-1}\boldsymbol{\Sigma}) > \frac{1}{2}(p+1)$, has several unnatural consequences. For instance, as $\mathbf{A} \to \mathbf{0}$, one would expect $\boldsymbol{\mu}^{\pi_p}(\mathbf{x}) \to \boldsymbol{\mu}$, but such is not the case (indeed $\boldsymbol{\mu}^{\pi_p}(\mathbf{x}) \to \mathbf{x} - h_p((\mathbf{x}-\boldsymbol{\mu})^t\boldsymbol{\Sigma}^{-1}(\mathbf{x}-\boldsymbol{\mu})) \times (\mathbf{x}-\boldsymbol{\mu})$). When $\mathbf{A}$ has very small characteristic roots, therefore, use of π_p is questionable, and the numerical approach at the end of this subsection may be preferable.

Lemma 4. *If* $p=1$,

$$\mu^{\pi_1}(x) = x - \frac{\sigma^2(x-\mu)}{(\sigma^2+A)}\left[\frac{2}{\|x\|^2} - \frac{2}{(e^{\|x\|^2}-1)}\right], \tag{4.117}$$

$$V^{\pi_1}(x) = \sigma^2 - \frac{\sigma^4}{(\sigma^2+A)}\left[\frac{2}{(e^{\|x\|^2}-1)}\left\{\frac{2\|x\|^2}{(1-e^{-\|x\|^2})} - 1\right\} - \frac{2}{\|x\|^2}\right], \tag{4.118}$$

and

$$m(x|\pi_1) = [4\pi(\sigma^2+A)]^{-1/2}(\|x\|^2)^{-1}[1-e^{-\|x\|^2}], \tag{4.119}$$

where $\|x\|^2 = (x-\mu)^2/(\sigma^2+A)$.

Proof. Noting that $\rho = \frac{2}{4}$ and that (from Appendix 2) $H_2(v) = v/(e^v - 1)$, the result follows by direct calculation from Lemma 3. □

To perform Bayesian analyses with respect to π_p, we will approximate the posterior distribution by a $\mathcal{N}_p(\boldsymbol{\mu}^{\pi_p}(\mathbf{x}), \mathbf{V}^{\pi_p}(\mathbf{x}))$ distribution. The natural estimate for $\boldsymbol{\theta}$ is, of course, $\boldsymbol{\mu}^{\pi_p}(\mathbf{x})$, and (approximate) credible sets for $\boldsymbol{\theta}$ can be constructed via (4.11). The relatively simple expression for $m(\mathbf{x}|\pi_p)$ also allows for testing of a point null hypothesis (see Subsection 4.3.3), and for implementation of the type of diagnostics discussed in Subsection 4.7.2.

Example 18 (continued). Here $p=1$, $X \sim \mathcal{N}(\theta, 1)$, and the prior median of 0 and quartiles of ± 1 led us to entertain the possibility of using a $\mathcal{N}(0, 2.19)$ prior. If, instead, the robust prior π_1 is used (with $\mu = 0$ and $A = 2.19$), and (say) $x=2$ is observed, Lemma 4 yields $\mu^{\pi_1}(2) = 1.50$, $V^{\pi_1}(2) = 0.872$, and $m(2|\pi_1) = 0.090$. An approximate 95% credible set for θ would thus be

$$1.50 \pm (1.96)\sqrt{0.872} = (-0.33, 3.33).$$

If the problem had been to test H_0: $\theta = 0$ versus H_1: $\theta \neq 0$, with $\pi_0 = 0.5$ being the prior probability that $\theta = 0$ and $0.5\pi_1$ being the prior density on

$\theta \neq 0$, then (from (4.14)) the posterior probability of the null hypothesis is

$$\alpha_0 = \left[1 + \frac{(0.5)m(2|\pi_1)}{(0.5)f(2|0)}\right]^{-1} = 0.37.$$

In Berger (1980b) various frequentist robustness properties of the robust Bayesian procedures are discussed. For instance, the following theorem provides a condition under which $\boldsymbol{\delta}^{\pi_p}(\mathbf{x}) \equiv \boldsymbol{\mu}^{\pi_p}(\mathbf{x})$ actually dominates the usual (least squares) estimator, $\boldsymbol{\delta}^0(\mathbf{x}) = \mathbf{x}$, in terms of frequentist risk under quadratic loss. The theorem is also related to that in Strawderman (1971), and the results in Subsection 5.4.3.

Theorem 6. *If* $p \geq 5$, $L(\boldsymbol{\theta}, \boldsymbol{\delta}) = (\boldsymbol{\theta} - \boldsymbol{\delta})'\mathbf{Q}(\boldsymbol{\theta} - \boldsymbol{\delta})$, *and*

$$(p+5)\, \mathrm{ch}_{\max}\{\boldsymbol{\Sigma}\mathbf{Q}\boldsymbol{\Sigma}(\boldsymbol{\Sigma} + \mathbf{A})^{-1}\} \leq 2\, \mathrm{tr}\{\boldsymbol{\Sigma}\mathbf{Q}\boldsymbol{\Sigma}(\boldsymbol{\Sigma} + \mathbf{A})^{-1}\}, \tag{4.120}$$

then

$$R(\boldsymbol{\theta}, \boldsymbol{\delta}^{\pi_p}) < R(\boldsymbol{\theta}, \boldsymbol{\delta}^0) = \mathrm{tr}(\mathbf{Q}\boldsymbol{\Sigma}) \quad \textit{for all } \boldsymbol{\theta},$$

and hence $r(\pi, \boldsymbol{\delta}^{\pi_p}) \leq r(\pi, \boldsymbol{\delta}^0)$ *for* <u>*all*</u> π. *If* $\boldsymbol{\Sigma} Q \boldsymbol{\Sigma}(\boldsymbol{\Sigma} + \mathbf{A})^{-1}$ *is a multiple of the identity,* (4.120) *is satisfied.*

It is also shown in Berger (1980b) that the confidence ellipsoids determined from $\boldsymbol{\mu}^{\pi_p}$ and $\mathbf{V}^{\pi_p}$ are typically much smaller than the classical confidence ellipsoids $\{\boldsymbol{\theta}: (\boldsymbol{\theta} - \mathbf{X})'\boldsymbol{\Sigma}^{-1}(\boldsymbol{\theta} - \mathbf{X}) \leq \chi_p^2(1-\alpha)\}$, and yet have *frequentist* coverage probability that is often close to $1 - \alpha$.

Partial Prior Information

It is often the case that only partial prior information about $\boldsymbol{\theta}$ is deemed to be available. For instance, it might be felt that the first k coordinates, $(\theta_1, \ldots, \theta_k)'$, of $\boldsymbol{\theta}$ have a given prior mean and covariance matrix, but that little or nothing is known about $(\theta_{k+1}, \ldots, \theta_p)$. More generally, it might be thought that $\mathbf{B}\boldsymbol{\theta}$ has prior mean $\mathbf{d}$ and covariance matrix $\mathbf{C}$, where $\mathbf{B}$, $\mathbf{d}$, and $\mathbf{C}$ are given $(k \times p)$, $(k \times 1)$, and $(k \times k)$ matrices, respectively, of ranks $k < p$, 1, and k.

EXAMPLE 38. Suppose $p = 5$, and that it is suspected that $\theta_1 = \theta_2$, $2\theta_1 + \theta_3 = \theta_5$, and $\theta_4 = 3$. Letting

$$\mathbf{B} = \begin{pmatrix} 1 & -1 & 0 & 0 & 0 \\ 2 & 0 & 1 & 0 & -1 \\ 0 & 0 & 0 & 1 & 0 \end{pmatrix} \quad \text{and} \quad \mathbf{d} = \begin{pmatrix} 0 \\ 0 \\ 3 \end{pmatrix},$$

these suspected relationships are indeed modelled by the equation $\mathbf{B}\boldsymbol{\theta} = \mathbf{d}$. Suppose further that the "variances" of $(\theta_1 - \theta_2)$, $(2\theta_1 + \theta_3 - \theta_5)$ and $(\theta_4 - 3)$

are thought to be 1, 3, and 4, respectively. We then might choose C to be diagonal with diagonal elements 1, 3, and 4.

In Appendix 2, it is argued that a reasonable robust Bayesian utilization of this partial prior information leads to the following modifications of (4.114) and (4.115):

$$\boldsymbol{\mu}^{\pi^*}(\mathbf{x}) = \mathbf{x} - h_k(\|\mathbf{x}\|_*^2)\boldsymbol{\Sigma}\mathbf{B}'\mathbf{T}^{-1}(\mathbf{B}\mathbf{x}-\mathbf{d}), \tag{4.121}$$

$$\begin{aligned}\mathbf{V}^{\pi^*}(\mathbf{x}) = {} & \boldsymbol{\Sigma} - h_k(\|\mathbf{x}\|_*^2)\boldsymbol{\Sigma}\mathbf{B}'\mathbf{T}^{-1}\mathbf{B}\boldsymbol{\Sigma} \\ & + g_k(\|\mathbf{x}\|_*^2)\boldsymbol{\Sigma}\mathbf{B}'\mathbf{T}^{-1}(\mathbf{B}\mathbf{x}-\mathbf{d})(\mathbf{B}\mathbf{x}-\mathbf{d})'\mathbf{T}^{-1}\mathbf{B}\boldsymbol{\Sigma},\end{aligned} \tag{4.122}$$

where $\mathbf{T} = (\mathbf{B}\boldsymbol{\Sigma}\mathbf{B}' + \mathbf{C})$ and $\|\mathbf{x}\|_*^2 = (\mathbf{B}\mathbf{x}-\mathbf{d})'\mathbf{T}^{-1}(\mathbf{B}\mathbf{x}-\mathbf{d})$.

Example 38 (continued). Suppose $\boldsymbol{\Sigma} = \mathbf{I}$ and $\mathbf{x} = (1, 0, 3, 4, 1)'$. Then

$$\mathbf{T} = \left[\begin{pmatrix} 2 & 2 & 0 \\ 2 & 6 & 0 \\ 0 & 0 & 1 \end{pmatrix} + \begin{pmatrix} 1 & 0 & 0 \\ 0 & 3 & 0 \\ 0 & 0 & 4 \end{pmatrix}\right] = \begin{pmatrix} 3 & 2 & 0 \\ 2 & 9 & 0 \\ 0 & 0 & 5 \end{pmatrix},$$

and

$$\|\mathbf{x}\|_*^2 = ((1,4,4)-(0,0,3))\mathbf{T}^{-1}((1,4,4)-(0,0,3))' = 1.98.$$

From the formula in Appendix 2,

$$H_4\left(\frac{\|\mathbf{x}\|_*^2}{2\rho}\right) = H_4\left(\frac{1.98}{2(\frac{4}{6})}\right) = H_4(1.49) = \frac{(1.49)^{4/2}}{[\frac{4}{2}]!\{e^{1.49} - (1+1.49)\}} = 0.57,$$

so that

$$h_3(\|\mathbf{x}\|_*^2) = \frac{4}{(1.98)}[1 - 0.57] = 0.87,$$

and

$$g_3(\|\mathbf{x}\|_*^2) = \frac{2(4)}{(1.98)^2}[1 + \{(1.49)[\tfrac{4}{6}(0.87) - 1] - 1\}(0.57)] = 0.15.$$

Plugging everything into (4.121) and (4.122) yields

$$\begin{aligned}\boldsymbol{\mu}^{\pi^*}(\mathbf{x}) &= (1, 0, 3, 4, 1)' - (0.87)(\tfrac{21}{23}, -\tfrac{1}{23}, \tfrac{10}{23}, \tfrac{1}{5}, -\tfrac{10}{23})' \\ &= (0.21, 0.04, 2.62, 3.83, 1.38)',\end{aligned}$$

and

$$\mathbf{V}^{\pi^*}(\mathbf{x}) = \begin{pmatrix} 0.44 & -0.20 & -0.09 & 0.03 & 0.09 \\ 0.18 & 0.66 & -0.08 & 0.00 & 0.08 \\ -0.09 & -0.08 & 0.91 & 0.01 & 0.09 \\ 0.03 & 0.00 & 0.01 & 0.83 & -0.01 \\ 0.09 & 0.08 & 0.09 & -0.01 & 0.91 \end{pmatrix}.$$

Partially Unknown $\boldsymbol{\Sigma}$

Consider the situation in which $\boldsymbol{\Sigma}=\sigma^2\boldsymbol{\Sigma}_0$, σ^2 unknown and $\boldsymbol{\Sigma}_0$ known. (As discussed at the beginning of the subsection, this is often the situation in regression analyses.) Assume there is available a statistic S^2, with S^2/σ^2 having a chi-square distribution with m degrees of freedom. (In the regression scenario, S^2 would be the residual sum of squares.)

A very *ad hoc* way of proceeding would be to simply replace $\boldsymbol{\Sigma}$ in (4.114) through (4.122) by $\hat{\sigma}^2\boldsymbol{\Sigma}_0$, where $\hat{\sigma}^2$ is an estimate of σ^2 based on the observed s^2. This, however, does not take into account the additional inaccuracy incurred by the estimation of σ^2. A fully Bayesian analysis would involve placing a prior distribution on σ^2 and working with the marginal posterior of $\boldsymbol{\theta}$, but closed form calculation (which we are striving for here) becomes impossible for suitably robust priors. It is, however, possible to approximate a fully Bayesian answer, and this is done in Appendix 2 for the prior

$$\pi_p^*(\boldsymbol{\theta}, \sigma^2)=\pi_p(\boldsymbol{\theta}|\sigma^2)\sigma^{-2},$$

where $\pi_p(\boldsymbol{\theta}|\sigma^2)$ is the robust prior defined earlier for known σ^2 (and $\boldsymbol{\Sigma}=\sigma^2\boldsymbol{\Sigma}_0$). Note that we are effectively giving σ^2 the (marginal) noninformative prior density $\pi(\sigma^2)=1/\sigma^2$, which is reasonable because σ^2 is merely a nuisance parameter here.

It is argued, in Appendix 2, that a reasonable approximation to the marginal posterior distribution, $\pi_p^*(\boldsymbol{\theta}|\mathbf{x}, s^2)$, is

$$\hat{\pi}_p^*(\boldsymbol{\theta}|\mathbf{x}, s^2)\sim \mathcal{T}_p(m, \boldsymbol{\mu}^*(\mathbf{x}, s^2), (1+2/m)\mathbf{V}^*(\mathbf{x}, s^2)), \tag{4.123}$$

where $\boldsymbol{\mu}^*(\mathbf{x}, s^2)$ and $\mathbf{V}^*(\mathbf{x}, s^2)$ are given by (4.114) and (4.115) (or (4.117) and (4.118), or (4.121) and (4.122), if applicable), with $\boldsymbol{\Sigma}$ replaced by

$$\hat{\boldsymbol{\Sigma}}=\frac{s^2}{(m+2)}\boldsymbol{\Sigma}_0. \tag{4.124}$$

It is almost as easy to work with a t-posterior as with a normal posterior. The natural estimate of $\boldsymbol{\theta}$ is still the posterior mean $\boldsymbol{\mu}^*(\mathbf{x}, s^2)$, and an (approximate) $100(1-\alpha)\%$ HPD credible set for $\boldsymbol{\theta}$ is the ellipsoid

$$C(\mathbf{x}, s^2)=\{\boldsymbol{\theta}\colon [\boldsymbol{\theta}-\boldsymbol{\mu}^*]^t\mathbf{V}^{*-1}[\boldsymbol{\theta}-\boldsymbol{\mu}^*]\le p(1+2/m)F_{p,m}(1-\alpha)\}, \tag{4.125}$$

where $F_{p,m}(1-\alpha)$ is the $(1-\alpha)$-fractile of the $\mathcal{F}(p, m)$ distribution. For evidence as to the robustness of these procedures (beyond their obvious similarity to the earlier robust procedures), see Berger (1980b).

EXAMPLE 39. A recently purchased truck obtained mileages of 8.5, 10.1, 9.0, 8.0, and 8.9 miles per gallon on five different 300-mile trips. It is assumed that these five observations, $X_1, \ldots, X_5$, are independent and from a common $\mathcal{N}(\theta, \sigma_0^2)$ distribution, both θ and σ_0^2 unknown. The sufficient statistics for (θ, σ_0^2) are $\bar{X}$ and $S_0^2=\sum_{i=1}^5(X_i-\bar{X})^2$. To write this in terms of our current notation, let $X = \bar{X}$ and $\sigma^2 = \sigma_0^2/5$ (so that $X \sim \mathcal{N}(\theta, \sigma^2)$), and let $S^2 = S_0^2/5$ (so that $S^2/\sigma^2 = S_0^2/\sigma_0^2$ is $\chi^2(4)$). Calculation yields $x = \bar{x} = 8.9$ and $s^2 = 0.48$. Note also that $m = 4$.

Suppose that information is available from the truck manufacturer to the effect that, among similar trucks, the average gas mileage is distributed according to a distribution with mean 9.14 and standard deviation 0.3. It is suspected, however, that this distribution is quite heavy tailed, so that a robust Bayesian analysis is desired.

Setting $\mu = 9.14$, $A = (0.3)^2 = 0.09$, and noting that (4.124) becomes (since $p = 1$ and so $\mathbf{\Sigma}_0 = 1$)

$$\hat{\sigma}^2 = \frac{s^2}{(m+2)} = 0.081,$$

Lemma 4 yields

$$\|x\|^2 = \frac{(8.9 - 9.14)^2}{(0.081 + 0.09)} = 0.337,$$

$$\mu^*(x) = 8.9 - \frac{(0.081)(8.9 - 9.14)}{(0.081 + 0.09)} \left[\frac{2}{(0.337)} - \frac{2}{(\exp\{0.337\} - 1)} \right] = 9.01,$$

$$V^*(x) = (0.081) - \frac{(0.081)^2}{(0.081 + 0.09)}$$

$$\times \left[\frac{2}{(e^{(0.337)} - 1)} \left\{ \frac{2(0.337)}{(1 - e^{-(0.337)})} - 1 \right\} - \frac{2}{(0.337)} \right] = 0.049.$$

Thus, using (4.125), an (approximate) 90% HPD credible set for θ is

$$\mu^* \pm [p(1 + 2/m) V^* F_{p,m}(0.90)]^{1/2} = (9.01) \pm [(1 + \tfrac{2}{4})(0.049) F_{1,4}(0.90)]^{1/2}$$
$$= (9.01) \pm 0.58 = (8.43, 9.59).$$

Independent Prior Coordinates

We have chosen, largely for technical convenience, robust priors that depend on some quadratic form in $\boldsymbol{\theta}$. Such priors incorporate dependence among the coordinates of $\boldsymbol{\theta}$, dependence which may not exist or be desirable. As an indication of why such dependence may be undesirable, observe, from say the expressions in Lemma 3, that if even one of the $(x_i - \mu_i)$ is extremely large, then $\boldsymbol{\mu}^{\pi_p}(\mathbf{x}) \cong \mathbf{x}$ and $\mathbf{V}^{\pi_p}(\mathbf{x}) \cong \mathbf{\Sigma}$. In other words, the Bayesian gains can be destroyed by imprecise prior specification of even one coordinate of $\boldsymbol{\theta}$. (Note, however, that we will be no worse off than the classical analysis which uses $\mathbf{x}$ and $\mathbf{\Sigma}$; this is, again, an indication of the robustness of $\boldsymbol{\mu}^{\pi_p}$ and $\mathbf{V}^{\pi_p}$.) One possibility, for eliminating this undesirable feature of the analysis, is to use independent robust priors for coordinates or groups of coordinates (cf. Berger and Dey (1983)). Unfortunately, the calculation of posterior means and variances can then be prohibitively difficult.

An *ad hoc*, but quite successful, solution to the difficulty is described in Berger and Dey (1985) (see also Dey and Berger (1983)), based on an idea in Stein (1981). The idea is to limit the allowable deviation of any coordinate from the m.l.e., x_i.

A reasonable way of limiting the deviations of the x_i is to write $\boldsymbol{\mu}^{\pi_p}$ as

$$\boldsymbol{\mu}^{\pi_p}(\mathbf{x}) = \mathbf{x} - \boldsymbol{\Sigma}(\boldsymbol{\Sigma}+\mathbf{A})^{-1/2} h_p(|\mathbf{y}|^2)\mathbf{y}, \tag{4.126}$$

where $|\mathbf{y}|^2 = \mathbf{y}^t\mathbf{y}$ and $\mathbf{y} = (y_1, \ldots, y_p)^t = (\boldsymbol{\Sigma}+\mathbf{A})^{-1/2}(\mathbf{x}-\boldsymbol{\mu})$, so that the Y_i have marginal mean zero and variance one ($\mathbf{X}$ has marginal mean $\boldsymbol{\mu}$ and covariance matrix $(\boldsymbol{\Sigma}+\mathbf{A})$). Then a large $|y_i|$ would indicate a possible prior misspecification problem in that coordinate. A reasonable method of truncating such coordinates would be to consider the variables

$$Z_i = (\text{sgn } Y_i)\min\{|Y_i|, |Y|_{(l)}\},$$

where $|Y|_{(l)}$ is the lth order statistic of $(|Y_1|, |Y_2|, \ldots, |Y_p|)$ (i.e., the lth entry in the sequence arising from arranging the $|Y_i|$ in increasing order). The idea is to replace the y_i in (4.126) by the z_i, limiting the effect of very extreme y_i.

Truncating the Y_i at the lth absolute order statistic reduces the effective dimension of the analysis from p to l (see Stein (1981), Dey and Berger (1983), and Berger and Dey (1985)). The recommended estimator is thus given by

$$\boldsymbol{\mu}^{T}(\mathbf{x}) = \mathbf{x} - \boldsymbol{\Sigma}(\boldsymbol{\Sigma}+\mathbf{A})^{-1/2} h_l(|\mathbf{z}|^2)\mathbf{z},$$

where $\mathbf{z} = (z_1, \ldots, z_p)^t$. It is suggested in Berger and Dey (1985) that l be chosen to be equal to p for $p \le 5$, and otherwise equal to

$$l = 3 + [(0.8)(p-3)],$$

where $[w]$ is the closest positive integer to w. The natural analog of $\mathbf{V}^{\pi_p}(\mathbf{x})$ (in (4.115)), for the estimate $\boldsymbol{\mu}^T$, is

$$\mathbf{V}^T(\mathbf{x}) = \boldsymbol{\Sigma} - h_l(|\mathbf{z}|^2)\boldsymbol{\Sigma}(\boldsymbol{\Sigma}+\mathbf{A})^{-1}\boldsymbol{\Sigma} + g_l(|\mathbf{z}|^2)\boldsymbol{\Sigma}(\boldsymbol{\Sigma}+\mathbf{A})^{-1/2}\mathbf{z}\mathbf{z}^t(\boldsymbol{\Sigma}+\mathbf{A})^{-1/2}\boldsymbol{\Sigma}.$$

A side benefit of this truncation is additional coordinatewise robustification. While $\boldsymbol{\mu}^{\pi_p}$ is robust as an estimate of the entire $\boldsymbol{\theta}$, certain coordinates could be estimated rather nonrobustly. By limiting the possible deviations from the x_i, $\boldsymbol{\mu}^T$ increases the robustness of the individual coordinates. As an indication of this, consider estimation of $\boldsymbol{\theta}$ under sum-of-squares error loss when $\boldsymbol{\Sigma}$ and $\mathbf{A}$ are multiples of the identity. The maximum component frequentist risk of an estimator $\boldsymbol{\delta}$ is

$$\max_{1 \le i \le p} \sup_{\boldsymbol{\theta}} E_{\theta_i}(\theta_i - \delta_i(\mathbf{X}))^2.$$

For $\boldsymbol{\mu}^{\pi_p}$, this maximum can be shown to be approximately $(p/4)$ for large p (see Efron and Morris (1972) for a related argument), while for $\boldsymbol{\mu}^T$ this maximum is only about 3.3 for large p.

Exact Calculation with t-Priors

The robust priors, π_p, are rather peculiar in their dependence on $\mathbf{\Sigma}$, and were indeed argued to be inadequate when improper (i.e., when (4.113) is violated). It is appealing, in such situations, to instead work with more standard flat-tailed priors, such as t-priors. We briefly discuss calculational features of such an approach here.

Suppose, once again, that $\mathbf{X} \sim \mathcal{N}_p(\boldsymbol{\theta}, \sigma^2 \mathbf{\Sigma}_0)$, and that σ^2 is unknown. (The calculational problem is actually sometimes easier when σ^2 is unknown.) Available to estimate σ^2 is a statistic S^2, where S^2/σ^2 has a $\chi^2(m)$ distribution. We again have prior inputs $\boldsymbol{\mu}$ and $\mathbf{A}$ for $\boldsymbol{\theta}$, but now select the prior density

$$\pi(\boldsymbol{\theta}, \sigma^2) = \pi^*(\boldsymbol{\theta}) \pi^{**}(\sigma^2),$$

where π^* is $\mathcal{T}_p(\alpha, \boldsymbol{\mu}, \mathbf{A})$ and $\pi^{**}(\sigma^2) = \sigma^{-2}$. For small α, π^* is indeed a flat-tailed density. Other (informative) choices for π^{**} are possible, and result in essentially the same calculational problem.

When $p = 1$, it is convenient to choose $\alpha = 1$, so that π^* is $\mathcal{C}(\mu, A^{1/2})$. One can then interpret μ as the prior median and $\mu \pm A^{1/2}$ as the prior quartiles; this makes elicitation of μ and A particularly simple. For $p > 1$, one could either relate A to marginal fractiles, or to a "subjective covariance matrix," using the fact that $[\alpha/(\alpha - 2)]\mathbf{A}$ is the covariance matrix of $\boldsymbol{\theta}$ under π^*.

At first sight, calculation of the posterior distribution or features thereof will involve high-dimensional integration, an unpleasant undertaking. It turns out, however, that a hierarchical representation for π^* exists, which allows easy numerical calculation of many features of the posterior. Indeed, one can write

$$\pi^*(\boldsymbol{\theta}) = \int_0^\infty \pi_1^*(\boldsymbol{\theta}|\lambda) \pi_2(\lambda) d\lambda, \tag{4.127}$$

where

$$\pi_1^*(\boldsymbol{\theta}|\lambda) \text{ is } \mathcal{N}_p(\boldsymbol{\mu}, \lambda^{-1}\mathbf{A}), \quad \text{and} \quad \pi_2(\lambda) \text{ is } \mathcal{G}\left(\frac{\alpha}{2}, \frac{2}{\alpha}\right)$$

(the proof is left as an exercise). Thus, as in Section 4.6, one can write, letting $f(\mathbf{x}|\boldsymbol{\theta}, \sigma^2)$ denote the normal density of $\mathbf{x}$ and $g(s^2|\sigma^2)$ denote the density of S^2,

$$E^{\pi(\boldsymbol{\theta}, \sigma^2|\mathbf{x}, s^2)}[\psi(\boldsymbol{\theta}, \sigma^2)] = E^{\tilde{\pi}(\sigma^2, \lambda|\mathbf{x}, s^2)} E^{\pi_1^*(\boldsymbol{\theta}|\lambda, \sigma^2, \mathbf{x}, s^2)}[\psi(\boldsymbol{\theta}, \sigma^2)]. \tag{4.128}$$

Now, since $\pi_1^*(\boldsymbol{\theta}|\lambda)$ is $\mathcal{N}_p(\boldsymbol{\mu}, \lambda^{-1}\mathbf{A})$ and $f(\mathbf{x}|\boldsymbol{\theta}, \sigma^2)$ is $\mathcal{N}_p(\boldsymbol{\theta}, \sigma^2\mathbf{\Sigma}_0)$, it is clear that

$$\pi_1^*(\boldsymbol{\theta}|\lambda, \sigma^2, \mathbf{x}, s^2) \text{ is } \mathcal{N}_p(\boldsymbol{\mu}^1(\mathbf{x}, \lambda, \sigma^2), \mathbf{V}^1(\mathbf{x}, \lambda, \sigma^2)),$$

where

$$\boldsymbol{\mu}^1(\mathbf{x}, \lambda, \sigma^2) = \mathbf{x} - \sigma^2\mathbf{\Sigma}_0(\sigma^2\mathbf{\Sigma}_0 + \lambda^{-1}\mathbf{A})^{-1}(\mathbf{x} - \boldsymbol{\mu})$$

and

$$\mathbf{V}^1(\mathbf{x}, \lambda, \sigma^2) = \sigma^2\boldsymbol{\Sigma}_0 - \sigma^4\boldsymbol{\Sigma}_0(\sigma^2\boldsymbol{\Sigma}_0 + \lambda^{-1}\mathbf{A})^{-1}\boldsymbol{\Sigma}_0.$$

Furthermore,

$$\tilde{\pi}(\sigma^2, \lambda | \mathbf{x}, s^2) \propto \pi^{**}(\sigma^2)\pi_2(\lambda)g(s^2|\sigma^2)m_1(\mathbf{x}|\lambda, \sigma^2),$$

where

$$m_1(\mathbf{x}|\lambda, \sigma^2) \text{ is } \mathcal{N}_p(\boldsymbol{\mu}, \sigma^2\boldsymbol{\Sigma}_0 + \lambda^{-1}\mathbf{A}).$$

Thus, an expression for the posterior mean of $\boldsymbol{\theta}$ is (setting $\psi(\boldsymbol{\theta}, \sigma^2) = \boldsymbol{\theta}$ in (4.128))

$$\boldsymbol{\mu}^\pi(\mathbf{x}) = \frac{\iint \boldsymbol{\mu}^1(\mathbf{x}, \lambda, \sigma^2)\pi^{**}(\sigma^2)\pi_2(\lambda)g(s^2|\sigma^2)m_1(\mathbf{x}|\lambda, \sigma^2)d\lambda\, d\sigma^2}{\iint \pi^{**}(\sigma^2)\pi_2(\lambda)g(s^2|\sigma^2)m_1(\mathbf{x}|\lambda, \sigma^2)d\lambda\, d\sigma^2}. \quad (4.129)$$

Replacing $\boldsymbol{\mu}^1$ by $[\mathbf{V}^1 + (\boldsymbol{\mu}^1 - \boldsymbol{\mu}^\pi)(\boldsymbol{\mu}^1 - \boldsymbol{\mu}^\pi)^t]$ will yield an expression for the posterior covariance matrix, $\mathbf{V}^\pi$.

The point here is that the analysis has been reduced from the initial $(p+1)$-dimensional integration (over $\boldsymbol{\theta}$ and σ^2) to a two-dimensional integration (over λ and σ^2). The calculation of $\boldsymbol{\mu}^\pi$ and $\mathbf{V}^\pi$ can thus be done by numerical integration without too much trouble. If σ^2 were known, the integral over σ^2 would be absent, so that only a one-dimensional numerical integration would be needed. Among the articles discussing or elaborating on these points are Tiao and Zellner (1964) and Dickey (1968, 1973, 1974).

An alternative approach can be taken when $p = 1$. For simplicity, suppose that $\pi^{**}(\sigma^2) = \sigma^{-2}$ (any inverted gamma distribution for σ^2 could be handled in the same way), and that $\pi^*(\theta)$ is $\mathcal{C}(\mu, A^{1/2})$. Also, let $\boldsymbol{\Sigma}_0 = n^{-1}$, corresponding to the situation in which X is a sample mean from a sample of size n. Then the posterior density of (θ, σ^2) is

$$\pi(\theta, \sigma^2 | x, s^2) = \frac{f(x|\theta, \sigma^2)g(s^2|\sigma^2)\pi^*(\theta)\sigma^{-2}}{m(x, s^2)},$$

and the marginal posterior of θ, given (x, s^2), is

$$\begin{aligned}
\pi^*(\theta|x, s^2) &= \int \pi(\theta, \sigma^2|x, s^2)d\sigma^2 \\
&= \frac{\pi^*(\theta)}{m(x, s^2)} \int_0^\infty \frac{1}{(2\pi\sigma^2/n)^{1/2}} \exp\left\{-\frac{n}{2\sigma^2}(x-\theta)^2\right\} \\
&\quad \times \frac{1}{\Gamma(m/2)(2\sigma^2)^{m/2}}(s^2)^{(m-2)/2} \exp\left\{-\frac{s^2}{2\sigma^2}\right\}\sigma^{-2}\, d\sigma^2 \\
&= K\pi^*(\theta) \int_0^\infty (\sigma^2)^{-(m+3)/2} \exp\left\{-\frac{1}{\sigma^2}\left[\frac{n}{2}(x-\theta)^2 + \frac{s^2}{2}\right]\right\} d\sigma^2 \\
&= K'\pi^*(\theta)\left[1 + \frac{(x-\theta)^2}{m(s^2/mn)}\right]^{-(m+1)/2}, \qquad (4.130)
\end{aligned}$$

where K and K' are the appropriate normalizing constants. Notice that the effect of having σ^2 unknown is to cause the effective likelihood function for θ to be $\mathcal{T}(m, x, s^2/(mn))$. (Notice also that $s^2/(mn)$ is the usual unbiased estimator of σ^2/n, the variance of $\bar{X}$.) This same reduction can also be effected in higher dimensions, providing $\mathbf{\Sigma} = \sigma^2\mathbf{\Sigma}_0$, $\mathbf{\Sigma}_0$ known.

One could now attempt to work directly with (4.130). For instance, the posterior mean is

$$\mu^\pi(x, s^2) = \frac{\displaystyle\int \theta\left[1+\frac{(\theta-\mu)^2}{A}\right]^{-1}\left[1+\frac{(x-\theta)^2}{m(s^2/mn)}\right]^{-(m+1)/2} d\theta}{\displaystyle\int \left[1+\frac{(\theta-\mu)^2}{A}\right]^{-1}\left[1+\frac{(x-\theta)^2}{m(s^2/mn)}\right]^{-(m+1)/2} d\theta}, \quad (4.131)$$

and a similar expression can be given for the posterior variance. (Actually, it is possible to express the posterior variance in terms of the numerator and denominator above.) Here it is clearly preferable to attempt evaluation of these one-dimensional integrals, rather than the two-dimensional integrals such as (4.129). Interestingly, when m is odd, the integrals in (4.131) can be evaluated in *closed form* using contour integration. The resultant formulas are given in Berger and Das Gupta (1985), and are particularly useful for preparation of graphs of Bayesian outputs (such as $\mu^\pi(x, s^2)$), with respect to the inputs μ and A. (Such graphs typically require hundreds or thousands of integrations.) The entire enterprise of robust Bayesian analysis of a one-dimensional normal mean is extensively discussed in Berger and Das Gupta (1985).

4.7.11. Other Issues in Robustness

Identification of Sensitivity

In high-dimensional problems, there are substantial difficulties in specifying any reasonable joint prior distribution for the parameters, much less doing a careful robustness study. We have discussed one possible type of solution: attempt to use an inherently robust prior. Development of high-dimensional robust priors for which calculation is feasible can be difficult, however. An alternative approach is to use conjugate priors (for which calculations can generally be carried out comparatively easily), and then investigate the sensitivity of the answers to choice of the conjugate prior inputs. This is typically done by differentiating the answers with respect to the prior inputs, and looking for inputs that sharply change the answers.

EXAMPLE 40. Suppose $\mathbf{X} = (X_1, \ldots, X_p)^t \sim \mathcal{N}_p(\boldsymbol{\theta}, \mathbf{I})$ and $\boldsymbol{\theta} \sim \mathcal{N}_p(\mathbf{0}, \mathbf{A})$, where $\mathbf{A}$ is the diagonal $(p \times p)$ matrix with diagonal elements A_i. Then the posterior mean is given, coordinatewise, by

$$\mu_i^{\pi}(\mathbf{x}) = [1 - (1 + A_i)^{-1}]x_i,$$

and

$$\frac{d}{dA_i}\mu_i^{\pi}(\mathbf{x}) = \frac{x_i}{(1 + A_i)^2}. \tag{4.132}$$

When x_i and A_i are such that this derivative is large, then small variations in A_i have a substantial effect on the answer, and robustness is a concern.

Sensitivity theory of the above type has been extensively developed in Leamer (1978, 1982) and Polasek (1982, 1984, 1985). They call it *local sensitivity* (as opposed to *global sensitivity*, which consists of the development of bounds on the range of posterior features as π ranges over a conjugate class of priors, as in Subsection 4.7.3).

The comparative simplicity of determining local sensitivity is its most attractive feature. Its major limitation is the restriction to conjugate priors, which may cause one to miss important causes of nonrobustness. For instance, when $A_i = 2.19$ and $x_i = 6$ in Example 40, the derivative in (4.132) is 0.59, which does not seem large enough to cause great concern about robustness. We saw, however, many indications (in the related Example 18 of Subsections 4.7.1 and 4.7.2) that use of a normal tail is very nonrobust in this situation. While thus not being curealls for all robustness concerns, these sensitivity detection methods can be useful tools in the robust Bayesian toolkit.

Model Robustness

In our discussion of robustness, we have assumed that the likelihood, $f(x|\theta)$, is known, but that the prior is uncertain. It will clearly often be the case, however, that f is itself uncertain. Indeed, there is a huge literature on the development of statistical methods that are robust with respect to possible misspecification of f. For developments from a classical perspective, see Huber (1981).

At a fundamental Bayesian level, model robustness is not different than prior robustness. This is because, as mentioned previously, choice of a model can be considered to be a particular (indeed, drastic) type of prior specification. (From the initially vast class, $\mathscr{P}$, of all possible probability distributions for $\mathbf{X}$, one selects a small-dimensional subset $\mathscr{P}_f = \{f(x|\theta), \theta \in \Theta\}$, i.e., one restricts oneself to priors on $\mathscr{P}$ which give probability 1 to $\mathscr{P}_f$.) Indeed, many of the robustness techniques we have discussed apply equally well to investigations of model robustness. For instance, $m(x)$, the marginal density of X, depends just as strongly on f as it does on π;

indeed, it is probably good to recognize this dependence and write

$$m(x|\pi, f) = \int_{\Theta} f(x|\theta)\pi(\theta)d\theta. \tag{4.133}$$

One can thus use m, as in Subsection 4.7.2, to assist in robustness investigations concerning f.

EXAMPLE 41. Suppose we observe $X_i = \theta + \varepsilon_i$, $i = 1, \ldots, n$, where $\theta \in \mathbb{R}^1$ is unknown and the ε_i are thought to be independent random errors with a common bell-shaped density, f_0. It is, of course, common to assume that f_0 is normal, but also common to worry about heavy-tailed departures from normality. A useful class of heavy-tailed densities is the $\mathcal{T}(\alpha, 0, \sigma^2)$ class, of which the normal distribution is the limiting case as $\alpha \to \infty$. For such an error density, f_0, the density of $\mathbf{X} = (X_1, \ldots, X_n)$ is

$$f(\mathbf{x}|\theta, \sigma^2, \alpha) = \prod_{i=1}^{n} f_0(x_i|\theta, \sigma^2, \alpha),$$

$f_0(x_i|\theta, \sigma^2, \alpha)$ denoting the $\mathcal{T}(\alpha, \theta, \sigma^2)$ density. If, now, one has some plausible prior distribution, $\pi_0(\theta, \sigma^2)$, for θ and σ^2 (even a noninformative prior can give interesting results), then (4.133) can be written

$$m(\mathbf{x}|\pi_0, \alpha) = \int\int f(\mathbf{x}|\theta, \sigma^2, \alpha)\pi_0(\theta, \sigma^2)d\theta\, d\sigma^2.$$

As in Subsection 4.7.2, this can be used to judge whether or not $\alpha = \infty$ (the normal distribution) is reasonable in light of the data, and even to suggest a choice of α (via the ML-II technique, say) when flatter tails seem indicated by the data. (Of course, one could also put a prior distribution on α and proceed in a complete Bayesian fashion.)

For specific examples and discussion of the above type of Bayesian model robustness investigation, along with other references, see Box and Tiao (1973) and Box (1980). (Other general discussions of Bayesian model robustness can be found in Jeffreys (1961), Dempster (1975), Geisser and Eddy (1979), O'Hagan (1979), Good (1983a), and Smith (1983).)

Of course, it is also possible to apply other Bayesian robustness techniques to the problem of model robustness. We make no attempt to do so here, for two reasons. The first is that, while formal definitions and concepts translate easily into the model robustness situation (by considering f to be part of π, so that the uncertainty in f translates into uncertainty in π), the analytical problems encountered are usually different enough to require separate study; there is not enough space for such separate study here. The second reason for our preoccupation with prior robustness is that it is *perceived*, especially by non-Bayesians, to be the bigger problem. It is often felt that models will typically have some external validity, not possessed by priors on the parameters of the model, implying that f will be known much more accurately than π. Substantial arguments can be presented on both sides of this issue. We will not discuss the arguments because, in a sense, they serve no purpose; if the model is securely known—fine—but, in the many cases where it is uncertain, model robustness should be investigated along with prior robustness (and the effect of incorrect model choice will often have a much larger impact than the effect of incorrect choice of a prior on the model parameters).

One aspect of the interaction of model and prior robustness that should be discussed is what to do when the "sample" and "prior" information clash. In particular, we repeatedly recommended use of priors with flat tails, so that the likelihood function could dominate the prior in the case of such conflict. In so arguing, we were implicitly assuming that: (i) a "robust" model is being used, so that the likelihood is plausible; and (ii) the likelihood will have reasonably sharp tails.

EXAMPLE 41 (continued). In this situation, it would not necessarily be reasonable to say that, if $\bar{X}$ is far from the prior median, then one should trust $\bar{X}$. There is not great confidence that f is normal, and the normal likelihood (and $\bar{x}$ as an estimate of θ) could be highly implausible in light of the actual data. If, however, a t-model with moderate or small α were used, one would find considerable "robustness," and a corresponding estimate, of θ, say the m.l.e., $\hat{\theta}$, *should* probably be trusted in the face of a conflict with the prior median. The reason is that the likelihood function will have tails that behave like polynomials in θ to the $-n(\alpha+1)$ power, which, for even moderate n, will cause the likelihood function to appear substantially sharper than a reasonably robust prior.

As a final comment on model robustness, it should be emphasized that, as with all types of Bayesian robustness, it is typically data dependent. If the data in Example 41 provides a perfect fit to normality, there is no real reason to use other than a normal analysis. If, on the other hand, the data supports the supposition of a heavier tail, then a more robust treatment of the problem is in order. (Another reference stressing the desirability of conditional robustness investigations is Hinkley (1983).)

Loss Robustness

In decision-theoretic problems, the difficulty in specifying a loss function can be even more severe than that of specifying a prior. Indeed, in our experience, practitioners are often far more ready to state "what they know about θ" than to quantify a loss reflecting the actual use that will be made of the results of their investigation; this loss is often to be determined at a later date by a third party. Of course, reporting the entire posterior distribution avoids this problem, since the posterior can be combined with any loss developed at a later date to yield the optimal action.

A robustness theory for loss functions can be set up, in a manner analogous to that for priors. We will not formally do so, partly because of space limitations, and partly because the robustness problems for loss functions are often not as severe as those for priors. In choosing a prior distribution, it is frequently necessary to worry about hard-to-specify features such as the prior tail, but such features of the loss function will not tend to be relevant. This is because *statistical* decision problems tend to involve only small or moderate errors in the determination of θ; the specification of the loss for large errors (the "tail" of the loss) will thus usually be irrelevant. (This is especially so when realistic bounded loss functions are used.) As a simple example, quadratic and absolute error estimation loss are quite different, and yet they result in estimating θ by the posterior mean and posterior median, respectively (see Subsection 4.4.2); these two estimates are typically reasonably similar.

A decision problem will, of course, generally be sensitive to the specification of the loss function for small or moderate errors, but this is as it should be. Also, if a weighted loss such as $w(\theta)\ (\theta-a)^2$ is used, the analysis can be sensitive to $w(\theta)$, in the same way that it is sensitive to the prior. (We saw in Subsection 4.4.2 that the "effective" prior is then $w(\theta)\pi(\theta)$). The literature on robustness with respect to the loss function includes Brown (1973), Kadane and Chuang (1978), Ramsay and Novick (1980), Shinozaki (1980), Kadane (1984), and Hwang (1985).

Ignoring Data

A sometimes useful solution to robustness problems is to ignore data whose modelling causes the nonrobustness. Consider, for instance, a survey sampling situation, in which all sorts of records (of classifying characteristics) are available concerning the members of the population, but in which most of the characteristics seem irrelevant to the attribute of interest. Constructing a general Bayesian (superpopulation) model for all this available data could be enormously difficult, and might not be trustworthy because of nonrobustness. Hence it might be reasonable to restrict consideration only to the subset of the characteristics deemed likely to be relevant.

Ignoring data causes no real problem to a Bayesian if the data seems unlikely to have an effect on the posterior distribution of the parameters of interest. Often, of course, this can only be ascertained through, at least informal, Bayesian reasoning. Consider the following examples.

EXAMPLE 42 (Fraser and Mackay (1976)). Suppose independent observations $X_1, \ldots, X_n$ from a $\mathcal{N}(\theta, \sigma^2)$ distribution are observed, where it is desired to estimate θ but σ^2 is also unknown. Independent observations $Y_1, \ldots, Y_m$ are also available, where Y_i is $\mathcal{N}(\mu_i, \sigma^2)$, μ_i unknown, $i=1, \ldots, m$. If virtually nothing is known *a priori* about the μ_i (and they are in no way related to θ), it is certainly reasonable to ignore the Y_i when estimating θ. (A formal Bayesian analysis would certainly show that the Y_i had almost negligible influence on the posterior distribution of θ.)

EXAMPLE 43. In a medical trial comparing two surgical techniques, a significant relationship was found between the time of the day in which the surgery was performed and the success of the surgery. Suppose the relationship was one of the following: (i) the later in the day the surgery occurred the less successful it was; (ii) when surgery began on even hours it was more successful than when it began on odd hours; (iii) when surgery ended on an even minute it was more successful than when it ended on an odd minute. The question before us is: Can we ignore the data "time of day"? The answer in case (i) is almost certainly no, and we better hope that the two treatment groups were not unbalanced concerning this covariate. The answer in case (iii) is almost certainly yes; it is hard to believe that this relationship is anything more than a coincidence. In case (ii) the answer is not so certain and indeed some investigation is called for. (Did certain surgeons work at certain times, etc.?)

The decision about ignoring data in Example 43 clearly involves prior opinions. The point, however, is that it *may* be possible to informally reason that certain data can be ignored, without having to go through a full-blown Bayesian analysis. This is not really a violation of Bayesian principles either, since the posteriors obtained by ignoring part of the data are felt to be the same as what would have been obtained by a sound Bayesian analysis with all of the data.

The real difficulty arises when it is necessary to throw away potentially relevant data. The reason for doing this would be an inability to carry out a (robust) Bayesian analysis involving everything. Hill (1975) considers a nonparametric problem of this nature in which a trustworthy complete Bayesian analysis seems almost impossible. Hill says,

> "When such a formal analysis simply cannot be made, or even when it is merely very difficult and of dubious validity, then there is little choice but to condition on that part of the data that can be effectively dealt with, and rely upon some form of stable estimation argument."

The last part of the comment can be interpreted to mean that, if you must ignore data, at least convince yourself that there is not reason to expect it to have a large effect on the posterior (or, more properly, the final conclusion). Other discussions of this issue can be found in Pratt (1965) (in the discussion of "insufficient statistics"), Dempster (1975), Good (1976, 1983) (in the discussion of the "Statistician's Stooge"), and Hill (1980).

4.8. Admissibility of Bayes Rules and Long Run Evaluations

As briefly mentioned in Section 4.1, there is a deep relationship between frequentist admissibility theory and Bayesian analysis. We begin the exploration of this relationship here (and develop it more fully in Chapter 8). Subsection 4.8.1 discusses the reasons that Bayes rules are almost always admissible. The more delicate subject of admissibility of generalized Bayes rules is introduced in Subsection 4.8.2, where it is shown that standard noninformative prior Bayes rules can, rather surprisingly, be inadmissible. The reaction of many Bayesians to such inadmissibility is:

> "So what? After all, admissibility is a frequentist criterion, and is hence suspect."

While agreeing with this reply at a certain philosophical level, we argue in Subsection 4.8.3 that inadmissibility can be of concern to Bayesians for various practical reasons. We also connect admissibility with more traditional Bayesian *Dutch Book* arguments in Subsection 4.8.3.

4.8.1. Admissibility of Bayes Rules

The basic reason that a Bayes rule is virtually always admissible is that, if a rule with better risk $R(\theta, \delta)$ existed, that rule would also have better Bayes risk $r(\pi, \delta) = E^{\pi}[R(\theta, \delta)]$. (Recall that we are assuming in this chapter that the Bayes risk of the problem is finite, unless indicated otherwise.) Three specific theorems in this regard follow.

Theorem 7. *Assume that Θ is discrete (say $\Theta = \{\theta_1, \theta_2, \ldots\}$) and that the prior, π, gives positive probability to each $\theta_i \in \Theta$. A Bayes rule δ^{π}, with respect to π, is then admissible.*

Proof. If δ^{π} is inadmissible, then there exists a rule δ with $R(\theta_i, \delta) \le R(\theta_i, \delta^{\pi})$ for all i, with strict inequality for, say, θ_k. Hence

$$r(\pi, \delta) = \sum_{i=1}^{\infty} R(\theta_i, \delta)\pi(\theta_i) < \sum_{i=1}^{\infty} R(\theta_i, \delta^{\pi})\pi(\theta_i) = r(\pi, \delta^{\pi}),$$

the inequality being strict since $R(\theta_k, \delta) < R(\theta_k, \delta^{\pi})$, $\pi(\theta_k) > 0$, and $r(\pi, \delta^{\pi}) < \infty$. This contradicts the fact that δ^{π} is Bayes. Therefore, δ^{π} must be admissible. □

The proofs of the next two theorems are similar proofs by contradiction, and will be left as exercises.

Theorem 8. *If a Bayes rule is unique, it is admissible.*

Theorem 9. *Assume that the risk functions $R(\theta, \delta)$ are continuous in θ for all decision rules δ. Assume also that the prior π gives positive probability to any open subset of Θ. Then a Bayes rule with respect to π is admissible.* (Conditions under which risk functions are continuous are given in Chapter 8.)

In some situations (such as the case of finite Θ) even more can be said; namely that all admissible rules must be Bayes rules. Discussion of this will be delayed until Chapters 5 and 8.

Formal Bayes rules need not be admissible if their Bayes risks are infinite. The following example demonstrates this.

Example 44. Assume $X \sim \mathcal{N}(\theta, 1)$, $\theta \sim \mathcal{N}(0, 1)$, and

$$L(\theta, a) = \exp\left\{\frac{3\theta^2}{4}\right\}(\theta - a)^2.$$

From Example 1, it is clear that $\pi(\theta|x)$ is a $\mathcal{N}(x/2, \frac{1}{2})$ density. Using Result 4, an easy calculation shows that the formal Bayes rule (i.e., the rule which minimizes the posterior expected loss) is given by $\delta^\pi(x) = 2x$. As in Example 4 of Section 1.3, a calculation then shows that

$$R(\theta, \delta^\pi) = \exp\left\{\frac{3\theta^2}{4}\right\}(4 + \theta^2) > \exp\left\{\frac{3\theta^2}{4}\right\}(1) = R(\theta, \delta_1),$$

where $\delta_1(x) = x$. Thus δ^π is seriously inadmissible. Note that $r(\pi, \delta^\pi) = E^\pi[R(\theta, \delta^\pi)] = \infty$, and indeed it can be shown that $r(\pi, \delta^*) = \infty$ for all $\delta^* \in \mathcal{D}^*$.

The possibility indicated in the above example is unsettling. The post-experimental Bayesian reasoning is unassailable. If the true prior and loss really are as given, and L was developed through utility theory, then one is forced (by utility analysis) to evaluate an action through posterior expected loss. But this results, from a pre-experimental basis, in a seriously inadmissible decision rule. If such a situation were to really occur, it would be best to trust the posterior analysis, reasoning that admissibility, being based on a measure of initial precision, is a suspect criterion. Of course, one can never determine π and L exactly, and, furthermore, standard utility axioms imply that a loss must be bounded. (See Chapter 2.) For a bounded loss the Bayes risk is clearly finite, and the above difficulty cannot arise.

4.8.2. Admissibility of Generalized Bayes Rules

As with formal Bayes rules, generalized Bayes rules need not be admissible. Unfortunately, the verification of admissibility (or inadmissibility) can be very difficult.

One situation in which a generalized Bayes rule, δ, can be easily shown to be admissible, is when the loss is positive and

$$r(\pi, \delta) = \int_{\Theta} R(\theta, \delta) dF^{\pi}(\theta) < \infty.$$

If π was proper, $r(\pi, \delta)$ would, of course, be the Bayes risk, but for improper π the meaning of $r(\pi, \delta)$ is unclear. Nevertheless, an argument identical to that leading to Result 1 of Subsection 4.4.1 then shows that the δ minimizing $r(\pi, \delta)$ is found by minimizing the posterior expected loss. Since this is exactly the way in which a generalized Bayes rule is defined, it follows that a generalized Bayes rule minimizes $r(\pi, \delta)$. Arguments, as in the previous subsection, can then be used to show that a generalized Bayes rule must be admissible under suitable conditions.

It is unfortunately rather rare to have $r(\pi, \delta) < \infty$ for improper π. When $r(\pi, \delta) = \infty$, even "natural" generalized Bayes rules can be inadmissible, as the following examples show.

EXAMPLE 45. Assume $X \sim \mathcal{G}(\alpha, \beta)$ ($\alpha > 1$ known) is observed, and it is desired to estimate β under squared-error loss. Since β is a scale parameter, it is felt that the noninformative prior density $\pi(\beta) = \beta^{-1}$ should be used. The (formal) posterior density of β given x is then

$$\pi(\beta|x) = \frac{f(x|\beta)\pi(\beta)}{\int_0^{\infty} f(x|\beta)\pi(\beta) d\beta} = \frac{\beta^{-\alpha} e^{-x/\beta} \beta^{-1}}{\int_0^{\infty} \beta^{-\alpha} e^{-x/\beta} \beta^{-1} \, d\beta},$$

which is clearly recognizable as an $\mathcal{IG}(\alpha, x^{-1})$ density. (Recall that, here, x is a fixed constant and β is the random variable.) Since the loss is squared-error loss, the generalized Bayes estimator, δ^0, of β is the mean of the posterior. Using Appendix 1, this is $\delta^0(x) = x/(\alpha - 1)$.

Consider now the risk of the estimator $\delta_c(x) = cx$. Clearly (since X has mean $\alpha\beta$ and variance $\alpha\beta^2$)

$$\begin{aligned} R(\beta, \delta_c) &= E_{\beta}^{X}[cX - \beta]^2 \\ &= E^{X}[c(X - \alpha\beta) + (c\alpha - 1)\beta]^2 \\ &= c^2\alpha\beta^2 + (c\alpha - 1)^2\beta^2 \\ &= \beta^2[c^2\alpha + (c\alpha - 1)^2]. \end{aligned}$$

Differentiating with respect to c and setting equal to zero shows that the value of c minimizing this expression is unique and is given by $c_0 = (\alpha + 1)^{-1}$. It follows that if $c \neq c_0$, then $R(\beta, \delta_{c_0}) < R(\beta, \delta_c)$ for all β, showing in particular that δ^0 (which is δ_c with $c = (\alpha - 1)^{-1}$) is inadmissible. Indeed the ratio of risks of δ^0 and δ_{c_0} is

$$\frac{R(\beta, \delta^0)}{R(\beta, \delta_{c_0})} = \frac{\alpha(\alpha - 1)^{-2} + (\alpha/(\alpha - 1) - 1)^2}{\alpha(\alpha + 1)^{-2} + (\alpha/(\alpha + 1) - 1)^2} = \frac{(\alpha + 1)^2}{(\alpha - 1)^2}.$$

For small α, δ^0 has significantly worse risk than δ_{c_0}.

EXAMPLE 46. Assume $\mathbf{X}=(X_1, X_2, \ldots, X_p)^t \sim \mathcal{N}_p(\boldsymbol{\theta}, \mathbf{I}_p)$, where $\boldsymbol{\theta}=(\theta_1, \ldots, \theta_p)^t$ and $\mathbf{I}_p$ is the $(p \times p)$ identity matrix. It is desired to estimate $\boldsymbol{\theta}$ under sum-of-squares error loss $(L(\boldsymbol{\theta}, \mathbf{a})=\sum_{i=1}^p (\theta_i - a_i)^2)$. (This could equivalently be stated as the problem of trying to simultaneously estimate p normal means from independent problems.) Since $\boldsymbol{\theta}$ is a location parameter, the noninformative prior density $\pi(\boldsymbol{\theta})=1$ is deemed appropriate. It is easy to see that the (formal) posterior density of $\boldsymbol{\theta}$ given $\mathbf{x}$ is then a $\mathcal{N}_p(\mathbf{x}, \mathbf{I}_p)$ density. The generalized Bayes estimator of $\boldsymbol{\theta}$ is the mean of the posterior (under sum-of-squares error loss, or indeed any quadratic loss), so $\boldsymbol{\delta}^0(\mathbf{x})=\mathbf{x}=(x_1, \ldots, x_p)^t$ is the generalized Bayes estimator. (If a sample of vectors $\mathbf{X}^1, \ldots, \mathbf{X}^n$ was taken, the generalized Bayes estimator would just be the vector of sample means.)

This most standard of estimators is admissible for $p=1$ or 2 (see Chapter 8), but surprisingly is inadmissible for $p \geq 3$. Indeed James and Stein (1960) showed that

$$\boldsymbol{\delta}^{\mathrm{JS}}(\mathbf{x})=\left(1-\frac{(p-2)}{\sum_{i=1}^p x_i^2}\right)\mathbf{x}$$

has $R(\boldsymbol{\theta}, \boldsymbol{\delta}^{\mathrm{JS}})<R(\boldsymbol{\theta}, \boldsymbol{\delta}^0)$ for all $\boldsymbol{\theta}$, if $p \geq 3$. (The proof is outlined in a more general setting in Subsection 5.4.3, where additional references are also given.)

It should be mentioned that the inadmissibility in this example is, in some sense, less serious than that in Example 45. In fact, the ratio of $R(\boldsymbol{\theta}, \boldsymbol{\delta}^{\mathrm{JS}})$ to $R(\boldsymbol{\theta}, \boldsymbol{\delta}^0)$ is very close to one over most of the parameter space. Only in a small region near zero (several standard deviations wide) will the ratio of risks be significantly smaller than one. This is in contrast to the situation of Example 45, in which the ratio of risks can be uniformly bad.

The estimator $\boldsymbol{\delta}^{\mathrm{JS}}$ can be modified so as to adjust the region of significant improvement to coincide with prior knowledge concerning $\boldsymbol{\theta}$. Indeed, one attractive such modification is the estimator $\boldsymbol{\delta}^{\mathrm{S}}$ in (4.102), which not only satisfies $R(\boldsymbol{\theta}, \boldsymbol{\delta}^{\mathrm{S}})<R(\boldsymbol{\theta}, \boldsymbol{\delta}^0)$, but also was shown to have excellent Bayesian performance when prior knowledge about $\boldsymbol{\theta}$ is available. (Such modifications also eliminate the conditionally silly feature of $\boldsymbol{\delta}^{\mathrm{JS}}$ that, if say $|\mathbf{x}|^2=(p-2)/1000$, then $\boldsymbol{\delta}^{\mathrm{JS}}(\mathbf{x})=-999\mathbf{x}$.) Nevertheless, if essentially *no* prior information about $\boldsymbol{\theta}$ is available, then use of $\boldsymbol{\delta}^{\mathrm{JS}}$ (or some modification) will not be significantly beneficial, and $\boldsymbol{\delta}^0$ might as well be used. (See Subsection 5.4.3 for further discussion.)

A version of this example, where use of $\pi(\boldsymbol{\theta}) \equiv 1$ does result in a uniformly inadmissible estimator, was given in Example 35 in Subsection 4.7.9. There it was desired to estimate $\eta=|\boldsymbol{\theta}|^2$, and the generalized Bayes estimator under squared-error loss was shown to be $\delta^\pi(\mathbf{x})=|\mathbf{x}|^2+p$. But it can be shown that the estimator $\delta^c(\mathbf{x})=|\mathbf{x}|^2-c$ has

$$R(\boldsymbol{\theta}, \delta^c)=2p+(p-c)^2+4|\boldsymbol{\theta}|^2, \tag{4.134}$$

so that $R(\boldsymbol{\theta}, \delta^\pi)-R(\boldsymbol{\theta}, \delta^p)=4p^2$, a substantial uniform difference (although again $R(\boldsymbol{\theta}, \delta^\pi)/R(\boldsymbol{\theta}, \delta^p) \to 1$ as $|\boldsymbol{\theta}|^2 \to \infty$).

4.8.3. Inadmissibility and Long Run Evaluations

Inadmissibility is a frequentist concept, and hence its relevance to a Bayesian can be debated. Indeed, it would be hard to quarrel with the conditional noninformative prior Bayesian who, in say Example 46, decided *after careful thought* that his prior was essentially uniform in the region of concentration of the likelihood function, and so used $\pi(\boldsymbol{\theta}) \equiv 1$. It is quite a different matter, however, to recommend *automatic* use of a given noninformative prior when prior information is vague (or, even worse, when it is not). The point is that, in recommending repeated use of a particular prior, a Bayesian has entered into the frequentist domain; it is then perfectly reasonable to investigate how repeated use of this prior actually performs.

The natural method of evaluating long run performance of a procedure δ (here the procedure determined by repeated use of a particular noninformative prior) was discussed in Subsection 1.6.2; consider a sequence of independent problems $\{(\theta^{(1)}, X^{(1)}), (\theta^{(2)}, X^{(2)}), \ldots\}$ and a loss (or criterion function) L that measures the performance of the procedure in each problem, and look at the long run performance of δ. One reasonable method of *comparing* two procedures, δ_1 and δ_2, would be to look at

$$S_N = \sum_{i=1}^{N} \{L(\theta^{(i)}, \delta_1(X^{(i)})) - L(\theta^{(i)}, \delta_2(X^{(i)}))\},$$

and consider the limiting behavior (in some sense) of S_N. We stress that this compares actual performance in repeated use, and is not some arbitrary frequentist comparison. Not surprisingly, however, the limiting behavior of S_N is very related to risk comparisons between δ_1 and δ_2. Indeed, the following theorem establishes two such results.

Theorem 10. *Consider* $\boldsymbol{\theta} = (\theta^{(1)}, \theta^{(2)}, \ldots)$ *to be any fixed sequence of parameters* $(\theta^{(i)} \in \Theta)$, *and suppose random variables* $X^{(i)} \in \mathscr{X}$ *are independently generated from the densities* $f(x^{(i)}|\theta^{(i)})$, $i = 1, 2, \ldots$ *(here f is the same for the entire sequence). Define the random variables*

$$Z_i = L(\theta^{(i)}, \delta_1(X^{(i)})) - L(\theta^{(i)}, \delta_2(X^{(i)})),$$

and assume that $E_{\theta^{(i)}}[Z_i - E_{\theta^{(i)}}(Z_i)]^2 < \infty$ *for all i.*

(a) *If* $R(\theta, \delta_1) - R(\theta, \delta_2) > \varepsilon > 0$ *for* <u>*all*</u> $\theta \in \Theta$, *then*

$$P_{\boldsymbol{\theta}}\left(\liminf_{N\to\infty} \frac{1}{N} S_N > \varepsilon\right) = 1 \tag{4.135}$$

for any sequence $\boldsymbol{\theta}$.

(b) *If* $R(\theta, \delta_1) - R(\theta, \delta_2) > 0$ *for all* θ, Θ *is closed, and* $R(\theta, \delta_1)$ *and* $R(\theta, \delta_2)$ *are continuous in* θ, *then* (4.135) *is valid for any* <u>*bounded*</u> *sequence* $\boldsymbol{\theta}$ *(although $\varepsilon > 0$ could depend on the bound).*

PROOF. Observe that

$$\psi(\theta^{(i)}) = E_{\theta^{(i)}}[Z_i] = R(\theta^{(i)}, \delta_1) - R(\theta^{(i)}, \delta_2).$$

Thus, by the strong law of large numbers,

$$\frac{1}{N}\sum_{i=1}^{N} [Z_i - \psi(\theta^{(i)})] \to 0$$

with probability one. Under the condition on the risks in part (a), $\psi(\theta^{(i)}) > \varepsilon$ for all i, and the result is immediate. The proof of part (b) is left as an exercise.

□

The moment condition in the theorem is trivially satisfied for bounded losses, and usually holds even for unbounded losses. Knowing that (4.135) holds would seem to be a serious indictment of δ_1, since this indicates that δ_1 will be inferior to δ_2 in actual practical use. The indictment in part (a) is particularly serious, since the conclusion holds for *any* sequence $\boldsymbol{\theta}$. Even the part (b) conclusion, that (4.135) holds for any bounded $\boldsymbol{\theta}$, is disturbing since, in practice, the sequence $\boldsymbol{\theta}$ probably will be bounded; we may not *know* the bound (the common reason for using an unbounded Θ in the first place), but δ_2 will beat δ_1 in repeated use *regardless of knowledge of the bound.*

EXAMPLE 46 (continued). It can be shown that the moment condition on the Z_i and the continuity condition on the risks are satisfied for either the problem of estimating means, $\boldsymbol{\theta}^{(i)}$, or of estimating $\eta^{(i)} = |\boldsymbol{\theta}^{(i)}|^2$. For estimating means, we saw that $R(\boldsymbol{\theta}, \boldsymbol{\delta}^0) - R(\boldsymbol{\theta}, \boldsymbol{\delta}^{\text{JS}}) > 0$ for all $\boldsymbol{\theta}$, so that the conclusion in part (b) of Theorem 10 applies. For estimating η, we saw that $R(\boldsymbol{\theta}, \delta^{\pi}) - R(\boldsymbol{\theta}, \delta^{p}) = 4p^2$, so that the conclusion in part (a) of Theorem 10 applies.

It is important to keep inadmissibility results, such as these, in proper perspective. Unless ε in (4.135) is quite large, δ_1 might be perfectly satisfactory in practice (as was stated to be the case for problems of estimating $\boldsymbol{\theta}$ in Example 46 when only very vague prior information is available). And, again, inadmissibility applies only to automated use of δ_1, in say a computer package. The need to at least consider admissibility when developing automated generalized Bayes rules seems strong, however. Other discussions, similar to that in this subsection, can be found in Hill (1974), Heath and Sudderth (1978), and Berger (1984d).

A repetitive but better-safe-than-sorry comment: the last two subsections should not be interpreted as demonstrating that noninformative prior Bayesian analysis is bad. Throughout the book, we have argued that such analysis is very powerful and almost always gives excellent results. An attempt has been made to expose the pitfalls of the noninformative prior Bayesian approach, but we feel that these pitfalls are far less frequent and less deep than those for other competing methods of "automatic" analysis.

Dutch Book Arguments

The use of (4.135) to compare δ_1 and δ_2 is reminiscent of so-called "Dutch book" or "betting coherency" arguments. The typical Dutch book scenario deals with evaluation of methods (usually inference methods) which produce, for each x, either a probability distribution for θ, say $q_x(\theta)$ (which could be a posterior distribution, a fiducial distribution, etc.), or a system of confidence statements $\{C(x), \alpha(x)\}$ with the interpretation that θ is felt to be in $C(x)$ with probability $\alpha(x)$. (Note that frequentist "confidence" theory is excludable from this scenario, in that it does not claim to yield anything resembling the probability that θ is in $C(x)$.) For simplicity, we will restrict ourselves to the confidence statement framework; any $\{q_x(\theta)\}$ can be at least partially evaluated through confidence statements, by choosing $\{C(x)\}$ and letting $\alpha(x)$ be the probability (with respect to q_x) that θ is in $C(x)$.

The assumption is then made (more on this later) that, since $\alpha(x)$ is thought to be the probability that θ is in $C(x)$, the statistician who proposes $\{C(x), \alpha(x)\}$ should be willing to make both the bet that θ is in $C(x)$ at odds of $(1-\alpha(x))$ to $\alpha(x)$, and the bet that θ is not in $C(x)$ at odds of $\alpha(x)$ to $(1-\alpha(x))$. One can then set up a long run evaluation scheme, where a sequence of problems $\{(\theta^{(1)}, X^{(1)}), (\theta^{(2)}, X^{(2)}), \ldots\}$ is again considered, and in which the statistician must accept any bets in each problem according to his stated odds.

Suppose the statistician's opponent bets according to the betting function $s(x)$, where (following Robinson (1979a, b)) $s(x)=0$ means that no bet is offered; $s(x)>0$ means that an amount $s(x)$ is bet that $\theta \in C(x)$; and $s(x)<0$ means that the amount $|s(x)|$ is bet that $\theta \notin C(x)$. The loss to the statistician in the ith problem can then be shown to be

$$W_i = [I_{C(x^{(i)})}(\theta^{(i)}) - \alpha(x^{(i)})]s(x^{(i)}),$$

and of interest is again the limiting behavior of $S_N = \sum_{i=1}^N W_i$. If (4.135) holds for all $\boldsymbol{\theta}$, the statistician is called *incoherent* (or, alternatively, $s(x)$ is said to be a *super relevant* betting strategy), while, if (4.135) holds only for bounded $\boldsymbol{\theta}$, the statistician is called *weakly incoherent* (or $s(x)$ is *weakly relevant*). These concepts can be found in this or related form in such works as Buehler (1959, 1976), Wallace (1959), Freedman and Purves (1969), Bondar (1977), Heath and Sudderth (1978), Robinson (1979a, 1979b), and Lane and Sudderth (1983).

If $\{C(x), \alpha(x)\}$ is incoherent or weakly incoherent, then the statistician will for sure lose money in the repeated betting scenario, which certainly casts doubt on the validity of the probabilities $\{\alpha(x)\}$. A number of objections to this scenario have been raised, however, and careful examination of these objections is worthwhile.

Objection 1. The statistician will have no incentive to bet, unless he perceives the odds as slightly favorable. This turns out to be no problem if incoherence is present, since the odds can be adjusted by $\varepsilon/2$ in the statistician's favor, and he will still lose. If only weak incoherence is present, it is still often possible to adjust the odds by a function $g(x)$, so that the statistician perceives the bets to be in his favor, and yet he will lose in the long run.

Objection 2. The situation is unfair to the statistician, since his opponent gets to choose when, how much, and which way to bet. Various proposals have been made to "even things up." The possibility mentioned in Objection 1 is one such proposal, but does not change the conclusions much. A more radical possibility, suggested by Fraser (1977), is to allow the statistician to decline bets. This can have a drastic effect, but strikes us as too radical, in that it gives the statistician license to state completely silly $\alpha(x)$ for some x. It is after all the $\{\alpha(x)\}$ that are being tested, and testing should be allowed for all x.

Objection 3. The most serious objection to the betting scenario is that $\{\alpha(x)\}$ is generally not selected for use in betting, but rather to communicate information about θ. It may be that there is no *better* choice of $\{\alpha(x)\}$ for communicating the desired information. Consider the following example, which can be found in Buehler (1971), and is essentially successive modifications by Buehler and H. Rubin of an earlier example of D. Blackwell.

EXAMPLE 47. Suppose X has density $f(\theta+1|\theta)=f(\theta-1|\theta)=\frac{1}{2}$, and $\theta\in\Theta=$ {integers}. We are to evaluate the confidence we attach to the sets $C(x)=\{x+1\}$ (the point $(x+1)$), and a natural choice is $\alpha(x)=\frac{1}{2}$ (since θ is either $x-1$ or $x+1$, and in the absence of fairly strong prior information about θ, either choice seems equally plausible). This choice can be beaten in the betting scenario, by betting that θ is not in $C(x)$ with probability $g(x)$, where $0<g(x)<1$ is a continuous increasing function. (Allowing a randomized betting strategy does not seem unreasonable.) Indeed, the expected gain per bet of one unit, for any bounded $\boldsymbol{\theta}$, is $\sup_{\theta_i\in\boldsymbol{\theta}}[g(\theta_i+1)-g(\theta_i-1)]>0$, so that $\alpha(x)=\frac{1}{2}$ is weakly incoherent. (A continuous version of this example, mentioned in Robinson (1979a), has $X\sim\mathcal{N}(\theta,1)$, $\Theta=\mathbb{R}^1$, $C(x)=(-\infty,x)$, and $\alpha(x)=\frac{1}{2}$.)

In this and other examples where $\{\alpha(x)\}$ loses in betting, one can ask the crucial question—Is there a better α that could be used? The question has no clear answer, because the purpose of α is not clearly defined. One possible justification for $\alpha(x)=\frac{1}{2}$, in the above example, is that it is the unique limiting probability of $C(x)$ for sequences of what could be called increasingly vague prior distributions (cf. Stone (1970)). (A more formal Bayesian justification along these lines would be a robust Bayesian justification, to the effect that the class of possible priors is so large that the range of possible posterior probabilities for $(-\infty,x)$ will include $\frac{1}{2}$ for all x.) An alternative justification can be found by retreating to decision theory, attempting to quantify how well $\alpha(x)$ performs as an indicator of whether or not θ is in $C(x)$, and then seeing if there is any better α. For instance, using the quadratic scoring function (see Subsection 2.4.3) as an indicator of how well $\alpha(x)$ performs, would mean considering the loss function

$$L_C(\theta, \alpha(x)) = (I_{C(x)}(\theta) - \alpha(x))^2.$$

(For the moment, we are considering $\{C(x)\}$ as given, and worrying only about the choice of α. Note that, for any posterior distribution on θ, the optimal choice of $\alpha(x)$ for L_C is the posterior probability of $C(x)$, so that L_C is a natural measure of the accuracy of α.) One can then ask if there is a better α, employing usual decision-theoretic ideas. The answer in the case of Example 47 is—no. It can be shown that $\alpha(x) = \frac{1}{2}$ is admissible for this loss, and hence no improvement is possible. (The same cannot necessarily be said, however, if choice of $C(x)$ is brought into the picture. For instance, a reasonable overall loss for $\{C(x), \alpha(x)\}$ is

$$L(\theta, C(x), \alpha(x)) = c_1(I_{C(x)}(\theta) - \alpha(x))^2 + c_2(1 - I_{C(x)}(\theta)) + c_3\,\mu(C(x)), \tag{4.136}$$

where c_i are constants and μ is a measure of the size of $C(x)$. It can be shown in Example 47 that $\{C^*(x), \alpha^*(x)\}$, with $\alpha^*(x) \equiv \frac{1}{2}$ and

$$C^*(x) = \begin{cases} \{x-1\} & \text{with probability } g(x), \\ \{x+1\} & \text{with probability } 1-g(x), \end{cases} \tag{4.137}$$

is a better procedure than the given $\{C(x), \alpha(x)\}$.)

Decision-theoretic inadmissibility, with respect to losses such as L_C, can be related to incoherency, and seems to be a criterion somewhere between weak incoherency and incoherency (cf. Robinson (1979a)). This supports the feeling that it may be a more valid criterion than the betting criterion. This is not to say that the betting scenarios are not important. Buehler, in discussion of Fraser (1977), makes the important point that, at the very least, betting scenarios show when quantities such as $\alpha(x)$ "behave differently from ordinary probabilities." And as Hill (1974) says,

> "... the desire for coherence ... is not primarily because he fears being made a sure loser by an intelligent opponent who chooses a judicious sequence of gambles ... but rather because he feels that incoherence is symptomatic of something basically unsound in his attitudes."

Nevertheless, Objection 3 often prevents betting incoherency from having a conclusive impact, and so decision-theoretic inadmissibility (with respect to an agreed upon criterion) is more often convincing.

Decision-theoretic methods of evaluating "inferences" such as $q_x(\theta)$ (i.e., distributions for θ given x) have also been proposed (cf. Eaton (1982) and Gatsonis (1984)). For the most part, however, little attention has been directed to these matters.

4.9. Bayesian Calculation

Conceptually, the Bayesian paradigm is easy to implement. For any prior, π, one need only calculate the desired posterior feature of interest, say

$$E^{\pi(\theta|x)}[g(\theta)]=\frac{\int_\Theta g(\theta)f(x|\theta)\pi(\theta)d\theta}{\int_\Theta f(x|\theta)\pi(\theta)d\theta} \tag{4.138}$$

(assuming a continuous situation). Here g could be an indicator function (if calculating the posterior probability of a credible set or a hypothesis); $g(\theta)=\theta$ would yield the posterior mean, μ^π; $g(\theta)=(\theta-\mu^\pi)^2$ would yield the posterior variance; and $g(\theta)=L(\theta, a)$ would yield the posterior expected loss of a.

Evaluating the integrals in (4.138) can be quite difficult, however, especially when Θ is high dimensional. For this reason, Bayesian theory is frequently concerned with choosing π so as to reduce the difficulty of the calculation, while retaining essential or desirable prior features (cf. the analyses in Section 4.6 and Subsection 4.7.10). Some numerical calculation is frequently inevitable, however.

In this section, we discuss three methods of calculating expressions such as that in (4.138): numerical integration, Monte Carlo integration, and analytic approximation. We can, unfortunately, only briefly discuss these three topics, and must completely ignore many other important areas in numerical analysis (such as solving likelihood equations, which can also be very important to a Bayesian as discussed in Subsection 3.5.4 and Section 4.5). The practicing Bayesian is well advised to become friends with as many numerical analysts as possible.

4.9.1. Numerical Integration

There are a wide variety of methods for numerical integration of quantities such as those in (4.138). These include Gaussian or Gauss–Hermite quadrature and use of Simpson's rule. As these methods belong almost entirely to the domain of numerical analysis, we will not discuss them here, except to mention two important points:

1. Do not assume that any given numerical integration method will work on a particular problem. Especially troublesome will be problems where $\pi(\theta)f(x|\theta)$ has sharp peaks and valleys (as a function of θ, of course). It is often necessary, in such problems, to guide the numerical integration process to ensure that no important peaks are missed. Indeed, some approximating analytical work may be needed to ensure sufficient accuracy or efficiency of the numerical integration (cf. Naylor and Smith (1982, 1983)).
2. The difficulties increase rapidly with the dimension of Θ. Indeed, numerical integration is rarely optimal in three or more dimensions; Monte Carlo methods become preferable.

4.9.2. Monte Carlo Integration

Suppose it is possible to generate an i.i.d. sequence of random variables $\{\theta_1, \theta_2, \ldots\}$, having common density $h(\theta) > 0$ on Θ. Note that

$$E^h\left[\frac{g(\theta)f(x|\theta)\pi(\theta)}{h(\theta)}\right] = \int_\Theta g(\theta)f(x|\theta)\pi(\theta)d\theta.$$

It follows (using the strong law of large numbers) that, under mild conditions,

$$\lim_{m\to\infty} \frac{1}{m}\sum_{i=1}^{m}\left[\frac{g(\theta_i)f(x|\theta_i)\pi(\theta_i)}{h(\theta_i)}\right] = \int_\Theta g(\theta)f(x|\theta)\pi(\theta)d\theta.$$

This can be used to approximate the numerator and denominator (setting $g(\theta)\equiv 1$) in (4.138), resulting in the approximation

$$E^{\pi(\theta|x)}[g(\theta)] \cong \frac{\sum_{i=1}^{m} g(\theta_i)w(\theta_i)}{\sum_{i=1}^{m} w(\theta_i)}, \tag{4.139}$$

where $w(\theta_i) = f(x|\theta_i)\pi(\theta_i)/h(\theta_i)$. The density $h(\theta)$ is called the *importance function*, and the process of generating θ_i according to h is often called *importance sampling.* (One could actually use different importance functions for the numerator and denominator in (4.138), but there are certain dangers in doing so.)

The key issue in Monte Carlo integration is that of finding a suitable h. On the one hand, it is desirable to choose h so that generation of the random $\{\theta_i\}$ is inexpensive. For instance, choosing $h(\theta)\equiv 1$ on $\Theta = (0, 1)$ will result in cheaper random number generation than will choosing h to be $\mathcal{B}e(10, 10)$ (it being cheaper to generate uniform random variables than beta random variables). This consideration must, however, be balanced against the desire to choose h so that the approximation in (4.139) is accurate for as small a choice of m as possible. Although the issue of choosing h to minimize m is very complex, a rough rule of thumb is that h should try to mimic the posterior, $\pi(\theta|x)$, as closely as possible. One wants to avoid having a large proportion of the $w(\theta_i)$ near zero (since such $w(\theta_i)$ will have little effect on (4.139) and will hence be "wasted" in terms of reducing the variability of (4.139)), or having some very large $w(\theta_i)$ (since these will increase the variability of (4.139)). Choosing $h(\theta)$ to be close to $\pi(\theta|x)$ will make $w(\theta)$ nearly constant, avoiding both problems.

Balancing the (often competing) goals in choosing h is something of an art, and no easy recipes can be given. The following comments will often be helpful, however.

1. When the likelihood, $l(\theta) = f(x|\theta)$, is itself a density *in* θ (after suitable normalization) it may make a good choice for h (since $\pi(\theta|x)$ will tend to be proportional to $l(\theta)$ for moderate or large samples). The normalizing constant for l will cancel in the numerator and denominator of

(4.139), so the approximation will become

$$E^{\pi(\theta|x)}[g(\theta)]\cong\frac{\sum_{i=1}^{m}g(\theta_i)\pi(\theta_i)}{\sum_{i=1}^{m}\pi(\theta_i)}. \tag{4.140}$$

When $f(x|\theta)$ is from an exponential family, $l(\theta)$ will indeed tend to be proportional to a standard density (in θ). For instance, if $\mathbf{X}=(X_1,\ldots,X_n)$ is an i.i.d. $\mathcal{N}(\theta,1)$ sample, then $l(\theta)=f(\mathbf{x}|\theta)$ is proportional to a $\mathcal{N}(\bar{x},n^{-1})$ density; this would often be a good choice for h.

2. We saw, in Subsection 4.7.8, that $\pi(\boldsymbol{\theta}|\mathbf{x})$ will tend to be approximately $\mathcal{N}_p(\hat{\boldsymbol{\theta}},[\hat{\mathbf{I}}(\mathbf{x})]^{-1})$ for large sample sizes, where $\hat{\boldsymbol{\theta}}$ is the m.l.e. for $\boldsymbol{\theta}=(\theta_1,\ldots,\theta_p)^t$ and $\hat{\mathbf{I}}(\mathbf{x})$ is the observed Fisher information matrix. It may thus be reasonable to choose $h(\boldsymbol{\theta})$ to equal this normal density and use (4.139). There is a possible danger, however, as evidenced by the following example from Subsection 4.7.8.

Example 10 (continued). The data consisted of five independent $\mathscr{C}(\theta,1)$ observations, and yielded $\hat{\theta}=4.45$ and $\hat{I}(\mathbf{x})=2.66$; one could thus consider choosing $h(\theta)$ to be the $\mathcal{N}(4.45,0.38)$ density. Note, however, that then (since $\mathbf{x}=(4.0,5.5,7.5,4.5,3.0)^t$ and $\pi(\theta)\equiv 1$)

$$\begin{aligned}w(\theta)&=\frac{f(\mathbf{x}|\theta)\pi(\theta)}{h(\theta)}\\&=\frac{[(0.76)\pi]^{1/2}\exp\{(\theta-4.45)^2/(0.76)\}}{\pi^5[1+(\theta-4)^2][1+(\theta-5.5)^2][1+(\theta-7.5)^2][1+(\theta-4.5)^2][1+(\theta-3)^2]},\end{aligned}$$

which is extremely large for large θ. Indeed

$$E^h\{[w(\theta)]^2\}=\int[f(\mathbf{x}|\theta)]^2[h(\theta)]^{-1}\,d\theta=\infty,$$

so that one cannot even be sure that the right-hand side of (4.139) converges to the correct answer!

The above example indicates that the tails of h should be no sharper than the tails of $[f(x|\theta)\pi(\theta)]$. This can sometimes be circumvented, however, for an otherwise suitable h, through the use of various tricks. One such trick is to break up the original integrals in (4.138) into integrals over various regions, using different importance functions over each region. Another possibility is to alter h by generating, say, the $\mathcal{N}_p(\hat{\boldsymbol{\theta}},[\hat{\mathbf{I}}(\mathbf{x})]^{-1})$ random $\boldsymbol{\theta}$, but replacing it by an independent $\boldsymbol{\theta}^*$ (having an appropriate large-tailed density) whenever $(\boldsymbol{\theta}-\hat{\boldsymbol{\theta}})^t[\hat{\mathbf{I}}(\mathbf{x})](\boldsymbol{\theta}-\hat{\boldsymbol{\theta}})$ is too large. (See Stein (1984) for some specific suggestions.) This process would correspond to a different importance function, of course.

3. The calculation in (4.139) depends on being able to calculate the $w(\theta_i)$ fairly easily. It may, however, be the case that either $f(x|\theta)$ or $\pi(\theta)$ is difficult to calculate. For instance, if π is a hierarchical prior, it may not have a simple closed-form density. It may then pay to choose $h(\theta)=\pi(\theta)$ (assuming the θ_i can be easily generated according to π, as may well be

the case in a hierarchical scenario), since the approximation then becomes

$$E^{\pi(\theta|x)}[g(\theta)] \cong \frac{\sum_{i=1}^{m} g(\theta_i) f(x|\theta_i)}{\sum_{i=1}^{m} f(x|\theta_i)},$$

avoiding the calculation of π.

4. The accuracy of the Monte Carlo approximation can sometimes be estimated theoretically, through determination of the variances of the $w(\theta_i)$. More often, however, it must be found empirically, through repetition of the calculation (with independent sets of random variables, of course), and calculation of a sample variance for (4.139).
5. Frequently, one must evaluate a large number of integrals of the same type. In statistical decision theory, for instance, it is necessary to evaluate $\rho(\pi(\theta|x), a) = E^{\pi(\theta|x)}[L(\theta, a)]$ for a range of a (so as to find the minimizing action). In such situations, it is efficient to use the *same* random $\{\theta_i\}$ to calculate the Monte Carlo approximations to $\rho(\pi(\theta|x), a)$ for all needed a. The generation of the $\{\theta_i\}$ is often the most expensive part of the Monte Carlo process, and efficiency savings in this generation can be very important.

   Another advantage of using the same $\{\theta_i\}$ for each a is that the $\rho(\pi(\theta|x), a)$ will then tend to have similar biases (i.e., the errors in the Monte Carlo approximations will all tend to be in the same direction for the various a). This makes it much easier to locate the approximate minimizing action. (Indeed, it can be impossible to numerically locate the minimum if significant biases, fluctuating in sign, were encountered.) Recall, of course, that one need only calculate and minimize the numerator of (4.138) (with $g(\theta) = L(\theta, a)$) to find the optimal action.
6. It pays to be somewhat wary of random number generators, especially pure pseudo-random number generators (as opposed to generators that utilize, at least partly, lists of physically generated random numbers). Problems with random number generators can include "nonrandomness" and extreme inefficiency (cf. Rubin (1976)).

There is a large and growing literature on Monte Carlo integration. A few references to Monte Carlo integration in Bayesian analysis are Kloek and van Dijk (1978), Stewart (1979, 1983), Stewart and Johnson (1972), and van Dijk and Kloek (1980, 1984). General references on Monte Carlo integration include Hammersley and Handscomb (1964), Simon (1976), and Rubinstein (1981). Johnstone and Velleman (1984) gives an extensive discussion of Monte Carlo estimation of variance.

4.9.3. Analytic Approximations

The expectation in (4.138) can be approximated by a number of analytic methods involving Taylor's series expansions. One such approximation, implicitly discussed in part (ii) of Result 8 (in Subsection 4.7.8), is based on approximating $\pi(\boldsymbol{\theta}|\mathbf{x})$ by a $\mathcal{N}_p(\hat{\boldsymbol{\theta}}^\pi, [\mathbf{I}^\pi(\mathbf{x})]^{-1}]$ distribution. The expectation of $g(\boldsymbol{\theta})$, with respect to this normal distribution, may be known. Although such approximations are basically "large n" approximations, they are often quite accurate for smaller n. This will especially be true if higher order terms of the Taylor series are taken into account. (See Lindley (1980), Tierney and Kadane (1984), and Berger and DasGupta (1985), for examples.)

The "art" in Taylor's series approximation is in selecting, from among the various possible expansions, one which achieves high accuracy with few terms. In this regard, Tierney and Kadane (1984) recommend the following modification of the standard asymptotic approximation discussed above:

Result 9. *Let* $L(\boldsymbol{\theta}) = \log[f(\mathbf{x}|\boldsymbol{\theta})\pi(\boldsymbol{\theta})]$ *and* $L^*(\boldsymbol{\theta}) = \log[g(\boldsymbol{\theta})f(\mathbf{x}|\boldsymbol{\theta})\pi(\boldsymbol{\theta})]$, *assuming that* f, π, *and* g *are positive. Let* $\hat{\boldsymbol{\theta}}$ *and* $\hat{\boldsymbol{\theta}}^*$ *denote the parameter values maximizing* L *and* L^*, *respectively. (We assume such exist.) Define* $\mathbf{I}^L(\mathbf{x})$ *and* $\mathbf{I}^{L^*}(\mathbf{x})$ *to be the* $(p \times p)$ *matrices with* (i, j) *elements, respectively,*

$$I_{ij}^L(\mathbf{x}) = -\left[\frac{\partial^2}{\partial\theta_i\,\partial\theta_j} L(\boldsymbol{\theta})\right]_{\boldsymbol{\theta}=\hat{\boldsymbol{\theta}}}, \qquad I_{ij}^{L^*}(\mathbf{x}) = -\left[\frac{\partial^2}{\partial\theta_i\,\partial\theta_j} L^*(\boldsymbol{\theta})\right]_{\boldsymbol{\theta}=\hat{\boldsymbol{\theta}}^*}.$$

Then, under suitable conditions,

$$E^{\pi(\boldsymbol{\theta}|\mathbf{x})}[g(\boldsymbol{\theta})] \cong \left(\frac{\det[\mathbf{I}^L(\mathbf{x})]}{\det[\mathbf{I}^{L^*}(\mathbf{x})]}\right)^{1/2} \frac{g(\hat{\boldsymbol{\theta}}^*)f(\mathbf{x}|\hat{\boldsymbol{\theta}}^*)\pi(\hat{\boldsymbol{\theta}}^*)}{f(\mathbf{x}|\hat{\boldsymbol{\theta}})\pi(\hat{\boldsymbol{\theta}})}. \tag{4.141}$$

PROOF. The heuristics of the result are essentially identical to that of Result 8; one writes

$$E^{\pi(\boldsymbol{\theta}|\mathbf{x})}[g(\boldsymbol{\theta})] = \frac{\int \exp\{L^*(\boldsymbol{\theta})\}d\boldsymbol{\theta}}{\int \exp\{L(\boldsymbol{\theta})\}d\boldsymbol{\theta}},$$

and expands L^* and L in Taylor's series about $\hat{\boldsymbol{\theta}}^*$ and $\hat{\boldsymbol{\theta}}$, respectively. The approximation tends to be increasingly accurate when $f(\mathbf{x}|\boldsymbol{\theta}) = \prod_{i=1}^n f_0(x_i|\boldsymbol{\theta})$ and $n \to \infty$. □

EXAMPLE 10 (continued). Recall that $X_1, \ldots, X_5$ were i.i.d. $\mathcal{C}(\theta, 1)$, $\pi(\theta) \equiv 1$, and $\mathbf{x} = (4.0, 5.5, 7.5, 4.5, 3.0)^t$ was observed. It was calculated, in Subsection 4.7.8, that $\hat{\theta}^L = 4.45$ (called $\hat{\theta}$ there) and $I^L(\mathbf{x}) = 2.66$ (called I^π there). To calculate the posterior mean, let $g(\theta) = \theta$. To strictly apply Result 9, one needs positive g, but the approximation is reasonable when, as here, $L(\boldsymbol{\theta})$ gives essentially no weight to $\boldsymbol{\theta}$ for which $g(\boldsymbol{\theta}) < 0$. (This can always be achieved by adding a constant to g, and then subtracting the constant from the final result.) Numerical maximization of $L^*(\theta)$ yields $\hat{\theta}^* = 4.54$, and calculation gives $I^{L^*}(\mathbf{x}) = 2.53$. Plugging these values into (4.141) yields $E^{\pi(\theta|\mathbf{x})}[\theta] \cong 4.61$.

To approximate the posterior variance, we could set $g(\theta) = (\theta - 4.61)^2$, but it works better (see Tierney and Kadane (1984)) to set $g(\theta) = \theta^2$, and use the fact that $E[(\theta-\mu^\pi)^2]=E[\theta^2]-[E\theta]^2$. For $g(\theta)=\theta^2$, maximization of $L^*(\theta)$ yields $\hat{\theta}^*=4.63$, and calculation of (4.141) yields $E^{\pi(\theta|x)}[\theta^2]\cong 21.92$. Thus the approximation to the posterior variance is $(21.92)-(4.61)^2=0.7$ (more digits would have had to be carried for two significant figure accuracy). These approximations to the posterior mean and variance are reasonably close to the exact values of 4.55 and 0.56, respectively, especially considering the small n.

When doing expansions, it can also sometimes pay to expand in terms of various special series of functions. An example can be found in Dickey (1967).

4.10. Bayesian Communication

4.10.1. Introduction

The majority of statistical analyses will not be solely for personal use by the analyst. Most will, instead, have the primary goal of communicating information to a variety of others. How is a Bayesian to undertake such communication?

We have previously discussed two possible answers. The first is to report the likelihood function, or statistics from which the likelihood function (or an approximation to it) can be easily reconstructed. Anyone seeking to utilize this information can simply process the likelihood function with their own prior. It may be advantageous to report the posterior, $\pi_0(\theta|x)$, from a noninformative prior Bayesian analysis instead of the likelihood function, catering to those who do not have or do not want to utilize their own prior information. Of course, anyone with a prior, π, can determine their own posterior from $\pi_0(\theta|x)$ by the formula (where π_0 is the noninformative prior)

$$\pi(\theta|x)=\frac{\pi_0(\theta|x)[\pi(\theta)/\pi_0(\theta)]}{\int \pi(\theta|x)[\pi(\theta)/\pi_0(\theta)]d\theta}; \tag{4.142}$$

note the simplicity that would result from conventionally using $\pi_0(\theta)\equiv 1$ as the noninformative prior.

The second type of Bayesian report for general consumption that we have discussed is the report that could arise when extreme robustness is present. Several situations were discussed in Section 4.7 in which the same (or essentially the same) Bayesian conclusion would follow from almost any reasonable prior distribution. Large sample theory was one such situation. Others occurred in various testing scenarios. Some hierarchical Bayes analyses were also judged to possess a good deal of such inherent robustness.

In general, however, it will not be the case that essentially unique answers occur. Indeed, Smith (1984) states,

> "... one of the most attractive features of the Bayesian approach is its recognition of the legitimacy of a plurality of (coherently constrained) responses to data. Any approach to scientific inference which seeks to legitimize *an* answer in response to complex uncertainty is, for me, a totalitarian parody of a would-be rational learning process."

The keys to effective Bayesian communication, when there are a multitude of possible answers, are:

(i) Keep the number and complexity of prior inputs that must be specified as low as possible.
(ii) Present the answers in a format that allows easy processing of prior information.

The first point, though seemingly obvious, goes somewhat contrary to natural Bayesian instincts. A Bayesian, in developing methodology, often attempts to accommodate as wide a variety of priors as possible. There is, of course, nothing inherently wrong with this. The result, however, can appear bewildering to the practitioner, if unaccompanied by specific practical advice about which inputs to worry about, or how to determine the inputs. And the second point is almost as clear; the interested reader of an applied statistical analysis will tend to become very uninterested if he is told that he must go to the computer and do a numerical integration to obtain the answer. Thus, either simple formulas or simple graphical displays would be highly desirable.

There is a moderate amount of literature on this issue of Bayesian communication, including Hildreth (1963), Dickey (1973), Dickey and Freeman (1975), Smith (1978), and Berger and DasGupta (1985). Ultimately desirable, of course, are computer packages capable of semi-automatic implementation of these communication techniques.

4.10.2. An Illustration: Testing a Point Null Hypothesis

The situation of testing a point null hypothesis provides a good illustration of the above ideas. The example is particularly appropriate because, as discussed in Subsection 4.3.3, there are not good "objective" methods of testing a point null hypothesis; subjective Bayesian analysis, with its multitude of possible answers, thus becomes almost necessary.

We will consider here the specific situation of observing i.i.d. $\mathcal{N}(\theta, \sigma^2)$ random variables, $X_1, \ldots, X_n$, where both θ and σ^2 are unknown, and in which it is desired to test H_0: $\theta = \theta_0$ versus H_1: $\theta \neq \theta_0$. It is reasonable to suppose that θ and σ^2 are independent, *a priori*; let π^* denote the distribution of θ and π^{**} denote that of σ^2. As in Subsection 4.3.3, we assume that

π^* gives probability π_0 to $\theta = \theta_0$, while spreading out the probability of $(1-\pi_0)$ on $\theta \neq \theta_0$ according to the density $g(\theta)$. Writing $\mathbf{X} = (X_1, \ldots, X_n)$ and letting $f(\mathbf{x}|\theta, \sigma^2)$ denote the density of $\mathbf{X}$, it follows, as in (4.14), that the posterior probability of H_0 is

$$\pi(\theta_0|\mathbf{x}) = \left[1 + \frac{(1-\pi_0)}{\pi_0} \frac{m(\mathbf{x}|g, \pi^{**})}{m(\mathbf{x}|\theta_0, \pi^{**})}\right]^{-1}, \tag{4.143}$$

where

$$m(\mathbf{x}|g, \pi^{**}) = \int\int f(\mathbf{x}|\theta, \sigma^2) g(\theta) \pi^{**}(\sigma^2) d\sigma^2 \, d\theta, \tag{4.144}$$

$$m(\mathbf{x}|\theta_0, \pi^{**}) = \int f(\mathbf{x}|\theta_0, \sigma^2) \pi^{**}(\sigma^2) d\sigma^2.$$

We now consider a series of specific choices for g and π^{**}, choices designed to allow reasonably flexible modelling of prior information, while allowing simplified Bayesian communication. The first issue concerns σ^2. This is a nuisance parameter, and may well be a parameter for which little prior information is available. If such is the case, substantial simplification can be achieved by choosing $\pi^{**}(\sigma^2) \propto \sigma^{-2}$, i.e., giving σ^2 a noninformative prior density. (Any inverted gamma prior for σ^2 could also be dealt with in the following fashion, but we are trying to avoid all but the most necessary prior inputs.) The integrations over σ^2 in (4.144) can then be carried out explicitly, yielding

$$m(\mathbf{x}|g, \pi^{**}) = k(n, s^2) \int \left[1 + \frac{n(\bar{x}-\theta)^2}{(n-1)s^2}\right]^{-n/2} g(\theta) d\theta \tag{4.145}$$

and

$$m(\mathbf{x}|\theta_0, \pi^{**}) = k(n, s^2)[1 + t^2/(n-1)]^{-n/2}, \tag{4.146}$$

where $s^2 = \sum_{i=1}^n (x_i - \bar{x})/(n-1)$, $t = \sqrt{n}(\bar{x} - \theta_0)/s$, and $k(n, s^2)$ is *chosen* for convenience to be

$$k(n, s^2) = \Gamma\left(\frac{n}{2}\right) \Big/ \left\{ s\Gamma\left(\frac{n-1}{2}\right)\left[\frac{(n-1)\pi}{n}\right]^{1/2} \right\}. \tag{4.147}$$

(Any multiple of σ^{-2} serves equally well as a noninformative prior for σ^2, so the choice of k is quite arbitrary. The reason for the choice in (4.147) is that integrating $f(\mathbf{x}|\theta, \sigma^2)$ with respect to π^{**} gave an *effective likelihood function of* $[1 + n(\bar{x}-\theta)^2/\{(n-1)s^2\}]^{-n/2}$, which is proportional to a $\mathcal{T}(n-1, \bar{x}, s^2/n)$ density (in θ); $k(n, s^2)$ is the appropriate normalizing constant for this density.)

Consider, next, the choice of g. One important point is that, if g is moderately flat (as is likely to be the case) and n is moderately large, then a quite good approximation to (4.145) is

$$m(\mathbf{x}|g, \pi^{**}) \cong g(\bar{x}). \tag{4.148}$$

Together with (4.143) and (4.146), this yields

$$\pi(\theta_0|\mathbf{x}) \cong \left[1 + \frac{(1-\pi_0)}{\pi_0} \times \frac{g(\bar{x})[1+t^2/(n-1)]^{n/2}}{k(n, s^2)}\right]^{-1}. \tag{4.149}$$

In terms of simplicity, this formula would appear to be an adequate Bayesian communication device. It is crucial to ask, however, if the prior inputs that are needed will be reasonably accessible to a user. Specification of π_0 should pose no real problem and is, in any case, an essential prior input (unless the users have been trained in interpretation of Bayes factors, which would be highly desirable). Is prior specification of $g(\bar{x})$ also feasible? The answer to this question depends on who is doing the specifying. Someone with a moderately sophisticated knowledge of probability might reasonably be asked to specify $g(\bar{x})$, but others would be hard pressed to do so. For communicating with nonexperts, therefore, an approach along the following lines might well be preferable.

For this situation, the functional form of g does not have a great effect (unless $\bar{x}$ is far from θ_0, in which case it is clear that H_0 should be rejected), so one could restrict attention to (say) the class of $\mathscr{C}(\mu, \tau)$ priors. The parameter μ could be interpreted as the "best guess" for θ, assuming H_0 to be false, while τ could be determined from the fact that $(\mu-\tau, \mu+\tau)$ should have probability $\frac{1}{2}$, assuming H_0 to be false. Specification of μ and τ should be possible, even by users with only rudimentary statistical knowledge. For such a choice of g, (4.149) becomes

$$\pi(\theta_0|\mathbf{x}) \cong \left[1 + \frac{(1-\pi_0)}{\pi_0} \cdot \frac{1}{k(n, s^2)\pi\tau} \cdot \frac{(1+t^2/(n-1))^{n/2}}{(1+(\bar{x}-\mu)^2/\tau^2)}\right]^{-1}. \tag{4.150}$$

(An even better approximation to $\pi(\theta_0|\mathbf{x})$ follows from Exercise 144.) To summarize the suggested procedure, the reader of a statistical analysis is required to specify: (i) π_0, his prior probability of H_0; (ii) μ, his best guess for θ if H_0 is false; and (iii) τ, the perceived accuracy of the guess μ. The conclusion for the reader is then given by (4.150).

Are the inputs π_0, μ, and τ all necessary? The importance of π_0 has already been discussed, and the effects of μ and τ in (4.150) are substantial. Very often it may be the case that $\mu = \theta_0$ (i.e., g is deemed to be nonincreasing in $|\theta-\theta_0|$) and it may then be reasonable to present (4.150) only as a function of π_0 and τ. Elimination of τ is impossible, however. Indeed, the effect of τ in (4.150) is almost as pronounced as the effect of π_0. (We would thus argue against the conventional choice in Jeffreys (1961) of $\tau=\sigma$, except that, if a conventional choice must be made so as to have an "objective" test, then choosing $\pi_0=\frac{1}{2}$ and $\tau=\sigma$ is probably the least of evils.)

As an alternative to a simple formula such as (4.150) (or when the approximation is too crude and numerical calculation of $\pi(\theta_0|\mathbf{x})$ is necessary), one can consider graphical displays of $\pi(\theta_0|\mathbf{x})$ as a function of the inputs π_0, μ, and τ. Figure 4.4 presents one such display, for the situation

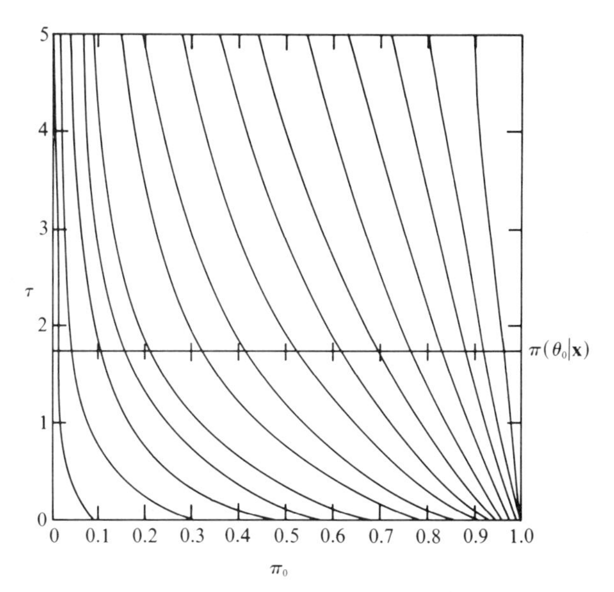

Figure 4.4. Posterior probabilities of H_0.

$n = 15$, $t = 2.145$, and $\mu = \bar{x}$. (The choice $\mu = \theta_0$ would be more natural in practice. We present this situation because it again corresponds to the choice of μ which is least favorable to H_0, and yet the posterior probabilities in Figure 4.4 indicate much less evidence against H_0 than would the classical P-value of 0.05, corresponding to $t = 2.145$.) The graph gives contours of $\pi(\theta_0|\mathbf{x})$ as π_0 and τ vary. The value of a contour is given by the point at which the contour crosses the axis labelled $\pi(\theta_0|\mathbf{x})$. For instance, the prior inputs $\pi_0 = 0.7$ and $\tau = 2.5$ yield the contour which crosses the $\pi(\theta_0|\mathbf{x})$ axis at about 0.76, which is thus the posterior probability of H_0 for the given prior inputs. For some readers of the statistical analysis, even a graph such as Figure 4.4 might be too complicated. It might pay to include, for such readers, a graph of (say) $\pi(\theta_0|\mathbf{x})$ versus τ, with π_0 fixed at $\frac{1}{2}$.

One of the advantages of graphical displays is that they allow easy sensitivity investigations. Thus, if $0.4 \le \pi_0 \le 0.6$ and $2 \le \tau \le 3$, it is clear from Figure 4.4 that the minimum and maximum of $\pi(\theta_0|\mathbf{x})$ are obtainable from the contours corresponding to (0.4, 2) and (0.6, 3), respectively: the result is (approximately) $0.44 \le \pi(\theta_0|\mathbf{x}) \le 0.72$. Calculational methods for easily determining various graphs in this situation can be found in Berger and Das Gupta (1985) (even for the general case of arbitrary π_0, μ, and τ).

4.11. Combining Evidence and Group Decisions

The Bayesian paradigm, as it tends to arise in statistics, is deceptively simple. One simply takes the likelihood ($l(\theta) = f(x|\theta)$) arising from the observed data x, multiplies it times the prior density $\pi(\theta)$, and normalizes to obtain the posterior probability distribution of θ. The simplicity with which the experimental and prior information can be combined is due to the fact that the likelihood arises as the conditional distribution of X given θ, whereas the prior can be interpreted as the "marginal" distribution of θ; multiplying the two densities gives the joint density of X and θ, from which the posterior can be directly calculated.

There are many situations, however, in which one has two or more information sources about θ, but for which the information does not naturally occur in the above easily utilizable form. Subsection 4.11.1 and 4.11.2 consider the case of a single analyst needing to process various such information sources. Subsection 4.11.3 considers the even more involved group decision scenario, in which there is a group of individuals with differing information (i.e., different posteriors and losses), and the group must reach a collective decision.

It should be stated, at the outset, that these are very hard problems. No attempt will be made to provide general answers, it being impossible to give broad prescriptions (analogous to Bayes's formula). Instead we concentrate on clearly exposing the problems, and arguing that many of the proposed *ad hoc* methods of combining evidence are not always viable. Our position will be that one can best deal with a hard problem in uncertainty by attempting honest probabilistic modelling of the situation, and processing probabilistic inputs with the machinery of probability (of which Bayes's theorem is one aspect).

4.11.1. Combining Probabilistic Evidence

The situation considered in this section is that of having available m sources of information about θ, to be denoted $\pi_1, \ldots, \pi_m$. These could be likelihoods arising from experiments, posterior distributions for θ from "experts," "prior" distributions arising from nonexperimental information sources, etc.

Example 48. A collection of experts is consulted, and each produces a probability distribution $(\pi_i(\theta))$ for θ, which summarizes the expert's judgement concerning θ. These could be considered to be the experts' posterior distributions for θ.

Example 49. Two different medical diagnostic tests are performed, resulting in observations x_1 and x_2, respectively, and likelihood functions (keeping with the above π_i notation) $\pi_1(\theta)=f_1(x_1|\theta)$ and $\pi_2(\theta)=f_2(x_2|\theta)$ (f_i being the density for the ith experiment).

Example 50. A satellite picture of a certain area yields spectral information, X, having density $f(x|\theta)$. (Usually Θ will be a finite set of possible classifications for the area, such as corn, forest, grass, dirt, etc.) Thus the experimental source of information can be summarized by the likelihood function $\pi_1(\theta) = f(x|\theta)$ (for the observed x, of course). Also available is a (prior) distribution, π_2, for θ that arises from knowing the overall distribution of θ at the known elevation of the area under consideration. (An area at 10,000 feet in altitude is unlikely to be a farm crop.) Also available is a prior distribution, π_3, from the preceding year, which arose as the empirical distribution of θ for the entire region under study. (The area in the satellite picture is a small subset of this region, and π_3 consists of the total proportions of corn, forest, grass, dirt, etc. that were found in the entire region during last year's study.)

How should one process these different sources of information? Many *ad hoc* systems have been proposed. Typical are the following two, based on first normalizing the $\pi_i(\theta)$ so that they are actual probability distributions (which we will henceforth assume):

Linear Opinion Pool. Assign a positive weight w_i (where $\sum_{i=1}^m w_i = 1$) to each information source π_i (supposedly to reflect the confidence in that information source), and then use

$$\pi(\theta) = \sum_{i=1}^{m} w_i \pi_i(\theta), \tag{4.151}$$

as the overall probability distribution for θ.

Independent Opinion Pool. When the information sources seem "independent," use, as the overall probability distribution for θ,

$$\pi(\theta) = k\left[\prod_{i=1}^{m} \pi_i(\theta)\right], \tag{4.152}$$

where k is the appropriate normalizing constant. (Note that independent likelihoods from statistical experiments are indeed combined in this way.) A frequently considered generalization of (4.152) is the *logarithmic opinion pool*, $\pi(\theta) = k[\prod_{i=1}^m \pi_i^{\alpha_i}]$, where the α_i are positive constants.

The inadequacies of such *ad hoc* formulas in general use are not difficult to expose. Consider the following examples, dealing first with the additive, and then the multiplicative, formulas.

EXAMPLE 51. Two paleontologists, P_1 and P_2, are asked to classify a fossil as belonging to time period T_1, T_2, or T_3. The first paleontologist, after considering all evidence, provides the probability distribution $\pi_1 = (0.1, 0.7, 0.2)$ (i.e., the fossil has probability 0.1 of belonging to period T_1, etc.). The second paleontologist provides the distribution $\pi_2 = (0, 0.6, 0.4)$.

Consider now the assignment, by P_2, of zero probability to T_1. Such a definitive assignment is likely to be based on some clearcut evidence missed by (or unavailable to) P_1; it would thus seem reasonable to conclude that T_1 has nearly zero probability of being the correct period. But this would rule out use of the linear opinion pool (4.151), assuming that both experts are to be given significant weight.

As an even more explicit version of this example, suppose P_2 is a carbon-dating specialist, and that his conclusion was based only on this evidence (i.e., was simply the normalized likelihood function from the carbon-dating experiment). If P_1 was unaware of this piece of evidence, then π_1 could be treated as the prior and π_2 as the likelihood; Bayes's theorem then yields, as the posterior probability vector for the time period,

$$\pi = \frac{(0.1\times 0, 0.7\times 0.6, 0.2\times 0.4)}{[0.1\times 0+0.7\times 0.6+0.2\times 0.4]} = (0, 0.84, 0.16). \tag{4.153}$$

Note that the experts' opinions have reinforced each other. A deficiency of the linear opinion pool is that it does not allow for such reinforcement of opinion.

The distribution in (4.153) is actually that which would result from application of the independent opinion pool. Thus the independent opinion pool does allow for reinforcement of opinion. While partial reinforcement is often desirable, the independent opinion pool can be too extreme in this regard. Outside of purely experimental statistical problems, it will be rare for the information sources to be independent, in the sense needed to apply (4.152). This is especially true of expert opinions, since the experts will often be basing their opinions on partially common data and training. Any such common background will introduce dependence among experts and this dependence can be very severe.

Example 52. As an artificial, but illustrative, example, suppose we train "experts" in a certain field. The experts are all taught to make judgements in the following way: if approached with evidence x, a realization of $X \sim f(x|\theta)$, they are to produce the final probability distribution

$$\pi_E(\theta) = f(x|\theta)\pi^*(\theta) \Big/ \int f(x|\theta)\pi^*(\theta)d\theta.$$

(Of course, these "experts" are being taught nothing more than to be Bayesians with the "expert" prior $\pi^*(\theta)$.) Suppose now that 100 such experts are each given identical evidence, x, and they each report back that $\pi(\theta)$ is a $\mathcal{N}(0, 1)$ distribution. The independent opinion pool would say to conclude that the final distribution for θ is $\pi(\theta) \propto [\pi_E(\theta)]^{100}$, which clearly yields the $\mathcal{N}(0, 1/100)$ distribution. The absurdity of this is apparent, the complete dependence among the experts implying there should be no reinforcement of opinion, whatsoever. The true final distribution for θ should be simply π_E (assuming, of course, that π^* is deemed to be reasonable). Although this is an extreme example, the effect of dependence among experts will often be too serious to ignore.

Note that, even if the experts obtained independent data, x_i (from experiments with densities $f_i(x_i|\theta)$), it would still be incorrect to use the independent opinion pool. Expert i would report

$$\pi_{E_i}(\theta) \propto f_i(x_i|\theta)\pi^*(\theta),$$

but the true posterior would be

$$\pi(\theta) \propto \left[\prod_{i=1}^{100} f_i(x_i|\theta)\right]\pi^*(\theta) \propto \left[\prod_{i=1}^{100} \pi_{E_i}(\theta)\right][\pi^*(\theta)]^{-99},$$

rather than $\pi(\theta) \propto \prod_{i=1}^{100} \pi_{E_i}(\theta)$ as the independent opinion pool would suggest.

The alternative to use of *ad hoc* rules for combining evidence is simply careful probabilistic modelling of the situation, combined with probabilistic processing (of which Bayes's theorem is one aspect). By probabilistic modelling, we mean obtaining the joint distribution of all random observables and unknown parameters of interest or, at least, determining enough to calculate the conditional (posterior) distribution of the desired θ given the observables (which may well include the expert opinions, etc.).

This is sometimes called the *super Bayesian* approach, to emphasize that it is a single decision maker (the super Bayesian) who is trying to process all the information to arrive at a distribution for θ which is consistent with probabilistic reasoning. It is, in general, hard to carry out extensive probabilistic modelling in complicated situations, and many (crude) approximations may be needed. We feel, however, that it is usually better to *approximate* a "correct answer," than to adopt a completely *ad hoc* approach.

Probabilistic modelling is often highly problem-specific. The following example presents, for illustrative purposes, a fairly complete probabilistic analysis in a simple case. Many related analyses have been done (cf. French (1980) and Lindley (1982b)).

EXAMPLE 53. Suppose $\Theta = \{\theta_1, \theta_2\}$, where θ_1 denotes the occurrence of some event (say, rain tomorrow) and θ_2 denotes nonoccurrence of the event. An expert is consulted, and his opinion is that θ_1 has probability $\pi_1(\theta_1) = p$ (and hence the probability of θ_2 is $\pi_1(\theta_2) = (1-p)$). The decision maker has his own estimate, $\pi_2(\theta_1)$, of the probability of θ_1. Thus the problem is simply to process these two pieces of information.

In attempting to construct an overall probability model for this situation, it is natural (and sensible) to focus on modelling the accuracy of the expert. In other words, we can seek to determine densities $f(p|\theta_1)$ and $f(p|\theta_2)$, reflecting the p that the expert would be likely to provide under each situation. For instance, suppose that we review the expert's predictions in similar situations in the past, and find that, when events analogous to θ_1 occurred (i.e., when it rained), the expert's predictions, p, roughly followed a $\mathcal{B}e(2, 1)$ distribution; and when the analogues to θ_2 occurred, the expert's p followed a $\mathcal{B}e(1, 2)$ distribution.

If no past records were available, one might intuitively reason something to the effect that, if θ_1 were true, then the expert would tend to report $p \cong \frac{2}{3}$, although this report could vary by as much as $\frac{1}{4}$ (interpreted as a standard deviation, say); and, if θ_2 were true, the expert would report $p \cong \frac{1}{3}$, with a standard deviation of again $\frac{1}{4}$. Reasonable distributions with roughly the implied means and variances are the $\mathcal{B}e(2, 1)$ and $\mathcal{B}e(1, 2)$ distributions, respectively.

Note that a good expert will tend to have p close to 1 or 0 as θ_1 or θ_2 is true, while a bad expert will have p much more spread out. In general, it might be sensible to specify a mean (p_0) and standard deviation (σ) for the expert's report (p) when θ_1 is true, and then model this by choosing $f(p|\theta_1)$ to be the $\mathcal{B}e(\alpha, \beta)$ density, where

$$p_0 = \alpha(\alpha+\beta)^{-1} \quad \text{and} \quad \sigma^2 = \alpha\beta(\alpha+\beta)^{-2}(\alpha+\beta+1)^{-1}$$

(see Appendix 1). The same modelling process could be undertaken for the expert's report when θ_2 is true, but it might be reasonable (for reasons of symmetry) to assume that $f(p|\theta_2) = f(1-p|\theta_1)$ (so that $f(p|\theta_2)$ will be $\mathcal{B}e(\beta, \alpha)$ when $f(p|\theta_1)$ is $\mathcal{B}e(\alpha, \beta)$).

Once $f(p|\theta_1)$ and $f(p|\theta_2)$ have been specified, Bayes's theorem can be applied to the problem to obtain

$$\pi(\theta_1|p) = \frac{f(p|\theta_1)\pi_2(\theta_1)}{f(p|\theta_1)\pi_2(\theta_1) + f(p|\theta_2)\pi_2(\theta_2)}.$$

Under the symmetry and beta density assumptions discussed above, this can be written in the simple form

$$\pi(\theta_1|p) = \left[1 + \left(\frac{1-p}{p}\right)^{\alpha-\beta} \frac{\pi_2(\theta_2)}{\pi_2(\theta_1)}\right]^{-1}. \tag{4.154}$$

As a specific instance, suppose $\alpha = 2$ and $\beta = 1$, as above, and that the expert reports $p = 0.7$ while the decision maker has $\pi_2(\theta_1) = 0.4$. Then $\pi(\theta_1|p) = 0.61$ (and $\pi(\theta_2|p) = 0.39$).

The modelling process undertaken in Example 53 is admittedly difficult, but we would argue that it is the only sound way to proceed. The crucial factor in evaluating the expert's advice is the skill of the expert, and anything short of probabilistic modelling of this skill (which then allows probabilistic processing of the information) is likely to be generally inadequate. If one must develop simplified formulas (for unsophisticated users), then at least the formulas should be based on some reasonable probability model, rather than sheer *ad hoc*-ery. Thus, the assumptions leading to formula (4.154) were not unreasonable, and this formula might represent a plausible simplified way of combining the two information sources. There is, furthermore, a certain appeal in choosing $(\alpha-\beta) = 1$, since then $\pi(\theta_1|p) = p$ when $\pi_2(\theta_1) = \frac{1}{2}$ (i.e., one just follows the expert's advice when nothing else is known about θ_1), but situations can certainly be imagined where other choices of $(\alpha - \beta)$ are desirable.

There have been a number of papers discussing the concerns mentioned here, and presenting useful probabilistic models in such situations as the combining of dependent expert opinions. A few of these references, from which others can be obtained, are Raiffa (1968), Winkler (1968, 1981), Hogarth (1975), Morris (1977, 1983), Lindley, Tversky, and Brown (1979), Dickey (1980), French (1980, 1981, 1984), McConway (1981), Bordley (1982), Genest and Schervish (1983), Schervish (1983), Barlow, Mensing, and Smiriga (1984), Genest (1984), Lindley (1982b, 1984), Lindley and Singpurwalla (1984), Zidek (1984) (which contains a general review), and Spiegelhalter and Knill-Jones (1984) (which includes a review of many nonprobabilistic expert systems). Among the many useful general works on probabilistic modelling, are Bernardo and Bermudez (1984), Howard and Matheson (1984) and Schacter (1984). It should be mentioned that probabilistic processing is possible through means other than Bayes's theorem (cf. Jeffrey (1968) and Diaconis and Zabell (1982)). Finally, there are general nonprobabilistic methods of combining evidence (cf. Dempster (1968) and Shafer (1976, 1982)), but we view these as either unnecessary elaborations on robust probabilistic analysis, or as insufficiently complicated representations of reality (see Subsection 4.12.2).

4.11.2. Combining Decision-Theoretic Evidence

The decision-theoretic issues involved in combining evidence are also quite interesting. It might, at first sight, seem that a partial solution to the problem of combining evidence in a decision problem would be to find an action that would be suitable for each source of information separately. That this need not be reasonable is indicated by the following example.

EXAMPLE 54. It is desired to decide between a_1 and a_2 under the loss function

$$L(\theta, a_1) = 1 - L(\theta, a_2) = \begin{cases} 0 & \text{if } |\theta| \le 1, \\ 1 & \text{if } |\theta| > 1. \end{cases}$$

Two experts provide posterior distributions for θ of $\pi_1 = \mathcal{N}(1, 2)$ and $\pi_2 = \mathcal{N}(0, 3)$, respectively. Calculation shows that $P^{\pi_1}(|\theta| \le 1) = 0.42$ and $P^{\pi_2}(|\theta| \le 1) = 0.44$; these are clearly the posterior expected losses of a_2 (and one minus the posterior expected losses of a_1) for π_1 and π_2. Thus, either expert would recommend a_2 on the basis of his own posterior distribution.

Suppose now that π_1 and π_2 actually arose as posteriors from normal experiments with noninformative priors. In other words, suppose Expert 1 performed the experiment of observing $X_1 \sim \mathcal{N}(\theta, 2)$, obtained the data $x_1 = 1$, and developed π_1 as the posterior distribution resulting from the use of the noninformative prior for θ. Similarly, suppose Expert 2 performed the experiment of observing $X_2 \sim \mathcal{N}(\theta, 3)$, obtained the data $x_2 = 0$, and did a noninformative prior Bayesian analysis. Well, if X_1 and X_2 were independent, then the combined evidence would clearly be given by the joint likelihood of x_1 and x_2. Indeed, a noninformative prior Bayesian analysis (for the data $x_1 = 1$ and $x_2 = 0$) would then yield, as the overall posterior,

$$\pi(\theta) \propto e^{-(\theta-1)^2/4} e^{-\theta^2/6} \propto e^{-(\theta-0.6)^2/(2.4)};$$

thus $\pi(\theta)$ would be a $\mathcal{N}(0.6, 1.2)$ density. But $P^{\pi}(|\theta| \leq 1) = 0.57$, so that the optimal action would be a_1, in spite of the fact that each expert, individually, would have recommended a_2. Such a result can hold even if there is substantial dependence between the experts. Thus there is no decision-theoretic shortcut that avoids careful probabilistic combination of evidence.

4.11.3. Group Decision Making

Suppose there are m members of a committee, to be identified simply as $\{1, 2, \ldots, m\}$, and that each member of the committee has a loss function, L_i, and prior, π_i, for a given decision problem involving θ. The job of the committee is to arrive at a consensus action, a, in the decision problem. We make the crucial simplifying assumption that sharing of information (i.e., discussion of losses and priors among the committee members) has already taken place, so that the L_i and π_i are the individuals' composite summaries of all available information. Without this assumption, the problem is not even well defined, in that the opinions would undoubtedly be changing as the committee members interacted. (See DeGroot (1974), Press (1978), and R. Berger (1981) for results and other references.)

If each committee member calculates his optimal Bayes action, a_i, and the actions turn out to be the same, then this common action will clearly be the group consensus action. (See Example 54 for an illustration of why it is crucial that information sharing occur before comparing perceived optimal actions; all committee members could begin deliberations believing that the same action is appropriate, but end up choosing quite a different action after sharing information.) An important generalization of this obvious conclusion arises upon recognition that the committee members should be robust Bayesian decision-theorists, and may well then have sets of actions they consider acceptable. If the intersection of the m members' sets is nonempty, it certainly seems natural to consider any action in the intersection to be a suitable consensus decision.

If no consensus decision emerges from the above considerations, what is to be done? In practice, it may happen that the final decision is referred to a single higher authority, but this would place the problem in the domain of Subsections 4.11.1 and 4.11.2 (a single decision maker trying to combine various sources of information to make a decision). Or the committee could take "no action," and seek further information, but this is itself really a type of decision (and one that should have been included as an option in the original decision-theoretic formulation of the problem). Or the committee could decide by some type of vote, but this introduces political considerations that vastly complicate the decision problem. We will restrict consideration to the special situation in which *all* committee members must agree on a particular action.

As a final aside, it is useful to recognize *why* there may be a lack of a consensus opinion. There are three main possibilities. First, the committee members may have different information which they will not reveal (secrets) or cannot adequately describe (personal experience). Second, certain of the committee members may be incoherent (i.e., non-Bayesians). Third, it may be impossible for the committee members to adequately process all available information, because of the problems discussed in Subsection 4.11.1, leaving them with no recourse but to use (possibly different) *ad hoc* methods of combining evidence to obtain the L_i and π_i.

Having clarified the problem, we can now state that it has no solution. There is *no* clearcut method by which the committee members should proceed to reach a consensus conclusion. One can, however, reduce the class of actions that need to be considered, through characterizations of *admissible* actions (i.e., those for which there exists no action which is considered as good or better by all committee members, and strictly better by at least one member). One such result follows.

Result 10. *Under certain (restrictive) conditions, the only actions that need to be considered (the admissible actions) are those which minimize*

$$\bar{\rho}(a) = \sum_{i=1}^{m} w_i \int L_i(\theta, a)\pi_i(\theta)d\theta, \tag{4.155}$$

for some set of constants $\{w_i\}$, *where* $0 \le w_i \le 1$ *and* $\sum_{i=1}^{m} w_i = 1$.

EXAMPLE 55. Three doctors in consultation must recommend a drug dosage level for a patient. There is an unknown minimal dosage, $0 \le \theta \le 1$, which will effect a cure, and the drug has serious deleterious side effects. Letting a denote the dosage level, the loss functions of the three doctors are of the form

$$L_i(\theta, a) = A_i I_{(a,\infty)}(\theta) + B_i a.$$

(There is a cost, A_i, for failure to cure, and a cost, $B_i a$, due to side effects, which is linearly increasing with the dosage level.) The doctors choose the pairs (A_i, B_i) to be $(1, 1)$, $(1, 1)$, and $(2, 1)$, respectively. The opinions of the three doctors concerning θ (after weighing all evidence and discussing the situation), are reflected by the densities $\pi_1(\theta) = \pi_2(\theta) = 6\theta(1-\theta)$ and $\pi_3(\theta) = 2(1-\theta)$ (on $(0, 1)$, of course).

Result 10 can be shown to apply in this situation. A calculation yields, for (4.155),

$$\bar{\rho}(a)=2(1-w_3)a^3+(5w_3-3)a^2+(1-4w_3)a+(1+w_3).$$

Differentiating and setting equal to zero yields the equation

$$0=\frac{d}{da}\bar{\rho}(a)=6(1-w_3)a^2+2(5w_3-3)a+(1-4w_3),$$

the largest root of which is

$$a^*(w_3)=\frac{[3-5w_3+(w_3^2+3)^{1/2}]}{[6(1-w_3)]};$$

this can be shown to minimize $\bar{\rho}(a)$. Furthermore, it is easy to check that $a^*(w_3)$ is decreasing in w_3, so that the range of possible actions is given by the interval $(a^*(1), a^*(0))=[0.75, 0.79]$. (Calculate $a^*(1)$ as $a^*(1)=\lim_{w_3\to 1} a^*(w_3)$.)

The main condition needed for the validity of Result 10 is that the nonrandomized actions form a complete class (i.e., any randomized action is inadmissible). The point is that it may well be "optimal" (if all committee members believe in personal expected utility) to choose a randomized action, say a 50–50 random choice between a_1 and a_2. The reasons for this can be found in Section 5.2. Indeed, the entire scenario, here, can be identified with the risk set formulation of Subsection 5.2.4; with θ_i identified with committee member i, $L(\theta_i, a)$ identified with $\int L_i(\theta, a)\pi_i(\theta)d\theta$, and π_i identified with w_i. It is shown in Subsection 5.2.4 (cf. Theorem 11), that Result 10 will hold under very mild conditions, if randomized actions are allowed. There are, however, practical objections to working with randomized actions: typical committees do not contain people that are completely comfortable with expected utility ideas, and the thought of letting a random mechanism select the final action will be unappealing to many such committees. Unfortunately, Result 10 will frequently fail if randomized actions are not allowed, so that determination of the admissible actions (admissible, now, within the class of nonrandomized actions) would have to proceed by some other mechanism. Conditions under which the nonrandomized actions *are* a complete class can be found in Weerahandi and Zidek (1983) and de Waal, Groenewald, van Zyl, and Zidek (1985). The latter paper contains a general discussion of complete class results for this problem.

Even if Result 10 does apply, the actual selection of an action from the admissible class remains difficult. Equation (4.155) suggests a Bayesian approach to selection: if the w_i are interpreted as the prior probabilities that are to be assigned to the committee members, then $\bar{\rho}(a)$ might be thought of as the actual Bayes risk, to be minimized over a. Unfortunately, even if it were possible to decide upon some prior probabilities (that the committee members were "correct," say), any interpretation of $\bar{\rho}(a)$ as the

Bayes risk is dependent on the L_i being constructed on a common utility scale. (As an extreme illustration, suppose the $w_i = 1/m$, but L_1 is measured on a scale that is 100 times larger than that of L_2; then $\bar{\rho}(a)$ will be influenced much more by committee member 1 than by 2.) One could attempt to have the members use a common utility scale, but it turns out that this is effectively impossible (cf. Luce and Raiffa (1957), Arrow (1966), Bacharach (1975), and Jones (1980)). It may be reasonable to *try* to put the L_i on a common scale, select the w_i in a mutually agreeable way, and determine a (or a randomized action, if necessary) by minimizing $\bar{\rho}(a)$ (cf. Bacharach (1975)), but such an approach is not clearly optimal.

An alternative approach to selecting an action from the admissible actions is via the idea of Nash equilibrium (originating in Nash (1950), for a slightly different problem). The key feature of this approach is that it does *not* require utility comparisons. For extensive development and examples of the approach, as it applies to the group decision problem, see Weerahandi and Zidek (1981, 1983), de Waal, Groenewald, van Zyl, and Zidek (1985), and Zidek (1984).

We have given only a very brief introduction to group decision making. References, in addition to those already mentioned, include Savage (1954), Winkler (1968), Hogarth (1975), French (1981, 1983), Hylland and Zeckhauser (1981), Wagner (1982), and Genest (1984).

4.12. Criticisms

We have discussed criticisms of the Bayesian viewpoint and given responses to these criticisms throughout this (and earlier) chapters. Nevertheless, for review and summary purposes, it is useful to list some of the major issues. The criticisms will be divided into two categories, the non-Bayesian criticisms and the Foundational criticisms. Under the heading of non-Bayesian criticisms, are placed those criticisms that tend to come from classical statisticians. Under the Foundational heading, are placed those criticisms that arise from beliefs that the Bayesian paradigm does not go far enough (into subjectivity) to provide a general foundation for statistics.

4.12.1. Non-Bayesian Criticisms

Objectivity

This issue was discussed in Sections 2.5, 3.7, 4.1, and 4.3, and in Subsection 4.7.11. The basic conclusion was that no method of analysis is fundamentally objective; the method with the greatest claim to being objective is probably noninformative prior Bayesian analysis, since it explicitly tries to model objectivity. Many non-Bayesian methods share the *appearance* of objectivity with noninformative prior Bayesian analysis, however, and it cannot be denied that such an appearance is important for certain audiences (due to our educational failures). The lack of objectivity of subjective Bayesian analysis is thus often a pragmatic consideration in applications. Of course, strenuous efforts should be directed towards disabusing users of the perceived magical objectivity of statistics.

Frequentist Justification

The argument that only frequentist probabilities and frequentist justifications are meaningful has been addressed in Sections 1.6, 3.1, 4.1, 4.3, 4.7, and 4.8. Frequentist justification (in the somewhat nonstandard sense discussed in Sections 1.6 and 4.8) was argued to be valuable if obtainable, but mainly in the role of a supporting player to the (conditional, robust) Bayesian star. We will not repeat these arguments.

Communicating Evidence

A naive view of the Bayesian position is that the statistician will choose his own prior and then report his personal Bayesian conclusion, leaving future users of the result (who might question the statistician's prior) in a quandary. The real goal of Bayesian analysis, however, is to assist the future (and present) users in conducting their own Bayesian analyses and drawing their own conclusions. Section 4.10 discussed this issue.

Prior Uncertainty

It is often very hard to determine prior distributions. This issue has been discussed extensively in Sections 3.7 and 4.7, and is unquestionably a serious problem. A large variety of approximation methods, compromises (some even based on frequentist answers), sensitivity studies, etc., were introduced as ways of helping to cope with this problem. The major contention we have made, however, is that this additional apparent complication of statistics is *unavoidable*. Also, the complications introduced by having to consider priors are usually more than compensated for by the simplifications discussed in Section 4.1. The issue of prior (and model and loss function) uncertainty should never be dismissed lightly, however.

Technical Difficulties

Bayesian calculations can be very hard, especially in high-dimensional problems. Again, however, we view this as an unavoidable difficulty, which often necessitates approximations and compromises (cf. Sections 4.3, 4.7, and 4.9). Much of the ongoing Bayesian research is concerned with overcoming such difficulties.

4.12.2. Foundational Criticisms

Model-Prior Decompositions

For inputs, Bayes's theorem formally requires $f(x|\theta)$ and $\pi(\theta)$, the likelihood (from a model for the experiment) and the prior for θ. Often, however, information does not arise in such a convenient form. This was discussed in Section 3.7 and extensively illustrated in Section 4.11. Bayesian analysis has been criticized because of this occasional inapplicability of Bayes's theorem.

The error in this criticism lies in the notion that Bayesian analysis is simply the application of Bayes's theorem. Most Bayesian analyses involving statistical experiments will indeed be based on Bayes's theorem, and in teaching Bayesian analysis it is convenient to emphasize Bayes's theorem. The essential feature of Bayesian analysis, however, is that of "determining the conditional probability distribution of all unknown quantities of interest (θ) given the values of the known quantities (x), to paraphrase comments to this effect that we have heard from Dennis Lindley. Thus the goal is to determine $\pi(\theta|x)$, but it is wrong to assume that only Bayes's theorem can achieve this end. One might well be able to determine the joint distribution of θ and x, or certain features of this joint distribution which allow determination of $\pi(\theta|x)$, by other means (cf. Jeffrey (1968) and Diaconis and Zabell (1982)). When viewed in this light, Bayesian analysis seems completely natural; one is just attempting to probabilistically model all random or uncertain quantities, and then condition on what becomes known. (See also Section 4.11.)

Observables Versus Parameters

In Subsections 2.4.4 and 4.3.4, it was indicated that prediction of future observable random quantities can be argued to be the primary goal of statistics. Model parameters play, at best, an intermediate role in such predictions. Our concern, here, is not in adjudicating this issue, since either the "predictive" or "parametric" views fit well within the Bayesian structure, but rather to mention the fundamental related view that models and parameters should be bypassed altogether. The motivation for this view can be found in works such as deFinetti (1974, 1975), which considers only observables to be real, and indeed defines models only as useful idealizations corresponding to certain types of prior beliefs about future observables (cf. Subsection 3.5.7). Thus, the argument goes, one should be concerned only with prior beliefs about future observables, and methods should be found to update these beliefs directly (as some of the observables become known) without passing through the intermediate stage of conventional parametric modelling. For arguments and developments of such a view see Goldstein (1981, 1984) and Dickey and Chong-Hong (1984).

Actually, such views are not necessarily anti-Bayesian. For instance, if one did know the joint distribution of all observables, and conditioned on those that become known, there would be no violation of Bayesian philosophy. Some of the recommended techniques in, say, Goldstein (1981, 1984) are not always compatible with probabilistic Bayesianism, however, and so leave room for debate.

Although a few brief words will be said about alternative subjective theories in general at the end of the subsection, this is not the place for extensive comparisons. We do feel, however, that models should be defended. *Anything* one does in a statistical analysis is an approximation in some manner; the key is to find approximations that are (i) reasonable, and (ii) *intelligence amplifiers* (to borrow a term used in other contexts by I. J. Good). Models *are* often intuitively reasonable (especially if rich enough), and they can definitely lead to sound conclusions that sheer intuition (without models) cannot perceive. And models are often much easier to construct in simple stages, than in one grand gestalt. For instance, in dealing with sequences of coin flips, we feel that one will tend to do much better by first constructing the intermediate Bernoulli trial model (with perhaps "fictitious" probability of heads, θ), and trying to develop a prior for θ (though perhaps by considering prior information about the marginal), than by attempting to directly ascertain a joint marginal distribution for coin flips (see also Subsection 3.5.7).

Data and Experiment Dependent Priors

In several places (Sections 3.7 and 4.7, for instance), we argued that selection of prior distributions will rarely follow the idealized scenario of being done without reference to the data or experimental structure. After all, models are often selected only after a careful examination of the data, so how could a prior on model parameters have been selected beforehand? But, if one has already seen the data, how can a legitimate prior be chosen?

Our answers to this criticism have been: (i) such problems beset all approaches to statistics (use of a model selected after seeing the data being already a serious problem); and (ii) it is the robust Bayesian approach that can actually best deal with this problem, by seeking to show that a proposed answer actually holds for the range of priors (and models) deemed acceptable. In elaboration of this last point, Bayesian model selection can temper the "desire" of the data to be overfitted, by bring in prior weights that can be assigned to models, with the more complex models receiving smaller weights (since there are so many possible complex models that each should receive smaller weight).

Prior Uncertainty

Whereas the classical reaction to the problem of prior uncertainty is to reject analysis which seeks such specifications, some subjectivists react by seeking alternatives to Bayesian subjectivism (cf. Shafer (1976, 1981, 1982a, b)). The typical argument is of the form "one cannot hope to exactly ascertain the prior, so Bayesian analysis should be abandoned in favor of a method which allows expression of uncertainty in beliefs." The trouble with this argument is that it ignores the possibility of dealing with prior uncertainty while staying within the Bayesian framework, through use of the robust Bayesian methodologies we have discussed. At the very best, therefore, these alternative theories become little more than alternative possible ways of dealing with uncertainty in beliefs. Our personal view of these alternative subjectivist theories is that they are either too simple or too complex. We have tried to demonstrate that the complexity of Bayesian analysis (or, alternatively, probabilistic treatment of uncertainty) is needed to deal with statistical problems; difficulties that arise, such as prior uncertainty, may call for small modifications (or enrichments) of the paradigm, but the basic language of probability need not be abandoned. Alternative theories attempt to replace probability by more complex constructs. We would argue that introducing such additional complexity is neither needed nor helpful. We know how to process probabilities, and probability is the simplest method of describing uncertainty that has been developed. Why leave the domain for less intuitive constructs that have no clear rules for processing?

It can, of course, be argued that probability is *not* a rich enough concept to describe uncertainty. The following example presents one such argument, and a reply.

Example 56. Suppose you pull a coin from your pocket and, without looking at it, are interested in the event A that it will come up heads when flipped. Suppose you (reasonably) judge the subjective probability of this event to be close to $\frac{1}{2}$. Next, you contemplate an experiment in which two drugs, about which you know nothing, will be tested, and are interested in the event B that Drug 1 is better than Drug 2. You (reasonably) judge your subjective probability of event B to also be $\frac{1}{2}$. The argument now proceeds:

"Even though both probabilities were $\frac{1}{2}$, you have a stronger 'belief' in the probability specified for event A, in that if you were told that five flips of the coin were all heads your opinion about the fairness of the coin would probably change very little, while if you were told that in tests on five patients Drug 1 worked better than Drug 2 you would probably change your opinion substantially about the worth of Drug 1." Thus, the argument goes, it is necessary to go beyond probability distributions and have measures of the "strength of belief" in probabilities.

It is easy to see the flaw in this reasoning. Before getting any data one *would be* equally secure in probabilities of $\frac{1}{2}$ for each A and B, in that one would be indifferent between placing a single bet on either event. The knowledge about the *events* A and B is well described by a probability of $\frac{1}{2}$. However, the knowledge about the overall phenomena being investigated in each case would be quite different. A description of overall knowledge about the situations is more fully described by defining the unknown (and perhaps fictitious) quantities p_C and p_M, reflecting the "true" proportion of heads and "true" proportion of patients for which Drug 1 would work better than Drug 2, respectively, and then quantifying prior distributions (or classes thereof) for p_C and p_M. The prior distribution for p_C would most likely be much more tightly concentrated about $\frac{1}{2}$, than would the prior distribution for p_M. Note that the subjective probabilities of events A and B are just the means of the respective prior distributions. (A very similar example can be found in the discussion of D. Lindley in Shafer (1982c).)

We would generally argue that prior distributions prove to be rich enough to reflect whatever is reasonably desired. Also interesting is the observation that, in taking account of experimental evidence, one is almost forced to think in the correct fashion. Thus, in Example 56, if one began by merely quantifying the probabilities of A or B, it would be impossible to directly update these probabilities, via Bayes's theorem, upon observing the data. One would be forced to the deeper level of prior specifications concerning p_C and p_M.

A second argument, to the effect that probability is not a rich enough concept, can be developed from the observation, made initially by Kraft, Pratt, and Seidenberg (1959) (see also Fine (1973)), that there may exist "likelihood orderings" of events that are internally consistent and yet which are not consistent with *any* probability distribution. Although surprised by this fact, we would question its relevance. The consistent modes of behavior seem to be those induced by probability distributions, so we would be distrustful of any likelihood ordering that is not consistent with some probability distribution. Note that we are not proposing Bayesian analysis as a model of how the mind can or does work. We are, instead, arguing that it is the most effective tool by which the mind can deal with uncertainty.

Exercises

Section 4.2

1. Prove Lemma 1.

2. There are three coins in a box. One is a two-headed coin, another is a two-tailed coin, and the third is a fair coin. When one of the three coins is selected at random and flipped, it shows heads. What is the probability that it is the two-headed coin?

3. (DeGroot (1970)) Suppose that, with probability $\frac{1}{10}$, a signal is present in a certain system at any given time, and that, with probability $\frac{9}{10}$, no signal is present. A measurement made on the system when a signal is present is normally distributed with mean 50 and variance 1, and a measurement made on the system when no signal is present is normally distributed with mean 52 and variance 1. Suppose that a measurement made on the system at a certain time has the value x. Show that the posterior probability that a signal is present is greater than the posterior probability that no signal is present, if $x < 51 - \frac{1}{2}\log 9$.

4. A scientific journal, in an attempt to maintain experimental standards, insists that all reported statistical results have (classical) error probability of α_0 (or better). To consider a very simple model of this situation, assume that all statistical tests conducted are of the form H_0: $\theta = \theta_0$ versus H_1: $\theta = \theta_1$, where θ_0 represents the standard and θ_1 the new proposal. Experimental results are reported in the journal only if the new proposal is verified with an error probability of $\alpha \le \alpha_0$. (Note that $\alpha = P_{\theta_0}$ (accepting H_1).) Let β denote the power of the test (i.e., $\beta = P_{\theta_1}$ (accepting H_1)). Assume further that α and β are fixed for all experiments conducted, with α being the specified value α_0. Let π_0 denote the proportion of all experiments conducted in which θ_0 is correct, and π_1 denote the proportion of experiments in which θ_1 is correct.
   (a) Show that the proportion of articles published in the journal that have correct results (i.e., $P(\theta = \theta_1 | \text{the test accepts } H_1)$) is $\pi_1\beta/[\alpha_0 + \pi_1(\beta - \alpha_0)]$. (Note that many people naively believe that the journal is guaranteeing a proportion of $(1-\alpha_0)$ of correct articles.)
   (b) Show that the proportion of correct published results is never less than π_1. (Note that $\beta \ge \alpha$ for reasonable tests.)

5. Suppose that X is $\mathcal{B}(n, \theta)$. Suppose also that θ has a $\mathcal{B}e(\alpha, \beta)$ prior distribution. Show that the posterior distribution of θ given x is $\mathcal{B}e(\alpha + x, \beta + n - x)$. What is the natural conjugate family for the binomial distribution?

6. Suppose that $\mathbf{X} = (X_1, \ldots, X_n)$ is a random sample from an exponential distribution. Thus $X_i \sim \mathcal{E}(\theta)$ (independently). Suppose also that the prior distribution of θ is $\mathcal{IG}(\alpha, \beta)$. Show that the posterior distribution of θ given $\mathbf{x}$ is
$$\mathcal{IG}\left(n + \alpha, \left[\left(\sum_{i=1}^{n} x_i\right) + \beta^{-1}\right]^{-1}\right).$$

7. Suppose that $\mathbf{X} = (X_1, \ldots, X_n)$ is a random sample from a $\mathcal{U}(0, \theta)$ distribution. Let θ have a $\mathcal{P}a(\theta_0, \alpha)$ distribution. Show that the posterior distribution of θ given $\mathbf{x}$ is $\mathcal{P}a(\max\{\theta_0, x_1, \ldots, x_n\}, \alpha + n)$.

8. Suppose that X is $\mathcal{G}(n/2, 2\theta)$ (so that X/θ is χ^2_n), while θ has an $\mathcal{IG}(\alpha, \beta)$ distribution. Show that the posterior distribution of θ given x is $\mathcal{IG}(n/2 + \alpha, [x/2 + \beta^{-1}]^{-1})$.

9. Suppose that $\mathbf{X} = (X_1, \ldots, X_k)^t \sim \mathcal{M}(n, \boldsymbol{\theta})$, and that $\boldsymbol{\theta} = (\theta_1, \ldots, \theta_k)^t$ has a $\mathcal{D}(\boldsymbol{\alpha})$ prior distribution ($\boldsymbol{\alpha} = (\alpha_1, \ldots, \alpha_k)^t$). Show that the posterior distribution of $\boldsymbol{\theta}$ given $\mathbf{x}$ is $\mathcal{D}((\boldsymbol{\alpha} + \mathbf{x}))$.

10. Suppose that $\mathbf{X} = (X_1, \ldots, X_n)$ is a sample from a $\mathcal{NB}(m, \theta)$ distribution, and that θ has a $\mathcal{B}e(\alpha, \beta)$ prior distribution. Show that the posterior distribution of θ given $\mathbf{x}$ is $\mathcal{B}e(\alpha + mn, (\sum_{i=1}^{n} x_i) + \beta)$.

11. Suppose that $\mathbf{X}=(X_1,\ldots,X_p)'\sim\mathcal{N}_p(\boldsymbol{\theta},\boldsymbol{\Sigma})$ and that $\boldsymbol{\theta}$ has a $\mathcal{N}_p(\boldsymbol{\mu},\mathbf{A})$ prior distribution. (Here $\boldsymbol{\theta}$ and $\boldsymbol{\mu}$ are p-vectors, while $\boldsymbol{\Sigma}$ and $\mathbf{A}$ are $(p\times p)$ positive definite matrices.) Also, $\boldsymbol{\Sigma}$, $\boldsymbol{\mu}$, and $\mathbf{A}$ are assumed known. Show that the posterior distribution of $\boldsymbol{\theta}$ given $\mathbf{x}$ is a p-variate normal distribution with mean

$$\mathbf{x}-\boldsymbol{\Sigma}(\boldsymbol{\Sigma}+\mathbf{A})^{-1}(\mathbf{x}-\boldsymbol{\mu})$$

and covariance matrix

$$(\mathbf{A}^{-1}+\boldsymbol{\Sigma}^{-1})^{-1}.$$

(*Hint*: Unless familiar with matrix algebra, it is probably easiest to simultaneously diagonalize $\boldsymbol{\Sigma}$ and $\mathbf{A}$, do the calculation for the transformed distributions, and then transform back to the original coordinates.)

12. Suppose that $\mathbf{X}=(X_1,\ldots,X_n)$ is a sample from a $\mathcal{N}(\theta,\sigma^2)$ distribution, where both θ and σ^2 are unknown. The prior density of θ and σ^2 is

$$\pi(\theta,\sigma^2)=\pi_1(\theta\,|\,\sigma^2)\pi_2(\sigma^2),$$

where $\pi_1(\theta\,|\,\sigma^2)$ is a $\mathcal{N}(\mu,\tau\sigma^2)$ density and $\pi_2(\sigma^2)$ is an $\mathcal{IG}(\alpha,\beta)$ density.

(a) Show that the joint posterior density of θ and σ^2 given $\mathbf{x}$ is

$$\pi(\theta,\sigma^2\,|\,\mathbf{x})=\pi_1(\theta\,|\,\sigma^2,\mathbf{x})\pi_2(\sigma^2\,|\,\mathbf{x})$$

where $\pi_1(\theta\,|\,\sigma^2,\mathbf{x})$ is a normal density with mean $\mu(\mathbf{x})=(\mu+n\tau\bar{x})/(n\tau+1)$ (here $\bar{x}=(1/n)\sum_{i=1}^n x_i$) and variance $(\tau^{-1}+n)^{-1}\sigma^2$, and $\pi_2(\sigma^2\,|\,\mathbf{x})$ is an inverted gamma density with parameters $\alpha+n/2$ and β', where

$$\beta'=\left[\beta^{-1}+\frac{1}{2}\sum_{i=1}^n(x_i-\bar{x})^2+\frac{n(\bar{x}-\mu)^2}{2(1+n\tau)}\right]^{-1}.$$

(b) Show that the marginal posterior density of σ^2 given $\mathbf{x}$ is $\mathcal{IG}(\alpha+n/2,\beta')$. (To find the marginal posterior density of σ^2 given $\mathbf{x}$, just integrate out over θ in the joint posterior density.)

(c) Show that the marginal posterior density of θ given $\mathbf{x}$ is a

$$\mathcal{T}(2\alpha+n,\mu(\mathbf{x}),[(\tau^{-1}+n)(\alpha+n/2)\beta']^{-1})$$

density.

(d) State why the joint prior density in this problem is from a conjugate family for the distribution of $\mathbf{X}$. (Note that, for this prior, the conditional prior variance of θ given σ^2 is proportional to σ^2. This is unattractive, in that prior knowledge of θ should generally not depend on the sample variance. If, for example, the X_i are measurements of some real quantity θ, it seems silly to base prior beliefs about θ on the accuracy of the measuring instrument. It must be admitted that the robust prior used in Subsection 4.7.10 suffers from a similar intuitive inadequacy. The excellent results obtained from its use, however, indicate that the above intuitive objection is not necessarily damning.)

13. General Motors wants to forecast new car sales for the next year. The number of cars sold in a year is known to be a random variable with a $\mathcal{N}((10^8)\theta,(10^6)^2)$ distribution, where θ is the unemployment rate during the year. The prior density for θ next year is thought to be (approximately) $\mathcal{N}(0.06,(0.01)^2)$. What is the distribution of car sales for next year? (That is, find the marginal or predictive distribution of car sales next year.)

14. Assume $\mathbf{X}=(X_1,\ldots,X_n)$ is a sample from a $\mathcal{P}(\theta)$ distribution. The improper noninformative prior $\pi(\theta)=\theta^{-1}I_{(0,\infty)}(\theta)$ is to be used. Find the (formal) posterior density of θ given $\mathbf{x}$, for $\mathbf{x}\neq(0,0,\ldots,0)$. (If $\mathbf{x}=(0,\ldots,0)$, the (formal) posterior does not exist.)

15. Assume X is $\mathcal{B}(n,\theta)$.
    (a) If the improper prior density $\pi(\theta)=[\theta(1-\theta)]^{-1}I_{(0,1)}(\theta)$ is used, find the (formal) posterior density of θ given x, for $1\leq x\leq n-1$.
    (b) Find the posterior density of θ given x, when $\pi(\theta)=I_{(0,1)}(\theta)$.

16. Assume $\mathbf{X}=(X_1,\ldots,X_n)$ is a sample from a $\mathcal{N}(\theta,\sigma^2)$ distribution, where θ and σ^2 are unknown. Let θ and σ^2 have the joint improper noninformative prior density

$$\pi(\theta,\sigma^2)=\sigma^{-2}I_{(0,\infty)}(\sigma^2).$$

(In Subsection 3.3.3 it was stated that a reasonable noninformative prior for (θ,σ) is $\pi(\theta,\sigma)=\sigma^{-1}I_{(0,\infty)}(\sigma)$. This transforms into the above prior for (θ,σ^2).)
    (a) Show that the (formal) posterior density of θ and σ^2 given $\mathbf{x}$ is

$$\pi(\theta,\sigma^2|\mathbf{x})=\pi_1(\theta|\sigma^2,\mathbf{x})\pi_2(\sigma^2|\mathbf{x}),$$

where $\pi_1(\theta|\sigma^2,\mathbf{x})$ is a $\mathcal{N}(\bar{x},\sigma^2/n)$ density and $\pi_2(\sigma^2|\mathbf{x})$ is an $\mathcal{IG}((n-1)/2,$ $[\frac{1}{2}\sum_{i=1}^n(x_i-\bar{x})^2]^{-1})$ density.
    (b) Show that the marginal posterior density of σ^2 given $\mathbf{x}$ is an $\mathcal{IG}((n-1)/2,$ $[\frac{1}{2}\sum_{i=1}^n(x_i-\bar{x})^2]^{-1})$ density.
    (c) Show that the marginal posterior density of θ given $\mathbf{x}$ is a $\mathcal{T}(n-1,\bar{x},$ $\sum_{i=1}^n(x_i-\bar{x})^2/n(n-1))$ density.

17. Show that mixtures of natural conjugate priors, as defined in (4.1), form a conjugate class of priors.

18. In the situation of Subsection 4.2.3, show that, if π_n is a $\mathcal{N}(0,n)$ prior density for θ, then $\pi_n(\theta|x)$:
    (a) converges pointwise to $\pi(\theta|x)$;
    (b) converges in probability to $\pi(\theta|x)$.

Subsections 4.3.1 and 4.3.2

19. Find the posterior mean and posterior variance or covariance matrix in Exercise (a) 5, (b) 6, (c) 7, (d) 8, (e) 9, (f) 10, (g) 11, (h) 12(b), (i) 12(c), (j) 14, (k) 15, (l) 16(b), (m) 16(c), and (n) in Example 5.

20. Find the generalized maximum likelihood estimate of θ and the posterior variance or covariance matrix of the estimate in Exercise (a) 5, (b) 6, (c) 7, (d) 8, (e) 9, (f) 10, (g) 11, (h) 12(c), (i) 14, (j) 15, (k) 16(c), and (l) in Example 5.

21. Find the median of the posterior distribution and the posterior variance of the median in Exercise (a) 5 (when $\alpha=\beta=n=x=1$), (b) 6 (when $\alpha=n=1$), (c) 7, (d) 12(b) (when $\alpha=1$, $n=2$), (e) 12(c), (f) 14, (g) 16(b) (when $n=2$), (h) 16(c).

22. In Example 7 (continued), show that $V^{\pi}(x)$ is increasing in x, with $V^{\pi}(0)=(2/\pi)\sigma^2$ and $V^{\pi}(\infty)=\sigma^2$.

23. A production lot of five electronic components is to be tested to determine θ, the mean lifetime. A sample of five components is drawn, and the lifetimes $X_1, \ldots, X_5$ are observed. It is known that $X_i \sim \mathscr{E}(\theta)$. From past records it is known that, among production lots, θ is distributed according to an $\mathscr{I}\mathscr{G}(10, 0.01)$ distribution. The five observations are 15, 12, 14, 10, 12. Find the generalized maximum likelihood estimate of θ, the posterior mean of θ, and their respective posterior variances.

24. Electronic components I and II have lifetimes X_1 and X_2 which have $\mathscr{E}(\theta_1)$ and $\mathscr{E}(\theta_2)$ densities, respectively. It is desired to estimate the mean lifetimes θ_1 and θ_2. Component I contains component II as a subcomponent, and will fail if the subcomponent does (or if something else goes wrong). Hence $\theta_1 < \theta_2$. Two independent observations, X_1^1 and X_1^2, of the lifetimes of component I are taken. Likewise, two observations X_2^1 and X_2^2 of the lifetimes of component II are taken. The $X_2^i (i=1, 2)$ are independent of each other, and also of the $X_1^i (i=1, 2)$. (The four observations are taken from different components.) It is decided that a reasonable noninformative prior for the situation is

$$\pi(\theta_1, \theta_2) = \theta_1^{-1}\theta_2^{-1} I_{(0,\theta_2)}(\theta_1) I_{(0,\infty)}(\theta_2).$$

Find reasonable Bayesian estimates of θ_1 and θ_2.

25. Find the $100(1-\alpha)\%$ HPD credible set in Exercise
(a) 7, (b) 11, (c) 12(c), (d) 14 (when $\sum_{i=1}^n x_i = 1$), (e) 15(a) (when $n=3$, $x=1$), (f) 16(b) (when $n=2$), (g) 16(c).

26. A large shipment of parts is received, out of which five are tested for defects. The number of defective parts, X, is assumed to have a $\mathscr{B}(5, \theta)$ distribution. From past shipments, it is known that θ has a $\mathscr{B}e(1, 9)$ prior distribution. Find the 95% HPD credible set for θ, if $x=0$ is observed.

27. The weekly number of fires, X, in a town has a $\mathscr{P}(\theta)$ distribution. It is desired to find a 90% HPD credible set for θ. Nothing is known *a priori* about θ, so the noninformative prior $\pi(\theta) = \theta^{-1} I_{(0,\infty)}(\theta)$ is deemed appropriate. The number of fires observed for five weekly periods was 0, 1, 1, 0, 0. What is the desired credible set?

28. From path perturbations of a nearby sun, the mass θ of a neutron star is to be determined. Five observations 1.2, 1.6, 1.3, 1.4, and 1.4 are obtained. Each observation is (independently) normally distributed with mean θ and unknown variance σ^2. *A priori* nothing is known about θ and σ^2, so the noninformative prior density $\pi(\theta, \sigma^2) = \sigma^{-2}$ is used. Find a 90% HPD credible set for θ.

29. Suppose $X_1, \ldots, X_n$ are an i.i.d. sample from the $\mathscr{U}(\theta - \frac{1}{2}, \theta + \frac{1}{2})$ density.
(a) For the noninformative prior $\pi(\theta) = 1$, show that the posterior probability, given $\mathbf{x} = (x_1, \ldots, x_n)$, that θ is in a set C is given by

$$[1 + x_{\min} - x_{\max}]^{-1} \int_C I_A(\theta) d\theta,$$

where $x_{\min} = \min\{x_i\}$, $x_{\max} = \max\{x_i\}$, and $A = (x_{\max} - \frac{1}{2}, x_{\min} + \frac{1}{2})$.

(b) Show that the confidence procedure $C(\mathbf{x}) = (\tilde{x} - 0.056, \tilde{x} + 0.056)$, where $\tilde{x} = (x_{\min} + x_{\max})/2$, is a frequentist 95% confidence procedure, and calculate its posterior probability.

30. For two independent $\mathscr{C}(\theta, 1)$ observations, X_1 and X_2, give the 95% HPD credible set with respect to the noninformative prior $\pi(\theta) = 1$. (Note that it will sometimes be an interval, and sometimes the union of two intervals; numerical work is required here.)

31. Suppose $X \sim \mathcal{N}(\theta, 1)$, and that a 95% HPD credible set for θ is desired. The prior information is that θ has a symmetric unimodal density with median 0 and quartiles ± 1. The observation is $x = 6$.
    (a) If the prior information is modelled as a $\mathcal{N}(0, 2.19)$ prior, find the 90% HPD credible set.
    (b) If the prior information is modelled as a $\mathscr{C}(0, 1)$ prior, find the 90% HPD credible set.

32. Find the approximate 90% HPD credible set, using the normal approximation to the posterior, in the situations of (a) Exercise 23; (b) Exercise 27; (c) Exercise 9.

33. Suppose S^2/σ^2 is $\chi^2(n)$ and that σ is given the noninformative prior $\pi(\sigma) = \sigma^{-1}$. Prove that the corresponding noninformative prior for σ^2 is $\pi^*(\sigma^2) = \sigma^{-2}$, and verify, when $n = 2$ and $s^2 = 2$ is observed, that the 95% HPD credible sets for σ and for σ^2 are not consistent with each other.

34. In the situation of Exercise 33, suppose that the size of a credible set, C, for σ is measured by

$$S(C) = \int_C \sigma^{-1}\, d\sigma.$$

    (a) If $n = 2$, find the S-optimal 90% credible set for σ.
    (b) If $n = 10$ and $s^2 = 2$, find the S-optimal 90% credible set for σ.
    (c) Show that the interval $C = (as, bs)$ has the same size under $S(C)$ for *any* s.
    (d) Show that, if $S(C)$ measures the size of a set for σ, then

$$S^*(C^*) = \int_{C^*} \sigma^{-2}\, d\sigma^2$$

    should be used to measure the size of a set for σ^2.

35. Show, in the continuous case, that the S-optimal $100(1-\alpha)\%$ credible set is given by (4.12), providing that $\{\theta\colon \pi(\theta|x) = ks(\theta)\}$ has measure (or size) zero.

Subsections 4.3.3 and 4.3.4

36. The waiting time for a bus at a given corner at a certain time of day is known to have a $\mathscr{U}(0, \theta)$ distribution. It is desired to test $H_0\colon 0 \le \theta \le 15$ versus $H_1\colon \theta > 15$. From other similar routes, it is known that θ has a $\mathscr{P}a(5, 3)$ distribution. If waiting times of 10, 3, 2, 5, and 14 are observed at the given corner, calculate the posterior probability of each hypothesis, the posterior odds ratio, and the Bayes factor.

37. In the situation of Exercise 26, it is desired to test H_0: $\theta \leq 0.1$ versus H_1: $\theta > 0.1$. Find the posterior probabilities of the two hypotheses, the posterior odds ratio, and the Bayes factor.

38. In the situation of Exercise 28, it is desired to test H_0: $\theta \leq 1$ versus H_1: $\theta > 1$. Find the posterior probabilities of the two hypotheses, the posterior odds ratio, and the Bayes factor.

39. (DeGroot (1970)) Consider two boxes A and B, each of which contains both red balls and green balls. It is known that, in one of the boxes, $\frac{1}{2}$ of the balls are red and $\frac{1}{2}$ are green, and that, in the other box, $\frac{1}{4}$ of the balls are red and $\frac{3}{4}$ are green. Let the box in which $\frac{1}{2}$ are red be denoted box W, and suppose $P(W=A)=\xi$ and $P(W=B)=1-\xi$. Suppose that the statistician may select one ball at random from either box A or box B and that, after observing its color, he must decide whether $W=A$ or $W=B$. Prove that if $\frac{1}{2}<\xi<\frac{2}{3}$, then in order to maximize the probability of making a correct decision, he should select the ball from box B. Prove also that if $\frac{2}{3}\leq\xi\leq 1$, then it does not matter from which box the ball is selected.

40. Theory predicts that θ, the melting point of a particular substance under a pressure of 10^6 atmospheres, is 4.01. The procedure for measuring this melting point is fairly inaccurate, due to the high pressure. Indeed it is known that an observation X has a $\mathcal{N}(\theta, 1)$ distribution. Five independent experiments give observations of 4.9, 5.6, 5.1, 4.6, and 3.6. The prior probability that $\theta=4.01$ is 0.5. The remaining values of θ are given the density $(0.5)\, g_1(\theta)$, where g_1 is a $\mathcal{N}(4.01, 1)$ density. Formulate and conduct a Bayesian test of the proposed theory.

41. In the situation of Exercise 40:
   (a) Calculate the P-value against H_0: $\theta=4.01$.
   (b) Calculate the lower bound on the posterior probability of H_0 for *any* g_1, and find the corresponding bound on the Bayes factor.
   (c) Calculate the lower bound on the posterior probability of H_0 for any g_1 in (4.22), and find the corresponding lower bound on the Bayes factor.
   (d) Assume that g_1 is $\mathcal{N}(4.01, \tau^2)$, and graph the posterior probability of H_0 as a function of τ^2.
   (e) Compare and discuss the answers obtained in parts (a) through (d).

42. Suppose that $X \sim \mathcal{B}(n, \theta)$, and that it is desired to test H_0: $\theta=\theta_0$ versus H_1: $\theta \neq \theta_0$.
   (a) Find lower bounds on the posterior probability of H_0 and on the Bayes factor for H_0 versus H_1, bounds which are valid for *any* g_1 (using the notation of (4.14)).
   (b) If $n=20$, $\theta_0=0.5$, and $x=15$ is observed, calculate the (two-tailed) P-value and the lower bound on the posterior probability when $\pi_0=0.5$.

43. Find the value of z for which the lower bound on B in (4.24) and the P-value of $(z-1)^+$ are equal.

44. Prove Theorem 1.

45. Consider $\mathscr{G}_N = \{g_1: g_1 \text{ is } \mathcal{N}(\theta_0, \tau^2), \tau^2 \geq 0\}$.
   (a) Show, in the situation of Example 11, that lower bounds on α_0 and B for this class of g_1 are given, respectively, by π_0 and 1 when $z \leq 1$, and by
$$\alpha_0 \geq \left[1 + \frac{(1-\pi_0)}{\pi_0} \cdot \frac{\exp\{z^2/2\}}{z\sqrt{e}}\right]^{-1}$$
   and
$$B \geq \sqrt{e}z \exp\{-z^2/2\}$$
   if $z > 1$.
   (b) Compare these bounds with those in Table 4.4.

46. Suppose $X \sim \mathscr{C}(\theta, 1)$, and that it is desired to test H_0: $\theta = 0$ versus H_1: $\theta \neq 0$. For *any* prior giving positive probability, π_0, to $\theta = 0$, show that α_0, the posterior probability of H_0, converges to π_0 as $|x| \to \infty$.

47. Suppose $\Theta_0 = (\theta_0 - b, \theta_0 + b)$ and $\Theta_1 = \Theta_0^c$. It is desired to test H_0: $\theta \in \Theta_0$ versus H_1: $\theta \in \Theta_1$, with a prior as in (4.13). Suppose the likelihood function, $f(x|\theta)$, satisfies $|f(x|\theta) - \gamma| \leq \varepsilon$ on Θ_0 (for the observed x, of course). Letting
$$m_1(x) = \int_{\Theta_1} f(x|\theta) dF^{g_1}(\theta),$$
defining α_0 to be the posterior probability of Θ_0, and letting α_0^* denote the posterior probability that would arise from replacing Θ_0 by the point $\{\theta_0\}$ and giving it probability π_0, while leaving the prior specification on Θ_1 unchanged (note that points in Θ_0 other than θ_0 are thus given zero density), show that
$$|\alpha_0 - \alpha_0^*| \leq \frac{2\varepsilon\pi_1 m_1(x)}{\pi_0[(\gamma + \pi_1 m_1(x)/\pi_0)^2 - \varepsilon^2]}.$$

48. Verify, in Example 11, that the likelihood function varies by no more than 5% on $(\theta_0 - b, \theta_0 + b)$ under the given condition.

49. In the situation of Example 12, find a number z_0 such that Z has probability 0.1 of being less than z_0.

Section 4.4

50. Prove Result 2 of Subsection 4.4.1 for randomized rules. (You may assume that all interchanges in orders of integration are legal.)

51. Prove Result 4 of Subsection 4.4.2. (You may assume that all integrals exist.)

52. Show that, in estimation of a vector $\boldsymbol{\theta} = (\theta_1, \theta_2, \ldots, \theta_p)^t$ by $\mathbf{a} = (a_1, \ldots, a_p)^t$ under a quadratic loss
$$L(\boldsymbol{\theta}, \mathbf{a}) = (\boldsymbol{\theta} - \mathbf{a})^t \mathbf{Q} (\boldsymbol{\theta} - \mathbf{a}),$$
where $\mathbf{Q}$ is a $(p \times p)$ positive definite matrix, the Bayes estimator of $\boldsymbol{\theta}$ is
$$\boldsymbol{\delta}^\pi(x) = E^{\pi(\boldsymbol{\theta}|x)}[\boldsymbol{\theta}].$$
(You may assume that $\pi(\boldsymbol{\theta}|x)$ and all integrals involved exist.)

53. Prove Result 6 of Subsection 4.4.2. (You may assume that $\pi(\theta|x)$ exists and that there is an action with finite posterior expected loss.)

54. If $X \sim \mathcal{B}(n, \theta)$ and $\theta \sim \mathcal{B}e(\alpha, \beta)$, find the Bayes estimator of θ under loss

$$L(\theta, a) = \frac{(\theta - a)^2}{\theta(1-\theta)}.$$

(Be careful about the treatment of $x = 0$ and $x = n$.)

55. If $X \sim \mathcal{G}(n/2, 2\theta)$ and $\theta \sim \mathcal{IG}(\alpha, \beta)$, find the Bayes estimator of θ under loss

$$L(\theta, a) = \left(\frac{a}{\theta} - 1\right)^2 = \frac{1}{\theta^2}(a - \theta)^2.$$

56. Assume θ, x, and a are real, $\pi(\theta|x)$ is symmetric and unimodal, and L is an increasing function of $|\theta - a|$. Show that the Bayes rule is then the mode of $\pi(\theta|x)$. (You may assume that all risk integrals exist.)

57. In the situation of Exercise 26, find the Bayes estimate of θ under loss
(a) $L(\theta, a) = (\theta - a)^2$,
(b) $L(\theta, a) = |\theta - a|$,
(c) $L(\theta, a) = (\theta - a)^2/\theta(1-\theta)$,
(d) $L(\theta, a) = (\theta - a)$ if $\theta > a$; $L(\theta, a) = 2(a - \theta)$ if $\theta \leq a$.

58. Suppose $\mathbf{X} \sim \mathcal{N}_2(\boldsymbol{\theta}, \mathbf{I}_2)$, $L(\boldsymbol{\theta}, \mathbf{a}) = (\boldsymbol{\theta}'\mathbf{a} - 1)^2$, $\mathcal{A} = \{(a_1, a_2)' \colon a_1 \geq 0,\ a_2 \geq 0$, and $a_1 + a_2 = 1\}$, and $\boldsymbol{\theta} \sim \mathcal{N}_2(\boldsymbol{\mu}, \mathbf{B})$, $\boldsymbol{\mu}$ and $\mathbf{B}$ known. Find the Bayes estimator of $\boldsymbol{\theta}$.

59. In the IQ example, where $X \sim \mathcal{N}(\theta, 100)$ and $\theta \sim \mathcal{N}(100, 225)$, assume it is important to detect particularly high or low IQs. Indeed the weighted loss

$$L(\theta, a) = (\theta - a)^2 e^{(\theta - 100)^2/900}$$

is deemed appropriate. (Note that this means that detecting an IQ of 145 (or 55) is about nine times as important as detecting an IQ of 100.) Find the Bayes estimator of θ.

60. In the situation of Exercise 9, find the Bayes estimator of $\boldsymbol{\theta}$ under loss

$$L(\boldsymbol{\theta}, \mathbf{a}) = \sum_{i=1}^{k} (\theta_i - a_i)^2.$$

Show that the Bayes risk of the estimator is

$$\frac{\alpha_0^2 - \sum_{i=1}^k \alpha_i^2}{\alpha_0(\alpha_0 + 1)(\alpha_0 + n)},$$

where $\alpha_0 = \sum_{i=1}^k \alpha_i$.

61. In the situation of Exercise 26, let a_0 denote the action "decide $0 \leq \theta \leq 0.15$," and a_1 denote the action "decide $\theta > 0.15$." Conduct the Bayes test under the loss
(a) "0–1" loss,

(b) $$L(\theta, a_0) = \begin{cases} 1 & \text{if } \theta > 0.15, \\ 0 & \text{if } \theta \leq 0.15. \end{cases}$$

$$L(\theta, a_1) = \begin{cases} 2 & \text{if } \theta \leq 0.15, \\ 0 & \text{if } \theta > 0.15. \end{cases}$$

(c) $L(\theta, a_0) = \begin{cases} 1 & \text{if } \theta > 0.15, \\ 0 & \text{if } \theta \le 0.15. \end{cases}$

$L(\theta, a_1) = \begin{cases} 0.15 - \theta & \text{if } \theta \le 0.15, \\ 0 & \text{if } \theta > 0.15. \end{cases}$

62. A company periodically samples products coming off a production line, in order to make sure the production process is running smoothly. They choose a sample of size 5 and observe the number of defectives. Past records show that the proportion of defectives, θ, varies according to a $\mathcal{B}e(1, 9)$ distribution. The loss in letting the production process run is 10θ, while the loss in stopping the production line, recalibrating, and starting up again is 1. What is the Bayes decision if one defective is observed in a sample?

63. A missile can travel at either a high or a low trajectory. The missile's effectiveness decreases linearly with the distance by which it misses its target, up to a distance of 2 miles at which it is totally ineffective.

If a low trajectory is used, the missile is safe from antimissile fire. However, its accuracy is subject to the proportion of cloud cover (θ). Indeed the distance, d, by which is misses its target is uniformly distributed on $(0, \theta)$. For the target area, θ is known to have the probability density

$$\pi_1(\theta) = 6\theta(1-\theta) I_{(0,1)}(\theta).$$

If a high trajectory is used, the missile will hit the target exactly, unless it is first destroyed by antimissile fire. From previous experience, the probability, ξ, of the missile being destroyed is thought to have the prior density $\pi_2(\xi) = 2(1-\xi) I_{(0,1)}(\xi)$. An experiment is conducted to provide further information about ξ. Two missiles are launched using a high trajectory, out of which none are shot down.

(a) Give the loss incurred in having the missile miss by a distance d. (Let 0 stand for a perfect hit and 1 for a total miss.)
(b) What is the optimal Bayes trajectory?

64. A wildcat oilman must decide how to finance the drilling of a well. It costs \$100,000 to drill the well. The oilman has available three options: a_1—finance the drilling himself (retaining all the profits); a_2—accept \$70,000 from investors in return for paying them 50% of the oil profits; a_3—accept \$120,000 from investors in return for paying them 90% of the oil profits. The oil profits will be \$$3\theta$, where θ is the number of barrels of oil in the well. From past data, it is believed that $\theta = 0$ with probability 0.9, while the $\theta > 0$ have density

$$\pi(\theta) = \frac{0.1}{300{,}000} e^{-\theta/300{,}000} I_{(0,\infty)}(\theta).$$

A seismic test is performed to determine the likelihood of oil in the given area. The test tells which type of geological structure, x_1, x_2, or x_3, is present. It is known that the probabilities of the x_i given θ are $f(x_1|\theta) = 0.8e^{-\theta/100{,}000}$, $f(x_2|\theta) = 0.2$, $f(x_3|\theta) = 0.8(1 - e^{-\theta/100{,}000})$.

(a) For monetary loss, what is the Bayes action if x_1 is observed?
(b) For monetary loss, what are the Bayes actions if x_2 and x_3 are observed?

(c) If the oilman has the utility function

$$U(z) = \begin{cases} 1 - e^{-(10^{-6})z} & \text{if } z \geq 0, \\ -(1 - e^{2(10^{-6})z}) & \text{if } z < 0, \end{cases}$$

where z is monetary gain or loss, what is the Bayes action if x_1 is observed?

65. A device has been created which can supposedly classify blood as type A, B, AB, or O. The device measures a quantity X, which has density

$$f(x|\theta) = e^{-(x-\theta)} I_{(\theta,\infty)}(x).$$

If $0 < \theta < 1$, the blood is of type AB; if $1 < \theta < 2$, the blood is of type A; if $2 < \theta < 3$, the blood is of type B; and if $\theta > 3$, the blood is of type O. In the population as a whole, θ is distributed according to the density

$$\pi(\theta) = e^{-\theta} I_{(0,\infty)}(\theta).$$

The loss in misclassifying the blood is given in the following table.

		Classified As			
		AB	A	B	O
True Blood Type	AB	0	1	1	2
	A	1	0	2	2
	B	1	2	0	2
	O	3	3	3	0

If $x = 4$ is observed, what is the Bayes action?

66. Verify the statement in Subsection 4.4.4, that the optimal Bayes choice of $\alpha(x)$ is the posterior probability of $C(x)$.

67. Solve the decision problem of choosing a credible set, $C(x)$, for θ and an associated accuracy measure, $\alpha(x)$, when $X \sim \mathcal{N}(\theta, 1)$ is to be observed, θ is given the noninformative prior $\pi(\theta) \equiv 1$, and the loss is given in (4.136) with $c_1 = 1$, $c_2 = 5$, $c_3 = 1$, and $\mu(C(x))$ being the length of $C(x)$.

Section 4.5

68. At a certain stage of an industrial process, the concentration, θ, of a chemical must be estimated. The loss in estimating θ is reasonably approximated by squared-error loss. The measurement, X, of θ has a $\mathcal{N}(\theta, 0.02)^2)$ distribution. Five measurements are available from determinations of the chemical concentration when the process was run in the past. These measurements are 3.29, 3.31, 3.35, 3.34, and 3.33. Each arose from (independent) $\mathcal{N}(\theta_i, (0.02)^2)$ distributions and it is believed that the θ_i are i.i.d. If $x = 3.30$ is the current measurement, estimate the current θ and find a 95% credible set for it, using the following methods:
(a) Naive empirical Bayes (i.e., pretend that $\theta \sim \mathcal{N}(\hat{\mu}_\pi, \hat{\sigma}^2_\pi)$).
(b) The Morris empirical Bayes formulas.

69. A steel mill casts p large steel beams. It is necessary to determine, for each beam, the average number of defects or impurities per cubic foot. (For the ith beam, denote this quantity θ_i.) On each beam, n sample cubic-foot regions are examined, and the number of defects in each region is determined. (For the ith beam, denote these $X_i^1, X_i^2, \ldots, X_i^n$.) The X_i^j have $\mathscr{P}(\theta_i)$ distributions.
(a) Show that a (naive) empirical Bayes estimate of each θ_i is

$$\delta_i^{\mathrm{EB}}(\mathbf{x}) = \begin{cases} \dfrac{\dfrac{\bar{x}^2}{(ns^2-\bar{x})}+n\bar{x}_i}{\dfrac{\bar{x}}{(ns^2-\bar{x})}+n} & \text{if } ns^2 > \bar{x}, \\ \bar{x} & \text{if } ns^2 \le \bar{x}, \end{cases}$$

where $\bar{x}_i = (1/n)\sum_{j=1}^n x_i^j$, $\bar{x} = (1/p)\sum_{i=1}^p \bar{x}_i$, and $s^2 = (1/(p-1))\sum_{i=1}^p \times (\bar{x}_i - \bar{x})^2$. (*Hint*: Proceed as in Exercises 23 and 29 from Chapter 3.)
(b) Find (naive) estimates of the accuracy of the $\delta_i^{\mathrm{EB}}(\mathbf{x})$.
(c) Under what circumstances are the answers in (a) and (b) likely to be good?

70. In the situation of Exercise 4, the journal editors decide to investigate the problem. They survey all of the recent contributors to the journal, and determine the total number of experiments, N, that were conducted by the contributors. Let x denote the number of such experiments that were reported in the journal (i.e., the experiments where H_1 was accepted by a size α_0 test). Again assume all experiments had fixed $\alpha(=\alpha_0)$ and $\beta(>\alpha_0)$.
(a) Show that a reasonable empirical Bayes estimate of π_1, is (defining $\hat{p} = x/N$)

$$\hat{\pi}_1 = \frac{\hat{p}-\alpha_0}{\beta-\alpha_0},$$

(truncated at zero and one to ensure it is a probability).
(b) If $\alpha_0 = 0.05$, $\beta = 0.8$, $N = 500$, and $x = 75$, what is the estimate of the proportion of correct articles in the journal?

71. Show that the V_i^{EB} in (4.34) can be larger than σ_f^2. Is this reasonable or unreasonable?

72. Suppose, in Exercise 68, that the measurements had common unknown variance σ_f^2, and that $(0.02)^2$ was the usual unbiased estimate of σ_f^2 based on 10 degrees of freedom. Find a reasonable empirical Bayes estimate and 95% credible set for θ.

73. Repeat Exercise 68, supposing that the first three past measurements were $\mathscr{N}(\theta_i, (0.03)^2)$ (less accurate measurements were taken), and that the subsequent measurements (including the current one) were $\mathscr{N}(\theta_i, (0.02)^2)$.

74. In Exercise 68, suppose that the θ_i are thought to possibly have a quadratic relationship of the form

$$\theta_i = \beta_1 + \beta_2 i + \beta_3 i^2 + \varepsilon_i,$$

where the β_j are unknown and the ε_i are independently $\mathscr{N}(0, \sigma_\pi^2), \sigma_\pi^2$ also unknown. (The current θ, being the sixth measurement, has index $i=6$.) Find the Morris empirical Bayes estimate and a 95% credible set for θ, and compare with the answers from Exercise 68(b).

75. In Example 15, graph the likelihood function for σ_π^2, (i.e., (4.39) for the given data), choosing $\boldsymbol{\beta}$ to be $\hat{\boldsymbol{\beta}}$ in (4.40) (replacing $\hat{\sigma}_\pi^2$ by σ_π^2). Does this strongly indicate that σ_π^2 is very close to zero?

76. In the two situations of Exercise 28 of Chapter 3, determine the empirical Bayes estimates of θ_1 and θ_2 that result from use of the nonparametric ML-II prior.

77. For $i=1,\ldots,p$, suppose that the X_i are independent $\mathcal{N}(\theta_i, 1)$ random variables, and that the θ_i are i.i.d. from a common prior π_0. Define m_0 by

$$m_0(y) = \int f(y|\theta)\pi_0(\theta)d\theta,$$

where f is the $\mathcal{N}(\theta, 1)$ density.
(a) Show that the posterior mean for θ_i, with given π_0, can be written

$$\mu^{\pi_0}(x_i) = x_i + \frac{m_0'(x_i)}{m_0(x_i)},$$

where m_0' is the derivative of m_0.
(b) Suggest a related nonparametric empirical Bayes estimator for θ_i.

Section 4.6

78. Prove Result 7.

79. Repeat Exercise 68 from the hierarchical Bayes perspective:
(a) With a constant second stage prior for (μ_π, σ_π^2).
(b) With a $\mathcal{N}(3.30, (0.03)^2)$ second stage prior for μ_π, and (independent) constant second stage prior for σ_π^2.
(c) Compare the answers obtained with each other and with those from Exercise 68, and try to explain any differences.

80. In the situation of Example 14 (continued) in Subsection 4.6.2, find the volume of the approximate $100(1-\alpha)\%$ credible ellipsoid for $\boldsymbol{\theta}$, and compare it with the volume of the classical confidence ellipsoid

$$C_0(\mathbf{x}) = \{\boldsymbol{\theta}: |\boldsymbol{\theta}-\mathbf{x}|^2/100 \le \chi_{20}^2(1-\alpha)\}.$$

81. Repeat Exercise 79(a) and (b), supposing that the measurements in Exercise 68 had common unknown variance σ_f^2. Available to estimate σ_f^2 was an (independent) random variable S^2, S^2/σ_f^2 having a chi-squared distribution with 10 degrees of freedom. Observed was $s^2=0.0036$, and σ_f^2 is to be given the noninformative prior $\pi^*(\sigma_f^2)=1/\sigma_f^2$.

82. Repeat Exercise 73 from the hierarchical Bayes perspective:
(a) With a constant second stage prior for (μ_π, σ_π^2).
(b) With a $\mathcal{N}(3.30, (0.03)^2)$ second stage prior for μ_π, and (independent) constant second stage prior for σ_π^2.
(c) Compare these answers with those from Exercise 73.

83. Repeat Exercise 74 from the hierarchical Bayes perspective, assuming a constant second stage prior for $(\beta_1, \beta_2, \beta_3, \sigma_\pi^2)$. Compare the answers so obtained with those from Exercise 74.

Subsections 4.7.1 through 4.7.4

84. Suppose $\mathbf{X} \sim \mathcal{N}_p(\boldsymbol{\theta}, \boldsymbol{\Sigma})$ and $\boldsymbol{\theta} \sim \mathcal{N}_p(\boldsymbol{\mu}, \mathbf{A})$, where $\boldsymbol{\mu}$, $\boldsymbol{\Sigma}$, and $\mathbf{A}$ are known. Develop analogs of (4.78) and (4.79) as measures of "surprising" data. What will tend to happen to these measures for large p? Are these measures likely to be useful for large p?

85. Consider the situation of Exercise 68. One prior that is entertained for the i.i.d. θ_i is that they are $\mathcal{N}(3.30, \sigma_\pi^2)$, σ_π^2 unknown. A second prior entertained is that they are i.i.d. $\mathcal{N}(\mu_\pi, \sigma_\pi^2)$, μ_π having a second stage $\mathcal{N}(3.30, (0.03)^2)$ distribution. In both cases, assume that nothing is known about σ_π^2, and that it is given a constant second stage density. Find and compare the marginals under the two prior models. (Note that an improper prior is reasonable to use in such an endeavor only when it is the same parameters being given the improper prior in the cases being compared.)

86. Assume $X \sim \mathcal{N}(\theta, 1)$ is observed, and that it is desired to estimate θ under squared-error loss. It is felt that θ has a $\mathcal{N}(2, 3)$ prior distribution, but the estimated mean and variance could each be in error by 1 unit. Hence the class of plausible priors is

$$\Gamma = \{\pi : \pi \text{ is a } \mathcal{N}(\mu, \tau^2) \text{ density with } 1 \le \mu \le 3 \text{ and } 2 \le \tau^2 \le 4\}.$$

   (a) Find the range of posterior means, as π ranges over Γ, when $x = 2$ is observed.
   (b) Find the Γ-posterior expected loss of the Bayes action with respect to the $\mathcal{N}(2, 3)$ prior, when $x = 2$ is observed.
   (c) For what ε is the Bayes action in (b) ε-posterior robust?
   (d) Repeat parts (a), (b), and (c) for $x = 10$.
   (e) Compare the posterior robustness present for $x = 2$ and $x = 10$.

87. In the situation of Exercise 86, suppose it is desired to test H_0: $\theta \le 0$ versus H_1: $\theta > 0$ under "0–1" loss.
   (a) Find the range of the posterior probability of H_0, as π ranges over Γ, when $x = 0$ is observed.
   (b) Find the Γ-posterior expected loss of the Bayes action with respect to the $\mathcal{N}(2, 3)$ prior, when $x = 2$ is observed.
   (c) For what ε is the Bayes action in (b) ε-posterior robust?
   (d) Repeat parts (a), (b), and (c), for $x = 2$.
   (e) Compare the posterior robustness present for $x = 0$ and $x = 2$.

88. Consider the ε-contamination class in (4.75), and suppose that π_0 is a $\mathcal{N}(0, 2.19)$ density. How large must ε be for the $\mathcal{C}(0, 1)$ density to also be in this class?

89. Prove Lemma 2.

90. Suppose $X \sim \mathcal{N}(\theta, 1)$ and $\pi = (0.9)\pi_0 + (0.1)q$, where π_0 is $\mathcal{N}(0, 2)$ and q is $\mathcal{N}(0, 10)$.
   (a) If $x = 1$ is observed, find $\pi(\theta | x)$ and the posterior mean and variance.
   (b) If $x = 7$ is observed, find $\pi(\theta | x)$ and the posterior mean and variance.

91. Derive formulas analogous to (4.85) and (4.86) when $\pi(\theta) = \sum_{i=1}^{l} \varepsilon_i \pi_i(\theta)$, where $\varepsilon_i \ge 0$, $\sum_{i=1}^{l} \varepsilon_i = 1$, and the π_i are densities.

92. In Example 1 (continued) in Subsection 4.3.2, the 95% HPD credible set, for $x=115$, was found to be $C=(94.08, 126.70)$. Let Γ be the ε-contamination class of priors with π_0 equal to the $\mathcal{N}(100, 225)$ distribution, $\varepsilon=0.1$, and $\mathcal{Q}=\{\text{all distributions}\}$. Find the range of the posterior probability of C as π ranges over Γ.

93. In the situation of Exercise 92, suppose that it is desired to test H_0: $\theta \leq 100$ versus H_1: $\theta > 100$. With the same Γ and $x=115$, find the range of the posterior probability of H_0.

94. It is desired to investigate the robustness of the analysis in Exercise 40 with respect to the ε-contamination class of priors, with π_0 being the prior described in Exercise 40, $\varepsilon=0.1$, and $\mathcal{Q}=\{\text{all distributions}\}$. Find the range of the posterior probability of H_0: $\theta=4.01$.

95. In the situation of Exercise 26, find the range of the posterior probability of the 95% HPD credible set, as π ranges over the ε-contamination class with $\varepsilon=0.05$, π_0 being the $\mathcal{Be}(1, 9)$ prior, and $\mathcal{Q}=\{\text{all distributions}\}$.

96. Using the ε-contamination class described in Example 26, with $\varepsilon=0.1$, repeat the analysis, and compare the resulting indicated robustness with the answer for the ε-contamination class with $\mathcal{Q}=\{\text{all distributions}\}$, in (a) Exercise 92; (b) Exercise 93; (c) Exercise 94.

Subsections 4.7.5 through 4.7.7

97. Let $X \sim \mathcal{N}(\theta, 1)$ be observed, and assume it is desired to estimate θ under squared-error loss. The class of possible priors is considered to be $\Gamma=\{\pi: \pi$ is a $\mathcal{N}(0, \tau^2)$ density, with $0<\tau^2<\infty\}$.
    (a) Let δ be any Bayes rule from a prior in Γ. Find the Γ-Bayes risk of any such δ.
    (b) Repeat part (a) with Γ replaced by $\Gamma'=\{\pi: \pi$ is a $\mathcal{N}(\mu, 1)$ density, with $-\infty<\mu<\infty\}$.
    (c) Find a decision rule which has reasonable Γ-Bayes risk for both Γ and Γ'.

98. Assume that $R(\theta, \delta^*)$ is continuous in θ.
    (a) Prove that, if $\Gamma=\{\text{all prior distributions}\}$, then
$$\sup_{\pi \in \Gamma} r(\pi, \delta^*) = \sup_{\theta \in \Theta} R(\theta, \delta^*).$$
    (b) Prove that, if $\Theta=(-\infty, \infty)$ and $\Gamma=\{\pi: \pi$ is a $\mathcal{N}(\mu, \tau^2)$ density, with $-\infty< \mu<\infty$ and $0<\tau^2<\infty\}$, then
$$\sup_{\pi \in \Gamma} r(\pi, \delta^*) = \sup_{\theta \in \Theta} R(\theta, \delta^*).$$
    (c) Suppose that $\Theta=\mathbb{R}^p$, that $R(\boldsymbol{\theta}, \delta^*)$ is nondecreasing in $|\boldsymbol{\theta}|$, and that, for any ball $B_c=\{\boldsymbol{\theta}: |\boldsymbol{\theta}| \leq c\}$ and $\varepsilon>0$, Γ contains a prior giving probability at most ε to B_c. Prove that, then,
$$\sup_{\pi \in \Gamma} r(\pi, \delta^*) = \sup_{\boldsymbol{\theta} \in \Theta} R(\boldsymbol{\theta}, \delta^*).$$

99. Prove that, if a decision rule δ^* has constant risk ($R(\theta, \delta^*)$) and is Bayes with respect to some proper prior $\pi \in \Gamma$, then δ^* is Γ-minimax.

100. Assume $X \sim \mathcal{B}(n, \theta)$ is observed, and that it is desired to estimate θ under squared-error loss. Let Γ be the class of all symmetric proper prior distributions. Find a Γ-minimax estimator. (*Hint*: Use Exercise 99, considering Bayes rules for conjugate priors.)

101. Assume $X \sim \mathcal{B}(1, \theta)$ is observed, and that it is desired to estimate θ under squared-error loss. It is felt that the prior mean is $\mu = \frac{3}{4}$. Letting Γ be the class of prior distributions satisfying this constraint, find the Γ-minimax estimator of θ. (Show first that only nonrandomized rules need be considered. Then show that $0 \le E^{\pi}(\theta - \mu)^2 \le \mu(1 - \mu)$.)

102. Assume $X \sim \mathcal{E}(\theta)$ is observed, where $\Theta = \{1, 2\}$. It is desired to test H_0: $\theta = 1$ versus H_1: $\theta = 2$ under "0–1" loss. It is known that $a \le \pi_0 \le b$, where π_0 is the prior probability of H_0. You may assume that it is only necessary to consider the most powerful tests, which have rejection regions of the form $\{x: x > c\}$.
    (a) Find the Γ-minimax test when $a = 0$, $b = 1$.
    (b) Find the Γ-minimax regret test when $a = 0$, $b = 1$.
    (c) Find the Γ-minimax test for arbitrary a and b ($0 \le a < b \le 1$).

103. Assume $X \sim \mathcal{U}(0, \theta)$ is observed, and that it is desired to estimate θ under squared-error loss. Let $\Gamma = \{\pi_1, \pi_2\}$, where π_1 is a $\mathcal{G}(2, \beta)$ prior density and π_2 is the prior density

$$\pi_2(\theta) = 6\alpha^{-2}\theta\left(1 + \frac{\theta}{\alpha}\right)^{-4} I_{(0,\infty)}(\theta).$$

The median of the prior is felt to be 6.
    (a) Determine α and β.
    (b) Calculate the Bayes estimators of θ with respect to π_1 and π_2.
    (c) Calculate $R(\theta, \delta)$ for each of the Bayes estimators.
    (d) Which of the two rules is more appealing if large θ are thought possible?

104. In Example 30,
    (a) verify (4.100);
    (b) verify (4.101).

105. In Example 1, suppose it is desired to estimate θ under squared-error loss, and that, for robustness reasons, it is deemed desirable to use a limited translation estimator, instead of the posterior mean $\mu(x) = (400 + 9x)/13$. Suppose that an increase by 20% over the (minimax) risk of $\sigma^2 = 100$ is deemed to be allowable. Find the limited translation estimator which achieves this limit, determine its Bayes risk, and evaluate the estimate for $x = 115$.

106. Archaeological digs discovered Hominid remains at three different sites. The remains were clearly of the same species. The geological structures within which they were found were similar, coming from about the same period. Geologists estimate this period to be about 8 million years ago, plus or minus 1 million years. A radiocarbon dating technique, used on the three remains, gave ages of 8.5 million, 9 million, and 7.8 million years. The technique is known to give a result which is normally distributed, with mean θ = true age,

and standard deviation 600,000 years. It is desired to estimate $\boldsymbol{\theta} = (\theta_1, \theta_2, \theta_3)^t (\theta_i$ is the true age of the ith remains) under a quadratic loss. The radiocarbon dating technique is considered much more reliable than the geological information, so the geological information is to be considered vague prior information, namely the specification of a prior mean and standard deviation for each θ_i. Estimate $\boldsymbol{\theta}$ using (4.102), and give the percentage increase in Bayes risk that would be suffered, compared to using the conjugate prior Bayes estimator, if a normal prior were correct. Also, calculate the conjugate prior Bayes estimate, and judge if concern for robustness in using (4.102) has caused any decline in Bayesian performance from a conditional perspective.

107. In Example 30 (continued), show that (4.105) holds for $\delta = \delta^M$, if $\mathcal{Q}$ contains all $\mathcal{U}(-k, k)$ distributions for $k \geq k_0$, k_0 an arbitrary constant.

108. In Example 1, suppose it is desired to estimate θ under squared-error loss, but that an approximate Γ-minimax estimator is desired for robustness reasons. Suppose Γ is the ε-contamination class, with $\mathcal{Q} = \{\text{all distributions}\}$, $\varepsilon = 0.1$, and π_0 being the $\mathcal{N}(100, 225)$ distribution. Find the appropriate estimator and give its Γ-minimax risk. Compare the estimate for $x = 115$ with the conjugate prior Bayes estimate and that from Exercise 105.

Subsections 4.7.8 and 4.7.9

109. Suppose $X_1, \ldots, X_n$ is an i.i.d. sample from the $\mathcal{E}(\theta)$ density. The prior density for θ is $\mathcal{IG}(10, 0.02)$.
   (a) Determine the exact posterior density for θ, given $x_1, \ldots, x_n$.
   (b) Determine the approximations to this posterior density from parts (i) through (iv) of Result 8.
   (c) Suppose $n = 5$ and $\sum_{i=1}^n x_i = 25$. Sketch the exact posterior and the four approximate posteriors, and comment on the adequacy of the approximations.
   (d) Suppose $n = 100$ and $\sum_{i=1}^n x_i = 500$. Sketch the exact posterior and the four approximate posteriors and comment on the adequacy of the approximations.

110. Suppose that $\mathbf{X} = (X_1, \ldots, X_k)^t \sim \mathcal{M}(n, \boldsymbol{\theta})$, and that $\boldsymbol{\theta}$ has a $\mathcal{D}(\boldsymbol{\alpha})$ prior distribution. Determine the four approximations to the posterior distribution of $\boldsymbol{\theta}$ given in parts (i) through (iv) of Result 8. (Note that $\mathbf{X}$ is the sufficient statistic from n i.i.d. $\mathcal{M}(1, \boldsymbol{\theta})$ random variables.)

111. Suppose that X_1, X_2, X_3, and X_4 are i.i.d. $\mathcal{C}(\theta, 1)$ random variables. The noninformative prior $\pi(\theta) = 1$ is to be used. If the data is $\mathbf{x} = (1.0, 6.7, 0.5, 7.1)$, show that *none* of the approximations to the posterior in Result 8 will be reasonable.

112. Repeat Exercise 16, using the "alternative Jeffreys noninformative prior" (see Subsection 3.3.3)
$$\pi(\theta, \sigma^2) = \sigma^{-3} I_{(0,\infty)}(\sigma^2).$$
(This corresponds to $\pi(\theta, \sigma) = \sigma^{-2} I_{(0,\infty)}(\sigma)$ for σ.) Do the two noninformative priors give similar answers?

113. In Example 36, suppose that $X \sim \mathcal{N}(\theta, \sigma^2)$ and π_0 is $\mathcal{N}(\mu, \tau^2)$, where σ^2, μ, and τ^2 are all assumed to be known. Verify that, as $|x-\mu| \to \infty$, it will happen that $\hat{\pi}(\theta|x)$ converges to $\langle\hat{\theta}\rangle$ (in probability, say; to be more precise, show that the posterior distribution of $(\theta - x)$ converges to a point mass at zero.)

114. In Example 1, where $X \sim \mathcal{N}(\theta, 100)$ and θ has the $\mathcal{N}(100, 225)$ prior density, π_0, suppose it is desired to "robustify" π_0 by using the ML-II prior from the ε-contamination class, with $\mathcal{Q}$ of the form (4.110) (where $\theta_0 = 100$) and $\varepsilon = 0.1$.
   (a) If $x = 115$ is observed, determine the ML-II prior and the resulting posterior, and calculate the corresponding posterior mean and variance.
   (b) Repeat (a) for $x = 150$.
   (c) Compare the answers in parts (a) and (b) with the corresponding Bayes answers with respect to π_0, and discuss the "robustness" of the ML-II prior.

115. Suppose $X \sim \mathcal{N}(\theta, \sigma^2)$ and π_0 is $\mathcal{N}(\mu, \tau^2)$, where σ^2, μ, and τ^2 are known. Let Γ be the ε-contamination class of priors with $\mathcal{Q}$ of the form (4.110) (where $\theta_0 = \mu$). Show that, as $|x-\mu| \to \infty$, the ML-II prior yields a posterior, $\hat{\pi}(\theta|x)$, which converges (in probability, say) to the $\mathcal{N}(x, \sigma^2)$ posterior. (To be more precise, show that the ML-II posterior distribution of $(\theta - x)$ converges to a $\mathcal{N}(0, \sigma^2)$ distribution.)

116. Using Theorem 5, find the ML-II prior for the indicated Γ when

$$f(x|\theta) = \frac{1}{2\sigma} \exp\left\{ -\frac{|x-\theta|}{\sigma} \right\},$$

$\sigma > 0$ known.

Subsection 4.7.10

117. Show that (4.113) is necessary and sufficient for π_p to be a proper prior.

118. In the situation of Example 1, where $X \sim \mathcal{N}(\theta, 100)$, suppose the robust prior, π_1 (in Subsection 4.7.10), is to be used, with $\mu = 100$ and $A = 225$, instead of the conjugate $\mathcal{N}(100, 225)$ prior.
   (a) If $x = 115$, calculate the robust posterior mean and posterior variance, and find an approximate 95% credible set for θ.
   (b) Repeat (a) for $x = 150$.
   (c) Compare the "robust Bayes" answers in (a) and (b) with those obtained from use of the $\mathcal{N}(100, 225)$ prior.

119. Conduct a Bayesian test in Exercise 40, with the $\mathcal{N}(4.01, 1)$ choice for g_1 replaced by the "robust" π_1 from Subsection 4.7.10, with $\mu = 4.01$ and $A = 1$.

120. In the situation of Exercise 106:
   (a) Find the posterior mean and covariance matrix for $\boldsymbol{\theta}$, using the relevant robust prior, π_3, from Section 4.7.10.
   (b) Determine an approximate 95% credible ellipsoid for $\boldsymbol{\theta}$.

121. The sufficient statistic from a regression study in econometrics is the least squares estimator $\mathbf{x} = (1.1, 0.3, -0.4, 2.2)'$, which can be considered to be an observation from a $\mathcal{N}_4(\boldsymbol{\theta}, 0.3\mathbf{I}_4)$ distribution. Economic theory suggests that $\theta_1 - \theta_2 = 1$, and $\theta_3 = -1$. These linear restrictions are felt to hold with accuracies

of 0.3 and 0.5, respectively. (Consider these standard deviations.) *A priori*, nothing is known about θ_4, and there is thought to be no relationship between $\theta_1 - \theta_2$ and θ_3.

(a) Find the posterior mean and covariance matrix for $\boldsymbol{\theta}$, using the relevant robust prior from Section 4.7.10.

(b) Determine an approximate 95% credible ellipsoid for $\boldsymbol{\theta}$.

122. In the situation of Exercise 28, suppose that past records show that neutron stars average about 1.3 in mass, with a standard deviation of 0.2. This prior information is rather vague, so the robust Bayesian analysis with π_1^* in Subsection 4.7.10 (for unknown σ^2) is to be used. Calculate the robust Bayes estimate of θ, and determine an approximate 90% credible set for θ.

123. After one month of the season, eleven baseball players have batting averages of 0.240, 0.310, 0.290, 0.180, 0.285, 0.240, 0.370, 0.255, 0.290, 0.260, and 0.210. It is desired to estimate the year-end batting average, θ_i, of each player; the one month batting averages, X_i, can be assumed to be $\mathcal{N}(\theta_i, (0.03)^2)$ random variables. The previous year's batting averages of the eleven players were 0.220, 0.280, 0.320, 0.280, 0.260, 0.280, 0.300, 0.240, 0.310, 0.240, and 0.240, respectively. These can be considered to be prior means for this year's θ_i, and it is known that the yearly variation of a player's average tends to be about 0.02 (which can be considered to be the prior standard deviations).

(a) Assuming a conjugate normal prior, calculate the posterior means and variances for the θ_i.

(b) Using the robust prior of Subsection 4.7.10, calculate the posterior means and variances for the θ_i.

(c) Suppose a few extreme conflicts, between the X_i and the prior means, will sometimes occur in this scenario. Perform a type of robust Bayes analysis which protects against this possibility.

124. Verify (4.127).

125. Find the posterior variance for the posterior in (4.130), expressing the answer solely in terms of the numerator and denominator of (4.131).

126. Repeat Exercise 122, assuming that the prior is $\pi(\theta, \sigma^2) = \pi^*(\theta)\sigma^{-2}$, where $\pi^*(\theta)$ is $\mathscr{C}(1.3, 0.14)$. (The quartiles of π^* match up with the quartiles of a $\mathcal{N}(1.3, 0.04)$ prior.)

Subsection 4.7.11

127. In Example 41, suppose that π_0 is chosen to be the noninformative prior $\pi_0(\theta, \sigma^2) = \sigma^{-2}$. Seven observations are taken, the result being $\mathbf{x} = (2.2, 3.1, 0.1, 2.7, 3.5, 4.5, 3.8)$. Graph $m(\mathbf{x} | \pi_0, \alpha)$ as a function of α, and comment on whether or not it seems reasonable, here, to assume that f_0 is normal.

128. In Example 42, find the marginal posterior distribution of θ, given $(x_1, \ldots, x_n, y_1, \ldots, y_m)$, for the noninformative prior $\pi(\theta, \sigma^2, \mu_1, \ldots, \mu_m) = \sigma^{-2}$.

Section 4.8

129. Prove Theorem 8.

130. Prove Theorem 9.

131. If $X \sim \mathscr{B}(n, \theta)$ and it is desired to estimate θ under squared-error loss, show that $\delta(x) = x/n$ is admissible. (*Hint*: Consider the improper prior density $\pi(\theta) = \theta^{-1}(1-\theta)^{-1}$. You may assume that all risk functions $R(\theta, \delta)$ are continuous.)

132. In Example 46, for estimating $\eta = |\boldsymbol{\theta}|^2$,
   (a) verify (4.134);
   (b) show that, on any bounded set of $\mathbf{x}$, there exists an admissible estimator which is arbitrarily close to $\delta^{\pi}(\mathbf{x}) = |\mathbf{x}|^2 + p$.

133. Assume $X \sim \mathscr{G}(\alpha, \beta)$ (α known) is observed, and that it is desired to estimate β under loss

$$L(\beta, a) = \left(1 - \frac{a}{\beta}\right)^2.$$

   It is decided to use the improper prior density $\pi(\beta) = \beta^{-2}$.
   (a) Show that the generalized Bayes estimator of β is

$$\delta^0(x) = \frac{x}{\alpha + 2}.$$

   (b) Show that δ^0 is inadmissible.
   (c) Show that δ^0 can be beaten in long-run use, in the sense of (4.135), for any sequence $\boldsymbol{\theta} = (\beta^{(1)}, \beta^{(2)}, \ldots)$.

134. Prove part (b) of Theorem 10.

135. In the situation of Example 45, show that δ^0 can be beaten in long-run use, in the sense of (4.135), for any sequence $\boldsymbol{\theta} = (\beta^{(1)}, \beta^{(2)}, \ldots)$ that is bounded away from zero.

136. In Example 47, verify that $\alpha(x) = \frac{1}{2}$ is weakly incoherent. (The loss for a randomized betting strategy is expected loss, as usual.)

137. Suppose $X \sim \mathcal{N}(\theta, 1)$, $\Theta = \mathbb{R}^1$, $C(x) = (-\infty, x)$, $\alpha(x) = \frac{1}{2}$, and $\{C(x), \alpha(x)\}$ is the system of confidence statements that is to be used. Show that this choice is weakly incoherent.

138. For the situation of Example 47, show that the given $\{C(x), \alpha(x)\}$ is an inadmissible procedure under loss (4.136).

Section 4.9

139. Suppose that $\bar{X} \sim \mathcal{N}(\theta, 0.2)$ is observed, and that the prior for θ is $\mathscr{C}(0, 1)$.
   (a) If it is possible to generate either normal or Cauchy random variables, describe two methods for calculating the posterior mean by Monte Carlo integration.
   (b) Which method in (a) is likely to give an accurate answer with the fewest repetitions?

(c) Carry out the two simulations in (a) for $\bar{x}=1$, using 5, 50, and 500 repetitions.
(d) For each of the repetition numbers and simulations in part (c), estimate the variance of the Monte Carlo analysis by repeating the analysis 10 times and calculating the sample variance. (Note that an independent set of random variables must be used each time the analysis is repeated.)
(e) Use the normal approximation to the posterior, from part (ii) of Result 8 (in Subsection 4.7.8), to suggest an importance function when $\bar{x}=1$.
(f) For the importance function from (e), repeat parts (c) and (d).

140. Suppose $\pi(\theta)$ is a hierarchical prior of the form

$$\pi(\theta)=\int_{\Lambda}\pi_1(\theta\mid\lambda)\pi_2(\lambda)d\lambda.$$

Imagine that it is possible to easily generate random variables having densities π_2 and π_1 (for each λ). Give a Monte Carlo approximation to $E^{\pi(\theta|x)}[g(\theta)]$ which does not involve evaluation of $\pi(\theta)$.

141. Using Result 9, approximate the posterior mean and variance in the situation of Exercise 109(c).

142. Using Result 9, approximate the posterior mean and variance in the situation of Exercise 139, when $\bar{x}=1$. (Calculate the posterior mean by defining $g(\theta)=\theta+2$, so that $E^{\pi(\theta|x)}[\theta]=E^{\pi(\theta|x)}[g(\theta)]-2$; this ensures that g is effectively positive for the calculation.)

Section 4.10

143. Develop versions of formulas (4.145) through (4.150) when π^{**} is an $\mathcal{IG}(\alpha,\beta)$ distribution.

144. If g in (4.145) is a $\mathcal{C}(\mu,\tau)$ density, show, via a Taylors series argument, that

$$m(\mathbf{x}\mid g,\pi^{**})\cong\frac{\tau}{\pi[\tau^2+(\bar{x}-\mu)^2]}\left(1+\frac{(n-1)s^2[3(\bar{x}-\mu)^2-\tau^2]}{n(n-3)[\tau^2+(\bar{x}-\mu)^2]^2}\right).$$

(This is very accurate for moderate or large n, $(\bar{x}-\mu)^2$, or τ.)

145. Repeat Exercise 40, assuming that the observations are $\mathcal{N}(\theta,\sigma^2)$, σ^2 unknown (instead of $\sigma^2=1$), and that $g_1(\theta)$ is $\mathcal{C}(4.01, 0.34)$ and σ^2 has (independently) the noninformative prior $\pi^{**}(\sigma^2)=\sigma^{-2}$:
(a) Use the (4.150) approximation.
(b) Use the approximation to $m(\mathbf{x}\mid g_1,\pi^{**})$ in Exercise 144.

146. Utilize Figure 4.4 to sketch a graph of $\pi(\theta_0\mid\mathbf{x})$ versus τ, when π_0 is fixed at $\frac{1}{2}$.

147. For the situation of Figure 4.4, suppose that it is known only that $1\le\tau\le 2$ and $0.3\le\pi_0\le 0.5$. Find the range of possible $\pi(\theta_0\mid\mathbf{x})$.

Section 4.11

148. In Example 50, suppose the four classifications under consideration are (corn, forest, grass, dirt). The likelihood, for the observed x, is the vector $\mathbf{l}=(f(x\mid\theta_1),$

$f(x|\theta_2)$, $f(x|\theta_3)$, $f(x|\theta_4)) = (0.3, 0.4, 0.6, 0)$. The topographical information yields (in similar vector form) $\boldsymbol{\pi}_2 = (0, 0.2, 0.6, 0.2)$. The information from the preceding year is $\boldsymbol{\pi}_3 = (0.1, 0.3, 0.3, 0.3)$. Making sure all inputs are converted to probability vectors, calculate the overall probability distribution for θ via the linear opinion pool, and via the independent opinion pool. Criticize each answer.

149. Experts $1, \ldots, m$ report estimates, $X_1, \ldots, X_m$, for θ. Suppose $\mathbf{X} = (X_1, \ldots, X_m)^t$ is $\mathcal{N}_m(\theta\mathbf{1}, \mathbf{\Sigma})$, where $\mathbf{1} = (1, \ldots, 1)^t$ and $\mathbf{\Sigma}$ has diagonal elements 1 and known off-diagonal elements $\rho > 0$. (Thus ρ reflects the fact that there is a dependence among experts.) Suppose θ is given the noninformative prior $\pi(\theta) = 1$.
    (a) Show that the posterior distribution of θ, given $\mathbf{x}$, is $\mathcal{N}(\bar{x}, m^{-1} + (1 - m^{-1})\rho)$.
    (b) Show that an infinite number of dependent experts, with $\rho = 0.2$, can convey no more accurate information about θ than five independent experts ($\rho = 0$).

150. In Example 53, suppose the expert would tend to report $p \cong \frac{1}{3}$ with a standard deviation of $\frac{1}{4}$, if θ_1 were true, and would tend to report $p \cong \frac{2}{3}$ with a standard deviation of $\frac{1}{4}$, if θ_2 were true. (The expert is mixed up in his reports.)
    (a) Determine the analog of (4.154) for this situation.
    (b) Which "expert" would you rather get an opinion from?

151. In Example 55, verify that any randomized action can be dominated by an action in the interval [0.75, 0.79].

CHAPTER 5
Minimax Analysis

5.1. Introduction

This chapter is devoted to the implementation and evaluation of decision-theoretic analysis based on the minimax principle introduced in Section 1.5. We began Chapter 4 with a discussion of axioms of rational behavior, and observed that they lead to a justification of Bayesian analysis. It would be nice to be able to say something similar about minimax analysis, but the unfortunate fact is that minimax analysis is not consistent with such sets of axioms. We are left in the uncomfortable position of asking why this chapter is of any interest. (Indeed many Bayesians will deny that it is of any interest.) It thus behooves us to start with a discussion of when minimax analysis can be useful.

Recall from Section 1.5 that the essence of the minimax principle is to try and protect against the worst possible state of nature. The one situation in which this is clearly appropriate is when the state of nature is determined by an intelligent opponent who desires to maximize your loss (see Section 1.4). You can then expect the worst possible state of nature to occur, and should plan accordingly. The study of this situation is called game theory, and is the subject of Section 5.2. (It should be mentioned that the axioms of rational behavior, alluded to earlier, are not valid in situations involving intelligent opponents.)

Statistical problems, on the other hand, involve a "neutral" nature, and it is then not clear why the minimax principle is useful. The most frequently given justification for minimax analysis in such problems is a possible desire for conservative behavior. A very simple example of this is Example 3 of Chapter 1. In a choice between a risky bond and a safe bond, a conservative attitude may well lead to the choice of the safe bond, even if its expected yield is lower than that of the risky bond. This consideration gets somewhat obviated by the fact that a utility function (if constructed) will naturally incorporate the desired conservatism, leaving no further need for explicit conservative action. This will be discussed more fully in Section 5.5. It can at least be said that, if a decision is to be made without the use of a utility approach (say loss is expressed solely in monetary terms), then a conservative principle may be appealing to some.

Perhaps the greatest use of the minimax principle is in situations for which no prior information is available. There is then no natural decision principle by which to proceed, and the minimax principle is often suggested as a good choice. Frequently the minimax principle will prove reasonable in such problems, but two notes of caution should be sounded. First, the minimax principle can lead to bad decision rules. This will be indicated in Section 5.5. Second, the minimax principle can be devilishly hard to implement. The alternative approach, discussed in Chapter 4, of using noninformative priors and Bayesian methodology is far simpler and usually gives as good or better results. Indeed, for problems in which the minimax approach is feasible, the two approaches will frequently give the same result. (Reasons for this will be indicated in Chapter 6, where a tie-in with the invariance principle will also be discussed.) It can be argued, however, that, due to the possible inadequacies of rules based on noninformative priors, it is wise to investigate such problems from a minimax viewpoint also. This is especially true when "standard" statistical procedures are being recommended for nonspecialists to use. Due to their importance, they should be examined from many directions; the Bayesian with noninformative prior and minimax approaches being two of the most appealing.

The final use of minimax rules, that should be mentioned, is that of providing a yardstick for Bayesian robustness investigations. In Subsections 4.7.5 through 4.7.7, the value of frequentist measures to Bayesians was discussed. Of specific interest here is the discussion of Γ-minimaxity in Subsection 4.7.6; this was introduced as an aid in the study of robustness with respect to specification of the prior. It was pointed out that, if Γ is the class of all priors, then Γ-minimaxity is simply minimaxity. In some sense, the implication is that a minimax rule is the most robust rule with respect to specification of the prior. (Essentially, it protects against the "prior" being concentrated at the "worst" state of nature.) The robustness of proposed Bayesian rules can thus be partially indicated by comparison of the rules with the minimax rule.

We feel that the above reasons provide some justification for studying the minimax approach. It is probably true, however, that the considerable *theoretical* popularity of the approach is not so much due to these justifications as to the fact that minimax theory is much richer mathematically than Bayesian theory. (Research involving difficult mathematics tends to be more prestigious.) This richness is partially indicated in this chapter, though only certain special cases can be considered.

Section 5.2 discusses game theory, and lays the mathematical foundation for the application of the minimax principle. This mathematical foundation (particularly the S-game formulation of Subsection 5.2.4) is of interest in its own right, being useful in a variety of other problems (such as the group decision problem of Subsection 4.11.3). Section 5.3 relates the general theory to statistical problems. Section 5.4 discusses the interesting situation of Stein estimation, where a multitude of minimax estimators can exist. Section 5.5 gives a critical discussion of minimax analysis.

5.2. Game Theory

Game theory originated with work of Borel (1921) and von Neumann (1928), and was extensively developed in von Neumann and Morgenstern (1944). For a general development and recent references, see Thomas (1984). The goal of game theory is the development of optimal strategies for action in competitive situations involving two or more intelligent antagonists. Such problems can be encountered in business situations involving competitors, in military or political settings of conflict between groups, and of course in games.

In Subsection 5.2.1 the basic elements of game theory will be discussed. Subsections 5.2.2 and 5.2.3 deal with explicit methods of solving games. Subsections 5.2.4, and especially 5.2.5 and 5.2.6, are more theoretically oriented, giving a fairly complete mathematical development for the case of finite Θ.

5.2.1. Basic Elements

We shall consider only games involving two opponents, henceforth to be called *players*. The statistician will be identified with player II (II for short), and his opponent will be called player I (or just I).

I. *Strategies*

In game theory, actions of players are usually called *strategies*, the terminology being designed to indicate that a strategy will involve a complete description of the "moves" that will be made in the game. In chess, for example, a strategy consists of a complete specification of all moves you will make throughout the game, including the response you will have to any given move of your opponent. To put this another way, a strategy can be thought of as a recipe (or computer program) which can be followed mechanically (say by a computer) to play the game. Clearly, good strategies can be very complicated, and in games such as chess seem almost unobtainable. (On the other hand, the computer programs that have recently been developed to play chess are nothing but strategies in the above sense, and indeed are very good strategies. Already, only extremely good chess players can beat the best computer strategies, and, in the near future, computers may become unbeatable.)

The set of strategies available to I will be denoted Θ, while $\mathscr{A}$ will stand for the set of possible strategies for II. (Hopefully this change in the interpretation of Θ will not cause confusion.)

II. *Two-Person Zero-Sum Games*

If player I chooses strategy $\theta \in \Theta$ and II chooses strategy $a \in \mathscr{A}$, it will be assumed that I *gains*, and II *loses*, an amount $L(\theta, a)$. We will continue to refer to L as the loss (to player II), though in game theory it is usually referred to as the *payoff function* (for I). For technical convenience, it will be assumed that $L(\theta, a) \geq -K$ for some finite constant K.

A game, such as defined above, is called a *two-person zero-sum game.* The reason for calling it a two-person game is obvious. The "zero sum" refers to the fact that the sum of the losses (or the sum of the gains) for the two players is zero, since a loss for one is a gain for the other. This introduces an enormous simplification of the theory, since the two players can then gain nothing by cooperation. Cooperation is senseless (feelings of altruism aside), since any extra gain to your opponent (through cooperation) comes directly from your own pocket. It can thus be assumed that each player is trying to maximize his gain *and* his opponent's loss. This makes the mathematical analysis relatively easy.

Zero-sum games are, of course, rather rare. In business settings, for example, competing companies are rarely in zero-sum situations. Cooperation (such as price fixing) will often be mutually beneficial to both. Nevertheless, zero-sum games serve as a useful introduction to game theory. Several examples of zero-sum games follow.

EXAMPLE 1. Refer again to the game of matching pennies discussed in Section 1.4 of Chapter 1. Each player has two strategies, and the loss matrix is

		Player II	
		a_1	a_2
Player I	θ_1	-1	1
	θ_2	1	-1

This is clearly a zero-sum game, since one player's loss is the other's gain.

EXAMPLE 2. Two countries, I and II, are at war. II has two airfields, and can defend one but not both. Let a_1 and a_2 denote defending airfield 1 and airfield 2 respectively. Country I can attack only one of the airfields. Denote the two possible strategies θ_1 and θ_2. If I attacks a defended airfield, it will immediately withdraw with no loss to either side. If I attacks an undefended airfield, the airfield will be destroyed. Airfield 1 is twice as valuable as airfield 2 to II. Letting 1 denote the value (to II) of airfield 2, it follows that the loss matrix is

	a_1	a_2
θ_1	0	2
θ_2	1	0

This is a zero-sum game, providing the values to I of the destruction of the airfields are the same as the values of the airfields to II.

EXAMPLE 3. Each player simultaneously extends one or two fingers on each hand. The number of fingers extended on the right hand of each player is the player's "choice," while the number of fingers extended on the left hand is a guess of the opponent's choice. If only *one* player guesses his opponent's choice correctly, he receives a payoff (from the other player) consisting of the total number of extended fingers (for both players). Otherwise there is no payoff.

For each player, let (i, j) denote the strategy "extend i fingers on your left hand and j fingers on your right hand." The loss matrix is then

		Player II			
		(1, 1)	(1, 2)	(2, 1)	(2, 2)
Player I	(1, 1)	0	−5	5	0
	(1, 2)	5	0	0	−7
	(2, 1)	−5	0	0	7
	(2, 2)	0	7	−7	0

The game is clearly zero sum.

EXAMPLE 4. Player I chooses a number $\theta \in \Theta = [0, 1]$, and II chooses a number $a \in \mathscr{A} = [0, 1]$. The loss to II (and gain to I) is $L(\theta, a) = |\theta - a|$.

EXAMPLE 5. Two companies, I and II, are the sole producers of widgets. In two of the firms which purchase widgets, the employees who decide whether to buy from I or II take bribes. Label these two firms A and B. Firm A buys 2000 widgets per year, while firm B buys 3000 widgets. Companies I and II set aside \$10,000 per year for bribery purposes. Each company must decide how much of the bribe money to allocate to A, and how much to allocate to B. Each company can expect an order from a bribed firm in proportion to the bribe ratio. In other words, if I and II bribe A with θ and a amounts of money respectively, I will obtain an order from A of $2000 \times \theta/(\theta + a)$ widgets. Since the bribes for B will then be $10{,}000 - \theta$ and $10{,}000 - a$, I will receive an order from B of

$$\frac{3000(10{,}000 - \theta)}{20{,}000 - [a + \theta]}$$

widgets. (If both x and y are zero, define $x/(x+y)$ as $\frac{1}{2}$.) Letting the loss to II be represented by the total sales of I to A and B, the loss function is

$$L(\theta, a) = \frac{2000\,\theta}{\theta + a} + \frac{3000(10{,}000 - \theta)}{20{,}000 - [a + \theta]}.$$

This can be considered a zero-sum game, since any sale made by I is a sale lost by II. Having the \$10,000 bribery cost go to someone outside the game may seem to violate the zero-sum condition. This is not the case, however, since the \$10,000 will *always* be spent. The money, therefore, has no effect on the choice of strategy. If, instead, the companies were trying to decide how much total bribe money to spend, the game would not be zero sum for the above reason. Note how delicate the zero-sum property really is. Because of this, it is extremely difficult to find realistic zero-sum games in business settings.

One facet of the above example deserves further emphasis, namely the fact that adding a constant to the loss function of II (or gain function of I) does not change the problem in terms of which strategies are optimal. This has two useful consequences. First, in verifying the zero-sum property, one can separately calculate the loss function for II and the gain function for I, and check that their difference is constant. Either function will then serve for the purpose of determining optimal strategies. Secondly, it will sometimes be of use to add a constant to the loss function (say, to make it positive). Again, this will not affect the conclusions.

So far, we have ignored the role of utility theory in the formulation of a game. As discussed in Chapter 2, the "values" of outcomes of a game should generally be measured in terms of utilities. This, however, poses a serious problem for game theory, in that the utility of a loss of z by player II will usually not be the negative of the utility of a gain of z by I. (Even if the utility functions of the two players are the same, the utility of a loss is frequently not the negative of the utility of the corresponding gain.) The game will then not be zero sum. (An obvious example of this would be playing matching pennies with a $1000 payoff.) Of course, if the outcome, z, of a game is always a fairly small amount of money, then the utility functions $U_i(z)$ $(i=1,2)$ of the two players will tend to be approximately equal to z. The game can then be considered to be approximately zero sum.

The reader may by now be quite discouraged as to the range of applicability of two-person zero-sum games. The reasons for studying them, however, transcend their limited usefulness. In the first place, there is a considerable theory on non-zero-sum games (cf. Jones (1980), Coleman (1982), Shubik (1982), and Thomas (1984)), for which zero-sum games can be considered an introduction. Secondly, the theory of two-person zero-sum games will be seen to apply directly to minimax analysis in strictly statistical settings. The reason the above problems are not encountered in such situations is that one just pretends that the statistician's loss is nature's gain and that nature is intelligent (i.e., out to maximize the loss). The theory of two-person zero-sum games then clearly applies. Of course serious questions can be raised concerning the desirability of so pretending. This issue will be discussed in Section 5.5.

In the remainder of the section it will be *assumed* that the loss is developed from utility considerations. (For problems that are stated in terms of money, this will entail assuming that the utility functions of the players are roughly the identity functions $U_i(z)=z$.) The reason for assuming this (besides the basic desire to accurately represent "value") is that, in game theory, it is of crucial importance to allow randomized strategies, and to evaluate random situations in terms of expected loss. As discussed in Chapter 2, it is precisely when losses are given in terms of utilities that this is justified.

In light of the above assumption, the loss function in a random situation will be *defined* as the expected loss to player II. In other words, if, for given strategies θ and a, the loss to player II is the random quantity Z, then $L(\theta, a)=E_{\theta,a}[Z]$.

EXAMPLE 6. Consider the game of matching pennies, but assume that, after I chooses θ_i, a spy reports the choice to II. Unfortunately, the reliability of the spy (i.e., the probability of his being correct) is only $p>\frac{1}{2}$. The possible strategies for II can be represented by (1, 1), (1, 2), (2, 1), and (2, 2), where (i, j) indicates that II chooses a_i if the spy announces θ_1, and a_j if the spy announces θ_2. (Recall that a strategy, by definition, consists of a complete description of the actions to be taken in light of all possible contingencies.) If, say, I chooses θ_1 and II chooses (1, 2), the expected loss to II is clearly

$$(-1)P(\text{spy says } \theta_1)+(1)P(\text{spy says } \theta_2)=-p+(1-p)=1-2p.$$

Similarly, the entire loss matrix can be calculated to be

	(1, 1)	(1, 2)	(2, 1)	(2, 2)
θ_1	-1	$1-2p$	$2p-1$	1
θ_2	1	$1-2p$	$2p-1$	-1

EXAMPLE 7. Ann arrives home from work at a random time between 4 and 5. She always goes out to dinner at 5. Mark and Jim are dating Ann, and would each like to take her to dinner. They must call her at home, before she leaves, in order to arrange to meet her for dinner. Unfortunately, they each have only one coin for a phone call. Whoever calls first, while Ann is at home, will get the dinner date. The utility function for both Mark and Jim is

U(getting the date) = 1,
U(other man getting the date) = −1,
U(neither getting the date) = 0.

The game will clearly be zero sum, since one man's gain is the other's loss.

Let Mark be player I and Jim be player II, and let θ and a denote the times they choose to call. We can clearly assume that $\Theta = \mathscr{A} = [4, 5]$. Let T denote the time at which Ann arrives home from work. We know that $T \sim \mathscr{U}(4, 5)$. If $T = t$, the utility of (θ, a) to I is

$$U_t(\theta, a) = \begin{cases} 1 & \text{if } t < \theta < a \quad \text{or} \quad a < t < \theta, \\ -1 & \text{if } t < a < \theta \quad \text{or} \quad \theta < t < a, \\ 0 & \text{if } a < t \quad \text{and} \quad \theta < t. \end{cases}$$

The gain to I (or loss to II) is thus

$$L(\theta, a) = E^T[U_T(\theta, a)] = \begin{cases} (1)P(T < \theta) + (-1)P(\theta < T < a) & \text{if } \theta < a, \\ (1)P(a < T < \theta) + (-1)P(T < a) & \text{if } \theta > a, \end{cases}$$

$$= \begin{cases} 2\theta - 4 - a & \text{if } \theta < a, \\ \theta + 4 - 2a & \text{if } \theta > a. \end{cases}$$

III. *Randomized Strategies*

As indicated in Section 1.4, it will be crucial to allow the use of randomized strategies. By a *randomized strategy* will be meant a probability distribution on the strategy space. As in Section 1.4, δ^* will be used to represent a randomized strategy for player II. For player I, a randomized strategy is a probability distribution on Θ, which for consistency we will denote by π. (Note the change in emphasis, however. Player II does not necessarily think that θ is distributed according to π. Instead, π is chosen by, and perhaps known only to, I.)

The loss when randomized strategies are used will again just be expected loss. Thus if I chooses π and II chooses δ^*, define

$$L(\pi, \delta^*) = E^{\pi}E^{\delta^*}L(\theta, a) = \int_{\Theta}\int_{\mathscr{A}} L(\theta, a)dF^{\delta^*}(a)dF^{\pi}(\theta). \qquad (5.1)$$

(By Fubini's theorem and the assumption that $L(\theta, a) \geq -K$, it does not matter in which order the expectation is taken.)

As in Section 1.4, a (or θ) are to be identified with $\langle a \rangle$ (or $\langle \theta \rangle$), which are defined to be the probability distributions which choose a (or θ) with probability one. Hence $L(\theta, \delta^*) = E^{\delta^*}[L(\theta, a)]$ and $L(\pi, a) = E^{\pi}[L(\theta, a)]$. Hopefully no confusion will result from this multiple use of L. Note that, in statistical problems, the quantity $L(\pi, \delta^*)$ has been called $r(\pi, \delta^*)$.

Only randomized strategies with finite loss will be considered. Let $\mathscr{A}^*$ denote the set of all δ^* for which $L(\theta, \delta^*) < \infty$ for all $\theta \in \Theta$, and let Θ^* denote the set of all π for which $L(\pi, a) < \infty$ for all $a \in \mathscr{A}$.

Example 2 (continued). For randomized strategies π and δ^*, (5.1) becomes

$$\begin{aligned} L(\pi, \delta^*) &= \sum_{i=1}^{2}\sum_{j=1}^{2} L(\theta_i, a_j)\pi(\theta_i)\delta^*(a_j) \\ &= (0)\pi(\theta_1)\delta^*(a_1) + (2)\pi(\theta_1)\delta^*(a_2) + (1)\pi(\theta_2)\delta^*(a_1) \\ &\quad + (0)\pi(\theta_2)\delta^*(a_2) \\ &= 2\pi(\theta_1) + \delta^*(a_1) - 3\pi(\theta_1)\delta^*(a_1) \end{aligned}$$

(using the facts that $\pi(\theta_2) = 1 - \pi(\theta_1)$ and $\delta^*(a_2) = 1 - \delta^*(a_1)$).

Example 4 (continued). Randomized strategies are probability distributions on [0, 1]. If $\pi(\theta)$ and $\delta^*(a)$ are probability *densities* on [0, 1], then

$$L(\pi, \delta^*) = \int_0^1 \int_0^1 |\theta - a|\pi(\theta)\delta^*(a)da\, d\theta.$$

IV. *Admissibility and Bayes Strategies*

The concepts of admissible strategies and Bayes strategies will prove useful. A strategy δ_1^* for II is *admissible* if there is no strategy $\delta_2^* \in \mathscr{A}^*$ for which $L(\theta, \delta_2^*) \leq L(\theta, \delta_1^*)$ for all $\theta \in \Theta$, with strict inequality for some θ. A strategy is *inadmissible* if such a δ_2^* exists. The definition of admissible and inadmissible strategies for I is analogous. A strategy δ_1^* for II is *Bayes with respect to* π, if $L(\pi, \delta_1^*) = \inf_{a \in \mathscr{A}} L(\pi, a)$. The quantity $L(\pi) = \inf_{a \in \mathscr{A}} L(\pi, a)$ is called the *Bayes loss* of π. These concepts are all, of course, merely translations into the game-theoretic setting of the corresponding statistical ideas.

Example 6 (continued). Since $p>\frac{1}{2}$, the strategy $(2, 1)$ is inadmissible, being dominated by $(1, 2)$. If π is given by $\pi(\theta_1)=\pi(\theta_2)=\frac{1}{2}$, then $L(\pi, \delta^*)$ can be calculated to be

$$L(\pi, \delta^*)=(1-2p)\delta^*((1, 2))+(2p-1)\delta^*((2, 1)).$$

Furthermore,

$$\inf_{a\in\mathscr{A}} L(\pi, a)=\min\{0, (1-2p), (2p-1), 0\}=1-2p.$$

Since the Bayes strategies with respect to π are those for which $L(\pi, \delta^*)= 1-2p$, it is clear that the only Bayes strategy is $\delta^*((1, 2))=1$ (i.e., choose $(1, 2)$ with probability one).

V. *Optimal Strategies and Value*

In choosing a strategy, it will be seen that, in some sense, player II should use the minimax principle. For clarity, we restate the definition of a minimax strategy, as it applies to this setting.

Definition 1. A *minimax strategy* for player II is a (randomized) strategy δ^{*M} which minimizes $\sup_{\theta\in\Theta} L(\theta, \delta^*)$, i.e., a strategy for which

$$\sup_{\theta\in\Theta} L(\theta, \delta^{*M})=\inf_{\delta^*\in\mathscr{A}^*}\sup_{\theta\in\Theta} L(\theta, \delta^*).$$

The quantity on the right-hand side of the above equation is called the *minimax value* of the game (many call it the *upper value*), and will be denoted $\bar{V}$.

Player I may also desire to act according to the minimax principle. Since L is his gain, however, the goal should be phrased in terms of maximizing the minimum possible gain.

Definition 2. A *maximin strategy* for player I is a (randomized) strategy π^M which maximizes $\inf_{a\in\mathscr{A}} L(\pi, a)$, i.e., a strategy for which

$$\inf_{a\in\mathscr{A}} L(\pi^M, a)=\sup_{\pi\in\Theta^*}\inf_{a\in\mathscr{A}} L(\pi, a).$$

The quantity on the right-hand side of the above equation is called the *maximin value* of the game (many call it the *lower value*), and will be denoted $\underline{V}$.

The following lemma will be needed later, and shows that $\inf_{a\in\mathscr{A}}$ and $\sup_{\theta\in\Theta}$ can be replaced by $\inf_{\delta^*\in\mathscr{A}^*}$ and $\sup_{\pi\in\Theta^*}$ in the above definitions, proving that minimax and maximin strategies also protect against the worst possible randomized strategy of the opponent. The proof of the lemma is left as an exercise.

Lemma 1. *For any strategies* δ_0^* *and* π_0,
(a) $\sup_{\pi \in \Theta^*} L(\pi, \delta_0^*) = \sup_{\theta \in \Theta} L(\theta, \delta_0^*)$,
(b) $\inf_{\delta^* \in \mathscr{A}^*} L(\pi_0, \delta^*) = \inf_{a \in \mathscr{A}} L(\pi_0, a)$.

Of considerable interest is the relationship between $\bar{V}$ and $\underline{V}$. The following lemma gives one important fact. The proof is easy and is also left as an exercise.

Lemma 2. $\underline{V} \le \bar{V}$.

In many games it so happens that $\underline{V} = \bar{V}$. This is so noteworthy that it deserves its own terminology.

Definition 3. If $\underline{V} = \bar{V}$, the game is said to have *value* $V = \underline{V} = \bar{V}$. If, in addition, the players have minimax and maximin strategies, the game is said to be *strictly determined.*

EXAMPLE 1 (continued). In Section 1.5, it was seen that a minimax strategy for the game of matching pennies is $\delta^*(a_1) = \delta^*(a_2) = \frac{1}{2}$, and that $\bar{V} = 0$. By symmetry, it is clear that a maximin strategy for I is $\pi(\theta_1) = \pi(\theta_2) = \frac{1}{2}$, and that $\underline{V} = 0$. The game thus has value $V = 0$ and is strictly determined.

In a strictly determined game, it is, in some sense, irrational for the players to use anything but their minimax (or maximin) strategies, providing they are playing against an intelligent opponent. Consider player II, for example. By using δ^{*M}, his loss will never be *more* than V $(=\bar{V})$. On the other hand, it seems plausible to believe that I will be using his maximin strategy, in which case II's loss can never be *less* than V $(=\underline{V})$. Using δ^{*M} to guarantee at most a loss of V thus seems natural. A nonminimax rule can never gain (and can incur considerable additional loss) if indeed I is using his maximin rule. Player I, by similar reasoning, should decide to use his maximin rule against an intelligent opponent. Note that in a strictly determined game there is no advantage in *knowing* that your opponent will be using a particular minimax or maximin strategy. This, in some sense, explains the phrase "strictly determined." Before the game even begins, the optimal strategies and (expected) outcome can be known.

It will be seen in Subsection 5.2.6 that a wide variety of two-person zero-sum games have values and are strictly determined. In particular, we will explicitly prove that if Θ is finite (and certain technical conditions hold), then the game has a value and is strictly determined. (This is the minimax theorem.)

5.2.2. General Techniques for Solving Games

By *solving* a game, we will mean finding the value and the minimax strategy. (The maximin strategy may also be of interest, though we tend to identify with player II.) Occasionally, it will be possible to explicitly solve a game by directly calculating the maximin and minimax strategies. Some situations in which this can be done are mentioned first.

I. *The Direct Method of Solving Games*

It is sometimes possible to find a minimax or maximin strategy directly from the definitions. To find a minimax strategy, for example, one can attempt to calculate the function $\bar{L}(\delta^*) = \sup_\theta L(\theta, \delta^*)$, and try to directly minimize it over all $\delta^* \in \mathscr{A}^*$. Usually this is a hopeless task, but in some interesting problems it can be done. The problems in which this approach will prove useful are certain statistical problems of Section 5.3 in which, by elimination of inadmissible strategies, the set $\mathscr{A}^*$ can be reduced to a low-dimensional space. If $\bar{L}(\delta^*)$ is not too complicated, the minimization can then be carried out.

Likewise, it will sometimes be possible to directly calculate $\underline{L}(\pi)$ (the Bayes loss of π), and maximize it over all $\pi \in \Theta^*$. Even if this can be done, however, one is still left with the problem of finding a minimax strategy. The following theorem can prove helpful in this regard. The proof of the theorem follows directly from the definitions, and is left as an exercise.

Theorem 1. *Assume that π^M is a maximin strategy for* I *and that the game has a value. Then any minimax strategy is a Bayes strategy with respect to π^M.*

In most games, the brute force approach described above will prove to be too difficult. Unfortunately, there is, in general, no other deterministic technique known for solving games. (Certain deterministic techniques are known for special cases, however, such as that discussed in the next subsection.)

The most useful general method for solving games is to try and guess a solution. This method will be discussed in the remainder of the subsection. The first problem which must be dealt with is that of determining whether a guessed optimal strategy really is minimax or maximin.

II. *Verification of Minimaxity*

The following theorem provides the basic tool for verifying minimaxity.

Theorem 2. *Let π_0 and δ_0^* be strategies for* I *and* II *respectively, and assume that for all $\theta \in \Theta$ and $a \in \mathscr{A}$,*

$$L(\theta, \delta_0^*) \le L(\pi_0, a). \tag{5.2}$$

Then the game has value $V = L(\pi_0, \delta_0^*)$ and is strictly determined, with π_0 and δ_0^* being maximin and minimax strategies.

PROOF. It is clear that

$$\bar{V} = \inf_{\delta^*} \sup_{\theta} L(\theta, \delta^*) \le \sup_{\theta} L(\theta, \delta_0^*),$$

and

$$\inf_{a} L(\pi_0, a) \le \sup_{\pi} \inf_{a} L(\pi, a) = \underline{V}.$$

Together with (5.2), these inequalities show that

$$\bar{V} \le \sup_{\theta} L(\theta, \delta_0^*) \le \inf_{a} L(\pi_0, a) \le \underline{V}.$$

By Lemma 2, $\underline{V} \le \bar{V}$, implying that

$$\bar{V} = \sup_{\theta} L(\theta, \delta_0^*) = \inf_{a} L(\pi_0, a) = \underline{V}. \tag{5.3}$$

It follows, by definition, that the game has value $V = \underline{V} = \bar{V}$, and that δ_0^* and π_0 are minimax and maximin strategies.

To prove that $L(\pi_0, \delta_0^*) = V$, note that

$$\inf_{\delta^*} L(\pi_0, \delta^*) \le L(\pi_0, \delta_0^*) \le \sup_{\pi} L(\pi, \delta_0^*).$$

Together with Lemma 1 and (5.3), this gives the desired result. □

EXAMPLE 8. Consider a game with the following loss matrix. Appended to the table are the row minima and column maxima.

	a_1	a_2	a_3	a_4	$\inf_a L(\theta_i, a)$
θ_1	7	2	5	1	1
θ_2	2	2	3	4	2
θ_3	5	3	4	4	3
θ_4	3	2	1	6	1
$\sup_\theta L(\theta, a_i)$	7	3	5	6	

It is clear that θ_3 and a_2 are best among nonrandomized strategies, according to the maximin and minimax principles respectively. Furthermore, since

$$L(\theta, a_2) \le L(\theta_3, a)$$

for all θ and a, Theorem 2 implies that θ_3 and a_2 are maximin and minimax among all strategies, and that the game has value $V = L(\theta_3, a_2) = 3$. (It was rather lucky to find nonrandomized strategies that are minimax and maximin.)

The following theorem is a useful generalization of Theorem 2. Its proof is similar and will be left as an exercise.

Theorem 3

(a) *Let* $\{\pi_n\}$ *and* δ_0^* *be strategies such that*

$$\sup_\theta L(\theta, \delta_0^*) \le \lim_{n\to\infty} \inf_a L(\pi_n, a).$$

Then the game has a value and δ_0^* *is a minimax strategy for* II.

(b) *Let* π_0 *and* $\{\delta_n^*\}$ *be strategies such that*

$$\lim_{n\to\infty} \sup_\theta L(\theta, \delta_n^*) \le \inf_a L(\pi_0, a).$$

Then the game has a value and π_0 *is a maximin strategy for* I.

III. *The Guess Method of Solving Games*

The main methods of guessing solutions are based on the following theorem.

Theorem 4. *Assume the game has a value V, and that* π^M *and* δ^{*M} *are maximin and minimax strategies. Define*

$$\Theta_V = \{\theta \in \Theta: L(\theta, \delta^{*M}) = V\},$$

and

$$\mathscr{A}_V = \{a \in \mathscr{A}: L(\pi^M, a) = V\}.$$

Then $\pi^M(\Theta_V) = 1$ *and* $\delta^{*M}(\mathscr{A}_V) = 1$.

Proof. We prove only that $\pi^M(\Theta_V) = 1$. The proof that $\delta^{*M}(\mathscr{A}_V) = 1$ is identical.

Since π^M and δ^{*M} are maximin and minimax strategies and the game has value V, it can be shown, as in Theorem 2, that $L(\pi^M, \delta^{*M}) = V$. Hence

$$\begin{aligned} 0 &= V - L(\pi^M, \delta^{*M}) = E^{\pi^M}[V - L(\theta, \delta^{*M})] \\ &= \int_{\Theta_V^c} [V - L(\theta, \delta^{*M})] dF^{\pi^M}(\theta). \end{aligned} \tag{5.4}$$

Since δ^{*M} is minimax, we know that $L(\theta, \delta^{*M}) \le V$ for all $\theta \in \Theta$. By the definition of Θ_V, it follows that $V - L(\theta, \delta^{*M}) > 0$ for $\theta \in \Theta_V^c$. Together with (5.4), this implies that $\pi^M(\Theta_V^c) = 0$, giving the desired conclusion. □

The implications of this theorem are twofold. First, it shows that, if either π^M or δ^{*M} can be guessed, then to find the other optimum strategy (needed for the verification of the solution through Theorem 1) it suffices to look at randomized strategies on Θ_V or $\mathscr{A}_V$. This is frequently a considerable simplification. (Note that any randomized strategy on $\mathscr{A}_V$ is Bayes with respect to π^M. Theorem 4 can thus be considered a more explicit version of Theorem 1.)

The second implication of Theorem 4 is that minimax or maximin strategies will often have constant loss, namely when $\Theta_V = \Theta$ or $\mathscr{A}_V = \mathscr{A}$. This concept is of enough importance to deserve its own name.

Definition 4. A strategy π_0 is an *equalizer strategy* for I if $L(\pi_0, a) = C$ (some constant) for all $a \in \mathscr{A}$. A strategy δ_0^* is an *equalizer strategy* for II if $L(\theta, \delta_0^*) = C'$ (some constant) for all $\theta \in \Theta$.

Theorem 4 thus suggests that an equalizer strategy might be a good guess for a minimax or maximin strategy. The guess method of solving games can now be stated as follows.

Step 1. Eliminate from the game as many inadmissible strategies as possible. (This is always the way to begin any method of solution.)

Step 2. Guess either π^M or δ^{*M}. If no obvious guess is available, look for an equalizer strategy.

Step 3. If δ^* is the guess from Step 2, calculate $\hat{V} = \sup_\theta L(\theta, \delta^*)$. If π is the guess, calculate $\hat{V} = \inf_a L(\pi, a)$.

Step 4. Determine $\Theta_{\hat{V}}$ or $\mathscr{A}_{\hat{V}}$ (depending on whether δ^* or π was guessed), and examine randomized strategies on this set to obtain a guess for the remaining optimum strategy. Again, looking for an equalizer strategy on the set may prove helpful.

Step 5. Use Theorem 2 to determine if the guesses for π^M and δ^{*M} are indeed maximin and minimax.

Sometimes Step 4 will fail because the reasonable guess for the other optimal strategy is not a proper strategy (i.e., is not a proper probability distribution). In such cases, however, a sequence of strategies can often be found which converge, in some sense, to the improper strategy. Theorem 3 can be used, in such situations, to draw the desired conclusions. Examples of this will be seen in Section 5.3. We now give some examples of application of the basic guess method.

EXAMPLE 4 (continued). The loss is $L(\theta, a) = |\theta - a|$. Note that, for $a \in [0, 1]$,

$$\sup_{\theta \in [0,1]} |\theta - a| = \max\{a, 1 - a\},$$

Since

$$\inf_{a\in[0,1]} \sup_{\theta\in[0,1]} |\theta-a| = \inf_{a\in[0,1]} \max\{a, 1-a\} = \tfrac{1}{2},$$

the minimum being attained at $a_0=\frac{1}{2}$, reasonable guesses for the minimax strategy and the value are a_0 and $\frac{1}{2}$. (It is hard to see how a randomized strategy could help II.)

Observe next that

$$\Theta_{1/2}=\{\theta\in[0,1]: L(\theta, a_0)=|\theta-\tfrac{1}{2}|=\tfrac{1}{2}\}=\{0,1\}.$$

Thus the guess for π^M should give probability one to this set. The obvious choice is

$$\pi_0=\tfrac{1}{2}\langle 0\rangle+\tfrac{1}{2}\langle 1\rangle$$

(i.e., the probability distribution which gives probability $\frac{1}{2}$ to each $\theta=0$ and $\theta=1$).

It remains only to verify, by Theorem 2, that a_0 and π_0 are indeed minimax and maximin. For all $a\in[0,1]$,

$$L(\pi_0, a)=\tfrac{1}{2}L(0,a)+\tfrac{1}{2}L(1,a)=\tfrac{1}{2}a+\tfrac{1}{2}(1-a)=\tfrac{1}{2}.$$

Since $L(\theta, a_0)\le\frac{1}{2}$, it follows that the condition of Theorem 2 is satisfied. Therefore, π_0 is maximin, a_0 is minimax, and the value is $V=\frac{1}{2}$.

EXAMPLE 6 (continued). It has already been observed that the strategy (2, 1) for II is inadmissible. Eliminating this strategy gives the reduced game

	(1, 1)	(1, 2)	(2, 2)
θ_1	-1	$1-2p$	1
θ_2	1	$1-2p$	-1

Equalizer strategies could be calculated, but symmetry suggests that the maximin strategy for I might be

$$\pi_0=\tfrac{1}{2}\langle\theta_1\rangle+\tfrac{1}{2}\langle\theta_2\rangle.$$

It was earlier calculated that $\inf_a L(\pi_0, a)=1-2p$, and that $\Theta_{1-2p}=\{(1,2)\}$. Since $L(\theta,(1,2))=1-2p$ for both θ_1 and θ_2, we can conclude from Theorem 2 that π_0 is maximin, (1, 2) is minimax, and $1-2p$ is the value.

EXAMPLE 7 (continued). There is no obvious maximin or minimax strategy. Hence, let us look for an equalizer strategy. It is clear that it will be necessary to consider randomized strategies. (The opponent could easily beat a non-randomized strategy.) Hence consider densities $\pi(\theta)$ and $\delta^*(a)$ on $\Theta=\mathscr{A}=[4,5]$. By symmetry, both optimal strategies should be the same (and the value should be zero), so we confine our attention to finding an equalizer strategy π for I.

Note first that

$$L(\pi, a)=\int_4^5 L(\theta, a)\pi(\theta)d\theta$$
$$=\int_4^a [2\theta-4-a]\pi(\theta)d\theta+\int_a^5 [\theta+4-2a]\pi(\theta)d\theta.$$

If this is to be an equalizer strategy, $L(\pi, a)$ must be constant for $a\in[4, 5]$, so that the derivative of L must be zero. Clearly

$$\frac{d}{da}L(\pi, a)=[2a-4-a]\pi(a)-\int_4^a \pi(\theta)d\theta$$
$$-[a+4-2a]\pi(a)-2\int_a^5 \pi(\theta)d\theta$$
$$=[2a-8]\pi(a)-\int_4^a \pi(\theta)d\theta-2\int_a^5 \pi(\theta)d\theta.$$

This is still rather complicated, so consider the second derivative, which must also be zero. Calculation gives (letting π' denote the derivative of π)

$$\frac{d^2}{da^2}L(\pi, a)=\pi(a)\left[3+(2a-8)\frac{\pi'(a)}{\pi(a)}\right]$$
$$=\pi(a)\left[3+(2a-8)\frac{d}{da}(\log \pi(a))\right].$$

Setting equal to zero and solving gives that

$$\frac{d}{da}\log \pi(a)=\frac{-3}{2a-8}.$$

Integrating gives

$$\log \pi(a)=c+-\tfrac{3}{2}\log(2a-8),$$

or $\pi(a)=K(2a-8)^{-3/2}$. Unfortunately, $\int_4^5 (2a-8)^{-3/2}\,da=\infty$, so this can not be normalized to give a proper density. Rather than giving up, however, it seems reasonable to try and modify the density to eliminate the offending portion. This suggests considering (for $4<\alpha<5$)

$$\pi_\alpha(\theta)=\begin{cases} K_\alpha(2\theta-8)^{-3/2} & \text{if } \alpha\le\theta\le 5,\\ 0 & \text{if } \theta<\alpha.\end{cases}$$

A calculation shows that the proper normalizing constant is $K_\alpha=\{(2\alpha-8)^{-1/2}-2^{-1/2}\}^{-1}$. Further calculation gives

$$L(\pi_\alpha, a)=\begin{cases} 8+(\alpha-4)^{1/2}-2a & \text{if } a\le\alpha,\\ [(\alpha-4)^{-1/2}-1]^{-1}\{a[2-(\alpha-4)^{-1/2}]-7 \\ \qquad +4(\alpha-4)^{-1/2}-2(\alpha-4)^{1/2}\} & \text{if } a\ge\alpha.\end{cases}$$

Choosing $\alpha = \frac{17}{4}$ gives

$$L(\pi_{17/4}, a) = \begin{cases} \frac{17}{2} - 2a & \text{if } a \leq \frac{17}{4}, \\ 0 & \text{if } a \geq \frac{17}{4}. \end{cases}$$

Clearly $L(\pi_{17/4}, a) \geq 0$. (Note that $\pi_{17/4}$ is an equalizer strategy for $a \geq \frac{17}{4}$, with larger gain for $a < \frac{17}{4}$.)

By symmetry, we try, as our guess for δ^{*M},

$$\delta^*_{17/4}(a) = \begin{cases} K_{17/4}(2a-8)^{-3/2} & \text{if } \frac{17}{4} \leq a \leq 5, \\ 0 & \text{if } a < \frac{17}{4}. \end{cases}$$

Also by symmetry, $L(\theta, \delta^*_{17/4}) \leq 0$ for $4 \leq \theta \leq 5$. Hence we can conclude from Theorem 2 that $\pi_{17/4}$ and $\delta^*_{17/4}$ are maximin and minimax strategies, and that the value of the game is zero.

The methods of solving games that have been discussed in this subsection are by no means foolproof. If direct calculation fails, if δ^{*M} or π^M cannot be guessed, or if equalizer strategies cannot be found or are not solutions, then we are out of luck. The next subsection discusses an important special case of game theory in which an explicit solution can always be obtained.

5.2.3. Finite Games

A *finite game* is a (two-person zero-sum) game in which the sets of (nonrandomized) strategies, Θ and $\mathscr{A}$, are finite. Finite games are important in game theory, since, in many games, each player will have only a finite set of options available to him. They are less important in statistical games, however, since the set of strategies available to the statistician will be the set of all decision rules, which is rarely finite.

It will be shown in Subsection 5.2.6 that all finite games have values and are strictly determined. Of even greater interest is that explicit procedures exist for solving finite games. One such procedure will be presented in this subsection.

Assume $\Theta = \{\theta_1, \theta_2, \ldots, \theta_m\}$ and $\mathscr{A} = \{a_1, a_2, \ldots, a_n\}$. The loss function can then be represented by the $(m \times n)$ loss matrix $\mathbf{L}$, whose elements are $l_{ij} = L(\theta_i, a_j)$. We will identify randomized strategies π and δ^* with probability vectors

$$\boldsymbol{\pi} = (\pi_1, \pi_2, \ldots, \pi_m)' \quad \text{and} \quad \boldsymbol{\delta}^* = (\delta_1^*, \delta_2^*, \ldots, \delta_n^*)',$$

where $\pi_i = \pi(\theta_i)$ and $\delta_i^* = \delta^*(a_i)$.

The following theorem provides a starting point for the investigation of finite games. It is valid for all games, but is mainly useful in solving finite games.

Theorem 5. *If both* I *and* II *have equalizer strategies, then the game has a value and the equalizer strategies are maximin and minimax strategies.*

PROOF. If π and δ^* are the equalizer strategies, then $L(\theta, \delta^*) = K_1$ for all $\theta \in \Theta$ and $L(\pi, a) = K_2$ for all $a \in \mathscr{A}$. But clearly

$$L(\pi, \delta^*) = E^{\pi} L(\theta, \delta^*) = E^{\pi} K_1 = K_1$$

and

$$L(\pi, \delta^*) = E^{\delta^*} L(\pi, a) = E^{\delta^*} K_2 = K_2.$$

Hence $K_1 = K_2$ and Theorem 2 gives the desired result. □

Definition 5. If, in a finite game, $\boldsymbol{\pi}$ and $\boldsymbol{\delta}^*$ are equalizer strategies for I and II, then the pair $(\boldsymbol{\pi}, \boldsymbol{\delta}^*)$ is called a *simple solution* of the game.

As a first step in solving finite games, the problem of finding simple solutions (for those games that have them) must be addressed.

I. *Finding a Simple Solution*

Assume that $m = n$, and that $\mathbf{L}$ is a nonsingular matrix. We search for strategies $\boldsymbol{\pi}$ and $\boldsymbol{\delta}^*$ such that $L(\boldsymbol{\pi}, a) = L(\theta, \boldsymbol{\delta}^*) = K$ for all $\theta \in \Theta$ and $a \in \mathscr{A}$. Note that

$$L(\boldsymbol{\pi}, a_j) = \sum_{i=1}^{m} \pi_i L(\theta_i, a_j) = \sum_{i=1}^{m} \pi_i l_{ij}.$$

Hence if $L(\boldsymbol{\pi}, a_j) = K$ for $j = 1, \ldots, m$, then

$$\boldsymbol{\pi}'\mathbf{L} = K\mathbf{1},$$

where $\mathbf{1}$ is the row vector of all ones. Similarly, if $L(\theta_i, \boldsymbol{\delta}^*) = K$ for $i = 1, \ldots, m$, then

$$\mathbf{L}\boldsymbol{\delta}^* = K\mathbf{1}^t.$$

Since $\mathbf{L}$ is nonsingular, solutions to these equations must be of the form

$$\boldsymbol{\pi} = K(\mathbf{L}^{-1})^t\mathbf{1}^t \quad \text{and} \quad \boldsymbol{\delta}^* = K\mathbf{L}^{-1}\mathbf{1}^t. \tag{5.5}$$

Since $\boldsymbol{\pi}$ and $\boldsymbol{\delta}^*$ must be probability vectors, it is clear that the following condition must be satisfied.

Condition 1. *The components of* $(\mathbf{L}^{-1})^t\mathbf{1}^t$ *and* $\mathbf{L}^{-1}\mathbf{1}^t$ *(i.e., the row and column sums of* $\mathbf{L}^{-1}$*) must all have the same sign.*

If Condition 1 holds, the components of $\boldsymbol{\pi}$ (and also the components of $\boldsymbol{\delta}^*$) will sum to one if $K = 1/(\mathbf{1}\mathbf{L}^{-1}\mathbf{1}^t)$. This analysis is summarized as

Theorem 6. *If* $\mathbf{L}$ *is an* $(m \times m)$ *nonsingular matrix satisfying Condition* 1, *then a simple solution to the game is given by* $(\boldsymbol{\pi}, \boldsymbol{\delta}^*)$, *where*

$$\boldsymbol{\pi} = (\mathbf{L}^{-1})^t\mathbf{1}^t/(\mathbf{1}\mathbf{L}^{-1}\mathbf{1}^t) \quad \textit{and} \quad \boldsymbol{\delta}^* = \mathbf{L}^{-1}\mathbf{1}^t/(\mathbf{1}\mathbf{L}^{-1}\mathbf{1}^t).$$

Furthermore, if Condition 1 *is violated, there is no simple solution.*

EXAMPLE 9. Assume $\mathbf{L}$ is

	a_1	a_2	a_3
θ_1	6	0	6
θ_2	8	−2	0
θ_3	4	6	5

A calculation shows that

$$\mathbf{L}^{-1} = \frac{1}{276}\begin{pmatrix} -10 & 36 & 12 \\ -40 & 6 & 48 \\ 56 & -36 & -12 \end{pmatrix}.$$

It is clear that the row and column sums of $\mathbf{L}^{-1}$ are all positive, so that Condition 1 is satisfied. A simple calculation using Theorem 6 then gives that a simple solution to the game is

$$\boldsymbol{\pi} = (\tfrac{1}{10}, \tfrac{1}{10}, \tfrac{8}{10})^t \quad \text{and} \quad \boldsymbol{\delta}^* = (\tfrac{19}{30}, \tfrac{7}{30}, \tfrac{4}{30})^t.$$

II. *Solving Finite Games*

A simple solution to a finite game may not exist. If such is the case, however, there will exist simple solutions of *subgames* that are solutions for the original game. Indeed the following procedure will locate all minimax and maximin solutions to the game.

Step 1. Ensure that the value of the game is nonzero. If by no other method, this can be achieved by adding a fixed constant to each element of $\mathbf{L}$ to make all elements positive. (Recall that adding a constant to the loss function does not affect the solutions.)

Step 2. Using Condition 1, find all nonsingular square submatrices of $\mathbf{L}$ which have simple solutions. Calculate these simple solutions, using Theorem 6.

Step 3. Write the simple solutions found in step 2 as vectors in the original game, by filling in zeros in the coordinates corresponding to the strategies deleted to arrive at the subgames. Then check, using Theorem 2, whether or not these vectors are maximin and minimax in the original game.

Step 4. Form the convex hulls of the minimax vectors and maximin vectors found in step 3. These consist of all minimax and maximin strategies.

We will not prove here that the above procedure works. The interested reader can find such a proof in Blackwell and Girshick (1954) (see Theorem 2.7.3).

Several observations should be made about the above procedure. First, at all stages of the investigation inadmissible strategies should be eliminated. (A game or subgame having an inadmissible strategy can not have a simple solution.) Second, frequently only one minimax strategy is wanted. If this is the case, it is usually easiest to begin by checking for nonrandomized minimax and maximin strategies, as in Example 8. (This corresponds to looking at the (1×1) subgames, using the above procedure.) If this fails, it is probably best to next see if a simple solution exists for the entire game (or for one of the largest square subgames). If this also fails to yield a solution, it is necessary to just start checking all the square subgames. In one sense, it is best to start with the (2×2) subgames, and work your way up. The advantage of this is that the calculation of inverses, made at each stage, can be useful at the next stage. Let us look at some examples.

EXAMPLE 10. Assume **L** is

	a_1	a_2	a_3	a_4	a_5
θ_1	3	5	−2	2	1
θ_2	3	6	−1	2	4
θ_3	4	3	6	7	8

We first note that θ_1 is inadmissible, being dominated by θ_2. Hence we can eliminate θ_1, reducing the game to

	a_1	a_2	a_3	a_4	a_5
θ_2	3	6	−1	2	4
θ_3	4	3	6	7	8

In this reduced game, a_4 and a_5 are inadmissible, being dominated by a_3. (Aren't we lucky though?) This leaves us with

	a_1	a_2	a_3
θ_2	3	6	−1
θ_3	4	3	6

Next we check to see if a pair of nonrandomized strategies might happen to work. The row minima and column maxima are

3	6	−1	−1
4	3	6	3
4	6	6	

Thus, the best nonrandomized strategy for II is a_1, and $L(\theta, a_1) \leq 4$, but the best nonrandomized strategy for I is θ_2, and $L(\theta_2, a) \geq 3$. Alas, Theorem 2 does not apply, so we have to look at the (2×2) subgames.

The (2×2) subgame

	a_1	a_2
θ_2	3	6
θ_3	4	3

can be checked to have the simple solution

$$\boldsymbol{\pi} = (\tfrac{1}{4}, \tfrac{3}{4})^t \quad \text{and} \quad \boldsymbol{\delta}^* = (\tfrac{3}{4}, \tfrac{1}{4})^t.$$

This corresponds in the original (reduced) game to

$$\boldsymbol{\pi} = (\tfrac{1}{4}, \tfrac{3}{4})^t \quad \text{and} \quad \boldsymbol{\delta}^* = (\tfrac{3}{4}, \tfrac{1}{4}, 0)^t.$$

To see if this is a solution, note that $L(\boldsymbol{\pi}, a_1) = \frac{1}{4}(3) + \frac{3}{4}(4) = \frac{15}{4}$, $L(\boldsymbol{\pi}, a_2) = \frac{15}{4}$, $L(\boldsymbol{\pi}, a_3) = \frac{17}{4}$, $L(\theta_2, \boldsymbol{\delta}^*) = \frac{15}{4}$, and $L(\theta_3, \boldsymbol{\delta}^*) = \frac{15}{4}$. Since $L(\theta, \boldsymbol{\delta}^*) \leq L(\boldsymbol{\pi}, a)$, Theorem 2 gives that it is indeed a solution. The other two (2×2) subgames have solutions which are not solutions to the original (reduced) game. There is thus only one minimax and one maximin strategy. In the original (nonreduced) game, these strategies are

$$\boldsymbol{\pi} = (0, \tfrac{1}{4}, \tfrac{3}{4})^t \quad \text{and} \quad \boldsymbol{\delta}^* = (\tfrac{3}{4}, \tfrac{1}{4}, 0, 0, 0)^t.$$

In the above example, we did not worry about step 1 of the procedure, namely ensuring that the game had a nonzero value. It turned out, of course, to have value $\frac{15}{4}$, so that all was well. If, say, a computer program was to be used to implement the procedure, it would undoubtedly be beneficial to always incorporate step 1. When solving games by hand, however, there are often reasons to ignore the step. The following example demonstrates this.

Example 3 (continued). From the symmetry of the game, it is clear that the value will be zero, so there is a temptation to add, say, 1 to each entry, and proceed with the calculation. Of course, we then have a (4×4) matrix, sixteen (3×3) matrices, thirty-four (2×2) matrices, and the nonrandomized strategies to check. Clearly some inspiration is needed instead, and looking at all those zeros in the original matrix (with zero being the value of the game) provides some. If we just concentrate on the nonzero entries, two (2×2) subgames stand out, namely

	(1, 1)	(2, 2)
(1, 2)	5	−7
(2, 1)	−5	7

and

	(1, 2)	(2, 1)
(1, 1)	−5	5
(2, 2)	7	−7

.

Let us solve the first of these. There is no nonrandomized solution, so we must find a simple solution. Since $|\mathbf{L}|=35-35=0$, we must first add, say, 1 to each element of the matrix, before applying Theorem 6. The simple solution can then be calculated to be

$$\boldsymbol{\pi}=(\tfrac{1}{2},\tfrac{1}{2})^t \quad \text{and} \quad \boldsymbol{\delta}^*=(\tfrac{7}{12},\tfrac{5}{12})^t.$$

In the original problem, this corresponds to

$$\boldsymbol{\pi}^1=(0,\tfrac{1}{2},\tfrac{1}{2},0)^t \quad \text{and} \quad \boldsymbol{\delta}^{*1}=(\tfrac{7}{12},0,0,\tfrac{5}{12})^t,$$

which can easily be checked to be maximin and minimax.

The second subgame is really just the first subgame with the roles of I and II reversed. Hence it will give rise to the maximin and minimax solutions

$$\boldsymbol{\pi}^2=(\tfrac{7}{12},0,0,\tfrac{5}{12})^t \quad \text{and} \quad \boldsymbol{\delta}^{*2}=(0,\tfrac{1}{2},\tfrac{1}{2},0)^t$$

in the original game.

Finally, any convex combination of these two maximin or two minimax strategies will also be maximin or minimax. Hence any strategy of the form

$$(\tfrac{7}{12}(1-\alpha),\tfrac{1}{2}\alpha,\tfrac{1}{2}\alpha,\tfrac{5}{12}(1-\alpha))^t$$

is minimax or maximin for $0\le\alpha\le 1$. Of course, there may be other maximin or minimax strategies, but these should be enough.

There are other methods of solving finite games based on linear programming methods. Of particular note is the simplex method, which is actually a more commonly used method of solving large games than the method discussed here. Unfortunately, the simplex method requires considerably more mathematical development.

For further discussion and examples of finite games the reader is referred to the delightfully amusing book *The Compleat Strategyst* by Williams (1954).

5.2.4. Games with Finite Θ

Assume $\Theta = \{\theta_1, \theta_2, \ldots, \theta_m\}$. The game can then be reformulated geometrically, so that geometric techniques can be brought to bear in obtaining a solution. As a method of solving games, this geometric approach is generally less useful than the approaches discussed in the two previous subsections. Theoretically, however, the approach is of great interest, since minimax theorems (i.e., theorems which state that the game has a value and that minimax rules exist) can be easily developed through geometric arguments. This subsection discusses the reformulation rather informally, leaving the explicit minimax theorems for the next two subsections.

I. *Reformulation as an S-Game*

The key to the reformulation is to transform the (perhaps complicated) set of all possible randomized strategies for player II into a simple geometric set in R^m.

Definition 6. For each randomized strategy $\delta^* \in \mathscr{A}^*$, let

$$R_i(\delta^*) = L(\theta_i, \delta^*),$$

and define

$$\mathbf{R}(\delta^*) = (R_1(\delta^*), R_2(\delta^*), \ldots, R_m(\delta^*))^t.$$

The point $\mathbf{R}(\delta^*)$ in m-dimensional Euclidean space is called the *risk point* of δ^*. The set

$$S = \{\mathbf{R}(\delta^*): \delta^* \in \mathscr{A}^*\}$$

is called the *risk set* of the game. Also, define $S_{\mathscr{A}}$ as the set of all risk points $\mathbf{R}(a)$ corresponding to nonrandomized actions $a \in \mathscr{A}$.

The following lemma shows that the risk set is convex, and indicates how it can be determined.

Lemma 3. *The risk set S is the convex hull* (*see Section* 1.8) *of $S_{\mathscr{A}}$.*

Proof. Consider a randomized strategy $\delta^* \in \mathscr{A}^*$. Since $L(\theta_i, \delta^*) = E^{\delta^*}L(\theta_i, a)$, it is clear that

$$\mathbf{R}(\delta^*) = (E^{\delta^*}L(\theta_1, a), \ldots, E^{\delta^*}L(\theta_m, a))^t = E^{\delta^*}\mathbf{R}(a),$$

the expectation being taken componentwise. Choosing δ^* to be a randomized strategy of the form

$$\delta^* = \sum_{i=1}^{\infty} p_i\langle a_i\rangle,$$

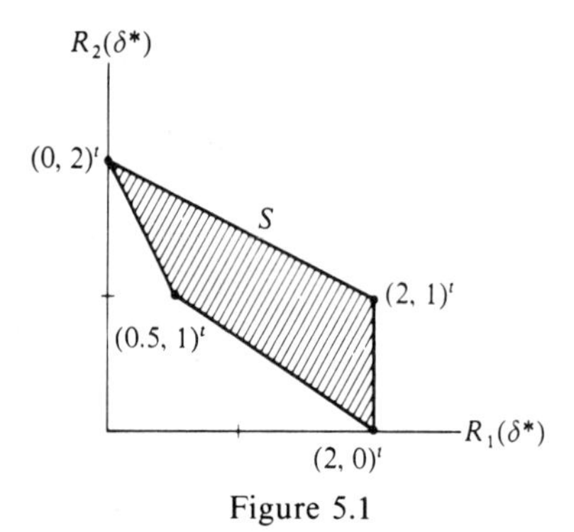

Figure 5.1

where $0 \le p_i \le 1$, $\sum_{i=1}^{\infty} p_i = 1$, and $a_i \in \mathscr{A}$, it follows that

$$\mathbf{R}(\delta^*) = \sum_{i=1}^{\infty} p_i \mathbf{R}(a_i).$$

From Definition 16 of Section 1.8, it follows that the convex hull of $S_{\mathscr{A}}$ is a subset of S.

Consider now any randomized strategy $\delta^* \in \mathscr{A}^*$, and let $\mathbf{X} = \mathbf{R}(a)$ be a random vector with the distribution induced by δ^*. Since δ^* has finite loss (by the definition of $\mathscr{A}^*$) and $\mathbf{X} = \mathbf{R}(a)$ is in the convex hull of $S_{\mathscr{A}}$, it follows from Lemma 2 of Section 1.8 that $E[\mathbf{X}] = E^{\delta^*}[\mathbf{R}(a)] = \mathbf{R}(\delta^*)$ is in the convex hull of $S_{\mathscr{A}}$. Thus S is a subset of the convex hull of $S_{\mathscr{A}}$. The conclusion follows. □

EXAMPLE 11. Assume the loss matrix for a two-person zero-sum game is

	a_1	a_2	a_3	a_4
θ_1	2	2	0	0.5
θ_2	0	1	2	1

Since $\mathscr{A}$ consists of a_1, a_2, a_3, and a_4, the set $S_{\mathscr{A}}$ consists of the four points $(2, 0)^t$, $(2, 1)^t$, $(0, 2)^t$, and $(0.5, 1)^t$. Graphing these and forming their convex hull gives the risk set S shown in Figure 5.1.

To complete the geometric reformulation of the game, geometric meanings must be given to the strategies of I and to the loss. Nonrandomized strategies for I are easy. They correspond simply to the choice of a coordinate of the risk point. In other words, saying that I chooses strategy θ_i is equivalent (from the viewpoint of loss) to saying that I chooses the ith coordinate of any risk point. As in the previous subsection, a randomized strategy for I can be thought of as a probability vector

$$\boldsymbol{\pi} = (\pi_1, \pi_2, \ldots, \pi_m)^t,$$

where π_i is the probability of choosing coordinate i (or the probability of choosing θ_i). The set of all such probability vectors will be denoted Ω_m, and is called the *simplex* in R^m.

The loss incurred if II chooses strategy δ^* and I chooses strategy π is clearly

$$L(\boldsymbol{\pi}, \delta^*) = \sum_{i=1}^{m} \pi_i L(\theta_i, \delta^*) = \pi' \mathbf{R}(\delta^*).$$

This completes the geometric description of the game. Instead of considering Θ^*, $\mathscr{A}^*$, and L, it suffices to deal with S, Ω_m, and $L^*(\boldsymbol{\pi}, \mathbf{s}) = \boldsymbol{\pi}'\mathbf{s}$, with the interpretation that II chooses a point $\mathbf{s} = (s_1, \ldots, s_m)' \in S$ (i.e., some $\mathbf{R}(\delta^*)$), I chooses $\boldsymbol{\pi} \in \Omega_m$, and the loss (to II) is then $L^*(\boldsymbol{\pi}, \mathbf{s}) = \boldsymbol{\pi}'\mathbf{s}$. (Note that $\boldsymbol{\pi}'\mathbf{s}$ is just the dot product of the vectors $\boldsymbol{\pi}$ and $\mathbf{s}$.)

Part of the simplicity of the above geometric formulation of the game is that only S need be specified to define the game. The other elements, Ω_m and $L^*(\boldsymbol{\pi}, \mathbf{s}) = \boldsymbol{\pi}'\mathbf{s}$, are automatically associated with the game. Such a geometric game is, therefore, simply called an *S-game*.

It will be seen in the next subsection that S-games are usually strictly determined. The remainder of this subsection is devoted to geometric techniques of actually solving such games. The techniques discussed will also form the basis of the theoretical arguments of the next subsection.

II. *Finding Minimax Strategies in S-games*

From the definition of a minimax strategy (and Lemma 1), it is clear that a minimax strategy in an S-game is a point $\mathbf{s}^M \in S$ such that

$$\sup_{\boldsymbol{\pi} \in \Omega_m} \boldsymbol{\pi}'\mathbf{s}^M = \max_i s_i^M = \inf_{\mathbf{s} \in S} \max_i s_i = \inf_{\mathbf{s} \in S} \sup_{\boldsymbol{\pi} \in \Omega_m} \boldsymbol{\pi}'\mathbf{s}. \tag{5.6}$$

In an S-game, we will call a minimax strategy a *minimax risk point.* In locating a minimax risk point, the concept of a lower quantant proves useful.

Definition 7. The *α-lower quantant in R^m*, to be denoted Q_α, is the set

$$Q_\alpha = \{\mathbf{z} = (z_1, \ldots, z_m)' \in R^m : z_i \le \alpha \quad \text{for } i = 1, \ldots, m\}.$$

The *corner* of Q_α is the point $(\alpha, \alpha, \ldots, \alpha)'$. (The set Q_α is thus the set of points whose maximum coordinate is less than or equal to α.)

EXAMPLE 12. For $m = 2$ and $\alpha > 0$, Q_α is the shaded region in Figure 5.2.

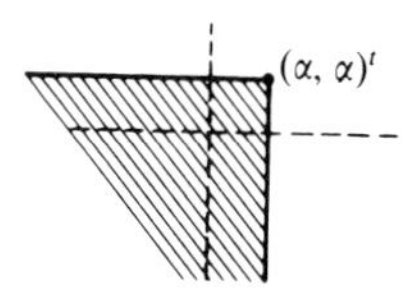

Figure 5.2

Definition 8. If S is the risk set of an S-game, define $S_\alpha = S \cap Q_\alpha$.

Note that S_α contains those points in S whose maximum coordinate is less than or equal to α. Therefore, if we can find a smallest α for which Q_α intersects S, line (5.6) implies that we will have found minimax risk points. The following theorem makes this precise. The proof is left as an exercise.

Theorem 7. *Let*

$$\alpha_M = \inf\{\alpha\colon S_\alpha \text{ is not empty}\}.$$

If $\alpha_M > -\infty$ and S_{α_M} is nonempty, then S_{α_M} consists of all minimax risk points in S.

EXAMPLE 11 (continued). The risk set is reproduced in Figure 5.3, along with Q_{α_M}. Clearly Q_{α_M} intersects S at the corner $(\alpha_M, \alpha_M)^t$ of Q_{α_M}. Hence S_{α_M} is the point on the line joining $(0.5, 1)^t$ and $(2, 0)^t$ which has equal coordinates. The line between $(0.5, 1)^t$ and $(2, 0)^t$ can be written

$$l = \{\tau(0.5, 1)^t + (1-\tau)(2, 0)^t = (2 - 1.5\tau, \tau)^t\colon 0 \le \tau \le 1\}.$$

The point with equal coordinates is clearly the point for which $2 - 1.5\tau = \tau$, or $\tau = 0.8$. Thus S_{α_M} is the single point $(0.8, 0.8)^t$. This is, by Theorem 7, the sole minimax risk point.

Note that this game is a finite game, and could have been solved by the method of the preceding subsection. The advantage of the geometric approach is that it immediately indicates the proper subgame to consider (namely that involving a_1 and a_4).

At this point, it is worthwhile to pause and consider how Q_{α_M} can intersect S. The following possibilities exist:

(i) S_{α_M} is empty. This will happen, for example, if S is an open set. Usually, however, S is closed.

(ii) S_{α_M} consists of the single point $(\alpha_M, \alpha_M, \ldots, \alpha_M)^t$. This was the situation in Example 11 (continued), and is the most frequently occurring case.

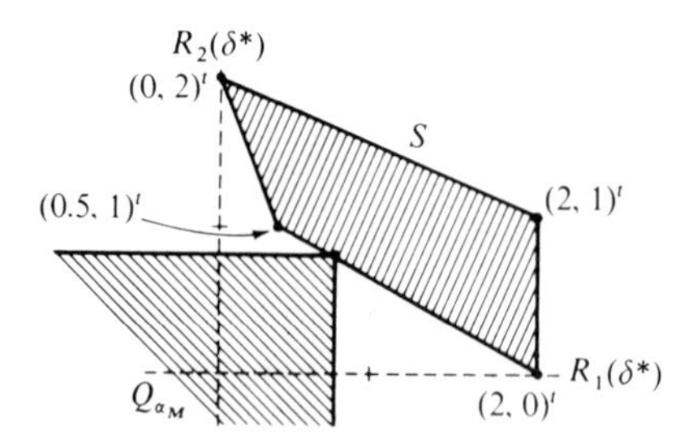

Figure 5.3

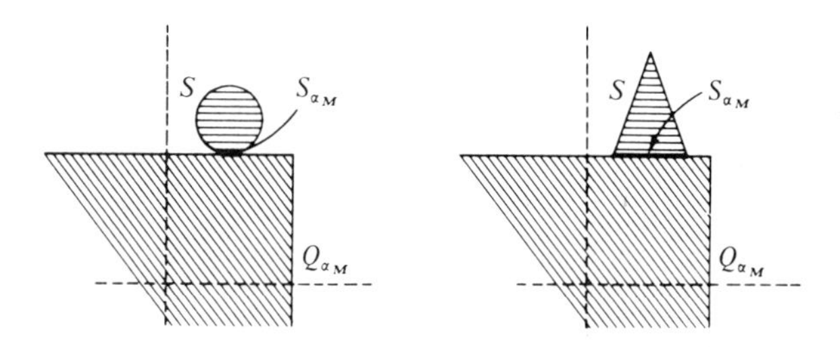

Figure 5.4

(iii) S_{α_M} is a set along the "side" of Q_{α_M}. It could be either a point or a convex subset of the side of Q_{α_M}, as the two parts of Figure 5.4 show.

In obtaining the minimax risk point in complicated situations, the problem can be considerably simplified by noting that, roughly speaking, only the "lower left edge" of S need be considered. This idea is made precise in the following definitions and theorems. The symbol $\bar{S}$ will be used to denote the *closure* of S (i.e., the union of S and all its limit points, or alternatively the smallest closed set containing S).

Definition 9. A point $\mathbf{z} \in R^m$ is a *lower boundary point* of S if $\mathbf{z} \in \bar{S}$ and there is no other point $\mathbf{y} \in \bar{S}$ such that $y_i \le z_i$ for $i = 1, \ldots, m$. The set of all lower boundary points of S will be denoted $\lambda(S)$.

In R^2, $\lambda(S)$ consists of all lower left or southwest points of $\bar{S}$. In Figure 5.5, $\lambda(S)$ is indicated for several sets S.

Intuitively, risk points not in $\lambda(S)$ can be improved upon by points in $\lambda(S)$. In looking for a minimax strategy, therefore, it often suffices to confine the search to $\lambda(S)$. The conditions needed for this to be true are that $\lambda(S)$ must exist and must be a subset of S. The following concepts are needed in verifying these conditions.

Definition 10. A set $S \subset R^m$ is *bounded from below* if, for some finite number K and all $\mathbf{s} = (s_1, \ldots, s_m)^t \in S$, $s_i \ge -K$ for $i = 1, \ldots, m$.

It is necessary to have S bounded from below to ensure the existence of $\lambda(S)$. Recall that we are assuming in this chapter that $L(\theta, a) \ge -K > -\infty$.

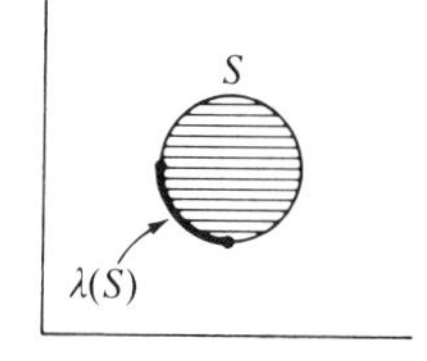

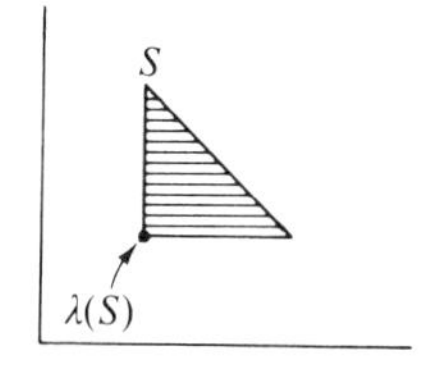

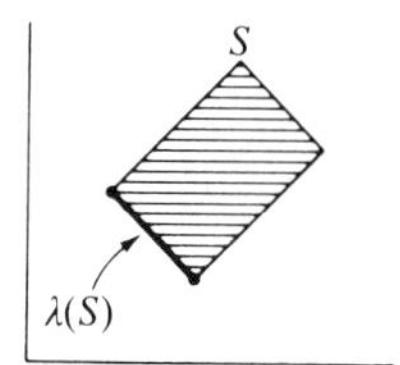

Figure 5.5

The following lemma shows that this implies that S is indeed bounded from below.

Lemma 4. *A risk set $S \subset R^m$ is bounded from below if $L(\theta_i, a) \geq -K > -\infty$ for all $a \in \mathscr{A}$ and $i = 1, \ldots, m$.*

PROOF. Clearly $s_i = L(\theta_i, \delta^*) = E^{\delta^*} L(\theta_i, a) \geq -K$. □

If S is closed, then $\lambda(S)$ is a subset of S (since then $S = \bar{S}$). More generally, we define the concept of closed from below.

Definition 11. A convex set $S \subset R^m$ is *closed from below* if $\lambda(S) \subset S$.

Usually, one just verifies that S is closed from below by showing that it is closed. The following lemma is an example. More interesting statistical examples will be seen in Section 5.3.

Lemma 5. *If $\mathscr{A}$ is finite, then the risk set S is closed.*

PROOF. Assume $\mathscr{A} = \{a_1, \ldots, a_n\}$. By Lemma 3, S is the convex hull of $\{\mathbf{R}(a_1), \ldots, \mathbf{R}(a_n)\}$. It is a straightforward exercise in limits and convexity to show that the convex hull of a finite set of points is closed. □

The following theorem, based on Lemma 7 of the next subsection, shows that frequently we need consider only points in $\lambda(S)$.

Theorem 8. *Assume that the risk set S is bounded from below and closed from below. Then any risk point $\mathbf{s} \notin \lambda(S)$ corresponds to an inadmissible strategy for* II. *Indeed, there exists a strategy corresponding to a point $\mathbf{s}' \in \lambda(S)$ for which $s'_i \leq s_i$ for $i = 1, \ldots, m$, with strict inequality for some i.*

PROOF. Lemma 7 of Subsection 5.2.5 shows that there exists a $\mathbf{y} \in \lambda(S)$ for which $y_i \leq s_i$, with strict inequality for some i. Since S is closed from below, $\mathbf{y} \in S$, giving the desired conclusion. □

Under the conditions of the above theorem, there is no reason to consider strategies other than those corresponding to points in $\lambda(S)$. In particular, the theorem shows that, if there is a minimax risk point not in $\lambda(S)$, then there is a better minimax risk point in $\lambda(S)$. It is, therefore, of interest to discuss how $\lambda(S)$ can be obtained. In simple situations such as Example 11 (continued), $\lambda(S)$ can be easily determined geometrically. In more complicated situations, however, an alternate approach is needed. The following theorems lead to such an approach.

Theorem 9. *Any strategy whose risk point is in $\lambda(S)$ is admissible.*

Theorem 10. *If a strategy in an S game is admissible, then it is Bayes with respect to some $\boldsymbol{\pi} \in \Omega_m$.*

Theorem 11. *If the risk set S is closed from below and bounded from below, then $\lambda(S)$ is a subset of the risk points arising from Bayes strategies.*

The proofs of Theorems 9 and 10 are left as exercises. (The proof of Theorem 10 requires results from the next subsection.) Theorem 11 is an immediate consequence of Theorems 9 and 10. Note that not all Bayes risk points need be in $\lambda(S)$. [The second graph in Figure 5.5 is an example of this. Any point on the left edge of S is Bayes with respect to $\boldsymbol{\pi}^0=(1,0)^t$. ($L(\boldsymbol{\pi}^0,\mathbf{s})=s_1$, which is minimized at any point along this left edge.) The only point which is in $\lambda(S)$, however, is the lower left corner of S.] When $\boldsymbol{\pi}$ has no zero coordinates, however, Theorem 7 of Section 4.8 shows that any Bayes strategy is admissible, implying (via Theorem 8) that the corresponding risk point is in $\lambda(S)$.

Theorem 11 provides a useful tool for calculating $\lambda(S)$. Frequently it is possible to explicitly determine all Bayes strategies for a game. Under the conditions of Theorem 11, $\lambda(S)$ is then a subset of the set of risk points corresponding to the Bayes strategies. Indeed, except possibly for risk points corresponding to Bayes strategies arising from $\boldsymbol{\pi}$ having zero coordinates, the two sets will be identical.

The important applications of this Bayesian technique for obtaining $\lambda(S)$ are to statistical games, and will be illustrated in Section 5.3. Indeed, it is generally unnecessary to even explicitly calculate $\lambda(S)$. The reduction of the problem to consideration of only Bayes strategies is often a great enough simplification to allow easy application of the techniques of Subsection 5.2.2.

When m is large, finding the minimax strategy can be difficult, even if $\lambda(S)$ has been obtained. In trying to solve such games, it is useful to recall the ways in which Q_{α_M} can intersect $\lambda(S)$. First, it can intersect at the corner $(\alpha_M,\ldots,\alpha_M)^t$. If this is the case, then the minimax risk point will be from an equalizer strategy. It is often quite easy to find an equalizer strategy with risk point in $\lambda(S)$, providing one exists. (An exercise will be to show that an admissible equalizer strategy is minimax. Hence an equalizer strategy with risk point in $\lambda(S)$ is always minimax by Theorem 9.)

Unfortunately, an equalizer strategy with risk point in $\lambda(S)$ will not always exist. This happens when $\lambda(S)$ intersects Q_{α_M} in a side. Note that any surface or edge of Q_{α_M} can be written as a set of the form

$$W=\{\mathbf{z}\in R^m\colon z_i=\alpha \quad \text{for} \quad i\in I,\, z_i\leq\alpha \quad \text{for} \quad i\notin I\},$$

where $\mathrm{I}=\{i(1),\ldots,i(k)\}$ is some set of integers. The problem in finding an equalizer strategy lies with the coordinates $i\notin \mathrm{I}$. An obvious idea is to ignore these coordinates and consider

$$\lambda_I(S)=\{(s_{i(1)},s_{i(2)},\ldots,s_{i(k)})^t\colon \mathbf{s}\in\lambda(S)\}.$$

(Essentially, this corresponds to projecting $\lambda(S)$ onto the k-dimensional subspace determined by the coordinates $i\in\mathrm{I}$.) This will determine a new and smaller S game, one in which the minimax risk point will have equal coordinates. This suggests the following procedure for determining a minimax strategy in a complicated S game:

(i) Determine $\lambda(S)$, and make sure that $\lambda(S) \subset S$.
(ii) Look for an equalizer strategy with risk point in $\lambda(S)$. If there isn't one, try to use the techniques of Subsection 5.2.2.
(iii) If (ii) fails, consider subproblems formed by deleting coordinates from the points $\mathbf{s} \in \lambda(S)$. Find the equalizer strategies for all such subgames, and determine the corresponding strategies in the original game.
(iv) Find the best strategy in this set according to the minimax principle. This strategy will be minimax.

III. *Finding Maximin Strategies in S-Games*

We now address the problem of determining a maximin strategy in an S-game. From the definition of a maximin strategy, it follows that $\boldsymbol{\pi}^M \in \Omega_m$ is maximin in an S-game if

$$\inf_{\mathbf{s} \in S} (\boldsymbol{\pi}^M)^t \mathbf{s} = \sup_{\boldsymbol{\pi} \in \Omega_m} \inf_{\mathbf{s} \in S} \boldsymbol{\pi}^t \mathbf{s}. \tag{5.7}$$

It will be seen that a maximin strategy has an interesting and useful interpretation in terms of a "tangent plane" to S at a minimax risk point. The following definition presents the needed concept.

Definition 12. A *hyperplane* in R^m is a set of the form

$$H(\boldsymbol{\xi}, k) = \left\{ \mathbf{z} \in R^m : \boldsymbol{\xi}^t \mathbf{z} = \sum_{i=1}^{m} \xi_i z_i = k \right\},$$

where k is some real number, $\boldsymbol{\xi} \in R^m$, and $\boldsymbol{\xi} \neq \mathbf{0}$ (the zero vector).

The following properties of a hyperplane are easy to check and aid in understanding its nature:

(i) A hyperplane is an $(m-1)$-dimensional plane (a line in R^2, a plane in R^3, etc.).
(ii) The hyperplane $H(\boldsymbol{\xi}, k)$ is perpendicular to the line $l = \{\lambda \boldsymbol{\xi}: -\infty < \lambda < \infty\}$. To see this, observe that if $\mathbf{x} \in H(\boldsymbol{\xi}, k)$, $\mathbf{y} \in H(\boldsymbol{\xi}, k)$, and $\lambda \boldsymbol{\xi} \in l$, then

$$(\lambda \boldsymbol{\xi})^t (\mathbf{x} - \mathbf{y}) = \lambda (\boldsymbol{\xi}^t \mathbf{x} - \boldsymbol{\xi}^t \mathbf{y}) = \lambda (k - k) = 0.$$

(This is what "perpendicular" means mathematically.)
(iii) If l is a line in R^m and $\mathbf{z}^0 \in R^m$, the hyperplane passing through $\mathbf{z}^0$ and perpendicular to l is $H(\boldsymbol{\xi}, k)$, where $\boldsymbol{\xi} = \mathbf{y}^1 - \mathbf{y}^2$ and $k = (\mathbf{z}^0)^t (\mathbf{y}^1 - \mathbf{y}^2)$, $\mathbf{y}^1$ and $\mathbf{y}^2$ being any distinct points of l.
(iv) Hyperplanes $H(\boldsymbol{\xi}, k)$, for fixed $\boldsymbol{\xi}$ and varying k, are parallel.

(v) If $\boldsymbol{\pi} \in \Omega_m$, then $H(\boldsymbol{\pi}, k)$ passes through the point $(k, k, \ldots, k)^t$ (since $\sum_{i=1}^m k\pi_i = k$), and has "nonpositive slope." (In R^2, for example, the slope of the hyperplane (a line) is $-\pi_1/\pi_2$.)

The reason for introducing hyperplanes is that if $\boldsymbol{\pi} \in \Omega_m$ is player I's strategy, then $H(\boldsymbol{\pi}, k) \cap S$ is the set of risk points in S with loss k. (The loss in an S-game is $\boldsymbol{\pi}^t\mathbf{s}$, and $H(\boldsymbol{\pi}, k)$ is defined as those points $\mathbf{z}$ for which $\boldsymbol{\pi}^t\mathbf{z} = k$.) Hence the hyperplanes $H(\boldsymbol{\pi}, k)$, for varying k, partition S into sets of constant loss. This will be seen in Subsection 5.2.6 to lead to the following result, which will here be stated informally.

Result 1. *Assume that* $\mathbf{s}^M$ *is a minimax risk point for* II. *Let* $H(\boldsymbol{\pi}^0, k)$ *be a hyperplane passing through* $\mathbf{s}^M$, *and "tangent" to* S, *where* $\boldsymbol{\pi}^0 \in \Omega_m$. *Then* $\boldsymbol{\pi}^0$ *is a maximin strategy for* I, *and the game has value* $(\boldsymbol{\pi}^0)^t\mathbf{s}^M = k$.

EXAMPLE 11 (continued). The hyperplane tangent to S at $\mathbf{s}^M = (0.8, 0.8)^t$ is clearly the line passing through $(0.5, 1)^t$ and $(2, 0)^t$. This line can be written as

$$l = \{\mathbf{z} \in R^2\colon (0.4, 0.6)\mathbf{z} = 0.8\},$$

which is the hyperplane $H((0.4, 0.6)^t, 0.8)$. By the above result, $(0.4, 0.6)^t$ is a maximin strategy for I and the game has value 0.8.

5.2.5. The Supporting and Separating Hyperplane Theorems

To make further theoretical progress towards establishing the existence of minimax strategies, it is necessary to develop two famous mathematical theorems: the supporting hyperplane theorem and the separating hyperplane theorem. We begin with some needed definitions.

Definition 13. A *boundary point* $\mathbf{s}^0$ of a set $S \subset R^m$ is a point such that every sphere about $\mathbf{s}^0$ contains points of both S and S^c.

Definition 14. A *supporting hyperplane* to a set $S \subset R^m$, at a boundary point $\mathbf{s}^0$ of S, is a hyperplane, $H(\boldsymbol{\xi}, k)$ $(\boldsymbol{\xi} \neq \mathbf{0})$, which contains $\mathbf{s}^0$ (i.e., $\boldsymbol{\xi}^t\mathbf{s}^0 = k$) and for which $\boldsymbol{\xi}^t\mathbf{s} \geq k$ when $\mathbf{s} \in S$. A *separating hyperplane* for sets S_1 and S_2 is a hyperplane, $H(\boldsymbol{\xi}, k)$ $(\boldsymbol{\xi} \neq \mathbf{0})$, such that $\boldsymbol{\xi}^t\mathbf{s}^1 \geq k$ for $\mathbf{s}^1 \in S_1$ and $\boldsymbol{\xi}^t\mathbf{s}^2 \leq k$ for $\mathbf{s}^2 \in S_2$.

The set of points $\{\mathbf{z} \in R^m\colon \boldsymbol{\xi}^t\mathbf{z} \geq k\}$ is the half-space lying on one side of the hyperplane $H(\boldsymbol{\xi}, k)$. Hence a supporting hyperplane to a set is a tangent hyperplane for which the set lies entirely on one side of the hyperplane. A separating hyperplane for two sets is a hyperplane for which one set lies on one side of the hyperplane, and the other set lies on the other side. Our goal will be to prove that a convex set has a supporting hyperplane at every boundary point, and that any two disjoint convex sets have a separating hyperplane. These results can be obtained from the following lemma.

Lemma 6. *If S is a closed convex set in R^m and $\mathbf{x}^0 \notin S$, then there is a hyperplane separating S and $\mathbf{x}^0$.*

PROOF. The proof will be geometric and involves several steps.

Step 1. We first show that there is a unique $\mathbf{s}^0 \in S$ nearest to $\mathbf{x}^0$, in the sense that $|\mathbf{s}^0 - \mathbf{x}^0| = \inf_{\mathbf{s}\in S} |\mathbf{s} - \mathbf{x}^0|$. To see this, choose $\mathbf{s}^n \in S$ so that $|\mathbf{s}^n - \mathbf{x}^0| \to \inf_{\mathbf{s}\in S} |\mathbf{s} - \mathbf{x}^0|$. It is easy to show that $\{\mathbf{s}^n\}$ can be chosen to be a bounded sequence, so that, by the Bolzano–Weierstrass theorem, $\{\mathbf{s}^n\}$ has a convergent subsequence $\{\mathbf{s}^{n(i)}\}$ with a limit point $\mathbf{s}^0$. Since S is closed, $\mathbf{s}^0 \in S$. Clearly

$$|\mathbf{s}^0 - \mathbf{x}^0| = \lim_{n\to\infty} |\mathbf{s}^{n(i)} - \mathbf{x}^0| = \inf_{\mathbf{s}\in S} |\mathbf{s} - \mathbf{x}^0|.$$

To see that $\mathbf{s}^0$ is unique, assume that there exists an $\mathbf{s}' \in S$ such that $\mathbf{s}' \neq \mathbf{s}^0$ and $|\mathbf{s}' - \mathbf{x}^0| = |\mathbf{s}^0 - \mathbf{x}^0|$. Then the points $\mathbf{x}^0$, $\mathbf{s}^0$, and $\mathbf{s}'$ form an isosceles triangle, and the midpoint of the line segment $\overline{\mathbf{s}'\mathbf{s}^0}$ joining $\mathbf{s}'$ and $\mathbf{s}^0$ is closer to $\mathbf{x}^0$ than are either $\mathbf{s}'$ or $\mathbf{s}^0$. But since S is convex, this midpoint is also in S, contradicting the assumption that $\mathbf{s}^0$ is closest to $\mathbf{x}^0$. Hence $\mathbf{s}^0$ must be unique.

Step 2. Define $\boldsymbol{\xi} = (\mathbf{s}^0 - \mathbf{x}^0)$ and $k = (|\mathbf{s}^0|^2 - |\mathbf{x}^0|^2)/2$. From property (iii) of hyperplanes, discussed in the previous subsection, it is clear that $H(\boldsymbol{\xi}, k)$ is the hyperplane perpendicular to the line segment $\overline{\mathbf{x}^0\mathbf{s}^0}$ and passing through the midpoint $(\mathbf{x}^0 + \mathbf{s}^0)/2$ of the line segment. Also, since

$$0 < \tfrac{1}{2}|\mathbf{s}^0 - \mathbf{x}^0|^2 = \tfrac{1}{2}(\boldsymbol{\xi}'\mathbf{s}^0 - \boldsymbol{\xi}'\mathbf{x}^0),$$

it follows that $\boldsymbol{\xi}'\mathbf{x}^0 < \boldsymbol{\xi}'\mathbf{s}^0$. We can conclude that

$$\boldsymbol{\xi}'\mathbf{x}^0 < \tfrac{1}{2}(\boldsymbol{\xi}'\mathbf{x}^0 + \boldsymbol{\xi}'\mathbf{s}^0)(=k) < \boldsymbol{\xi}'\mathbf{s}^0,$$

so that $H(\boldsymbol{\xi}, k)$ separates $\mathbf{x}^0$ and $\mathbf{s}^0$.

Step 3. We next show that $H(\boldsymbol{\xi}, k) \cap S$ is empty. To prove this, assume the contrary, namely that $\mathbf{s}^1 \in S$ and $\boldsymbol{\xi}'\mathbf{s}^1 = k$. Because $H(\boldsymbol{\xi}, k)$ is the perpendicular bisector of $\overline{\mathbf{x}^0\mathbf{s}^0}$, the triangle joining the points $\mathbf{x}^0$, $\mathbf{s}^0$, and $\mathbf{s}^1$ is isosceles, with $\overline{\mathbf{x}^0\mathbf{s}^0}$ as base. Let $\mathbf{s}^2$ be the point on $\overline{\mathbf{s}^0\mathbf{s}^1}$ for which $\overline{\mathbf{x}^0\mathbf{s}^2}$ is perpendicular to $\overline{\mathbf{s}^0\mathbf{s}^1}$ (see Figure 5.6). Then $|\mathbf{x}^0 - \mathbf{s}^2| < |\mathbf{x}^0 - \mathbf{s}^0|$, and since

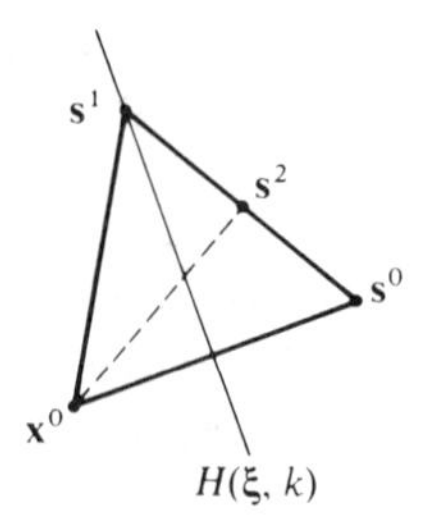

Figure 5.6

$\mathbf{s}^2 \in S$ (by convexity), Step 1 is again contradicted. It follows that $H(\boldsymbol{\xi}, k) \cap S$ must be empty.

Step 4. We conclude the proof by showing that $\boldsymbol{\xi}^t\mathbf{s} > k$ for $\mathbf{s} \in S$, so that $H(\boldsymbol{\xi}, k)$ separates $\mathbf{x}^0$ and S. To see this, assume that $\mathbf{s}^1 \in S$ and that $\boldsymbol{\xi}^t\mathbf{s}^1 = \alpha_1 \leq k$. From Step 2, we know that $\boldsymbol{\xi}^t\mathbf{s}^0 = \alpha_0 > k$. Let $\lambda = (k - \alpha_1)/(\alpha_0 - \alpha_1)$, and note that $0 \leq \lambda < 1$. By convexity, $\mathbf{s}^2 = (\lambda\mathbf{s}^0 + (1-\lambda)\mathbf{s}^1) \in S$, but

$$\boldsymbol{\xi}^t\mathbf{s}^2 = \lambda\alpha_0 + (1-\lambda)\alpha_1 = \alpha_1 + \lambda(\alpha_0 - \alpha_1) = k.$$

This contradicts the result of Step 3, yielding the desired conclusion. □

Theorem 12 (The Supporting Hyperplane Theorem). *If* $\mathbf{s}^0$ *is a boundary point of a convex set* S, *then there is a supporting hyperplane to* S *at* $\mathbf{s}^0$.

Proof. It can be checked that S and $\bar{S}$ (the closure of S) have the same boundary points. From the definition of a boundary point, it follows that there are points $\mathbf{x}^m \notin \bar{S}$ such that $\mathbf{x}^m \to \mathbf{s}^0$ (i.e., $\lim_{m\to\infty} |\mathbf{x}^m - \mathbf{s}^0| = 0$). By Lemma 6, there are hyperplanes $H(\boldsymbol{\xi}^m, k_m)$ such that

$$(\boldsymbol{\xi}^m)^t\mathbf{s} \geq k_m \quad \text{for } \mathbf{s} \in \bar{S}, \quad \text{and} \quad (\boldsymbol{\xi}^m)^t\mathbf{x}^m \leq k_m. \tag{5.8}$$

The hyperplanes and inequalities in (5.8) are not affected if $\boldsymbol{\xi}^m$ and k_m are replaced by $\boldsymbol{\tau}^m = \boldsymbol{\xi}^m/|\boldsymbol{\xi}^m|$ and $k'_m = k_m/|\boldsymbol{\xi}^m|$. Thus for $H(\boldsymbol{\tau}^m, k'_m)$,

$$(\boldsymbol{\tau}^m)^t\mathbf{s} \geq k'_m \quad \text{for } \mathbf{s} \in S, \quad \text{and} \quad (\boldsymbol{\tau}^m)^t\mathbf{x}^m \leq k'_m. \tag{5.9}$$

Observe that since $|\boldsymbol{\tau}^m| = 1$ for all m, the Bolzano–Weierstrass theorem implies that the sequence $\{\boldsymbol{\tau}^m\}$ has a convergent subsequence $\{\boldsymbol{\tau}^{m(i)}\}$. Let $\boldsymbol{\tau}^0 = \lim_{i\to\infty} \boldsymbol{\tau}^{m(i)}$. We must now show that the sequence $\{k'_{m(i)}\}$ has a convergent subsequence. Note first that for any fixed $\mathbf{s} \in S$, a use of Schwarz's inequality in (5.9) gives

$$k'_m \leq (\boldsymbol{\tau}^m)^t\mathbf{s} \leq |\boldsymbol{\tau}^m||\mathbf{s}| = |\mathbf{s}|.$$

Similarly,

$$k'_m \geq (\boldsymbol{\tau}^m)^t\mathbf{x}^m \geq -|\boldsymbol{\tau}^m||\mathbf{x}^m| = -|\mathbf{x}^m|.$$

Since $\mathbf{x}^m \to \mathbf{s}^0$, these inequalities imply that $\{k'_{m(i)}\}$ is a bounded sequence. Another application of the Bolzano–Weierstrass theorem thus yields a convergent subsequence $\{k'_{m(i(j))}\}$, with say limit k_0.

Since (by (5.9)) $k'_{m(i(j))} \leq (\boldsymbol{\tau}^{m(i(j))})^t\mathbf{s}$ for all $\mathbf{s} \in \bar{S}$, it follows that

$$k_0 \leq (\boldsymbol{\tau}^0)^t\mathbf{s} \quad \text{for all } \mathbf{s} \in \bar{S}. \tag{5.10}$$

Likewise, since $k'_{m(i(j))} \geq (\boldsymbol{\tau}^{m(i(j))})^t\mathbf{x}^{m(i(j))}$, it follows that $k_0 \geq (\boldsymbol{\tau}^0)^t\mathbf{s}^0$. Together with (5.10), this shows that

$$(\boldsymbol{\tau}^0)^t\mathbf{s}^0 = k_0. \tag{5.11}$$

From (5.10) and (5.11), we can conclude that $H(\boldsymbol{\tau}^0, k_0)$ is a supporting hyperplane to S at $\mathbf{s}^0$. □

Theorem 13 (The Separating Hyperplane Theorem). *If S_1 and S_2 are disjoint convex subsets of R^m, then there exists a vector $\boldsymbol{\xi} \in R^m$ ($\boldsymbol{\xi} \neq \mathbf{0}$) such that*

$$\boldsymbol{\xi}^t \mathbf{s}^1 \geq \boldsymbol{\xi}^t \mathbf{s}^2$$

for all $\mathbf{s}^1 \in S_1$ and $\mathbf{s}^2 \in S_2$. Indeed defining

$$k = \sup_{\mathbf{s}^2 \in S_2} \boldsymbol{\xi}^t \mathbf{s}^2,$$

the hyperplane $H(\boldsymbol{\xi}, k)$ separates S_1 and S_2.

Proof. Let

$$A = \{\mathbf{x} = \mathbf{s}^1 - \mathbf{s}^2 : \mathbf{s}^1 \in S_1 \quad \text{and} \quad \mathbf{s}^2 \in S_2\}.$$

It is straightforward to check that A is convex. Also, $\mathbf{0} \notin A$, since S_1 and S_2 are disjoint. Two cases must now be considered.

Case 1. Assume $\mathbf{0} \in \bar{A}$ (i.e., $\mathbf{0}$ is a boundary point of A). Then, by Theorem 12, there is a supporting hyperplane, $H(\boldsymbol{\xi}, k)$, to A at $\mathbf{0}$. By the definition of a supporting hyperplane, it is clear that $k = \boldsymbol{\xi}^t \mathbf{0} = 0$, and $\boldsymbol{\xi}^t \mathbf{x} \geq k = 0$ for $\mathbf{x} \in A$. Hence, for $\mathbf{s}^1 \in S_1$ and $\mathbf{s}^2 \in S_2$, it follows that $\boldsymbol{\xi}^t(\mathbf{s}^1 - \mathbf{s}^2) \geq 0$. This gives the first conclusion of the theorem. The final conclusion follows from the observation that

$$\boldsymbol{\xi}^t \mathbf{s}^1 \geq \sup_{\mathbf{s}^2 \in S_2} \boldsymbol{\xi}^t \mathbf{s}^2 \geq \boldsymbol{\xi}^t \mathbf{s}^2.$$

Case 2. Assume $\mathbf{0} \notin \bar{A}$. Then, by Lemma 6, there exists a hyperplane separating $\mathbf{0}$ and $\bar{A}$. The remainder of the argument proceeds as in Case 1. □

The main application of the above theorems will be in proving the minimax theorem in the next subsection. First, however, we will use them to prove three other results: Lemma 2 of Chapter 1 and Jensen's inequality (both discussed in Section 1.8), and Lemma 7 (alluded to in the proof of Theorem 8 in the previous subsection).

Lemma 2 of Chapter 1. *Let $\mathbf{X}$ be an m-variate random vector such that $E[|\mathbf{X}|] < \infty$ and $P(\mathbf{X} \in S) = 1$, where S is a convex subset of R^m. Then $E[\mathbf{X}] \in S$.*

Proof. Define $\mathbf{Y} = \mathbf{X} - E[\mathbf{X}]$, and let

$$S' = S - E[\mathbf{X}] = \{\mathbf{y} : \mathbf{y} = \mathbf{x} - E[\mathbf{X}] \quad \text{for some } \mathbf{x} \in S\}.$$

Note that S' is convex, $P(\mathbf{Y} \in S') = 1$, and $E[\mathbf{Y}] = \mathbf{0}$. Showing that $E[\mathbf{X}] \in S$ is clearly equivalent to showing that $\mathbf{0} \in S'$. We will establish this by induction on m.

When $m = 0$, $\mathbf{Y}$ is degenerate (a point), so that $E[\mathbf{Y}] = \mathbf{Y} \in S'$. Now suppose that the result holds for all dimensions up to and including $m-1$. We must show that the result is then true for dimension m. This will be established by contradiction. Thus assume that $\mathbf{0} \notin S'$. Then, by Theorem 13, there exists a vector $\boldsymbol{\xi} \neq \mathbf{0}$ in R^m such that $\boldsymbol{\xi}^t \mathbf{y} \geq \boldsymbol{\xi}^t \mathbf{0} = 0$ for all $\mathbf{y} \in S'$. Defining $Z = \boldsymbol{\xi}^t \mathbf{Y}$, it follows that $P(Z \geq 0) = 1$. However, $E[Z] = \boldsymbol{\xi}' E[\mathbf{Y}] = 0$, so that it must be true that $P(Z = 0) = 1$. Hence, with probability one, $\mathbf{Y}$ lies in the hyperplane defined by $\boldsymbol{\xi}'\mathbf{y} = 0$. Now define $S'' = S' \cap \{\mathbf{y}: \boldsymbol{\xi}'\mathbf{y} = 0\}$, and observe that S'' is a convex subset of an $(m-1)$-dimensional Euclidean space, and that $P(\mathbf{Y} \in S'') = 1$ and $E[\mathbf{Y}] = \mathbf{0}$. By the induction hypothesis, $\mathbf{0} \in S''$. Since $S'' \subset S'$, this contradicts the supposition that $\mathbf{0} \notin S'$, completing the proof. □

Theorem 14 (Jensen's Inequality). *Let $g(\mathbf{x})$ be a convex real valued function defined on a convex subset S of R^m, and let $\mathbf{X}$ be an m-variate random vector which has finite expectation $E[\mathbf{X}]$. Suppose, also, that $P(\mathbf{X} \in S) = 1$. Then*

$$g(E[\mathbf{X}]) \leq E[g(\mathbf{X})],$$

with strict inequality if g is strictly convex and $\mathbf{X}$ is not concentrated at a point.

Proof. The proof will be done by induction on m. For $m = 0$ the theorem is trivially satisfied, since S is a single point. Assume next that the theorem holds for all dimensions up to and including $(m-1)$. It must now be shown that the result holds for dimension m. To this end, define

$$B = \{(\mathbf{x}', y)' \in R^{m+1}: \mathbf{x} \in S, y \in R^1, \quad \text{and } y \geq g(\mathbf{x})\}.$$

Step 1. We first show that B is convex in R^{m+1}. If $(\mathbf{x}', y_1)'$ and $(\mathbf{z}', y_2)'$ are two points in B, then

$$\lambda(\mathbf{x}', y_1) + (1-\lambda)(\mathbf{z}', y_2) = ([\lambda\mathbf{x} + (1-\lambda)\mathbf{z}]', \lambda y_1 + (1-\lambda)y_2).$$

But, since S is convex, $[\lambda\mathbf{x} + (1-\lambda)\mathbf{z}] \in S$ for $0 \leq \lambda \leq 1$. Also, since $y_1 \geq g(\mathbf{x})$, $y_2 \geq g(\mathbf{z})$, and g is convex,

$$\lambda y_1 + (1-\lambda)y_2 \geq \lambda g(\mathbf{x}) + (1-\lambda)g(\mathbf{z}) \geq g(\lambda\mathbf{x} + (1-\lambda)\mathbf{z}).$$

Therefore, $[\lambda(\mathbf{x}', y_1) + (1-\lambda)(\mathbf{z}', y_2)]' \in B$ for $0 \leq \lambda \leq 1$.

Step 2. By Lemma 2 of Chapter 1, $E[\mathbf{X}] \in S$. It follows that $\mathbf{b}^0 = (E[\mathbf{X}]', g(E[\mathbf{X}]))'$ is a boundary point of B.

Step 3. Let $H(\boldsymbol{\xi}, k)$ be the supporting hyperplane to B at $\mathbf{b}^0$. Writing $\boldsymbol{\xi} = (\boldsymbol{\tau}', \rho)'$, where $\boldsymbol{\tau} \in R^m$ and $\rho \in R^1$, it follows that

$$\boldsymbol{\xi}'\mathbf{b}^0 = \boldsymbol{\tau}' E[\mathbf{X}] + \rho g(E[\mathbf{X}]) = k, \tag{5.12}$$

and

$$\boldsymbol{\xi}'(\mathbf{x}', y)' = \boldsymbol{\tau}'\mathbf{x} + \rho y \geq k \quad \text{for } \mathbf{x} \in S \quad \text{and} \quad y \geq g(\mathbf{x}). \tag{5.13}$$

Letting $\mathbf{x} = \mathbf{X}$ and $y = g(\mathbf{X})$ in (5.12) and (5.13), it follows that, with probability one,

$$\boldsymbol{\tau}'\mathbf{X} + \rho g(\mathbf{X}) \geq \boldsymbol{\tau}' E[\mathbf{X}] + \rho g(E[\mathbf{X}]). \tag{5.14}$$

Step 4. Observe that $\rho \geq 0$. (If $\rho < 0$, line (5.13) will be contradicted by sending y to infinity.) If $\rho > 0$, taking expectations in (5.14) and cancelling common terms gives the desired result. If $\rho = 0$, it follows from (5.14) that

$$h(\mathbf{X}) = \boldsymbol{\tau}'(\mathbf{X} - E[\mathbf{X}]) \geq 0.$$

On the other hand, clearly $E[h(\mathbf{X})] = 0$. It follows that $h(\mathbf{X}) = 0$ with probability one, or equivalently that

$$P(\boldsymbol{\tau}'\mathbf{X} = k) = 1.$$

This means that $\mathbf{X}$ is concentrated on the hyperplane $H(\boldsymbol{\tau}, k)$ with probability one. Let $S_H = S \cap H(\boldsymbol{\tau}, k)$. Since the intersection of convex sets is convex, S_H is convex. Also, $P(\mathbf{X} \in S_H) = 1$. Since S_H is an $(m-1)$-dimensional set, the induction hypothesis can be applied to give the desired conclusion. (The conclusion about strict inequality follows by noting that, for $\rho \neq 0$, the inequality in (5.14) is strict when g is strictly convex. Since $\mathbf{X}$ is not concentrated at a point, ρ must be nonzero for some nonzero dimension.) □

Lemma 7. *In an S-game, assume that the risk set S is bounded from below. If* $\mathbf{s} \in S$ *but* $\mathbf{s} \notin \lambda(S)$ *($\lambda(S)$ is the lower boundary of S), then there exists a point* $\mathbf{y} \in \lambda(S)$ *such that* $y_i \leq s_i$, *the inequality being strict for some i.*

Proof. The proof will proceed by induction on m. For $m = 0$, S contains only one point, and the result is trivially satisfied. Assume next that the result holds for all dimensions up to and including $m-1$. It must now be shown that the lemma holds for m dimensions. To this end, define

$$T = \{\mathbf{x} \in \bar{S}: x_i \leq s_i \quad \text{for } i = 1, \ldots, m\}.$$

Step 1. It is easy to check that T is convex, bounded, and closed. As a consequence of T being bounded, note that $t_m = \inf_{\mathbf{x} \in T} x_m$ is finite.

Step 2. Define $T_m = \{\mathbf{x} \in T: x_m = t_m\}$. We must show that T_m is not empty. To this end, let $\{\mathbf{x}^n\}$ be a sequence of points in T such that $x_m^n \to t_m$. Since T is bounded, the Bolzano–Weierstrass theorem implies the existence of a subsequence $\{\mathbf{x}^{n(i)}\}$ converging to a point $\mathbf{x}^0$. Since T is closed, $\mathbf{x}^0 \in T$. But clearly $x_m^0 = t_m$, so that $\mathbf{x}^0 \in T_m$.

Step 3. Let $T'_m = \{(x_1, \ldots, x_{m-1})': \mathbf{x} \in T_m\}$. It is easy to verify that T'_m is bounded, closed, and convex. It thus satisfies the induction hypothesis, from which it can be concluded that there exists a point $\mathbf{z} = (z_1, \ldots, z_{m-1})' \in \lambda(T'_m)$ such that $z_i \leq x_i^0$ for $i = 1, \ldots, m-1$.

Step 4. Define $\mathbf{y} = (z_1, \ldots, z_{m-1}, t_m)'$. Since T'_m and T are closed, $\mathbf{y} \in T$. Hence $y_i \leq s_i$ for $i = 1, \ldots, m$. Also, $\mathbf{y} \in \lambda(S)$. If it were not, then for some $\mathbf{y}' \in \bar{S}$, it would follow that $y'_i \leq y_i$ with strict inequality for some i. But the inequality can't be strict for $i = m$ by the definition of t_m, while, for $i < m$, strict inequality would violate the fact that $\mathbf{z} \in \lambda(T'_m)$. (Clearly y' would also be in T_m, and $(y'_1, \ldots, y'_{m-1})$ would be in T'_m.)

Step 5. At least one of the inequalities $y_i \leq s_i$ must be strict, since $\mathbf{y}$ and $\mathbf{s}$ are different points. (One is in $\lambda(S)$, while the other is not.) □

5.2.6. The Minimax Theorem

We are ready to prove the fundamental theorem concerning the existence of minimax strategies. The theorem will be proven only for finite Θ. Generalizations to other situations will be discussed, however.

Theorem 15 (The Minimax Theorem). *Consider a (two-person zero-sum) game in which $\Theta=\{\theta_1, \theta_2, \ldots, \theta_m\}$. Assume that the risk set S is bounded from below. Then the game has value*

$$V=\inf_{\delta^*\in\mathscr{A}^*}\sup_{\pi\in\Theta^*} L(\pi,\delta^*)=\sup_{\pi\in\Theta^*}\inf_{\delta^*\in\mathscr{A}^*} L(\pi,\delta^*),$$

and a maximin strategy π^M exists. Moreover, if S is closed from below, then a minimax strategy δ^{*M} exists, and $L(\pi^M, \delta^{*M})=V$.

Proof. Consider the problem in terms of the S-game formulation, as discussed in Subsection 5.2.4. The strategy spaces are then S and Ω_m, and the loss is $\boldsymbol{\pi}^t\mathbf{s}$. As in Theorem 7 of Subsection 5.2.4, consider $S_\alpha=Q_\alpha\cap S$ and $\alpha_M=\inf\{\alpha: S_\alpha \text{ is not empty}\}$. Note that α_M is finite, since S is bounded from below.

The basic idea of the proof will be to find a separating hyperplane between S and the interior of Q_{α_M}. This hyperplane will give rise to a maximin strategy. Also, it will follow that $V=\alpha_M$ and that S_{α_M} will consist of the minimax risk points.

Step 1. Construction of the hyperplane: Let

$$T=\{\mathbf{x}\in R^m: x_i<\alpha_M \quad \text{for } i=1,\ldots,m\}.$$

(Thus T is the interior of Q_{α_M}.) It is easy to check that T is a convex set, and S and T are obviously disjoint. Since S is also convex, Theorem 13 can be used to conclude that there is a hyperplane $H(\boldsymbol{\xi}, k)$ separating S and T, such that

$$\boldsymbol{\xi}^t\mathbf{s}\geq k \quad \text{for } \mathbf{s}\in S, \qquad \boldsymbol{\xi}^t\mathbf{x}\leq k \quad \text{for } \mathbf{x}\in T, \quad \text{and} \quad k=\sup_{\mathbf{x}\in T}\boldsymbol{\xi}^t\mathbf{x}. \tag{5.15}$$

Step 2. The vector $\boldsymbol{\xi}$ has $\xi_i\geq 0$ for $i=1,\ldots,m$. This follows from the observation that, if $\xi_j<0$, letting $x_j\to-\infty$ in (5.15) reveals a contradiction. Note, as a consequence, that

$$k=\sup_{\mathbf{x}\in T}\boldsymbol{\xi}^t\mathbf{x}=\boldsymbol{\xi}^t(\alpha_M,\ldots,\alpha_M)=\alpha_M\sum_{i=1}^m \xi_i.$$

Step 3. Dividing $\boldsymbol{\xi}$ and k by $(\sum_{i=1}^m \xi_i)$ shows that the hyperplane $H(\boldsymbol{\xi}, k)$ can be rewritten $H(\boldsymbol{\pi}, \alpha_M)$, where $\boldsymbol{\pi}\in\Omega_m$ is given by $\pi_i=\xi_i/(\sum_{j=1}^m \xi_j)$. Note also that (5.15) can be rewritten

$$\boldsymbol{\pi}^t\mathbf{s}\geq\alpha_M \quad \text{for } \mathbf{s}\in S, \quad \text{and} \quad \boldsymbol{\pi}^t\mathbf{x}\leq\alpha_M \quad \text{for } \mathbf{x}\in T. \tag{5.16}$$

Step 4. From (5.16), it is clear that $\inf_{\mathbf{s}\in S}\boldsymbol{\pi}^t\mathbf{s}\geq\alpha_M$.

Step 5. There exists a sequence $\{\mathbf{s}^n\}$ of risk points in S such that

$$\lim_{n\to\infty} \max_i \{s_i^n\} = \alpha_M.$$

This follows from the fact that S_α is nonempty for each $\alpha > \alpha_M$. In particular, choosing $\mathbf{s}^n \in S_{(\alpha_M+1/n)}$ gives the desired result.

Step 6. By Steps 4 and 5, together with Lemma 1 in Subsection 5.2.1, the hypothesis of Theorem 3(b) (Subsection 5.2.2) is satisfied for the S-game. Hence the game has value α_M and $\boldsymbol{\pi}$ is a maximin strategy.

Step 7. Assume that S is closed from below. Consider the sequence $\{\mathbf{s}^n\}$ found in Step 5. This is clearly a bounded sequence, since S is bounded from below and the $Q_{(\alpha_M+1/n)}$ all have a common bound from above. Hence, by the Bolzano-Weierstrass theorem, a subsequence $\{\mathbf{s}^{n(i)}\}$ exists which converges to a limit point $\mathbf{s}^0$. Clearly $\mathbf{s}^0 \in \bar{S}$ and $\max_i \{\mathbf{s}_i^0\} = \alpha_M$.

If $\mathbf{s}^0 \in \lambda(S)$, then, since S is closed from below, $\mathbf{s}^0 \in S$. Since $\max_i \{s_i^0\} = \alpha_M$, which is the value of the game, it follows that $\mathbf{s}^0$ is a minimax risk point. If $\mathbf{s}^0 \notin \lambda(S)$, then by Lemma 7 there exists an $\mathbf{s}' \in \lambda(S)$ such that $s_i' \le s_i^0 \le \alpha_M$. By the previous argument, $\mathbf{s}' \in S$ and is a minimax risk point. This completes the proof. □

Versions of the minimax theorem hold in settings other than that considered above. As an example, we state the following theorem, which can be found in Blackwell and Girshick (1954).

Theorem 16. *If $\mathscr{A}$ is a closed, bounded, and convex subset of R^n, and $L(\theta, a)$ is, for each θ, a continuous convex function of a, then the game has a value and player* II *has a minimax strategy which is nonrandomized. If, in addition, Θ is a closed bounded subset of R^m and $L(\theta, a)$ is continuous in θ for each a, then player* I *has a maximin strategy which randomizes among at most $n+1$ nonrandomized strategies.*

There are quite general minimax theorems that hold in a wide variety of settings. Unfortunately, these theorems involve advanced topological and measure-theoretic concepts. Some simple versions of such theorems can be found in Ferguson (1967). For more powerful advanced theorems see Le Cam (1955), Brown (1976), and Diaconis and Stein (1983).

It is, of course, not always true that minimax and maximin strategies exist, or that the game has a value. Indeed the basic assumptions in the minimax theorem, that the risk set be closed from below and bounded from below, are quite essential. For example, if the risk set is open, then there is no minimax strategy. (The concept of an *ε-minimax* strategy has been developed for such a situation. This is a strategy δ^* for which $\sup_\theta L(\theta, \delta^*) \le \bar{V} + \varepsilon$; in other words, a strategy within ε of being optimal according to the minimax principle.)

For unbounded strategy spaces, even more serious difficulties can be encountered. The following is an example in which the game doesn't even have a value.

EXAMPLE 13. Consider the game in which each player chooses a positive integer, with the player who chooses the largest integer winning 1 from the other player. Here $\Theta = \mathscr{A} = \{1, 2, \ldots\}$, and

$$L(\theta, a) = \begin{cases} 1 & \text{if } \theta > a, \\ 0 & \text{if } \theta = a, \\ -1 & \text{if } \theta < a. \end{cases}$$

A randomized strategy for player I is a probability vector $\boldsymbol{\pi} = (\pi_1, \pi_2, \ldots)^t$, where π_i is the probability of selecting integer i. Clearly

$$L(\boldsymbol{\pi}, a) = \left(\sum_{\{i>a\}} \pi_i\right) - \left(\sum_{\{i<a\}} \pi_i\right).$$

For any fixed $\boldsymbol{\pi}$, it follows that

$$\inf_a L(\boldsymbol{\pi}, a) = \lim_{a \to \infty} L(\boldsymbol{\pi}, a) = -1.$$

Hence

$$\underline{V} = \sup_{\boldsymbol{\pi}} \inf_a L(\boldsymbol{\pi}, a) = -1.$$

It can similarly be calculated that $\bar{V} = 1$. Since $\underline{V} \neq \bar{V}$, the game does not have a value. There are clearly no optimum strategies in this game.

5.3. Statistical Games

5.3.1. Introduction

A statistical decision problem (with action space $\mathscr{A}$, state-of-nature space Θ, and loss L) can be viewed as a two-person zero-sum game in which the statistician is player II and nature is player I. The strategy space for the statistician will be assumed to be $\mathscr{D}^*$, the space of all randomized decision rules, while the strategy space for nature is Θ. The minimax approach views the problem as that of choosing among rules $\delta^* \in \mathscr{D}^*$, before experimentation. (Some criticisms of this viewpoint will be presented in the next section.) Since the loss to the statistician will be $L(\theta, \delta^*(X, \cdot))$, and X (the sample observation) is random, the proper measure of loss for this game is the risk function

$$R(\theta, \delta^*) = E^X L(\theta, \delta^*(X, \cdot)).$$

It is, of course, necessary to consider randomized strategies. A randomized strategy for the statistician is, by definition, a probability distribution on $\mathscr{D}^*$. Since $\mathscr{D}^*$ is the space of randomized decision rules, the obvious question arises as to whether another randomization is necessary. The answer is no, as the following lemma shows.

Lemma 8. *Let P be a probability distribution on $\mathscr{D}^*$. Define $\delta_P^*(x, A)$, for all $x \in \mathscr{X}$ and $A \subset \mathscr{A}$, by*

$$\delta_P^*(x, A) = E^P[\delta^*(x, A)].$$

Then δ_P^* is equivalent to P, in the sense that

$$R(\theta, \delta_P^*) = E^P[R(\theta, \delta^*)].$$

PROOF. By definition,

$$\begin{aligned} L(\theta, \delta_P^*(x, \cdot)) &= E^{\delta_P^*(x, \cdot)}[L(\theta, a)] \\ &= E^P E^{\delta^*(x, \cdot)}[L(\theta, a)]. \end{aligned}$$

Hence

$$\begin{aligned} R(\theta, \delta_P^*) &= E^X L(\theta, \delta_P^*(X, \cdot)) \\ &= E^X E^P E^{\delta^*(X, \cdot)}[L(\theta, a)] \\ &= E^P E^X E^{\delta^*(X, \cdot)}[L(\theta, a)] \\ &= E^P[R(\theta, \delta^*)]. \end{aligned}$$

The interchange in the order of integration above is justified by Fubini's theorem, since $L(\theta, a) \geq -K > -\infty$ by the original assumption on L. □

The above lemma is not quite correct technically, the problem being that δ_P^* need not necessarily be measurable, and hence need not be in $\mathscr{D}^*$. This problem does not arise in reasonable statistical situations, however, and need not concern us.

The import of Lemma 8 is that it is unnecessary to consider randomized strategies on $\mathscr{D}^*$, since the risk function of any randomized strategy can be duplicated by that of a randomized decision rule in $\mathscr{D}^*$. Therefore, in applying to statistical games the game-theoretic results of Section 5.2, both $\mathscr{A}$ and $\mathscr{A}^*$ will be taken to be $\mathscr{D}^*$. Note, in particular, that the definition of a minimax strategy then coincides with that of a minimax decision rule as given in Section 1.5.

Randomized strategies for nature are again probability distributions $\pi \in \Theta^*$. The expected loss if nature uses $\pi \in \Theta^*$ and the statistician uses $\delta^* \in \mathscr{D}^*$ is clearly $r(\pi, \delta^*) = E^\pi[R(\theta, \delta^*)]$, the Bayes risk of δ^* with respect to π. In statistical games, a maximin strategy for nature is more commonly called a *least favorable prior distribution.* The reason is that nature is no longer an intelligent opponent, and the statistician is really just trying, through the minimax approach, to protect against the worst or least favorable possible values of θ.

In the remainder of the chapter, we will tend to use the statistical, as opposed to game-theoretic, terminology. Subsection 5.3.2 deals with general techniques for solving statistical games. Subsection 5.3.3 discusses the situation of finite Θ.

5.3.2. General Techniques for Solving Statistical Games

The game-theoretic techniques of Subsection 5.2.2 can be readily adapted to solve statistical games. We discuss three basic methods of solution here. A fourth method will be mentioned in Chapter 6, after invariance has been discussed.

I. *The Direct Method of Solving Statistical Games*

If the class $\mathscr{D}^*$ of decision rules is small enough, or can be reduced by admissibility considerations to a small enough set, the minimax rule can be calculated directly from the definition. Chapter 8 is devoted to the use of admissibility in reducing $\mathscr{D}^*$. Here we will content ourselves with an example in a standard situation.

EXAMPLE 14. Assume $X \sim \mathcal{N}(\theta, \sigma^2)$, σ^2 known, is observed, and that it is desired to test $H_0: \theta \leq \theta_0$ versus $H_1: \theta > \theta_0$, under "0-K_i" loss. A randomized decision rule in a testing situation is usually represented by a test function $\phi(x)$, where $\phi(x)$ is the probability of rejecting the null hypothesis when $X = x$ is observed. (Thus $\phi(x) = \delta^*(x, a_1) = 1 - \delta^*(x, a_0)$, where a_i denotes accepting H_i.)

It will be seen in Chapter 8 that only the "uniformly most powerful" tests $\phi_c(x) = I_{(c,\infty)}(x)$ need be considered. (Any other test can be improved upon by a test ϕ_c, for some c.) Clearly

$$R(\theta, \phi_c) = \begin{cases} K_1 P_\theta(X > c) & \text{if } \theta \leq \theta_0, \\ K_0 P_\theta(X < c) & \text{if } \theta > \theta_0. \end{cases}$$

Since $\beta(\theta) = P_\theta(X > c)$ is increasing in θ, it follows that

$$\begin{aligned} \sup_\theta R(\theta, \phi_c) &= \max\left\{ \sup_{\theta \leq \theta_0} K_1 \beta(\theta), \sup_{\theta > \theta_0} K_0[1 - \beta(\theta)] \right\} \\ &= \max\{K_1 \beta(\theta_0), K_0[1 - \beta(\theta_0)]\}. \end{aligned}$$

Letting $z = \beta(\theta_0)$, and graphing the functions $K_1 z$ and $K_0(1 - z)$, it becomes clear that $\max\{K_1 z, K_0(1 - z)\}$ is minimized when $K_1 z = K_0(1 - z)$, or $z = K_0/(K_1 + K_0)$. But if

$$\beta(\theta_0) = P_{\theta_0}(X > c) = K_0/(K_1 + K_0),$$

it follows that c is the $1 - K_0/(K_1 + K_0) = K_1/(K_1 + K_0)$-fractile of the

$\mathcal{N}(\theta_0, \sigma^2)$ distribution. Denoting this by $z(K_1/(K_1+K_0))$, it can be concluded that

$$\inf_{\delta^* \in \mathscr{D}^*} \sup_\theta R(\theta, \delta^*) = \inf_c \sup_\theta R(\theta, \phi_c) = \sup_\theta R(\theta, \phi_{z(K_1/(K_1+K_0))}).$$

Hence $\phi_{z(K_1/(K_1+K_0))}$ is minimax. (See Hald (1971) for general results.)

II. *Guessing a Least Favorable Prior*

A very useful method of determining a minimax rule is to guess a least favorable prior distribution, determine the resulting Bayes rule, and check to see if this rule is minimax. In verifying minimaxity, the following consequence of Theorem 2 of Subsection 5.2.2 proves to be very useful. (Its proof will be left as an exercise.)

Theorem 17. *If $\delta_0^* \in \mathscr{D}^*$ is Bayes with respect to $\pi_0 \in \Theta^*$, and*

$$R(\theta, \delta_0^*) \leq r(\pi_0, \delta_0^*)$$

for all $\theta \in \Theta$, then δ_0^* is minimax and π_0 is least favorable.

Often, the obvious guess for a least favorable prior is an improper prior π. The generalized Bayes rule, δ^π, is then a good guess for the minimax rule, but if $r(\pi, \delta^\pi)$ is infinite, Theorem 17 will not apply. Frequently, however, a sequence of proper priors $\{\pi_n\}$ can be found, such that $[c_n \pi_n(\theta)] \rightarrow \pi(\theta)$ for all θ, where the c_n are appropriate constants. The following immediate consequence of Theorem 3(a) can then be useful.

Theorem 18. *Assume that $\{\pi_n\}$ is a sequence of proper priors and δ_0^* is a decision rule such that*

$$R(\theta, \delta_0^*) \leq \lim_{n \rightarrow \infty} r(\pi_n) < \infty,$$

for all $\theta \in \Theta$. Then δ_0^* is minimax.

The sequence $\{\pi_n\}$ can usually be chosen to be any convenient sequence such that constants c_n can be found for which $c_n \pi_n \rightarrow \pi$. Choosing an appropriate sequence of conjugate prior densities often succeeds, and is calculationally simple. See Brown (1976) for some general theorems for situations in which conjugate priors do not work.

Example 15. Assume $X \sim \mathcal{N}(\theta, 1)$ is observed, and that it is desired to estimate θ under squared-error loss. The obvious guess for a least favorable prior is the noninformative prior density $\pi(\theta) = 1$. (It contains no helpful information about θ.) The generalized Bayes rule, with respect to π, was seen in Chapter 4 to be $\delta_0(x) = x$.

Since π is improper, we must look for an approximating sequence $\{\pi_n\}$ of proper priors. A reasonable choice for π_n is a $\mathcal{N}(0, n)$ density, since $(2\pi n)^{1/2}\pi_n(\theta) \to \pi(\theta) = 1$ for all θ. (Intuitively, the π_n are very flat for large n, and approximate a uniform density.)

From previous calculations, we know that the Bayes risk of π_n is $r(\pi_n) = n/(n+1)$. Since

$$R(\theta, \delta_0) = E_\theta(X-\theta)^2 = 1 = \lim_{n\to\infty} r(\pi_n)$$

for all θ, Theorem 18 can be used to conclude that δ_0 is minimax.

EXAMPLE 16. Assume $X \sim \mathscr{E}(\theta)$ is observed. The parameter space is $\Theta = (0, 1] \cup [2, \infty)$, and it is desired to test $H_0: 0 < \theta \leq 1$ versus $H_1: 2 \leq \theta < \infty$. Let a_i denote accepting H_i. The loss function is given by

$$L(\theta, a_0) = \begin{cases} 0 & \text{if } 0 < \theta \leq 1, \\ 4 & \text{if } \theta \geq 2, \end{cases}$$

$$L(\theta, a_1) = \begin{cases} 0 & \text{if } \theta \geq 2, \\ 5-\theta & \text{if } 0 < \theta \leq 1. \end{cases}$$

It seems plausible that the least favorable prior is one which makes H_0 and H_1 as hard to distinguish as possible, namely one which gives positive probability only to the points $\theta = 1$ and $\theta = 2$. Assume that π is such a prior, and let $\pi_1 = \pi(1) = 1 - \pi(2)$. We still must find the least favorable choice of π_1.

To find the Bayes rule with respect to π, note that the posterior expected losses of a_0 and a_1 are (letting $m(x)$ denote the marginal density of X)

$$E^{\pi(\theta|x)}L(\theta, a_0) = \frac{4(1-\pi_1)(\frac{1}{2}e^{-x/2})}{m(x)}$$

and

$$E^{\pi(\theta|x)}L(\theta, a_1) = \frac{4\pi_1 e^{-x}}{m(x)}.$$

Thus the Bayes rule is to choose a_0 if

$$\frac{(1-\pi_1)(e^{-x/2})}{m(x)} < \frac{2\pi_1 e^{-x}}{m(x)},$$

which can be rewritten as

$$x < 2\log\left[\frac{2\pi_1}{(1-\pi_1)}\right] = c.$$

We will, therefore, consider the class of rules defined by the test functions $\phi_c(x) = I_{(c,\infty)}(x)$. (This reduction could also have been established using results on uniformly most powerful tests from Chapter 8.)

To find the least favorable π_1, or alternatively the least favorable c, note that

$$R(\theta, \phi_c) = \begin{cases} \displaystyle\int_c^\infty (5-\theta)\theta^{-1}e^{-x/\theta}\,dx & \text{if } 0<\theta\le 1, \\ \displaystyle\int_0^c 4\theta^{-1}e^{-x/\theta}\,dx & \text{if } \theta\ge 2, \end{cases}$$

$$= \begin{cases} (5-\theta)e^{-c/\theta} & \text{if } 0<\theta\le 1, \\ 4(1-e^{-c/\theta}) & \text{if } \theta\ge 2. \end{cases}$$

Define

$$h(c) = \sup_{0<\theta\le 1} (5-\theta)e^{-c/\theta} \quad \text{and} \quad g(c) = \sup_{2\le\theta<\infty} 4(1-e^{-c/\theta}).$$

Clearly

$$\bar{R}(c) = \sup_{\theta\in\Theta} R(\theta, \phi_c) = \max\{h(c), g(c)\}.$$

Note that $h(c)$ is strictly decreasing in c, with $h(0)=5$ and $\lim_{c\to\infty} h(c)=0$. Also, $g(c)$ is strictly increasing in c, with $g(0)=0$ and $\lim_{c\to\infty} g(c)=4$. Hence $h(c)$ and $g(c)$ are equal for just one value of c, call it c_0, and $\bar{R}(c_0)=\inf_c \bar{R}(c)$. The test ϕ_{c_0} is a good candidate for a minimax test.

To find c_0, note first that $4(1-e^{-c/\theta})$ is decreasing in θ, so that $g(c)=4(1-e^{-c/2})$. Next observe that

$$\frac{d}{d\theta}[(5-\theta)e^{-c/\theta}] = -e^{-c/\theta} + (5-\theta)c\theta^{-2}e^{-c/\theta}$$
$$= \theta^{-2}e^{-c/\theta}(-\theta^2 - \theta c + 5c).$$

This derivative is positive for $0<\theta\le 1$, providing $\theta^2+\theta c-5c<0$. The roots of the equation $\theta^2+\theta c-5c=0$ are $\frac{1}{2}(-c\pm[c^2+20c]^{1/2})$, one of which is negative, while the other is larger than 1 for $c>\frac{1}{4}$. Hence if $c>\frac{1}{4}$ and $0<\theta\le 1$, it follows that $\theta^2+\theta c-5c<0$, and $(5-\theta)e^{-c/\theta}$ is maximized at $\theta=1$. Thus $h(c)=4e^{-c}$ for $c>\frac{1}{4}$.

Let's assume the solution to $h(c)=g(c)$ is some $c>\frac{1}{4}$. Then we want to solve the equation

$$4e^{-c} = 4(1-e^{-c/2}).$$

Letting $z=e^{-c/2}$, this is equivalent to solving $4z^2=4(1-z)$. The positive solution of this latter equation is $z\cong 0.618$, which corresponds to $c\cong 0.96$ in the original equation. Since $c>\frac{1}{4}$, this is indeed the unique solution. Note that $c=0.96$ corresponds to $\pi_1\cong 0.45$.

Observe finally that for $\pi_1\cong 0.45$,

$$\begin{aligned} r(\pi) = r(\pi, \phi_{0.96}) &= \pi_1 R(1, \phi_{0.96}) + (1-\pi_1)R(2, \phi_{0.96}) \\ &= \pi_1 h(0.96) + (1-\pi_1)g(0.96) \\ &= h(0.96) \end{aligned}$$

(since $h(0.96) = g(0.96)$). Also, as shown earlier, $R(\theta, \phi_{0.96}) \le h(0.96)$ for all $\theta \in \Theta$. Hence Theorem 17 can be used to conclude that $\phi_{0.96}$ is a minimax test, and π is least favorable.

In a variety of situations, the least favorable prior, π_0, will be concentrated on a finite set of points. The following theorem describes one such class of problems.

Theorem 19. *Suppose the statistical game has a value and that a (proper) least favorable prior, π_0, and minimax rule δ^{π_0}* (see Theorem 1) *exist. Suppose further that Θ is a compact subset of $\mathbb{R}^1$, and that $R(\theta, \delta^{\pi_0})$ is an analytic function of θ in some open neighborhood of Θ in the complex plane* (cf. Lehmann (1959) for conditions under which this is so). *Then either π_0 is supported on a finite set of points, or $R(\theta, \delta^{\pi_0})$ is constant on Θ.*

Proof. Theorem 4 implies that π_0 is supported on the set

$$\Theta_v = \{\theta: R(\theta, \delta^{\pi_0}) = r(\pi_0, \delta^{\pi_0})\}.$$

If this set has a limit point, $R(\theta, \delta^{\pi_0})$ must be a constant function on Θ. (If any level set of an analytic function contains a limit point, the function must be constant.) If Θ_v does not contain a limit point, then the compactness of Θ guarantees the finiteness of Θ_v. □

Examples in which the conditions of Theorem 19 can be verified include Exercise 40 (and generalizations), estimating a bounded normal mean under squared-error loss (cf. Casella and Strawderman (1981) and Bickel (1981)), and estimating a bounded normal mean under zero-one loss (cf. Zeytinoglu and Mintz (1984); note that this effectively yields optimal α-level confidence procedures for a bounded normal mean). The major difficulty in problems with a finitely supported least favorable prior is in determining the support points and the masses they are to be given. Some general numerical approaches to such determination can be found in Nelson (1966) and Kempthorne (1984).

III. *Equalizer Rules*

The third method of determining a minimax decision rule is to search for an equalizer rule δ_0^*, i.e., a rule for which $R(\theta, \delta_0^*) = C$ for all θ. If δ_0^* can be shown to be Bayes with respect to a prior π, then since $r(\pi, \delta_0^*) = E^{\pi}[C] = C$, Theorem 17 can be used to conclude that δ_0^* is minimax and π is least favorable.

Example 17. Assume $X \sim \mathcal{B}(n, \theta)$ is observed, and that it is desired to estimate θ under squared-error loss. Let us try to find an equalizer rule of the form $\delta(x) = ax + b$. Clearly

$$\begin{aligned} R(\theta, \delta) &= E_\theta[aX + b - \theta]^2 = E_\theta[a(X - n\theta) + \{b + (an - 1)\theta\}]^2 \\ &= a^2 n\theta(1-\theta) + \{b + (an-1)\theta\}^2 \\ &= \theta^2[-a^2 n + (an-1)^2] + \theta[a^2 n + 2b(an-1)] + b^2. \end{aligned}$$

For the risk to be constant in θ, we must have $-a^2 n + (an-1)^2 = 0$ and $a^2 n + 2b(an-1) = 0$. Solving these equations for a and b gives $a = (n + \sqrt{n})^{-1}$ and $b = \sqrt{n}/[2(n+\sqrt{n})]$. Thus

$$\delta_0(x) = ax + b = \frac{x + \sqrt{n}/2}{n + \sqrt{n}}$$

is an equalizer rule.

To complete the argument, we must show that δ_0 is Bayes. From Chapter 4 (Exercise 5), we know that the mean of the posterior distribution arising from a $\mathscr{B}e(\alpha, \beta)$ prior is $(x+\alpha)/(\alpha+\beta+n)$. The equalizer rule δ_0 is clearly of this form with $\alpha = \beta = \sqrt{n}/2$. Hence δ_0 is Bayes, and must, by Theorem 17, be minimax.

Often an equalizer rule δ_0^* will not be Bayes itself, but will be Bayes in the limit, in the sense that there will exist a sequence of priors $\{\pi_n\}$ such that

$$\lim_{n\to\infty} r(\pi_n, \delta_0^*) = \lim_{n\to\infty} r(\pi_n).$$

Such a δ_0^* is called *extended Bayes.* If an extended Bayes equalizer rule can be found, it is minimax by Theorem 18. Example 15 illustrates such a situation.

When dealing with finite Θ, it frequently happens that a minimax rule can be found by looking for an equalizer rule corresponding to a point with equal coordinates on the lower boundary of the risk set. This was indicated in Subsection 5.2.4, and will be discussed further in the next subsection.

It should be emphasized that the three methods discussed above for finding a minimax rule are not really distinct in the sense that only one can apply. Frequently, two or all three of the techniques will be useful in the determination of a minimax rule. The methods can just be thought of as three helpful guidelines to employ in tackling a statistical game.

5.3.3. Statistical Games with Finite Θ

When $\Theta = \{\theta_1, \theta_2, \ldots, \theta_m\}$, the statistical problem can be converted to an S-game. The risk set will then be

$$S = \{\mathbf{R}(\delta^*) = (R(\theta_1, \delta^*), \ldots, R(\theta_m, \delta^*))' : \delta^* \in \mathscr{D}^*\}.$$

When S can be explicitly constructed, the methods of Subsection 5.2.4 can be used to find a minimax rule. Unfortunately, the space $\mathscr{D}^*$ of decision rules is usually quite complicated, so that obtaining S in its entirety is difficult. Frequently, however, the lower boundary, $\lambda(S)$, can be obtained, and a minimax risk point found for this much reduced set. Theorem 11 can be very helpful in obtaining $\lambda(S)$, since it shows that $\lambda(S)$ is contained in the set of risk points corresponding to Bayes rules. In order to apply the theorem, however, it is necessary to verify that S is closed from below and bounded from below. These conditions are also needed to apply the minimax theorem, which ensures the existence of a minimax rule. We, therefore, give a brief discussion of the verification of these conditions in statistical games.

Under the assumption that $L(\theta, a) \geq -K > -\infty$ (which we are making), it is easy to check that $R(\theta, \delta^*) \geq -K$. Hence S will clearly be bounded from below.

It is considerably more difficult to determine if S is closed from below. Usually this is done by showing that S is closed. To show that S is closed, it is useful to consider

$$W = \{\mathbf{w}(a) = (w_1(a), \ldots, w_m(a))^t : w_i(a) = L(\theta_i, a) \text{ and } a \in \mathscr{A}\}.$$

This is the set of "loss points" in R^m. The following lemma gives a simple condition, in terms of W, under which S is closed and bounded.

Lemma 9. *If $W \subset R^m$ is closed and bounded, then so is the risk set S.*

Unfortunately, the proof of the above lemma involves measure theory and so is beyond the scope of the book. A non-measure-theoretic proof for discrete distributions can be found in Blackwell and Girshick (1954).

The following lemma, in turn, states simple conditions under which W is closed and bounded. Its proof will also be omitted.

Lemma 10. *If* (i) *$\mathscr{A}$ is finite, or* (ii) *$\mathscr{A}$ is a bounded closed subset of R^m with $L(\theta_i, a)$ continuous in a for $i = 1, \ldots, m$, then W is closed and bounded, and hence S is closed and bounded.*

Of course, S will be closed under far more general conditions than those of Lemma 10. Indeed only very rarely will S not be closed in statistical problems. The theory is, therefore, mainly of academic interest.

We next discuss two typical situations in which Θ is finite: testing simple hypotheses and classification problems.

I. *Testing Simple Hypotheses*

Assume it is desired to test $H_0: \theta = \theta_0$ versus $H_1: \theta = \theta_1$, based on the sample observation X. Let a_i denote accepting $H_i (i = 0, 1)$, and assume "$0 - K_i$" loss is used. As usual, a rule $\delta^* \in \mathscr{D}^*$ will be represented by a test function

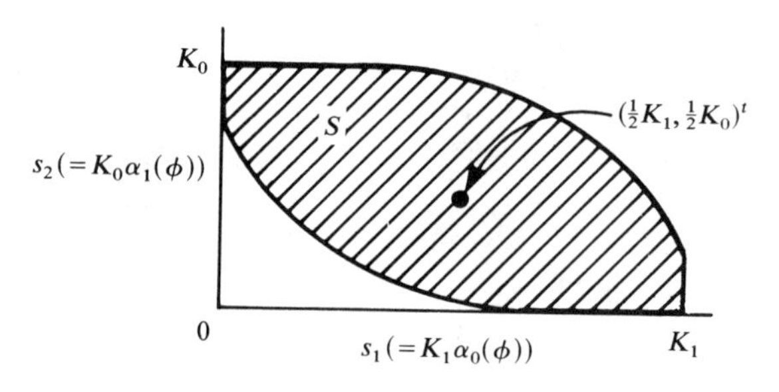

Figure 5.7

$\phi(x)$, where $\phi(x) = \delta^*(x, a_1)$ is the probability of rejecting when x is observed. Define

$$\alpha_0(\phi) = E_{\theta_0}[\phi(X)] = \text{probability of type I error},$$

$$\alpha_1(\phi) = E_{\theta_1}[1 - \phi(X)] = \text{probability of type II error}.$$

It is easy to check that $R(\theta_0, \phi) = K_1\alpha_0(\phi)$ and $R(\theta_1, \phi) = K_0\alpha_1(\phi)$. Thus the risk set is

$$S = \{(K_1\alpha_0(\phi), K_0\alpha_1(\phi))^t : \phi \in \mathscr{D}^*\}.$$

Since $\mathscr{A}$ is finite, Lemma 10 can be used to conclude that S is closed and bounded. We thus know that a minimax rule exists, and that $\lambda(S)$ is a subset of the Bayes risk points.

In this situation, we can actually say a good deal about the shape of S. Indeed S must have a form similar to that shown in Figure 5.7. This follows from noting that (i) $0 \le \alpha_i(\phi) \le 1$, (ii) $(0, K_0)^t$ and $(K_1, 0)^t$ are in S (arising from $\phi(x) \equiv 0$ and $\phi(x) \equiv 1$, respectively), and (iii) S is symmetric about $(\frac{1}{2}K_1, \frac{1}{2}K_0)^t$ (if $(s_1, s_2)^t$ arises from ϕ, then $(K_1 - s_1, K_0 - s_2)^t$ arises from $(1 - \phi)$).

From Figure 5.7, it is clear that the minimax risk point is the point in $\lambda(S)$ with equal coordinates. Furthermore, this minimax point is the unique minimax point. (S can intersect Q_{α_M} solely at the corner $(\alpha_M, \alpha_M)^t$.) Since any point in $\lambda(S)$ is a Bayes risk point, it follows that the minimax rule is the Bayes rule which has constant risk. (Theorem 17 could have been used directly to show that such a rule is minimax, but we now know that the minimax risk point is unique, and more importantly that this approach will always succeed in finding a minimax rule.)

Bayes rules for this situation are very simple. Indeed, in Subsection 4.4.3 it was observed that they are of the form

$$\phi(x) = \begin{cases} 1 & \text{if } K_0\pi(\theta_1|x) > K_1\pi(\theta_0|x), \\ \gamma(x) & \text{if } K_0\pi(\theta_1|x) = K_1\pi(\theta_0|x), \\ 0 & \text{if } K_0\pi(\theta_1|x) < K_1\pi(\theta_0|x), \end{cases}$$

where $0 \leq \gamma(x) \leq 1$ is arbitrary and $\pi(\theta_i | x)$ is the posterior probability of θ_i given x. From the definition of posterior probability, it follows that ϕ can be written

$$\phi(x) = \begin{cases} 1 & \text{if } K_0\pi_1 f(x|\theta_1) > K_1\pi_0 f(x|\theta_0), \\ \gamma(x) & \text{if } K_0\pi_1 f(x|\theta_1) = K_1\pi_0 f(x|\theta_0), \\ 0 & \text{if } K_0\pi_1 f(x|\theta_1) < K_1\pi_0 f(x|\theta_0), \end{cases}$$

where π_i is the prior probability of $\theta_i (i=0, 1)$. These tests are simply the usual most powerful tests of H_0 versus H_1. The test of this form for which $K_1\alpha_0(\phi) = K_0\alpha_1(\phi)$ will be minimax.

EXAMPLE 18. Assume $X \sim \mathscr{B}(n, \theta)$ is observed, and that it is desired to test $H_0: \theta = \theta_0$ versus $H_1: \theta = \theta_1$, where $\theta_0 < \theta_1$. Note that $f(x|\theta)$ has monotone likelihood ratio (i.e., if $\theta < \theta'$, the likelihood ratio $f(x|\theta')/f(x|\theta)$ is a nondecreasing function of x). Hence the Bayes tests can be rewritten

$$\phi(x) = \begin{cases} 1 & \text{if } x > j, \\ \gamma & \text{if } x = j, \\ 0 & \text{if } x < j, \end{cases}$$

where j is some integer (depending on the K_i and π_i). For a test of this form,

$$\alpha_0(\phi) = P_{\theta_0}(X > j) + \gamma P_{\theta_0}(X = j),$$

and

$$\alpha_1(\phi) = P_{\theta_1}(X < j) + (1-\gamma) P_{\theta_1}(X = j).$$

We seek j and γ for which $K_1\alpha_0(\phi) = K_0\alpha_1(\phi)$.

As an explicit example, assume that $n = 15$, $\theta_0 = \frac{1}{4}$, $\theta_1 = \frac{1}{2}$, $K_1 = 1$, and $K_0 = 2$. A table of binomial probabilities shows that only $j = 5$ can possibly work. For this value of j,

$$K_1\alpha_0(\phi) = 0.1484 + \gamma(0.1651),$$
$$K_0\alpha_1(\phi) = 2[0.0592 + (1-\gamma)(0.0916)].$$

Setting these expressions equal to each other and solving for γ gives $\gamma \cong 0.44$. The minimax test is thus

$$\phi(x) = \begin{cases} 1 & \text{if } x > 5, \\ 0.44 & \text{if } x = 5, \\ 0 & \text{if } x < 5. \end{cases}$$

II. *Classification Problems*

It is desired to classify an observation X as belonging to one of m populations. If X belongs to the ith population, X occurs according to the density $f(x|\theta_i)$. Thus the problem is to decide among $\{\theta_1, \theta_2, \ldots, \theta_m\}$. Let a_i denote the decision that θ_i is the true state of nature, and assume that the loss for an incorrect decision is 1, while the loss for a correct decision is 0. Thus

$$L(\theta_i, a_j) = \begin{cases} 1 & \text{if } i \neq j, \\ 0 & \text{if } i = j. \end{cases} \tag{5.17}$$

For simplicity, let $\phi_i(x) = \delta^*(x, a_i)$ denote the probability of taking action a_i when x is observed, and let $\boldsymbol{\phi} = (\phi_1, \phi_2, \ldots, \phi_m)$. Clearly

$$\begin{aligned} R(\theta_i, \boldsymbol{\phi}) &= \sum_{j=1}^{m} L(\theta_i, a_j) E_{\theta_i}[\phi_j(X)] \\ &= 1 - E_{\theta_i}[\phi_i(X)]. \end{aligned} \tag{5.18}$$

Since $\mathscr{A}$ and Θ are finite, it follows from Lemma 10 that the risk set is closed and bounded. This in turn implies that there exists a minimax rule among the class of all Bayes rules. It can even be shown, as in the situation of testing simple hypotheses, that this minimax rule is unique and has constant risk. Hence we again seek a constant risk Bayes rule.

The posterior expected loss of action a_i is simply

$$E^{\pi(\theta|x)}[L(\theta, a_i)] = \sum_{j \neq i} \pi(\theta_j | x) = 1 - \pi(\theta_i | x).$$

A Bayes action is an action that minimizes this posterior expected loss, or, equivalently, one that maximizes $\pi(\theta_i | x)$. Since $\pi(\theta_i | x) = \pi_i f(x | \theta_i)/m(x)$, where π_i is the prior probability of θ_i and $m(x)$ is the marginal density of X, it is clear that $\pi(\theta_i | x)$ is maximized for those i for which $\pi_i f(x | \theta_i)$ is maximized. A Bayes rule is thus a rule for which $\phi_i(x) > 0$ only if $\pi_i f(x | \theta_i) = \max_j \pi_j f(x | \theta_j)$. Any minimax rule will be a constant risk rule of this form.

EXAMPLE 19. Assume $X \sim \mathscr{E}(\theta)$ and $\Theta = \{\theta_1, \theta_2, \ldots, \theta_m\}$, where $\theta_1 < \theta_2 < \cdots < \theta_m$. It is desired to classify X as arising from the distribution indexed by $\theta_1, \theta_2, \ldots$, or θ_m, with the loss as given in (5.17).

To find the minimax rule for this problem, we will seek a constant risk Bayes rule. In determining the form of a Bayes rule, note that for $i < j$,

$$\pi_i f(x | \theta_i) = \pi_i \theta_i^{-1} e^{-x/\theta_i} > \pi_j \theta_j^{-1} e^{-x/\theta_j} = \pi_j f(x | \theta_j) \tag{5.19}$$

if and only if

$$x < c_{ij} = \frac{\theta_i \theta_j}{\theta_j - \theta_i} \left(\log \frac{\pi_i}{\pi_j} + \log \frac{\theta_j}{\theta_i} \right). \tag{5.20}$$

On the other hand, for $j < i$,

$$\pi_i f(x | \theta_i) > \pi_j f(x | \theta_j) \quad \text{if and only if } x > c_{ji}. \tag{5.21}$$

We will show that these relationships imply the existence of constants $0 = a_0 \leq a_1 \leq \cdots \leq a_m = \infty$, such that

$$\phi_i(x) = 1 \quad \text{for } a_{i-1} < x < a_i, \qquad i = 1, \ldots, m. \tag{5.22}$$

Note first that $\pi_i f(x|\theta_i) = \pi_j f(x|\theta_j)$ only if $x = c_{ij}$. Since $\{x: x = c_{ij}, 1 \le i \le m, 1 \le j \le m\}$ is finite, it follows that, with probability one, $\phi_i(X)$ is zero or one. But (5.19), (5.20), and (5.21) imply that $\phi_i(x) = 1$ if and only if

$$\max_{1 \le j < i} \{c_{ji}\} < x < \min_{i < j \le m} \{c_{ij}\}.$$

This shows that $\phi_i(x) = 1$ for x in some interval. The fact that the intervals must be ordered according to i follows directly from (5.19) and (5.20), noting that if $i < j$, $\phi_i(x_i) = 1$, and $\phi_j(x_j) = 1$ (so that $\pi_i f(x_i|\theta_i) > \pi_j f(x_i|\theta_j)$ and $\pi_i f(x_j|\theta_i) < \pi_j f(x_j|\theta_j)$), then $x_i < c_{ij} < x_j$.

For a decision rule of the form (5.22), it is clear from (5.18) that

$$R(\theta_i, \phi) = 1 - \int_{\alpha_{i-1}}^{\alpha_i} \theta_i^{-1} \exp\left\{\frac{-x}{\theta_i}\right\} dx = 1 - \exp\left\{\frac{-\alpha_{i-1}}{\theta_i}\right\} + \exp\left\{\frac{-\alpha_i}{\theta_i}\right\}. \quad (5.23)$$

To find a constant risk rule of this form, simply set the $R(\theta_i, \boldsymbol{\phi})$ equal to each other and solve for the α_i.

As an explicit example, assume that $m = 3$, $\theta_1 = 1$, $\theta_2 = 2$, and $\theta_3 = 4$. Setting the risks in (5.23) equal to each other, results in the equations

$$e^{-\alpha_1} = 1 - e^{-\alpha_1/2} + e^{-\alpha_2/2} = 1 - e^{-\alpha_2/4}.$$

Letting $z = e^{-\alpha_2/4}$ and $y = e^{-\alpha_1/2}$, these equations become

$$y^2 = 1 - y + z^2 = 1 - z.$$

Clearly $z = 1 - y^2$, so that the equation $y^2 = 1 - y + z^2$ can be rewritten $y^2 = 1 - y + (1 - y^2)^2$, or

$$y^4 - 3y^2 - y + 2 = 0.$$

The appropriate solution to this is $y \cong 0.74$, from which it can be calculated that $z \cong 1 - (0.74)^2 = 0.45$, $\alpha_1 \cong 0.60$, and $\alpha_2 \cong 1.60$. (These can be checked to correspond to the prior with $\pi_1 = 0.23$, $\pi_2 = 0.33$, and $\pi_3 = 0.44$.) The minimax rule is thus to decide a_1 (i.e., classify X as arising from $f(x|\theta_1)$) if $0 < X < 0.6$, decide α_2 if $0.6 < X < 1.6$, and decide a_3 if $X > 1.6$.

5.4. Classes of Minimax Estimators

5.4.1. Introduction

Usually, application of the minimax approach will lead to a unique minimax decision rule. In multivariate estimation problems, however, this turns out not to be the case; large classes of minimax rules can often be constructed, raising several complex and interesting statistical issues. We first encountered the existence of alternative minimax estimators in Subsection 4.7.7

(see (4.102)), Subsection 4.7.10 (see Theorem 6), and Subsection 4.8.2 (see Example 46), where alternative minimax estimators to the usual estimator of a multivariate normal mean were given. The subject will be treated more systematically here.

The basic fact that motivates this section is that, in many multivariate estimation problems, the "standard" minimax estimator, $\boldsymbol{\delta}^0$, is inadmissible. (The first discovery of this phenomenon was in Stein (1955a); indeed Stein is recognized as the founder of this area of study, and the general phenomenon has become known as the *Stein phenomenon.*)

EXAMPLE 20. Suppose $\mathbf{X} \sim \mathcal{N}_p(\boldsymbol{\theta}, \boldsymbol{\Sigma})$, $\boldsymbol{\Sigma}$ known, and that $L(\boldsymbol{\theta}, \boldsymbol{\delta}) = (\boldsymbol{\theta} - \boldsymbol{\delta})^t \mathbf{Q}(\boldsymbol{\theta} - \boldsymbol{\delta})$, $\mathbf{Q}$ a known $(p \times p)$ positive definite matrix. The standard minimax estimator is $\boldsymbol{\delta}^0(\mathbf{x}) = \mathbf{x}$ (the least squares estimator in standard regression settings, see Subsection 4.7.10). This estimator is "standard," in the sense that it is a minimax equalizer rule (see Exercise 34). For $p = 1$ or 2, this estimator is admissible (see Section 8.9) and is indeed thus the unique minimax estimator (see Exercise 32). But $\boldsymbol{\delta}^0$ is inadmissible for $p \geq 3$, so that other minimax estimators then exist. (Indeed, Brown (1966) shows this to be the case when $p \geq 3$ for essentially any location problem.)

EXAMPLE 21. Suppose $\mathbf{X} = (X_1, \ldots, X_p)^t$, where the X_i are independent $\mathcal{P}(\theta_i)$ random variables, and that it is desired to estimate $\boldsymbol{\theta} = (\theta_1, \ldots, \theta_p)^t$ using $\boldsymbol{\delta}(\mathbf{x})$ under the loss

$$L(\boldsymbol{\theta}, \boldsymbol{\delta}) = \sum_{i=1}^{p} \theta_i^{-1}(\delta_i - \theta_i)^2. \tag{5.24}$$

Then the estimator $\boldsymbol{\delta}^0(\mathbf{x}) = \mathbf{x}$ is a minimax equalizer rule (see Exercise 36), and is admissible and the unique minimax rule when $p = 1$. But Clevenson and Zidek (1975) show that $\boldsymbol{\delta}^0$ is inadmissible if $p \geq 2$, so that other minimax estimators then exist.

When the standard estimator, $\boldsymbol{\delta}^0$, is a minimax equalizer rule with constant risk $R(\boldsymbol{\theta}, \boldsymbol{\delta}^0) \equiv R_0$, but is inadmissible, it follows that

$$\mathcal{D}_0 = \{\boldsymbol{\delta}^* \in \mathcal{D}^*: R(\boldsymbol{\theta}, \boldsymbol{\delta}^*) \leq R_0\}$$

is the class of all minimax rules for the situation. Unfortunately, it is very difficult to determine which estimators are in $\mathcal{D}_0$. In the next subsection, therefore, a relatively simple subclass of $\mathcal{D}_0$ is found that is much easier to work with. This is followed, in Subsections 5.4.3 and 5.4.4, by fairly careful treatments of Examples 20 and 21, with special emphasis placed on the basic statistical issues encountered in dealing with classes of minimax estimators.

Examples 20 and 21 were selected for discussion, here, because they are representative illustrations of the general phenomenon in a continuous and a discrete setting, respectively. Many other multivariate estimation problems have also been investigated from this viewpoint. There has been work on nonnormal continuous exponential families in general (cf. Hudson (1978), Chen (1983), Chou (1984), Johnson (1984)), and on important special cases, such as estimation of variances or more general scale parameters (cf. Berger (1980a), DasGupta (1984a, b), Ghosh, Hwang, and Tsui (1984)), and such as estimating a covariance matrix (cf. James and Stein (1960), Stein (1975), Efron and Morris (1976b), Haff (1977), Olkin and Selliah (1977), Muirhead (1982), Dey and Srinivasan (1985)). There is also a substantial body of work on minimax multivariate estimation in location problems (cf. Strawderman (1974), Berger (1975), Brandwein and Strawderman (1980), Shinozaki (1984), Sen and Saleh (1985)), and some work on estimation in Pearson curve families (cf. Haff and Johnson (1984), Johnson (1984)). The extensive literature dealing with multivariate estimation in discrete settings includes Peng (1975), Hudson (1978), Tsui (1979), Tsui and Press (1982), Hwang (1982a), Ghosh, Hwang, and Tsui (1983), and Chou (1984). Some situations in which the Stein phenomenon does *not* hold are discussed in Gutmann (1982a, b) and Cheng (1982). A general review of the entire area can be found in Berger (1986).

Classes of multivariate estimators also arise in many other statistical situations, such as in quantile estimation (cf. Zidek (1971), Rukhin and Strawderman (1982), and Rukhin (1983, 1984b, 1985a, b)).

5.4.2. The Unbiased Estimator of Risk

This subsection is moderately technical, and can be skipped by those interested only in the minimax estimators themselves. We will restrict consideration to $\boldsymbol{\delta} \in \mathscr{D}$, the class of nonrandomized rules. In situations such as Examples 20 and 21, this is a natural restriction, since the convex loss ensures that any randomized estimator can be beaten by a nonrandomized estimator. Further restriction to "nice" estimators will also be made. (The conditions defining "nice" will usually be moment and differentiability conditions.) We will let $\mathscr{D}_1$ refer, generically, to this subclass of "nice" nonrandomized estimators.

Definition 15. Suppose there exists an operator $\mathscr{R}$ on $\mathscr{D}_1$ such that, for any $\boldsymbol{\delta} \in \mathscr{D}_1$,

$$R(\boldsymbol{\theta}, \boldsymbol{\delta}) = E_{\boldsymbol{\theta}}[\mathscr{R}\boldsymbol{\delta}(\mathbf{X})]. \tag{5.25}$$

Then $\mathscr{R}\boldsymbol{\delta}(\mathbf{x})$ is called an *unbiased estimator of the risk* of $\boldsymbol{\delta}$ (for obvious reasons).

The primary use of an unbiased estimator of risk is to allow characterization of many minimax estimators (though see Diaconis and Stein (1983) and Berger (1986) for other uses). Indeed, the class

$$\tilde{\mathscr{D}}_0 = \{\boldsymbol{\delta} \in \mathscr{D}_1 \colon \mathscr{R}\boldsymbol{\delta}(\mathbf{x}) \le R_0 \text{ for all } \mathbf{x}\}$$

is a subset of $\mathscr{D}_0$ (i.e., a subclass of minimax estimators), as can be seen by simply taking expectations on both sides of the inequality. Furthermore, $\tilde{\mathscr{D}}_0$ usually contains the most important minimax estimators, in which case little is lost by replacing $\mathscr{D}_0$ by the much more tractable $\tilde{\mathscr{D}}_0$.

EXAMPLE 20 (continued). We must first specify $\mathscr{D}_1$, a class of appropriately "nice" estimators for which an unbiased estimator of risk exists. It is convenient to write

$$\boldsymbol{\delta}(\mathbf{x}) = \mathbf{x} + \boldsymbol{\Sigma}\boldsymbol{\gamma}(\mathbf{x}),$$

where $\boldsymbol{\gamma} = (\gamma_1, \ldots, \gamma_p^t)$, and then *define* $\mathscr{D}_1$ as the class of all such estimators for which the following two conditions are satisfied for all $\boldsymbol{\theta}$, i, and j:

Condition 1. Except possibly for $\mathbf{x}_{(i)} = (x_1, \ldots, x_{i-1}, x_{i+1}, \ldots, x_p)$ in a set of probability zero, $\gamma_j(\mathbf{x})$ is a continuous piecewise differentiable function of x_i, and

$$\lim_{|x_i| \to \infty} |\boldsymbol{\gamma}(\mathbf{x})| \exp\{-\tfrac{1}{2}(\mathbf{x} - \boldsymbol{\theta})^t \boldsymbol{\Sigma}^{-1}(\mathbf{x} - \boldsymbol{\theta})\} = 0.$$

Condition 2. $E_{\boldsymbol{\theta}}[|\boldsymbol{\gamma}(\mathbf{X})|^2] < \infty$ and $E_{\boldsymbol{\theta}}[|\nabla \gamma_i(\mathbf{X})|^2] < \infty$, where

$$\nabla \gamma_i(\mathbf{x}) = \left(\frac{\partial}{\partial x_1} \gamma_i(\mathbf{x}), \ldots, \frac{\partial}{\partial x_p} \gamma_i(\mathbf{x})\right)^t.$$

Under these conditions, Chen (1983) shows that, for $\boldsymbol{\delta} \in \mathscr{D}_1$,

$$\mathscr{R}\boldsymbol{\delta}(\mathbf{x}) = (\operatorname{tr} \mathbf{Q}\boldsymbol{\Sigma}) + 2 \operatorname{tr}[\mathbf{J}_{\boldsymbol{\gamma}}(\mathbf{x})\mathbf{Q}^*] + \boldsymbol{\gamma}^t(\mathbf{x})\mathbf{Q}^*\boldsymbol{\gamma}(\mathbf{x}) \tag{5.26}$$

satisfies (5.25), where $\mathbf{Q}^* = \boldsymbol{\Sigma}\mathbf{Q}\boldsymbol{\Sigma}$ and $\mathbf{J}_{\boldsymbol{\gamma}}(\mathbf{x})$ is the $(p \times p)$ matrix with (i, j) element $(\partial/\partial x_i)\gamma_j(x)$. Hence (5.26) defines an unbiased estimator of risk. Note that, in the special case when $\mathbf{Q}^* = \mathbf{I}$ (which can always be achieved without loss of generality by a linear transformation of the problem, see Exercise 52),

$$\mathscr{R}\boldsymbol{\delta}(\mathbf{x}) = \operatorname{tr}(\mathbf{Q}\boldsymbol{\Sigma}) + 2 \sum_{i=1}^{p} \frac{\partial}{\partial x_i} \gamma_i(\mathbf{x}) + |\boldsymbol{\gamma}(\mathbf{x})|^2. \tag{5.27}$$

EXAMPLE 21 (continued). Write

$$\boldsymbol{\delta}(\mathbf{x}) = \mathbf{x} + \boldsymbol{\gamma}(\mathbf{x}),$$

and *define* $\mathscr{D}_1$ as the class of all such estimators which satisfy the conditions:

Condition 1. $\gamma_i(\mathbf{x}) = 0$ whenever $x_i = 0$, $i = 1, \ldots, p$.

Condition 2. $E_\theta|\boldsymbol{\gamma}(\mathbf{x})|^2 < \infty$ for all $\boldsymbol{\theta}$.

Then it can be shown that

$$\mathscr{R}\boldsymbol{\delta}(\mathbf{x}) = p + \sum_{i=1}^{p} \left\{ \frac{1}{(x_i+1)} \gamma_i^2(\mathbf{x}+\mathbf{e}_i) + 2[\gamma_i(\mathbf{x}+\mathbf{e}_i) - \gamma_i(\mathbf{x})] \right\} \tag{5.28}$$

is an unbiased estimator of risk for $\boldsymbol{\delta} \in \mathscr{D}_1$; here $\mathbf{e}_i$ is the unit vector with a 1 in the ith coordinate.

We will use these unbiased estimators of risk to construct classes of minimax estimators in the next two subsections. The utilization of unbiased estimators of risk to prove minimaxity began with Stein (1973, 1981). Versions of the unbiased estimators of risk in (5.26) and (5.28) can, in addition, be found in Peng (1975), Hudson (1978), Tsui (1979), Berger (1976b, c, 1980a, 1982c), Hwang (1982a), and Chou (1984). Note that an unbiased estimator of risk can be used to attempt to construct estimators improving on *any* inadmissible estimator $\boldsymbol{\delta}^0$; simply try to find an estimator $\boldsymbol{\delta}$ such that $\mathscr{R}\boldsymbol{\delta}(\mathbf{x}) \le \mathscr{R}\boldsymbol{\delta}^0(\mathbf{x})$ (guaranteeing that $R(\boldsymbol{\theta}, \boldsymbol{\delta}) \le R(\boldsymbol{\theta}, \boldsymbol{\delta}^0)$). This is discussed further in Subsection 8.9.3.

5.4.3. Minimax Estimators of a Normal Mean Vector

We continue with Example 20 here, providing various minimax estimators of a multivariate normal mean. The general class of minimax estimators developed in the preceding subsection was (noting that $R_0 = \text{tr}(\mathbf{Q}\boldsymbol{\Sigma})$ is the minimax risk)

$$\tilde{\mathscr{D}}_0 = \{\boldsymbol{\delta} = \mathbf{x} + \boldsymbol{\Sigma}\boldsymbol{\gamma}(\mathbf{x}) \in \mathscr{D}_1 \colon 2\,\text{tr}[\mathbf{J}_{\boldsymbol{\gamma}}(\mathbf{x})Q^*] + \boldsymbol{\gamma}^t(\mathbf{x})\mathbf{Q}^*\boldsymbol{\gamma}(\mathbf{x}) \le 0\}. \tag{5.29}$$

(Versions of this class were given in Berger (1976) and Chen (1983).) A particularly simple subclass of $\tilde{\mathscr{D}}_0$ arises from taking $\boldsymbol{\gamma}$ to be of the form

$$\boldsymbol{\gamma}(\mathbf{x}) = -\frac{r(\|\mathbf{x}-\boldsymbol{\mu}\|^2)}{\|\mathbf{x}-\boldsymbol{\mu}\|^2} \mathbf{Q}^{*-1}(\mathbf{x}-\boldsymbol{\mu}), \tag{5.30}$$

where $\|\mathbf{x}-\boldsymbol{\mu}\|^2 = (\mathbf{x}-\boldsymbol{\mu})^t\mathbf{Q}^{*-1}(\mathbf{x}-\boldsymbol{\mu}) = (\mathbf{x}-\boldsymbol{\mu})^t\boldsymbol{\Sigma}^{-1}\mathbf{Q}^{-1}\boldsymbol{\Sigma}^{-1}(\mathbf{x}-\boldsymbol{\mu})$.

Theorem 20. *Suppose that* $\boldsymbol{\gamma}$ *is of the form* (5.30), *where r is a continuous, positive, piecewise differentiable function which is bounded by* $2(p-2)$. *Then*

$$\boldsymbol{\delta}(\mathbf{x}) = \mathbf{x} + \boldsymbol{\Sigma}\boldsymbol{\gamma}(\mathbf{x}) = \mathbf{x} - \frac{r(\|\mathbf{x}-\boldsymbol{\mu}\|^2)}{\|\mathbf{x}-\boldsymbol{\mu}\|^2}\mathbf{Q}^{-1}\boldsymbol{\Sigma}^{-1}(\mathbf{x}-\boldsymbol{\mu}) \tag{5.31}$$

is minimax.

PROOF. It is straightforward to verify Conditions 1 and 2 of Example 20 in Subsection 5.4.2, so that $\boldsymbol{\delta} \in \mathscr{D}_1$. Calculation then gives

$$\mathbf{J}_\gamma(\mathbf{x}) = \left(-\frac{r}{\|\mathbf{x}-\boldsymbol{\mu}\|^2}\right)\mathbf{Q}^{*-1}$$

$$+2\left(-\frac{r'}{\|\mathbf{x}-\boldsymbol{\mu}\|^2}+\frac{r}{\|\mathbf{x}-\boldsymbol{\mu}\|^4}\right)\mathbf{Q}^{*-1}(\mathbf{x}-\boldsymbol{\mu})(\mathbf{x}-\boldsymbol{\mu})^t\mathbf{Q}^{*-1},$$

where $r'(z) = (d/dz)r(z)$, so that

$$2\,\mathrm{tr}[\mathbf{J}_\gamma(\mathbf{x})\mathbf{Q}^*] = -\frac{2pr}{\|\mathbf{x}-\boldsymbol{\mu}\|^2}+4\left(-\frac{r'}{\|\mathbf{x}-\boldsymbol{\mu}\|^2}+\frac{r}{\|\mathbf{x}-\boldsymbol{\mu}\|^4}\right)(\mathbf{x}-\boldsymbol{\mu})^t\mathbf{Q}^{*-1}(\mathbf{x}-\boldsymbol{\mu}).$$

Also,

$$\boldsymbol{\gamma}^t(\mathbf{x})\mathbf{Q}^*\boldsymbol{\gamma}(\mathbf{x}) = \frac{r^2}{\|\mathbf{x}-\boldsymbol{\mu}\|^4}(\mathbf{x}-\boldsymbol{\mu})^t\mathbf{Q}^{*-1}(\mathbf{x}-\boldsymbol{\mu}) = \frac{r^2}{\|\mathbf{x}-\boldsymbol{\mu}\|^2},$$

so that

$$2\,\mathrm{tr}[\mathbf{J}_\gamma(\mathbf{x})\mathbf{Q}^*]+\boldsymbol{\gamma}^t\mathbf{Q}^*\boldsymbol{\gamma} = -\frac{r}{\|\mathbf{x}-\boldsymbol{\mu}\|^2}[2p-4-r]-\frac{4r'}{\|\mathbf{x}-\boldsymbol{\mu}\|^2}.$$

Since $0 \le r \le 2(p-2)$ and $r' \ge 0$, this is negative, and so $\boldsymbol{\delta} \in \tilde{\mathscr{D}}_0$. □

Among the interesting examples of minimax $\boldsymbol{\delta}$ of the form (5.31) are

$$\boldsymbol{\delta}(\mathbf{x}) = \mathbf{x} - \frac{(p-2)}{\|\mathbf{x}-\boldsymbol{\mu}\|^2}\mathbf{Q}^{-1}\boldsymbol{\Sigma}^{-1}(\mathbf{x}-\boldsymbol{\mu}),$$

which reduces to the James–Stein estimator when $\boldsymbol{\mu} = \mathbf{0}$ and $\boldsymbol{\Sigma} = \mathbf{Q} = \mathbf{I}$ (see Example 46 in Subsection 4.8.2), and

$$\boldsymbol{\delta}(\mathbf{x}) = \mathbf{x} - \min\left\{1, \frac{(p-2)}{\|\mathbf{x}-\boldsymbol{\mu}\|^2}\right\}\mathbf{Q}^{-1}\boldsymbol{\Sigma}^{-1}(\mathbf{x}-\boldsymbol{\mu}), \tag{5.32}$$

a positive part version. (These estimators were first developed in Berger (1976c) and Hudson (1974).)

The existence of large classes ($\mathscr{D}_0$ or $\tilde{\mathscr{D}}_0$) of minimax estimators in this problem raises a troubling issue: which minimax estimator should be selected for actual use? There is no satisfactory resolution to this question, unless one resorts to at least partial Bayesian reasoning (see Berger (1982a, 1986) and Berger and Berliner (1984); Smith and Campbell (1980) raise related issues involving another "class" of possible estimators, the ridge regression class). The key facts are that: (i) the admissible minimax estimators in $\mathscr{D}_0$ (or $\tilde{\mathscr{D}}_0$) will have risk functions which cross; and (ii) each such risk function will be substantially smaller than the minimax risk, $R_0 = \mathrm{tr}(\mathbf{Q}\boldsymbol{\Sigma})$, only in some relatively small region (or subspace) of $\Theta = \mathbb{R}^p$. The most sensible way of selecting a minimax estimator would, therefore, seem to be to decide, *a priori*, where $\boldsymbol{\theta}$ is felt likely to be, and then to choose a minimax estimator which performs well in the selected region. (Alternatively, one could specify a prior distribution π_0 for $\boldsymbol{\theta}$, and then seek a minimax estimator which does well with respect to π_0; see Subsection 4.7.7 for discussion of this approach.) Note that use of prior information in this fashion is essentially unavoidable; if one is determined to make no use of prior information, then one may as well just use $\boldsymbol{\delta}^0(\mathbf{x}) = \mathbf{x}$ (even though it is inadmissible), since any specific alternative minimax estimator chosen will have an essentially negligible chance of improving significantly on $\boldsymbol{\delta}^0$. Being forced (by the duo of minimaxity and admissibility) to introduce prior information may be troubling to some frequentists.

We finish this subsection with the presentation of a minimax estimator which is adapted to do well in a prespecified region of the parameter space. (More elaborate and somewhat better such estimators can be found in Chen (1983).) Although a variety of types of regions could be considered, we will suppose here that one selects an ellipse

$$\{\boldsymbol{\theta}: (\boldsymbol{\theta} - \boldsymbol{\mu})^t \mathbf{A}^{-1}(\boldsymbol{\theta} - \boldsymbol{\mu}) \le p - 0.6\}, \tag{5.33}$$

which is felt to have a 50% chance (*a priori*) of containing $\boldsymbol{\theta}$. Construction of $\boldsymbol{\mu}$ and $\mathbf{A}$ is discussed in Subsection 4.7.10, where it is also noted that $\boldsymbol{\mu}$ and $\mathbf{A}$ can be roughly interpreted as the prior mean and covariance matrix of $\boldsymbol{\theta}$, respectively. (Indeed, for a $\mathcal{N}_p(\boldsymbol{\mu}, \mathbf{A})$ prior, the probability of the ellipse in (5.33) is very nearly $\frac{1}{2}$.) Note that this is *not* the empirical Bayes situation in which $\boldsymbol{\mu}$ and $\mathbf{A}$ have a certain structure (but are otherwise unknown) and are to be estimated from the data. (Minimaxity results can be established for such situations—cf. Efron and Morris (1972a, b, 1973, 1976b), Haff (1976, 1978), Rao (1976, 1977), Morris (1977), Reinsel (1983, 1984), and Exercise 56—but space precludes extensive discussion.) Instead we are assuming that the θ_i are not related in such structural ways, but that $\boldsymbol{\mu}$ and $\mathbf{A}$ can be completely specified to reflect available knowledge about the location of $\boldsymbol{\theta}$. (It may help to think of the regression scenario, in which $\boldsymbol{\theta}$ is the vector of unknown regression coefficients; often the θ_i are highly different in character, precluding empirical Bayes types of analyses, and yet intelligent guesses, μ_i, and associated accuracies, $\sqrt{A_{ii}}$, can be specified concerning the location of the θ_i.)

For simplicity, attention will be restricted to the case where $\mathbf{Q}$, $\boldsymbol{\Sigma}$, and $\mathbf{A}$ are diagonal matrices, with diagonal elements $\{q_i\}$, $\{\sigma_i^2\}$, and $\{A_i\}$, respectively. (See Berger (1982a) for discussion of the general case.) Also, suppose that a relabelling has occurred (if necessary) so that the $q_i^* = q_i\sigma_i^4/(\sigma_i^2 + A_i)$ satisfy $q_1^* \ge q_2^* \ge \cdots \ge q_p^* > 0$ (and define $q_{p+1}^* = 0$). Finally, define $(j-2)^+$ as the positive part of $(j-2)$, and

$$\|\mathbf{x} - \boldsymbol{\mu}\|_j^2 = \sum_{l=1}^{j} \frac{(x_l - \mu_l)^2}{(\sigma_l^2 + A_l)}.$$

Theorem 21. *If $p \geq 3$, the estimator $\boldsymbol{\delta}^M$, given componentwise by*

$$\delta_i^M(\mathbf{x}) = x_i - \frac{\sigma_i^2}{(\sigma_i^2 + A_i)}(x_i - \mu_i)\left[\frac{1}{q_i^*}\sum_{j=i}^{p}(q_j^* - q_{j+1}^*)\min\left\{1, \frac{2(j-2)^+}{\|\mathbf{x}-\boldsymbol{\mu}\|_j^2}\right\}\right], \tag{5.34}$$

is a minimax estimator of $\boldsymbol{\theta}$.

Proof. The proof is given in Berger (1982a), and is based on modification of the estimators in (5.31) according to techniques discussed in Bhattacharya (1966) and Berger (1979). □

Example 22. Suppose $p = 5$, $\mathbf{Q} = \mathbf{I}$, $\boldsymbol{\Sigma} = \text{diag.}\{10, 1, 1, 1, 0.1\}$ and $\boldsymbol{\theta}$ is thought likely to lie in the ellipse with center $\boldsymbol{\mu}$ and $\mathbf{A} = \mathbf{I}$. Then $q_1^* = q_1\sigma_1^4/(\sigma_1^2 + A_1) = \frac{100}{11}$, $q_2^* = q_3^* = q_4^* = \frac{1}{2}$, and $q_5^* = \frac{1}{110}$. (Note that these are decreasingly ordered, so that no relabelling is necessary.) For simplicity of notation, define

$$r_j(\mathbf{x}) = \min\left\{1, \frac{2(j-2)^+}{\|\mathbf{x}-\boldsymbol{\mu}\|_j^2}\right\}.$$

Then

$$\delta_i^M(\mathbf{x}) = \begin{cases} x_1 - \dfrac{\sigma_1^2}{(\sigma_1^2 + A_1)}(x_1 - \mu_1)[\frac{11}{100}\{(\frac{1}{2} - \frac{1}{110})r_4(\mathbf{x}) + \frac{1}{110}r_5(\mathbf{x})\}] & \text{if } i = 1, \\ x_i - \dfrac{\sigma_i^2}{(\sigma_i^2 + A_i)}(x_i - \mu_i)[2\{(\frac{1}{2} - \frac{1}{110})r_4(\mathbf{x}) + \frac{1}{110}r_5(\mathbf{x})\}] & \text{if } 2 \leq i \leq 4, \\ x_5 - \dfrac{\sigma_5^2}{(\sigma_5^2 + A_5)}(x_5 - \mu_5)r_5(\mathbf{x}) & \text{if } i = 5. \end{cases}$$

Note that we "expect" $\|\mathbf{x}-\boldsymbol{\mu}\|_j^2$ to be about j (since, marginally, $\mathbf{X}$ would have mean $\boldsymbol{\mu}$ and covariance matrix $\boldsymbol{\Sigma} + \mathbf{A}$, in which case the expected value of $\|\mathbf{X}-\boldsymbol{\mu}\|_j^2$ is j), which is less than or equal to $2(j-2)^+$ for $j \geq 4$. Hence, we "expect" $r_4(\mathbf{x})$ and $r_5(\mathbf{x})$ to be one, in which case δ_i^M becomes

$$\delta_i^M(\mathbf{x}) = \begin{cases} x_1 - \dfrac{\sigma_1^2}{(\sigma_1^2 + A_1)}(x_1 - \mu_1)[\frac{11}{200}] = x_1 - (0.05)(x_1 - \mu_1) & \text{if } i = 1, \\ x_i - \dfrac{\sigma_i^2}{(\sigma_i^2 + A_i)}(x_i - \mu_i) = x_i - (0.5)(x_i - \mu_i) & \text{if } 2 \leq i \leq 4, \\ x_5 - \dfrac{\sigma_5^2}{(\sigma_5^2 + A_5)}(x_5 - \mu_5) = x_5 - (0.09)(x_5 - \mu_5) & \text{if } i = 5. \end{cases}$$

Thus, for $i \geq 2$, δ_i^M is expected to behave like the conjugate prior Bayes estimator (with respect to the $\mathcal{N}_p(\boldsymbol{\mu}, \mathbf{A})$ prior), which indicates that $\boldsymbol{\delta}^M$ is likely to perform well for $\boldsymbol{\theta}$ in the ellipse (5.33).

Contrast the reasonable behavior of $\boldsymbol{\delta}^M$ with that of $\boldsymbol{\delta}$ in (5.32). Noting that the "marginal" distribution of $\mathbf{X}$ has mean $\boldsymbol{\mu}$ and covariance matrix

$\boldsymbol{\Sigma}+\mathbf{A}$, calculation yields

$$\begin{aligned}E[\|\mathbf{X}-\boldsymbol{\mu}\|^2] &= E[(\mathbf{X}-\boldsymbol{\mu})'\boldsymbol{\Sigma}^{-1}\mathbf{Q}^{-1}\boldsymbol{\Sigma}^{-1}(\mathbf{X}-\boldsymbol{\mu})] \\ &= \text{tr}[\{E(\mathbf{X}-\boldsymbol{\mu})(\mathbf{X}-\boldsymbol{\mu})'\}\boldsymbol{\Sigma}^{-1}\mathbf{Q}^{-1}\boldsymbol{\Sigma}^{-1}] \\ &= \text{tr}[\{\boldsymbol{\Sigma}+\mathbf{A}\}\boldsymbol{\Sigma}^{-1}\mathbf{Q}^{-1}\boldsymbol{\Sigma}^{-1}] = 116.11.\end{aligned}$$

Hence we would "expect" that

$$\boldsymbol{\delta}(\mathbf{x}) \cong \mathbf{x} - \frac{3}{116.11}(\text{diag}\{0.1, 1, 1, 1, 10\})(\mathbf{x}-\boldsymbol{\mu}).$$

This estimator is quite unsatisfactory, with only x_5 (the coordinate with *small* variance) being shrunk by a nonnegligible amount. And note that we even "guided" $\boldsymbol{\delta}$ to the extent of specifying $\boldsymbol{\mu}$. This reinforces the necessity of carefully involving prior information (including specification of $\mathbf{A}$) in the selection of a minimax estimator.

For more formal indications (involving actual Bayes risk calculations) of the success of $\boldsymbol{\delta}^M$ in the specified region, see Berger (1982a) and Chen (1983). It should be noted, however, that the success of $\boldsymbol{\delta}^M$ is only a *comparative* success among minimax estimators. It is not the case that $\boldsymbol{\delta}^M$ will necessarily do well compared to, say, the robust Bayesian estimators discussed in Subsection 4.7.10. The reason is indicated by the behavior of δ_1^M in Example 22; the percent of shrinkage towards μ_1 is only 5%, while the conjugate prior Bayes estimator or the robust Bayes estimators suggest a shrinkage factor on the order of $\sigma_1^2/(\sigma_1^2+A_1)=\frac{10}{11}$, or 91%. It will quite generally be true (exchangeable or nearly exchangeable situations being the main exception) that the first two coordinates of $\boldsymbol{\delta}^M$ will shrink towards the respective μ_i much less than will, say, robust Bayes or empirical Bayes estimators; and $\boldsymbol{\delta}^M$ tries about as hard as is possible to shrink towards the μ_i, subject to the minimaxity constraint. At first sight, it may not seem bad that only two coordinates will be seriously "undershrunk," but, unfortunately, the first two coordinates are the *most important* ones in which to shrink. As an indication of this, observe that, in estimating θ_i, the difference in the Bayes risk (with respect to the conjugate $\mathcal{N}(\mu_i, A_i)$ prior, π_i) of the usual estimator $\delta^0(x_i)=x_i$ and the Bayes estimator $\delta^{\pi_i}(x_i)=x_i-[\sigma_i^2/(\sigma_i^2+A_i)](x_i-\mu_i)$ is

$$r(\pi_i, \delta^0)-r(\pi_i, \delta^{\pi_i}) = \frac{q_i\sigma_i^4}{(\sigma_i^2+A_i)} = q_i^*. \tag{5.35}$$

Thus it is the coordinates with large q_i^* that offer the greatest potential for improvement through utilization of the prior information, and it is the first two coordinates that have the largest q_i^*; serious undershrinking of these coordinates can thus lead to the sacrifice of most of the overall potential risk improvement available.

The unfortunate conclusion is that minimaxity is simply not generally compatible with a desire for substantial risk improvement in a specified region, or, equivalently, good Bayesian performance. If faced with a need to choose between good Bayesian performance and minimaxity, we would opt for good Bayesian performance, and use an appropriate (robust) Bayesian estimator.

It should be mentioned that the same conflict occurs in empirical or hierarchical Bayes situations. Except for the exchangeable case, empirical Bayes or hierarchical Bayes estimators will not typically be minimax, and insistence on minimaxity can lead to sacrifice of most of the possible gains.

Another advantage of using the robust, empirical, or hierarchical Bayes estimators is that they do not depend on $\mathbf{Q}$, the weighting matrix of the quadratic loss. In applications, determination of $\mathbf{Q}$ is typically very difficult, because of uncertainty as to the future uses of the results of the study. (Prior knowledge, such as $\boldsymbol{\mu}$ and $\mathbf{A}$, tends to be much more readily available.) Often, "convenient" choices of $\mathbf{Q}$, such as $\mathbf{Q}=\mathbf{I}$ or $\mathbf{Q}=\boldsymbol{\Sigma}^{-1}$, are made. While we do not seriously recommend "convenience" choices, a very interesting $\mathbf{Q}$ is given by

$$\mathbf{Q}=\boldsymbol{\Sigma}^{-1}(\boldsymbol{\Sigma}+\mathbf{A})\boldsymbol{\Sigma}^{-1}=\boldsymbol{\Sigma}^{-1}+\boldsymbol{\Sigma}^{-1}\mathbf{A}\boldsymbol{\Sigma}^{-1}.$$

The fascinating feature of this choice is that the q_i^* are then all equal, in which case $\boldsymbol{\delta}^M$ reduces to

$$\boldsymbol{\delta}^M(\mathbf{x})=\mathbf{x}-\min\left\{1, \frac{2(p-2)}{\|\mathbf{x}-\boldsymbol{\mu}\|_p^2}\right\}\boldsymbol{\Sigma}(\boldsymbol{\Sigma}+\mathbf{A})^{-1}(\mathbf{x}-\boldsymbol{\mu}). \tag{5.36}$$

This estimator can be shown to be nearly *optimal* in terms of Bayesian performance, while preserving minimaxity! (The argument is analogous to that given in Subsection 4.7.7 for the exchangeable case; see Chen (1983) and Berger (1986) for deails.) Indeed, for any $\mathbf{Q}$ for which the q_i^* are nearly equal (or, at least for which the first four or five of the q_i^* are nearly equal), it can be seen that $\boldsymbol{\delta}^M$ will be excellent from a robust Bayesian perspective. Hence, minimaxity *is compatible* with good Bayesian performance for such $\mathbf{Q}$. Unfortunately, the only natural situations in which such $\mathbf{Q}$ arise are: (i) the obvious symmetric case, where $\mathbf{Q}$, $\boldsymbol{\Sigma}$, and $\mathbf{A}$ are all multiples of the identity; and (ii) certain "prediction" cases where $\mathbf{Q}=\boldsymbol{\Sigma}^{-1}$ and $\mathbf{A}$ is proportional to $\boldsymbol{\Sigma}$ (see Berger (1986)).

We have drifted somewhat far afield, in that we started out being concerned only with selection of a minimax estimator. If minimaxity is unequivocally demanded, use of $\boldsymbol{\delta}^M$ will, at least, offer some improvement over $\boldsymbol{\delta}^0(\mathbf{x})=\mathbf{x}$. And, since $R(\boldsymbol{\theta}, \boldsymbol{\delta}^M)<R(\boldsymbol{\theta}, \boldsymbol{\delta}^0)$ for all $\boldsymbol{\theta}$, use of $\boldsymbol{\delta}^M$ would have the sociological advantage (over, say, a nonminimax robust Bayes estimator) of guaranteeing (from a frequentist perspective) that the prior information, $\boldsymbol{\mu}$ and $\mathbf{A}$, has been utilized with no possibility of harm (assuming, of course, that the loss function is generally acceptable—a rather doubtful assumption). This guarantee may be attractive to some.

Many other issues could be raised. For instance, $\boldsymbol{\delta}^M$ can be modified according to the ideas in the sections "Partial Prior Information" and

"Independent Prior Coordinates" in Subsection 4.7.10, while preserving minimaxity (cf. Berger (1982a, 1986)). These modifications are very desirable from a practical perspective, for essentially the same reasons as discussed in Subsection 4.7.10. Generalizations to unknown $\mathbf{\Sigma}$ are also of obvious interest, and a number of preliminary steps in this direction have been made (cf. Berger, Bock, Brown, Casella, and Gleser (1977), Gleser (1979), and Berger and Haff (1982)).

There is a huge literature, for this normal mean problem, on development of various alternative classes of minimax estimators, and exploration of the properties thereof. A scattering of references includes Stein (1966, 1973, 1981), Sclove (1968), Baranchik (1970), Strawderman (1971, 1978), Bock (1975, 1982), Efron and Morris (1976a), Bibby and Toutenburg (1977), Bunke (1977), Faith (1978), Zidek (1978), Judge and Bock (1978), Alam and Hawkes (1979), Takada (1979), Brown and Zidek (1980), Casella (1980), Shinozaki (1980), Arnold (1981), Vinod and Ullah (1981), Thisted (1982), Zheng (1982), Berger and Dey (1983), Berger and Wolpert (1983), Berliner (1983, 1984b), Lin and Mousa (1983), Oman (1983), Hwang (1984), and George (1985). A general review of the area can be found in Berger (1986).

5.4.4. Minimax Estimators of Poisson Means

We continue with Example 21 in this subsection, developing various minimax estimators of $\boldsymbol{\theta}$, the vector of Poisson means. The general class of minimax estimators developed in Subsection 5.4.2 was

$$\tilde{\mathscr{D}}_0=\left\{\boldsymbol{\delta}=\mathbf{x}+\boldsymbol{\gamma}(\mathbf{x})\in\mathscr{D}_1\colon \sum_{i=1}^{p}\left(\frac{1}{(x_i+1)}\gamma_i^2(\mathbf{x}+\mathbf{e}_i)+2[\gamma_i(\mathbf{x}+\mathbf{e}_i)-\gamma_i(\mathbf{x})]\right)\leq 0\right\}. \tag{5.37}$$

A particularly simple subclass of $\tilde{\mathscr{D}}_0$ arises from taking $\boldsymbol{\gamma}$ to be of the form

$$\boldsymbol{\gamma}(\mathbf{x})=-h\left(\sum_{i=1}^{p}x_i\right)\mathbf{x}, \tag{5.38}$$

where h is a real valued, bounded function. (Note that Conditions 1 and 2 in Example 21 (continued) in Subsection 5.4.2 are satisfied by any such $\boldsymbol{\gamma}$.)

Theorem 22. *If $\boldsymbol{\gamma}$ is of the form* (5.38), *and*

$$h^2(z+1)(z+p)-2[h(z+1)(z+p)-h(z)z]\leq 0 \tag{5.39}$$

for all $z\geq 0$, then

$$\boldsymbol{\delta}(\mathbf{x})=\mathbf{x}+\boldsymbol{\gamma}(\mathbf{x})=\left(1-h\left(\sum_{i=1}^{p}x_i\right)\right)\mathbf{x}$$

is a minimax estimator of $\boldsymbol{\theta}$.

PROOF. The inequality in the definition of $\tilde{\mathscr{D}}_0$ reduces to (5.39) for $\boldsymbol{\gamma}$ of this form. □

Corollary 1. *Choosing* $h(z)=c/(b+z)$ *for* $0\le c\le 2(p-1)$ *and* $b\ge (p-2+p^{-1})$, *yields the minimax estimator*

$$\boldsymbol{\delta}(\mathbf{x})=\left(1-\frac{c}{b+\sum_{i=1}^{p}x_i}\right)\mathbf{x}. \tag{5.40}$$

(Such estimators were first shown to be minimax, by a different argument, in Clevenson and Zidek (1975).)

PROOF. Verification of (5.39) will be left as an exercise. □

Several features of the analysis here deserve comment. First, note that only $p=2$ is needed to ensure inadmissibility of $\boldsymbol{\delta}^0$ (and the existence of a multitude of minimax estimators). This need for only $p=2$ seems to be the rule in multivariate estimation problems, with the $p=3$ needed for the normal mean problem being somewhat of an exception (see Berger (1980a) and Brown (1980b) for more discussion).

A second point concerns the loss function $L(\boldsymbol{\theta},\boldsymbol{\delta})=\sum_{i=1}^{p}\theta_i^{-1}(\delta_i-\theta_i)^2$. The presence of the weights θ_i^{-1} may seem somewhat strange, but they can be argued to yield a very reasonable loss (cf. Clevenson and Zidek (1975)). Results of a related nature can be obtained for $L(\boldsymbol{\theta},\boldsymbol{\delta})=(\sum_{i=1}^{p}\theta_i)^{-1}\sum_{i=1}^{p}(\delta_i-\theta_i)^2$, and even for other weighted losses (cf. Peng (1975), Tsui and Press (1982), Hwang (1982a), Ghosh, Hwang, and Tsui (1983), and Chou (1984)). The troubling feature of these results is that the classes of minimax estimators are very different for the different losses, and it is very difficult to decide which loss is most reasonable. The same phenomenon, of heavy dependence of the class of minimax estimators on the choice from among reasonable and hard to distinguish loss functions, can be found for estimating normal variances (or general scale parameters) in Berger (1980a) and DasGupta (1984a, b). Such a heavy dependence leaves the frequentist decision-theorist in the uncomfortable position of not being able to make firm recommendations concerning estimator choice.

Even for a given loss function, one can again raise the issue of selection from $\tilde{\mathscr{D}}_0$ of a particular minimax estimator for actual use. Unfortunately, the theory for such selection is much less developed in this situation, than in the situation of estimating a multivariate normal mean, in that methods of utilizing general prior information in the selection are not known.

5.5. Evaluation of the Minimax Principle

In this section, we will evaluate the minimax principle from a statistical viewpoint. Thus we will be discussing the consequences of treating nature as an intelligent opponent. Since we are simply pretending to act as if this were true, the criticisms given in Section 5.2 concerning the applicability of game theory do not apply. Instead, the concern is whether or not the minimax principle gives good results.

We will begin with a discussion of admissibility of minimax rules. Subsection 5.5.2 then considers the minimax approach from the viewpoint of rationality. In Subsection 5.5.3, the minimax approach is compared with the Bayesian approach, and it is argued that the minimax approach is successful only if it corresponds to a reasonable Bayesian approach. Subsection 5.5.4 discusses the relationship between the minimax approach and a desire for conservative action. Subsection 5.5.5 introduces a criterion called *minimax regret*, which, if used, alleviates some of the problems of the minimax approach.

5.5.1. Admissibility of Minimax Rules

Minimax rules are usually, but not always, admissible. Many of the minimax rules encountered in this chapter were also proper Bayes rules with finite Bayes risks. The results of Subsection 4.8.1 can be used to prove admissibility for such rules.

Certain admissibility results have been implicitly derived for finite Θ. Indeed if the risk set S is closed from below and bounded from below, then, from the minimax theorem and Theorems 8 and 9 it follows that an admissible minimax rule exists. Note, however, that not all minimax rules are admissible. Indeed, all points in $Q_{\alpha_M} \cap S$ are minimax, yet only those in $Q_{\alpha_M} \cap \lambda(S)$ can be both minimax and admissible. (The second graph in Figure 5.4 is an example in which $Q_{\alpha_M} \cap S$ is not equal to $Q_{\alpha_M} \cap \lambda(S)$.)

A situation in which inadmissible minimax rules abound is when all risk functions are unbounded. Then, since $\sup_\theta R(\theta, \delta^*) = \infty$ for all $\delta^* \in \mathscr{D}^*$, all rules are minimax. Of course, no one would seriously propose using the minimax principle in such a situation.

When a minimax rule corresponds to a generalized Bayes rule, as in Example 15, admissibility becomes hard to verify. A useful technique does exist, however, for proving admissibility in such situations. The technique involves consideration of sequences of proper priors, as in Subsection 5.3.2, and will be discussed in Chapter 8. Of course, we saw in Section 5.4 that even very standard generalized Bayes minimax estimators can be inadmissible.

5.5.2. Rationality and the Minimax Principle

At the beginning of the chapter, it was stated that the minimax principle violates the axioms of rational behavior which were discussed in Chapters 2 and 4. The problem is caused not only by the use of the minimax principle, but also by the use of $R(\theta, \delta^*)$ as a measure of loss. We discuss the latter point first.

In Chapter 1 it was pointed out that the use of $R(\theta, \delta^*)$ violates the Likelihood Principle (and hence the axioms of rational behavior), since it involves an average over the sample space. The risk function was, to an extent, redeemed in Chapter 4, where it was seen to be of some use in robustness studies. The fact remains, however, that the risk function is not necessarily a good measure of loss for a statistical problem. In a statistical situation, the statistician will observe X before making a decision. Using, as the loss, an average over all X seems to contradict common sense, as well as the basic ideas of game theory. (In game theory, one averages over all unknown random quantities, but not over something known.) Of course, if faced with a sequence of experiments or a problem of experimental design, averages over $\mathscr{X}$ are certainly appropriate.

The above objection to the minimax principle is somewhat philosophical. A more significant objection is that the minimax principle, by considering $\sup_\theta R(\theta, \delta^*)$, can violate the rationality principles in serious ways. As an example of this, consider Figure 5.8. The first graph in this figure contains the risk functions of two rules, δ^1 and δ^2. The rule δ^2 is clearly preferred according to the minimax principle, and would probably be preferred by any principle (providing θ is not felt to be concentrated near 1.5). Now imagine we are told that for $1<\theta<2$, we must incur an additional fixed loss of 1. The new risk functions are then those in the second graph of Figure 5.8, and δ^1 is now preferred to δ^2 according to the minimax principle. This is a clear violation of the axioms of rational behavior (in particular Axiom 3 in Section 2.2 is violated), and seems rather silly. Why should we change our minds about the best rule just because of an additional unavoidable loss that will be incurred for both rules? (To avoid nitpicking, let's assume that the utility function for the problem is linear.) This example is still rather artificial, but similar problems will be seen for more realistic situations in the next subsection.

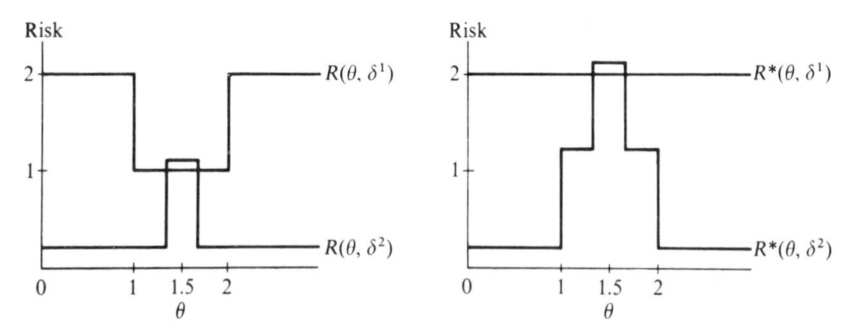

Figure 5.8

5.5.3. Comparison with the Bayesian Approach

When considered from a Bayesian viewpoint, it is clear that the minimax approach can be unreasonable. Consider the second graph in Figure 5.8 of the preceding subsection, for example. Unless the prior information is concentrated near 1.5, it would be absurd to prefer δ^1 to δ^2 (as suggested by the minimax principle). The following discrete example also illustrates this problem.

EXAMPLE 23. Assume $\Theta = \{\theta_1, \theta_2\}$, $\mathscr{A} = \{a_1, a_2\}$, and the loss matrix is

	a_1	a_2
θ_1	10	10.01
θ_2	8	-8

Since $\sup_\theta L(\theta, a_1) = 10$ and $\sup_\theta L(\theta, a_2) = 10.01$, the best nonrandomized action (according to the minimax principle) is a_1. Observe also that $L(\theta_1, a) \geq 10$ for all $a \in \mathscr{A}$. Hence $L(\theta, a_1) \leq L(\theta_1, a)$, and Theorem 2 implies that a_1 is a minimax rule. It can also be shown to be the unique minimax rule.

From a practical viewpoint, action a_1 is unappealing when compared to a_2. If θ_1 occurs, a_1 is only slightly better than a_2, while for θ_2 it is considerably worse. (Recall that θ is not being determined by an intelligent opponent, so there is no reason to think that θ_1 is certain to occur.) This point is emphasized by a Bayesian analysis, which shows that a_1 is preferred to a_2 only if the prior probability of θ_1 is greater than 0.9994. This probability seems unrealistically large, indicating that a_2 will generally be best.

The above two examples are, of course, extreme, but they indicate the problem. A minimax rule can be very bad if it does not correspond to reasonable prior beliefs. Let's look at further examples of a more realistic nature.

EXAMPLE 24. A physician diagnoses a disease as being one of three possibilities, θ_1, θ_2, or θ_3, and can prescribe medicine a_1, a_2, or a_3. Through consideration of the seriousness of each disease and the effectiveness of the various medicines, he arrives at the following loss matrix.

	a_1	a_2	a_3
θ_1	7	1	3
θ_2	0	1	6
θ_3	1	2	0

The minimax rule turns out to be

$$\delta^* = \tfrac{3}{49}\langle a_1\rangle + \tfrac{39}{49}\langle a_2\rangle + \tfrac{7}{49}\langle a_3\rangle.$$

(The proof is left as an exercise.) This does not seem unreasonable, but neither does it appear to be obviously optimal. More insight can be obtained by observing that this rule is Bayes with respect to the prior $\boldsymbol{\pi} = (\frac{7}{49}, \frac{10}{49}, \frac{32}{49})^t$ (which is least favorable). If this prior is in rough agreement with the physician's beliefs concerning the likelihood of each disease, then the minimax rule is fine. Otherwise, it may not be good.

Turning to true statistical problems, consider first the testing of simple hypotheses, as discussed in Subsection 5.3.3. It was seen there that a minimax test is always a Bayes rule with respect to certain prior probabilities of the hypotheses. If these prior probabilities correspond to true prior beliefs, or are at least close, use of the minimax rule is reasonable. Otherwise, however, use of the minimax rule seems ill advised. A similar observation can be made concerning many of the other examples in the chapter, such as Examples 16 and 19.

It can be argued that the above examples are decision problems, and that in inference problems where objectivity is desired the minimax approach will fare better. This is to an extent true, but the following examples indicate that caution should still be used.

Example 25 (Due to Herman Rubin). Assume $X \sim \mathscr{B}(n, \theta)$ is observed, where $\Theta = (0, 1]$, $\mathscr{A} = [0, 1]$, and that it is desired to estimate θ under loss

$$L(\theta, a) = \min\left\{2, \left(1 - \frac{a}{\theta}\right)^2\right\}.$$

(The loss is truncated at 2, in order to allay suspicions concerning the unrealistic nature of an unbounded loss.)

It turns out that the rule $\delta_0(x) \equiv 0$ is the unique minimax rule. We will only show here that it is the unique minimax rule among the class of all nonrandomized rules, leaving the proof for randomized rules as an exercise.

Note first that $R(\theta, \delta_0) = 1$ for all $\theta \in \Theta$. Now consider a nonrandomized rule δ which is not identically zero. Then

$$B = \{x \in \mathscr{X}: \delta(x) > 0\}$$

is nonempty. Define $c = \min_{x \in B}\{\delta(x)\}$, and note that $L(\theta, \delta(x)) = 2$ for $\theta < c/(1+\sqrt{2})$ and $x \in B$. Thus if $\theta < c/(1+\sqrt{2})$, it follows that

$$\begin{aligned} R(\theta, \delta) = E^X L(\theta, \delta(X)) &= \sum_{x \in B} (2) f(x \mid \theta) + \sum_{x \notin B} (1) f(x \mid \theta) \\ &= 1 + \sum_{x \in B} f(x \mid \theta) > 1. \end{aligned}$$

Hence no nonrandomized rule other than δ_0 has maximum risk of 1 or less, and the conclusion follows.

The rule δ_0 does not seem particularly sensible or objective. The loss function is very reasonable, yet the minimax principle says to ignore the data and estimate θ to be zero. This would only be appropriate if there was a very strong prior belief that θ was near zero.

EXAMPLE 17 (continued). This is an example in which, at first sight, the minimax rule

$$\delta_0(x) = \frac{x+\sqrt{n}/2}{n+\sqrt{n}}$$

seems very reasonable. It is enlightening, however, to investigate the rule from a Bayesian viewpoint.

We saw that δ_0 corresponds to a Bayes rule from a $\mathcal{B}e(\sqrt{n}/2, \sqrt{n}/2)$ prior. The mean of this prior is $\frac{1}{2}$, and the variance is $1/[4(1+\sqrt{n})]$. The prior mean seems natural (if one is trying to be objective), but the prior variance does not. Indeed if n is large, the use of δ_0 clearly corresponds to having strong prior beliefs that θ is concentrated near $\frac{1}{2}$. When n is small, the prior seems more reasonable. Indeed for $n = 1$, the prior is the Jeffreys noninformative prior (see Example 8 of Chapter 3), while, for $n = 4$, the prior is the uniform prior on $(0, 1)$. Both seem very reasonable. The Bayesian viewpoint thus suggests that δ_0 is attractive for small n (at least when objectivity is desired), but unattractive for large n. This conclusion is strongly supported by a visual examination of the risk function of δ_0 for various n. A calculation shows that

$$R(\theta, \delta_0) = \frac{1}{[4(1+\sqrt{n})^2]}.$$

For various n, the normalized risk functions $[nR(\theta, \delta_0)]$ are graphed in Figure 5.9. (These are, of course, the constant lines.) For purposes of comparison, the normalized risk of the usual estimator $\delta(x) = x/n$ is also graphed. This is given by

$$nR(\theta, \delta) = nE_\theta[\theta - X/n]^2 = \theta(1-\theta).$$

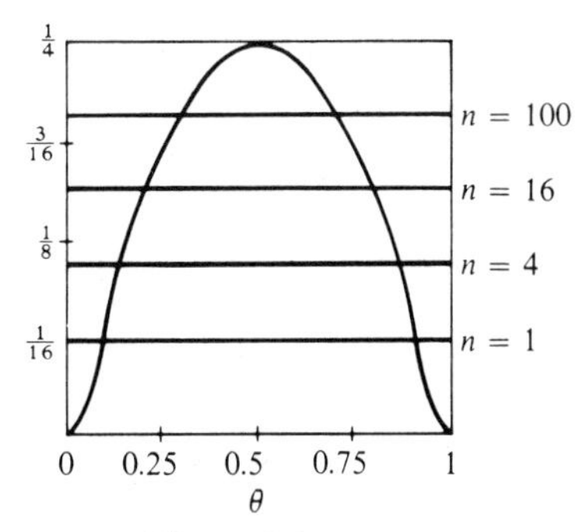

Figure 5.9

The minimax rule compares very favorably with $\delta(x) = x/n$ for small n, but rather unfavorably for large n.

The purpose of this subsection has not just been to point out inadequacies of minimax rules, but also to indicate that, to properly evaluate decision rules, Bayesian reasoning is almost indispensable. When a decision rule is derived in a non-Bayesian fashion, one should check that it corresponds to a reasonable prior distribution.

5.5.4. The Desire to Act Conservatively

The last bastion of defense for the minimax principle is the argument that someone might really want to act conservatively no matter how silly the result seems to be. The obvious counter-argument has already been mentioned, namely that conservatism will naturally be built into the utility function, and hence into the loss and risk function. Further attempts at conservatism would be overkill. The possible loophole in this argument is the need to develop the loss through a utility analysis. We saw in Chapter 2 that doing this is very hard, and a supporter of the minimax principle might argue that it is far simpler to assume, say, a linear utility (if money is involved) and to incorporate the desired conservatism through the minimax principle. We could, of course, argue against this on "rationality" grounds and give examples where it doesn't work, but instead we will hoist the minimax man by his own petard. If this is the way he desires to act, there can then be no justification in statistical problems for randomization. If several actions are available, the conservative action is obviously that with the smallest maximum loss. Randomizing among actions is taking a gamble; risking a larger than necessary loss. To put this another way, the justification for evaluating randomized rules through expected loss was utility theory. If one argues that utility theory doesn't apply, then randomized rules can no longer be evaluated through expected loss, and much of minimax theory is inapplicable. But if utility theory does apply, it is wrong to evaluate a rule by its worst possible loss or risk.

5.5.5. Minimax Regret

In many situations, the worst inadequacies of the minimax approach can be partially alleviated through use of what is called the minimax regret approach. The key idea can be seen by considering Example 23. Note that if θ_1 is the state of nature, the loss is bound to be at least 10. The minimax strategy ignores this, and tries to protect against 10.01. An obvious way to remedy the situation is to consider

$$L^*(\theta, a) = L(\theta, a) - \inf_{a \in \mathscr{A}} L(\theta, a).$$

For a given θ, this will result in evaluating actions on the basis of how they compare with the best possible action. In other words, it will be the *differences* in losses between the actions that matter, not the absolute magnitudes of the losses. The function $L^*(\theta, a)$ is called *regret loss* (since it gives the regret we have for not using the best action) and is reputed to have originated with L. J. Savage.

Regret loss can be dealt with exactly as is a usual loss. Indeed defining

$$R^*(\theta, \delta^*) = E_\theta L^*(\theta, \delta^*(X, \cdot)),$$

where for a randomized rule δ^*

$$L^*(\theta, \delta^*(x, \cdot)) = E^{\delta^*(x, \cdot)}[L^*(\theta, a)],$$

one can formally define the following decision principle.

Minimax Regret Principle. *A rule* δ_1^* *is preferred to* δ_2^* *if*

$$\sup_\theta R^*(\theta, \delta_1^*) < \sup_\theta R^*(\theta, \delta_2^*).$$

Definition 16. A *minimax regret* decision rule δ^{*MR} is a rule which is minimax for the loss L^*.

EXAMPLE 23 (continued). The regret loss matrix is easily calculated to be

	a_1	a_2
θ_1	0	0.01
θ_2	16	0

The minimax regret rule is simply the minimax rule for this loss. This can be calculated to be

$$\delta^{*MR} = \tfrac{1}{1601}\langle a_1\rangle + \tfrac{1600}{1601}\langle a_2\rangle.$$

Hence $a_2$is selected with probability nearly equal to one, exactly as intuition would suggest. Contrast this with the opposite behavior of δ^{*M} for the original loss matrix.

In many of the statistical problems we have encountered, $\inf_{a \in \mathscr{A}} L(\theta, a) = 0$. The "standard" losses of Subsection 2.4.2 all satisfy this condition, for example. In such a situation it is clear that $L^*(\theta, a) = L(\theta, a)$ and $R^*(\theta, \delta^*) = R(\theta, \delta^*)$. The minimax and minimax regret approaches then coincide. This is probably why minimax rules in standard statistical situations are usually much more reasonable than the minimax rule in Example 23.

It is interesting to note that analysis according to the Bayes risk principle is unaffected by the use of L^* instead of L. For example, defining $r(\theta) = \inf_a L(\theta, a)$, the Bayes risk of a rule δ under L^* is

$$\begin{aligned} r^*(\pi, \delta) &= E^{\pi} R^*(\theta, \delta) = E^{\pi} E_{\theta}^{X} L^*(\theta, \delta(X)) \\ &= E^{\pi} E_{\theta}^{X}[L(\theta, \delta(X)) - r(\theta)] \\ &= r(\pi, \delta) - E^{\pi}[r(\theta)]. \end{aligned}$$

Providing $E^{\pi}[r(\theta)]$ is finite, minimizing $r^*(\pi, \delta)$ (over δ) is thus equivalent to minimizing $r(\pi, \delta)$. The same result can be demonstrated for minimizing the posterior expected loss. It is thus as rational to be basing a decision on L^* as on L, and the former seems definitely more attractive for the minimax approach (at least when L is developed through utility theory).

5.5.6. Conclusions

In actually making decisions, the use of the minimax principle is definitely suspect. It may be of some use when dealing with an intelligent opponent, but the conditions needed for a strict application of the principle are quite unrealistic and restrictive. In problems of statistical decision theory, the principle can lead to very bad results, and works well only when it happens to coincide with reasonable prior information. (The minimax regret principle works somewhat better in statistical situations.)

When dealing with problems of statistical inference, the minimax principle tends to be more successful, especially when used with standard losses. The reason for this success, however, is that it will frequently correspond to an objective Bayesian approach (in which, say, a noninformative prior is used). Indeed the best method of evaluating a minimax rule is to determine the prior distribution to which it corresponds, and to observe whether or not this prior information is reasonable. Being as the minimax approach tends to be much more difficult than a Bayesian approach (using, say, a noninformative prior), the Bayesian approach seems superior. Of course, we have seen that the Bayesian approach has problems of its own, particularly that of robustness. Because of this, it is often useful to compare the risk of a Bayes rule with that of the minimax rule, in an attempt to detect unappealing features of the Bayes rule. The concern here is that, through misspecification of the prior information, a dangerously nonconservative rule could result. The minimax rule, being naturally conservative, can, in comparison, point out such a problem.

Exercises

Subsection 5.2.1

1. Prove Lemma 1.

2. Prove Lemma 2.

3. Prove that a unique minimax strategy is admissible.

4. For a two-person zero-sum game in which $\mathscr{A}$ is a closed bounded convex subset of R^m and L is convex in a for each $\theta \in \Theta$, prove that there exists an $a_0 \in \mathscr{A}$ such that

$$\sup_{\theta \in \Theta} L(\theta, a_0) = \bar{V}.$$

Subsection 5.2.2

5. Prove Theorem 1.

6. Prove Theorem 3.

7. In Example 2, find the minimax and maximin strategies and the value of the game.

8. Prove that, if an equalizer strategy is admissible, then it is minimax (or maximin).

9. Assume $\Theta = \mathscr{A} = [0, 1]$. Find minimax and maximin strategies and the value of the game for each of the following losses:
   (a) $\theta - 2\theta a + a/2$,
   (b) $(\theta - a)^2$,
   (c) $\theta^2 - \theta + a - a^2$,
   (d) $a - 2\theta a + 1$,
   (e) $|\theta - a| - (\theta - a)^2$.

10. (Blackwell and Girshick (1954).) Consider the following game: I chooses a number θ, where $0 \le \theta \le 1$. II consists of two partners II_1 and II_2. II_1 observes θ, then chooses a number z, where $0 \le z \le 1$. The number z is then told to II_2 who proceeds to choose a number a, $0 \le a \le 1$, and pays to I the amount $[z + |\theta - a|]$. II_2 and II_1 can agree, before the game, on the manner in which II_1 will choose z.
    (a) What is the strategy space for II?
    (b) Show that the game has value zero.
    (c) Find the class of all nonrandomized maximin stategies.
    (d) Show that no minimax strategy exists.

Subsection 5.2.3

In the following exercises, solving a game will mean finding a minimax strategy, a maximin strategy, and the value of the game.

11. Solve the game in Example 24 (Subsection 5.5.3).

12. Solve the game of *scissors-paper-stone*, in which each player can choose between the strategies scissors (s), paper (p), and stone (r), and the loss matrix is

	s	p	r
s	0	1	-1
p	-1	0	1
r	1	-1	0

13. Solve the game with loss matrix

	a_1	a_2	a_3
θ_1	l_1	0	0
θ_2	0	l_2	0
θ_3	0	0	l_3

14. Let $\Theta = \mathscr{A} = \{1, 2, \ldots, m\}$. Assume the loss matrix is given by

$$l_{ij} = \begin{cases} 1 & \text{if } |i-j| = 0 \quad \text{or } 1, \\ 0 & \text{otherwise.} \end{cases}$$

Solve the game.

15. (Blackwell and Girshick (1954).) Solve the following game of *hide and seek*. II can hide at location A, B, C, or D. Hiding at A is free, hiding at B or C costs 1 unit, and hiding at D costs 2 units. (The cost is to be payed to I at the end of the game.) I can choose to look for II in only one of A, B, or C. If he finds II, he will be paid an additional 3 units by II.

16. The good guys (army G) are engaged in a war with the bad guys (army B). B (player I) must go through mountain pass 1 or pass 2. G (player II) has a choice of three strategies to defend the passes: a_1—use all available men to defend pass 1; a_2—use all available men to defend pass 2; a_3—use half of the men at each pass. The loss to G is the difference between the number of G companies destroyed and the number of B companies destroyed. In defending pass 1 with all available men, G will lose 2 of their companies while destroying 10 of B's. In defending pass 2 with all available men, G will lose 2 compared with 12 for B. If G splits its force and B goes through pass 1, the losses will be 8 for B and 5 for G, while if B goes through pass 2, the losses will be 10 for B and 5 for G.
    (a) Solve the game.
    (b) A spy reports that the probability that B will choose pass 1 is 0.8. What strategy should G now use?

17. Tom and Burgess together inherit an antique pipe, valued at \$400. They agree to decide ownership by the method of sealed bids. They each write down a bid

and put it in an envelope. They open the envelopes together, and the higher bidder receives the pipe, while paying the amount of his bid to the other player. If the two bids are equal, a coin flip decides who gets the pipe, the winner receiving the pipe and paying the other the amount of the bid.

(a) Solve the game, assuming any bid from \$0 to \$400 is allowed.

(b) Solve the game (finding at least two minimax and two maximin strategies) if only bids in \$100 increments are allowed.

18. There is an interesting card game called *liar's poker* (also known by a number of less refined names). Consider the following very simplified version of the game. A large deck of cards contains only 2s, 3s, and 4s, in equal proportions. (The deck is so large that, at all stages of the game, the probability of drawing either a 2, 3, or 4 can be considered to be $\frac{1}{3}$.)

Player I draws a card. He claims that its denomination is i ($i=2$, 3, or 4). Player II can either believe I (call this action b) or call him a liar (action l). If II takes action l, and I's card is really an i, II loses 1. If II takes action l and I's card is not an i, II wins 1. If II takes action b, I gives him the drawn card. II can now either keep the card, or discard it and draw a new one from the deck. He must then claim that the card is a j, where j is an integer larger than i. I can then take action b or action l. If he says "l," he wins or loses according to whether or not II was lying. If he says "b," he then gets II's card, and can either keep it or take another, after which he must claim $k>j$. If the game has reached this stage, it is clear that $k=4$ (and $i=2$, $j=3$). Whenever a claim of $k=4$ is made, the other player cannot improve the hand, since 4 is the largest possible hand. Hence the only sensible action, and the only one we will allow, is for a player to say l if his opponent claims 4.

Find the minimax and maximin strategies for the game, and show that the value of the game is 4/9.

19. Prove that, in a finite game, the sets of maximin and minimax strategies are bounded and convex.

Subsection 5.2.4

20. Prove Theorem 7.

21. Prove Theorem 9.

22. Consider the finite game with loss matrix

	a_1	a_2	a_3	a_4	a_5
θ_1	4	5	8	2	6
θ_2	1	8	5	6	6

(a) Graph the risk set S.

(b) Find the minimax strategy and the value of the game.

(c) Find the maximin strategy, and determine the tangent line to S at the minimax point.

23. For the following S games find: (i) the value; (ii) the minimax risk point; (iii) the maximin strategy and the tangent line to S at the minimax point; and (iv) the Bayes risk point with respect to $\pi = (\frac{1}{3}, \frac{2}{3})^t$.
    (a) $S = \{\mathbf{x} \in R^2: (x_1 - 8)^2 + (x_2 - 3)^3 \le 9\}$.
    (b) $S = \{\mathbf{x} \in R^2: (x_1 - 10)^2 + (x_2 - 10)^2 \le 400\}$.
    (c) $S = \{\mathbf{x} \in R^2: x_1 > 0 \text{ and } x_2 \ge 1/(2x_1)\}$.

24. Let $K = \{\mathbf{x} \in R^2: (x_1 - 8)^2 + (x_2 - 8)^2 \le 100\}$, $Q = \{\mathbf{x} \in R^2: x_1 \ge 0 \text{ and } x_2 \ge 0\}$, and $P = \{(0, 2)^t\}$. Consider the S-game in which $S = K \cap Q - P$.
    (a) Graph the set S, and describe the set
        (i) of admissible risk points,
        (ii) of Bayes risk points,
        (iii) of Bayes risk points against $\boldsymbol{\pi}$ which have no zero coordinates.
    (b) Show that the last conclusion in Theorem 8 is false for this situation.
    (c) Use S to demonstrate that a Bayes risk point need not always exist.
    (d) Which Bayes risk points are inadmissible?

25. (a) Prove that, if S is closed and bounded, a Bayes risk point always exists.
    (b) Give an example in which S is closed and bounded from below, and yet a Bayes risk point does not exist for at least one $\boldsymbol{\pi}$.

26. Give an example of a finite game in which a nonrandomized strategy a_0 is a minimax strategy and is Bayes with respect to $\boldsymbol{\pi}$, yet $\boldsymbol{\pi}$ is not maximin.

Subsections 5.2.5 and 5.2.6

27. Prove Theorem 10 of Subsection 5.2.4.

28. (a) Prove that if S_1 and S_2 are closed disjoint convex subsets of R^m, and at least one of them is bounded, then there exists a vector $\boldsymbol{\xi} \in R^m$ such that

$$\sup_{\mathbf{s}^2 \in S_2} \boldsymbol{\xi}^t \mathbf{s}^2 < \inf_{\mathbf{s}^1 \in S_1} \boldsymbol{\xi}^t \mathbf{s}^1.$$

    (b) Find a counterexample to the above result if both sets are unbounded.

29. For the situation of Exercise 10, find ε-minimax strategies for II.

Subsection 5.3.2

30. Prove Theorem 17.

31. Prove Theorem 18.

32. Suppose that $L(\theta, a)$ is strictly convex in a for each θ. If δ_0 is an equalizer rule which is also admissible, prove that δ_0 is the unique minimax rule.

33. An IQ test score $X \sim \mathcal{N}(\theta, 100)$ is to be observed, on the basis of which it is desired to test H_0: $\theta \le 100$ versus H_1: $\theta > 100$. The loss in incorrectly concluding that $\theta \le 100$ is 3, while the loss in incorrectly deciding that $\theta > 100$ is 1. A correct decision loses zero. What is the minimax decision if $x = 90$ is observed?

34. Assume that the waiting time, X, for a bus has a $\mathcal{U}(0, \theta)$ distribution. It is desired to test H_0: $\theta \le 10$ versus H_1: $\theta > 10$. The loss in incorrectly deciding that

$\theta \leq 10$ is $(\theta - 10)^2$, while the loss in incorrectly concluding that $\theta > 10$ is 10. The loss of a correct decision is zero.

(a) If n independent observations $X_1, \ldots, X_n$ are taken, determine the minimax test. (You may assume that, from admissibility considerations, only tests which reject whan $z = \max\{x_i\} > c$ (where $c \leq 10$) need be considered, and furthermore, that any such test is admissible.)

(b) If five observations, 1, 5, 9, 3, and 4, are taken, what is the minimax decision.

35. Assume $\mathbf{X} \sim \mathcal{N}_p(\boldsymbol{\theta}, \boldsymbol{\Sigma})$, $\boldsymbol{\Sigma}$ known, and that it is desired to estimate $\boldsymbol{\theta}$ under a quadratic loss. Prove that $\boldsymbol{\delta}^0(\mathbf{x}) = \mathbf{x}$ is minimax, and that it is an equalizer rule.

36. Let $\Theta = (0, \infty)$, $\mathscr{A} = [0, \infty)$, $X \sim \mathscr{P}(\theta)$, and $L(\theta, a) = (\theta - a)^2/\theta$.
(a) Show that $\delta_0(x) = x$ is an equalizer rule.
(b) Show that δ_0 is generalized Bayes with respect to $\pi(\theta) = 1$ on Θ.
(c) Show that δ_0 is minimax.
(d) Verify minimaxity of $\boldsymbol{\delta}^0(\mathbf{x}) = \mathbf{x}$ in Example 21.

37. Let $\Theta = (0, 1)$, $\mathscr{A} = [0, 1]$, $X \sim \mathscr{B}(n, \theta)$, and $L(\theta, a) = (\theta - a)^2/[\theta(1-\theta)]$. Show that $\delta(x) = x/n$ is a minimax estimator of θ, and find the least favorable prior distribution.

38. (Ferguson (1967).) Let $\Theta = [0, 1)$, $\mathscr{A} = [0, 1]$, $X \sim \mathscr{G}e(1-\theta)$, and $L(\theta, a) = (\theta - a)^2/(1-\theta)$.
(a) Write the risk function, $R(\theta, \delta)$, of a nonrandomized estimator δ as a power series in θ.
(b) Show that the only nonrandomized equalizer rule is

$$\delta_0(i) = \begin{cases} \frac{1}{2} & \text{if } i = 0. \\ 1 & \text{if } i \geq 1. \end{cases}$$

(c) Show that a nonrandomized rule is Bayes with respect to a prior π if and only if $\delta(i) = \mu_{i+1}/\mu_i (i = 0, 1, 2, \ldots)$, where $\mu_i = E^{\pi}[\theta^i]$.
(d) Show that δ_0 is minimax. (*Hint*: Observe that the $\mu_i (i \geq 1)$ can be equal for a distribution π on $[0, 1]$ only if π is concentrated on $\theta = 0$ and $\theta = 1$. Note, however, that $\Theta = [0, 1)$ for this problem.)

39. (Ferguson (1967).) Let Θ be the set of all distributions over $[0, 1]$, let $\mathscr{A} = [0, 1]$, and let $L(\theta, a) = (\mu_1 - a)^2$, where μ_1 is the mean of the distribution θ. (Note that θ refers to the entire distribution, not just a parameter value.) Let $X_1, \ldots, X_n$ be a sample of size n from the distribution θ, and let $X = X_1 + X_2 + \cdots + X_n$.
(a) Show that $\delta_0(x) = [x + \sqrt{n}/2]/(n + \sqrt{n})$ has risk function

$$R(\theta, \delta_0) = \frac{n(\mu_2 - \mu_1 + \frac{1}{4})}{(n + \sqrt{n})^2},$$

where $\mu_2 = E_\theta[X^2]$.
(b) Show that $(\mu_2 - \mu_1)$ (and hence $R(\theta, \delta_0)$) is maximized if and only if θ is concentrated on zero and one.
(c) Prove that δ_0 is minimax.

40. Assume $X \sim \mathscr{B}(1, \theta)$, and that it is desired to estimate θ under loss $L(\theta, a) = |\theta - a|$. Find the minimax rule, the least favorable prior distribution, and the value of the game.

41. (Blackwell and Girshick (1954).) Let $X_1, \ldots, X_n$ be a sample from a $\mathcal{N}(\theta, \sigma^2)$ distribution. It is desired to test

$$H_0\colon \sigma^2 = \sigma_0^2, \qquad -\infty < \theta < \infty,$$

versus

$$H_1\colon \sigma^2 = \sigma_1^2, \qquad \theta = \theta_1 \quad (\text{where } \sigma_0^2 < \sigma_1^2).$$

The loss is "0-K_i" loss.

(a) Show the following:

(i) There exists a minimax rule of the form: accept H_0 when $\sum_{i=1}^n (x_i - \bar{x})^2 \leq k$ and accept H_1 when $\sum_{i=1}^n (x_i - \bar{x})^2 > k$.

(ii) The following is a least favorable prior distribution: give the point (θ_1, σ_1^2) prior probability ξ; and give the set $\Omega_0 = \{(\theta, \sigma_0^2)\colon \theta \in R^1\}$ prior probability $1 - \xi$, where, furthermore, the density of θ in Ω_0 is chosen to be $\pi_1(\theta) = [n/2\pi(\sigma_1^2 - \sigma_0^2)]^{1/2} \exp\{-n(\theta - \theta_1)^2/[2(\sigma_1^2 - \sigma_0^2)]\}$.

(*Hint*: Show that, for any k, there exists a value of ξ for which the proposed decision rule is Bayes against the above prior, and that there exists a value of ξ, and hence a value of k, which makes the maximum risk under H_0 and H_1 equal. Note that, if nature employs the above prior, then the marginal distributions of $\bar{X}$ under H_0 and H_1 (i.e., with respect to the conditional priors on Ω_0 and (θ_1, σ_1^2) separately) are the same. This makes $\bar{X}$ useless for discriminating between H_0 and H_1.)

(b) Given an equation from which k can be determined.

Subsection 5.3.3

42. Give an example, for finite Θ, in which the set of loss points W is closed but not bounded, and in which the risk set S is not closed. (*Hint*: Find a set W which is closed but unbounded, for which the convex hull of W is not closed.)

43. For the following situations involving the testing of simple hypotheses, sketch the risk set S, and find the minimax rule and the least favorable prior distribution. Assume the loss is "0–1" loss.

(a) $H_0\colon X \sim \mathcal{U}(0, 1)$ versus $H_1\colon X \sim \mathcal{U}(\frac{1}{2}, \frac{3}{2})$.

(b) $H_0\colon X \sim \mathcal{B}(2, \frac{1}{2})$ versus $H_1\colon X \sim \mathcal{B}(2, \frac{2}{3})$.

(c) $H_0\colon X \sim \mathcal{G}(1, 1)$ versus $H_1\colon X \sim \mathcal{G}(1, 2)$.

44. Assume $X \sim \mathcal{B}(10, \theta)$ and that it is desired to test $H_0\colon \theta = 0.4$ versus $H_1\colon \theta = 0.6$ under "0–1" loss. Obtain the minimax procedure and compute the least favorable prior distribution.

45. Assume X has density

$$f(x|\theta) = 2^{-(x+\theta)}, \quad \text{for } x = 1-\theta, 2-\theta, \ldots.$$

It is desired to test $H_0\colon \theta = 0$ versus $H_1\colon \theta = 1$ under "0–1" loss.

(a) Sketch the risk set S.

(b) Find a minimax decision rule.

(c) Find a least favorable prior distribution.

(d) Find a nonrandomized minimax decision rule.

46. Let $(X_1, \ldots, X_{100})$ be a sample from a $\mathcal{N}(\theta, 25)$ distribution. It is desired to test H_0: $\theta = 0$ versus H_1: $\theta = 2$ under "0-K_i" loss, where $K_0 = 10$ and $K_1 = 25$. Obtain the minimax procedure and compute the least favorable prior distribution.

47. We are given two coins and are told that one of them is fair but that the other is biased, having a *known* probability $p > \frac{1}{2}$ of falling heads when flipped. The problem is to decide which coin is biased on the basis of n tosses of each coin. Assuming "0–1" loss, determine the minimax procedure.

48. In Exercise 47, assume that we are allowed a third possible action, namely deciding that the experiment is inconclusive. If the loss for this decision is l (where $l < \frac{1}{2}$), find a minimax procedure.

49. Let $X \sim \mathcal{P}(\theta)$, and assume that it is desired to test H_0: $\theta = 1$ versus H_1: $\theta = 2$ under "0–1" loss. Find the form of the Bayes tests for this problem. Using these Bayes tests, determine an adequate number of points in $\lambda(S)$ (the lower boundary of the risk set) and sketch this lower boundary. Find the minimax test and the least favorable prior distribution.

50. Let $X \sim \mathcal{P}(\theta)$, where $\Theta = \{1, 2\}$, $\mathcal{A} = \{a_1, a_2, a_3\}$, and the loss matrix is

		a_1	a_2	a_3
θ	1	0	20	10
	2	50	0	20

(a) Show that the Bayes rules are of the following form: decide a_1 if $x < k - (\log 3)/\log 2$, decide a_3 if $k - (\log 3)/(\log 2) < x < k - 1$, and decide a_2 if $x > k - 1$. (On the boundaries of these regions, randomization is, of course, allowed.)

(b) Sketch $\lambda(S)$ (the lower boundary of the risk set), by finding an adequate number of Bayes risk points. (It suffices to look at $k = \frac{1}{2}i + \frac{1}{3}$ for various integers i. This choice of k eliminates the need to worry about the boundaries of the acceptance regions.)

(c) Find the minimax rule and the least favorable prior distribution.

51. Assume that X is an observation from the density

$$f(x \mid \theta) = e^{-(x-\theta)} I_{(\theta,\infty)}(x),$$

and that the parameter space is $\Theta = \{1, 2, 3\}$. It is desired to classify X as arising from $f(x \mid 1)$, $f(x \mid 2)$, or $f(x \mid 3)$, under a "0–1" loss (zero for the correct decision, one for an incorrect decision).

(a) Find the form of the Bayes rules for this problem.

(b) Find the minimax rule and the least favorable prior distribution.

Subsection 5.4.2

52. Suppose that $\mathbf{X} \sim \mathcal{N}_p(\boldsymbol{\theta}, \mathbf{\Sigma})$ and $L(\boldsymbol{\theta}, \boldsymbol{\delta}) = (\boldsymbol{\theta} - \boldsymbol{\delta})^t \mathbf{Q}(\boldsymbol{\theta} - \boldsymbol{\delta})$, where $\mathbf{\Sigma}$ and $\mathbf{Q}$ are known $(p \times p)$ positive definite matrices. Consider the linearly transformed problem defined by $\mathbf{X}^* = \mathbf{B}\mathbf{X}$, $\boldsymbol{\theta}^* = \mathbf{B}\boldsymbol{\theta}$, $\boldsymbol{\delta}^* = \mathbf{B}\boldsymbol{\delta}$, and $L^*(\boldsymbol{\theta}^*, \boldsymbol{\delta}^*) =$

$(\boldsymbol{\theta}^* - \boldsymbol{\delta}^*)^t\mathbf{Q}^*(\boldsymbol{\theta}^* - \boldsymbol{\delta}^*)$, where $\mathbf{B}$ is a $(p \times p)$ nonsingular matrix and $\mathbf{Q}^* = (\mathbf{B}^t)^{-1}\mathbf{Q}\mathbf{B}^{-1}$. Show that:

(a) If $\boldsymbol{\delta}^1$ is R-better than $\boldsymbol{\delta}^2$, then $\boldsymbol{\delta}^{1*}$ is R-better than $\boldsymbol{\delta}^{2*}$ in the transformed problem.

(b) If $\boldsymbol{\delta}$ is minimax, then $\boldsymbol{\delta}^*$ is minimax in the transformed problem.

53. Verify that (5.26) is an unbiased estimator of risk:

(a) In the special case where $\mathbf{Q} = \boldsymbol{\Sigma} = \mathbf{I}$.

(b) In the general case. (*Hint*: First make a linear transformation, so that $\boldsymbol{\Sigma}^* = \mathbf{I}$. Then expand the risk as

$$R(\boldsymbol{\theta}^*, \boldsymbol{\delta}^*) = E_{\boldsymbol{\theta}^*}\{(\mathbf{X}^* - \boldsymbol{\theta}^*)^t\mathbf{Q}^*(\mathbf{X}^* - \boldsymbol{\theta}^*) + 2\boldsymbol{\gamma}^{*t}(\mathbf{X}^*)\mathbf{Q}^*(\mathbf{X}^* - \boldsymbol{\theta}^*)$$
$$+ \boldsymbol{\gamma}^*(\mathbf{X}^*)^t\mathbf{Q}^*\boldsymbol{\gamma}^*(X^*)\}.$$

Finally, verify and use the fact that

$$\int \gamma_j^*(\mathbf{x}^*)(x_i^* - \theta_i^*)e^{-\frac{1}{2}(x_i^* - \theta_i^*)^2}\,dx_i^* = \int \left[\frac{\partial}{\partial x_i^*}\gamma_j^*(\mathbf{x}^*)\right]e^{-\frac{1}{2}(x_i^* - \theta_i^*)^2}\,dx_i^*.)$$

54. Verify that (5.28) is an unbiased estimator of risk. (*Hint*: Follow the hint in Exercise 53, but replace the integration by parts with an appropriate summation by parts.)

Subsections 5.4.3 and 5.4.4

55. Verify that the estimator in (5.32) is minimax.

56. Suppose that $\mathbf{X} \sim \mathcal{N}_p(\boldsymbol{\theta}, \sigma_f^2\mathbf{I})$, σ_f^2 known and $p \geq 4$, and that $L(\boldsymbol{\theta}, \boldsymbol{\delta}) = |\boldsymbol{\theta} - \boldsymbol{\delta}|^2$ (i.e., $\mathbf{Q} = \mathbf{I}$).

(a) Show that the empirical Bayes estimator defined in (4.33) through (4.35), namely

$$\boldsymbol{\delta}(\mathbf{x}) = \mathbf{x} - \min\left\{\frac{p-3}{p-1}, \frac{(p-3)\sigma_f^2}{\sum_{i=1}^p (x_i - \bar{x})^2}\right\}(\mathbf{x} - \bar{x}\mathbf{1})$$

is minimax.

(b) Show that the hierarchical Bayes estimator defined in (4.60) and (4.63), namely

$$\boldsymbol{\delta}(\mathbf{x}) = \mathbf{x} - \frac{(p-3)\sigma_f^2}{\sum (x_i - \bar{x})^2}\left[1 - H_{p-3}\left(\frac{\sum (x_i - \bar{x})^2}{2\sigma_f^2}\right)\right](\mathbf{x} - \bar{x}\mathbf{1})$$

is minimax. (*Hint*: Note, from Appendix 2, that H_{p-3} is a decreasing function.)

57. Suppose $p = 4$, $\boldsymbol{\Sigma} = \text{diag}\{16, 8, 4, 2\}$, $\mathbf{Q} = \mathbf{I}$, $\boldsymbol{\mu} = \mathbf{0}$, $\mathbf{A} = \text{diag}\{1, 24, 1, 1\}$, and that $\mathbf{x} = (3, -12, 1, -1)^t$ is observed.

(a) Calculate $\boldsymbol{\delta}^M$. (Be careful about the order of the indices.)

(b) Calculate $\boldsymbol{\delta}$ in (5.32).

(c) Discuss the shrinkage behavior of $\boldsymbol{\delta}^M$ and $\boldsymbol{\delta}$.

(d) Suppose that $\mathbf{Q}$ were $\boldsymbol{\Sigma}^{-1}(\boldsymbol{\Sigma} + \mathbf{A})\boldsymbol{\Sigma}^{-1}$. Calculate and discuss the shrinkage of $\boldsymbol{\delta}^M$ in this case.

58. Verify Corollary 1.

Section 5.5

59. For the situation of Example 25, show that $\delta_0(x) \equiv 0$ is the unique minimax rule (among the class $\mathscr{D}^*$ of all randomized rules).

60. In Exercise 65 of Chapter 4, assume that no data, x, is available.
    (a) Using the given loss matrix, find the minimax rule for blood classification, and find the least favorable prior distribution.
    (*Hint*: Consider the subgame involving only the choices A, B, and O.)
    (b) Find the Bayes rule (for this no-data situation), and evaluate the performance of the minimax rule.

61. Discuss whether or not the minimax rule in the situation of Exercise 38 is reasonable from a Bayesian viewpoint.

62. Discuss whether or not the minimax rule in the situation of Exercise 40 is reasonable from a Bayesian viewpoint.

63. Assume that an S game has risk set
$$S = \{\mathbf{x} \in R^2: (x_1 - 10)^2 + (x_2 - 1)^2 \le 4\}.$$
    (a) Find the minimax strategy.
    (b) Convert S into S^*, the corresponding risk set if regret loss is used. Find the minimax regret strategy.

64. Prove that, if $\Theta = \{\theta_1, \theta_2\}$ and the risk set S is bounded from below and closed from below, then the minimax regret rule is an equalizer rule and is unique.

CHAPTER 6
Invariance

6.1. Introduction

The invariance principle is an intuitively appealing decision principle which is frequently used, even in classical statistics. It is interesting not only in its own right, but also because of its strong relationship with several other proposed approaches to statistics, including the *fiducial inference* of Fisher (1935), the *structural inference* of Fraser (1968, 1979), and the use of noninformative priors. Unfortunately, space precludes discussion of fiducial inference and structural inference. Many of the key ideas in these approaches will, however, be brought out in the discussion of invariance and its relationship to the use of noninformative priors. The basic idea of invariance is best conveyed through an example.

EXAMPLE 1. It is known that X, the decay time of a certain atomic particle, is exponentially distributed with density

$$f(x|\theta)=\theta^{-1}\exp\left\{\frac{-x}{\theta}\right\}I_{(0,\infty)}(x),$$

where $\theta>0$ is unknown. It is desired to estimate θ on the basis of one observation X, under loss $L(\theta, a)=(1-a/\theta)^2$. Imagine that X is measured in terms of seconds, and that a certain decision rule, $\delta_0(x)$, is proposed.

Consider now the decision problem that would result from measuring the decay time in minutes, instead of seconds. The observation would then be $Y=X/60$. Defining $\eta=\theta/60$, it is easy to check that the density of Y is

$$f(y|\eta)=\eta^{-1}\exp\left\{\frac{-y}{\eta}\right\}I_{(0,\infty)}(y),$$

where $\eta > 0$. If actions (to be denoted a^*) in this new problem are also expressed in terms of minutes, so that $a^* = a/60$, then it is clear that

$$L(\theta, a) = \left(1 - \frac{a}{\theta}\right)^2 = \left(1 - \frac{[a/60]}{[\theta/60]}\right)^2 = \left(1 - \frac{a^*}{\eta}\right)^2 = L(\eta, a^*).$$

It follows that the formal structure of the problem in terms of minutes (i.e., the class of densities, the parameter space, and the loss) is exactly the same as the structure of the problem in terms of seconds. It thus seems reasonable to use the same decision rule for the two formulations. Letting $\delta_0^*(y)$ denote the proposed decision rule for the transformed problem, this means that $\delta_0^*(y)$ should equal $\delta_0(y)$.

From another viewpoint, it seems reasonable to insist that the same decision be made no matter what unit of measurement is used. This implies that $\delta_0^*(y)$ should satisfy

$$\delta_0^*(y) = \frac{\delta_0(x)}{60}.$$

Combining this with the earlier conclusion results in the relationship

$$\delta_0(x) = 60\delta_0^*(y) = 60\delta_0(y) = 60\delta_0\left(\frac{x}{60}\right).$$

The above reasoning holds for any transformation of the form $Y = cX$, where $c > 0$. It follows that δ_0 should satisfy

$$\delta_0(x) = c^{-1}\delta_0(cx) \tag{6.1}$$

for all $c > 0$. The functional equation in (6.1) can be easily solved by setting $c = 1/x$, the result being

$$\delta_0(x) = x\delta_0(1).$$

The only decision rules consistent with the above intuitive reasoning are thus rules of the form $\delta_0(x) = Kx$, where K is a positive constant. Such rules are said to be *invariant* (for the given problem of course), and the invariance principle states that one should only use invariant rules. It will indeed be seen that there is a best invariant rule for this problem (i.e., a choice of K which minimizes the risk), so that the invariance principle completely determines the rule to be used.

In understanding the invariance principle, it is useful to consider the following variation of the above example.

Example 2. Consider the situation of Example 1, but now assume that theoretical considerations imply that θ must be at least 120 seconds. The problem in terms of seconds then has the parameter space $\Theta = (120, \infty)$, while the corresponding problem in terms of minutes has the parameter space $(2, \infty)$. The structures of the original and transformed problems thus differ, and there is no reason to think that $\delta_0^*(y)$ should equal $\delta_0(y)$. It is still reasonable to expect that

$$\delta_0(x) = 60\delta_0^*(y) = 60\delta_0^*\left(\frac{x}{60}\right)$$

(the action taken should not depend on the unit of measurement), but, being unable to conclude that $\delta_0^*(y) = \delta_0(y)$, no further progress can be made.

Examples 1 and 2 clearly delineate the two intuitive arguments upon which the invariance approach is based. These two arguments can be paraphrased as follows:

Principle of Rational Invariance. *The action taken in a decision problem should not depend on the unit of measurement used, or other such arbitrarily chosen incidentals.*

Invariance Principle. *If two decision problems have the same formal structure (in terms of $\mathscr{X}, \Theta, f(x|\theta)$, and L), then the same decision rule should be used in each problem.*

The relationship $\delta_0^*(y) = \delta_0(y)$, in Example 1, follows from the invariance principle. The principle of rational invariance, on the other hand, implies that, in both Examples 1 and 2, the relationship $\delta_0(x) = 60\delta_0^*(y)$ should hold.

The principle of rational invariance is so intuitively sensible that it merits little discussion. The invariance principle, on the other hand, though at first sight appealing, is in actuality rather unreasonable. This is because it ignores what we have seen to be the other crucial component of a decision problem, the prior information π. The invariance principle would be sensible if the prior, π, were included in the list of the formal structure of the problem, but it is not customary to do so. It should be kept in mind, therefore, that application of the invariance principle will only be sound for priors which are naturally "invariant." In Example 1, for instance, it is easy to see that a prior density, $\pi(\theta)$, in the original problem, transforms into

$$\pi^*(\eta) = c^{-1}\pi\left(\frac{\eta}{c}\right)$$

under the transformation $\eta = c\theta$. The original and transformed decision problems thus really have the same structure only if $\pi^*(\eta) = \pi(\eta)$, which, by the reasoning used for δ_0 in Example 1, implies that $\pi(\theta) = K/\theta$ for some positive constant K. Note that this prior density is the noninformative prior density discussed in Subsection 3.3.2 (since θ is a scale parameter). Hence the indication is that the invariance principle is suitable only when no prior information is available, and indeed that analysis by invariance will correspond to Bayesian analysis with a noninformative prior. Note that the inapplicability of the invariance principle in Example 2 can be reinterpreted in this light.

If invariance analysis is equivalent to the use of noninformative priors, a natural question is—Why should we bother with a chapter on invariance? There are basically three reasons. First, people who don't like to talk about noninformative priors are welcome to do the same thing using invariance. Second, it is not strictly true that the two approaches always correspond, although when they don't, invariance is probably suspect. Third, and most importantly, it was seen in Section 3.3 that there are many possible choices for a noninformative prior. For example, the argument above, that an invariant prior in Example 1 should be $\pi(\theta)=K/\theta$, is not completely sound, since the resulting prior is improper. Indeed, as in Section 3.3, it seems that all one can logically conclude is that π, when improper, need only be "relatively invariant." (See Section 3.3 for the meaning of this.) It is unnecessary to examine this concept in detail, since it will be seen (in Section 6.6) that invariance suggests one particular noninformative prior for use, namely that which is called the right invariant Haar measure on the group of transformations.

A proper understanding of invariance can be obtained only through the study of groups of transformations of a problem. No knowledge of group theory will be assumed in this chapter, but the reader with no previous exposure to groups should be warned that the chapter will not be light reading. More thorough and deeper study of invariance can be found in Eaton (1983).

6.2. Formulation

In this section, the notation and structure needed to apply invariance are discussed.

6.2.1. Groups of Transformations

As in Example 1, the important concept will be that of transformations of the problem. For the moment, let $\mathscr{X}$ denote an arbitrary space (assumed to be a subset of R^n), and consider transformations of $\mathscr{X}$ into itself. We will be concerned only with transformations that are one-to-one and onto. (A transformation g is *one-to-one* if $g(x_1)=g(x_2)\Rightarrow x_1=x_2$, and it is *onto* if the range of g is all of $\mathscr{X}$.)

If g_1 and g_2 are two transformations, it will be important to consider the *composition* of g_2 and g_1, which is the transformation, to be denoted g_2g_1, which is defined by

$$g_2g_1(x)=g_2(g_1(x)).$$

We are now ready to define a group of transformations.

Definition 1. A *group of transformations* of $\mathscr{X}$, to be denoted $\mathscr{G}$, is a set of (measurable) one-to-one and onto transformations of $\mathscr{X}$ into itself, which satisfies the following conditions:

(i) If $g_1 \in \mathscr{G}$ and $g_2 \in \mathscr{G}$, then $g_2 g_1 \in \mathscr{G}$.
(ii) If $g \in \mathscr{G}$, then g^{-1}, the inverse transformation defined by the relation $g^{-1}(g(x)) = x$, is in $\mathscr{G}$.
(iii) The identity transformation e, defined by $e(x) = x$, is in $\mathscr{G}$. (This actually follows from (i) and (ii) if $\mathscr{G}$ is nonempty.)

EXAMPLE 3. Let $\mathscr{X} = R^1$ or $\mathscr{X} = (0, \infty)$, and consider the group of transformations $\mathscr{G} = \{g_c\colon c > 0\}$, where $g_c(x) = cx$. This will be called the *multiplicative group* or the group of *scale transformations.* Clearly the functions g_c are one-to-one and onto. Note that

$$g_{c_2} g_{c_1}(x) = g_{c_2}(g_{c_1}(x)) = g_{c_2}(c_1 x) = c_2 c_1 x = g_{c_2 c_1}(x),$$

so that

$$g_{c_2} g_{c_1} = g_{c_2 c_1} \in \mathscr{G}.$$

It can similarly be checked that $g_c^{-1} = g_{c^{-1}} \in \mathscr{G}$ and $e = g_1 \in \mathscr{G}$, so that $\mathscr{G}$ is indeed a group of transformations. Note that this was the group considered in Example 1.

EXAMPLE 4. Let $\mathscr{X} = R^1$, and consider the group of transformations $\mathscr{G} = \{g_c\colon c \in R^1\}$, where $g_c(x) = x + c$. Clearly the functions g_c are one-to-one and onto, $g_{c_2} g_{c_1} = g_{(c_2 + c_1)} \in \mathscr{G}$, $g_c^{-1} = g_{(-c)} \in \mathscr{G}$, and $e = g_0 \in \mathscr{G}$. This group will be called the *additive group* or *location group* (on R^1).

EXAMPLE 5. Let $\mathscr{X} = R^1$, and consider the group of transformations

$$\mathscr{G} = \{g_{b,c}\colon -\infty < b < \infty, 0 < c < \infty\},$$

where $g_{b,c}(x) = cx + b$. Clearly

$$g_{b_2,c_2} g_{b_1,c_1}(x) = g_{b_2,c_2}(c_1 x + b_1) = c_2(c_1 x + b_1) + b_2 = g_{(c_2 b_1 + b_2), c_2 c_1}(x),$$

so that $g_{b_2,c_2} g_{b_1,c_1} = g_{(c_2 b_1 + b_2), c_2 c_1} \in \mathscr{G}$. It can similarly be checked that $g_{b,c}^{-1} = g_{-b/c, 1/c} \in \mathscr{G}$, and that $e = g_{0,1} \in \mathscr{G}$. This group will be called the *affine group* (on R^1).

EXAMPLE 6. Let $\mathbf{x} = (x_1, x_2, \ldots, x_n)$, where each $x_i \in \mathscr{X}_0 \subset R^1$. (Formally, $\mathscr{X} = \mathscr{X}_0 \times \mathscr{X}_0 \times \cdots \times \mathscr{X}_0$.) Consider the group of transformations $\mathscr{G} = \{g_{\mathbf{i}}\colon \mathbf{i} = (i_1, i_2, \ldots, i_n)$ is some ordering of the integers $(1, 2, \ldots, n)\}$, where $g_{\mathbf{i}}(\mathbf{x}) = (x_{i_1}, x_{i_2}, \ldots, x_{i_n})$. It will be left as an exercise to check that this is indeed a group of transformations. This group is called the *permutation group.*

6.2.2. Invariant Decision Problems

In this subsection, we formally define the concept that a group of transformations can result in transformed problems of equivalent structure. As usual, X will denote a random variable (or vector) with sample space $\mathscr{X}$ and density $f(x|\theta)$. Also, $\mathscr{F}$ will denote the class of all densities $f(x|\theta)$ for $\theta \in \Theta$. If $\mathscr{G}$ is a group of transformations of $\mathscr{X}$ (which we will also call a group of transformations of X), we want to consider the problems based on observation of the random variables $g(X)$.

Definition 2. The family of densities $\mathscr{F}$ is said to be *invariant under the group* $\mathscr{G}$ if, for every $g \in \mathscr{G}$ and $\theta \in \Theta$, there exists a unique $\theta^* \in \Theta$ such that $Y = g(X)$ has density $f(y|\theta^*)$. In such a situation, θ^* will be denoted $\bar{g}(\theta)$.

The following equations (for $A \subset \mathscr{X}$ and integrable function h) follow immediately from Definition 2, and will be frequently used:

$$P_\theta(g(X) \in A) = P_{\bar{g}(\theta)}(X \in A), \tag{6.2}$$

and

$$E_\theta[h(g(X))] = E_{\bar{g}(\theta)}[h(X)]. \tag{6.3}$$

For a given g, the transformation $\theta \to \bar{g}(\theta)$ is a transformation of Θ into itself. It will be left as an exercise to show that, if $\mathscr{F}$ is invariant under $\mathscr{G}$, then

$$\bar{\mathscr{G}} = \{\bar{g} \colon g \in \mathscr{G}\}$$

is a group of transformations of Θ. We will henceforth use this fact.

To be invariant, a decision problem must also have a loss function which is unchanged by the relevant transformations. To avoid complications, we will assume that no two actions have identical loss for all θ.

Definition 3. If $\mathscr{F}$ is invariant under the group $\mathscr{G}$, a loss function $L(\theta, a)$ is said to be *invariant under* $\mathscr{G}$ if, for every $g \in \mathscr{G}$ and $a \in \mathscr{A}$, there exists an $a^* \in \mathscr{A}$ such that $L(\theta, a) = L(\bar{g}(\theta), a^*)$ for all $\theta \in \Theta$. In such a situation, the action a^* will be denoted $\tilde{g}(a)$, and the decision problem will itself be said to be *invariant under* $\mathscr{G}$.

The action a^*, in the above definition, can be shown to be unique, because of the assumption that no two actions have identical loss and because of the fact that $\bar{g}$ is onto. It is also true that

$$\tilde{\mathscr{G}} = \{\tilde{g} \colon g \in \mathscr{G}\}$$

is a group of transformations of $\mathscr{A}$ into itself. The proof of this will be left as an exercise.

Instead of beginning with the group $\mathscr{G}$ of transformations on $\mathscr{X}$ and deriving from it the groups $\bar{\mathscr{G}}$ and $\tilde{\mathscr{G}}$, we could have just started with one big composite group with vector elements of the form (g_1, g_2, g_3), where g_1, g_2, and g_3 operate on $\mathscr{X}$, Θ, and $\mathscr{A}$, respectively. This has a certain theoretical advantage in manipulation and is slightly more general, but is also somewhat obscure at first sight. We, therefore, opted for the $\mathscr{G}$, $\bar{\mathscr{G}}$, $\tilde{\mathscr{G}}$ approach. One will encounter both sets of notations in the literature.

EXAMPLE 1 (continued). Consider the group of scale transformations defined in Example 3. Clearly $Y = g_c(X) = cX$ has density

$$c^{-1}f\left(\frac{y}{c}\,|\,\theta\right) = (c\theta)^{-1}\exp\left\{\frac{-y}{c\theta}\right\} I_{(0,\infty)}(y) = f(y|\,c\theta).$$

Hence $\bar{g}_c(\theta) = c\theta$ is the induced transformation of Θ, and the class $\mathscr{F}$ of exponential densities is clearly invariant under $\mathscr{G}$. It is also easy to see that $\tilde{g}(a) = ca$, since

$$L(\theta, a) = \left(1 - \frac{a}{\theta}\right)^2 = \left(1 - \frac{\tilde{g}(a)}{\bar{g}(\theta)}\right)^2 = L(\bar{g}(\theta), \tilde{g}(a)).$$

Hence the loss and decision problem are invariant under $\mathscr{G}$. Note that $\mathscr{G}$, $\bar{\mathscr{G}}$, and $\tilde{\mathscr{G}}$ are the same group of transformations.

EXAMPLE 7. Assume that $X \sim \mathscr{N}(\theta, 1)$, and that it is desired to estimate θ under a loss of the form $L(\theta, a) = W(|\theta - a|)$. Let $\mathscr{G}$ be the group of additive transformations introduced in Example 4. Clearly $Y = g_c(X) = X + c$ has a $\mathscr{N}((\theta + c), 1)$ density. Hence $\bar{g}_c(\theta) = \theta + c$, and $\mathscr{F}$ is invariant under $\mathscr{G}$. Also, if $\tilde{g}_c(a) = a + c$, then

$$L(\theta, a) = W(|\theta - a|) = W(|\bar{g}_c(\theta) - \tilde{g}_c(a)|) = L(\bar{g}_c(\theta), \tilde{g}_c(a)),$$

so that the loss and problem are invariant under $\mathscr{G}$. Note that $\mathscr{G}$, $\bar{\mathscr{G}}$, and $\tilde{\mathscr{G}}$ are the same group.

EXAMPLE 8. Assume that $X \sim \mathscr{B}(n, \theta)$, and that it is desired to estimate θ under squared-error loss. Let $\mathscr{G} = \{e, g^*\}$, where e is the identity transformation and g^* is the transformation $g^*(x) = n - x$. Clearly $Y = g^*(X) = n - X$ has a $\mathscr{B}(n, 1-\theta)$ density, so that $\mathscr{F}$ is invariant under $\mathscr{G}$, with $\bar{g}^*(\theta) = 1 - \theta$. The loss and decision problem are also invariant, as can be seen by defining $\tilde{g}^*(a) = 1 - a$.

EXAMPLE 9. Assume that $X \sim \mathscr{N}(\theta, \sigma^2)$, both θ and σ^2 unknown, and that it is desired to test H_0: $\theta \le 0$ versus H_1: $\theta > 0$ under "0–1" loss. Here $\mathscr{A} = \{a_0, a_1\}$, where a_i denotes accepting H_i. Let $\mathscr{G}$ be the group of scale transformations given in Example 3. Clearly $Y = g_c(X) = cX$ has a $\mathscr{N}(c\theta, c^2\sigma^2)$ density, so that $\mathscr{F}$ is invariant under $\mathscr{G}$, with $\bar{g}_c((\theta, \sigma^2)) = (c\theta, c^2\sigma^2)$. (Note that $\bar{g}_c$ is a transformation on $\Theta = R^1 \times (0, \infty)$.) Note also that if $\theta < 0$, then $c\theta < 0$, while if $\theta > 0$, then $c\theta > 0$. Hence, defining $\tilde{g}_c(a) = \tilde{e}(a) = a$ (the identity transformation), it is clear that the loss is invariant under $\mathscr{G}$. Interestingly enough, $\tilde{\mathscr{G}}$ consists only of $\tilde{e}$.

As in the above example, it is frequently the case in hypothesis testing that $\tilde{\mathscr{G}}$ consists only of the identity transformation. (The group operation will usually not cause a change in the hypothesis accepted.) When this is the case, an invariant loss must satisfy

$$L(\theta, a_i) = L(\bar{g}(\theta), \tilde{g}(a_i)) = L(\bar{g}(\theta), a_i).$$

Of the various standard testing losses we have considered, only "0-K_i" loss will generally satisfy this condition, and then only if Θ_0 and Θ_1 (the parameter regions under the null and alternative hypotheses) are invariant subspaces, in the sense that, for all g, $\bar{g}(\theta) \in \Theta_i$ when $\theta \in \Theta_i$ (for $i = 0, 1$). (In other words, Θ_0 and Θ_1 must be mapped back into themselves.) Thus invariance in hypothesis testing has come to mean essentially that the loss is "0-K_i" loss, and that Θ_0 and Θ_1 are invariant subspaces. (It is possible for invariance in hypothesis testing to take other forms. In Example 9, for instance, one could enlarge $\mathscr{G}$ to include the transformations $g_c(x) = cx$ for $c < 0$. These transformations would have the effect of interchanging Θ_0 and Θ_1, so that $\tilde{\mathscr{G}}$ would also include the transformation that switches actions.)

6.2.3. Invariant Decision Rules

In an invariant decision problem, the formal structures of the problems involving X and $Y = g(X)$ are identical. Hence the invariance principle states that δ and δ^*, the decision rules used in the X and Y problems, respectively, should be identical. The principle of rational invariance, in addition, states that the actions taken in the two problems should correspond, or that

$$\tilde{g}(\delta(x)) = \delta^*(y).$$

Combining these principles, it follows that a decision rule should be invariant in the sense of the following definition.

Definition 4. If a decision problem is invariant under a group $\mathscr{G}$ of transformations, a (nonrandomized) decision rule, $\delta(x)$, is *invariant under* $\mathscr{G}$ if, for all $x \in \mathscr{X}$ and $g \in \mathscr{G}$,

$$\delta(g(x)) = \tilde{g}(\delta(x)). \tag{6.4}$$

Let $\mathscr{D}_I$ denote the set of all (nonrandomized) invariant decision rules.

In Example 1, the invariant decision rules were explicitly found. We content ourselves with one further example here.

EXAMPLE 9 (continued). Using the results in Example 9, Equation (6.4) becomes

$$\delta(cx)=\delta(x). \tag{6.5}$$

Imagine now that $\delta(x_0)=a_i$ for some $x_0>0$. Then (6.5) implies that $\delta(x)=a_i$ for all $x>0$. A similar result holds for $x<0$. The conclusion is that the only invariant (nonrandomized) decision rules are $\delta_{0,0}$, $\delta_{0,1}$, $\delta_{1,0}$, and $\delta_{1,1}$, where

$$\delta_{i,j}(x)=\begin{cases} a_i & \text{if } x<0,\\ a_j & \text{if } x>0.\end{cases}$$

The crucial property of invariant decision rules is that their risk functions are constant on "orbits" of Θ.

Definition 5. Two points, θ_1 and θ_2, in Θ are said to be *equivalent* if $\theta_2=\bar{g}(\theta_1)$ for some $\bar{g}\in\bar{\mathscr{G}}$. An *orbit* in Θ is an equivalence class of such points. Thus the θ_0-*orbit* in Θ, to be denoted $\Theta(\theta_0)$, is the set

$$\Theta(\theta_0)=\{\bar{g}(\theta_0):\ \bar{g}\in\bar{\mathscr{G}}\}.$$

Theorem 1. *The risk function of an invariant decision rule δ is constant on orbits of Θ, or, equivalently,*

$$R(\theta,\delta)=R(\bar{g}(\theta),\delta)$$

for all $\theta\in\Theta$ and $\bar{g}\in\bar{\mathscr{G}}$.

PROOF. Clearly

$$\begin{aligned} R(\theta,\delta) &= E_\theta L(\theta,\delta(X))\\ &= E_\theta L(\bar{g}(\theta),\tilde{g}(\delta(X))) \quad \text{(invariance of loss)}\\ &= E_\theta L(\bar{g}(\theta),\delta(g(X))) \quad \text{(invariance of } \delta)\\ &= E_{\bar{g}(\theta)} L(\bar{g}(\theta),\delta(X)) \quad \text{(invariance of distributions and (6.3))}\\ &= R(\bar{g}(\theta),\delta(X)), \end{aligned}$$

completing the proof. □

EXAMPLE 1 (continued). Recalling that $\bar{g}_c(\theta)=c\theta$, it is clear that all points in Θ are equivalent ($\theta_1=c\theta_2$ for some c), so that Θ is itself an orbit. It follows from Theorem 1 that any invariant decision rule has constant risk for all θ. This can be verified by the direct calculation (letting $\delta_K(x)=Kx$)

$$\begin{aligned} R(\theta,\delta_K) &= E_\theta\left(1-\frac{\delta_K(X)}{\theta}\right)^2\\ &= \int_0^\infty \theta^{-1}\left(1-\frac{Kx}{\theta}\right)^2 \exp\left\{\frac{-x}{\theta}\right\}dx\\ &= 1-2K+2K^2. \end{aligned}$$

The special situation in which the risk function of an invariant rule is constant is of great importance and deserves further investigation.

Definition 6. A group $\bar{\mathscr{G}}$ of transformations of Θ is said to be *transitive* if Θ consists of a single orbit, or equivalently if, for any θ_1 and θ_2 in Θ, there exists some $\bar{g} \in \bar{\mathscr{G}}$ for which $\theta_2 = \bar{g}(\theta_1)$.

If $\bar{\mathscr{G}}$ is transitive, then from Theorem 1 it is clear that any invariant decision rule has a constant risk. An invariant decision rule which minimizes this constant risk will be called a *best invariant decision rule.*

EXAMPLE 1 (continued). All that remains to be done is to find the optimal value of K. Clearly

$$\frac{d}{dK} R(\theta, \delta_K) = -2 + 4K,$$

so that $K = \frac{1}{2}$ is the minimizing value of K. Hence $\delta_{1/2}(x) = x/2$ is the best invariant decision rule.

Even when $\bar{\mathscr{G}}$ is not transitive on Θ, the class $\mathscr{D}_I$ of invariant procedures is often small enough so that an "optimal" invariant rule can easily be selected according to some other principle. We will not actively pursue such situations here.

Invariant randomized decision rules can also be considered, but for simplicity we will not do so. Randomization is sometimes useful in testing problems, but is seldom needed in estimation problems. The justification for the latter statement is that $\bar{\mathscr{G}}$ will usually be transitive in estimation problems to which invariance is applied, so that a best invariant rule will typically exist. It can be shown for such a situation that, providing the group $\mathscr{G}$ is not too large (see Section 6.9), a nonrandomized best invariant rule will always exist. (The proof is related to the proof that a nonrandomized Bayes rule exists, and can be found in Kiefer (1957).) This will be true, in particular, when $\bar{\mathscr{G}}$ can be identified with Θ, as will be the case in virtually all of our estimation examples.

As a final comment, it is important to note that, if possible, sufficiency should always be used to reduce the complexity of a problem before applying invariance. This, of course, is true for all decision principles.

6.3. Location Parameter Problems

One of the simplest and most natural applications of invariance is to location parameter problems. The concept of a location parameter was defined in Subsection 3.3.2. Recall that θ is a location parameter (or a location vector in higher dimensional settings) if the density of X is of the form $f(x-\theta)$.

Consider first the one-dimensional estimation problem, in which $\mathscr{X}=\Theta=\mathscr{A}=R^1$, and θ is a location parameter. Assume also that the loss is a function of $a-\theta$, say $L(a-\theta)$. As in Example 7, it is easy to check that this problem is invariant under the group $\mathscr{G}=\{g_c\colon g_c(x)=x+c\}$, and that $\bar{\mathscr{G}}=\tilde{\mathscr{G}}=\mathscr{G}$. Hence (6.4), the determining equation for invariant rules, becomes

$$\delta(x+c)=\delta(x)+c.$$

Choosing $c=-x$, it follows that $\delta(0)=\delta(x)-x$, or

$$\delta(x)=x+K \tag{6.6}$$

(where $K=\delta(0)$). Any invariant rule must thus be of this form.

It is easy to check, in this situation, that $\bar{\mathscr{G}}$ is transitive on Θ, so that any rule of the form (6.6) has constant risk

$$R(\theta,\delta)=R(0,\delta)=E_0L(X+K). \tag{6.7}$$

The best invariant decision rule (assuming one exists) is simply that rule of the form (6.6) for which K minimizes (6.7).

The minimization of (6.7) is a problem we encountered in Bayesian analysis with a noninformative prior. The analogy is of enough interest to recall that development. For a location parameter, the noninformative prior is the constant prior $\pi(\theta)=1$. The posterior density for this prior is

$$\pi(\theta|x)=\frac{\pi(\theta)f(x-\theta)}{\int f(x-\theta)d\theta}=f(x-\theta).$$

Hence the posterior expected loss is

$$\begin{aligned} E^{\pi(\theta|x)}[L(a-\theta)]&=\int_{-\infty}^{\infty}L(a-\theta)f(x-\theta)d\theta\\ &=\int_{-\infty}^{\infty}L(y+a-x)f(y)dy \quad (\text{letting } y=x-\theta)\\ &=E_0L(X+K) \qquad\qquad\qquad (\text{letting } a=x+K).\end{aligned}$$

The generalized Bayes rule is thus also $\delta(x)=a=x+K$, where K minimizes (6.7). The results of Chapter 4 can thus be applied to the minimization of (6.7). If, for example, L is squared-error loss, then we know that the generalized Bayes rule is the posterior mean

$$\int_{-\infty}^{\infty}\theta f(x-\theta)d\theta=x-\int_{-\infty}^{\infty}yf(y)dy.$$

Equating this with $\delta(x)=x+K$, it is clear that the optimum choice of K is $-E_0[X]$. If L is absolute error loss, it can similarly be shown that the optimal choice of K is $-m$, where m is the median of X when $\theta=0$. Two other useful results concerning the optimum value of K are given in Exercises 10 and 11. The equivalence shown in the above situation between the best invariant estimator and the generalized Bayes rule will be seen in Section 6.6 to be part of a very general phenomenon.

EXAMPLE 10. Assume that $X_1, X_2, \ldots, X_n$ is a sample from a $\mathcal{N}(\theta, \sigma^2)$ distribution, σ^2 known. The sample mean, $\bar{X}$, is sufficient for θ, and has a $\mathcal{N}(\theta, \sigma^2/n)$ distribution. Clearly θ is a location parameter for this distribution. Hence, by the above results, the best invariant estimator of θ under squared-error loss is

$$\delta(\bar{x}) = \bar{x} - E_0[\bar{X}] = \bar{x}.$$

Since the median of $\bar{X}$ when $\theta = 0$ is $m = 0$, it also follows that the above estimator is best invariant under absolute error loss. Indeed, it can be shown, using Exercise 11, that this estimator is best invariant for any loss which is increasing in $|\theta - a|$.

For location vector problems (i.e., problems in which $\mathbf{X} = (X_1, \ldots, X_p)^t$ has density $f(\mathbf{x} - \boldsymbol{\theta})$, the location vector $\boldsymbol{\theta} = (\theta_1, \ldots, \theta_p)^t$ being unknown), exactly the same kinds of results can be obtained. The appropriate group of transformations is the additive group on R^p, defined by

$$\mathscr{G} = \{g_{\mathbf{c}} \colon \mathbf{c} = (c_1, \ldots, c_p)^t \in R^p\},$$

where $g_{\mathbf{c}}(\mathbf{x}) = \mathbf{x} + \mathbf{c}$. The analysis and conclusions in this situation are virtually identical to those in the one-dimensional case.

Pitman's Estimator

A somewhat different problem arises when a sample $X_1, \ldots, X_n$ is taken from a one-dimensional location density $f(x - \theta)$. This is a special case of the more general situation in which $\mathbf{X} = (X_1, \ldots, X_n)$ has density $f(x_1 - \theta, \ldots, x_n - \theta)$. The relevant group of transformations for this problem is

$$\mathscr{G} = \{g_c \colon g_c(\mathbf{x}) = (x_1 + c, \ldots, x_n + c), \text{ where } c \in R^1\}.$$

It is easy to check that $\bar{\mathscr{G}}$ is simply the additive group on $\Theta = R^1$, which is transitive. It follows that if the loss is of the form $L(a - \theta)$, then a best invariant rule can be sought. Unfortunately, the derivation of such a rule is considerably more difficult here than in the single observational case, and will be delayed until Section 6.5. It turns out, however (see Section 6.6), that the best invariant rule again corresponds to the generalized Bayes rule with respect to the noninformative prior $\pi(\theta) = 1$. It follows that the best invariant estimator is that action which minimizes

$$E^{\pi(\theta|\mathbf{x})}[L(a - \theta)] = \frac{\int_{-\infty}^{\infty} L(a - \theta) f(x_1 - \theta, \ldots, x_n - \theta) d\theta}{\int_{-\infty}^{\infty} f(x_1 - \theta, \ldots, x_n - \theta) d\theta}.$$

When L is squared-error loss, the minimizing action is the posterior mean, so that the best invariant estimator is

$$\delta(\mathbf{x}) = \frac{\int_{-\infty}^{\infty} \theta f(x_1 - \theta, \ldots, x_n - \theta) d\theta}{\int_{-\infty}^{\infty} f(x_1 - \theta, \ldots, x_n - \theta) d\theta}.$$

This estimator is called Pitman's estimator, and was derived in Pitman (1939).

6.4. Other Examples of Invariance

I. *Scale Parameter Problems*

In Subsection 3.3.2 it was stated that $\theta(>0)$ is a scale parameter if $X \in R^1$ or $X \in (0, \infty)$ has a density of the form $\theta^{-1} f(x/\theta)$. If the loss is of the form $L(a/\theta)$, it can be checked, as in Example 1, that the problem is invariant under the group of scale transformations.

Rather than carrying out a general analysis in this situation, we simply note that the problem can be transformed into an invariant location parameter problem. To see this, assume for simplicity that $X > 0$, and consider the transformation $Y = \log X$ and $\eta = \log \theta$. The density of Y given η is clearly $\exp\{y - \eta\} f(\exp\{y - \eta\})$, which is a location density with location parameter η. Defining $a^* = \log a$, the loss in estimating η by a^* is clearly

$$L\left(\frac{a}{\theta}\right) = L\left(\frac{\exp\{a^*\}}{\exp\{\eta\}}\right) = L(\exp\{a^* - \eta\}),$$

which is invariant in the location parameter problem. One can thus use the results of the previous section to obtain the best invariant rule, $\delta^*(y)$, for the transformed problem. Converting back to the original problem results in the decision rule

$$\delta(x) = \exp\{\delta^*(\log x)\}.$$

It can easily be checked that this is indeed a best invariant rule for the original scale parameter problem.

EXAMPLE 11. Assume that $X_1, X_2, \ldots, X_n$ is a sample from a $\mathcal{N}(0, \sigma^2)$ distribution, and that it is desired to estimate σ^2 under loss

$$L(\sigma^2, a) = (\log a - \log \sigma^2)^2 = \left(\log\left[\frac{a}{\sigma^2}\right]\right)^2.$$

It is easy to check that $Z = \sum_{i=1}^n X_i^2 \sim \mathcal{G}(n/2, 2\sigma^2)$ is sufficient for σ^2. For this density, σ^2 is a scale parameter, and the group of scale transformations leaves the problem invariant.

Transforming to a location parameter problem results in the consideration of $Y = \log Z$, $\eta = \log \sigma^2$, and $L^*(\eta, a^*) = (a^* - \eta)^2$. From Section 6.3, it follows that the best invariant estimator in the transformed problem is

$$\delta^*(y) = y - E_0[Y].$$

Noting that $E_{\eta=0}[Y] = E_{\sigma^2=1}[\log Z]$, it can be concluded that the best invariant estimator for the original scale parameter problem is

$$\delta(z) = \exp\{\delta^*(\log z)\} = \exp\{(\log z) - E_{\sigma^2=1}[\log Z]\}$$
$$= \frac{z}{\exp\{E_{\sigma^2=1}[\log Z]\}}.$$

The quantity $\exp\{E_{\sigma^2=1}[\log Z]\}$ can be numerically calculated, and is very close to $(n-1)$ for $n \geq 2$.

II. *Location-Scale Parameter Problems*

In Subsection 3.3.3 a location-scale density was defined to be a density of the form $\sigma^{-1}f([x-\theta]/\sigma)$, where $x \in R^1$ and both $\theta \in R^1$ and $\sigma > 0$ are unknown. The class of such densities is invariant under the group of affine transformations $g_{b,c}(x) = cx + b$ (see Example 5), since $Y = g_{b,c}(X) = cX + b$ has density $(c\sigma)^{-1}f([y-(b+c\theta)]/(c\sigma))$. The group, $\bar{\mathscr{G}}$, of induced transformations of

$$\Theta = \{(\theta, \sigma): \theta \in R^1 \text{ and } \sigma > 0\}$$

thus consists of the transformations defined by

$$\bar{g}_{b,c}((\theta, \sigma)) = (c\theta + b, c\sigma).$$

Clearly $\bar{\mathscr{G}}$ is transitive, so that a best invariant rule can be sought when the loss is invariant. Some specific examples will be given in the Exercises and in Section 6.6.

III. *A Nontransitive Example*

Consider the binomial estimation problem of Example 8. The only restriction imposed by (6.4) is that

$$\delta(n-x) = 1 - \delta(x),$$

which is a symmetry restriction. Any rule satisfying this restriction is invariant. Note that there are a large number of such procedures. Also, $\bar{\mathscr{G}} = \{\bar{e}, \bar{g}^*\}$ is not transitive on $\Theta = (0, 1)$, and indeed the orbits in Θ are simply the pairs of points of the form $\{\theta, 1-\theta\}$. There is no best invariant decision rule for this problem, and invariance is not of much help in choosing a procedure.

6.5. Maximal Invariants

In dealing with statistical problems in which a sample of size greater than one is taken, it can be fairly difficult to apply invariance arguments. The concept of a maximal invariant can be very useful in simplifying such problems.

Definition 7. Let $\mathcal{G}$ be a group of transformations on a space $\mathcal{X}$. A function $T(x)$ on $\mathcal{X}$ is said to be *invariant* with respect to $\mathcal{G}$ if $T(g(x)) = T(x)$ for all $x \in \mathcal{X}$ and $g \in \mathcal{G}$. A function $T(x)$ is said to be *maximal invariant* with respect to $\mathcal{G}$ if it is invariant and satisfies

$$T(x_1) = T(x_2) \quad \text{implies} \quad x_1 = g(x_2) \quad \text{for some } g \in \mathcal{G}. \tag{6.8}$$

These concepts are best pictured by recalling that $\mathcal{X}$ can be divided up into orbits of points that are equivalent under $\mathcal{G}$. An invariant function is a function which is constant on orbits. A maximal invariant is a function which is constant on orbits, and which assigns different values to different orbits. Note that if $\mathcal{G}$ is transitive on $\mathcal{X}$, so that $\mathcal{X}$ itself is an orbit, then the only maximal invariants are constant functions.

Example 12 (Location Invariance). Let $\mathcal{X} = R^n$ and let $\mathcal{G}$ consist of the transformations

$$g_c((x_1, \ldots, x_n)) = (x_1 + c, \ldots, x_n + c).$$

(This would arise in a statistical problem in which a sample of size n was taken from a location density.) The $(n-1)$-vector

$$\mathbf{T}(\mathbf{x}) = (x_1 - x_n, \ldots, x_{n-1} - x_n)$$

is then a maximal invariant. To see this, note that $\mathbf{T}(g_c(\mathbf{x})) = \mathbf{T}(\mathbf{x})$ (since $(x_i + c) - (x_n + c) = (x_i - x_n)$), implying that $\mathbf{T}$ is invariant. To check (6.8), observe that, if $\mathbf{T}(\mathbf{x}) = \mathbf{T}(\mathbf{x}')$, then $(x_i - x_n) = (x_i' - x_n')$ for all i. Hence $g_c(\mathbf{x}') = \mathbf{x}$ for $c = (x_n - x_n')$.

Example 13 (Scale Invariance). Let $\mathcal{X} = R^n$ and let $\mathcal{G}$ consist of the transformations

$$g_c((x_1, \ldots, x_n)) = (cx_1, \ldots, cx_n),$$

where $c > 0$. (This would arise in a statistical problem in which a sample of size n was taken from a scale density.) Defining

$$z^2 = \sum_{i=1}^{n} x_i^2,$$

the function

$$\mathbf{T}(\mathbf{x}) = \begin{cases} \mathbf{0} & \text{if } z = 0, \\ \left(\dfrac{x_1}{z}, \ldots, \dfrac{x_n}{z}\right) & \text{if } z \neq 0, \end{cases}$$

is a maximal invariant. The proof will be left as an exercise.

EXAMPLE 14 (Location-Scale Invariance). Let $\mathscr{X} = R^n$ and let $\mathscr{G}$ consist of the transformations

$$g_{b,c}((x_1, \ldots, x_n)) = (cx_1 + b, \ldots, cx_n + b),$$

where $c > 0$. Defining

$$\bar{x} = \frac{1}{n} \sum_{i=1}^{n} x_i \quad \text{and} \quad s^2 = \frac{1}{n} \sum_{i=1}^{n} (x_i - \bar{x})^2,$$

the function

$$\mathbf{T}(\mathbf{x}) = \begin{cases} \mathbf{0} & \text{if } s = 0, \\ \left(\dfrac{x_1 - \bar{x}}{s}, \ldots, \dfrac{x_n - \bar{x}}{s}\right) & \text{if } s \neq 0, \end{cases}$$

is a maximal invariant. The proof is left as an exercise.

The major uses of maximal invariants are based on the following theorems.

Theorem 2. *Let $\mathscr{G}$ be a group of transformations of a space $\mathscr{X}$, and assume that $T(x)$ is a maximal invariant. Then a function $h(x)$ is invariant with respect to $\mathscr{G}$ if and only if h is a function of $T(x)$.*

PROOF. If $h(x) = s(T(x))$, say, then

$$h(g(x)) = s(T(g(x))) = s(T(x)) = h(x).$$

Hence h is invariant if it is a function of T.

To establish the "only if" part of the theorem, assume that h is invariant and that $T(x_1) = T(x_2)$. Then $x_1 = g(x_2)$ for some $g \in \mathscr{G}$, so that

$$h(x_1) = h(g(x_2)) = h(x_2).$$

This implies that h is a function of T. □

Theorem 3. *Consider a statistical decision problem that is invariant under a group $\mathscr{G}$, and let $v(\theta)$ be a maximal invariant on Θ, with respect to $\bar{\mathscr{G}}$. Then, if $h(x)$ is invariant under $\mathscr{G}$, the distribution of $h(X)$ depends only on $v(\theta)$.*

PROOF. Observe that, for any fixed set $A \subset \mathscr{X}$,

$$\begin{aligned} P_{\bar{g}(\theta)}(h(X) \in A) &= P_\theta(h(g(X)) \in A) \quad \text{(invariance of distribution)} \\ &= P_\theta(h(X) \in A) \quad \text{(invariance of } h\text{)}, \end{aligned}$$

so that $s(\theta) = P_\theta(h(X) \in A)$ is an invariant function on Θ. By Theorem 2 (with $\mathscr{X}$ and $\mathscr{G}$ replaced by Θ and $\bar{\mathscr{G}}$), it follows that $s(\theta)$ is a function of $v(\theta)$. Since this is true for all A, the theorem follows. □

We now briefly describe the application of the above results to testing and estimation problems.

Hypothesis Testing

In Example 9 and the subsequent discussion it was indicated that, for invariant hypothesis testing problems, it will typically be the case that an invariant decision rule satisfies $\delta(g(x)) = \delta(x)$ for all $g \in \mathscr{G}$ (i.e., the hypothesis accepted remains the same under transformation of the problem by g). But this means that $\delta(x)$ is an invariant function, which, by Theorem 2, implies that $\delta(x)$ is a function of the maximal invariant $T(x)$. Hence, for invariant hypothesis testing problems, it is frequently the case that the only invariant decision rules are those which are a function of the maximal invariant. Noting from Theorem 3 that the distribution of $T(X)$ depends only on the maximal invariant $v(\theta)$ on Θ, it follows that the invariance approach reduces the original problem to a test concerning $v(\theta)$, based on the observation $T(X)$. This can be a great simplification, as indicated in the following example.

EXAMPLE 15. Let $\mathbf{X} = (X_1, \ldots, X_n)$ have a density of the form $f(x_1 - \theta, \ldots, x_n - \theta)$ on R^n, where $\theta \in R^1$ is unknown. It is desired to test H_0: $f = f_0$ versus H_1: $f = f_1$, where f_0 and f_1 are known (up to θ). The loss is "0–1" loss. This problem is invariant under the group of transformations in Example 12, so that

$$\mathbf{T}(\mathbf{x}) = (x_1 - x_n, \ldots, x_{n-1} - x_n)$$

is a maximal invariant. Since $\bar{\mathscr{G}}$ is transitive on Θ, it is clear that the only maximal invariants on Θ are constant functions, which by Theorem 3 implies that the distribution of $\mathbf{Y} = \mathbf{T}(\mathbf{X})$ is independent of θ. (This, of course, is directly obvious here.) It follows that any invariant test will depend only on $\mathbf{Y}$, and hence that the problem can be reformulated as a test of H_0: $f^* = f_0^*$ versus H_1: $f^* = f_1^*$, where $*$ denotes a marginal density of $\mathbf{Y}$. Note that the joint density of $(Y_1, \ldots, Y_{n-1}, X_n)$ at, say, $\theta = 0$ is

$$f(y_1 + x_n, \ldots, y_{n-1} + x_n, x_n),$$

so that

$$f^*(\mathbf{y}) = \int_{-\infty}^{\infty} f(y_1 + x_n, \ldots, y_{n-1} + x_n, x_n)\, dx_n.$$

This new problem is just a test of a simple hypothesis against a simple alternative, and it is well known that the optimal or most powerful tests in such a situation (see Chapter 8) are to reject H_0 if $f_1^*(\mathbf{y}) \geq K f_0^*(\mathbf{y})$, where the constant K depends on the desired error probabilities (i.e., the desired risks under H_0 and H_1). In terms of the original problem, these tests are called *uniformly most powerful invariant* tests.

Estimation

The following considerations apply mainly to invariant estimation problems, but are potentially useful for any invariant decision problem in which $\bar{\mathscr{G}}$ is transitive on Θ. The technique that is discussed gives a conceptually simple way of determining a best invariant decision rule.

Assume that $\bar{\mathscr{G}}$ is transitive on Θ, and that $T(x)$ is a maximal invariant on $\mathscr{X}$. By Theorem 1, any invariant rule δ will have constant risk, so we need only consider, say, $\theta = \bar{e}$. Defining $Y = T(X)$ and, as usual, letting $E^{X|y}$ denote expectation with respect to the conditional distribution of X given $Y = y$, it is clear that

$$R(\bar{e}, \delta(X)) = E_{\bar{e}}^{X}[L(\bar{e}, \delta(X))] = E_{\bar{e}}^{Y} E_{\bar{e}}^{X|Y}[L(\bar{e}, \delta(X))].$$

It follows that an invariant rule which, for each y, minimizes

$$E_{\bar{e}}^{X|y}[L(\bar{e}, \delta(X))],$$

will be best invariant (measurability concerns aside). This last conditional problem is quite easy to solve. Note first that $\mathscr{X}(y) = \{x \in \mathscr{X}: T(x) = y\}$ is an orbit of $\mathscr{X}$, and that any $x \in \mathscr{X}(y)$ can be written (uniquely) as $x = g(x_y)$, where x_y is some fixed point in $\mathscr{X}(y)$. Now an invariant decision rule must satisfy

$$\delta(x) = \delta(g(x_y)) = \tilde{g}\delta(x_y),$$

and so is specified by the choice of the $\delta(x_y)$. In the conditional problem, therefore, the conditional risk of δ depends only on the choice of $\delta(x_y)$. Finding the best choice will usually be a straightforward minimization problem.

Example 16 (Pitman's Estimator). We now present the derivation of Pitman's estimator, which was briefly discussed in Section 6.3. The setup, recall, is that of observing $\mathbf{X} = (X_1, \ldots, X_n)$ which has density $f(x_1 - \theta, \ldots, x_n - \theta)$, and desiring to estimate θ under loss $L(a - \theta)$. From Example 12, we know that $\mathbf{T}(\mathbf{x}) = (x_1 - x_n, \ldots, x_{n-1} - x_n)$ is a maximal invariant for this problem. Setting $\mathbf{y} = \mathbf{T}(\mathbf{x})$ and observing that, on the $\mathscr{X}(\mathbf{y})$ orbit of $\mathscr{X}$, a point $\mathbf{x}$ can be written

$$\mathbf{x} = (y_1 + x_n, \ldots, y_{n-1} + x_n, x_n) = g_{x_n}((y_1, \ldots, y_{n-1}, 0)),$$

it is clear that, instead of working with the conditional distribution of $\mathbf{X}$ given $\mathbf{Y} = \mathbf{y}$, it will be more convenient (and equivalent) to work with the conditional distribution of X_n given $\mathbf{Y} = \mathbf{y}$. Choosing $\theta = \bar{e} = 0$ ($\bar{\mathscr{G}}$ is transitive so the risk is constant), it is easy to show (see Example 15) that the conditional density of X_n, given $\mathbf{Y} = \mathbf{y}$, is

$$\frac{f(y_1 + x_n, \ldots, y_{n-1} + x_n, x_n)}{\int_{-\infty}^{\infty} f(y_1 + x_n, \ldots, y_{n-1} + x_n, x_n) dx_n}.$$

Since an invariant rule must satisfy

$$\delta(\mathbf{x}) = \delta(g_{x_n}((y_1, \ldots, y_{n-1}, 0))) = \delta((y_1, \ldots, y_{n-1}, 0)) + x_n, \tag{6.9}$$

it follows that $\delta((y_1, \ldots, y_{n-1}, 0))$ must be chosen to minimize the conditional risk

$$E_0^{X_n|\mathbf{y}}[L(\delta(\mathbf{X}) - 0)]$$
$$= \frac{\int_{-\infty}^{\infty} L(\delta((y_1, \ldots, y_{n-1}, 0)) + x_n) f(y_1 + x_n, \ldots, y_{n-1} + x_n, x_n)\, dx_n}{\int_{-\infty}^{\infty} f(y_1 + x_n, \ldots, y_{n-1} + x_n, x_n)\, dx_n}.$$

The variable x_n in the above integrals is just a dummy variable of integration. Replace it by (say) z, and then make the change of variables $z = x_n - \theta$, where θ is the new variable of integration and x_n is the true observed value of X_n. Using (6.9), the conditional risk then becomes

$$\frac{\int_{-\infty}^{\infty} L(\delta(\mathbf{x}) - \theta) f(x_1 - \theta, \ldots, x_n - \theta)\, d\theta}{\int_{-\infty}^{\infty} f(x_1 - \theta, \ldots, x_n - \theta)\, d\theta}. \tag{6.10}$$

The best invariant estimator is thus that action, $\delta(\mathbf{x})$, which minimizes (6.10), as stated in Section 6.3.

6.6. Invariance and Noninformative Priors

In Section 6.3 it was seen that the best invariant estimator of a location parameter is also the generalized Bayes rule with respect to the noninformative prior $\pi(\theta) = 1$. A similar relationship is valid in many situations in which $\bar{\mathscr{G}}$ is transitive on Θ. Indeed it is then usually the case that the best invariant rule is (generalized) Bayes with respect to what is called the right invariant Haar measure on the group $\bar{\mathscr{G}}$. To obtain this result in a general setting requires measure theory and advanced group theory. We will, therefore, restrict ourselves to rather simple (yet frequently occurring) groups, which can be dealt with by calculus.

6.6.1. Right and Left Invariant Haar Densities

We assume, for the remainder of this section, that the group $\bar{\mathscr{G}}$ is a subset of R^p with positive Lebesgue measure (i.e., is not a discrete set or a set of lower dimension), and that for each fixed $\bar{g}_0 \in \bar{\mathscr{G}}$, the transformations

$$\bar{g} \to \bar{g}_0 \bar{g} \quad \text{and} \quad \bar{g} \to \bar{g} \bar{g}_0$$

have differentials and hence Jacobians. (It would actually be sufficient to require only that the transformations have differentials almost everywhere (i.e., except on a set of lower dimension). This would allow treatment of such groups as the orthogonal group (see Exercise 5). For simplicity, however, we will restrict ourselves to the everywhere differentiable case.)

Definition 8. Let $\mathbf{H}^r_{\bar{g}_0}(\bar{g})$ denote the differential (i.e., matrix of first partial derivatives) of the transformation $\bar{g} \to \bar{g}\bar{g}_0$, and let

$$J^r_{\bar{g}_0}(\bar{g}) = |\det(\mathbf{H}^r_{\bar{g}_0}(\bar{g}))|$$

denote the Jacobian of the transformation. Similarly, define $\mathbf{H}^l_{\bar{g}_0}(\bar{g})$ and $J^l_{\bar{g}_0}(\bar{g})$ as the differential and Jacobian of the transformation $\bar{g} \to \bar{g}_0\bar{g}$.

Example 17. Let $\bar{\mathscr{G}}$ be the affine group on R^1 (see Example 5). The transformation $\bar{g}_{b,c}$ can be considered to be the point $(b, c) \in R^2$, so that we can represent $\bar{\mathscr{G}}$ as

$$\bar{\mathscr{G}} = \{(b, c): -\infty < b < \infty \text{ and } 0 < c < \infty\}.$$

In this new notation, the group operation $\bar{g} \to \bar{g}\bar{g}_0$ can be written

$$(b, c) \to (b, c)(b_0, c_0) = ([cb_0 + b], cc_0).$$

The function

$$t((b, c)) = (t_1, t_2) = ([cb_0 + b], cc_0),$$

has differential

$$\mathbf{H}^r_{\bar{g}_0}((b, c)) = \begin{pmatrix} \dfrac{\partial t_1((b, c))}{\partial b} & \dfrac{\partial t_1((b, c))}{\partial c} \\ \dfrac{\partial t_2((b, c))}{\partial b} & \dfrac{\partial t_2((b, c))}{\partial c} \end{pmatrix} = \begin{pmatrix} 1 & b_0 \\ 0 & c_0 \end{pmatrix}.$$

The Jacobian of the transformation $\bar{g} \to \bar{g}\bar{g}_0$ is thus

$$J^r_{\bar{g}_0} = |\det(\mathbf{H}^r_{\bar{g}_0})| = c_0.$$

It will be left as an exercise to show that the group transformation $\bar{g} \to \bar{g}_0\bar{g}$ has Jacobian $J^l_{\bar{g}_0} = c_0^2$.

Definition 9. A *right invariant Haar density* on $\bar{\mathscr{G}}$, to be denoted $h^r(\bar{g})$, is a density (with respect to Lebesgue measure) which, for $A \subset \bar{\mathscr{G}}$ and all $g_0 \in \bar{\mathscr{G}}$, satisfies

$$\int_{A\bar{g}_0} h^r(y)dy = \int_A h^r(x)dx, \tag{6.11}$$

where $A\bar{g}_0 = \{\bar{g}\bar{g}_0: \bar{g} \in A\}$ and y and x are just dummy variables of integration.

Similarly, a *left invariant Haar density* on $\bar{\mathscr{G}}$, to be denoted $h^l(\bar{g})$, must satisfy

$$\int_{\bar{g}_0 A} h^l(y)dy = \int_A h^l(x)dx, \tag{6.12}$$

where $\bar{g}_0 A = \{\bar{g}_0\bar{g}: \bar{g} \in A\}$.

The idea behind invariant Haar densities is very similar to the idea behind invariant noninformative priors discussed in Subsection 3.3.2. Indeed, if h^r and h^l are probability densities, then (6.11) and (6.12) can be written

$$P^{h^r}(A\bar{g}_0) = P^{h^r}(A) \quad \text{and} \quad P^{h^l}(\bar{g}_0 A) = P^{h^l}(A),$$

implying that probabilities of sets are invariant under transformation of the sets by $\bar{g}_0$ on the right and left respectively. (Actually, when h^r and h^l are probability densities, it turns out that they must be equal, so that one can speak of the *invariant Haar density.*)

Note from (6.11) and (6.12) that h^r and h^l are not unique, since multiplying each by a constant will not affect the validity of the equations. The following is true, however.

Important Property. *The right and left invariant Haar densities, h^r and h^l, exist and are unique up to a multiplicative constant.*

For a proof of this property see Halmos (1950). Nachbin (1965) is also a useful reference.

Calculation of h^r and h^l

To determine h^r, simply make the change of variables $y = x\bar{g}_0$ on the left-hand side of (6.11). This integral then becomes

$$\int_{A\bar{g}_0} h^r(y)\,dy = \int_A h^r(x\bar{g}_0) J^r_{\bar{g}_0}(x)\,dx. \tag{6.13}$$

Using this in (6.11) gives

$$\int_A h^r(x\bar{g}_0) J^r_{\bar{g}_0}(x)\,dx = \int_A h^r(x)\,dx.$$

Since this must hold for all A, it follows that

$$h^r(x\bar{g}_0) J^r_{\bar{g}_0}(x) = h^r(x)$$

for all $x \in \bar{\mathscr{G}}$ and $\bar{g}_0 \in \bar{\mathscr{G}}$. Choosing $x = \bar{e}$ (the identity element in $\bar{\mathscr{G}}$), this implies that

$$h^r(\bar{g}_0) = \frac{h^r(\bar{e})}{J^r_{\bar{g}_0}(\bar{e})}.$$

Since $h^r(\bar{e})$ is just a multiplicative constant, it can be ignored, and the following result is established.

Result 1. *The right invariant Haar density on $\bar{\mathscr{G}}$ is*

$$h^r(\bar{g}) = \frac{1}{J^r_{\bar{g}}(\bar{e})},$$

where $J^r_{\bar{g}}(x)$ is the Jacobian of the transformation $x \to x\bar{g}$.

It can similarly be shown that

$$h^l(\bar{g}_0 x) J^l_{\bar{g}_0}(x) = h^l(x), \tag{6.14}$$

so that the following result holds.

Result 2. *The left invariant Haar density on* $\bar{\mathscr{G}}$ *is*

$$h^l(\bar{g}) = \frac{1}{J^l_{\bar{g}}(\bar{e})},$$

where $J^l_{\bar{g}}(x)$ *is the Jacobian of the transformation* $x \to \bar{g}x$.

Example 17 (continued). From Results 1 and 2 and the calculation in Example 17, it is clear that $h^r(\bar{g}) = 1/c$ and $h^l(\bar{g}) = 1/c^2$, where $\bar{g} = (b, c)$.

Left and Right Invariant (Generalized) Prior Densities

To relate the results on Haar densities to prior densities, we will assume that $\bar{\mathscr{G}}$ is *isomorphic* to Θ. This means that a one-to-one linear mapping between points in Θ and points in $\bar{\mathscr{G}}$ exists. In the cases considered here this will be satisfied because Θ and $\bar{\mathscr{G}}$ will be the same space. In the location-scale problem of Section 6.4, for instance,

$$\Theta = \{(\theta, \sigma): \theta \in R^1 \text{ and } \sigma > 0\},$$

while the group $\bar{\mathscr{G}} = \{(b, c): b \in R^1 \text{ and } c > 0\}$ (see Example 17). Hence the two spaces are the same. In such a situation we will denote the left and right invariant Haar densities on Θ (considered as a group) by $\pi^l(\theta)$ and $\pi^r(\theta)$, respectively, and will call them the *left invariant* and *right invariant* (generalized) prior densities. (The term "generalized" is used to indicate that these may be improper densities.) In the location-scale problem, for instance, it follows from Example 17 (continued) that $\pi^l((\theta, \sigma)) = 1/\sigma^2$ and $\pi^r((\theta, \sigma)) = 1/\sigma$. Note that these correspond to the two choices for a noninformative prior discussed in Example 7 in Subsection 3.3.3. Indeed, the general method discussed in that subsection of determining a noninformative prior can now be recognized as being simply the calculation of the left invariant Haar density.

6.6.2. The Best Invariant Rule

In this subsection, we establish the major result concerning the relationship of the best invariant rule to the (generalized) Bayes rule with respect to the right invariant prior density, $\pi^r(\theta)$. For simplicity, we will only consider the case in which $\mathscr{X}$, $\mathscr{G}$, $\bar{\mathscr{G}}$, and Θ are all isomorphic, although some comments about more general situations will be made. A good example to keep in mind is the following.

EXAMPLE 18. Assume that $X_1, \ldots, X_n$ is a sample from a $\mathcal{N}(\theta, \sigma^2)$ density, with θ and σ^2 both unknown. A sufficient statistic for (θ, σ) is $\mathbf{X} = (\bar{X}, S)$, where $\bar{X}$ is the sample mean and $S = [(1/n)\sum_{i=1}^{n}(X_i - \bar{X})^2]^{1/2}$. It is well known that $\bar{X}$ and S are independent, and that their joint density (for the appropriate constant K) is

$$f(\mathbf{x} | (\theta, \sigma)) = f((\bar{x}, s) | (\theta, \sigma))$$
$$= K\sigma^{-n} s^{(n-2)} \exp\left\{-\frac{n}{2\sigma^2}(\bar{x} - \theta)^2\right\} \exp\left\{-\frac{ns^2}{(2\sigma^2)}\right\}$$

on $\mathscr{X} = \{(\bar{x}, s): \bar{x} \in R^1 \text{ and } s > 0\}$. This density can easily be seen to be invariant under the group

$$\mathscr{G} = \{g_{b,c} = (b, c): b \in R^1 \text{ and } c > 0\},$$

where

$$g_{b,c}(\mathbf{x}) = (b, c)((\bar{x}, s)) = (c\bar{x} + b, cs).$$

Hence $\mathscr{X}$ and $\mathscr{G}$ are clearly the same space. It is also easy to check that Θ and $\bar{\mathscr{G}}$ are the same group as $\mathscr{X}$ and $\mathscr{G}$.

When $\mathscr{X}$, $\mathscr{G}$, $\bar{\mathscr{G}}$, and Θ are the same group, it is convenient to think of x and θ as group elements also. For instance, we can then write an invariant rule, $\delta(x)$, as $\delta(x) = \tilde{x}(\delta(e))$.

Result 3. *Consider an invariant decision problem in which $\mathscr{X}$, $\mathscr{G}$, $\bar{\mathscr{G}}$, and Θ are all isomorphic, with the common group being as in Subsection* 6.6.1. *Then, for an invariant decision rule $\delta(x) = \tilde{x}(a)$,*

$$E^{\pi^r(\theta|x)}[L(\theta, \tilde{x}(a))] = \int_{\mathscr{X}} L(\bar{e}, \tilde{y}(a)) f(y | \bar{e}) dy = R(\theta, \delta), \tag{6.15}$$

where $\pi^r(\theta | x)$ is the posterior distribution of θ given x with respect to the right invariant (generalized) prior density, $\pi^r(\theta)$. Also, the (generalized) Bayes rule, with respect to π^r, and the best invariant decision rule coincide (if they exist) and are equal to $\delta^*(x) = \tilde{x}(a^*)$, where a^* minimizes the middle expression in (6.15).

PROOF. The proof will be carried out in several steps. All integrals will be assumed to be over the common group, unless indicated otherwise.

Step 1. $f(x | \bar{g}) = f(g^{-1}(x) | \bar{e}) J^l_{g^{-1}}(x)$. To show this, note from (6.2) that

$$\int_{g^{-1}A} f(x | \theta) dx = \int_A f(x | \bar{g}(\theta)) dx.$$

Making the change of variables $x = g^{-1}(y)$ (the Jacobian of which is $J^l_{g^{-1}}(y)$) in the first integral above, results in the equality

$$\int_A f(g^{-1}(y) | \theta) J^l_{g^{-1}}(y) dy = \int_A f(x | \bar{g}(\theta)) dx.$$

Since this must be true for all A, it follows that (replacing y by x)

$$f(g^{-1}(x)|\theta)J^l_{g^{-1}}(x)=f(x|\bar{g}(\theta))$$

for all x, θ, and g. Choosing $\theta=\bar{e}$ gives the desired result.

Step 2. There exists a constant K such that, for any integrable function t,

$$\int t(x^{-1})h^r(x)dx=K\int t(y)h^l(y)dy, \tag{6.16}$$

where x^{-1} is the group inverse of x.

To see this, recall that

$$\int_A h^r(x)dx=\int_{Ag} h^r(x)dx. \tag{6.17}$$

Making the change of variables $x=y^{-1}$ in each integral, and defining $q(y)=h^r(y^{-1})J(y)$, where $J(y)$ is the Jacobian of the transformation, (6.17) becomes

$$\int_{A^{-1}} q(y)dy=\int_{(Ag)^{-1}} q(y)dy=\int_{g^{-1}A^{-1}} q(y)dy.$$

This implies that $q(y)$ is a left invariant Haar measure on the group. Since left invariant Haar measure is unique up to a multiplicative constant, it follows that $q(y)$ must equal $Kh^l(y)$ for some positive constant K. Performing the same change of variables on the left-hand side of (6.16) then gives the desired result.

Step 3. There exists a function $v(g)$ such that, for any integrable function t,

$$\int t(yg)h^l(y)dy=v(g)\int t(y)h^l(y)dy. \tag{6.18}$$

To see this, define

$$\mu(A)=\int_A h^l(y)dy$$

and

$$\mu_g(A)=\int_{Ag} h^l(y)dy.$$

Note that, for $g'\in\mathscr{G}$,

$$\mu_g(g'A)=\int_{g'Ag} h^l(y)dy=\int_{Ag} h^l(y)dy=\mu_g(A),$$

the middle equality following from the left invariance of h^l. Thus μ_g is what is called a *left invariant measure* on $\mathscr{G}$, which again must be unique up to a multiplicative constant. It follows that $\mu_g(A)=v(g)\mu(A)$ for some constant $v(g)$, i.e.,

$$\int_{Ag} h^l(y)dy = v(g)\int_A h^l(y)dy. \tag{6.19}$$

The change of variables $y = xg$, on the left-hand side of (6.19), results in the equivalent equation

$$\int_A h^l(xg)J_g^r(x)dx = v(g)\int_A h^l(y)dy.$$

Since this is true for all A, we have that

$$h^l(yg)J_g^r(y) = v(g)h^l(y).$$

Making the same change of variables, $y = xg$, on the left-hand side of (6.18) gives the desired result.

Step 4. Observe that

$$\begin{aligned}
&\int L(\theta, \tilde{x}(a))f(x|\theta)h^r(\theta)d\theta \\
&\quad = K\int L(\bar{y}^{-1}, \tilde{x}(a))f(x|\bar{y}^{-1})h^l(y)dy && \text{(step 2)} \\
&\quad = K\int L(\bar{e}, \tilde{y}(\tilde{x}(a)))f(x|\bar{y}^{-1})h^l(y)dy && \text{(invariance of loss)} \\
&\quad = K\int L(\bar{e}, \widetilde{yx}(a))f(yx|\bar{e})J_y^l(x)h^l(y)dy && \text{(step 1)} \qquad (6.20) \\
&\quad = K\int L(\bar{e}, \widetilde{yx}(a))f(yx|\bar{e})\frac{h^l(x)}{h^l(yx)}h^l(y)dy && \text{(from (6.14))} \\
&\quad = Kh^l(x)v(x)\int L(\bar{e}, \tilde{y}(a))f(y|\bar{e})\frac{1}{h^l(y)}h^l(y)dy && \text{(step 3)} \\
&\quad = Kh^l(x)v(x)\int L(\bar{e}, \tilde{y}(a))f(y|\bar{e})dy.
\end{aligned}$$

Choosing $L(\theta, a) \equiv 1$ in (6.20) establishes that

$$\int f(x|\theta)h^r(\theta)d\theta = Kh^l(x)v(x)\int f(y|\bar{e})dy = Kh^l(x)v(x).$$

Since $\pi^r(\theta) = h^r(\theta)$, this implies that

$$\pi^r(\theta|x) = \frac{f(x|\theta)h^r(\theta)}{\int f(x|\theta)h^r(\theta)d\theta} = \frac{f(x|\theta)h^r(\theta)}{Kh^l(x)v(x)}.$$

The first equality in Equation (6.15) follows immediately from this and (6.20). The second equality in (6.15) follows from the transitivity of $\bar{\mathscr{G}}$ (automatic when $\bar{\mathscr{G}}$ and Θ are the same group), which, by Theorem 1, implies that any invariant rule must have constant risk, say, $R(\bar{e}, \delta)$. The final conclusions of the result follow immediately from (6.15). □

EXAMPLE 18 (continued). In the situation of Example 18, assume that it is desired to estimate $(\alpha\theta+\beta\sigma)$ under a loss of the form

$$L((\theta, \sigma), a) = W\left(\frac{[\alpha\theta+\beta\sigma-a]}{\sigma}\right).$$

It can be checked that this loss is invariant when $\tilde{g}_{b,c}(a) = ca + \alpha b$, and hence that the decision problem is invariant.

We saw earlier that, for the group $\bar{\mathscr{G}}$,

$$\pi^r((\theta, \sigma)) = \frac{1}{\sigma}.$$

Result 3 thus implies that the best invariant estimator is given by

$$\delta(\mathbf{x}) = \tilde{\mathbf{x}}(a^*) = \tilde{g}_{\bar{x},s}(a^*) = sa^* + \alpha\bar{x},$$

where a^* is that a which minimizes

$$\begin{aligned}
&\int_{\mathscr{X}} L(\bar{e}, \tilde{y}(a)) f(y|\bar{e})dy \\
&\quad = \int_0^\infty \int_{-\infty}^\infty W\left(\frac{[0+\beta(1)-\tilde{g}_{y_1,y_2}(a)]}{1}\right) f((y_1, y_2)|(0, 1))dy_1\, dy_2 \\
&\quad = K \int_0^\infty \int_{-\infty}^\infty W(\beta - y_2 a - \alpha y_1) y_2^{n-2} \exp\left\{-\frac{n}{2}y_1^2\right\} \exp\left\{-\frac{n}{2}y_2^2\right\} dy_1\, dy_2.
\end{aligned}$$

Some explicit examples of this are given in the Exercises.

Observe that Result 3 settles, at least for invariant problems, the question of what is the correct choice for a noninformative prior. The right invariant Haar density is the correct choice. This is somewhat surprising, in light of the fact that the "natural" prior invariance argument of Subsection 3.3.2 led to the left invariant Haar density. (In Examples 5 and 6 of Chapter 3 the left and right invariant Haar densities are equal, but they differ for Example 7.) At the end of Subsection 3.3.2, however, it was pointed out that this natural invariance argument is not logically sound, and that all that can be concluded is that a noninformative prior should be what is called *relatively left invariant*. The choice of the right invariant Haar measure can be viewed as a sensible choice (for invariant problems) among this class of possible noninformative priors.

Analogues of Result 3 hold in much greater generality than discussed here. (See Kudo (1955), Stein (1965), Hora and Buehler (1965), and Zidek (1969) for indications of this.) The most important generalization is that to essentially arbitrary $\mathscr{X}$. This can be carried out by conditioning on the maximal invariant T (see Section 6.5), and noting that, in the conditional problem, the orbit $\mathscr{X}(t)$ will usually be isomorphic to $\mathscr{G}$. Providing everything else is satisfactory, it will follow that the best invariant rule in this conditional problem is the (generalized) Bayes rule with respect to the right invariant Haar density on $\bar{\mathscr{G}}$. Since this is true for all conditional problems, it can be concluded that the overall best invariant rule is simply this (generalized) Bayes rule. Finding the generalized Bayes rule can be much easier than finding the best invariant estimator directly, as is evidenced by the derivations of Pitman's estimator in Sections 6.3 and 6.5.

When $\bar{\mathscr{G}}$ is not isomorphic to Θ, results similar to Result 3 can still be obtained, but involve more detailed properties of groups and subgroups.

6.6.3. Confidence and Credible Sets

In Example 8 of Subsection 4.3.2, it was observed that the classical $100(1-\alpha)\%$ confidence interval for a normal mean (with known variance) is identical to the $100(1-\alpha)\%$ HPD credible set based on the noninformative prior $\pi(\theta)=1$. This observation was of considerable interest, since it showed that the classical interval has good "final precision" in this situation. It would be nice to show that such a correspondence holds in many situations, and indeed this can be done, using Result 3 of the previous subsection.

It is first necessary to decide exactly what a "classical $100(1-\alpha)\%$ confidence rule" is. A *confidence rule*, $C(\cdot)$, is a function from $\mathscr{X}$ into the collection of subsets of Θ, with the interpretation that $C(x)$ will be the suggested confidence set when x is observed. A *confidence level* of $100(1-\alpha)\%$ means that $P_\theta(\theta \in C(X)) \geq 1-\alpha$ for all θ. (Note that this is a probability over X, not θ.)

If now the family of densities of X is invariant under a group $\mathscr{G}$, then it seems reasonable to require C to be an invariant confidence rule. This has come to mean that C should satisfy

$$C(g(x)) = \bar{g}C(x), \tag{6.21}$$

where, as usual, $\bar{g}C(x)=\{\bar{g}(\theta)\colon \theta \in C(x)\}$. The classical justification for this definition of an invariant confidence rule is that the probability of coverage is then appropriately invariant. It can also be justified by a decision-theoretic invariance argument based on an appropriate invariant loss, such as

$$L(\theta, C(x)) = \alpha(1-I_{C(x)}(\theta)) + \beta S(C(x)),$$

where $S(C(x))$ is an invariant measure of the "size" of $C(x)$ (i.e., $S(\bar{g}C(x)) = S(C(x))$).

For simplicity, we will henceforth assume, as in the previous subsection, that $\mathscr{X}$, $\mathscr{G}$, $\bar{\mathscr{G}}$, and Θ are all isomorphic, and that the common group is as in Subsection 6.6.1. Note that only continuous densities are thus being considered.

The first results of interest are obtained by letting the decision loss (as a technical device) be

$$L(\theta, C(x)) = I_{C(x)}(\theta).$$

It is clear that this is an invariant loss, that

$$R(\theta, C) = E_\theta^X[L(\theta, C(X))] = P_\theta(\theta \in C(X)),$$

and that

$$E^{\pi^r(\theta|x)}[L(\theta, C(x))] = P^{\pi^r(\theta|x)}(\theta \in C(x)),$$

where, as before, $\pi^r(\theta|x)$ is the posterior distribution with respect to the right invariant (generalized) prior, π^r. Equation (6.15) becomes, for this loss,

$$P^{\pi^r(\theta|x)}(\theta \in C(x)) = P_{\bar{e}}(\bar{e} \in C(X)) = P_\theta(\theta \in C(X)). \tag{6.22}$$

The second equality states that an invariant confidence rule has a constant probability of coverage (which could, of course, have been obtained from Theorem 1). More importantly, (6.22) says that the probability of coverage of an invariant confidence rule is equal to the posterior probability that θ is in $C(x)$ for the right invariant prior, π^r. This basically means that the measures of initial and final precision for an invariant confidence rule coincide in such invariant situations, which is quite appealing.

It remains to find the optimal invariant confidence rule. This could, of course, be approached from a proper decision-theoretic viewpoint, with specification of the loss, etc., but let us instead examine the problem classically and from a Bayesian inference viewpoint. Thus we specify a desired probability of coverage, $(1-\alpha)$, and seek the optimal invariant confidence rule in the class, $\mathscr{C}$, of invariant confidence rules with probability of coverage at least $1-\alpha$. (By (6.22), this class will be the same whether we adopt the classical or Bayesian (with respect to π^r) interpretation of probability of coverage.)

The natural way of choosing among confidence rules in $\mathscr{C}$ is according to size (see also Subsection 4.3.2). For this purpose, we must first decide upon a natural measure of size. Following Kiefer (1966), one very appealing choice for the measure of the size of $C(x)$ is

$$S^l(C(x)) = \int_{C(x)} \pi^l(\theta)d\theta \tag{6.23}$$

(i.e., size with respect to the left invariant (generalized) prior density). For location parameter problems, this corresponds to the volume of $C(x)$. To see that this is natural in other situations, consider the following example.

Example 19. Assume that S^2 is to be observed, where $S^2/\sigma^2 \sim \chi^2(n)$, and that it is desired to construct a confidence interval, $C(s^2) = (L(s^2), U(s^2))$, for the scale parameter σ^2. Since $\pi^l(\sigma^2) = 1/\sigma^2$ for a scale parameter, (6.23) gives that

$$S^l(C(s^2)) = \int_{L(s^2)}^{U(s^2)} \sigma^{-2}\, d\sigma^2 = \log\left[\frac{U(s^2)}{L(s^2)}\right].$$

A more obvious measure of the size of $C(s^2)$ is just the length $[U(s^2) - L(s^2)]$. The decision as to which measure is the most reasonable is a subjective decision which must be made for each individual problem. The ratio measure $S^l(C(s^2))$ is very appealing for a scale parameter, however, and indeed corresponds to common practice. To see this, note that the usual form considered for $C(s^2)$ is (for an appropriate statistic $t(s^2)$)

$$C(s^2) = \left\{\sigma^2 : a < \frac{t(s^2)}{\sigma^2} < b\right\} = \left(\frac{t(s^2)}{b}, \frac{t(s^2)}{a}\right).$$

This interval is frequently felt to have the same size for all s^2, which is true when $S^l(C(s^2))$, and not length, is used to measure size.

Another appealing facet of $S^l(C(x))$, as a measure of size, is that it is constant for all x when C is invariant. To see this, note that, by choosing $x = e$ in (6.21) (recall we are assuming that $\mathscr{X}$ and $\mathscr{G}$ are the same group), it is clear that

$$C(g) = \bar{g}C(e). \tag{6.24}$$

Hence

$$S^l(C(x)) = \int_{C(x)} \pi^l(\theta)d\theta = \int_{\bar{x}C(e)} \pi^l(\theta)d\theta = \int_{C(e)} \pi^l(\theta)d\theta,$$

the last step following from the left invariance of π^l.

Since size (as measured by S^l) and probability of coverage (Bayesian and classical) are constant for an invariant confidence rule, we can define an optimal invariant rule as follows.

Definition 10. In the above situation, a *π^l-optimal* $100(1-\alpha)\%$ *invariant confidence rule*, $C(\cdot)$, is an invariant confidence rule which minimizes

$$S^l(C(x)) = \int_{C(e)} \pi^l(\theta)d\theta,$$

subject to

$$\begin{aligned} P^{\pi^r(\theta|x)}(\theta \in C(x)) &= P^{\pi^r(\theta|e)}(\theta \in C(e)) = P_{\bar{e}}^X(\bar{e} \in C(X)) \\ &= P_\theta^X(\theta \in C(X)) \geq 1-\alpha. \end{aligned}$$

Note that the π^l-optimal $100(1-\alpha)\%$ invariant confidence rule is, from the Bayesian viewpoint (with prior π^r), the $100(1-\alpha)\%$ credible set which minimizes $S^l(C(x))$. This follows easily from Definition 10 and the facts that $S^l(C(x))$ and $P^{\pi^r(\theta|x)}(\theta \in C(x))$ are constant for this rule.

Result 4. *In the above situation, the* π^l*-optimal* $100(1-\alpha)\%$ *invariant confidence rule is*

$$C(x) = \bar{x}C(e),$$

where

$$C(e) = \{\theta\colon \pi^r(\theta \mid e) > K\pi^l(\theta)\}$$
$$= \left\{\theta\colon \frac{f(e \mid \theta)\pi^r(\theta)}{\pi^l(\theta)} > K'\right\}, \tag{6.25}$$

the constants K or K' being chosen so that

$$P_{\bar{e}}^{X}(\bar{e} \in C(X)) = P^{\pi^r(\theta \mid e)}(\theta \in C(e)) = 1-\alpha. \tag{6.26}$$

Proof. Since we are in the continuous case, it can be assumed that $C(x)$ attains the desired probability of coverage, $1-\alpha$. (Having a probability of coverage larger than $1-\alpha$ will obviously lead to a larger size.) Hence it is only necessary to choose $C(e)$ to minimize

$$S^l(C(x)) = \int_{C(e)} \pi^l(\theta)d\theta,$$

subject to (6.26). It is easy to see that $C(e)$ should consist of those θ for which $\pi^l(\theta)$ is small compared to $\pi^r(\theta \mid e)$, which, together with the fact that $\pi^r(\theta \mid e) = f(e \mid \theta)\pi^r(\theta)/m(e)$, gives (6.25). □

Example 20. Assume that $S^2/\sigma^2 \sim \chi^2(2)$, and that it is desired to construct a π^l-optimal $100(1-\alpha)\%$ invariant confidence rule for σ^2. Clearly the problem is invariant, with σ^2 being a scale parameter, so that

$$\pi^l(\sigma^2) = \pi^r(\sigma^2) = \frac{1}{\sigma^2}.$$

It follows from (6.25) that $C(e) = C(1)$ is of the form

$$C(1) = \{\sigma^2\colon f(1 \mid \sigma^2) > K'\}$$
$$= \left\{\sigma^2\colon \sigma^{-2}\exp\left\{-\frac{1}{2\sigma^2}\right\} > K''\right\}$$
$$= (a, b),$$

where a and b satisfy

$$a^{-1}\exp\left\{-\frac{1}{2a}\right\} = b^{-1}\exp\left\{-\frac{1}{2b}\right\}. \tag{6.27}$$

An easy calculation shows that

$$\pi^r(\sigma^2 \mid 1) = \frac{1}{2}(\sigma^2)^{-2}\exp\left\{-\frac{1}{2\sigma^2}\right\},$$

so that (6.26) implies that a and b must also satisfy

$$\begin{aligned} 1-\alpha &= \int_a^b \frac{1}{2}(\sigma^2)^{-2} \exp\left\{-\frac{1}{2\sigma^2}\right\} d\sigma^2 \\ &= \int_{1/b}^{1/a} \frac{1}{2} \exp\left\{-\frac{1}{2} y\right\} dy \left(\text{letting } y = \frac{1}{\sigma^2}\right) \\ &= \exp\left\{-\frac{1}{2b}\right\} - \exp\left\{-\frac{1}{2a}\right\}. \end{aligned} \tag{6.28}$$

Equations (6.27) and (6.28) can be numerically solved for a and b, and the desired confidence rule is then

$$C(s^2) = \overline{s^2 C}(1) = (s^2 a, s^2 b).$$

Note that the π^l-optimal $100(1-\alpha)\%$ invariant confidence rule will not, in general, equal the $100(1-\alpha)\%$ HPD credible set with respect to π^r. This is because the HPD credible set is chosen to minimize the volume of $C(x)$, which as discussed earlier, may be less suitable than minimizing $S^l(C(x))$.

The correspondence indicated in (6.22) holds in much greater generality than that considered here. See Stein (1965) for a more general result. For further results involving size of confidence sets see Hooper (1982).

6.7. Invariance and Minimaxity

The class of invariant decision rules is often much smaller than the class of all decision rules. This being the case, finding a minimax rule within the class of invariant rules is usually much easier than finding a minimax rule within the class of all rules. For example, a best invariant rule (if one exists) is clearly minimax within the class of invariant rules. It would, therefore, be nice if such a minimax invariant rule turned out to be minimax in the overall problem. This is frequently the case, and the following theorem indicates why.

For simplicity, the theorem will be given only for a finite group of transformations, $\mathcal{G} = \{g_1, g_2, \ldots, g_m\}$, such as those in Examples 6 and 8. Also, to avoid having to worry about randomized rules, we will assume that the loss is convex and that the invariant problem is such that

$$\tilde{g}\left(\frac{1}{m}\sum_{i=1}^{m} a_i\right) = \frac{1}{m}\sum_{i=1}^{m} \tilde{g}(a_i) \tag{6.29}$$

for all $g \in \mathcal{G}$. Equation (6.29) will be satisfied by most of the groups we have encountered, providing the action space $\mathcal{A}$ is convex.

Theorem 4. *Consider a decision problem that is invariant under a finite group, $\mathcal{G} = \{g_1, g_2, \ldots, g_m\}$, for which $\tilde{\mathcal{G}}$ satisfies (6.29). Assume also that $\mathcal{A}$ is* *convex and that the loss, $L(\theta, a)$, is convex in a. Then, if there exists a minimax rule, there will exist a minimax rule which is also invariant. Conversely, if an invariant rule is minimax within the class of all invariant rules, then it is minimax.*

PROOF. Since the loss is convex, we can restrict consideration to nonrandomized rules. It will be shown that, for any decision rule δ, there exists an invariant rule δ^I such that

$$\sup_\theta R(\theta, \delta^I) \le \sup_\theta R(\theta, \delta). \tag{6.30}$$

The conclusions of the theorem then follow immediately.

Define, for a given decision rule δ, the new rule

$$\delta^I(x) = \frac{1}{m} \sum_{i=1}^m \tilde{g}_i^{-1} \delta(g_i(x)).$$

To see that δ^I is an invariant decision rule, observe that

$$\begin{aligned}
\delta^I(g(x)) &= \frac{1}{m} \sum_{i=1}^m \tilde{g}_i^{-1} \delta(g_i(g(x))) \\
&= \frac{1}{m} \sum_{i=1}^m \tilde{g}\tilde{g}^{-1} \tilde{g}_i^{-1} \delta(g_i(g(x))) \\
&= \tilde{g}\left(\frac{1}{m} \sum_{i=1}^m (\tilde{g}_i\tilde{g})^{-1} \delta((g_i g)(x))\right) \quad \text{(using (6.29))} \\
&= \tilde{g}\left(\frac{1}{m} \sum_{i=1}^m (\widetilde{g_i g})^{-1} \delta((g_i g)(x))\right) \\
&= \tilde{g}(\delta^I(x)),
\end{aligned}$$

the last step following from the fact that, for any $g \in \mathscr{G}$, $\{g_1 g, \ldots, g_m g\}$ will just be a reordering of $\{g_1, \ldots, g_m\}$.

To verify (6.30), let δ^i denote $\tilde{g}_i^{-1} \delta(g_i(\cdot))$, and observe that

$$\begin{aligned}
\sup_\theta R(\theta, \delta^I) &\le \sup_\theta \frac{1}{m} \sum_{i=1}^m R(\theta, \delta^i) \quad \text{(convexity of } L) \\
&= \sup_\theta \frac{1}{m} \sum_{i=1}^m E_\theta[L(\bar{g}_i(\theta), \delta(g_i(X)))] \quad \text{(invariance of loss)} \\
&= \sup_\theta \frac{1}{m} \sum_{i=1}^m R_{\bar{g}_i(\theta)}[L(\bar{g}_i(\theta), \delta(X))] \quad \text{(by (6.3))} \\
&\le \frac{1}{m} \sum_{i=1}^m \sup_\theta R(\bar{g}_i(\theta), \delta) \\
&= \frac{1}{m} \sum_{i=1}^m \sup_\theta R(\theta, \delta) = \sup_\theta R(\theta, \delta).
\end{aligned}$$

This completes the proof. □

EXAMPLE 21. Consider the situation of Example 8, where $X \sim \mathcal{B}(n, \theta)$, $\mathcal{A} = \Theta = [0, 1]$, $L(\theta, a) = (\theta - a)^2$, $\mathcal{G} = \{e, g^*\}$ (g^* being the transformation $g^*(x) = n - x$), $\bar{\mathcal{G}} = \{\bar{e}, \bar{g}^*\}$ ($\bar{g}^*$ being the transformation $\bar{g}^*(\theta) = 1 - \theta$), and $\tilde{\mathcal{G}}$ is the same as $\bar{\mathcal{G}}$. Note that

$$\begin{aligned}\tilde{g}^*(\tfrac{1}{2}(a_1 + a_2)) &= 1 - \tfrac{1}{2}(a_1 + a_2)\\ &= \tfrac{1}{2}[(1 - a_1) + (1 - a_2)]\\ &= \tfrac{1}{2}(\tilde{g}^*(a_1) + \tilde{g}^*(a_2)),\end{aligned}$$

so that (6.29) is satisfied. The other conditions of Theorem 4 are also clearly satisfied. It can be concluded that one need only search for a minimax rule within the class of invariant rules, i.e., those which satisfy

$$\delta(n - x) = 1 - \delta(x).$$

This is a helpful reduction of the problem.

It is quite generally true that a minimax rule within the class of invariant rules is an overall minimax rule. By reasoning similar to that in Theorem 4, this result can be established for any finite group or, more generally, for any group for which the Haar density (or Haar measure) has finite mass. (The average $(1/m) \sum_{i=1}^{m}$ in Theorem 4 gets replaced, in a sense, by an average over the Haar density, normalized to be a probability measure.) Even for groups which have infinite Haar measure (such as the groups in Examples 3, 4, and 5), it can often be shown that a minimax rule within the class of invariant rules is overall minimax. This result, known as the Hunt–Stein theorem, is unfortunately too difficult to be presented here. The crucial condition needed for the validity of the Hunt–Stein theorem concerns the nature of the group $\mathcal{G}$. For example, the condition is satisfied if the group is what is called *amenable*. (All groups considered so far have been amenable.) The interested reader is referred to Kiefer (1957) (a very general exposition), Lehmann (1959) (an exposition for certain testing situations), Kiefer (1966) (an heuristic discussion with other references), and Bondar and Milnes (1981) (a modern survey).

It is not always true that a minimax rule must be invariant. The following example (due to Stein) demonstrates this.

EXAMPLE 22. Assume that $\mathbf{X} \sim \mathcal{N}_p(\mathbf{0}, \mathbf{\Sigma})$ and that $\mathbf{Y} \sim \mathcal{N}_p(\mathbf{0}, \Delta\mathbf{\Sigma})$, where $\mathbf{X}$ and $\mathbf{Y}$ are independent and $p \geq 2$. Thus

$$\Theta = \{(\Delta, \mathbf{\Sigma}) \colon \Delta > 0 \text{ and } \mathbf{\Sigma} \text{ is a nonsingular covariance matrix}\}.$$

Let $\mathcal{A} = (0, \infty)$, and assume that it is desired to estimate Δ under loss

$$L((\Delta, \mathbf{\Sigma}), a) = \begin{cases} 0 & \text{if } |1 - a/\Delta| \leq \tfrac{1}{2},\\ 1 & \text{if } |1 - a/\Delta| > \tfrac{1}{2}.\end{cases}$$

Consider the group $\mathscr{G}$ of transformations

$$g_{\mathbf{B}}(\mathbf{x}, \mathbf{y}) = (\mathbf{Bx}, \mathbf{By}),$$

where $\mathbf{B}$ is any nonsingular $(p \times p)$ matrix, and $g_{\mathbf{B}_1} g_{\mathbf{B}_2} = g_{\mathbf{B}_1 \mathbf{B}_2}$. It can be checked that the decision problem is invariant under $\mathscr{G}$, and that $\bar{g}_{\mathbf{B}}((\Delta, \mathbf{\Sigma})) = (\Delta, \mathbf{B\Sigma B}')$ and $\tilde{g}_{\mathbf{B}}(a) = a$. It follows that an invariant decision rule must satisfy

$$\delta(\mathbf{Bx}, \mathbf{By}) = \delta(g_{\mathbf{B}}(\mathbf{x}, \mathbf{y})) = \tilde{g}_{\mathbf{B}}(\delta(\mathbf{x}, \mathbf{y})) = \delta(\mathbf{x}, \mathbf{y}), \tag{6.31}$$

for all $\mathbf{B}$, $\mathbf{x}$, and $\mathbf{y}$. Letting $\mathbf{x} = \mathbf{e}_1 \equiv (1, 0, \ldots, 0)'$, $\mathbf{y} = \mathbf{e}_p \equiv (0, \ldots, 0, 1)'$, $\mathbf{b}^1$ denote the first column of $\mathbf{B}$, and $\mathbf{b}^p$ denote the last column of $\mathbf{B}$, (6.31) becomes

$$\delta(\mathbf{b}^1, \mathbf{b}^p) = \delta(\mathbf{e}_1, \mathbf{e}_p).$$

The only restriction on $\mathbf{b}^1$ and $\mathbf{b}^p$ is that they not be multiples of each other ($\mathbf{B}$ is nonsingular). We can hence conclude that δ must be constant with probability one; say, $\delta(\mathbf{x}, \mathbf{y}) = K$. For any such δ,

$$R((\Delta, \mathbf{\Sigma}), \delta) = \begin{cases} 0 & \text{if } |1 - K/\Delta| \le \frac{1}{2}, \\ 1 & \text{if } |1 - K/\Delta| > \frac{1}{2}, \end{cases}$$

so that

$$\sup_{(\Delta, \mathbf{\Sigma})} R((\Delta, \mathbf{\Sigma}), \delta) = 1.$$

Any invariant rule thus has a minimax risk of 1. (This can be shown to be true for any randomized invariant rule as well.)

Consider now the decision rule

$$\delta(\mathbf{x}, \mathbf{y}) = \left| \frac{y_1}{x_1} \right|.$$

For this rule,

$$\begin{aligned} R((\Delta, \mathbf{\Sigma}), \delta) &= E_{\Delta, \mathbf{\Sigma}} L((\Delta, \mathbf{\Sigma}), \delta(\mathbf{X}, \mathbf{Y})) \\ &= P_{\Delta, \mathbf{\Sigma}}\left(\left| 1 - \frac{\delta(\mathbf{X}, \mathbf{Y})}{\Delta} \right| > \frac{1}{2} \right) \\ &= P_{\Delta, \mathbf{\Sigma}}\left(\left| 1 - \frac{|Y_1/(\Delta\sigma_{11})|}{|X_1/\sigma_{11}|} \right| > \frac{1}{2} \right) \\ &= P\left(\left| 1 - \left| \frac{Z}{Z'} \right| \right| > \frac{1}{2} \right), \end{aligned}$$

where σ_{11} is the standard deviation of X_1, and $Z = Y_1/\Delta\sigma_{11}$ and $Z' = X_1/\sigma_{11}$ are independently $\mathcal{N}(0, 1)$. Clearly this probability is a constant (independent of $(\Delta, \mathbf{\Sigma})$) which is less than one. Hence this rule has smaller minimax risk than any invariant rule.

The Hunt–Stein theorem fails in the above example because the group $\mathscr{G}$, called the full linear group, is too large. Frequently, in such problems, there are subgroups of $\mathscr{G}$ (such as the group of $(p \times p)$ lower triangular matrices) which do satisfy the hypotheses of the Hunt–Stein theorem, and can hence be used to find a minimax rule (cf. James and Stein (1960)).

6.8. Admissibility of Invariant Rules

When a best invariant decision rule exists, it is natural to ask if it is admissible. When the best invariant rule is a proper Bayes rule (for example, when the Haar density is a probability density), the answer is, of course, "yes." Usually, however, the best invariant rule will only be generalized Bayes (with respect to the right invariant Haar measure), and hence need not be admissible. Some typical situations in which the best invariant estimator is inadmissible follow.

Example 23. Assume that $\mathbf{X} \sim \mathcal{N}_p(\boldsymbol{\theta}, \mathbf{I}_p)$, and that it is desired to estimate $\boldsymbol{\theta}$ under loss $L(\boldsymbol{\theta}, \mathbf{a}) = \sum_{i=1}^{p} (\theta_i - a_i)^2$. This problem is invariant under the p-dimensional additive group, and it is easy to see that $\delta_0(\mathbf{x}) = \mathbf{x}$ is the best invariant estimator. When $p \geq 3$, however, it was seen in Example 46 of Section 4.8 that δ_0 is inadmissible.

Example 24. Assume that $X \sim \mathscr{C}(\theta, 1)$, and that it is desired to estimate θ under squared-error loss. This is a location invariant problem, so that the invariant estimators are of the form $\delta_c(x) = x + c$. Now

$$R(\theta, \delta_c) = E_\theta[(\theta - (X + c))^2] = \infty$$

for all c, while the noninvariant estimator $\delta^*(x) = 0$ has risk

$$R(\theta, \delta^*) = \theta^2 < R(\theta, \delta_c).$$

Hence all invariant estimators are inadmissible.

Example 25. In an invariant estimation problem for which the best invariant estimator is not unique, it will usually be the case that any invariant rule is inadmissible. A general proof of this for location densities can be found in Farrell (1964). The basic idea of this proof can be seen in the following specific example due to Blackwell (1951).

Assume that X is either $\theta + 1$ or $\theta - 1$, with probability $\frac{1}{2}$ each. The loss is

$$L(\theta, a) = \begin{cases} |\theta - a| & \text{if } |\theta - a| \leq 1, \\ 1 & \text{if } |\theta - a| > 1, \end{cases}$$

and $\Theta = \mathscr{A} = (-\infty, \infty)$. This is a location invariant problem, so that the invariant rules are of the form $\delta_c(x) = x + c$. The risk of an invariant rule in a location problem is constant, so we need only calculate $R(0, \delta_c)$. This can easily be seen to be

$$R(0, \delta_c) = E_0[L(0, \delta_c(X))] = \begin{cases} 1 - (\frac{1}{2})|c| & \text{if } |c| \leq 1, \\ (\frac{1}{2})|c| & \text{if } 1 \leq |c| \leq 2, \\ 1 & \text{if } |c| \geq 2. \end{cases}$$

This is minimized at $c = +1$ and $c = -1$. Hence $\delta_1(x) = x+1$ and $\delta_{-1}(x) = x-1$ are both best invariant estimators.

Consider now the estimator

$$\delta^*(x) = \begin{cases} \delta_1(x) = x+1 & \text{if } x < 0, \\ \delta_{-1}(x) = x-1 & \text{if } x \geq 0. \end{cases}$$

This is not invariant, but has a risk function which is easily calculated to be

$$R(\theta, \delta^*) = \begin{cases} 0 & \text{if } -1 \leq \theta < 1, \\ \frac{1}{2} & \text{otherwise.} \end{cases}$$

Hence δ^* is R-better than δ_1 or δ_{-1}.

General admissibility results for invariant rules are hard to obtain. Some of the most general results can be found in Brown (1966), Brown and Fox (1974a), Brown and Fox (1974b), and Zidek (1976). These articles show, under very weak conditions, that the best invariant decision rule in one- or two-dimensional location parameter or scale parameter problems is admissible. (The conditions do require that the best invariant rule be unique, and that the density have certain moments, precluding the situations in Examples 24 and 25.)

6.9. Conclusions

In this concluding discussion it will be assumed that no prior information is available. The invariance approach is then appealing, both intuitively and practically. The intuitive appeal was discussed in Section 6.1. The practical appeal lies in the relative ease of application of invariance. The major problem with invariance concerns the amount of invariance that can be used.

The first difficulty is that there may be too little invariance in a problem to lead to a significant simplification. In Section 6.4, for example, it was indicated that invariance considerations are of limited usefulness for the binomial distribution. Even worse are problems involving the Poisson distribution. It can be shown that no group of transformations (other than the identity transformation) leaves the family of Poisson densities invariant. Hence invariance is totally useless for Poisson problems. Finally, the invariance approach is limited by the necessity of having an invariant loss function.

In location parameter estimation problems, typical losses tend to be invariant, but this need not be the case for other types of problems. In scale or location-scale parameter estimation problems, for instance, there are many reasonable losses that are not invariant. Even worse is the situation for testing problems, in which the only loss that will typically be invariant is "$0-K_i$" loss (see Subsection 6.2.2). This is a very serious limitation.

The second difficulty with invariance is that too much invariance may be available, as in Example 22. Indeed a modification of this example will be given as an exercise in which an even larger group is used and there are *no* invariant procedures. A more natural example of the difficulty caused by too much invariance follows.

EXAMPLE 26. Consider a location parameter problem in which X has density $f(|x-\theta|)$ on R^1, and in which the loss is of the form $L(\theta, a) = W(|\theta - a|)$. This problem is not only invariant under the additive group, but is also invariant under the transformation $X \to -X$. Analysis of this new invariance problem will be left as an exercise, but it can be shown that the *only* nonrandomized invariant rule is $\delta_0(x) = x$. An example will be given in the Exercises in which δ_0 is worse than other translation invariant estimators of the form $\delta_c(x) = x + c$. Hence use of the additional transformation in the invariance argument is bad. (Actually, there are reasonable randomized invariant estimators for this problem, namely those that estimate $x+c$ and $x-c$ with probability $\frac{1}{2}$ each. Having to consider randomized rules can itself be considered to be an undesirable complication, however.)

The solution to the problem of having too much invariance is simply to not use it all. One should generally use only the smallest group of transformations, $\mathscr{G}_0$, for which $\bar{\mathscr{G}}_0$ is transitive on Θ. If $\bar{\mathscr{G}}_0$ is transitive on Θ, a best invariant rule will usually exist, which is why transitivity is desirable. The use of a group of transformations, $\mathscr{G}$, which is larger than $\mathscr{G}_0$, serves no real purpose, however, since it only reduces the class of invariant rules. Indeed the best invariant rule for $\mathscr{G}_0$ may not be invariant under $\mathscr{G}$, so that the use of $\mathscr{G}$ instead of $\mathscr{G}_0$ can cause a definite increase in the risk of the invariant rule chosen. In Example 26, for instance, the additive group

$$\mathscr{G}_0 = \{g_c\colon g_c(x) = x + c,\ c \in R^1\}$$

results in a $\bar{\mathscr{G}}_0$ which is transitive on Θ. The additional use of the transformation $X \to -X$ reduces the class of nonrandomized invariant rules to the single rule $\delta_0(x) = x$, which may be a harmful reduction.

As a concluding consideration, the relationship between the invariance approach and the Bayesian noninformative prior approach should be discussed. In Section 6.6 it was shown that the two approaches are often virtually equivalent (if the right invariant (generalized) prior density is used as the noninformative prior). To this author, it seems preferable to approach a problem from the noninformative prior viewpoint, rather than from the invariance viewpoint. The reasons are that:

(i) the noninformative prior viewpoint is much more widely applicable than the invariance viewpoint, applying in noninvariant situations;
(ii) the noninformative prior approach is much easier than the invariance approach (compare the respective derivations of Pitman's estimator);
(iii) the noninformative prior approach yields conditional measures of accuracy (which sometimes will differ from the frequentist risk);
(iv) the noninformative prior approach at least forces some consideration of relative likelihoods of various θ; and
(v) the noninformative prior approach will rarely lead to the difficulties encountered by the invariance approach when too much invariance is present. (In Example 26, for instance, the noninformative prior is $\pi(\theta)=1$, and the application of the Bayesian noninformative prior approach will lead directly to the best translation invariant rule. One need not be concerned with how much invariance should be used.)

Of course, the study of invariance has been very useful in suggesting a reasonable choice for the noninformative prior, namely the right invariant Haar density. Also, as always, it is good to examine a statistical problem from as many viewpoints as possible, and the invariance viewpoint can definitely be instructive.

Exercises

Section 6.2

1. Verify that the permutation group, given in Example 6, is indeed a group of transformations.

2. Verify that $\bar{\mathscr{G}}$ and $\tilde{\mathscr{G}}$ are groups of transformations of Θ and $\mathscr{A}$, respectively.

3. Assume that $\mathbf{X} \sim \mathcal{N}_p(\theta_1 \mathbf{1}, \theta_2^2 \mathbf{I}_p)$, where $\mathbf{1}=(1, 1, \ldots, 1)^t$. The parameter space is $\Theta=\{(\theta_1, \theta_2)\colon \theta_1 \in R^1 \text{ and } \theta_2 > 0\}$. Let $\mathscr{A}=R^1$ and $L(\theta, a)=(\theta_1 - a)^2/\theta_2^2$. Finally, let
$$\mathscr{G}=\{g_{b,c}\colon g_{b,c}(\mathbf{x})=b\mathbf{x}+c\mathbf{1}, \text{ where } c \in R^1 \text{ and } b \neq 0\}.$$
   (a) Show that $\mathscr{G}$ is a group of transformations.
   (b) Show that the decision problem is invariant under $\mathscr{G}$, and find $\bar{\mathscr{G}}$ and $\tilde{\mathscr{G}}$.
   (c) Determine the relationship which must be satisfied by an invariant decision rule.
   (d) Show that an invariant decision rule must have constant risk.

4. Assume that $\mathbf{X} \sim \mathcal{N}_p(\theta_1 \mathbf{1}, \theta_2^2 \mathbf{I}_p)$, where $\mathbf{1}=(1, 1, \ldots, 1)^t$. The parameter space is $\Theta=\{(\theta_1, \theta_2)\colon \theta_1 \in R^1 \text{ and } \theta_2 > 0\}$. It is desired to test $H_0\colon \theta_1 \leq 0$ versus $H_1\colon \theta_1 > 0$ under "0–1" loss. Let
$$\mathscr{G}=\{g_c\colon g_c(\mathbf{x})=c\mathbf{x}, \text{ where } c>0\}.$$
   (a) Show that the decision problem is invariant under $\mathscr{G}$, and find $\bar{\mathscr{G}}$ and $\tilde{\mathscr{G}}$.
   (b) Find the form of invariant decision rules.

5. Assume that $\mathbf{X} \sim \mathcal{N}_p(\boldsymbol{\theta}, \mathbf{I}_p)$, where $\Theta = R^p$. The loss function is of the form $L(\boldsymbol{\theta}, a) = W(|\boldsymbol{\theta}|, a)$, so that it depends on $\boldsymbol{\theta}$ only through its length. Let $\mathscr{G}$ be the group of orthogonal transformations of R^p. (Thus

$$\mathscr{G} = \{g_{\mathcal{O}}: g_{\mathcal{O}}(\mathbf{x}) = \mathcal{O}\mathbf{x}, \text{ where } \mathcal{O} \text{ is an orthogonal } (p \times p) \text{ matrix}\}.)$$

   (a) Verify that $\mathscr{G}$ is indeed a group of transformations.
   (b) Show that the decision problem is invariant under $\mathscr{G}$, and find $\bar{\mathscr{G}}$ and $\tilde{\mathscr{G}}$.
   (c) Show that the class of invariant decision rules is the class of rules based on $|\mathbf{x}|$.
   (d) Show that the risk function of an invariant decision rule depends only on $|\boldsymbol{\theta}|$.

Section 6.3

6. Assume that $X \sim \mathscr{C}(\theta, \beta)$, where β is known, and that it is desired to estimate θ under loss

$$L(\theta, a) = \begin{cases} 0 & \text{if } |\theta - a| \le c, \\ 1 & \text{if } |\theta - a| > c. \end{cases}$$

   Find the best invariant estimator of θ.

7. Assume that $X \sim \mathcal{N}(\theta + 2, 1)$, and that it is desired to estimate θ under squared-error loss. Find the best invariant estimator of θ.

8. Assume that $\mathbf{X} = (X_1, X_2)$ has a density (on R^2) of the form $f(x_1 - \theta_1, x_2 - \theta_2)$, and that it is desired to estimate $(\theta_1 + \theta_2)/2$ under loss

$$L((\theta_1, \theta_2), a) = (\tfrac{1}{2}(\theta_1 + \theta_2) - a)^2.$$

   Here $\mathscr{A} = R^1$ and $\Theta = R^2$.
   (a) Show that the problem is invariant under the group

$$\mathscr{G} = \{g_{\mathbf{c}}: g_{\mathbf{c}}(\mathbf{x}) = (x_1 + c_1, x_2 + c_2), \text{ where } \mathbf{c} = (c_1, c_2) \in R^2\},$$

   and find $\bar{\mathscr{G}}$ and $\tilde{\mathscr{G}}$.
   (b) Find the best invariant decision rule.

9. Assume that $X_1, \ldots, X_n$ is a sample of size n from the density

$$f(x|\theta) = \exp\{-(x - \theta)\} I_{(\theta,\infty)}(x).$$

   Find the best invariant estimator of θ for the loss function
   (a) $L(\theta, a) = (\theta - a)^2$.
   (b) $L(\theta, a) = |\theta - a|$.
   (c) $L(\theta, a) = 0$ if $|\theta - a| \le c$; $L(\theta, a) = 1$ if $|\theta - a| > c$.

10. Assume that X has a density of the form $f(|x - \theta|)$ on R^1, and that it is desired to estimate θ under a strictly convex loss which depends only on $|\theta - a|$. Prove that $\delta(x) = x$ is a best invariant estimator.

11. Assume that X has a unimodal density of the form $f(|x - \theta|)$ on R^1 (so that $f(z)$ is nonincreasing for $z \ge 0$) and that it is desired to estimate θ under a loss of the form $L(\theta, a) = W(|\theta - a|)$, where $W(z)$ is nondecreasing for $z \ge 0$. Prove that $\delta(x) = x$ is a best invariant estimator.

12. Assume that $X_1, \ldots, X_n$ is a sample from a $\mathcal{N}(\theta, 1)$ distribution, and that it is desired to estimate θ. Find the best invariant estimator of θ for the loss
    (a) $L(\theta, a) = |\theta - a|^r$, $r \geq 1$.
    (b) $L(\theta, a) = |\theta - a|^r$, $r < 1$.
    (c) $L(\theta, a) = 0$ if $|\theta - a| \leq c$; $L(\theta, a) = 1$ if $|\theta - a| > c$.

13. Assume that $X_1, \ldots, X_n$ is a sample from the $\mathcal{U}(\theta - \frac{1}{2}, \theta + \frac{1}{2})$ distribution. Find the best invariant estimator of θ for the loss
    (a) $L(\theta, a) = (\theta - a)^2$.
    (b) $L(\theta, a) = |\theta - a|$.
    (c) $L(\theta, a) = 0$ if $|\theta - a| \leq c$; $L(\theta, a) = 1$ if $|\theta - a| > c$.

14. Assume that $X_1, \ldots, X_n$ is a sample from the half-normal distribution, which has density

$$f(x|\theta) = \left(\frac{2}{\pi}\right)^{1/2} \exp\left\{-\frac{1}{2}(x - \theta)^2\right\} I_{(\theta, \infty)}(x).$$

Show that the best invariant estimator of θ under squared-error loss is

$$\delta(x) = \bar{x} - \frac{\exp\{-n[(\min x_i) - \bar{x}]^2/2\}}{(2n\pi)^{1/2} P(Z < n^{1/2}[(\min x_i) - \bar{x}])},$$

where Z is $\mathcal{N}(0, 1)$.

Section 6.4

15. Assume that $X_1, \ldots, X_n$ is a sample from the $\mathcal{G}(\alpha, \beta)$ distribution, with α known and β unknown. Find the best invariant estimator of β for the loss
    (a) $L(\beta, a) = (1 - a/\beta)^2$.
    (b) $L(\beta, a) = (a/\beta) - 1 - \log(a/\beta)$.

16. Assume that $X_1, \ldots, X_n$ is a sample from the $\mathcal{P}a(\theta, 1)$ distribution, where $\theta > 0$. Find the best invariant estimator of θ for the loss
    (a) $L(\theta, a) = (1 - a/\theta)^2$, when $n \geq 3$.
    (b) $L(\theta, a) = |\log a - \log \theta|$.
    (c) $L(\theta, a) = |1 - a/\theta|$, when $n \geq 2$.

17. Assume that $X_1, \ldots, X_n$ is a sample from the $\mathcal{U}(0, \theta)$ distribution where $\theta > 0$. Find the best invariant estimator of θ for the loss
    (a) $L(\theta, a) = (1 - a/\theta)^2$.
    (b) $L(\theta, a) = |1 - a/\theta|$.
    (c) $L(\theta, a) = 0$ if $c^{-1} \leq a/\theta \leq c$; $L(\theta, a) = 1$ otherwise.

18. Assume that $X_1, \ldots, X_n$ is a sample from the $\mathcal{N}(0, \sigma^2)$ distribution, where $\sigma > 0$.
    (a) Find the best invariant estimator of σ^2 for the loss $L(\sigma^2, a) = (1 - a/\sigma^2)^2$.
    (b) Find the best invariant estimator of σ for the loss $L(\sigma, a) = (1 - a/\sigma)^2$.

19. Assume that $X_1, \ldots, X_n$ are positive random variables with a joint density of the form $\theta^{-n} f(x_1/\theta, \ldots, x_n/\theta)$, where $\theta > 0$ is an unknown scale parameter. It is desired to estimate θ under a loss of the form $L(\theta, a) = W(a/\theta)$.

(a) Show that the best invariant estimator of θ is that action, a, which minimizes

$$\frac{\int_0^\infty \theta^{-(n+1)} W(a/\theta) f(x_1/\theta, \ldots, x_n/\theta) d\theta}{\int_0^\infty \theta^{-(n+1)} f(x_1/\theta, \ldots, x_n/\theta) d\theta}.$$

(You may use the corresponding result for the location parameter case.)

(b) If $L(\theta, a) = (1 - a/\theta)^2$, show that the best invariant estimator of θ is

$$\delta(\mathbf{x}) = \frac{\int_0^\infty \theta^{-(n+2)} f(x_1/\theta, \ldots, x_n/\theta) d\theta}{\int_0^\infty \theta^{-(n+3)} f(x_1/\theta, \ldots, x_n/\theta) d\theta}.$$

(c) If $X_1, \ldots, X_n$ are a sample from the $\mathcal{U}(\theta, 2\theta)$ distribution, where $\theta > 0$, and it is desired to estimate θ under loss $L(\theta, a) = (1 - a/\theta)^2$, show that the best invariant estimator of θ is

$$\delta(\mathbf{x}) = \frac{(n+2)[(V/2)^{-(n+1)} - U^{-(n+1)}]}{(n+1)[(V/2)^{-(n+2)} - U^{-(n+2)}]},$$

where $U = \min x_i$ and $V = \max x_i$.

Section 6.5

20. Verify that $\mathbf{T}(\mathbf{x})$ in Example 13 is indeed a maximal invariant.

21. Verify that $\mathbf{T}(\mathbf{x})$ in Example 14 is indeed a maximal invariant.

22. Assume that $\mathscr{X} = R^p$, and let $\mathscr{G}$ be the group of orthogonal transformations of $\mathscr{X}$ (see Exercise 5). Show that $T(\mathbf{x}) = \sum_{i=1}^p x_i^2$ is a maximal invariant.

23. Assume that X_1 and X_2 are independent observations from a common density, f. It is desired to test H_0: f is $\mathcal{N}(\theta, 1)$ versus H_1: f is $\mathscr{C}(\theta, 1)$ under "0–1" loss. Show that the uniformly most powerful invariant tests reject H_0 if $|X_1 - X_2| > K$ and accept H_0 otherwise, where K is a constant which depends on the desired error probabilities.

24. Assume that $X_1 \sim \mathcal{N}(\theta_1, 1)$ and (independently) $X_2 \sim \mathcal{N}(\theta_2, 1)$. It is desired to test H_0: $\theta_1 = \theta_2$ versus H_1: $\theta_1 < \theta_2$ under "0–1" loss. Let $\mathscr{G}$ be the group of transformations

$$\mathscr{G} = \{g_c: g_c((x_1, x_2)) = (x_1 + c, x_2 + c)\}.$$

(a) Show that the problem is invariant under $\mathscr{G}$, and find $\bar{\mathscr{G}}$ and $\tilde{\bar{\mathscr{G}}}$.
(b) Show that $T((x_1, x_2)) = (x_2 - x_1)$ is a maximal invariant.
(c) Show that the uniformly most powerful invariant tests reject H_0 if $Y = X_2 - X_1 > K$ and accept H_0 otherwise, where K is a constant which depends on the desired error probabilities.

25. Assume that $X_1, \ldots, X_n$ are a sample from the $\mathcal{N}(\theta, \sigma^2)$ distribution, both θ and σ^2 unknown. It is desired to test H_0: $\sigma^2 \geq 1$ versus H_1: $\sigma^2 < 1$ under "0–1" loss.
(a) After reducing to the sufficient statistics $\bar{X}$ (the sample mean) and S^2 (the sample variance), show that the problem is invariant under the location group

$$\mathscr{G} = \{g_c: g_c((\bar{x}, s^2)) = (\bar{x} + c, s^2), \text{ where } c \in R^1\}.$$

(b) Show that $T((\bar{x}, s^2)) = s^2$ is a maximal invariant.

(c) Show that the uniformly most powerful invariant tests reject H_0 if $S^2 < K$ and accept H_0 otherwise, where K is a constant which depends on the desired error probabilities.

Section 6.6

26. Calculate J_g^l in Example 17.

27. Prove Result 2.

28. Let $\mathscr{G}$ be the group of one-dimensional scale transformations, and identify each g_c with the point $c \in (0, \infty)$. Show that the left and right invariant Haar densities for this group are $h^l(c) = h^r(c) = 1/c$.

29. Let $\mathscr{G}$ be the p-dimensional location (or additive) group, and identify each $g_{\mathbf{c}}$ with the point $\mathbf{c} \in R^p$. Show that the left and right invariant Haar densities for this group are $h^l(\mathbf{c}) = h^r(\mathbf{c}) = 1$.

The following three problems deal with groups, $\mathscr{G}$, of matrix transformations of R^p. We will identify a transformation with the relevant matrix, so that $\mathbf{g} \in \mathscr{G}$ is a $(p \times p)$ matrix (with (i, j) element $g_{i,j}$), and the transformation is simply $\mathbf{x} \to \mathbf{gx}$. The composition of two transformations (or the group multiplication) corresponds simply to matrix multiplication. Thus to find Haar densities, we must be concerned with the transformations of $\mathscr{G}$ given by $\mathbf{g} \to \mathbf{g}^0\mathbf{g}$ and $\mathbf{g} \to \mathbf{g}\mathbf{g}^0$. To calculate the Jacobians of these transformations, write the matrices $\mathbf{g}$, $\mathbf{g}^0$, $\mathbf{g}^0\mathbf{g}$, and $\mathbf{g}\mathbf{g}^0$ as vectors, by stringing the rows of the matrices end to end. Thus a $(p \times p)$ matrix will be treated as a vector in R^{p^2}. Coordinates which are always zero can be ignored, however, possibly reducing the dimension of the vector. The densities calculated for the transformations are then densities with respect to Lebesgue measure on the Euclidean space spanned by the reduced vectors.

30. If $\mathscr{G}$ is the group of nonsingular $(p \times p)$ diagonal matrices, show that the left and right invariant Haar densities (w.r.t. $\prod_{i=1}^p dg_{ii}$) are
$$h^l(\mathbf{g}) = h^r(\mathbf{g}) = \frac{1}{|\det \mathbf{g}|} = \frac{1}{\prod_{i=1}^p |g_{ii}|}.$$

31. Let $\mathscr{G}$ be the group of all linear transformations of R^p or, equivalently, the group of all $(p \times p)$ nonsingular matrices. Show that the left and right invariant Haar densities (w.r.t. $\prod_{i=1}^p \prod_{j=1}^p dg_{ij}$) are $h^l(\mathbf{g}) = h^r(\mathbf{g}) = |\det \mathbf{g}|^{-p}$,
    (a) when $p = 2$,
    (b) for general p.

32. Let $\mathscr{G}$ be the group of all $(p \times p)$ nonsingular lower triangular matrices (i.e., $g_{ij} = 0$ for $j > i$). Show that the left and right invariant Haar densities (w.r.t. $\prod_{i=1}^p \prod_{j=1}^i dg_{ij}$) are
$$h^l(\mathbf{g}) = \frac{1}{\prod_{i=1}^p |g_{ii}|^i} \quad \text{and} \quad h^r(\mathbf{g}) = \frac{1}{\prod_{i=1}^p |g_{ii}|^{p+1-i}},$$
    (a) when $p = 2$,
    (b) for general p.

33. In the situation of Example 18, find the best invariant estimator for the loss
    (a) $L((\theta, \sigma), a) = (\alpha\theta + \beta\sigma - a)^2/\sigma^2$.
    (b) $L((\theta, \sigma), a) = (1 - a/\sigma)^2$.
    (c) $L((\theta, \sigma), a) = (\theta - a)^2/\sigma^2$.

34. Assume that $X_1, \ldots, X_n$ is a sample from the $\mathscr{U}(\theta - \sqrt{3}\sigma, \theta + \sqrt{3}\sigma)$ distribution, and that it is desired to estimate $\alpha\theta + \beta\sigma$ under the loss $L((\theta, \sigma), a) = (\alpha\theta + \beta\sigma - a)^2/\sigma^2$, where α and β are given constants. Find the generalized Bayes estimator with respect to the generalized prior density $\pi(\theta, \sigma) = 1/\sigma$ (on $\Theta = \{(\theta, \sigma): \theta \in R^1 \text{ and } \sigma > 0\}$), and show that this is also the best invariant estimator.

35. In the situation of Example 20, find the π^l-optimal 90% invariant confidence rule for σ^2, and compare it to the 90% HPD credible set with respect to the generalized prior $\pi(\sigma^2) = 1/\sigma^2$.

36. Assume that X has density (on $(0, \infty)$)

$$f(x|\beta) = \frac{2\beta}{\pi(\beta^2 + x^2)},$$

where $\beta > 0$. Find the π^l-optimal 90% invariant confidence rule for β.

37. For a one-dimensional location parameter problem, prove that the π^l-optimal $100(1-\alpha)\%$ invariant confidence rule coincides with the $100(1-\alpha)\%$ HPD credible set with respect to the noninformative prior $\pi(\theta) = 1$.

38. Assume that $X_1, \ldots, X_n$ is a sample from the $\mathscr{U}(0, \theta)$ distribution. Find the π^l-optimal $100(1-\alpha)\%$ invariant confidence rule, and show that it coincides with the $100(1-\alpha)\%$ HPD credible set with respect to the noninformative prior $\pi(\theta) = 1/\theta$.

Section 6.7

39. Assume that $X \sim \mathscr{B}(1, \theta)$ is observed, on the basis of which it is desired to estimate θ under loss $L(\theta, a) = |\theta - a|$. Find a minimax estimator of θ.

40. Assume that $X \in R^1$ has density

$$f(x|\theta) = \frac{2\theta}{\pi(e^{\theta x} + e^{-\theta x})},$$

where $\theta > 0$. It is desired to estimate θ under a convex loss. Show that, to determine a minimax estimator, one need only consider estimators which are a function of $|x|$.

Sections 6.8 and 6.9

41. In Example 25, show that, for any constant b, the estimator

$$\delta_b^*(x) = \begin{cases} \delta_1(x) = x + 1 & \text{if } x < b, \\ \delta_{-1}(x) = x - 1 & \text{if } x \geq b, \end{cases}$$

is R-better than the best invariant estimators.

42. Consider the situation of Example 22.
 (a) Show that the given decision problem is actually invariant under the larger group

$$\mathscr{G} = \{g_{\mathbf{B},c}\colon g_{\mathbf{B},c}(\mathbf{x}, \mathbf{y}) = (\mathbf{B}\mathbf{x}, c\mathbf{B}\mathbf{y}), \text{ where } \mathbf{B} \text{ is nonsingular and } c > 0\},$$

 and find $\bar{\mathscr{G}}$ and $\tilde{\mathscr{G}}$.
 (b) Show that there are *no* invariant decision rules for this problem. (Clearly, too much invariance has been used.)

43. In Example 26, let $\mathscr{G}$ be the group consisting of the transformations $g_{1,c}(x) = x + c$ and $g_{-1,c'}(x) = -x + c'$, where c and c' vary over R^1.
 (a) Show that the decision problem is invariant under $\mathscr{G}$, and find $\bar{\mathscr{G}}$ and $\tilde{\mathscr{G}}$.
 (b) Show that $\delta_0(x) = x$ is the only nonrandomized decision rule which is invariant under $\mathscr{G}$.
 (c) (Kiefer) Suppose that $f(|x-\theta|) = \frac{1}{2}I_{(1,2)}(|x-\theta|)$, and that $W(|\theta - a|) = |\theta - a|^{1/2}$. Consider the estimators $\delta_c(x) = x + c$, and show that, for $|c| < 1$, $R(\theta, \delta_c) = R(0, \delta_c)$ is a strictly concave and symmetric function of c. Use this to conclude that $\delta_0(x) = x$ is inadmissible. (Note that since the best invariant translation estimator cannot be unique, it too will be inadmissible, though not as seriously so as δ_0.)
 (d) Consider the randomized (invariant) estimators of the form

$$\delta_c^*(x, a) = \begin{cases} \frac{1}{2} & \text{if } a = x + c, \\ \frac{1}{2} & \text{if } a = x - c, \end{cases}$$

 (i.e., δ_c^* estimates $x + c$ and $x - c$ with probability $\frac{1}{2}$ each). Show that there are c for which δ_c^* is R-better than δ_0.
 (e) Prove that δ_c^* is inadmissible for any c.

CHAPTER 7
Preposterior and Sequential Analysis

7.1. Introduction

Until now, we have dealt only with the making of decisions or inferences. Another very important aspect of statistics is that of the choice of experiment, commonly called experimental design. Being as this choice must (usually) be made before the data (and hence the posterior distribution) can be obtained, the subject is frequently called *preposterior analysis* by Bayesians.

The goal of preposterior analysis is to choose the experiment or experimental design which minimizes overall cost. This overall cost consists of the decision loss *and* the cost of conducting and analyzing the experiment. We will essentially ignore the cost of analyzing the results (after all, statisticians are so underpaid that the cost of their labors is usually negligible), and thus consider only the decision loss and the cost of experimentation. Note that these last quantities are in opposition to each other. To lower the decision loss it will generally be necessary to run a larger experiment, whereby the experimental cost will be increased. In this chapter, we will be concerned with the balancing of these two costs.

The general problem of experimental design is very complicated, concerning choices among quite different experiments (say, a choice between a completely randomized design and a randomized block design, or maybe a choice of the values of the independent variables in a regression study). Though of considerable interest, investigation of such general questions is beyond the scope of this book. We will content ourselves with a study of the simplest design problem, that of deciding when to stop sampling. (Discussion and references for other problems in Bayesian design can be found in Pilz (1983) and Chaloner (1984).)

To get specific, assume random variables $X_1, X_2, \ldots$ are available for observation. Let $\mathscr{X}_i$ be the sample space of X_i, define $\mathbf{X}^j = (X_1, X_2, \ldots, X_j)$, and assume that $\mathbf{X}^j$ has density $f_j(\mathbf{x}^j|\theta)$ on $\mathscr{X}^j = \mathscr{X}_1 \times \cdots \times \mathscr{X}_j$. As usual, $\theta \in \Theta$ is the unknown state of nature, concerning which some inference or decision is to be made. Most of the examples we will consider deal with situations in which the X_i are independent observations from a common density $f(x|\theta)$. In such a situation,

$$f_j(\mathbf{x}^j|\theta) = \prod_{i=1}^{j} f(x_i|\theta),$$

and we will say that $X_1, X_2, \ldots$ is a *sequential sample* from the density $f(x|\theta)$. It will be assumed that the observations can be taken in stages, or *sequentially*. This means that after observing, say, $\mathbf{X}^j = (X_1, \ldots, X_j)$, the experimenter has the option of either making an immediate decision or taking further observations.

The experimental cost in this setup is simply the cost of taking observations. This cost can depend on many factors, two of the most crucial being the number of observations ultimately taken, and the way in which the observations are taken (i.e., one at a time, in groups, etc.). To quantify this, let n denote the number of observations ultimately taken, let s denote the manner in which the observations are taken, and, as usual, let $a \in \mathscr{A}$ denote the action taken. Then

$$L(\theta, a, n, s)$$

will denote the overall loss or cost when θ turns out to be the true state of nature.

Frequently, it will be the case that $L(\theta, a, n, s)$ can be considered to be the sum of the *decision loss*, $L(\theta, a)$, and the *cost of observation* (or *sampling cost*), $C(n, s)$. This will happen when the decision maker has a (nearly) linear utility function, so that the combined loss is just the sum of the individual losses. If the utility function U is nonlinear and the situation involves, say, monetary gain or loss, then it will generally happen, instead, that $L(\theta, a, n, s) = -U(G(\theta, a) - C(n, s))$, where $G(\theta, a)$ represents the monetary gain when the pair (θ, a) occurs and $C(n, s)$ is the sampling cost.

Unfortunately, even the sequential decision problem in this generality is too hard to handle. The difficulty occurs in trying to deal with all possible methods, s, of taking observations. We will, therefore, restrict ourselves to studying the two most common methods of taking observations. The first is the *fixed sample size* method, in which one preselects a sample size n, observes $\mathbf{X}^n = (X_1, \ldots, X_n)$, and makes a decision. The overall loss for this situation will be denoted $L^F(\theta, a, n)$. This problem will be considered in the next section.

The second common method of taking observations is that of *sequential analysis*, in which the observations are taken one at a time, with a decision being made, after each observation, to either cease sampling (and choose an action $a \in \mathscr{A}$) or take another observation. This is the situation that will be discussed in the bulk of the chapter (Sections 7.3 through 7.6), and so, for simplicity, the overall loss will just be written $L(\theta, a, n)$ in this case.

An important special case of the preceding situation is that in which the decision loss is $L(\theta, a)$, the ith observation costs c_i, and the utility function is approximately linear. Then $L(\theta, a, n)$ will be of the form

$$L(\theta, a, n) = L(\theta, a) + \sum_{i=1}^{n} c_i. \tag{7.1}$$

In such a situation, the phrases "loss," "risk," "Bayes risk," etc., will refer to the overall loss $L(\theta, a, n)$ and its derived risks, while "decision loss," "decision risk," etc., will refer to $L(\theta, a)$ and its derived risks.

When the fixed sample size loss, $L^F(\theta, a, n)$, is equal to the sequential loss, $L(\theta, a, n)$, (essentially a statement that it is no cheaper to take observations in a batch than one at a time) a sequential analysis of the problem will usually be considerably cheaper than a fixed sample size analysis. To intuitively see why, consider a lot inspection situation in which it is desired to test H_0: $\theta = 0.05$ versus H_1: $\theta = 0.15$, where θ is the proportion of defectives in the lot. The possible observations, $X_1, X_2, \ldots,$ are a sequential sample (i.e., are independent) from the $\mathscr{B}(1, \theta)$ distribution.

Suppose that the optimal fixed sample size experiment requires a sample of size $n = 100$, and imagine that, instead, we decide to observe the X_i sequentially. Now if it so happens that the first 50 items tested from the lot are all good, then there is clearly overwhelming evidence that H_0 is true. Taking another 50 observations (as required by the fixed sample size approach) would almost certainly be a waste of money.

At the other extreme in this example, imagine that all 100 observations are taken, out of which 10 defectives are observed. There is then very little evidence as to whether H_0 or H_1 is true, and if it is important to determine which is true (and it is known that θ can only be 0.05 or 0.15), then more observations are needed.

The advantage of sequential analysis should be clear. It allows one to gather exactly the correct amount of data needed for a decision of the desired accuracy.

A word of warning should be given. Sequential analysis is not easy. Much of the chapter may seem somewhat abstract and technical, but it is abstraction that cannot be avoided. A considerable amount of theory must be brought to bear to obtain reasonable answers to sequential problems. Because of the difficulty of preposterior and sequential analysis, we are also very limited as to topics that can be covered. Indeed, a serious effort is made only to introduce the ideas behind: (i) optimal fixed sample size selection (Section 7.2); (ii) sequential Bayes analysis (Sections 7.3 and 7.4); (iii) the sequential probability ratio test (Section 7.5); and (iv) the evidential relevance of the stopping rule (Section 7.7). For more general coverage of sequential analysis (particularly from a classical perspective) see Wald

(1947, 1950), Ghosh (1970), Govindarajulu (1981), and Siegmund (1985). (Note that Abraham Wald is generally considered to be the founder of sequential analysis.) The material on sequential Bayes analysis is very related to two other extensively studied areas: dynamic programming and optimal stopping problems (cf. DeGroot (1970), Chow, Robbins, and Siegmund (1971), Larson and Casti (1978, 1982), and Whittle (1982)); and bandit problems, where one can choose not only whether or not to take another observation, but also which experiment to perform (cf. Chernoff (1959, 1972), Gitting (1979), Keener (1980, 1984), and Berry and Fristedt (1985)).

7.2. Optimal Fixed Sample Size

In determining the optimal fixed sample size in a decision problem, the most sensible approach to use is the Bayesian approach. This is because, in balancing the decision loss and the sampling cost for any given sample size, the only reasonable pre-experimental measure of the expected decision loss is the Bayes decision risk. Formally, we define

Definition 1. Assume that $L^F(\theta, a, n)$ is the loss in observing $\mathbf{X}^n = (X_1, \ldots, X_n)$ and taking action a, and that θ has the prior density $\pi(\theta)$. Let δ_n^π denote a Bayes decision rule for this problem (if one exists), and let

$$r^n(\pi) = E^\pi E_\theta^{\mathbf{X}^n}[L^F(\theta, \delta_n^\pi(\mathbf{X}^n), n)]$$

denote the Bayes risk for the problem. (δ_0^π and $r^0(\pi)$ are the corresponding quantities for the no-observation problem.)

The optimal fixed sample size is clearly that n which minimizes $r^n(\pi)$. This can often be found by simply differentiating $r^n(\pi)$ with respect to n and setting equal to zero. Some examples follow. In all examples it is assumed that L^F is of the form

$$L^F(\theta, a, n) = L(\theta, a) + C(n). \tag{7.2}$$

It is clear, in such a situation, that δ_n^π will be the Bayes decision rule (based on n observations) for the loss $L(\theta, a)$. We will, as usual, use $r(\pi, \delta_n^\pi)$ to denote the Bayes decision risk of this rule.

Example 1. Assume that $X_1, X_2, \ldots$ is a sequential sample from a $\mathcal{N}(\theta, \sigma^2)$ density (σ^2 known), and that it is desired to estimate θ under a loss of the form (7.2) with $L(\theta, a) = (\theta - a)^2$. The parameter θ is thought to have a $\mathcal{N}(\mu, \tau^2)$ prior density.

If $\mathbf{X}^n = (X_1, \ldots, X_n)$ is to be observed, we know that a sufficient statistic for θ is

$$\bar{X}_n = \frac{1}{n} \sum_{i=1}^{n} X_i.$$

As in Chapter 4, it follows that the posterior distribution of θ given $\mathbf{x}^n$, to be denoted $\pi^n(\theta|\bar{x}_n)$, is $\mathcal{N}(\mu_n(\bar{x}_n), \rho_n)$, where

$$\mu_n(\bar{x}_n) = \frac{\sigma^2}{\sigma^2 + n\tau^2} \mu + \frac{n\tau^2}{\sigma^2 + n\tau^2} \bar{x}_n,$$

and (7.3)

$$\rho_n = \frac{\sigma^2 \tau^2}{\sigma^2 + n\tau^2}.$$

Since the decision loss is squared-error loss, it follows that $\delta_n^\pi(\mathbf{x}^n) = \mu_n(\bar{x}_n)$ is the Bayes decision rule, and that

$$r(\pi, \delta_n^\pi) = \rho_n = \frac{\sigma^2 \tau^2}{\sigma^2 + n\tau^2}.$$

Case 1. Assume that each observation costs c, so that $C(n) = nc$. Then

$$r^n(\pi) = r(\pi, \delta_n^\pi) + C(n) = \frac{\sigma^2 \tau^2}{\sigma^2 + n\tau^2} + nc.$$

Pretending that n is a continuous variable and differentiating with respect to n gives

$$\frac{d}{dn} r^n(\pi) = -\frac{\sigma^2 \tau^4}{(\sigma^2 + n\tau^2)^2} + c.$$

Setting equal to zero and solving gives

$$n^* = \sigma c^{-1/2} - \frac{\sigma^2}{\tau^2}$$

as the approximate minimizing value of n. Since the second derivative of $r^n(\pi)$ is positive, $r^n(\pi)$ is strictly convex in n. It follows that the n minimizing $r^n(\pi)$ is one (or both) of the integers closest to n^*, unless $n^* < 0$, in which case $n = 0$ minimizes $r^n(\pi)$. (Choosing $n = 0$ corresponds to making a decision without taking observations.) The smallest attainable Bayes risk is thus approximately

$$r^{n^*}(\pi) = 2\sigma c^{1/2} - \frac{c\sigma^2}{\tau^2}.$$

Case 2. Assume that $C(n) = \log(1 + n)$. (Thus it is more efficient, in terms of cost per observation, to take larger samples.) Then

$$r^n(\pi) = \frac{\sigma^2 \tau^2}{\sigma^2 + n\tau^2} + \log(1 + n),$$

and it can be checked that this is (approximately) minimized at

$$n^* = \frac{\sigma^2}{2} - \frac{\sigma^2}{\tau^2} + \frac{\sigma}{2\tau}[4(\tau^2 - \sigma^2) + \sigma^2\tau^2]^{1/2}.$$

EXAMPLE 2. Consider the situation of Example 1, Case 1, except assume that the decision loss is now $L(\theta, a) = |\theta - a|^k$ $(k > 0)$. From Exercise 56 of Chapter 4, it is clear that $\delta_n^\pi(\mathbf{x}^n) = \mu_n(\bar{x}_n)$ (see (7.3)) is still the Bayes rule. It follows, using Result 2 of Section 4.4, that

$$r(\pi, \delta_n^\pi) = E^{m_n} E^{\pi^n(\theta|\bar{X}_n)}[|\theta - \mu_n(\bar{X}_n)|^k],$$

where m_n is the marginal density of $\bar{X}_n$. But clearly

$$E^{\pi^n(\theta|\bar{X}_n)}[|\theta - \mu_n(\bar{X}_n)|^k] = v(k)(\rho_n)^{k/2} = v(k)\left(\frac{\sigma^2\tau^2}{\sigma^2 + n\tau^2}\right)^{k/2},$$

where $v(k)$ is the kth absolute moment of a $\mathcal{N}(0, 1)$ distribution. Hence

$$r(\pi, \delta_n^\pi) = v(k)\left(\frac{\sigma^2\tau^2}{\sigma^2 + n\tau^2}\right)^{k/2},$$

and

$$r^n(\pi) = v(k)\left(\frac{\sigma^2\tau^2}{\sigma^2 + n\tau^2}\right)^{k/2} + nc.$$

Differentiating and setting equal to zero gives an optimal n of (approximately)

$$n^* = \sigma^2\left(\frac{v(k)k}{2\sigma^2 c}\right)^{2/(k+2)} - \frac{\sigma^2}{\tau^2}.$$

Note that

$$r(\pi, \delta_{n^*}^\pi) = v(k)\left(\frac{\sigma^2\tau^2}{\sigma^2 + n^*\tau^2}\right)^{k/2} = v(k)\left(\frac{2c\sigma^2}{kv(k)}\right)^{k/(k+2)}$$

and

$$cn^* = c\sigma^2\left(\frac{kv(k)}{2c\sigma^2}\right)^{2/(k+2)} - \frac{c\sigma^2}{\tau^2} = \frac{kv(k)}{2}\left(\frac{2c\sigma^2}{kv(k)}\right)^{k/(k+2)} - \frac{c\sigma^2}{\tau^2}.$$

Hence $r^{n^*}(\pi)$ can be easily calculated. For later reference, observe that when c is very small,

$$r(\pi, \delta_{n^*}^\pi) \cong \frac{2cn^*}{k}$$

(since c is then much smaller than $c^{k/(k+2)}$). It follows that, for small c, the decision risk and the sampling cost are roughly proportional functions of c.

EXAMPLE 3. Let $X_1, X_2, \ldots$ be a sequential sample from a $\mathcal{N}(\theta, \sigma^2)$ density, σ^2 known. It is desired to test $H_0: \theta \in \Theta_0$ versus $H_1: \theta \in \Theta_1$, where Θ_0 is "less than" Θ_1 in the sense that, if $\theta_0 \in \Theta_0$ and $\theta_1 \in \Theta_1$ then $\theta_0 < \theta_1$. Let a_i denote accepting H_i, and assume that the loss is of the form (7.2) with $L(\theta, a_i)$ being zero if a correct decision is made and positive otherwise. Let $\pi(\theta)$ denote the prior density for θ.

If $\mathbf{X}^n = (X_1, \ldots, X_n)$ is to be observed, $\bar{X}_n$ will again be sufficient for θ. Letting $\pi^n(\theta|\bar{x}_n)$ again denote the posterior distribution of θ given $\mathbf{x}^n$, the Bayes decision rule is to select a_0 if

$$\int_{\Theta_1} L(\theta, a_0) dF^{\pi^n(\theta|\bar{x}_n)}(\theta) < \int_{\Theta_0} L(\theta, a_1) dF^{\pi^n(\theta|\bar{x}_n)}(\theta) \tag{7.4}$$

(i.e., if the posterior expected decision loss of a_0 is smaller than that of a_1). Multiplying through by $(2\pi\sigma^2/n)^{1/2} m_n(\bar{x}_n)$ in (7.4) (m_n is the marginal density) gives the equivalent inequality

$$\begin{aligned} &\int_{\Theta_1} L(\theta, a_0) \exp\left\{-\frac{n}{2\sigma^2}(\bar{x}_n - \theta)^2\right\} dF^{\pi}(\theta) \\ &\quad < \int_{\Theta_0} L(\theta, a_1) \exp\left\{-\frac{n}{2\sigma^2}(\bar{x}_n - \theta)^2\right\} dF^{\pi}(\theta). \end{aligned} \tag{7.5}$$

It will be seen in Chapter 8 that the inequality in (7.5) holds for $\bar{x}_n < K(n)$, where $K(n)$ is some constant. (This must be true for any admissible test.) The reverse inequality holds for $\bar{x}_n > K(n)$, and

$$\begin{aligned} &\int_{\Theta_1} L(\theta, a_0) \exp\left\{-\frac{n}{2\sigma^2}(K(n) - \theta)^2\right\} dF^{\pi}(\theta) \\ &\quad = \int_{\Theta_0} L(\theta, a_1) \exp\left\{-\frac{n}{2\sigma^2}(K(n) - \theta)^2\right\} dF^{\pi}(\theta). \end{aligned} \tag{7.6}$$

Thus the Bayes rule, δ_n^π, is to decide a_0 if $\bar{x}_n < K(n)$, decide a_1 if $\bar{x}_n > K(n)$, and do anything if $\bar{x}_n = K(n)$.

The decision risk of δ_n^π is clearly

$$R(\theta, \delta_n^\pi) = \begin{cases} L(\theta, a_0) P_\theta(\bar{X}_n < K(n)) & \text{if } \theta \in \Theta_1, \\ L(\theta, a_1)[1 - P_\theta(\bar{X}_n < K(n))] & \text{if } \theta \in \Theta_0, \end{cases}$$

and the Bayes decision risk is

$$\begin{aligned} r(\pi, \delta_n^\pi) = &\int_{\Theta_1} L(\theta, a_0) P_\theta(\bar{X}_n < K(n)) dF^{\pi}(\theta) \\ &+ \int_{\Theta_0} L(\theta, a_1)[1 - P_\theta(\bar{X}_n < K(n))] dF^{\pi}(\theta). \end{aligned} \tag{7.7}$$

Pretending that n is a continuous variable, it is shown in Appendix 3 that

$$\frac{d}{dn} r(\pi, \delta_n^\pi) = (8\pi n\sigma^2)^{-1/2} \left[\int_{\Theta_0} L(\theta, a_1)(\theta - K^*) \times \exp\left\{ -\frac{n}{2\sigma^2} (K(n) - \theta)^2 \right\} dF^\pi(\theta) - \int_{\Theta_1} L(\theta, a_0)(\theta - K^*) \exp\left\{ -\frac{n}{2\sigma^2} (K(n) - \theta)^2 \right\} dF^\pi(\theta) \right], \tag{7.8}$$

where K^* is any constant. (The calculation is often simplified by a convenient choice of K^*.)

Using (7.6) and (7.8), it is often easy to calculate

$$\frac{d}{dn} r^n(\pi) = \frac{d}{dn} [r(\pi, \delta_n^\pi) + C(n)]$$

and approximate the optimal n. Some specific cases follow.

Case 1. Assume $\Theta_0 = \{\theta_0\}$ and $\Theta_1 = \{\theta_1\}$. The situation is thus that of testing a simple hypothesis against a simple alternative. Using (7.6), an easy calculation shows that

$$K(n) = \tfrac{1}{2}(\theta_0 + \theta_1) + \frac{\sigma^2}{n(\theta_1 - \theta_0)} \log \left[\frac{L(\theta_0, a_1)\pi(\theta_0)}{L(\theta_1, a_0)\pi(\theta_1)} \right].$$

Also, (7.8) gives (letting $K^* = 0$)

$$\frac{d}{dn} r(\pi, \delta_n^\pi) = (8\pi n\sigma^2)^{-1/2} \left[L(\theta_0, a_1)\theta_0 \exp\left\{ -\frac{n}{2\sigma^2} (K(n) - \theta_0)^2 \right\} \pi(\theta_0) - L(\theta_1, a_0)\theta_1 \exp\left\{ -\frac{n}{2\sigma^2} (K(n) - \theta_1)^2 \right\} \pi(\theta_1) \right].$$

The above equations simplify further if it happens that $L(\theta_0, a_1)\pi(\theta_0) = L(\theta_1, a_0)\pi(\theta_1) = b$ (say). Then

$$K(n) = \frac{\theta_0 + \theta_1}{2}$$

and

$$\frac{d}{dn} r(\pi, \delta_n^\pi) = \frac{-b(\theta_1 - \theta_0)}{(8\pi n\sigma^2)^{1/2}} \exp\left\{ -\frac{n}{8\sigma^2} (\theta_1 - \theta_0)^2 \right\}.$$

If now $C(n) = nc$, then

$$\frac{d}{dn} r^n(\pi) = \frac{-b(\theta_1 - \theta_0)}{(8\pi n\sigma^2)^{1/2}} \exp\left\{ -\frac{n}{8\sigma^2} (\theta_1 - \theta_0)^2 \right\} + c. \tag{7.9}$$

This can be set equal to zero and numerically solved for the (approximate) optimal n.

It is interesting for later comparisons to note that when b/c is large (basically a statement that the decision loss is much larger than the cost of an observation), then it can be shown that the optimal n, above, is approximately

$$n^* = \frac{8\sigma^2}{(\theta_1-\theta_0)^2}\left[\log\frac{b}{c} - \frac{1}{2}\log\log\frac{b}{c} + \log\frac{(\theta_1-\theta_0)^2}{8\pi^{1/2}\sigma^2}\right].$$

(This approximation is good when $\log(b/c)$ is much larger than $8\sigma^2/(\theta_1-\theta_0)^2$ and $\log\log(b/c)$.) Hence the optimal n is of the order of $\log(b/c)$. It can also be checked, in this situation, that

$$r(\pi, \delta^\pi_{n^*}) \cong 8\sigma^2 c(\theta_1-\theta_0)^{-2},$$

while the sampling cost is

$$n^* c \cong 8\sigma^2 c(\theta_1-\theta_0)^{-2}\log\left(\frac{b}{c}\right).$$

Since $\log(b/c)$ is large, the sampling cost clearly dominates the overall Bayes risk, $r^{n^*}(\pi)$. Note that this is in sharp contrast to the situations of Examples 1 and 2.

Case 2. Let $\Theta_0=(-\infty, \theta_0]$, $\Theta_1=(\theta_0, \infty)$, $L(\theta, a_0)=W(|\theta-\theta_0|)$ for $\theta>\theta_0$, and $L(\theta, a_1)=W(|\theta-\theta_0|)$ for $\theta<\theta_0$. Assume also that π is a $\mathcal{N}(\mu, \tau^2)$ prior density.

An analysis, given in Appendix 3, shows that

$$\frac{d}{dn}r(\pi, \delta^\pi_n) = -\frac{\exp\{-\frac{1}{2}(1/\tau^2+\sigma^2/n\tau^4)(\theta_0-\mu)^2\}\sigma\tau}{2\pi(\sigma^2+n\tau^2)n^{1/2}}\int_0^\infty W(\rho_n^{1/2}y)ye^{-y^2/2}\,dy. \tag{7.10}$$

If, for instance, $W(y)=y^k$ $(k>0)$, then (7.10) becomes

$$\begin{aligned}\frac{d}{dn}r(\pi, \delta^\pi_n) &= -\frac{\exp\{-\frac{1}{2}(1/\tau^2+\sigma^2/n\tau^4)(\theta_0-\mu)^2\}\sigma\tau}{2\pi(\sigma^2+n\tau^2)n^{1/2}}\cdot\frac{(2\pi)^{1/2}\rho_n^{k/2}v(k+1)}{2} \\ &= -\frac{\exp\{-\frac{1}{2}(1/\tau^2+\sigma^2/n\tau^4)(\theta_0-\mu)^2\}v(k+1)(\sigma\tau)^{(k+1)}}{2(2\pi)^{1/2}(\sigma^2+n\tau^2)^{(k+2)/2}n^{1/2}},\end{aligned} \tag{7.11}$$

where $v(k+1)$, as before, is the $(k+1)$st absolute moment of a $\mathcal{N}(0, 1)$ distribution. Using this, it is an easy numerical task to minimize $r^n(\pi)$.

If $C(n)=nc$ and c is small, then n will be large. For large n, (7.11) is approximately

$$\frac{d}{dn}r(\pi, \delta^\pi_n) \cong -\frac{v(k+1)\sigma^{(k+1)}\exp\{-(1/2\tau^2)(\theta_0-\mu)^2\}}{2(2\pi)^{1/2}\tau n^{(k+3)/2}}.$$

In this case, the optimal n (approximately that for which $(d/dn)r^n(\pi)=(d/dn)r(\pi, \delta^\pi_n)+c=0$) is clearly

$$n^* = \left[\frac{v(k+1)\sigma^{(k+1)}\exp\{-(1/2\tau^2)(\theta_0-\mu)^2\}}{2(2\pi)^{1/2}\tau c}\right]^{2/(k+3)}.$$

This situation is considerably different than Case 1, in which n^* was of the order $\log(b/c)$. Also, the Bayes decision risk and the sampling cost will be of comparable magnitude here, rather than the sampling cost dominating as in Case 1. Note that this is very similar to the result obtained in Example 2 for estimation. Further discussion of this behavior will be given in Subsection 7.4.10.

It should be noted that, at least in the above examples, the optimal sample size, n^*, will be fairly robust with respect to the tail of the prior distribution. This is because the Bayes decision risk, $r(\pi, \delta_n^\pi)$, is not changed very much by small changes in the tail of the prior (although, as seen in Chapter 4, δ_n^π can change considerably). Hence for purposes of determining the optimal sample size, it is frequently reasonable to use simple priors, such as conjugate priors. In actually determining $\delta_{n^*}^\pi$, however, more attention must be paid to robustness.

7.3. Sequential Analysis—Notation

In Section 7.1 it was stated that the distinguishing feature of sequential analysis is that the observations are taken one at a time, with the experimenter having the option of stopping the experiment and making a decision at any time. The observations, as mentioned earlier, will be denoted $X_1, X_2, \ldots$, where $\mathbf{X}^n = (X_1, X_2, \ldots, X_n)$ has density $f_n(\mathbf{x}^n|\theta)$ (and distribution function $F_n(\mathbf{x}^n|\theta)$) on $\mathscr{X}^n = \mathscr{X}_1 \times \mathscr{X}_2 \times \cdots \times \mathscr{X}_n$. Also, let $\mathbf{X} = (X_1, X_2, \ldots)$, $\mathscr{X} = \mathscr{X}_1 \times \mathscr{X}_2 \times \cdots$, and, for notational convenience, let X_0 and $\mathbf{X}^0$ stand for "no observation taken." As usual, $\theta \in \Theta$ is the unknown state of nature, and $\pi(\theta)$ will denote a prior density on Θ.

If n observations are taken sequentially, at which point action $a \in \mathscr{A}$ is taken, then the loss when θ is the true state of nature will be denoted $L(\theta, a, n)$. It will be assumed that this loss is increasing in n. Also, losses of the special form (7.1) will often be considered.

A (nonrandomized) *sequential decision procedure* will be denoted

$$\mathbf{d} = (\boldsymbol{\tau}, \boldsymbol{\delta}),$$

and, as indicated, consists of two components. The first component, $\boldsymbol{\tau}$, is called the *stopping rule*, and consists of functions τ_0, $\tau_1(\mathbf{x}^1)$, $\tau_2(\mathbf{x}^2), \ldots$, where $\tau_i(\mathbf{x}^i)$ is the probability (zero or one for a nonrandomized procedure) of stopping sampling and making a decision after $\mathbf{x}^i$ is observed. (τ_0 is the probability of making an immediate decision without sampling.) The second component, $\boldsymbol{\delta}$, is called the *decision rule* and consists of a series of decision functions δ_0, $\delta_1(\mathbf{x}^1)$, $\delta_2(\mathbf{x}^2), \ldots$, where $\delta_i(\mathbf{x}^i)$ is the action to be taken if sampling has stopped after observing $\mathbf{x}^i$. (The definition of a randomized sequential decision procedure should be obvious. As we will, for the most part, be considering the problem from a Bayesian viewpoint, randomized procedures will not be needed.)

It is frequently convenient to talk in terms of the *stopping time*, N, rather than the stopping rule. The stopping time is simply the final sample size, i.e., the sample size at which $\boldsymbol{\tau}$ says to stop and make a decision. Formally, the stopping time is the random function of $\mathbf{X}$ given (for a nonrandomized sequential procedure) by

$$N(\mathbf{X}) = \min\{n \geq 0\colon \tau_n(\mathbf{X}^n) = 1\}.$$

For $n \geq 1$, let $\{N = n\}$ denote the set of all $\mathbf{x}^n \in \mathscr{X}^n$ for which $\tau_n(\mathbf{x}^n) = 1$ and $\tau_j(\mathbf{x}^j) = 0$ $(j < n)$. Clearly $\{N = n\}$ is the set of observations for which the sequential procedure stops at time n. Note that

$$\begin{aligned} P_\theta(N < \infty) &= P(N = 0) + \sum_{n=1}^{\infty} P_\theta(N = n) \\ &= P(N = 0) + \sum_{n=1}^{\infty} \int_{\{N=n\}} dF_n(\mathbf{x}^n | \theta). \end{aligned}$$

A sequential procedure will be called *proper* if $P_\theta(N < \infty) = 1$ for all $\theta \in \Theta$. (In a Bayesian setting, this need only hold for all θ in a set of probability one under the prior π.) We will restrict consideration to such procedures.

The *risk function* of a sequential procedure $\mathbf{d}$ is the expected loss

$$\begin{aligned} R(\theta, \mathbf{d}) &= E_\theta[L(\theta, \delta_N(\mathbf{X}^N), N)] \\ &= P(N = 0) L(\theta, \delta_0, 0) + \sum_{n=1}^{\infty} \int_{\{N=n\}} L(\theta, \delta_n(\mathbf{x}^n), n) dF_n(\mathbf{x}^n | \theta). \end{aligned} \tag{7.12}$$

When the loss is as in (7.1),

$$\begin{aligned} R(\theta, \mathbf{d}) = P(N = 0) L(\theta, \delta_0) &+ \sum_{n=1}^{\infty} \int_{\{N=n\}} L(\theta, \delta_n(\mathbf{x}^n)) dF_n(\mathbf{x}^n | \theta) \\ &+ \sum_{n=1}^{\infty} \left(\sum_{i=1}^{n} c_i \right) P_\theta(N = n). \end{aligned}$$

7.4. Bayesian Sequential Analysis

7.4.1. Introduction

While Bayesian analysis in fixed sample size problems is straightforward (robustness considerations aside), Bayesian sequential analysis is very difficult. A considerable amount of notation and machinery will be needed to deal with the problem, all of which tends to obscure the simple idea that is involved. This idea is that at every stage of the procedure (i.e., after every given observation) one should compare the (posterior) Bayes risk of making an immediate decision with the "expected" (posterior) Bayes risk that will be obtained if more observations are taken. If it is cheaper to stop and make a decision, that is what should be done. To clarify this idea, we begin with a very simple illustration.

EXAMPLE 4. A manufacturing firm is trying to decide whether to build a new plant in Ohio (action a_0) or in Alabama (action a_1). The plant would cost \$1,000,000 less to build at the site in Alabama, but there is perhaps a lack of skilled labor in the area. (The site in Ohio has an abundance of skilled labor.) A total of 700 skilled workers are needed, and the company feels that θ, the size of the available skilled labor force near the site in Alabama, has a $\mathcal{N}(350, (100)^2)$ prior density. (For convenience, we will treat θ as a continuous variable.) The company will have to train workers if skilled workers are not available, at a cost of \$3500 each. Assuming the company has an approximately linear utility function for money, the decision loss can be written as

$$L(\theta, a) = \begin{cases} 1{,}000{,}000 & \text{if } a = a_0, \\ 3500(700-\theta) & \text{if } a = a_1 \text{ and } 0 \le \theta \le 700, \\ 0 & \text{if } a = a_1 \text{ and } \theta > 700. \end{cases}$$

Suppose now that the company either can make an immediate decision, or can commission a survey to be conducted (at a cost of \$20,000), the result of which would be an estimate, X, of θ. It is known that the accuracy of the survey would be such that X would be $\mathcal{N}(\theta, (30)^2)$. The problem is to decide whether or not to commission the survey (i.e., to decide whether to make an immediate decision, or to take the observation and then make a decision).

The Bayes risk of an immediate decision is the smaller of $r(\pi, a_0) = 1{,}000{,}000$ and

$$\begin{aligned} r(\pi, a_1) &= 3500 \int_0^{700} (700-\theta)\pi(\theta)d\theta \\ &\cong 3500 \int_{-\infty}^{\infty} (700-\theta)\pi(\theta)d\theta \\ &= 3500(700-350) = 1{,}225{,}000. \end{aligned}$$

(The error in replacing the exact limits of integration above by ∞ and $-\infty$ is negligible.) Hence the Bayes risk of an immediate decision is 1,000,000.

If the survey is commissioned and x observed, the posterior density, $\pi(\theta|x)$, is $\mathcal{N}(\mu(x), \rho^{-1})$, where

$$\mu(x) = \frac{900}{900+10{,}000}(350) + \frac{10{,}000}{900+10{,}000}(x) \cong 28.90 + (0.9174)x,$$

$$\rho^{-1} = \frac{(900)(10{,}000)}{(900+10{,}000)} \cong 825.66.$$

Hence

$$r(\pi(\theta|x), a_1) = 3500 \int_0^{700} (700-\theta)\pi(\theta|x)\,d\theta,$$

which, if $100 < x < 600$ (so that $\mu(x)$ is over four standard deviations from 0 or 700), is approximately

$$3500 \int_{-\infty}^{\infty} (700-\theta)\pi(\theta|x)\,d\theta = 3500[700-\mu(x)]$$
$$= 3500[671.1-(0.9174)x].$$

Clearly $r(\pi(\theta|x), a_0) = 1{,}000{,}000$, so that the Bayes risk of taking the survey, observing x, and then making a decision is

$$\begin{aligned} r(x) &= \min\{r(\pi(\theta|x), a_0), r(\pi(\theta|x), a_1)\} + 20{,}000 \\ &\cong \min\{(1{,}020{,}000), 3500[676.8-(0.9174)x]\} \quad (\text{for } 100 < x < 600) \\ &= \begin{cases} 1{,}020{,}000 & \text{if } 100 < x < 420.08, \\ 3500[676.8-(0.9174)x] & \text{if } 420.08 < x < 600. \end{cases} \end{aligned}$$

Note, however, that we do not know which x will occur. Therefore we can only evaluate the Bayes risk, $r(x)$, through expected value over X. The relevant distribution for X is the "predictive" or marginal distribution, $m(x)$, which in this situation is $\mathcal{N}(350, (100)^2+(30)^2)$. Since $P^m(100 < X < 600) \cong 0.983$, it is clear that (approximately)

$$\begin{aligned} E^m[r(X)] &\cong \int_{-\infty}^{420.08} (1{,}020{,}000)m(x)\,dx \\ &\quad + \int_{420.08}^{\infty} 3500[676.8-(0.9174)x]m(x)\,dx \\ &= 763{,}980 + 205{,}745 = 969{,}725. \end{aligned}$$

This is less than the Bayes risk of an immediate decision, so the survey would be well worth the money. (Note that the calculations here were quite easy, partly because it just "happened" that replacing limits of integration by $-\infty$ and ∞ (when desired) gave reasonable approximations. May you always be so lucky.)

Often in decision problems, such as that above, there will be a number of possible stages of investigation, each stage corresponding to the commission of a more elaborate (and expensive) screening study. Of course, it is also typical to have a large number of possible stages (observations) in standard statistical settings. Unfortunately, determining the expected Bayes risk of continuing to sample becomes progressively harder as the number of possible stages increases. The remainder of this section is devoted to methods of dealing with this difficulty. Many of the basic methods discussed were first developed in Arrow, Blackwell, and Girshick (1949), and in Wald (1950).

7.4.2. Notation

The *Bayes risk* of a sequential procedure $\mathbf{d}$ is defined to be

$$r(\pi, \mathbf{d}) = E^{\pi}[R(\theta, \mathbf{d})].$$

A sequential procedure which minimizes $r(\pi, \mathbf{d})$ (over all proper procedures) is called a *Bayes sequential procedure*, and will be denoted

$$\mathbf{d}^{\pi} = (\tau^{\pi}, \boldsymbol{\delta}^{\pi}).$$

The *Bayes risk* of the problem is defined to be

$$r(\pi) = \inf_{\mathbf{d}} r(\pi, \mathbf{d}).$$

It is useful to introduce special notation for the marginal (predictive) and posterior densities. Thus define (for $n = 1, 2, \ldots$) the marginal densities

$$m_n(\mathbf{x}^n) = E^{\pi}[f_n(\mathbf{x}^n | \theta)] = \int_{\Theta} f_n(\mathbf{x}^n | \theta) dF^{\pi}(\theta),$$

and (assuming $m_n(\mathbf{x}^n) > 0$) the posterior densities

$$\pi^n(\theta) = \pi(\theta | \mathbf{x}^n) = \frac{f_n(\mathbf{x}^n | \theta) \pi(\theta)}{m_n(\mathbf{x}^n)}.$$

(It will have to be remembered that π^n depends on $\mathbf{x}^n$. This dependence is suppressed to simplify notation.)

It is important to think in a sequential fashion. Indeed, after each new observation is taken, it is useful to consider the new sequential problem which starts at that point. To be precise, assume $\mathbf{x}^n$ has been observed, and consider the new sequential experiment, to be denoted $\mathscr{E}_n(\mathbf{x}^n)$, for which the possible observations are $X_{n+1}, X_{n+2}, \ldots$, the loss is $L(\theta, a, j)$, and the prior is $\pi^n(\theta)$. The distributions of the X_i must, naturally, be considered to be the relevant conditional distributions given $\mathbf{x}^n$ and θ. Since we will usually deal with independent X_i, explicit discussion of these conditional distributions is unnecessary.

Definition 2. Let $\mathscr{D}^n$ denote the class of all proper (normalized) sequential procedures in the problem $\mathscr{E}_n(\mathbf{x}^n)$, which has loss $L(\theta, a, j)$, prior $\pi^n(\theta)$, and sequential observations $X_{n+1}, X_{n+2}, \ldots$. Denote the Bayes risk of a procedure $\mathbf{d} \in \mathscr{D}^n$ by $r(\pi^n, \mathbf{d}, n)$, and let

$$r(\pi^n, n) = \inf_{\mathbf{d} \in \mathscr{D}^n} r(\pi^n, \mathbf{d}, n).$$

We will be very concerned with the quantity $r(\pi^n, n)$. This represents the (conditional on $\mathbf{x}^n$) Bayes risk of proceeding in an optimal fashion at stage n. (It must be remembered that $r(\pi^n, n)$ can depend on $\mathbf{x}^n$, both through π^n and through the distributions of $X_{n+1}, X_{n+2}, \ldots$. When (as usual) the observations are independent, however, $r(\pi^n, n)$ will only depend on $\mathbf{x}^n$ through π^n. For this reason, $\mathbf{x}^n$ is suppressed.) It will be convenient to let π^0 correspond to the original prior π, so that $r(\pi^0, 0) = r(\pi)$, the Bayes risk for the original problem.

The quantity $r(\pi^n, n)$ represents the smallest Bayes risk that can be attained once $\mathbf{x}^n$ has been observed. To decide whether or not to make an immediate decision, therefore, it is intuitively obvious that one should compare $r(\pi^n, n)$ to the Bayes risk of an immediate decision, going on if $r(\pi^n, n)$ is smaller. Clearly, notation for the Bayes risk of an immediate decision will be needed. Thus let

$$r_0(\pi^n, a, n) = E^{\pi^n}[L(\theta, a, n)]$$

denote the posterior expected loss of action a at time n, and define

$$r_0(\pi^n, n) = \inf_{a \in \mathscr{A}} r_0(\pi^n, a, n).$$

This last quantity will be called the *posterior Bayes risk* at time n.

One other notational device will be frequently used. In a situation in which θ has prior π, and X has (conditional on θ) density $f(x|\theta)$ and marginal density $m^*(x)$, define, for any function $g(x)$,

$$E^*[g(X)] = E^{m^*}[g(X)] = E^{\pi}E_\theta^X[g(X)].$$

Thus the symbol E^* will stand for expectation over X, with respect to the implied marginal density of X.

7.4.3. The Bayes Decision Rule

Assuming, as usual, that $L(\theta, a, n) \geq -K$, it follows from interchanging orders of integration, as in Chapter 4, that

$$\begin{aligned}
r(\pi, \mathbf{d}) &= E^{\pi}[R(\theta, \mathbf{d})] \\
&= P(N=0)E^{\pi}[L(\theta, \delta_0, 0)] \\
&\quad + \sum_{n=1}^{\infty} \int_{\Theta} \int_{\{N=n\}} L(\theta, \delta_n(\mathbf{x}^n), n) dF_n(\mathbf{x}^n|\theta) dF^{\pi}(\theta) \\
&= P(N=0)E^{\pi}[L(\theta, \delta_0, 0)] \\
&\quad + \sum_{n=1}^{\infty} \int_{\{N=n\}} \int_{\Theta} L(\theta, \delta_n(\mathbf{x}^n), n) dF^{\pi^n}(\theta) dF^{m_n}(\mathbf{x}^n) \\
&= P(N=0) r_0(\pi, \delta_0, 0) + \sum_{n=1}^{\infty} \int_{\{N=n\}} r_0(\pi^n, \delta_n(\mathbf{x}^n), n) dF^{m_n}(\mathbf{x}^n).
\end{aligned}$$

From this expression, it is clear that $r(\pi, \mathbf{d})$ will be minimized if δ_0 and the δ_n are chosen to minimize (for each $\mathbf{x}^n$) the posterior expected loss $r_0(\pi^n, \delta_n(\mathbf{x}^n), n)$. This, of course, is exactly what is done in the fixed sample size situation, and can be summarized as

Result 1. *For $n = 0, 1, 2, \ldots$, assume that $\delta_n^\pi(\mathbf{x}^n)$ is a Bayes decision rule for the fixed sample size decision problem with observations $X_1, \ldots, X_n$ and loss $L(\theta, a, n)$. Then $\boldsymbol{\delta}^\pi = \{\delta_0^\pi, \delta_1^\pi, \ldots\}$ is a Bayes sequential decision rule.*

Half of the problem of determining the Bayes sequential procedure, $\mathbf{d}^\pi$, has thus been solved. Regardless of the stopping rule used, the optimal action, once one has stopped, is simply the Bayes action for the given observations. This is, of course, automatically called for if one adopts the post-experimental Bayesian viewpoint discussed in Chapter 4. Further discussion of the implications of this will be given in Section 7.7.

Unfortunately, the determination of the optimal stopping time, $\boldsymbol{\tau}^\pi$, is usually much harder than the determination of $\boldsymbol{\delta}^\pi$. Much of the remainder of the chapter is devoted to an analysis of this problem. We begin with a discussion of the one case in which the optimal stopping rule is easy to find.

7.4.4. Constant Posterior Bayes Risk

For each n, it may happen that $r_0(\pi^n, n)$, the posterior Bayes risk at time n, is a constant independent of the actual observation $\mathbf{x}^n$. Since (see Section 7.2)

$$r^n(\pi) = E^\pi E_\theta[L(\theta, \delta_n^\pi(\mathbf{X}^n), n)] = E^{m_n}[r_0(\pi^n, n)],$$

it is clear that if $r_0(\pi^n, n)$ is a constant, this constant must be $r^n(\pi)$.

Assuming a Bayes decision rule, $\boldsymbol{\delta}^\pi$, exists, we may, by Result 1, restrict the search for a Bayes procedure to procedures of the form $\mathbf{d} = (\boldsymbol{\tau}, \boldsymbol{\delta}^\pi)$. For such a procedure,

$$\begin{aligned} r(\pi, \mathbf{d}) &= P(N=0) r_0(\pi, \delta_0^\pi, 0) + \sum_{n=1}^{\infty} \int_{\{N=n\}} r_0(\pi^n, \delta_n^\pi(\mathbf{x}^n), n) dF^{m_n}(\mathbf{x}^n) \\ &= P(N=0) r^0(\pi) + \sum_{n=1}^{\infty} r^n(\pi) P^{m_n}(N=n). \end{aligned}$$

Letting $\lambda_0 = P(N=0)$, $\lambda_n = P^{m_n}(N=n)$ $(n \geq 1)$, and observing that $0 \leq \lambda_n \leq 1$ and $\sum_{n=0}^{\infty} \lambda_n = 1$, it is clear that the Bayes risk,

$$r(\pi, \mathbf{d}) = \sum_{n=0}^{\infty} r^n(\pi) \lambda_n,$$

is minimized when λ_n is nonzero only for those n for which $r^n(\pi)$ is minimized. Hence the choice of an optimal stopping rule corresponds exactly to the choice, before experimentation, of an optimal fixed sample size n. This is summarized as

Result 2. *If, for each n, $r_0(\pi^n, n)$ is a constant $(r^n(\pi))$ for all $\mathbf{x}^n$, then a Bayes sequential stopping rule is to stop after observing n^* observations, where n^* is a value of n which minimizes $r^n(\pi)$ (assuming such a value exists).*

EXAMPLE 5. Assume that $X_1, X_2, \ldots$ is a sequential random sample from a $\mathcal{N}(\theta, \sigma^2)$ distribution (σ^2 known), that it is desired to estimate θ under loss $L(\theta, a, n) = (\theta - a)^2 + \sum_{i=1}^n c_i$, and that θ has a $\mathcal{N}(\mu, \tau^2)$ prior distribution. As in Example 1, the posterior distribution of θ given $\mathbf{x}^n = (x_1, \ldots, x_n)$ is $\mathcal{N}(\mu_n(\bar{x}_n), \rho_n)$, where $\rho_n = \sigma^2\tau^2/(\sigma^2 + n\tau^2)$ does not depend on $\mathbf{x}^n$. Since, for squared-error decision loss, the posterior mean is the Bayes action and the posterior variance is its posterior expected decision loss, it is clear that

$$r_0(\pi^n, n) = \rho_n + \sum_{i=1}^n c_i,$$

which does not depend on $\mathbf{x}^n$. Hence Result 2 applies, and the optimal sequential stopping rule is simply to choose that n for which $\rho_n + \sum_{i=1}^n c_i$ is minimized.

Unfortunately, it is extremely rare for the posterior Bayes risk to be independent of the observations. When it is not, finding the Bayes stopping rule is very difficult. In principle, a technique does exist for at least approximating (arbitrarily closely) the optimal stopping rule. It is to this technique we now turn.

7.4.5. The Bayes Truncated Procedure

The difficulty in determining a Bayes sequential stopping rule is that there is an infinite possible future to consider. An obvious method of bypassing this problem is to restrict consideration to procedures for which the stopping time is bounded. As mentioned in the introduction, it is also important to consider the sequential problems starting at time $n+1$. These concepts are tied together in the following definition.

Definition 3. In the sequential experiment $\mathscr{E}_n(\mathbf{x}^n)$ (observations $X_{n+1}, X_{n+2}, \ldots$, loss $L(\theta, a, j)$, and prior π^n) with procedures $\mathscr{D}^n$, let $\mathscr{D}_m^n$ denote the subset of procedures in $\mathscr{D}^n$ which take at most m observations. These procedures will be called *m-truncated procedures.* Also, define

$$r_m(\pi^n, n) = \inf_{\mathbf{d} \in \mathscr{D}_m^n} r(\pi^n, \mathbf{d}, n),$$

which will be called the *m-truncated Bayes risk* for the sequential experiment starting at stage n.

The important concept to understand is that $r_m(\pi^n, n)$ represents the (conditional on $\mathbf{x}^n$) Bayes risk of performing in an optimal sequential fashion, given that stage n has been reached, and assuming that at most m additional observations can be taken. In particular, $r_m(\pi, 0)$ represents the Bayes risk in the original problem of the optimal m-truncated procedure. Note that the definition of $r_0(\pi^n, n)$ given here is consistent with the definition in Subsection 7.4.2. Note also that, since $\mathscr{D}_m^n \subset \mathscr{D}_{m+1}^n$, the functions $r_m(\pi^n, n)$ are clearly nonincreasing in m. (We can do no worse by allowing the option of taking more observations.)

A natural goal in the original sequential problem is to seek the optimal m-truncated procedure, in the hope that it will be close to $\mathbf{d}^\pi$ for suitable m. Formally we define

Definition 4. If there exists an m-truncated procedure, $\mathbf{d}^m = (\boldsymbol{\tau}^m, \boldsymbol{\delta}^m) \in \mathscr{D}_m^0$, for which $r(\pi, \mathbf{d}^m) = r_m(\pi, 0)$, then $\mathbf{d}^m$ will be called a *Bayes m-truncated procedure.*

Due to the finite future of truncated procedures, a Bayes m-truncated procedure turns out to be theoretically calculable. Indeed, such a procedure is given in the following theorem.

Theorem 1. *Assume that Bayes decision rules, δ_n^π, exist for all n, and that the functions $r_j(\pi^n, n)$ are finite for all $j \leq m$ and $n \leq m-j$. Then a Bayes m-truncated procedure, $\mathbf{d}^m$, is given by $\mathbf{d}^m = (\boldsymbol{\tau}^m, \boldsymbol{\delta}^\pi)$, where $\boldsymbol{\tau}^m$ is the stopping rule which says to stop sampling and make a decision for the first n ($n = 0, 1, 2, \ldots, m$) for which*

$$r_0(\pi^n, n) = r_{m-n}(\pi^n, n).$$

Proof. The fact that $\boldsymbol{\delta}^m$ must be $\boldsymbol{\delta}^\pi$ follows from Result 1 of Subsection 7.4.3. The conclusion that $\boldsymbol{\tau}^m$ should be as given follows from induction on m and Theorem 2 (which follows this theorem). The induction argument will be left as an exercise. □

The intuition behind $\mathbf{d}^m$ should be clear. Initially (at stage 0) one compares $r_0(\pi, 0)$, the Bayes risk of an immediate Bayes decision, with $r_m(\pi, 0)$, the overall m-truncated Bayes risk. If the two are equal, the risk of immediately stopping is equal to the overall optimal Bayes risk, so one should stop. If $r_0(\pi, 0)$ is greater than $r_m(\pi, 0)$, one can expect to do better by not immediately stopping, i.e., by observing X_1. After x_1 has been observed, compare $r_0(\pi^1, 1)$, the posterior Bayes risk of an immediate decision, with $r_{m-1}(\pi^1, 1)$, the $(m-1)$-truncated Bayes risk at stage 1. (After x_1 is observed, at most $m-1$ additional observations can be taken.) Again one should stop if $r_0(\pi^1, 1)$ and $r_{m-1}(\pi^1, 1)$ are equal, and continue sampling otherwise. Proceeding in the obvious fashion results in the stopping rule $\boldsymbol{\tau}^m$.

To find $\mathbf{d}^m$, it is thus only necessary to calculate the functions $r_j(\pi^n, n)$. In principle, this can always be done inductively, as shown in the following theorem. The proof of this theorem should be carefully read.

Theorem 2. *Assuming the quantities $r_j(\pi^n, n)$ are finite, they can be calculated inductively from the relationship*

$$r_j(\pi^n, n) = \min\{r_0(\pi^n, n), E^*[r_{j-1}(\pi^n(\theta|X_{n+1}), n+1)]\}.$$

Here $\pi^n(\theta|X_{n+1})$ *is the posterior after observing* X_{n+1} *(i.e., is* $\pi^{n+1}(\theta)$*), and* E^* *(as defined earlier) is expectation with respect to the predictive (or marginal) distribution of* X_{n+1} *given* $\mathbf{x}^n$. *This predictive distribution has density*

$$m^*(x_{n+1}) = \frac{m_{n+1}(\mathbf{x}^{n+1})}{m_n(\mathbf{x}^n)},$$

providing $m_n(\mathbf{x}^n) > 0$.

PROOF. The proof is given somewhat informally for ease of understanding, but can easily be made rigorous using an induction argument along the lines indicated. Note that the sequential experiment $\mathscr{E}_n(\mathbf{x}^n)$ is being considered (i.e., we imagine that the first n observations have already been taken).

Assume $j=1$ (i.e., at most one additional observation can be taken). The only two possible courses of action are to then make an immediate decision, incurring posterior Bayes risk $r_0(\pi^n, n)$, or to observe X_{n+1} and then make a decision. In the latter case, the posterior Bayes risk will be $r_0(\pi^n(\theta|X_{n+1}), n+1)$. Since X_{n+1} is random, the expected posterior Bayes risk

$$E^*[r_0(\pi^n(\theta|X_{n+1}), n+1)]$$

measures the Bayes risk of observing X_{n+1} and then making a decision. Clearly $r_1(\pi^n, n)$, the Bayes risk of the optimal way of acting when at most one observation is allowed, must be the smaller of $r_0(\pi^n, n)$ and $E^*[r_0(\pi^n(\theta|X_{n+1}), n+1)]$.

Exactly the same reasoning holds for any j. Either an immediate decision must be made, incurring posterior Bayes risk $r_0(\pi^n, n)$, or X_{n+1} must be observed, in which case at most $j-1$ more observations can be taken. The expected Bayes risk of this latter course of action is $E^*[r_{j-1}(\pi^n(\theta|X_{n+1}), n+1)]$, and so the smaller of this and $r_0(\pi^n, n)$ must be $r_j(\pi^n, n)$.

The predictive distribution of X_{n+1} given $\mathbf{x}^n$, with respect to which E^* is taken, can be most easily determined by taking the predictive distribution of $\mathbf{X}^{n+1}$ (i.e., $m_{n+1}(\mathbf{x}^{n+1})$) and conditioning on $\mathbf{x}^n$. The resulting density is (in the continuous case for simplicity)

$$\begin{aligned} m^*(x_{n+1}) &= \frac{m_{n+1}(\mathbf{x}^{n+1})}{\int_{\mathscr{X}_{n+1}} m_{n+1}(\mathbf{x}^{n+1})dx_{n+1}} \\ &= \frac{m_{n+1}(\mathbf{x}^{n+1})}{\int_{\mathscr{X}_{n+1}} E^{\pi}[f_{n+1}(\mathbf{x}^{n+1}|\theta)]dx_{n+1}} \\ &= \frac{m_{n+1}(\mathbf{x}^{n+1})}{E^{\pi}[f_n(\mathbf{x}^n|\theta)]} \\ &= \frac{m_{n+1}(\mathbf{x}^{n+1})}{m_n(\mathbf{x}^n)}. \end{aligned}$$

The theorem is thus established. □

Observe from Theorem 1, that the stopping rule is completely specified by the functions $r_0(\pi^n, n)$ and $r_{m-n}(\pi^n, n)$ for $n = 0, 1, 2, \ldots, m$. Furthermore, Theorem 2 shows that $r_{m-n}(\pi^n, n)$ can be determined solely from the $r_0(\pi^{j-1}, j-1)$ and $r_{m-j}(\pi^j, j)$ for $j = n+1, \ldots, m$. (Note that $\pi^n(\theta|X_{n+1})$ is just π^{n+1}.) Hence one need only be concerned with these special cases of the $r_m(\pi^n, n)$, a pleasant fact which substantially reduces calculation.

Note that, for all m,

$$r_m(\pi, 0) \geq r(\pi). \tag{7.13}$$

(Minimizing the Bayes risk over the subclass of procedures $\mathscr{D}_m^0$ results in a quantity at least as large as the minimum over all procedures.) Hence the computable quantities $r_m(\pi, 0)$ provide upper bounds on the true Bayes risk of the problem. More will be said about this in Subsection 7.4.8.

An important special case to consider is that in which $X_1, X_2, \ldots$ is a sequential sample from a common density $f(x|\theta)$, and $L(\theta, a, n)$ is of the form (7.1) with $c_i = c$. It is then easy to see that

$$r_0(\pi^n, n) = \rho_0(\pi^n) + nc,$$

where the quantity

$$\rho_0(\pi^n) = E^{\pi^n}[L(\theta, \delta_n^\pi(\mathbf{x}^n))]$$

is the posterior Bayes decision risk. It follows that

$$r_1(\pi^n, n) = \min\{\rho_0(\pi^n), E^*[\rho_0(\pi^n(\theta|X_{n+1}))] + c\} + nc,$$

and suggests defining inductively

$$\rho_j(\pi^n) = \min\{\rho_0(\pi^n), E^*[\rho_{j-1}(\pi^n(\theta|X_{n+1}))] + c\}. \tag{7.14}$$

It can be easily checked that

$$r_j(\pi^n, n) = \rho_j(\pi^n) + nc,$$

so that Theorem 1 can be restated as

Corollary 1. *Assume that $X_1, X_2, \ldots$ is a sequential sample, and that $L(\theta, a, n) = L(\theta, a) + nc$. Then the Bayes m-truncated stopping rule, $\boldsymbol{\tau}^m$, is to stop sampling and make a decision for the first n $(n = 0, 1, \ldots, m)$ for which*

$$\rho_0(\pi^n) = \rho_{m-n}(\pi^n),$$

providing these quantities are finite.

The advantage of using Corollary 1 (if it applies) is that sometimes the functions ρ_j can be explicitly computed for all possible posteriors. Note also that

$$m^*(x_{n+1}) = \frac{m_{n+1}(\mathbf{x}^{n+1})}{m_n(\mathbf{x}^n)} = E^{\pi^n}[f(x_{n+1}|\theta)],$$

so that, to calculate E^* in (7.14), one can use the often more convenient formula

$$E^*[h(X_{n+1})] = E^{\pi_n} E^{X_{n+1}|\theta}[h(X_{n+1})].$$

EXAMPLE 6. Assume that $X_1, X_2, \ldots$ is a sequential sample from a $\mathcal{B}(1, \theta)$ density, and that it is desired to test $H_0: \theta = \frac{1}{3}$ versus $H_1: \theta = \frac{2}{3}$. Let a_i denote accepting H_i $(i = 0, 1)$, and suppose that $L(\theta, a, n) = L(\theta, a) + nc$ with $c = 1$ and $L(\theta, a)$ being "0–20" loss (i.e., no decision loss for a correct decision, and a decision loss of 20 for an incorrect decision). Finally, let π_i^* $(i = 0, 1)$ denote the prior probability that H_i is true. (The symbols π_i $(i = 0, 1)$ will be reserved for use as arbitrary prior probabilities of H_i.)

Imagine that it is desired to find $\mathbf{d}^2$, the optimal procedure among all those taking at most two observations. To use Corollary 1, the functions ρ_0, ρ_1, and ρ_2 must be calculated. It is easiest to calculate ρ_0 and ρ_1 for arbitrary π, later specializing to the π_i^{*n}.

To calculate $\rho_0(\pi)$, note that the Bayes risks of immediately deciding a_0 and a_1 are $20(1 - \pi_0)$ and $20\pi_0$, respectively, so that the Bayes action is a_0 if $\pi_0 > \frac{1}{2}$ and a_1 if $\pi_0 \le \frac{1}{2}$. Hence

$$\rho_0(\pi) = \begin{cases} 20\pi_0 & \text{if } \pi_0 \le \frac{1}{2}, \\ 20(1 - \pi_0) & \text{if } \pi_0 > \frac{1}{2}. \end{cases}$$

To determine the function $\rho_1(\pi)$ using (7.14), it is first necessary to calculate $E^*[\rho_0(\pi(\theta|X)]$. This expectation is calculated with respect to

$$\begin{aligned} m^*(x) = E^{\pi}[f(x|\theta)] &= f(x|\tfrac{1}{3})\pi_0 + f(x|\tfrac{2}{3})\pi_1 \\ &= \begin{cases} (1 - \frac{1}{3})\pi_0 + (1 - \frac{2}{3})\pi_1 & \text{if } x = 0, \\ \frac{1}{3}\pi_0 + \frac{2}{3}\pi_1 & \text{if } x = 1, \end{cases} \\ &= \begin{cases} \frac{1}{3}(1 + \pi_0) & \text{if } x = 0, \\ \frac{1}{3}(2 - \pi_0) & \text{if } x = 1. \end{cases} \end{aligned}$$

Clearly $\pi(\theta|X)$ is determined by

$$\begin{aligned} \pi(\tfrac{1}{3}|0) &= \frac{\pi(\frac{1}{3})f(0|\frac{1}{3})}{m(0)} = \frac{2\pi_0}{1 + \pi_0}, \\ \pi(\tfrac{1}{3}|1) &= \frac{\pi(\frac{1}{3})f(1|\frac{1}{3})}{m(1)} = \frac{\pi_0}{2 - \pi_0}. \end{aligned} \tag{7.15}$$

Hence

$$\begin{aligned} \rho_0(\pi(\theta|0)) &= \begin{cases} 20\pi(\frac{1}{3}|0) & \text{if } \pi(\frac{1}{3}|0) \le \frac{1}{2}, \\ 20[1 - \pi(\frac{1}{3}|0)] & \text{if } \pi(\frac{1}{3}|0) > \frac{1}{2}, \end{cases} \\ &= \begin{cases} \dfrac{40\pi_0}{1 + \pi_0} & \text{if } \pi_0 \le \frac{1}{3}, \\ \dfrac{20(1 - \pi_0)}{1 + \pi_0} & \text{if } \pi_0 > \frac{1}{3}, \end{cases} \end{aligned}$$

and similarly

$$\rho_0(\pi(\theta|1)) = \begin{cases} \dfrac{20\pi_0}{2-\pi_0} & \text{if } \pi_0 \le \frac{2}{3}, \\ \dfrac{40(1-\pi_0)}{2-\pi_0} & \text{if } \pi_0 > \frac{2}{3}. \end{cases}$$

Considering separately the regions $\pi_0 \le \frac{1}{3}$, $\frac{1}{3} < \pi_0 \le \frac{2}{3}$, and $\pi_0 > \frac{2}{3}$, it follows that

$$\begin{aligned} &E^*[\rho_0(\pi(\theta|X))] \\ &\quad = \rho_0(\pi(\theta|0))m^*(0) + \rho_0(\pi(\theta|1))m^*(1) \\ &\quad = \begin{cases} \left(\dfrac{40\pi_0}{1+\pi_0}\right)\left(\dfrac{1+\pi_0}{3}\right) + \left(\dfrac{20\pi_0}{2-\pi_0}\right)\left(\dfrac{2-\pi_0}{3}\right) & \text{if } \pi_0 \le \frac{1}{3}, \\ \left(\dfrac{20(1-\pi_0)}{1+\pi_0}\right)\left(\dfrac{1+\pi_0}{3}\right) + \left(\dfrac{20\pi_0}{2-\pi_0}\right)\left(\dfrac{2-\pi_0}{3}\right) & \text{if } \frac{1}{3} < \pi_0 \le \frac{2}{3}, \\ \left(\dfrac{20(1-\pi_0)}{1+\pi_0}\right)\left(\dfrac{1+\pi_0}{3}\right) + \left(\dfrac{40(1-\pi_0)}{2-\pi_0}\right)\left(\dfrac{2-\pi_0}{3}\right) & \text{if } \frac{2}{3} < \pi_0, \end{cases} \\ &\quad = \begin{cases} 20\pi_0 & \text{if } \pi_0 \le \frac{1}{3}, \\ \frac{20}{3} & \text{if } \frac{1}{3} < \pi_0 \le \frac{2}{3}, \\ 20(1-\pi_0) & \text{if } \frac{2}{3} < \pi_0. \end{cases} \end{aligned}$$

Thus

$$\begin{aligned} \rho_1(\pi) &= \min\{\rho_0(\pi), E^*[\rho_0(\pi(\theta|X))] + 1\} \\ &= \begin{cases} 20\pi_0 & \text{if } \pi_0 \le \frac{23}{60}, \\ \frac{23}{3} & \text{if } \frac{23}{60} < \pi_0 \le \frac{37}{60}, \\ 20(1-\pi_0) & \text{if } \frac{37}{60} < \pi_0. \end{cases} \end{aligned} \tag{7.16}$$

The calculation of $\rho_2(\pi)$ is similar. Rather than going through it in gory detail for all π, we will make life simple and assume that the original prior has $\pi_0^* = \frac{2}{5}$. (Note that, to apply Corollary 1, it is only necessary to calculate $\rho_2(\pi^*)$.)

Note first, from (7.15), that $\pi^*(\frac{1}{3}|0) = \frac{4}{7}$ and $\pi^*(\frac{1}{3}|1) = \frac{1}{4}$. Hence (7.16) implies that

$$\rho_1(\pi^*(\theta|0)) = \tfrac{23}{3} \quad (\text{since } \pi^*(\tfrac{1}{3}|0) = \tfrac{4}{7} \text{ and } \tfrac{23}{60} < \tfrac{4}{7} \le \tfrac{37}{60})$$

and

$$\rho_1(\pi^*(\theta|1)) = \tfrac{20}{4} = 5 \quad (\text{since } \pi^*(\tfrac{1}{3}|1) = \tfrac{1}{4} < \tfrac{23}{60}).$$

Also, $m^*(0) = \frac{1}{3}(1+\frac{2}{5}) = \frac{7}{15}$ and $m^*(1) = 1 - m^*(0) = \frac{8}{15}$, so that

$$\begin{aligned} E^*[\rho_1(\pi^*(\theta|X))] &= \rho_1(\pi^*(\theta|0))m^*(0) + \rho_1(\pi^*(\theta|1))m^*(1) \\ &= (\tfrac{23}{3})(\tfrac{7}{15}) + 5(\tfrac{8}{15}) \cong 6.24. \end{aligned}$$

Therefore,

$$\rho_2(\pi^*) = \min\{\rho_0(\pi^*), E^*[\rho_1(\pi^*(\theta|X))] + 1\} = \min\{\tfrac{40}{5}, 7.24\} = 7.24.$$

The optimal procedure, $\mathbf{d}^2$, is now given immediately by Corollary 1. For $\pi_0^* = \frac{2}{5}$, note that

$$\rho_0(\pi^*) = 8 > 7.24 = \rho_2(\pi^*),$$

so that X_1 should be observed. After observing X_1, $\rho_0(\pi^{*1})$ and $\rho_1(\pi^{*1})$ must be compared. If $x_1 = 0$, then

$$\pi_0^{*1} = \pi^{*1}(\tfrac{1}{3}) = \pi^*(\tfrac{1}{3}|0) = \tfrac{4}{7},$$

so that $\rho_0(\pi^{*1}) = 20(1-\pi_0^{*1}) = \frac{60}{7}$ and $\rho_1(\pi^{*1}) = 20(1-\pi_0^{*1}) = \frac{60}{7}$. Since $\rho_0(\pi^{*1}) = \rho_1(\pi^{*1})$, it is optimal to stop and decide a_0 (since $\pi_0^{*1} > \frac{1}{2}$). If $x_1 = 1$, then $\pi_0^{*1} = \pi^*(\frac{1}{3}|1) = \frac{1}{4}$, so that $\rho_0(\pi^{*1}) = \frac{20}{4} = 5$ and $\rho_1(\pi^{*1}) = \frac{20}{4} = 5$. Hence it is again optimal to stop, but now action a_1 should be taken (since $\pi_0^{*1} < \frac{1}{2}$). Note that it never pays to take the second observation (although, of course, it might pay if more than two observations could be taken).

The above example indicates the difficulty of actually computing the $\rho_j(\pi^n)$ or, more generally, the $r_j(\pi^n, n)$. Indeed, except for special cases, the calculations generally become unmanageable for large j. Even with a computer, the calculations can generally only be done for conjugate priors (so that the π^n depend on $\mathbf{x}^n$ only through a few parameters), and even then can be very hard for large j. It is thus to be hoped that $\mathbf{d}^m$, the Bayes m-truncated procedure, will be close to $\mathbf{d}^\pi$ for small or moderate m. This issue will be discussed in Subsection 7.4.8, but for now it is worthwhile to note that it frequently happens that the Bayes sequential procedure is, in fact, a truncated procedure. The following theorem gives a condition under which this is so. For simplicity, we restrict attention to the special case in which $L(\theta, a, n)$ is of the form (7.1).

Theorem 3. *Assume that $L(\theta, a, n)$ is of the form (7.1), with $L(\theta, a)$ being nonnegative, and that the posterior Bayes decision risk at stage m satisfies $\rho_0(\pi^m) < c_{m+1}$ for all $\mathbf{x}^m$. Then a Bayes sequential procedure must be truncated at m.*

The proof of this theorem is obvious, since at stage m it costs more to take another observation than to make a decision. Often it is the case that, as $m \to \infty$, $\rho_0(\pi^m) \to 0$ uniformly in $\mathbf{x}^m$, in which case the condition of the theorem is sure to be satisfied for large enough m. An example follows.

EXAMPLE 7. Assume that $X_1, X_2, \ldots$ is a sequential sample from a $\mathcal{B}(1, \theta)$ distribution, and that it is desired to estimate θ under the loss $L(\theta, a, n) = (\theta - a)^2 + nc$. Assume that θ has a $\mathcal{U}(0, 1)$ prior distribution.

Letting $S_n = \sum_{i=1}^n X_i$ denote the sufficient statistic for θ, and noting that $S_n \sim \mathcal{B}(n, \theta)$, an easy calculation shows that $\pi^n(\theta) = \pi(\theta|s_n)$ (where $s_n = \sum_{i=1}^n x_i$) is a $\mathcal{B}e(s_n+1, n-s_n+1)$ density. Since the decision loss is squared-error loss, the posterior Bayes decision risk at time n is the variance of the posterior. From Appendix 1, this is given by

$$\rho_0(\pi^n) = \frac{(s_n+1)(n-s_n+1)}{(n+2)^2(n+3)}.$$

It can be checked that $(s_n+1)(n-s_n+1) \le (n+2)^2/4$, so that

$$\rho_0(\pi^n) \le \frac{1}{4(n+3)}.$$

Thus, letting m denote the smallest integer n for which $1/4(n+3) < c$, Theorem 3 implies that the Bayes sequential procedure must be truncated at m.

Conditions more delicate than that in Theorem 3 can be derived, under which a Bayes sequential procedure is truncated. See S. N. Ray (1965) for a discussion of this. We content ourselves with a theorem involving one quite strong, yet still fairly easy to use, condition. The proof of the theorem is left as an exercise.

Theorem 4. *Assume that $r_m(\pi, 0) \to r(\pi)$ as $m \to \infty$ (see Theorem 5), and that for all $j \ge M$ and all $\mathbf{x}^j$,*

$$r_0(\pi^j, j) \le E^*[r_0(\pi^j(\theta|X_{j+1}), j+1)].$$

Then the Bayes M-truncated procedure, $\mathbf{d}^M$, is a Bayes sequential procedure with respect to π.

7.4.6. Look Ahead Procedures

The Bayes m-truncated procedure, $\mathbf{d}^m$, has the obvious failing that m must be significantly larger than the expected stopping time of the true Bayes procedure, $\mathbf{d}^\pi$, before $\mathbf{d}^m$ can be a reasonable approximation to $\mathbf{d}^\pi$. The difficulty of calculating $\mathbf{d}^m$ for large m makes this a serious problem when the Bayes procedure will be likely to require a fairly large sample. One obvious method of circumventing this difficulty is to use what we will call the *m-step look ahead* procedure. The idea is to look m steps ahead at every stage, rather than only looking ahead up to stage m. Formally we define

Definition 5. Let $\boldsymbol{\tau}_{\mathrm{L}}^m$ be the stopping rule (assuming it is well defined and proper) which stops for the first n $(n = 0, 1, 2, \ldots)$ for which

$$r_0(\pi^n, n) = r_m(\pi^n, n).$$

The procedure $\mathbf{d}_{\mathrm{L}}^m = (\boldsymbol{\tau}_{\mathrm{L}}^m, \boldsymbol{\delta}^\pi)$ will be called the *m-step look ahead procedure.*

The m-step look ahead procedure is a very reasonable procedure. Since it looks ahead farther than $\mathbf{d}^m$, it will have a smaller Bayes risk than $\mathbf{d}^m$. It will also tend to be a good approximation to $\mathbf{d}^\pi$ for much smaller m than is needed for $\mathbf{d}^m$ to be a good approximation. Evidence of this will be given in Subsection 7.4.8. Finally, $\mathbf{d}_{\mathrm{L}}^m$ is usually not much harder to calculate than $\mathbf{d}^m$, since the difficulty in the calculation of either procedure is determined by the difficulty in calculation of the $r_m(\pi^n, n)$.

In the special situation of Corollary 1 of the preceding subsection, it is easy to see that the m-step look ahead procedure stops for the first n for which

$$\rho_0(\pi^n) = \rho_m(\pi^n).$$

An example follows.

EXAMPLE 6 (continued). We will find the 1-step look ahead procedure for this situation. It was calculated that, for all π,

$$\rho_0(\pi) = \begin{cases} 20\pi_0 & \text{if } \pi_0 \le \frac{1}{2}, \\ 20(1-\pi_0) & \text{if } \pi_0 > \frac{1}{2}, \end{cases}$$

and

$$\rho_1(\pi) = \begin{cases} 20\pi_0 & \text{if } \pi_0 \le \frac{23}{60}, \\ \frac{23}{3} & \text{if } \frac{23}{60} < \pi_0 \le \frac{37}{60}, \\ 20(1-\pi_0) & \text{if } \frac{37}{60} < \pi_0, \end{cases}$$

where π_0 is the prior probability that $\theta = \frac{1}{3}$. Hence the 1-step look ahead procedure, when the true prior is π^*, stops for the first n for which

$$\rho_0(\pi^{*n}) = \rho_1(\pi^{*n}),$$

or, equivalently, for the first n for which

$$\pi_0^{*n} \le \tfrac{23}{60} \quad \text{or} \quad \pi_0^{*n} \ge \tfrac{37}{60}. \tag{7.17}$$

Letting $s_n = \sum_{i=1}^n x_i$, it is clear that

$$\begin{aligned} \pi_0^{*n} = \pi^*(\tfrac{1}{3}|x_1, \ldots, x_n) &= \frac{\pi_0^*(\frac{1}{3})^{s_n}(\frac{2}{3})^{n-s_n}}{\pi_0^*(\frac{1}{3})^{s_n}(\frac{2}{3})^{n-s_n} + (1-\pi_0^*)(\frac{2}{3})^{s_n}(\frac{1}{3})^{n-s_n}} \\ &= \{1 + [(\pi_0^*)^{-1} - 1]2^{(2s_n - n)}\}^{-1}. \end{aligned}$$

A calculation then shows that (7.17) is equivalent to

$$|s_n - K^*| \ge \frac{\log(\frac{37}{23})}{2\log 2},$$

where

$$K^* = \frac{1}{2}\left\{n - \frac{\log[(\pi_0^*)^{-1} - 1]}{2\log 2}\right\}.$$

The 1-step look ahead procedure thus has a very simple form in this example.

In determining the m-step look ahead procedure, it is calculationally easiest to look ahead as few steps as necessary. The idea is that it may not be necessary to calculate $r_m(\pi^n, n)$ at every stage, since it may be the case that $r_0(\pi^n, n) > r_j(\pi^n, n)$ for some $j < m$, in which case one should certainly continue sampling. (If it is desirable to continue sampling when looking ahead j steps, it will also be desirable when looking ahead $m > j$ steps.) Thus, typically, the simplest way to proceed at stage n is to compare $r_0(\pi^n, n)$ with $r_1(\pi^n, n)$, going on if $r_0(\pi^n, n) > r_1(\pi^n, n)$. If the two quantities are equal, calculate $r_2(\pi^n, n)$ and compare with $r_0(\pi^n, n)$. Continue in the obvious manner, stopping whenever $r_m(\pi^n, n)$ is calculated and found to be equal to $r_0(\pi^n, n)$. Note that this stopping rule is exactly equivalent to $\boldsymbol{\tau}_L^m$, the stopping rule for the m-step look ahead procedure. This principle, of never looking ahead farther than necessary, can also be used in the calculation of $\mathbf{d}^m$ or any other truncated rule, and will usually save considerably on computation.

Another useful look ahead procedure has been proposed by Amster (1963). This procedure does not look ahead sequentially, but looks ahead with a fixed sample size rule. Formally, we define

Definition 6. Let $r^m(\pi^n, n)$ denote the Bayes risk for the fixed sample size problem with observations $X_{n+1}, \ldots, X_{n+m}$, prior π^n, and loss $L^F(\theta, a, m) = L(\theta, a, n+m)$. The *fixed sample size look ahead procedure* will be denoted $\mathbf{d}_{FS} = (\boldsymbol{\tau}_{FS}, \boldsymbol{\delta}^\pi)$, and $\boldsymbol{\tau}_{FS}$ is the stopping rule which stops for the first n ($n = 0, 1, 2, \ldots$) for which

$$r_0(\pi^n, n) \le \inf_{m \ge 1} r^m(\pi^n, n).$$

The rationale for this rule is that if, at stage n, there is a fixed sample size way of continuing, which has smaller Bayes risk than does making an immediate decision, then certainly another observation should be taken. The attractiveness of $\mathbf{d}_{FS}$ is that it is usually relatively easy to compute the fixed sample size Bayes risk $r^m(\pi^n, n)$, and hence $\mathbf{d}_{FS}$. An example follows.

Example 8. As in Example 7, assume that $X_1, X_2, \ldots$ is a sequential sample from a $\mathscr{B}(1, \theta)$ density, and that it is desired to estimate θ under loss $L(\theta, a, n) = (\theta - a)^2 + nc$. Assume now, however, that θ has a $\mathscr{B}e(\alpha_0, \beta_0)$ prior.

Note that $S_n = \sum_{i=1}^n X_i$ is a sufficient statistic for θ and has a $\mathscr{B}(n, \theta)$ distribution. An easy calculation then shows that $\pi^n(\theta) = \pi(\theta | s_n)$ is a $\mathscr{B}e(\alpha_0 + s_n, \beta_0 + n - s_n)$ density. Define, for simplicity, $\alpha_n = \alpha_0 + s_n$ and $\beta_n = \beta_0 + n - s_n$. Since the decision loss is squared-error loss, the variance of the posterior will be the posterior Bayes decision risk. Hence

$$r_0(\pi^n, n) = \frac{\alpha_n \beta_n}{(\alpha_n + \beta_n)^2(\alpha_n + \beta_n + 1)} + nc.$$

To compute $r^m(\pi^n, n)$, define $Y_m = \sum_{i=n+1}^{n+m} X_i$, note that $Y_m \sim \mathscr{B}(m, \theta)$, and conclude, as above, that the posterior at time $n + m$ will be $\mathscr{B}e(\alpha_n +$

$y_m, \beta_n+m-y_m)$. The Bayes decision rule will be the posterior mean, namely

$$\delta(y_m)=\frac{\alpha_n+y_m}{\alpha_n+\beta_n+m}.$$

The Bayes risk, for prior π^n and loss $L(\theta, a, n+m)$, will thus be

$$\begin{aligned}
r^m(\pi^n, n) &= E^{\pi^n}E_\theta^{Y_m}[L(\theta, \delta(Y_m), n+m)] \\
&= E^{\pi^n}E_\theta^{Y_m}[(\delta(Y_m)-\theta)^2]+(n+m)c \\
&= \frac{\alpha_n\beta_n}{(\alpha_n+\beta_n+m)(\alpha_n+\beta_n)(\alpha_n+\beta_n+1)}+(n+m)c,
\end{aligned}$$

the last step following from a fairly laborious calculation.

Now $\mathbf{d}_{\mathrm{FS}}$ stops when $r_0(\pi^n, n)\le r^m(\pi^n, n)$ for all $m\ge 1$, i.e., when

$$\begin{aligned}
&\frac{\alpha_n\beta_n}{(\alpha_n+\beta_n)^2(\alpha_n+\beta_n+1)}+nc \\
&\qquad\le\frac{\alpha_n\beta_n}{(\alpha_n+\beta_n+m)(\alpha_n+\beta_n)(\alpha_n+\beta_n+1)}+(n+m)c.
\end{aligned}$$

This inequality is equivalent to the inequality

$$\frac{\alpha_n\beta_n[(\alpha_n+\beta_n+m)-(\alpha_n+\beta_n)]}{(\alpha_n+\beta_n)^2(\alpha_n+\beta_n+1)(\alpha_n+\beta_n+m)}\le mc,$$

or

$$\frac{\alpha_n\beta_n}{(\alpha_n+\beta_n)^2(\alpha_n+\beta_n+1)(\alpha_n+\beta_n+m)}\le c.$$

This will hold for all $m\ge 1$ if and only if it holds for $m=1$. The procedure $\mathbf{d}_{\mathrm{FS}}$ thus stops for the first n for which

$$\frac{(\alpha_0+s_n)(\beta_0+n-s_n)}{(\alpha_0+\beta_0+n)^2(\alpha_0+\beta_0+n+1)^2}\le c.$$

This stopping rule, by the way, must also be that of $\mathbf{d}_{\mathrm{L}}^1$, the 1-step look ahead procedure. This is because the stopping condition depends only on what happens for $m=1$. But when only one more observation is considered, there is no difference between looking ahead in a sequential manner and looking ahead in a fixed sample size manner.

Recall, from Subsection 7.4.4, that it can happen that the optimal fixed sample size rule is the sequential Bayes rule; this in particular happened in several common exponential family-conjugate prior situations. In such a situation, the fixed sample size look ahead rule is likely to be an exceptionally good approximation to the optimal sequential rule, because it tends to correspond to the rule that would arise from *approximating* the actual posterior *at each time n* by a conjugate posterior; such an approximation is unlikely to significantly affect the stopping time.

7.4.7. Inner Truncation

It can sometimes happen that the functions $r_m(\pi^n, n)$ are infinite (so that $\mathbf{d}^m$, $\mathbf{d}_L^m$, and $\mathbf{d}_{FS}$ are not well defined), but yet there do exist sequential procedures with finite Bayes risk. An example follows.

EXAMPLE 9. Let $X_1, X_2, \ldots$ be a sequential sample from the distribution which gives probability $\frac{1}{2}$ to the point zero, and is $\mathcal{N}(\theta, 1)$ with probability $\frac{1}{2}$. (This can be interpreted as a situation in which there is only a 50% chance of an observation being valid, $X_i = 0$ corresponding, for instance, to nonresponse.) Assume that the prior π is $\mathcal{C}(0, 1)$, and that $L(\theta, a, n) = (\theta - a)^2 + nc$.

Consider now the point $\mathbf{x}^n = \mathbf{0}^n = (0, \ldots, 0)$. The probability that this point will occur is clearly 2^{-n}. Also, for this point,

$$\pi^n(\theta) = \pi(\theta|\mathbf{0}^n) = \frac{\pi(\theta)2^{-n}}{\int \pi(\theta)2^{-n}\,d\theta} = \pi(\theta),$$

and so

$$r_0(\pi^n, n) = \inf_a E^{\pi^n}[(\theta - a)^2 + nc] = \inf_a E^{\pi}[(\theta - a)^2 + nc] = \infty.$$

It follows, by a simple induction argument, that $r_j(\pi^n, n) = \infty$, for all j, when $\mathbf{x}^n = \mathbf{0}^n$. The procedures $\mathbf{d}^m$, $\mathbf{d}_L^m$, and $\mathbf{d}_{FS}$ are thus not well defined when $\mathbf{x}^n = \mathbf{0}^n$. It is, furthermore, clear that any m-truncated procedure must have infinite Bayes risk, since the probability of observing $\mathbf{0}^m$ (for which the posterior Bayes risk is infinite) is positive.

There are reasonable sequential procedures with finite Bayes risk for this situation. Consider, for instance, the stopping rule $\boldsymbol{\tau}^*$, which says to stop sampling the first time a nonzero observation is obtained. (The probability of eventually observing a nonzero observation is one, so that this is a proper stopping time.) Assume that upon stopping, the decision rule $\delta_n^*(\mathbf{x}^n) = x_n$ will be used. The risk of $\mathbf{d}^* = (\boldsymbol{\tau}^*, \boldsymbol{\delta}^*)$ is

$$\begin{aligned} R(\theta, \mathbf{d}^*) &= \sum_{n=1}^{\infty} \int_{\{N=n\}} [(\theta - x_n)^2 + nc]\,dF_n(\mathbf{x}^n|\theta) \\ &= \sum_{n=1}^{\infty} 2^{-n} \int_{-\infty}^{\infty} [(\theta - x_n)^2 + nc](2\pi)^{-1/2} \exp\{-\tfrac{1}{2}(x_n - \theta)^2\}\,dx_n \\ &= \sum_{n=1}^{\infty} 2^{-n}(1 + nc) = 1 + 2c. \end{aligned}$$

Hence

$$r(\pi, \mathbf{d}^*) = E^{\pi}[R(\theta, \mathbf{d}^*)] = 1 + 2c < \infty.$$

The point of the above example is that procedures based on truncation cannot always be counted upon to exist, or to be good approximations to the Bayes procedure, $\mathbf{d}^{\pi}$. There does exist, however, a modification of the truncation technique which circumvents the indicated difficulty. This modification is called *inner truncation* and was introduced by Herman Rubin. Besides providing useful approximations to the Bayes procedure, inner truncation leads to a relatively simple proof (given in Subsection 7.4.9) of the existence of a Bayes procedure.

Definition 7. For a given sequential problem with loss $L(\theta, a, n)$, the *m-inner truncated loss* is defined to be

$$L^m(\theta, a, n) = \begin{cases} L(\theta, a, n) & \text{if } n < m, \\ \inf_a L(\theta, a, m) & \text{if } n \geq m. \end{cases} \tag{7.18}$$

The sequential problem with L replaced by L^m is called the *m-inner truncated problem.* A Bayes procedure for this new problem (assuming one exists) will be denoted $\mathbf{d}_{\mathrm{I}}^m$, and will be called a *Bayes m-inner truncated procedure.* The Bayes risk of any procedure $\mathbf{d}$, in this problem, will be denoted $r_{\mathrm{I}}^m(\pi, \mathbf{d})$.

The important feature of $L^m(\theta, a, n)$ is that, when $n = m$, it is assumed that the optimal action will be taken. It is enlightening to consider a loss of the form (7.1) with $\inf_a L(\theta, a) = 0$. The m-inner truncated loss is then

$$L^m(\theta, a, n) = \begin{cases} L(\theta, a) + \sum_{i=1}^n c_i & \text{if } n < m, \\ \sum_{i=1}^m c_i & \text{if } n \geq m. \end{cases} \tag{7.19}$$

In this situation, inner truncation essentially means that the decision loss will be waived if the mth observation is taken.

To find a Bayes m-inner truncated procedure, $\mathbf{d}_{\mathrm{I}}^m$, note first that $L^m(\theta, a, n)$ is a constant (depending possibly upon θ) for $n \geq m$. Hence there can be no gain in taking more than m observations, and the investigation can be restricted to consideration of m-truncated procedures. The results of Subsection 7.4.5 are then immediately applicable; Theorems 1 and 2 giving the Bayes m-inner truncated procedure.

In interpreting Theorems 1 and 2 for this situation, it is helpful to indicate the dependence of the functions $r_j(\pi^n, n)$ on the inner truncation point m. (For the m-inner truncated problem, the $r_0(\pi^n, n)$ and hence the $r_j(\pi^n, n)$ must, of course, be calculated with respect to $L^m(\theta, a, n)$.) Thus let $r_j^m(\pi^n, n)$ denote the relevant quantities for the loss $L^m(\theta, a, n)$. The symbols $r_j(\pi^n, n)$ will be reserved for the Bayes risks for the original loss, $L(\theta, a, n)$. Note that

$$r_j^m(\pi^n, n) = r_j(\pi^n, n) \quad \text{if } n + j < m. \tag{7.20}$$

This is because the mth stage cannot be reached if $n + j < m$ (the $r_j^m(\pi^n, n)$ and $r_j(\pi^n, n)$ are the Bayes risks of optimally proceeding at most j steps beyond n), and $L^m(\theta, a, i) = L(\theta, a, i)$ for $i < m$.

Of particular interest is the quantity $r_m^m(\pi, 0)$, which, being the optimal Bayes risk among all procedures truncated at m for the loss $L^m(\theta, a, n)$, must be the Bayes risk of the m-inner truncated problem. (When $\mathbf{d}_{\mathrm{I}}^m$ exists, it follows that $r_{\mathrm{I}}^{m}(\pi, \mathbf{d}_{\mathrm{I}}^{m}) = r_{m}^{m}(\pi, 0)$.) Note that, since

$$L^{m}(\theta, a, n) \le L(\theta, a, n) \tag{7.21}$$

(recall that $L(\theta, a, j)$ is nondecreasing in j), it must be true that

$$r_{m}^{m}(\pi, 0) \le r(\pi), \tag{7.22}$$

where $r(\pi)$, as usual, stands for the Bayes risk for the loss $L(\theta, a, n)$. The constants $r_{m}^{m}(\pi, 0)$ thus provide lower bounds for $r(\pi)$, complementing the upper bounds given in (7.13) of Subsection 7.4.5. From (7.22), it is also apparent that the difficulties encountered in Example 9 for regular truncation cannot occur for inner truncation. If there is a sequential procedure with finite Bayes risk, one can always find a truncated procedure with finite inner truncated Bayes risk.

As in Subsection 7.4.6, it is possible to define a look ahead procedure for inner truncation. The definition is complicated, however, by the fact that the loss must be changed as one looks ahead. The idea is that in looking ahead (say) m steps, one should imagine that the decision loss will be waived if m more observations are actually taken. Formally, the procedure is defined as follows.

Definition 8. Let $\boldsymbol{\tau}_{\mathrm{IL}}^{m}$ be the stopping rule (assuming it is proper) which stops for the first n $(n = 0, 1, 2, \ldots)$ for which

$$r_0(\pi^n, n) = r_{m}^{m+n}(\pi^n, n),$$

and let $\mathbf{d}_{\mathrm{IL}}^{m} = (\boldsymbol{\tau}_{\mathrm{IL}}^{m}, \boldsymbol{\delta}^{\pi})$. This will be called the *m-step inner look ahead procedure.*

When $L(\theta, a, n)$ is of the form (7.1) with $\inf_a L(\theta, a) = 0$, the 1-step inner look ahead procedure, $\mathbf{d}_{\mathrm{IL}}^{1}$, is particularly interesting. Using (7.19), an easy calculation shows that

$$r_{1}^{n+1}(\pi^n, n) = \min\left\{ r_0(\pi^n, n), \sum_{i=1}^{n+1} c_i \right\}.$$

Hence the 1-step inner look ahead procedure stops for the first n for which

$$r_0(\pi^n, n) \le \sum_{i=1}^{n+1} c_i,$$

or, since $r_0(\pi^n, n) = \rho_0(\pi^n) + \sum_{i=1}^{n} c_i$, for the first n for which

$$\rho_0(\pi^n) \le c_{n+1}.$$

In other words, $\mathbf{d}_{\mathrm{IL}}^{1}$ stops sampling when the posterior Bayes decision risk is smaller than the cost of another observation. It is, of course, obvious that one would then want to stop.

7.4.8. Approximating the Bayes Procedure and the Bayes Risk

In this subsection, the accuracy of the various truncated and look ahead procedures (and their Bayes risks) as approximations to $\mathbf{d}^{\pi}$ (and $r(\pi)$) will be discussed. Considering first the Bayes risks, the following relationships were pointed out in previous subsections:

$$r_{\mathrm{I}}^{m}(\pi, \mathbf{d}_{\mathrm{I}}^{m}) \le r(\pi) \le r(\pi, \mathbf{d}_{\mathrm{L}}^{m}) \le r(\pi, \mathbf{d}^{m}). \tag{7.23}$$

Recall that $\mathbf{d}^{m}$ is the Bayes m-truncated procedure, $\mathbf{d}_{\mathrm{L}}^{m}$ is the m-step look ahead procedure, and $\mathbf{d}_{\mathrm{I}}^{m}$ is the Bayes m-inner truncated procedure. The bounds $r_{\mathrm{I}}^{m}(\pi, \mathbf{d}_{\mathrm{I}}^{m})$ $(= r_{m}^{m}(\pi, 0))$ and $r(\pi, \mathbf{d}^{m})$ $(= r_{m}(\pi, 0))$ are explicitly calculable using Theorem 2 in Subsection 7.4.5. Unfortunately, the bound $r(\pi, \mathbf{d}_{\mathrm{L}}^{m})$ is often too difficult to calculate.

It has been shown that the $r(\pi, \mathbf{d}^{m})$ are nonincreasing in m, and it can similarly be shown that the $r_{\mathrm{I}}^{m}(\pi, \mathbf{d}_{\mathrm{I}}^{m})$ are nondecreasing in m. It will, in fact, be shown in the next subsection that these quantities converge to $r(\pi)$ as $m \to \infty$ (under certain conditions). We can thus (theoretically) calculate arbitrarily close approximations to $r(\pi)$, and find procedures arbitrarily close to optimal, in terms of Bayes risk.

Another method of investigating the accuracy of the various procedures, as approximations to $\mathbf{d}^{\pi}$, is to consider the stopping times. Letting N^{m}, N_{L}^{m}, N^{π}, N_{IL}^{m} and N_{FS} denote the stopping times of $\mathbf{d}^{m}$, $\mathbf{d}_{\mathrm{L}}^{m}$, $\mathbf{d}^{\pi}$, $\mathbf{d}_{\mathrm{IL}}^{m}$, and $\mathbf{d}_{\mathrm{FS}}$, respectively, it can be shown that

$$N^{m} \le N_{\mathrm{L}}^{m} \le N^{\pi} \le N_{\mathrm{IL}}^{m} \quad \text{and} \quad N_{\mathrm{FS}} \le N^{\pi}. \tag{7.24}$$

(The proof will be left as an exercise.) These inequalities are useful in that if, in a given situation, (say) N_{L}^{m} and N_{IL}^{m} are close, then either stopping time (or one in between) should be nearly optimal.

To aid in understanding the behavior of all these Bayes risks and stopping times, we turn to some examples. Unfortunately, the behavior is quite dependent on the type of problem being dealt with. In particular, it is important to distinguish between testing and estimation problems.

I. *Testing*

The following example gives some indication of the behavior of the procedures in testing problems.

Example 10. Assume that $X_1, X_2, \ldots$ is a sequential sample from a $\mathscr{B}(1, \theta)$ distribution, and that it is desired to test $H_0: \theta = 0.4$ versus $H_1: \theta = 0.6$. The loss is of the form (7.1) with "0-K" decision loss (zero for a correct decision and K for an incorrect decision) and $c_i = 1$. The prior distribution is $\pi(0.4) = \pi(0.6) = \frac{1}{2}$.

Table 7.1. Look Ahead Procedures and Bayes Risks ($K = 10^6$).

m	$\mathbf{d}_{\mathrm{L}}^{m}$	$\mathbf{d}_{\mathrm{IL}}^{m}$	$r(\pi, \mathbf{d}_{\mathrm{L}}^{m})$	$r(\pi, \mathbf{d}_{\mathrm{IL}}^{m})$	$r(\pi, \mathbf{d}^{m})$	$r_{\mathrm{I}}^{m}(\pi, \mathbf{d}_{\mathrm{I}}^{m})$
1	1	35	400,001	175.7	400,001	1
2	2	33	307,696	166.6	400,001	2
3	3	32	228,580	162.3	352,002	3
4	4	31	164,962	158.48	352,002	4
5	5	31	116,383	158.48	317,444	5
6	6	31	80,731	158.48	317,444	6
7	7	30	55,322	155.21	289,798	7
8	8	30	37,590	155.21	289,798	8
9	9	30	25,396	155.21	266,575	9
10	10	30	17,094	155.21	266,575	10
15	13	29	5176.3	152.82	213,116	15
20	16	29	1599.9	152.82	186,108	20
25	17	29	1098.8	152.82	153,789	25
30	18	28	766.1	151.73	136,237	30
40	20	28	400.6	151.73	102,091	40
60	23	28	204.1	151.73	59,601	59.97
80	24	28	179.39	151.73	35,874	79.49
100	26	28	156.39	171.73	22,007	97.33
150	27	28	152.60	151.73	6851.4	128.67
200	28	28	151.73	151.73	2282.9	143.10
300	28	28	151.73	151.73	378.74	150.68
400	28	28	151.73	151.73	176.61	151.61
500	28	28	151.73	151.73	154.48	151.72
600	28	28	151.73	151.73	152.04	151.73
700	28	28	151.73	151.73	151.76	151.73

It can be shown, for this problem, that the procedures $\mathbf{d}^{\pi}$, $\mathbf{d}_{\mathrm{IL}}^{m}$, $\mathbf{d}_{\mathrm{L}}^{m}$, and $\mathbf{d}_{\mathrm{FS}}$ all have stopping rules of the form

$$\text{stop for the first } n \text{ for which } |2s_n - n| = k, \tag{7.25}$$

where $s_n = \sum_{i=1}^{n} x_i$ and k is some integer.

Case 1. $K = 10^6$. This is a situation in which observations are cheap (compared to the decision loss). The true Bayes procedure, $\mathbf{d}^{\pi}$, has a stopping rule of the form (7.25) with $k = 28$. The Bayes risk is approximately 151.73, and the expected number of observations for $\mathbf{d}^{\pi}$ is approximately 140. The values of k in (7.25) for the look ahead procedures $\mathbf{d}_{\mathrm{L}}^{m}$ and $\mathbf{d}_{\mathrm{IL}}^{m}$ depend on m, and are given in the second and third columns of Table 7.1. Finally, the Bayes risks of $\mathbf{d}_{\mathrm{L}}^{m}$, $\mathbf{d}_{\mathrm{IL}}^{m}$, $\mathbf{d}^{m}$, and $\mathbf{d}_{\mathrm{I}}^{m}$ (the last for the m-inner truncated loss $L^m(\theta, a, n)$) are given in the remaining columns of Table 7.1.

As expected, all Bayes risks converge to $r(\pi) = 151.73$, and the relationships in (7.23) and (7.24) are satisfied. Note that the truncated procedures $\mathbf{d}^{m}$ and $\mathbf{d}_{\mathrm{I}}^{m}$ have Bayes risks which converge very slowly to $r(\pi)$. The

Table 7.2. Look Ahead Procedures and Bayes Risks ($K = 100$).

m	$\mathbf{d}_{\mathrm{L}}^{m}$	$\mathbf{d}_{\mathrm{IL}}^{m}$	$r(\pi, \mathbf{d}_{\mathrm{L}}^{m})$	$r(\pi, \mathbf{d}_{\mathrm{IL}}^{m})$	$r(\pi, \mathbf{d}^{m})$	$r_{\mathrm{I}}^{m}(\pi, \mathbf{d}_{\mathrm{I}}^{m})$
1	1	12	41.00	59.85	41.00	1.00
2	2	10	34.62	50.00	41.00	2.00
3	3	9	31.00	45.25	37.68	3.00
4	3	9	31.00	45.25	37.68	4.00
5	3	8	31.00	40.75	35.81	5.00
6	3	8	31.00	40.75	35.81	6.00
7	3	7	31.00	36.66	34.46	7.00
8	3	7	31.00	36.66	34.46	8.00
9	3	7	31.00	36.66	33.49	9.00
10	4	7	29.90	36.66	33.49	10.00
15	4	6	29.90	33.23	31.84	14.89
20	4	5	29.90	30.82	31.20	19.16
28	4	5	29.90	30.82	30.48	24.09
36	4	4	29.90	29.90	30.16	26.99
50	4	4	29.90	29.90	29.96	29.17
70	4	4	29.90	29.90	29.91	29.80
100	4	4	29.90	29.90	29.90	29.89

truncation point, m, must be several hundred before the approximation is reasonably accurate. (This is to be expected, since the expected stopping time of $\mathbf{d}^{\pi}$ is about 140, and, unless the truncation point is considerably larger than this, there is not much hope of doing well.) The look ahead procedures do not suffer from this difficulty, and perform well for much smaller m. Indeed the look ahead procedures are exactly the Bayes procedure (i.e., $k = 28$) when $m = 200$ (for $\mathbf{d}_{\mathrm{L}}^{m}$) and when $m = 30$ (for $\mathbf{d}_{\mathrm{IL}}^{m}$). The m-step inner look ahead procedure, $\mathbf{d}_{\mathrm{IL}}^{m}$, is particularly attractive, having a Bayes risk close to optimal even for very small m.

The fixed sample size look ahead procedure, $\mathbf{d}_{\mathrm{FS}}$, has a stopping rule of the form (7.25) with $k = 21$, and has Bayes risk 305.4. This is moderately accurate as an approximation to $\mathbf{d}^{\pi}$.

Case 2. $K = 100$. This is a situation in which observations are fairly expensive. The true Bayes rule is now of the form (7.25) with $k = 4$. The Bayes risk is $r(\pi) = 29.90$, and the expected number of observations is 13.4. Table 7.2 gives the values of k and Bayes risks for the various procedures, in this situation.

Since the expected number of observations needed in this situation is small, it is not surprising that the truncated procedures do well for much smaller m than were needed in Case 1. Also in contrast to Case 1, $\mathbf{d}_{\mathrm{L}}^{m}$ performs better than $\mathbf{d}_{\mathrm{IL}}^{m}$ for small m. Indeed for m as small as three, $\mathbf{d}_{\mathrm{L}}^{m}$ has Bayes risk very close to optimal.

The stopping rule for $\mathbf{d}_{FS}$, in this case, is of the form (7.25) with $k=2$. The corresponding Bayes risk of 34.62 is quite good.

The above example suggests that for expensive observations (relative to the decision loss) or equivalently for situations in which only a small number of observations are likely to be needed, the m-step look ahead procedure is likely to do well for moderate or small m. When, on the other hand, observations are cheap (or a large sample size will be needed), the m-step inner look ahead procedure seems to be best. (The reason for the success of $\mathbf{d}_{IL}^m$, in such a situation, will be explained in Subsection 7.4.10.) In both cases, the performance of $\mathbf{d}_{FS}$ seems fair.

In case 1 of Example 10, the relative failure of $\mathbf{d}_L^m$ compared to $\mathbf{d}_{IL}^m$ may be somewhat surprising. The reason for this failure is indicated in the following example.

EXAMPLE 11. In the general sequential setup, assume that it is desired to test $H_0: \theta \in \Theta_0$ versus $H_1: \theta \in \Theta_1$. Let a_i denote accepting H_i $(i=0,1)$.

Suppose now that $\mathbf{x}^n$ has been observed, and that (say) $\delta_n(\mathbf{x}^n)=a_0$ is the Bayes action. This means that

$$E^{\pi^n}[L(\theta, a_0, n)] < E^{\pi^n}[L(\theta, a_1, n)],$$

and that the posterior Bayes risk is

$$r_0(\pi^n, n) = E^{\pi^n}[L(\theta, a_0, n)].$$

Now imagine that up to m more observations can be taken, but that, for all $j \leq m$ and all $x_{n+1}, \ldots, x_{n+j}$,

$$E^{\pi^{n+j}}[L(\theta, a_0, n+j)] < E^{\pi^{n+j}}[L(\theta, a_1, n+j)]. \tag{7.26}$$

This means that no matter what the observations turn out to be, a_0 will still be the Bayes action upon stopping. In other words, m additional observations cannot possibly counter the evidence already existing for H_0. In Example 10, for instance, the Bayes action (upon stopping) is a_0 if $2s_n - n \leq 0$, and is a_1 otherwise. If, at a certain stage, $2s_n - n = -20$, then clearly even 20 more observations cannot cause a change in action.

It is intuitively clear that, since $L(\theta, a, n)$ is increasing in n, the taking of more observations is wasteful if these observations cannot cause a change in action. This can be easily established rigorously, and the conclusion is that the m step look ahead procedure, $\mathbf{d}_L^m$, will stop at stage n when (7.26) holds.

The intuition to be gained here is that in order to expect to lower the posterior Bayes risk it is necessary to allow the possibility of taking enough observations so that our mind could be changed concerning the action to be taken. In testing, as fairly conclusive evidence for one hypothesis accumulates, $\mathbf{d}_L^m$ must, therefore, look quite far ahead before it will recognize that another observation might be desirable. This difficulty will be most pronounced when observations are cheap, in that a large number of observations will then probably be taken, and very conclusive evidence for one hypothesis will tend to accumulate.

II. *Estimation*

Estimation problems tend to be easier to deal with than testing problems. Indeed, for most estimation problems, the m-step look ahead procedure, $\mathbf{d}_{\mathrm{L}}^{m}$, will perform very well for quite small m. To see why, imagine that we knew the infinite sequence $\mathbf{x}=(x_1, x_2, \ldots)$ that would occur, and were allowed to stop at any time n, with the proviso that, upon stopping, the Bayes rule $\delta_n^{\pi}(\mathbf{x}^n)$ must be used. Writing the posterior Bayes risk at time n as

$$V(\mathbf{x}, n) = r_0(\pi^n, n),$$

it is clear that the optimal n would be that which minimized $V(\mathbf{x}, n)$. Now it happens that, in estimation problems, $V(\mathbf{x}, n)$ is usually a fairly smooth function of n, behaving as in Figure 7.1. In Example 7, for instance,

$$V(\mathbf{x}, n) = r_0(\pi^n, n) = \frac{(s_n+1)(n-s_n+1)}{(n+2)^2(n+3)} + nc.$$

Graphing this for various $\mathbf{x}$ will convince you that $V(\mathbf{x}, n)$ often behaves as in Figure 7.1. The minimum value of $V(\mathbf{x}, n)$ will thus tend to occur when the graph begins to swing upwards, or, equivalently, for the first n for which $V(\mathbf{x}, n) < V(\mathbf{x}, n+1)$. Of course, in reality, $V(\mathbf{x}, n+1)$ is not known at stage n, but an m-step look ahead procedure will usually detect when $V(\mathbf{x}, n)$ is likely to swing up, even when m is small. Indeed the 1-step look ahead procedure will often succeed very well.

The m-step inner look ahead procedure is generally not very good in estimation problems, because the posterior Bayes risk, based on the inner truncated loss $L^m(\theta, a, n)$, can be a poor approximation to the true posterior Bayes risk, and can hence result in poor estimates of upcoming values of $V(\mathbf{x}, n)$. Indeed, by pretending that the optimal action will be taken m steps ahead, $\mathbf{d}_{\mathrm{IL}}^{m}$ will tend to considerably overshoot the true turning point of $V(\mathbf{x}, n)$.

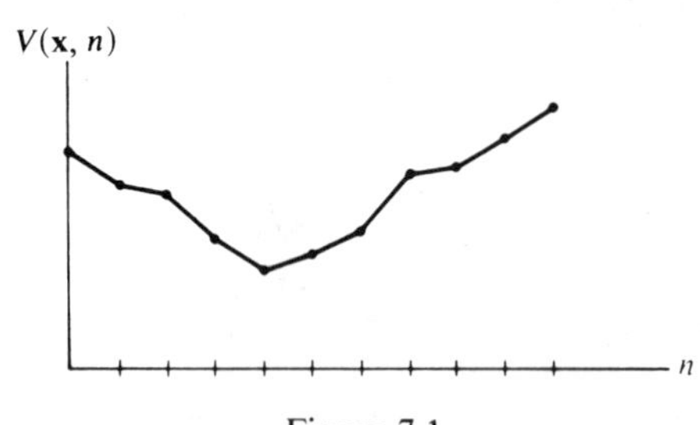

Figure 7.1

The fixed sample size look ahead procedure, $\mathbf{d}_{FS}$, will usually be roughly equivalent to $\mathbf{d}_L^m$ for m equal to one or two. It will hence perform reasonably well, but is advantageous only if it is easier to calculate than is $\mathbf{d}_L^m$.

Testing problems do not behave as nicely as estimation problems because $V(\mathbf{x}, n)$ for a testing problem will tend to have many ups and downs. The long term trend of $V(\mathbf{x}, n)$ will be as in Figure 7.1, but local oscillations can be quite large. A fairly large m is thus needed if $\mathbf{d}_L^m$ is to "look past" these local oscillations. This does suggest, however, that the reason $\mathbf{d}_{FS}$ performs fairly well in testing problems is that it can look quite far ahead, past the local oscillations.

7.4.9. Theoretical Results

As mentioned in the previous subsection, $r_m(\pi, 0) = r(\pi, \mathbf{d}^m)$ and $r_m^m(\pi, 0) = r_I^m(\pi, \mathbf{d}_I^m)$ will usually converge to $r(\pi)$ as $m \to \infty$. Several theorems establishing this are given below. As a side benefit, one of the theorems gives conditions under which a Bayes procedure is guaranteed to exist.

For simplicity, it will be assumed throughout the subsection that Bayes decision rules, δ_n^π, exist for all n. The first two theorems will deal only with losses of the form (7.1). (Generalizations to arbitrary loss are given in the Exercises.) In these two theorems, the usual notation,

$$r(\pi, \delta_n^\pi) = E^\pi E_\theta[L(\theta, \delta_n^\pi(\mathbf{X}^n))] = E^{m_n} E^{\pi^n}[L(\theta, \delta_n^\pi(\mathbf{X}^n))],$$

will be used to denote the Bayes decision risk of δ_n^π. Hopefully, the simultaneous use of $r(\pi, \delta_n^\pi)$ and $r(\pi, \mathbf{d})$ will cause no confusion.

Theorem 5. *Assume that $L(\theta, a, n)$ is of the form* (7.1) *with* $\inf_a L(\theta, a) = 0$, *and that* $\lim_{n\to\infty} r(\pi, \delta_n^\pi) = 0$. *Then*

$$\lim_{m\to\infty} r_m(\pi, 0) = r(\pi)$$

(*i.e., the Bayes risks of* $\mathbf{d}^m$ *converge to the true Bayes risk*).

Proof. Let $\boldsymbol{\tau}^\varepsilon$ be a stopping rule for which the risk of $\mathbf{d}^\varepsilon = (\boldsymbol{\tau}^\varepsilon, \boldsymbol{\delta}^\pi)$ satisfies $r(\pi, \mathbf{d}^\varepsilon) < r(\pi) + \varepsilon$. (Such an "$\varepsilon$-Bayes" procedure must exist, even though a true Bayes procedure need not.) Define $\boldsymbol{\tau}^{\varepsilon,m} = (\tau_0^{\varepsilon,m}, \tau_1^{\varepsilon,m}, \ldots)$ by

$$\tau_n^{\varepsilon,m} = \begin{cases} \tau_n^\varepsilon & \text{if } n < m, \\ 1 & \text{if } n = m. \end{cases}$$

The procedure $\mathbf{d}^{\varepsilon,m} = (\boldsymbol{\tau}^{\varepsilon,m}, \boldsymbol{\delta}^\pi)$ is thus the procedure $\mathbf{d}^\varepsilon$ truncated at m. Let N denote the stopping time of $\mathbf{d}^\varepsilon$, and define

$$\{N \geq m\} = \{\mathbf{x}^m \in \mathscr{X}^m : \tau_j^\varepsilon(\mathbf{x}^j) = 0 \text{ for all } j < m\}.$$

Clearly

$$r(\pi, \mathbf{d}^{\varepsilon}) - r(\pi, \mathbf{d}^{\varepsilon, m}) = \sum_{n=m}^{\infty} \int_{\{N=n\}} r_0(\pi^n, n) dF^{m_n}(\mathbf{x}^n)$$

$$- \int_{\{N \geq m\}} r_0(\pi^m, m) dF^{m_m}(\mathbf{x}^m)$$

$$= \sum_{n=m}^{\infty} \int_{\{N=n\}} \left[E^{\pi^n}[L(\theta, \delta_n^{\pi}(\mathbf{x}^n))] + \sum_{i=1}^{n} c_i \right] dF^{m_n}(\mathbf{x}^n)$$

$$- \int_{\{N \geq m\}} \left[E^{\pi^m}[L(\theta, \delta_m^{\pi}(\mathbf{x}^m))] + \sum_{i=1}^{m} c_i \right] dF^{m_m}(\mathbf{x}^m)$$

$$\geq - \int_{\{N \geq m\}} E^{\pi^m}[L(\theta, \delta_m^{\pi}(\mathbf{x}^m))] dF^{m_m}(\mathbf{x}^m)$$

(since the c_i are positive)

$$\geq -E^{m_m} E^{\pi^m}[L(\theta, \delta_m^{\pi}(\mathbf{X}^m))] = -r(\pi, \delta_m^{\pi}).$$

By assumption, $r(\pi, \delta_m^{\pi}) < \varepsilon$ for large enough m, so that, for large enough m,

$$r(\pi, \mathbf{d}^{\varepsilon, m}) < r(\pi, \mathbf{d}^{\varepsilon}) + \varepsilon < r(\pi) + 2\varepsilon.$$

Since $r_m(\pi, 0) \leq r(\pi, \mathbf{d}^{\varepsilon, m})$ ($r_m(\pi, 0)$ is the minimum Bayes risk over *all* m-truncated procedures), it follows that, for large enough m,

$$r_m(\pi, 0) < r(\pi) + 2\varepsilon.$$

But $r(\pi) \leq r_m(\pi, 0)$, so it can be concluded that, for large enough m,

$$|r_m(\pi, 0) - r(\pi)| < 2\varepsilon.$$

Since ε was arbitrary, the conclusion follows. □

Theorem 5 is actually true under conditions considerably weaker than $\lim_{n \to \infty} r(\pi, \delta_n^{\pi}) = 0$. This condition suffices for typical applications, however. Note that Theorem 5 and (7.23) (recall $r(\pi, \mathbf{d}^m) = r_m(\pi, 0)$) immediately imply that $\lim_{m \to \infty} r(\pi, \mathbf{d}_{\mathrm{L}}^m) = r(\pi)$.

The following theorem establishes that the m-inner truncated Bayes risks, $r_m^m(\pi, 0)$ $(= r_{\mathrm{I}}^m(\pi, \mathbf{d}_{\mathrm{I}}^m))$, converge to $r(\pi)$ as $m \to \infty$, and also that the true Bayes risk of $\mathbf{d}_{\mathrm{I}}^m$ converges to $r(\pi)$.

Theorem 6. *Assume that $L(\theta, a, n)$ is of the form* (7.1) *with $\inf_a L(\theta, a) = 0$, and that $\lim_{n \to \infty} r(\pi, \delta_n^{\pi}) = 0$. Then*

$$\lim_{m \to \infty} r_m^m(\pi, 0) = r(\pi) = \lim_{m \to \infty} r(\pi, \mathbf{d}_{\mathrm{I}}^m).$$

Proof. Let N denote the stopping time of $\mathbf{d}_{\mathrm{I}}^{m}$. For any $\varepsilon > 0$,

$$\begin{aligned} r_m^m(\pi, 0) &= r_{\mathrm{I}}^m(\pi, \mathbf{d}_{\mathrm{I}}^m) \\ &= r(\pi, \mathbf{d}_{\mathrm{I}}^m) - \int_{\{N=m\}} E^{\pi^m}[L(\theta, \delta_m^\pi(\mathbf{x}^m))]dF^{m_m}(\mathbf{x}^m) \\ &\geq r(\pi, \mathbf{d}_{\mathrm{I}}^m) - r(\pi, \delta_m^\pi) \\ &> r(\pi, \mathbf{d}_{\mathrm{I}}^m) - \varepsilon \end{aligned}$$

for large enough m (since $r(\pi, \delta_m^\pi) \to 0$). Hence

$$r(\pi) \leq r(\pi, \mathbf{d}_{\mathrm{I}}^m) < r_m^m(\pi, 0) + \varepsilon$$

for large enough m. Combining this with (7.22) (in Subsection 7.4.7), it follows that, for large enough m,

$$|r(\pi) - r_m^m(\pi, 0)| < \varepsilon \quad \text{and} \quad |r(m, \mathbf{d}_{\mathrm{I}}^m) - r(\pi)| < \varepsilon.$$

Since ε was arbitrary, the conclusion follows. □

Versions of Theorems 5 and 6 appeared in Hoeffding (1960).

In a situation such as that of Example 9, $r(\pi, \delta_n^\pi)$ and $E^{m_n}[r_0(\pi^n, n)]$ are infinite for all n. Hence Theorems 5 and 6 do not apply. It can still be true, however, that $r_m^m(\pi, 0) \to r(\pi)$ as $m \to \infty$. The following theorem establishes this, and also proves the existence of a Bayes procedure, under certain conditions. This theorem is a special case of that in Magwire (1953).

Theorem 7. *Assume that*

(a) *there is a sequential procedure,* $\mathbf{d}$, *with finite Bayes risk, and*
(b) *with probability one (under* π*),* $\inf_a L(\theta, a, m) \to \infty$ *as* $m \to \infty$ *(i.e., the sampling cost goes to infinity as* $m \to \infty$*).*

Then

$$\lim_{m \to \infty} r_m^m(\pi, 0) = r(\pi).$$

Also, a Bayes sequential procedure, $\mathbf{d}^\pi$, *exists, and* $\mathbf{d}^\pi$ *stops sampling for the first* n $(n = 0, 1, \ldots)$ *for which* $r_0(\pi^n, n)$ *is finite and*

$$r_0(\pi^n, n) = r^\infty(\pi^n, n),$$

where r^∞ *satisfies the equation*

$$r^\infty(\pi^n, n) = \min\{r_0(\pi^n, n), E^*[r^\infty(\pi^n(\theta|X_{n+1}), n+1)]\}.$$

Proof. Consider again the m-inner truncated loss

$$L^m(\theta, a, n) = \begin{cases} L(\theta, a, n) & \text{if } n < m, \\ \inf_a L(\theta, a, m) & \text{if } n \geq m. \end{cases}$$

Observe that

$$L^m(\theta, a, n) \text{ is nondecreasing in } m, \text{ and } \lim_{m\to\infty} L^m(\theta, a, n) = L(\theta, a, n). \quad (7.27)$$

Consider the sequential problem with prior π^n, loss $L^m(\theta, a, j)$, and possible observations $X_{n+1}, X_{n+2}, \ldots$. Let $r^m(\pi^n, \mathbf{d}, n)$ denote the Bayes risk of a procedure $\mathbf{d}$ in this problem, and let $r^m(\pi^n, n)$ denote the Bayes risk for the problem. Observe that, if $n < m$, then

$$r^m(\pi^n, n) = \min\{r_0(\pi^n, n), E^*[r^m(\pi^n(\theta|X_{n+1}), n+1)]\}. \quad (7.28)$$

This follows from applying Theorem 2 of Subsection 7.4.5 to the loss L^m, noting that $r^m(\pi^n, n) = r_{m-n}(\pi^n, n)$ in the notation of Theorem 2. (Recall from Subsection 7.4.7 that for the loss L^m one can restrict attention to m-truncated procedures.)

Since $L^m(\theta, a, n)$ and hence $r^m(\pi^n, n)$ are nondecreasing in m, it follows that

$$r^\infty(\pi^n, n) = \lim_{m\to\infty} r^m(\pi^n, n)$$

exists (though it could be infinite). The monotone convergence theorem hence implies that

$$\lim_{m\to\infty} E^*[r^m(\pi^n(\theta|X_{n+1}), n+1)] = E^*[r^\infty(\pi^n(\theta|X_{n+1}), n+1)].$$

Together with (7.28), these equalities imply that

$$r^\infty(\pi^n, n) = \min\{r_0(\pi^n, n), E^*[r^\infty(\pi^n(\theta|X_{n+1}), n+1)]\}. \quad (7.29)$$

Note also, from (7.27), that $r^m(\pi, 0) \le r(\pi)$ for all m, so that

$$r^\infty(\pi, 0) = \lim_{m\to\infty} r^m(\pi, 0) \le r(\pi). \quad (7.30)$$

Consider now (for the original sequential problem) the stopping rule, to be denoted $\boldsymbol{\tau}^*$, which stops sampling for the first n for which $r_0(\pi^n, n) < \infty$ and $r_0(\pi^n, n) = r^\infty(\pi^n, n)$. The pseudoprocedure $\mathbf{d}^* = (\boldsymbol{\tau}^*, \boldsymbol{\delta}^\pi)$ will be shown to be a Bayes procedure for the original problem. (At this point we merely call $\mathbf{d}^*$ a pseudoprocedure because it has not yet been shown to stop sampling with probability one.)

As a first step, note that, since

$$L^m(\theta, a, n) = \inf_a L(\theta, a, m)$$

for $n \ge m$, it is immaterial what happens for $n \ge m$. Thus, even for a pseudoprocedure $\mathbf{d}$, $r^m(\pi^n, \mathbf{d}, n)$ is well defined, providing we conventionally assign a loss of $\inf_a L(\theta, a, m)$ to not stopping.

Now define $\mathbf{d}^n = (\boldsymbol{\tau}^n, \boldsymbol{\delta}^n)$, where $\boldsymbol{\tau}^n = (\tau_n^*, \tau_{n+1}^*, \ldots)$ and $\boldsymbol{\delta}^n = (\delta_n^\pi, \delta_{n+1}^\pi, \ldots)$. Clearly $\mathbf{d}^n$ is the procedure $\mathbf{d}^*$ restricted to the sequential problem continuing from stage n. The crucial step of the proof is to show that, for $n \leq m$,

$$r^m(\pi^n, \mathbf{d}^n, n) \leq r^\infty(\pi^n, n). \tag{7.31}$$

This will be established by induction on $m-n$. Assume that $n \leq m$ in all of the following.

First, if $m-n=0$, then

$$L^m(\theta, a, n) = \inf_a L(\theta, a, m).$$

Hence

$$r^m(\pi^n, \mathbf{d}^n, n) = E^{\pi^m}_{\cdot}[\inf_a L(\theta, a, m)] = r^m(\pi^m, m) \leq r^\infty(\pi^m, m).$$

This establishes (7.31) for $m-n=0$.

Assume now that (7.31) holds for $m-n=j$. We must show that it then holds for $m-n=j+1$. Thus let $m-n=j+1$, which in particular implies that $n<m$. Note first, from the definition of $\boldsymbol{\tau}^*$, that if

$$r_0(\pi^n, n) = r^\infty(\pi^n, n) < \infty, \tag{7.32}$$

then $\mathbf{d}^n$ will make an immediate decision, incurring the posterior Bayes risk $r_0(\pi^n, n)$. Hence (7.31) is satisfied when (7.32) is true. If, instead, $r_0(\pi^n, n)$ is infinite or $r_0(\pi^n, n) > r^\infty(\pi^n, n)$, then the definition of $\boldsymbol{\tau}^*$ ensures that another observation will be taken. In this case, therefore, familiar arguments show that

$$r^m(\pi^n, \mathbf{d}^n, n) = E^*[r^m(\pi^n(\theta|X_{n+1}), \mathbf{d}^{n+1}, n+1)]. \tag{7.33}$$

But

$$r^m(\pi^n(\theta|X_{n+1}), \mathbf{d}^{n+1}, n+1) = r^m(\pi^{n+1}, \mathbf{d}^{n+1}, n+1),$$

and $m-(n+1)=j$. Hence, by the induction hypothesis,

$$r^m(\pi^n(\theta|X_{n+1}), \mathbf{d}^{n+1}, n+1) \leq r^\infty(\pi^n(\theta|X_{n+1}), n+1).$$

Combining this with (7.33) shows that

$$r^m(\pi^n, \mathbf{d}^n, n) \leq E^*[r^\infty(\pi^n(\theta|X_{n+1}), n+1)]. \tag{7.34}$$

Since we are considering the case in which $r_0(\pi^n, n)$ is infinite or $r_0(\pi^n, n) > r^\infty(\pi^n, n)$, (7.34) and (7.29) imply (7.31). Thus (7.31) is established.

Taking the limit in (7.31) for $n=0$ and using (7.30) shows that

$$\lim_{m\to\infty} r^m(\pi, \mathbf{d}^*, 0) \leq r^\infty(\pi, 0) \leq r(\pi). \tag{7.35}$$

Since $r(\pi) < \infty$ (By Condition (a) of the theorem), we can conclude that

$$\lim_{m\to\infty} r^m(\pi, \mathbf{d}^*, 0) < \infty. \tag{7.36}$$

Now, since $\inf_a L(\theta, a, m)$ was defined to be the loss if $\mathbf{d}^*$ does not stop sampling, it is clear that

$$r^m(\pi, \mathbf{d}^*, 0) \geq E^\pi[\{\inf_a L(\theta, a, m)\}\lambda(\theta)],$$

where $\lambda(\theta)$ is the probability (given θ) that $\mathbf{d}^*$ does not stop sampling. Hence (7.36) implies that

$$\lim_{m\to\infty} E^{\pi}[\{\inf_a L(\theta, a, m)\}\lambda(\theta)] < \infty.$$

A standard analysis argument shows that this is consistent with Condition (b) of the theorem only if $\lambda(\theta)=0$ with probability one (with respect to π). Hence $\mathbf{d}^*$ is indeed a proper procedure.

Finally, (7.27) and another application of the monotone convergence theorem show that

$$\lim_{m\to\infty} r^m(\pi, \mathbf{d}^*, 0) = r(\pi, \mathbf{d}^*).$$

Together with (7.35), this implies that

$$r(\pi, \mathbf{d}^*) = r(\pi),$$

and hence that $\mathbf{d}^*$ is a Bayes procedure. It also follows from (7.35) that $r^{\infty}(\pi, 0) = r(\pi)$, so that

$$r_m^m(\pi, 0) = r^m(\pi, 0) \to r(\pi)$$

as $m \to \infty$. The remaining conclusions of the theorem are merely restatements of the definition of $\boldsymbol{\tau}^*$ and (7.29). □

The equation

$$r^{\infty}(\pi^n, n) = \min\{r_0(\pi^n, n), E^*[r^{\infty}(\pi^n(\theta|X_{n+1}), n+1)]\} \tag{7.37}$$

has a simple intuitive explanation. The quantity $r^{\infty}(\pi^n, n)$ can be seen to be the Bayes risk for the sequential problem continuing from stage n, so that

$$r^*(\pi^n, n) = E^*[r^{\infty}(\pi^n(\theta|X_{n+1}), n+1)]$$

is the smallest Bayes risk among procedures which take at least one observation. Hence (7.37) merely states that the Bayes risk of the problem is the smaller of the posterior Bayes risk of an immediate decision and the minimum Bayes risk among procedures which take at least one observation. This, of course, seems intuitively obvious.

The following special case of Theorem 7 is of interest.

Corollary 2. *Suppose that*

(a) $X_1, X_2, \ldots$ *is a sequential sample,*
(b) $L(\theta, a, n) = L(\theta, a) + nc$, *where* $c > 0$, *and*
(c) *there is a sequential procedure,* $\mathbf{d}$, *with finite Bayes risk.*

Then a Bayes procedure, $\mathbf{d}^{\pi}$, *exists, and* $\mathbf{d}^{\pi}$ *stops sampling for the first n for which* $\rho_0(\pi^n)$ *is finite and*

$$\rho_0(\pi^n) = \rho^{\infty}(\pi^n),$$

where $\rho^{\infty}(\pi^n)$ *satisfies the equation*

$$\rho^{\infty}(\pi^n) = \min\{\rho_0(\pi^n), E^*[\rho^{\infty}(\pi^n(\theta|X))] + c\}.$$

PROOF. Recalling that $\rho_0(\pi^n) = \inf_a E^{\pi^n}[L(\theta, a)]$, it can easily be checked that

$$r_0(\pi^n, n) = \rho_0(\pi^n) + nc,$$

and that $r^\infty(\pi^n, n)$ can be written as

$$r^\infty(\pi^n, n) = \rho^\infty(\pi^n) + nc.$$

The result is immediate from Theorem 7, since (as always) we are assuming that $L(\theta, a) \geq K > -\infty$. □

A sequential Bayes procedure need not always exist, as the following example shows.

EXAMPLE 12. Assume that $X_1, X_2, \ldots$ is a sequential sample from a $\mathcal{N}(\theta, 1)$ density, and that it is desired to estimate θ under loss

$$L(\theta, a, n) = (\theta - a)^2 + \sum_{i=1}^{n} c_i,$$

where $c_i = [2i(i+1)]^{-1}$. It can easily be shown that

$$\sum_{i=1}^{n} c_i = \sum_{i=1}^{n} \frac{1}{2i(i+1)} = \frac{1}{2}\left(1 - \frac{1}{n+1}\right).$$

Hence

$$L(\theta, a, n) = (\theta - a)^2 + \frac{1}{2}\left(1 - \frac{1}{n+1}\right).$$

If now the prior is $\mathcal{N}(0, 1)$, so that π^n is $\mathcal{N}(\mu_n, (n+1)^{-1})$, then the posterior Bayes decision risk is

$$\rho_0(\pi^n) = E^{\pi^n}[(\theta - \mu_n)^2] = (n+1)^{-1}.$$

Hence

$$r_0(\pi^n, n) = \frac{1}{n+1} + \frac{1}{2}\left(1 - \frac{1}{n+1}\right) = \frac{1}{2}\left(1 + \frac{1}{n+1}\right).$$

This is decreasing in n, so that it never pays to stop sampling. Another observation always lowers the posterior Bayes risk. No proper Bayes procedure can, therefore, exist.

7.4.10. Other Techniques for Finding a Bayes Procedure

Although truncation methods are, in principle, capable of providing close approximations to the Bayes procedure, the considerable calculational difficulties often encountered with such methods suggest that consideration of other techniques is desirable. We briefly discuss some alternative techniques here.

I. *Solution of the Functional Equation*

One potentially useful method follows from Corollary 2 of the preceding subsection. This corollary shows (under the indicated conditions) that the Bayes procedure is determined by the function ρ^∞ which satisfies

$$\rho^\infty(\pi^n) = \min\{\rho_0(\pi^n), E^*[\rho^\infty(\pi^n(\theta|X))] + c\}. \tag{7.38}$$

One can thus attempt to find the Bayes procedure by solving (7.38) for ρ^∞. (The solution to (7.38) is usually unique. This is so, for example, if $r(\pi, \delta_n^\pi) = E^{m_n}[\rho_0(\pi^n)] \to 0$ as $n \to \infty$. See DeGroot (1970) for a proof of this.)

Unfortunately, the calculational problems encountered in solving (7.38) are usually as difficult as those involved in finding truncated or look ahead procedures. Therefore, we will not discuss general techniques of solution. (See DeGroot (1970) for an introduction to the subject.)

In some special cases, (7.38) can be solved explicitly, or at least the solution can be reduced to a fairly simple numerical problem. The next section considers, in detail, one such very important special case. Here we content ourselves with a single example (due to Wald (1950)).

Example 13. Assume that $X_1, X_2, \ldots$ is a sequential sample from a $\mathcal{U}(\theta - \frac{1}{2}, \theta + \frac{1}{2})$ density, and that it is desired to estimate θ under loss

$$L(\theta, a, n) = 12(\theta - a)^2 + nc.$$

The prior distribution is $\mathcal{U}(a, b)$. A straightforward calculation shows that $\pi^n = \pi(\theta|x_1, \ldots, x_n)$ is $\mathcal{U}(a_n, b_n)$, where $a_n = \max\{a, (\max x_i) - \frac{1}{2}\}$, and $b_n = \min\{b, (\min x_i) + \frac{1}{2}\}$. The Bayes estimator at stage n is the posterior mean,

$$\delta_n^\pi(\mathbf{x}^n) = \frac{a_n + b_n}{2},$$

and the posterior decision risk is twelve times the posterior variance, i.e.,

$$\rho_0(\pi^n) = (b_n - a_n)^2.$$

Note that this is the square of the posterior range.

It is easy to check, in this situation, that the conditions of Corollary 2 are satisfied. Hence a Bayes procedure can be determined if the solution to (7.38) can be found. Note that all π^n are uniform distributions. Hence it suffices to find a solution to (7.38) for the class of uniform priors. By the reasoning in the previous paragraph, if π is any prior with range r, then the posterior Bayes decision risk will be

$$\rho_0(\pi) = r^2.$$

It thus seems reasonable to believe that $\rho^\infty(\pi)$ will also be a function of r, say

$$\rho^\infty(\pi) = h(r).$$

If this is true, then (7.38) becomes

$$h(r) = \min\{r^2, E^*[h(r(X))] + c\}, \tag{7.39}$$

where $r(X)$ is the posterior range of $\pi(\theta|X)$.

Intuition also suggests that the Bayes procedure will stop sampling when the posterior range gets small enough. But, from Corollary 2, we know that the Bayes procedure stops sampling when $\rho_0(\pi^n) = \rho^\infty(\pi^n)$. The conclusion is that $h(r)$ is likely to be of the form

$$h(r) = \begin{cases} r^2 & \text{if } r \le r_0, \\ g(r) & \text{if } r > r_0, \end{cases} \tag{7.40}$$

where $g(r) < r^2$ for $r > r_0$.

To proceed further, it is necessary to find a specific representation for $E^*[h(r(X))]$. A moment's thought will make clear the fact that the exact interval specified by π is immaterial, as long as it has range r. It is convenient to choose π to be $\mathcal{U}(\frac{1}{2}, r + \frac{1}{2})$. An easy calculation then shows that the posterior range of $\pi(\theta|X)$ is

$$r(X) = \min\{r, X\} - \max\{0, X - 1\}, \tag{7.41}$$

and that the marginal density of X is

$$m(x) = r^{-1} r(x) I_{(0, r+1)}(x).$$

It is now necessary to distinguish between the two cases $r \le 1$ and $r > 1$. Consider first $r \le 1$. For such r, it follows from (7.41) that

$$r(x) = \begin{cases} x & \text{if } 0 < x < r, \\ r & \text{if } r \le x < 1, \\ r + 1 - x & \text{if } 1 \le x < r + 1. \end{cases}$$

Hence

$$\begin{aligned} E^*[h(r(X))] &= E^m[h(r(X))] \\ &= \frac{1}{r} \int_0^{r+1} r(x) h(r(x)) dx \\ &= \frac{1}{r} \left[\int_0^r x h(x) dx + \int_r^1 r h(r) dx \right. \\ &\quad \left. + \int_1^{r+1} (r + 1 - x) h(r + 1 - x) dx \right] \\ &= \frac{1}{r} \left[\int_0^r x h(x) dx + r h(r)(1 - r) + \int_0^r y h(y) dy \right] \\ &= \frac{2}{r} \int_0^r x h(x) dx + h(r)(1 - r). \end{aligned}$$

Using the form for h suggested in (7.40), this becomes (for $r > r_0$)

$$E^*[h(r(X))] = \frac{2}{r}\int_0^{r_0} x^3\,dx + \frac{2}{r}\int_{r_0}^{r} xg(x)dx + g(r)(1-r)$$
$$= \frac{r_0^4}{2r} + \frac{2}{r}\int_{r_0}^{r} xg(x)dx + g(r)(1-r).$$

Inserting this in (7.39) and again using (7.40) gives (for $r > r_0$)

$$g(r) = \min\{r^2, E^*[h(r(X))] + c\}$$
$$= E^*[h(r(X))] + c$$
$$= \frac{r_0^4}{2r} + \frac{2}{r}\int_{r_0}^{r} xg(x)dx + g(r)(1-r) + c.$$

Multiplying through by r and collecting terms results in the equation

$$0 = \tfrac{1}{2}r_0^4 + 2\int_{r_0}^{r} xg(x)dx - r^2 g(r) + cr. \tag{7.42}$$

Differentiating both sides with respect to r gives (letting $g'(r) = dg(r)/dr$)

$$0 = 2rg(r) - 2rg(r) - r^2 g'(r) + c$$
$$= -r^2 g'(r) + c.$$

Hence $g'(r) = c/r^2$, which implies that

$$g(r) = K - \frac{c}{r}$$

for some constant K. Now it is reasonable to assume that $g(r_0) = r_0^2$ ($h(r)$ should be continuous), so that $K = r_0^2 + c/r_0$. Plugging

$$g(r) = r_0^2 + c(r_0^{-1} - r^{-1})$$

back into (7.42) and solving for r_0 gives $r_0 = (2c)^{1/3}$. The suggested function h is thus

$$h^*(r) = \begin{cases} r^2 & \text{if } r \le (2c)^{1/3}, \\ \tfrac{3}{2}(2c)^{2/3} - c/r & \text{if } (2c)^{1/3} < r \le 1, \end{cases}$$

and it can be checked that this is indeed a valid solution to (7.39).

Consider next the case $r > 1$. Note that it will still be the case that $r(X) \le 1$. A calculation (left as an exercise) shows that

$$E^*[h^*(r(X))] + c = \frac{3}{2}(2c)^{2/3} - \frac{c}{r}.$$

Hence, extending the definition of h^* to all $r>0$ by

$$h^*(r)=\begin{cases} r^2 & \text{if } r\le(2c)^{1/3}, \\ \frac{3}{2}(2c)^{2/3}-c/r & \text{if } (2c)^{1/3}<r, \end{cases} \tag{7.43}$$

it follows that (7.39) is satisfied for all r.

It can also be shown that $r(\pi, \delta_n^\pi)\to 0$ as $n\to\infty$, so that the given solution to (7.38) is unique.

In conclusion, it has been shown that

$$\rho^\infty(\pi^n)=h^*(b_n-a_n).$$

Corollary 2 thus gives that the Bayes procedure is to stop sampling for the first n for which $\rho_0(\pi^n)=\rho^\infty(\pi^n)$, i.e., the first n for which

$$(b_n-a_n)^2\le(2c)^{1/3},$$

and then use the Bayes estimate $(a_n+b_n)/2$.

Sometimes, one can infer that the Bayes procedure is of a particular form, depending on, say, a small number of unknown constants. If nothing else, this reduces the problem to a very tractable numerical problem (minimizing a Bayes risk over a small number of constants). A very important example of this will be given in the next section. See also Section 8.6 (Chapter 8).

Due to the difficulty in finding explicit Bayes procedures, a number of interesting theoretical studies have been performed, with the goal of finding "asymptotic" Bayes procedures. Virtually all such works assume that $c_j=c$ for all j, and are asymptotic in that they consider the problem as $c\to 0$. They are, hence, potentially relevant when the cost per observation is much smaller than the decision loss. (Unfortunately, this is usually not the case, in that sequential analysis is most often used with expensive observations. When observations are cheap, it is usually wasteful to analyze the situation after each observation, and taking batches of observations is often cheaper.)

The simplest asymptotic results to state are those of Bickel and Yahav (1967). Besides being of interest in the asymptotic sense, these results provide insight into the behavior of certain previously discussed sequential procedures.

The results of Bickel and Yahav pertain to quite general settings, but for this brief discussion we will assume that we have a sequential sample X_1, $X_2, \ldots$, that $L(\theta, a, n)$ is of the form (7.1), and that each observation costs $c_i=c$. The procedures that will be considered are called *asymptotically pointwise optimal*, which basically means that, as $c\to 0$, the posterior Bayes risks of the procedures, evaluated at the stopping point, converge to the corresponding posterior Bayes risk of the Bayes procedure, for almost all samples. See Bickel and Yahav (1967) and Kiefer and Sacks (1963) for a discussion of this and related concepts. It is again convenient to consider testing and estimation separately.

II. *Asymptotically Pointwise Optimal Procedures in Testing Problems*

It frequently happens, in testing problems, that (with probability one for each θ)

$$n^{-1}\log[\rho_0(\pi^n)]\to Y(\theta),\quad\text{where}\quad P^{\pi}(-\infty<Y(\theta)<0)=1. \tag{7.44}$$

Consider, for instance, Example 3 in Section 7.2.

EXAMPLE 3 (revisited). The posterior Bayes decision risk is clearly the smaller of the posterior expected losses of a_0 and a_1, namely

$$\rho_0(\pi^n)=\min\left\{\int_{\Theta_1} L(\theta,a_0)dF^{\pi^n(\theta|\bar{X}_n)}(\theta),\int_{\Theta_0} L(\theta,a_1)dF^{\pi^n(\theta|\bar{X}_n)}(\theta)\right\}.$$

In Case 1, for which $\Theta_0=\{\theta_0\}$ and $\Theta_1=\{\theta_1\}$,

$$\begin{aligned}\rho_0(\pi^n)=\frac{n^{1/2}}{(2\pi)^{1/2}\sigma m_n(\bar{X}_n)}\min\Big\{&L(\theta_1,a_0)\pi(\theta_1)\exp\left[-\frac{n}{2\sigma^2}(\theta_1-\bar{X}_n)^2\right],\\ &L(\theta_0,a_1)\pi(\theta_0)\exp\left[-\frac{n}{2\sigma^2}(\theta_0-\bar{X}_n)^2\right]\Big\}.\end{aligned} \tag{7.45}$$

Imagine now that θ_1 is true, so that $\bar{X}_n\to\theta_1$ with probability one. Then, for large enough n, the minimum in (7.45) will be (with probability approaching one)

$$L(\theta_0,a_1)\pi(\theta_0)\exp\left[-\frac{n}{2\sigma^2}(\theta_0-\bar{X}_n)^2\right].$$

Hence, with probability one,

$$\begin{aligned}\lim_{n\to\infty}\frac{1}{n}\log[\rho_0(\pi^n)]&=\lim_{n\to\infty}\frac{1}{n}\log\left[\frac{n^{1/2}L(\theta_0,a_1)\pi(\theta_0)}{(2\pi)^{1/2}\sigma}\right]\\ &\quad-\lim_{n\to\infty}\frac{1}{n}\log[m_n(\bar{X}_n)]-\lim_{n\to\infty}\frac{1}{2\sigma^2}(\theta_0-\bar{X}_n)^2\\ &=-\lim_{n\to\infty}\frac{1}{n}\log[m_n(\bar{X}_n)]-\frac{1}{2\sigma^2}(\theta_0-\theta_1)^2.\end{aligned}$$

Now

$$\begin{aligned}m_n(\bar{X}_n)=\frac{n^{1/2}}{(2\pi)^{1/2}\sigma}\Big[&\pi(\theta_0)\exp\left\{-\frac{n}{2\sigma^2}(\theta_0-\bar{X}_n)^2\right\}\\ &+\pi(\theta_1)\exp\left\{-\frac{n}{2\sigma^2}(\theta_1-\bar{X}_n)^2\right\}\Big],\end{aligned}$$

for which it is easy to show that, with probability one,

$$\lim_{n\to\infty}\frac{1}{n}\log[m_n(\bar{X}_n)]=0.$$

In conclusion, it is true that, with probability one,

$$\lim_{n\to\infty} \frac{1}{n} \log[\rho_0(\pi^n)] = -\frac{1}{2\sigma^2}(\theta_0 - \theta_1)^2.$$

The same result holds if θ_0 is the true state of nature. Thus (7.44) is clearly satisfied.

In Case 2 of Example 3 (where $\Theta_0 = (-\infty, \theta_0]$, $\Theta_1 = (\theta_0, \infty)$, and $W(y) = y^k$), it can be shown that, if θ^* is the true value of θ, then, with probability one,

$$\lim_{n\to\infty} \frac{1}{n} \log[\rho_0(\pi^n)] = -\frac{1}{2\sigma^2}(\theta_0 - \theta^*)^2.$$

Hence (7.44) is satisfied for the normal priors considered.

When (7.44) is satisfied, an asymptotically pointwise optimal procedure (see Bickel and Yahav (1967)) is to stop sampling for the first n for which

$$\rho_0(\pi^n) \le c, \tag{7.46}$$

and then make the Bayes decision. As shown at the end of Subsection 7.4.4, this is exactly the 1-step inner look ahead procedure, $\mathbf{d}^1_{\mathrm{IL}}$. That this procedure turns out to be nearly optimal for very small c is somewhat surprising. It does explain, however, why, in Case 1 of Example 10 (Subsection 7.4.8), the procedure $\mathbf{d}^1_{\mathrm{IL}}$ performed so well.

Comparison of the stopping times of $\mathbf{d}^1_{\mathrm{IL}}$ and the optimal fixed sample size rule is instructive. Clearly (7.44) implies that, for large n,

$$\log[\rho_0(\pi^n)] \cong nY(\theta).$$

Since (7.46) is equivalent to

$$\log[\rho_0(\pi^n)] \le \log c,$$

it follows that $\mathbf{d}^1_{\mathrm{IL}}$ will stop (approximately) for the first n for which

$$n \ge \frac{\log c}{Y(\theta)} = \frac{\log c^{-1}}{|Y(\theta)|}.$$

Hence the stopping time of $\mathbf{d}^1_{\mathrm{IL}}$ will be of the order of $\log c^{-1}$. This was also observed for the optimal fixed sample size rule in Case 1 of Example 3, which dealt with separated hypotheses. In case 2 of Example 3, however, the hypotheses were connected, and it was then observed that the optimal fixed sample size was of the order $c^{-\lambda}$ (λ depending on the loss). Thus, for connected hypotheses, there appears to be a significant difference between the sample sizes needed for the fixed sample size approach and the sequential approach. The fixed sample size approach seems to take considerably more observations than are needed ($c^{-\lambda}$ compared with $\log c^{-1}$). The reason for this is that the fixed sample size approach gets very concerned about θ being near the joint boundary of the connected hypotheses, and chooses a very large sample size to protect against this possibility. The sequential procedure quickly realizes, however, when θ is away from the boundary, and can then make do with a fairly small sample size.

III. *Asymptotically Pointwise Optimal Procedures in Estimation Problems*

It frequently happens, in estimation problems, that (with probability one for each θ)

$$n^{\beta}\rho_0(\pi^n)\rightarrow Y(\theta), \quad \text{where} \quad P^{\pi}(0<Y(\theta)<\infty)=1, \tag{7.47}$$

and β is some positive constant. For example, when $L(\theta, a)=|\theta-a|^k$, this condition is usually satisfied with $\beta=k/2$. See Example 2 (Section 7.2) and Example 7 (Subsection 7.4.5) for illustrations of this.

When (7.47) is satisfied, an asymptotically pointwise optimal procedure (see Bickel and Yahav (1967)) is to stop sampling for the first n for which

$$\rho_0(\pi^n)\leq\frac{cn}{\beta}, \tag{7.48}$$

and then use the Bayes estimate. Note, from the discussion at the end of Example 2, that this is essentially the stopping rule for the optimal fixed sample size rule, except that $\rho_0(\pi^n)$ is used instead of

$$r(\pi, \delta_n^{\pi})=E^{m_n}[\rho_0(\pi^n)].$$

Frequently, $\rho_0(\pi^n)$ and $r(\pi, \delta_n^{\pi})$ will be similar functions. Indeed, (7.47) implies that, for large n and given θ,

$$\rho_0(\pi^n)\cong n^{-\beta}Y(\theta),$$

so that

$$r(\pi, \delta_n^{\pi})=E^{m_n}[\rho_0(\pi^n)]=E^{\pi}E_0^{\mathbf{X}^n}[\rho_0(\pi^n)]\cong n^{-\beta}E^{\pi}[Y(\theta)].$$

Hence the sample sizes for the asymptotically pointwise optimal procedure and optimal fixed sample size rule will tend to be comparable (in contrast to the testing situation with connected hypotheses).

It can also be shown that the asymptotically pointwise optimal procedure is (as $c\rightarrow 0$) usually equivalent to the 1-step look ahead procedure $\mathbf{d}_L^1$ (or, for that matter, any $\mathbf{d}_L^m$). All these procedures seek to minimize (see Subsection 7.4.8)

$$V(\mathbf{x}, n)=\rho_0(\pi^n)+nc\cong n^{-\beta}Y(\theta)+nc. \tag{7.49}$$

To see intuitively that the asymptotically pointwise optimal procedure minimizes this expression (asymptotically), differentiate with respect to n on the right-hand side of (7.49) and set equal to zero. The result is

$$0=-\beta n^{-(\beta+1)}Y(\theta)+c\cong-\beta n^{-1}\rho_0(\pi^n)+c.$$

Clearly, the n which satisfies this equality will essentially be that determined by (7.48). Likewise, as discussed in Subsection 7.4.8, the look ahead procedures seek the minimum of $V(\mathbf{x}, n)$. They succeed, in an asymptotic sense, as $c \to 0$ (see Shapiro and Wardrop (1980)). A third class of procedures, based on this same idea, has been proposed by Alvo (1977) for certain special cases. They are better (i.e., less dependent on small c) than the asymptotically pointwise optimal procedure, and yet are quite easy to use. The idea (see also Lorden (1977) and Woodroofe (1981)) is to also be concerned with the second-order term in the asymptotic risk.

IV. *Asymptotically Optimal Procedures: Concluding Remarks*

The major question concerning asymptotically optimal procedures is whether or not they are reasonable for moderately small c. The fact that they are based on the assumption that (7.44) or (7.47) give approximate expressions for $\log \rho_0(\pi^n)$ or $\rho_0(\pi^n)$, and that these expressions are independent of the prior, is cause for concern. It means that the procedures can be written in such a way that they do not depend on the prior. One wonders how successful they will then be, when c is such that the prior will matter. (When c is very small, enough observations will be taken to overwhelm the prior information.) The relevant look ahead procedures, as discussed in Subsection 7.4.8, seem much more attractive from this viewpoint.

Of course, to a classical statistician it may seem attractive to have a procedure which does not depend on the prior information. Thus work has been done on describing and evaluating, in a classical fashion, asymptotically optimal procedures such as those above. As an example, Schwarz (1962) has shown that, for testing H_0: $\theta \leq -\lambda$ versus H_1: $\theta \geq \lambda$ based on a sequential sample $X_1, X_2, \ldots$ from a $\mathcal{N}(\theta, 1)$ distribution, the asymptotically optimal test (for a quite general loss structure, with cost c per observation, and everywhere positive prior) is (approximately) to stop sampling for the first n for which

$$|s_n| = \left| \sum_{i=1}^{n} x_i \right| \geq [2n \log c^{-1}]^{1/2} - n\lambda,$$

deciding H_0 if s_n is negative and deciding H_1 if s_n in positive. This non-Bayesian procedure is evaluated classically in Woodroofe (1976), and is shown to have quite desirable properties. More general results of this nature can be found in Schwarz (1962), Schwarz (1968), and Woodroofe (1978). See also Kiefer and Sacks (1963), Chernoff (1968, 1972), Vardi (1979), Berry (1982), and Lorden (1984) for other asymptotic results.

7.5. The Sequential Probability Ratio Test

The most commonly used sequential procedure is the sequential probability ratio test (SPRT), introduced by Wald in the 1940s. The SPRT is designed for testing a simple null hypothesis against a simple alternative hypothesis when a sequential sample $X_1, X_2, \ldots$ is available. The common density of the (independent) X_i will, as usual, be denoted $f(x|\theta)$, and the two hypotheses of interest will be denoted H_0: $\theta = \theta_0$ and H_1: $\theta = \theta_1$. The densities $f(x|\theta_0)$ and $f(x|\theta_1)$ need not actually be from the same parametric family. When they are not, the θ_i can just be considered to be indices.

When considering the problem from a sequential decision-theoretic viewpoint, it will be assumed that the loss is

$$L(\theta, a, n) = L(\theta, a) + nc,$$

where $L(\theta, a)$ is "0-K_i" loss. Thus, letting a_i denote accepting H_i, the decision loss is $L(\theta_0, a_0) = L(\theta_1, a_1) = 0$, $L(\theta_0, a_1) = K_1$, and $L(\theta_1, a_0) = K_0$. The parameter space is formally $\Theta = \{\theta_0, \theta_1\}$, so that a prior π is specified by

$$\pi_0 \equiv \pi(\theta_0) = 1 - \pi(\theta_1) \equiv 1 - \pi_1.$$

We first derive the SPRT as a Bayes procedure, and in later sections discuss its classical properties.

7.5.1. The SPRT as a Bayes Procedure

The Bayes sequential problem, as described above, is of the type dealt with in Corollary 2 of Subsection 7.4.9. The conclusion of that corollary thus applies, namely that a Bayes procedure is given by $\mathbf{d}^\pi = (\boldsymbol{\tau}^*, \boldsymbol{\delta}^\pi)$, where $\boldsymbol{\delta}^\pi$ is the Bayes decision rule and $\boldsymbol{\tau}^*$ is the stopping rule which stops sampling for the first n ($n = 0, 1, 2, \ldots$) for which

$$\rho_0(\pi^n) = \rho^\infty(\pi^n). \tag{7.50}$$

Again $\rho_0(\pi^n)$ is the posterior Bayes decision risk at stage n, while $\rho^\infty(\pi^n)$ satisfies (for all n)

$$\rho^\infty(\pi^n) = \min\{\rho_0(\pi^n), \rho^*(\pi^n)\}, \tag{7.51}$$

where

$$\rho^*(\pi^n) = E^*[\rho^\infty(\pi^n(\theta|X))] + c, \tag{7.52}$$

the expectation being with respect to the predictive (or marginal) density of X induced by $\pi^n(\theta)$.

The posterior distributions, π^n, are determined by $\pi_0^n = \pi^n(\theta_0)$. Hence the functions ρ_0 and ρ^∞ can be considered to be functions of a single variable. For *any* prior π, clearly

$$\begin{aligned}\rho_0(\pi) &= \inf_a E^\pi[L(\theta, a)] = \min\{\pi_1 K_0, \pi_0 K_1\} \\ &= \min\{(1 - \pi_0)K_0, \pi_0 K_1\}.\end{aligned} \tag{7.53}$$

Also, note from the discussion at the end of Theorem 7 in Subsection 7.4.9 that

$$\rho^*(\pi) = \inf_{\mathbf{d}: N \geq 1} r(\pi, \mathbf{d}), \tag{7.54}$$

i.e., $\rho^*(\pi)$ is the minimum Bayes risk among procedures taking at least one observation. The following lemma describes the behavior of $\rho^*(\pi)$.

Lemma 1. *The function $\rho^*(\pi)$ is a concave continuous function of π_0, and is equal to c when $\pi_0 = 1$ or $\pi_0 = 0$.*

Proof. For convenience, write $\rho^*(\pi_0)$, $r(\pi_0, \mathbf{d})$, etc., for the relevant Bayes risks. (Recall that π is determined by π_0.) It is easy to check, for $0 \le \alpha \le 1$, $0 \le \pi_0 \le 1$, and $0 \le \pi_0' \le 1$, that

$$r(\alpha\pi_0 + (1-\alpha)\pi_0', \mathbf{d}) = \alpha r(\pi_0, \mathbf{d}) + (1-\alpha)(\pi_0', \mathbf{d}).$$

Hence

$$\begin{aligned}\rho^*(\alpha\pi_0 + (1-\alpha)\pi_0') &= \inf_{\mathbf{d}: N \ge 1} r(\alpha\pi_0 + (1-\alpha)\pi_0', \mathbf{d}) \\ &= \inf_{\mathbf{d}: N \ge 1} [\alpha r(\pi_0, \mathbf{d}) + (1-\alpha) r(\pi_0', \mathbf{d})] \\ &\ge \inf_{\mathbf{d}: N \ge 1} [\alpha r(\pi_0, \mathbf{d})] + \inf_{\mathbf{d}: N \ge 1} [(1-\alpha) r(\pi_0', \mathbf{d})] \\ &= \alpha\rho^*(\pi_0) + (1-\alpha)\rho^*(\pi_0').\end{aligned}$$

This establishes that ρ^* is concave. The continuity of $\rho^*(\pi_0)$ for $0 < \pi_0 < 1$ follows from the well known fact that a convex or concave function on a convex subset, Ω, of R^1 is continuous on the interior of Ω.

To prove the remainder of the lemma, note, from (7.54) and the fact that the decision loss is nonnegative, that $\rho^*(\pi_0) \ge c$. Finally, let $\mathbf{d}_0$ be the procedure which stops after taking one observation and makes the Bayes decision. Clearly

$$\rho^*(\pi_0) \le r(\pi_0, \mathbf{d}_0) = c + E^*[\rho_0(\pi(\theta | X))].$$

As $\pi_0 \to 0$ or $\pi_0 \to 1$, it is easy to check that $E^*[\rho_0(\pi(\theta | X))] \to 0$. Together with the fact that $\rho^*(\pi_0) \ge c$, this shows that $\rho^*(\pi_0)$ is continuous at $\pi_0 = 0$ and $\pi_0 = 1$, and is equal to c at these values. □

Two cases must now be distinguished.

Case 1. $\rho_0(\pi) \le \rho^*(\pi)$ for all π_0.

In this situation, the Bayes procedure is to immediately make the Bayes decision, taking no observations.

Case 2. $\rho_0(\pi) > \rho^*(\pi)$ for some π_0 $(0 < \pi_0 < 1)$.

In this situation, graphing $\rho_0(\pi)$ and $\rho^*(\pi)$ as functions of π_0 (using (7.53) and Lemma 1) results in Figure 7.2 below. From this figure it is clear that $\rho_0(\pi) > \rho^*(\pi)$ if $\pi_0' < \pi_0 < \pi_0''$. Together with (7.50), (7.51), and (7.53), this proves the following theorem. (From now on, π will again refer to the specific true prior for the problem.)

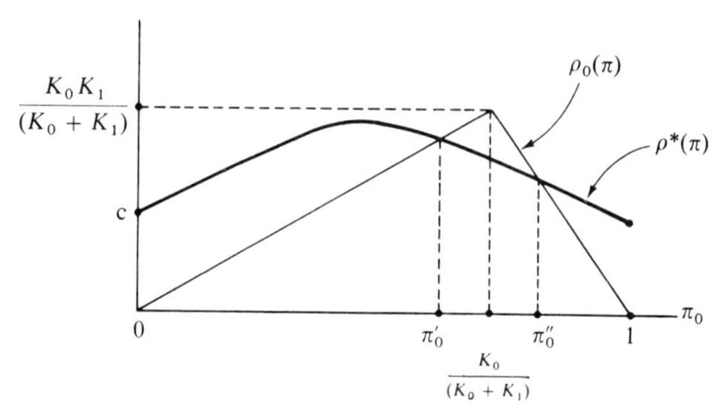

Figure 7.2

Theorem 8. *The Bayes sequential procedure, $\mathbf{d}^\pi$, stops sampling for the first n ($n = 0, 1, 2, \ldots$) for which $\pi_0^n \le \pi_0'$ or $\pi_0^n \ge \pi_0''$, deciding a_0 if $\pi_0^n \ge \pi_0''$ and a_1 if $\pi_0^n \le \pi_0'$. The constants π_0' and π_0'' satisfy $\pi_0' \le K_0/(K_0 + K_1) \le \pi_0''$.*

Note that this theorem is also valid for Case 1, as can be seen by choosing $\pi_0' = \pi_0'' = K_0/(K_0 + K_1)$.

The Bayes sequential procedure can be written in a more enlightening form by defining the *likelihood ratio* of θ_1 to θ_0 at stage n as

$$L_n = \frac{\prod_{i=1}^n f(x_i | \theta_1)}{\prod_{i=1}^n f(x_i | \theta_0)},$$

and noting that

$$\begin{aligned}
\pi_0^n = \pi(\theta_0 | \mathbf{x}^n) &= \frac{\pi(\theta_0) \prod_{i=1}^n f(x_i | \theta_0)}{\pi(\theta_0) \prod_{i=1}^n f(x_i | \theta_0) + \pi(\theta_1) \prod_{i=1}^n f(x_i | \theta_1)} \\
&= \frac{1}{1 + (\pi_1/\pi_0) \prod_{i=1}^n (f(x_i | \theta_1)/f(x_i | \theta_0))} \\
&= \frac{1}{1 + (\pi_1/\pi_0) L_n}.
\end{aligned}$$

(In this and subsequent expressions, define K/∞ as zero and $K/0$ as infinity.) Assuming $0 < \pi_0 < 1$ (which if not satisfied makes the problem trivial), it can easily be checked that $\pi_0^n \le \pi_0'$ if and only if $L_n \ge \pi_0(1 - \pi_0')/(\pi_1 \pi_0')$, while $\pi_0^n \ge \pi_0''$ if and only if $L_n \le \pi_0(1 - \pi_0'')/(\pi_1 \pi_0'')$. This establishes the following corollary to Theorem 8.

Corollary 3. *If $0 < \pi_0 < 1$, then the Bayes procedure, $\mathbf{d}^\pi$, is of the following form:*

$$\begin{aligned}
&\text{if } L_n \le A, && \text{stop sampling and decide } a_0; \\
&\text{if } L_n \ge B, && \text{stop sampling and decide } a_1; \qquad (7.55) \\
&\text{if } A < L_n < B, && \text{take another observation};
\end{aligned}$$

where $A = \pi_0(1 - \pi_0'')/(\pi_1 \pi_0'')$ and $B = \pi_0(1 - \pi_0')/(\pi_1 \pi_0')$. (Note that $A \le B$.)

From now on, we will assume that it is desirable to take at least one observation. This is, of course, equivalent to assuming that

$$\pi_0' < \pi_0 < \pi_0''. \tag{7.56}$$

The difficulty is that π_0' and π_0'' are not known, and the ensuing development, which is aimed at determining π_0' and π_0'', assumes (7.56). In applications, therefore, one must separately check whether or not the immediate Bayes decision has smaller Bayes risk than the optimal procedure derived by assuming (7.56). When (7.56) is satisfied, it is easy to check that $A < 1$ and $B > 1$.

Definition 9. The procedure defined by (7.55) with constants $A < 1$ and $B > 1$ is called the *sequential probability ratio test* (SPRT) with *stopping boundaries* A and B, and will be denoted $\mathbf{d}^{A,B}$.

The Bayesian problem can now be phrased as that of choosing A and B to minimize $r(\pi, \mathbf{d}^{A,B})$. To express the Bayes risk of $\mathbf{d}^{A,B}$ conveniently, let N denote the stopping time of $\mathbf{d}^{A,B}$, i.e.,

$$N = \min\{n: L_n \leq A \text{ or } L_n \geq B\},$$

and define the probabilities of Type I error and Type II error as

$$\alpha_0 = P_{\theta_0} (\text{deciding } a_1) = P_{\theta_0}(L_N \geq B),$$

$$\alpha_1 = P_{\theta_1} (\text{deciding } a_0) = P_{\theta_1}(L_N \leq A).$$

Also, let $E_{\theta_0}N$ and $E_{\theta_1}N$ denote the expected stopping times under θ_0 and θ_1 respectively. (It is being implicitly assumed that $P_{\theta_i}(N < \infty) = 1$ and $E_{\theta_i}N < \infty$. This will be verified later.) An easy calculation then gives

$$\begin{aligned} r(\pi, \mathbf{d}^{A,B}) &= \pi(\theta_0)R(\theta_0, \mathbf{d}^{A,B}) + \pi(\theta_1)R(\theta_1, \mathbf{d}^{A,B}) \\ &= \pi_0[\alpha_0 K_1 + cE_{\theta_0}N] + \pi_1[\alpha_1 K_0 + cE_{\theta_1}N]. \end{aligned} \tag{7.57}$$

The problem thus reduces to the calculation of α_0, α_1, $E_{\theta_0}N$, and $E_{\theta_1}N$, and the subsequent minimization of (7.57) over A and B. Unfortunately, only rarely can this program be analytically carried out. Two other options are available, however. First, numerical calculation and minimization of (7.57) is generally quite feasible, only two variables being involved. Second, reasonably accurate approximations to α_0, α_1, $E_{\theta_0}N$, and $E_{\theta_1}N$ exist, which simplify the calculation considerably. It is to these approximations that we now turn.

7.5.2. Approximating the Power Function and the Expected Sample Size

From a classical viewpoint, the relevant properties of an SPRT are the error probabilities, α_0 and α_1, and the expected sample sizes, $E_{\theta_0}N$ and $E_{\theta_1}N$. As these are also the quantities needed for a Bayesian analysis, it seems that nothing else need be considered. Unfortunately, this is too narrow a view in practice, in that it is unlikely that the simple versus simple hypothesis testing formulation of the problem is realistic. Far more common will be parametric situations in which, for example, it is desired to test H_0: $\theta \le \theta_0$ versus H_1: $\theta \ge \theta_1$. Here θ_0 and θ_1 represent the boundaries of regions it is important to distinguish between, and the simple versus simple formulation would seem to be a reasonable approximation to this more complicated situation. Even if an SPRT is then used, however, it is clear that one must be concerned with all possible values of θ, and not just θ_0 and θ_1. Hence it is important to investigate the error probabilities and the expected sample sizes of an SPRT for all values of θ.

Because of the above reasoning, we will, in the following, deal with the more general situation in which the X_i are a sequential sample from the density $f(x|\theta)$, the parameter θ possibly assuming values other than θ_0 or θ_1. It is then important to consider

$$\beta(\theta) = P_\theta \text{ (deciding } a_1) = P_\theta(L_N \ge B),$$

$$\alpha(\theta) = P_\theta \text{ (deciding } a_0) = P_\theta(L_N \le A),$$

and $E_\theta N$, for arbitrary values of θ. (The results obtained will, of course, also be valid for just θ_0 and θ_1. Note that $\alpha_0 = \beta(\theta_0)$ and $\alpha_1 = \alpha(\theta_1)$.) The function $\beta(\theta)$ is called the *power function*, $\alpha(\theta)$ is called the *operating characteristic* (OC) curve, and $E_\theta N$ is the *expected stopping time* (often called the average sample number (ASN)).

As mentioned earlier, it will be shown that $P_\theta(N < \infty) = 1$. Assuming this, it follows that

$$1 = P_\theta(N < \infty) = P_\theta(L_N \le A) + P_\theta(L_N \ge B) = \alpha(\theta) + \beta(\theta). \quad (7.58)$$

Hence $\alpha(\theta) = 1 - \beta(\theta)$, and it is only necessary to determine, say, $\beta(\theta)$.

To obtain approximations to $\beta(\theta)$ and $E_\theta(N)$, it is helpful to consider the following reformulation of the SPRT. (This reformulation and much of the following development is due to Wald (1947).) Define

$$Z_i = \log\left[\frac{f(X_i|\theta_1)}{f(X_i|\theta_0)}\right],$$

where Z_i is allowed to take on the values $\pm\infty$. Note that the Z_i are i.i.d. (since the X_i are), and that

$$S_n = \sum_{i=1}^n Z_i = \log\left[\prod_{i=1}^n \frac{f(X_i|\theta_1)}{f(X_i|\theta_0)}\right] = \log L_n.$$

Because log y is monotone in y, it follows from Definition 9 and (7.55) that the SPRT $\mathbf{d}^{A,B}$ can be rewritten as follows:

$$\begin{aligned} &\text{if } S_n \le a = \log A, && \text{stop sampling and decide } a_0;\\ &\text{if } S_n \ge b = \log B, && \text{stop sampling and decide } a_1;\\ &\text{if } a < S_n < b, && \text{take another observation.}\end{aligned} \quad (7.59)$$

In this formulation, we will denote the SPRT by $\mathbf{d}_{a,b}$. Note that $a<0$ and $b>0$, since $A<1$ and $B>1$.

The advantage of the above formulation of the problem is that S_n, being a sum of i.i.d. random variables, is much easier to deal with than is L_n. Indeed, $\mathbf{d}_{a,b}$ can be interpreted as a random walk, stopping when S_n reaches the "barriers" a and b. Though fruitful, this last interpretation is beyond the scope of the book.

Example 14. Suppose $X_1, X_2, \ldots$ is a sequential sample from a $\mathcal{N}(\theta, \sigma^2)$ density, σ^2 known. It is desired to test H_0: $\theta=\theta_0$ versus H_1: $\theta=\theta_1(\theta_0<\theta_1)$. Clearly

$$Z_i = \log\frac{f(X_i\,|\,\theta_1)}{f(X_i\,|\,\theta_0)} = -\frac{1}{2\sigma^2}[(X_i-\theta_1)^2-(X_i-\theta_0)^2]$$
$$=\frac{1}{\sigma^2}(\theta_1-\theta_0)X_i+\frac{1}{2\sigma^2}(\theta_0^2-\theta_1^2).$$

Hence

$$S_n=\sum_{i=1}^{n} Z_i=\frac{1}{\sigma^2}(\theta_1-\theta_0)\sum_{i=1}^{n} X_i+\frac{n}{2\sigma^2}(\theta_0^2-\theta_1^2)$$
$$=\frac{n}{\sigma^2}(\theta_1-\theta_0)\bar{X}_n+\frac{n}{2\sigma^2}(\theta_0^2-\theta_1^2).$$

The SPRT $\mathbf{d}_{a,b}$ can thus be written as follows:

if $\bar{X}_n<\dfrac{a\sigma^2}{n(\theta_1-\theta_0)}+\dfrac{1}{2}(\theta_0+\theta_1)$, stop sampling and decide a_0;

if $\bar{X}_n>\dfrac{b\sigma^2}{n(\theta_1-\theta_0)}+\dfrac{1}{2}(\theta_0+\theta_1)$, stop sampling and decide a_1;

and otherwise continue sampling.

In the remainder of the section, it will be assumed that we are considering an SPRT $\mathbf{d}_{a,b}$ with $a<0$ and $b>1$, and that N is the stopping time of the SPRT and $\beta(\theta)$ is its power function. We begin the development of the approximations to $\beta(\theta)$ and $E_\theta N$ by establishing, as promised, that $P_\theta(N<\infty)=1$.

Theorem 9. *Let N be the stopping time of the* SPRT $\mathbf{d}_{a,b}$. *If $P_\theta(Z_i=0)<1$, then $P_\theta(N<\infty)=1$ and all moments of N exist.*

Proof. The proof follows Stein (1946). The index θ is immaterial to the proof, and so will be omitted.

Since $P(Z_i=0)<1$, either $P(Z_i>\varepsilon)>0$ or $P(Z_i<-\varepsilon)>0$ for some $\varepsilon>0$. Assume that

$$P(Z_i>\varepsilon)=\lambda>0,$$

the other case being handled similarly. Note that N is defined by ($N=\infty$ formally being allowed)

$$N=\min\{n: S_n \le a \text{ or } S_n \ge b\}.$$

Letting m be an integer greater than $(b-a)/\varepsilon$, it follows from the independence of the Z_i that, for all j,

$$\begin{aligned} P(S_{j+m}-S_j > b-a) &= P\left(\sum_{i=j+1}^{j+m} Z_i > b-a\right) \\ &\ge P(Z_i > (b-a)/m, \text{for } i=j+1,\ldots,j+m) \\ &\ge P(Z_i > \varepsilon, \text{for } i=j+1,\ldots,j+m) = \lambda^m. \end{aligned}$$

Hence, again using the independence of the Z_i,

$$\begin{aligned} P(N \ge jm+1) &= P(a < S_i < b, \text{for } i=1,\ldots,jm) \\ &\le P(S_{im}-S_{(i-1)m} \le (b-a), \text{for } i=1,\ldots,j) \\ &= \prod_{i=1}^{j} P(S_{im}-S_{(i-1)m} \le b-a) \le (1-\lambda^m)^j. \end{aligned} \tag{7.60}$$

Define $\gamma=(1-\lambda^m)^{-1}$ and $\rho=(1-\lambda^m)^{1/m}$. For any given n, let j be that integer for which $jm < n \le (j+1)m$. It follows from (7.60) that

$$P(N \ge n) \le P(N \ge jm+1) \le (1-\lambda^m)^j = \gamma\rho^{(j+1)m} \le \gamma\rho^n.$$

Hence

$$P(N<\infty) \ge 1-P(N \ge n) \ge 1-\gamma\rho^n.$$

Since this is true for all n, it is clear that $P(N<\infty)=1$. Also,

$$E[N^k]=\sum_{n=0}^{\infty} n^k P(N=n) \le \sum_{n=0}^{\infty} n^k P(N \ge n) \le \sum_{n=0}^{\infty} n^k \gamma\rho^n < \infty,$$

so that all moments of N exist. The proof is complete. □

The condition $P_\theta(Z_i=0)<1$, in the above lemma, is quite innocuous, since $Z_i=0$ if and only if $f(X_i|\theta_1)=f(X_i|\theta_0)$. The condition thus ensures that $f(X_i|\theta_1)$ and $f(X_i|\theta_0)$ differ with positive probability, clearly a necessary condition to be able to distinguish between the two.

The basic tool that will be used in approximating $\beta(\theta)$ and $E_\theta N$ is the *fundamental identity of sequential analysis*, first established by Wald (1947). This identity involves consideration of

$$M_\theta(t)=E_\theta[e^{tZ_i}],$$

the moment generating function of Z_i. It will be necessary to assume that $M_\theta(t)$ is finite (at least for t near zero), and also that the Z_i are finite valued. Both of these conditions can be assumed, without loss of generality, by considering, in place of the Z_i,

$$Z_i^* = \begin{cases} (b-a) & \text{if } Z_i > (b-a), \\ Z_i & \text{if } |Z_i| \le (b-a), \\ -(b-a) & \text{if } Z_i < -(b-a). \end{cases} \tag{7.61}$$

The SPRT is unchanged if the Z_i are replaced by the Z_i^*. (Any Z_n, for which $|Z_n| > (b-a)$, would automatically cause S_n to jump outside the stopping boundaries. The same thing would happen, however, for $S_n^* = \sum_{i=1}^n Z_i^*$.) It is clear that the Z_i^* are finite valued, and that the moment generating function of the Z_i^* exists for all t. We will use the Z_i^* in place of the Z_i only if needed, however, since the Z_i are often calculationally easier to work with.

Theorem 10 (Fundamental Identity of Sequential Analysis). *If $P_\theta(Z_i = 0) < 1$ and $P_\theta(|Z_i| < \infty) = 1$, then*

$$E_\theta[\exp(tS_N) M_\theta(t)^{-N}] = 1, \tag{7.62}$$

for all t for which $M_\theta(t)$ is finite.

Proof. (The proof that will be given (due to Bahadur (1958)) is rather unintuitive. Equation (7.62) is intuitively reasonable, however, as can be seen by verifying the equation for fixed n.)

Define (assuming that $M_\theta(t) < \infty$)

$$g(z \mid t, \theta) = \frac{\exp(tz) f(z \mid \theta)}{M_\theta(t)}.$$

This is clearly a density in z, and does not give probability one to zero (since $\exp(tz)$ is positive, $M_\theta(t)$ is a constant, and $f(z \mid \theta)$ does not give probability one to zero). Pretending that the Z_i have density $g(z \mid t, \theta)$ and considering the SPRT based on the Z_i, it follows from Theorem 9 that

$$P_{t,\theta}(N_t < \infty) = 1,$$

where N_t is the stopping time for the SPRT in this new problem. Note, however, that in terms of the original Z_i (for simplicity only the continuous case is considered)

$$\begin{aligned} E_\theta[\exp(tS_N) M_\theta(t)^{-N}] &= \sum_{n=1}^{\infty} \int_{\{N=n\}} \cdots \int \exp(tS_n) M_\theta(t)^{-n} \prod_{i=1}^{n} [f(z_i \mid \theta) dz_i] \\ &= \sum_{n=1}^{\infty} \int_{\{N=n\}} \cdots \int \prod_{i=1}^{n} [\exp(tz_i) M_\theta(t)^{-1} f(z_i \mid \theta) dz_i] \\ &= \sum_{n=1}^{\infty} \int_{\{N_t=n\}} \cdots \int \prod_{i=1}^{n} [g(z_i \mid t, \theta) dz_i] = P_{t,\theta}(N_t < \infty) = 1. \end{aligned}$$

This completes the proof. □

The next theorem is a direct consequence of the fundamental identity, and will itself be crucial in obtaining the approximations to $\beta(\theta)$ and $E_\theta N$.

Theorem 11. *Assume that $P_\theta(Z_i=0)<1$, that $P_\theta(|Z_i|<\infty)=1$, and that $M_\theta(t)$ exists for t in a neighborhood of the origin. Then*

(i) $E_\theta[S_N]=\mu_\theta E_\theta N$, *and*
(ii) $E_\theta[S_N-N\mu_\theta]^2=\sigma_\theta^2 E_\theta N$,

where $\mu_\theta=E_\theta[Z_i]$ and $\sigma_\theta^2=E_\theta(Z_i-\mu_\theta)^2$.

Proof. It can be shown that derivatives with respect to t can be taken inside the expectation sign of the fundamental identity. Hence (7.62) implies that

$$\begin{aligned} 0=\frac{d}{dt}1&=E_\theta\left[\frac{d}{dt}\{\exp(tS_N)M_\theta(t)^{-N}\}\right] \\ &=E_\theta\left[S_N\exp(tS_N)M_\theta(t)^{-N}-N\exp(tS_N)M_\theta(t)^{-(N+1)}\left\{\frac{d}{dt}M_\theta(t)\right\}\right]. \end{aligned} \tag{7.63}$$

Notice that $M_\theta(0)=1$, and that

$$\begin{aligned} \frac{d}{dt}M_\theta(t)\bigg|_{t=0}&=E_\theta\left[\frac{d}{dt}\exp(tZ_i)\right]\bigg|_{t=0}=E_\theta[Z_i\exp(tZ_i)]\bigg|_{t=0} \\ &=E_\theta[Z_i]=\mu_\theta. \end{aligned} \tag{7.64}$$

Hence, evaluating (7.63) at $t=0$ gives

$$0=E_\theta[S_N-N\mu_\theta],$$

which proves part (i).

The proof of part (ii) is similar (based on taking the second derivative) and is left as an exercise. □

One final technical result is needed.

Lemma 2. *Assume that $P_\theta(|Z_i|<\infty)=1$, that $M_\theta(t)$ exists for all t, that $P_\theta(Z_i<0)>0$ and $P_\theta(Z_i>0)>0$, and that $\mu_\theta\neq 0$. Then there exists a unique nonzero number t_θ for which $M_\theta(t_\theta)=1$.*

Proof. The proof is based on showing that, under the given conditions, $M_\theta(t)$ will be a function behaving as in Figure 7.3. To see that this is correct, note that

$$\frac{d^2}{dt^2}M_\theta(t)=E_\theta\left[\frac{d^2}{dt^2}\exp(tZ_i)\right]=E_\theta[Z_i^2\exp(tZ_i)]>0,$$

the strict inequality following from the assumption that $P(Z_i=0)<1$. From Lemma 1 in Section 1.8, it follows that $M_\theta(t)$ is strictly convex.

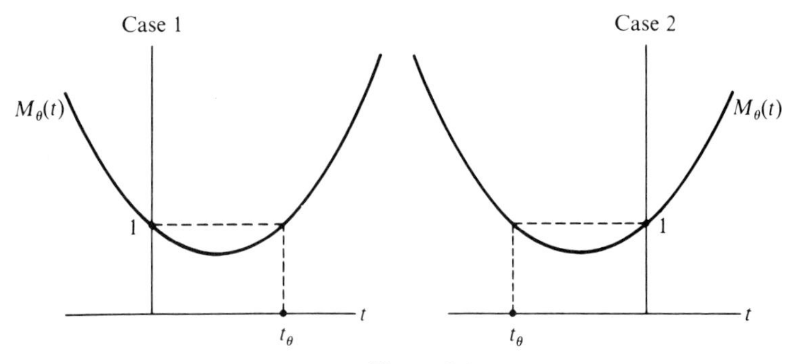

Figure 7.3

Note next that $M_\theta(0)=1$ and, by (7.64), that

$$\frac{d}{dt}M_\theta(t)\bigg|_{t=0} = \mu_\theta \neq 0.$$

Finally, let $\varepsilon>0$ and $\lambda>0$ be such that $P_\theta(Z_i>\varepsilon)>\lambda$. Clearly, for $t>0$,

$$M_\theta(t)=E_\theta[\exp(tZ_i)]\geq \lambda\exp(t\varepsilon).$$

Hence

$$\lim_{t\to\infty} M_\theta(t)\geq \lim_{t\to\infty}\lambda\exp(t\varepsilon)=\infty.$$

It can similarly be shown that

$$\lim_{t\to-\infty} M_\theta(t)=\infty.$$

These facts imply that $M_\theta(t)$ must behave as in Case 1 or Case 2 of Figure 7.3, giving the desired conclusion. □

The conditions of this lemma are not particularly stringent. It has already been mentioned that, using (7.61) if necessary, $M_\theta(t)$ can be guaranteed to exist for all t. The conditions $P_\theta(Z_i<0)>0$ and $P_\theta(Z_i>0)>0$ merely ensure that trivialities are avoided. If, for instance, $P_\theta(Z_i<0)=0$, then S_n will be sure to exit through the b boundary. (Recall, we are assuming that $a<0$ and $b>0$.) In such a situation, it is obviously silly to take any observations, since the action that will be taken is already known. The calculation of t_θ will be seen to be relatively easy in many common situations.

The approximations to $\beta(\theta)$ and $E_\theta N$ are based on pretending that S_N hits the boundaries a and b *exactly*. In other words, we will pretend that S_N has the two-point distribution, P_θ^*, defined by

$$P_\theta^*(S_N=a)=P_\theta(S_N\leq a) \text{ and } P_\theta^*(S_N=b)=P_\theta(S_N\geq b). \tag{7.65}$$

The Wald Approximation to $\beta(\theta)$

Case 1. $\mu_\theta \neq 0$.

Assume t_θ is such that $M_\theta(t_\theta)=1$. The fundamental identity then implies that

$$1 = E_\theta[\exp(t_\theta S_N) M_\theta(t_\theta)^{-N}] = E_\theta[\exp(t_\theta S_N)].$$

Pretending that S_N has the distribution P^*_θ defined in (7.65), it follows that

$$\begin{aligned} 1 &\cong \exp(t_\theta a) P^*_\theta(S_N = a) + \exp(t_\theta b) P^*_\theta(S_N = b) \\ &= \exp(t_\theta a) P_\theta(S_N \le a) + \exp(t_\theta b) P_\theta(S_N \ge b). \end{aligned} \tag{7.66}$$

Now (7.58) implies that

$$P_\theta(S_N \le a) + P_\theta(S_N \ge b) = 1, \tag{7.67}$$

so that (7.66) gives

$$\beta(\theta) = P_\theta(S_N \ge b) \cong \frac{1 - \exp(t_\theta a)}{\exp(t_\theta b) - \exp(t_\theta a)} \overset{\text{(defn.)}}{=} \tilde{\beta}(\theta). \tag{7.68}$$

Case 2. $\mu_\theta = 0$.

If $\mu_\theta = 0$, Theorem 11(i) implies that $E_\theta[S_N]=0$. Again pretending that S_N has the distribution P^*_θ, this becomes

$$a P_\theta(S_N \le a) + b P_\theta(S_N \ge b) \cong 0.$$

Together with (7.67), this implies that

$$\beta(\theta) = P_\theta(S_N \ge b) \cong \frac{-a}{b-a} \overset{\text{(defn.)}}{=} \tilde{\beta}(\theta). \tag{7.69}$$

It can be shown that this agrees with (7.68) in the sense that as $\mu_\theta \to 0$ (so that $t_\theta \to 0$) the expression in (7.68) converges to that in (7.69).

The Wald Approximation to $E_\theta N$

Case 1. $\mu_\theta \neq 0$.

Theorem 11(i) implies that

$$E_\theta N = (\mu_\theta)^{-1} E_\theta[S_N]. \tag{7.70}$$

Use of P^*_θ leads to the approximation

$$E_\theta S_N \cong a P_\theta(S_N \le a) + b P_\theta(S_N \ge b) = a + (b-a)\beta(\theta).$$

Replacing $\beta(\theta)$ by its approximation $\tilde{\beta}(\theta)$ and inserting this expression into (7.70) gives

$$E_\theta N \cong (\mu_\theta)^{-1}[a + (b-a)\tilde{\beta}(\theta)] \overset{\text{(defn.)}}{=} \tilde{E}_\theta N. \tag{7.71}$$

Case 2. $\mu_\theta = 0$.

It is now necessary to use Theorem 11(ii), which, for $\mu_\theta = 0$, states that

$$E_\theta N = (\sigma_\theta^2)^{-1} E_\theta(S_N^2).$$

Use of P_θ^* leads to the approximation

$$\begin{aligned} E_\theta[S_N^2] &\cong a^2 P_\theta(S_N \le a) + b^2 P_\theta(S_N \ge b) \\ &\cong a^2 + (b^2 - a^2)\tilde{\beta}(\theta) \\ &= a^2 + (b-a)(b+a)\left[\frac{-a}{b-a}\right] = -ab. \end{aligned}$$

Hence

$$E_\theta N \cong \frac{-ab}{\sigma_\theta^2} \overset{\text{(defn.)}}{=} \tilde{E}_\theta^N. \tag{7.72}$$

This can be shown to be the limiting case of (7.71) as μ_θ (and hence t_θ) approach zero.

It is of particular interest to consider $\tilde{\alpha}_0 = \tilde{\beta}(\theta_0)$, $\tilde{\alpha}_1 = 1 - \tilde{\beta}(\theta_1)$, $\tilde{E}_{\theta_0}N$, and $\tilde{E}_{\theta_1}N$. To this end, note that if $P_{\theta_j}(|Z_i| < \infty) = 1$ for $j = 0, 1$, then

$$t_{\theta_0} = 1 \quad \text{and} \quad t_{\theta_1} = -1. \tag{7.73}$$

(The proof will be left as an exercise.) It is then easy to calculate, from (7.68) and (7.72), that

$$\begin{aligned} \tilde{\alpha}_0 &= \frac{1-\exp(a)}{\exp(b)-\exp(a)} = \frac{1-A}{B-A}, \\ \tilde{\alpha}_1 &= 1 - \frac{1-\exp(-a)}{\exp(-b)-\exp(-a)} = \frac{\exp(-b)-1}{\exp(-b)-\exp(-a)} \\ &= \frac{B^{-1}-1}{B^{-1}-A^{-1}} = \frac{A(B-1)}{B-A}, \end{aligned} \tag{7.74}$$

$$\tilde{E}_{\theta_0}N = (\mu_{\theta_0})^{-1}[a + (b-a)\tilde{\alpha}_0],$$

and

$$\tilde{E}_{\theta_1}N = (\mu_{\theta_1})^{-1}[a + (b-a)(1-\tilde{\alpha}_1)].$$

The approximations $\tilde{\alpha}_0$ and $\tilde{\alpha}_1$ are particularly interesting, in that they are distribution free, depending only on A and B. Note also that the equations for $\tilde{\alpha}_0$ and $\tilde{\alpha}_1$ can be solved for A and B, the result being $A = \tilde{\alpha}_1/(1-\tilde{\alpha}_0)$ and $B = (1-\tilde{\alpha}_1)/\tilde{\alpha}_0$. This suggests a way of determining approximate stopping boundaries for the SPRT, if it is desired to attain certain specified error probabilities. Indeed, if error probabilities α_0 and α_1 are desired (this is a classical approach, of course), the suggested SPRT is that with stopping boundaries

$$A^* = \frac{\alpha_1}{1-\alpha_0} \quad \text{and} \quad B^* = \frac{1-\alpha_1}{\alpha_0}. \tag{7.75}$$

Example 14 (continued). To determine $\tilde{\beta}(\theta)$ and $\tilde{E}_\theta N$, note that

$$\mu_\theta = E_\theta[Z_i] = \frac{1}{\sigma^2}(\theta_1 - \theta_0)\theta + \frac{1}{2\sigma^2}(\theta_0^2 - \theta_1^2)$$

and

$$\sigma_\theta^2 = E_\theta(Z_i - \mu_\theta)^2 = E_\theta\left[\frac{1}{\sigma^2}(\theta_1 - \theta_0)(X_i - \theta)\right]^2 = \frac{1}{\sigma^2}(\theta_1 - \theta_0)^2.$$

A standard calculation also shows that

$$\begin{aligned} M_\theta(t) &= E_\theta[\exp(tZ_i)] \\ &= \exp\left\{-\frac{1}{2\sigma^2}t(\theta_1 - \theta_0)[\theta_1 + \theta_0 - 2\theta - t(\theta_1 - \theta_0)]\right\}. \end{aligned}$$

Hence $M_\theta(t_\theta) = 1$ (for $t_\theta \neq 0$) if and only if

$$\theta_1 + \theta_0 - 2\theta - t_\theta(\theta_1 - \theta_0) = 0,$$

or, equivalently,

$$t_\theta = \frac{\theta_1 + \theta_0 - 2\theta}{\theta_1 - \theta_0}.$$

Using these formulas, it is easy to calculate the Wald approximations $\tilde{\beta}(\theta)$ and $\tilde{E}_\theta N$.

As a specific example, suppose $\sigma^2 = 1$, $\theta_0 = -\frac{1}{2}$, $\theta_1 = \frac{1}{2}$, and assume an SPRT with error probabilities $\alpha_0 = \alpha_1 = 0.1$ is desired. The approximations to the stopping boundaries of the desired SPRT are, from (7.75), $A^* = \frac{1}{9}$ and $B^* = 9$, or, equivalently, $a^* = -\log 9$ and $b^* = \log 9$ (for the Z_i formulation). To find $\tilde{\beta}(\theta)$ and $\tilde{E}_\theta N$ for $\mathbf{d}_{a^*,b^*}$, note first that $\mu_\theta = \theta$, $\sigma_\theta^2 = 1$, and $t_\theta = -2\theta$.

Case 1. $\theta \neq 0$ (i.e., $\mu_\theta \neq 0$).

From (7.68) and (7.71) it is clear that

$$\tilde{\beta}(\theta) = \frac{1 - \exp(-2\theta a^*)}{\exp(-2\theta b^*) - \exp(-2\theta a^*)} = \frac{1 - 9^{2\theta}}{9^{-2\theta} - 9^{2\theta}}$$

and

$$\tilde{E}_\theta N = \theta^{-1}(\log 9)[-1 + 2\tilde{\beta}(\theta)].$$

In particular,

$$\tilde{E}_{\theta_1} N = \tilde{E}_{\theta_0} N = (-2)(\log 9)[-1 + 2(0.1)] \cong 3.52.$$

Case 2. $\theta = 0$ (i.e., $\mu_\theta = 0$).

For this case, (7.69) and (7.72) give that $\tilde{\beta}(0) = \frac{1}{2}$ and $\tilde{E}_0 N = (\log 9)^2 \cong 4.83$.

For purposes of comparison, it is interesting to note, in this example, that the optimal fixed sample size test with error probabilities $\alpha_0 = \alpha_1 = 0.1$ requires a sample of size $n \cong 6.55$. (Of course, n can't really be a fraction, but 6.55 corresponds, in some sense, to how much sample information is needed to achieve the desired error probabilities.) Thus the fixed sample size procedure requires, on the average, almost twice as many observations as does the SPRT. (Actually, as will be discussed in the next subsection, $\tilde{E}_{\theta_i}N$ is an underestimate of $E_{\theta_i}N$. The true value of $E_{\theta_i}N$ in this example is still, however, considerably smaller than 6.55.)

In certain situations the Wald approximations to $\beta(\theta)$ and $E_\theta N$ are exact. This is the case when S_N must hit a or b exactly, i.e., there is no "overshoot." Such a situation occurs when the Z_i can assume only the values $-v$, 0, and v, for some constant v. If a and b are then chosen to be integral multiples of v, it is clear that S_N must hit a or b exactly. (There is no sense in choosing a or b to be other than integral multiples of v, since S_n can only move in steps of size $\pm v$.) An example follows.

EXAMPLE 15. Assume that $X_1, X_2, \ldots$ is a sequential sample from a $\mathscr{B}(1, \theta)$ distribution, and that it is desired to test H_0: $\theta = \theta_0$ versus H_1: $\theta = 1 - \theta_0$, where $0 < \theta_0 < \frac{1}{2}$. Clearly

$$Z_i = \log \frac{f(X_i \mid 1 - \theta_0)}{f(X_i \mid \theta_0)}$$

$$= \log \frac{(1-\theta_0)^{X_i}(\theta_0)^{(1-X_i)}}{(\theta_0)^{X_i}(1-\theta_0)^{(1-X_i)}} = (2X_i - 1)\log(\theta_0^{-1} - 1).$$

Setting $v = \log(\theta_0^{-1} - 1)$ and noting that X_i is zero or one, it follows that Z_i is either $-v$ or v. Hence, choosing $a = -jv$ and $b = kv$, where j and k are positive integers, the Wald approximations to $\beta(\theta)$ and $E_\theta N$ will be exact. The calculational details are left as an exercise.

It should be noted that exact SPRTs can sometimes be obtained by methods other than that above (see Ghosh (1970)). Also, Le Page (1979) develops a modification of the SPRT which is guaranteed to always achieve the exact error probabilities.

7.5.3. Accuracy of the Wald Approximations

The accuracy of the Wald approximations to $\beta(\theta)$ and $E_\theta N$ is, to a large extent, determined by the amount that S_N will tend to "overshoot" a or b. If this overshoot tends to be small, the approximations will be quite good. If the overshoot tends to be large, the approximations can be bad. (In such a situation, the approximations can sometimes be improved by using the Z_i^* (see (7.61)) in place of the Z_i, since the overshoot is then reduced when $|Z_i| > b - a$.) As a typical intermediate case, consider the situation of Example 14, in which the X_i are $\mathcal{N}(\theta, 1)$ and it is desired to test H_0: $\theta = -\frac{1}{2}$

Table 7.3. Accuracy of the Wald Approximations.

	$a=-2.5, b=7.5$				$a=-5.0, b=5.0$			
θ	$\beta(\theta)$	$\tilde{\beta}(\theta)$	$E_\theta(N)$	$\tilde{E}_\theta N$	$\beta(\theta)$	$\tilde{\beta}(\theta)$	$E_\theta N$	$\tilde{E}_\theta N$
−1.000	0.0000	0.0000	3.37	2.50	0.0000	0.0000	5.87	5.00
−0.750	0.0000	0.0000	4.39	3.33	0.0003	0.0005	7.72	6.66
−0.500	0.0003	0.0005	6.43	4.99	0.0038	0.0067	11.35	9.87
−0.250	0.0139	0.0169	11.97	9.32	0.0579	0.0759	20.01	16.96
−0.125	0.0759	0.0776	18.14	13.79	0.1985	0.2227	27.18	22.18
0.000	0.2761	0.2500	25.17	18.75	0.5000	0.5000	31.42	25.00
0.125	0.5724	0.5063	26.71	20.50	0.8015	0.7773	27.18	22.18
0.250	0.7887	0.7183	23.16	18.73	0.9421	0.9241	20.01	16.96
0.500	0.9540	0.9180	15.41	13.36	0.9962	0.9933	11.35	9.87
0.750	0.9901	0.9765	10.91	9.69	0.9997	0.9995	7.72	6.66
1.000	0.9978	0.9933	8.35	7.43	1.0000	1.0000	5.87	5.00

versus H_1: $\theta=\frac{1}{2}$. In Table 7.3, the true values of $\beta(\theta)$ and $E_\theta N$ and the corresponding Wald approximations, $\tilde{\beta}(\theta)$ and $\tilde{E}_\theta N$, are given for two different SPRTs. (This table, in modified form, appeared in Ghosh (1970).)

An examination of the table shows that $\tilde{\beta}(\theta)$ and $\tilde{E}_\theta N$ are reasonably close to $\beta(\theta)$ and $E_\theta N$, and that $\tilde{E}_\theta N$ is consistently an underestimate of $E_\theta N$. Note also that $\tilde{\beta}(\theta)>\beta(\theta)$ for $\theta\leq-\frac{1}{2}$, while $\tilde{\alpha}(\theta)=1-\tilde{\beta}(\theta)<1-\beta(\theta)=\alpha(\theta)$ for $\theta\geq\frac{1}{2}$. This indicates that the Wald approximations tend to *overestimate* the true error probabilities. This is good from a conservative classical viewpoint, in that, if an SPRT is chosen using the Wald approximations so as to attain certain specified error probabilities, then the true error probabilities of the SPRT are probably smaller. The following theorem gives a theoretical justification for some of these comments.

Theorem 12. *Assume that the conditions of Lemma 2 are satisfied, and consider the* SPRT $\mathbf{d}^{A,B}$. *The following inequalities then hold*:

$$A\geq A^*=\frac{\alpha_1}{1-\alpha_0}\quad\text{and}\quad B\leq B^*=\frac{1-\alpha_1}{\alpha_0};\tag{7.76}$$

$$E_{\theta_0}N\geq(\mu_{\theta_0})^{-1}\left[(1-\alpha_0)\log\left(\frac{\alpha_1}{1-\alpha_0}\right)+\alpha_0\log\left(\frac{1-\alpha_1}{\alpha_0}\right)\right],$$

and (7.77)

$$E_{\theta_1}N\geq(\mu_{\theta_1})^{-1}\left[\alpha_1\log\left(\frac{\alpha_1}{1-\alpha_0}\right)+(1-\alpha_1)\log\left(\frac{1-\alpha_1}{\alpha_0}\right)\right].$$

Proof. For simplicity, we will deal only with the continuous case. Note that since $P_{\theta_j}(|Z_i|<\infty)=1$ for $j=0, 1$, the densities $f(x_i|\theta_0)$ and $f(x_i|\theta_1)$

must be mutually absolutely continuous. Clearly

$$\alpha_1 = P_{\theta_1}(L_N \le A) = \sum_{n=1}^{\infty} \int_{B_n} \cdots \int \prod_{i=1}^{n} [f(x_i \mid \theta_1) dx_i], \tag{7.78}$$

where $B_n = \{N = n\} \cap \{\mathbf{x}^n \colon L_n \le A\}$. Now $L_n \le A$ can be rewritten

$$\prod_{i=1}^{n} f(x_i \mid \theta_1) \le A \prod_{i=1}^{n} f(x_i \mid \theta_0),$$

which, when used in (7.78), gives

$$\alpha_1 \le \sum_{n=1}^{\infty} \int_{B_n} \cdots \int A \prod_{i=1}^{n} [f(x_i \mid \theta_0) dx_i] = A P_{\theta_0}(L_N \le A).$$

Using (7.58), it follows that

$$\alpha_1 \le A(1 - P_{\theta_0}(L_N \ge B)) = A(1 - \alpha_0),$$

proving the first inequality in (7.76). The second inequality follows similarly.

To prove the first inequality in (7.77), note that

$$\begin{aligned} E_{\theta_0}[S_N] &= E_{\theta_0}[S_N | S_N \le a] P_{\theta_0}(S_N \le a) + E_{\theta_0}[S_N | S_N \ge b] P_{\theta_0}(S_N \ge b) \\ &= E_{\theta_0}[S_N \mid S_N \le a](1 - \alpha_0) + E_{\theta_0}[S_N \mid S_N \ge b] \alpha_0. \end{aligned} \tag{7.79}$$

Jensen's inequality (applied to $(-\log)$) gives

$$\begin{aligned} E_{\theta_0}[S_N \mid S_N \le a] &= E_{\theta_0}[\log L_N \mid S_N \le a] \\ &\le \log E_{\theta_0}[L_N \mid S_N \le a] \\ &= \log\left[\{P_{\theta_0}(S_N \le a)\}^{-1} \left\{ \sum_{n=1}^{\infty} \int_{B_n} \cdots \int L_n \prod_{i=1}^{n} [f(x_i \mid \theta_0) dx_i] \right\} \right] \\ &= \log\left[\{P_{\theta_0}(S_N \le a)\}^{-1} \left\{ \sum_{n=1}^{\infty} \int_{B_n} \cdots \int \prod_{i=1}^{n} [f(x_i \mid \theta_1) dx_i] \right\} \right] \\ &= \log\left[\frac{P_{\theta_1}(S_N \le a)}{P_{\theta_0}(S_N \le a)} \right] = \log\left(\frac{\alpha_1}{1 - \alpha_0} \right). \end{aligned} \tag{7.80}$$

Similarly it can be shown that

$$E_{\theta_0}[S_N \mid S_N \ge b] \le \log\left(\frac{1 - \alpha_1}{\alpha_0} \right). \tag{7.81}$$

Note also that

$$\begin{aligned} \mu_{\theta_0} = E_{\theta_0}[Z_i] &= E_{\theta_0}\left[\log\left\{ \frac{f(X_i \mid \theta_1)}{f(X_i \mid \theta_0)} \right\} \right] \\ &< \log \int_{\mathscr{X}_i} \frac{f(x_i \mid \theta_1)}{f(x_i \mid \theta_0)} f(x_i \mid \theta_0) dx_i \\ &= \log 1 = 0, \end{aligned} \tag{7.82}$$

the strict inequality following from the strict convexity of $(-\log)$ and the fact that $P_{\theta_0}(Z_i=0)<1$. Finally, Theorem 11(i) implies that

$$E_{\theta_0}[N]=\frac{E_{\theta_0}[S_N]}{\mu_{\theta_0}},$$

which together with (7.79), (7.80), (7.81), and (7.82) gives the desired result. The last inequality in (7.77) can be established by similar reasoning. □

In interpreting this theorem, it is important to keep track of several different quantities. Imagine that we are in the standard situation (at least standard classically) of specifying desired α_0 and α_1. Let A and B denote the true stopping boundaries which give these error probabilities, let $A^*=\alpha_1/(1-\alpha_0)$ and $B^*=(1-\alpha_1)/\alpha_0$ be the Wald approximations to these stopping boundaries, and let α_0^* and α_1^* denote the error probabilities of $\mathbf{d}^{A^*,B^*}$. The inequalities in (7.76) show that $\mathbf{d}^{A^*,B^*}$ has more "distant" stopping boundaries than $\mathbf{d}^{A,B}$, and this will usually mean that α_0^* and α_1^* are smaller than α_0 and α_1. (This need not always be true, however. If it is absolutely necessary to guarantee error probabilities α_0 and α_1, one can use, as stopping boundaries, $A^{**}=\alpha_1$ and $B^{**}=1/\alpha_0$. The proof that these stopping boundaries result in error probabilities $\alpha_0^{**}\le\alpha_0$ and $\alpha_1^{**}\le\alpha_1$ will be left as an exercise.)

Consider next the stopping times N and N^* of $\mathbf{d}^{A,B}$ and $\mathbf{d}^{A^*,B^*}$. Clearly $N\le N^*$. (It takes S_n at least as long to get to A^* or B^* as to A or B.) Hence the use of the approximations A^* and B^*, while usually lowering the error probabilities, will cause an increase in the stopping time. Studies in Wald (1947) indicate, however, that this increase is relatively modest.

Note that the question of how N^* differs from N is not the same as the question discussed earlier concerning how close the Wald approximation $\tilde{E}_\theta N$ is to $E_\theta N$. The inequalities in (7.77) are interesting in this regard, because the right-hand sides are $\tilde{E}_{\theta_i}N^*$ (the Wald approximations to the expected stopping times for the procedure $\mathbf{d}^{A^*,B^*}$). Combining these inequalities with the fact that $N\le N^*$, we have (for $i=0,1$)

$$\tilde{E}_{\theta_i}N^*\le E_{\theta_i}N\le E_{\theta_i}N^*.$$

This proves that the Wald approximations to $E_{\theta_i}N^*$ $(i=0,1)$ are, indeed, underestimates.

7.5.4. Bayes Risk and Admissibility

Consider, once again, the Bayesian situation discussed in Subsection 7.5.1. The problem of finding a Bayes procedure was reduced to that of finding the SPRT which minimizes (7.57). An appealing simplification is to replace α_0, α_1, $E_{\theta_0}N$, and $E_{\theta_1}N$ in (7.57) by their Wald approximations. The approximate Bayes risk of $\mathbf{d}^{A,B}$ is then (using (7.74))

$$r(\pi, \mathbf{d}^{A,B}) \cong \pi_0 \left\{ \frac{(1-A)}{(B-A)} K_1 + c(\mu_{\theta_0})^{-1} \left[(\log A) + \left(\log \frac{B}{A} \right) \frac{(1-A)}{(B-A)} \right] \right\} \tag{7.83}$$

$$+ \pi_1 \left\{ \frac{A(B-1)}{(B-A)} K_0 + c(\mu_{\theta_1})^{-1} \left[(\log A) + \left(\log \frac{B}{A} \right) \frac{B(1-A)}{(B-A)} \right] \right\}.$$

This is numerically easy to minimize over A and B.

An even simpler approximation can be found when c is very small. For small c, it is likely that a large number of observations will be desired, which in turn implies that the optimal A must be small and the optimal B large. Assuming this in (7.83) gives

$$r(\pi, \mathbf{d}^{A,B}) \cong \pi_0 \left[\frac{1}{B} K_1 + c(\mu_{\theta_0})^{-1} \log A \right] + \pi_1 [A K_0 + c(\mu_{\theta_1})^{-1} \log B].$$

Differentiating with respect to A and B and setting equal to zero gives "optimal" A and B of

$$A = \frac{-c\pi_0}{\mu_{\theta_0} K_0 \pi_1} \quad \text{and} \quad B = \frac{\pi_0 K_1 \mu_{\theta_1}}{c\pi_1}. \tag{7.84}$$

Of course, these solutions are subject to the inaccuracies of the Wald approximations.

Being as Bayes procedures are generally admissible, one would expect SPRTs to be admissible. This is true in the rather strong sense that, among all sequential procedures with fixed error probabilities α_0 and α_1, the SPRT with these error probabilities simultaneously minimizes *both* $E_{\theta_0} N$ and $E_{\theta_1} N$. This result was established by Wald and Wolfowitz (1948). We formally state the theorem below and indicate the idea of the proof. For convenience, let $\alpha_i(\mathbf{d})$ and $N(\mathbf{d})$ denote the error probabilities and stopping time of a procedure $\mathbf{d}$.

Theorem 13. *Let* $\mathbf{d}^{A,B}$ *be an* SPRT, *and assume that* $\mathbf{d}$ *is any other sequential procedure for which*

$$\alpha_0(\mathbf{d}) \le \alpha_0(\mathbf{d}^{A,B}) \quad \textit{and} \quad \alpha_1(\mathbf{d}) \le \alpha_1(\mathbf{d}^{A,B}).$$

Then

$$E_{\theta_0}[N(\mathbf{d}^{A,B})] \le E_{\theta_0}[N(\mathbf{d})] \quad \textit{and} \quad E_{\theta_1}[N(\mathbf{d}^{A,B})] \le E_{\theta_1}[N(\mathbf{d})].$$

PROOF. The proof is based on the fact that, for any $0 < \pi_0 < 1$, there exist positive constants c, K_0, and K_1 such that the given $\mathbf{d}^{A,B}$ is a Bayes procedure for the sequential problem with these constants. (See Ferguson (1967) for a proof of this.) It follows that $r(\pi, \mathbf{d}^{A,B}) \le r(\pi, \mathbf{d})$ for any sequential procedure $\mathbf{d}$, or, equivalently,

$$\pi_0 K_1 [\alpha_0(\mathbf{d}^{A,B}) - \alpha_0(\mathbf{d})] + (1-\pi_0) K_0 [\alpha_1(\mathbf{d}^{A,B}) - \alpha_1(\mathbf{d})]$$
$$\le \pi_0 c \{ E_{\theta_0}[N(\mathbf{d})] - E_{\theta_0}[N(\mathbf{d}^{A,B})] \}$$
$$+ (1-\pi_0) c \{ E_{\theta_1}[N(\mathbf{d})] - E_{\theta_1}[N(\mathbf{d}^{A,B})] \}.$$

By assumption, the left-hand side of this inequality is nonnegative, and so the right-hand side must be nonnegative. Dividing by c and letting $\pi_0 \to 0$ and $\pi_0 \to 1$ gives the desired result. □

7.5.5. Other Uses of the SPRT

As mentioned at the beginning of Subsection 7.5.2, SPRTs are frequently used for testing problems which are more complicated than just testing simple against simple hypotheses. The most common such use is in testing H_0: $\theta \le \theta_0$ versus H_1: $\theta \ge \theta_1 (\theta_0 < \theta_1)$, where one is supposedly indifferent when $\theta_0 < \theta < \theta_1$. It is fairly natural, in this situation, to pretend that the problem is that of testing H_0: $\theta = \theta_0$ versus H_1: $\theta = \theta_1$, and to use the relevant SPRT. It can indeed be shown that, if the X_i have a density with monotone likelihood ratio in θ (see Section 8.3), then the SPRT with error probabilities α_0 and α_1 gives error probabilities $\alpha_0(\theta) \le \alpha_0$ for $\theta < \theta_0$ and $\alpha_1(\theta) \le \alpha_1$ for $\theta > \theta_1$. In a classical sense, therefore, an SPRT is very reasonable in this situation, though not necessarily optimal (see Brown, Cohen and Samuel-Cahn (1983) and Siegmund (1985)).

In the above situation, one could attempt to find the optimal SPRT from a Bayesian viewpoint. More generally, in any testing situation for which the true Bayes procedure is too hard to calculate, it may be reasonable to simply find the best SPRT (if the form of the SPRT seems natural for the problem). Thus, if it is desired to test H_0: $\theta \in \Theta_0$ versus H_1: $\theta \in \Theta_1$ (where $\theta_0 < \theta_1$ for $\theta_0 \in \Theta_0$ and $\theta_1 \in \Theta_1$) under decision loss $L(\theta, a_i)(i = 0, 1)$ for a wrong decision and zero loss for a correct decision, then the Bayes risk of the SPRT $\mathbf{d}^{A,B}$ will be

$$r(\pi, \mathbf{d}^{A,B}) = \int_{\Theta_0} L(\theta, a_1)\beta(\theta) dF^{\pi}(\theta) + \int_{\Theta_1} L(\theta, a_0)[1 - \beta(\theta)] dF^{\pi}(\theta) + cE^{\pi}[E_{\theta}N]. \tag{7.85}$$

This can be numerically minimized over A and B (perhaps using the Wald approximations to $\beta(\theta)$ and $E_{\theta}N$), giving the optimal SPRT. Note that one must choose parameter values $\theta_0 \in \Theta_0$ and $\theta_1 \in \Theta_1$ from which to develop the SPRT. Sometimes there is a natural choice, as when testing H_0: $\theta \le \theta_0$ versus H_1: $\theta \ge \theta_1$. In some situations, the choice does not affect the form of the SPRT. In Example 14, for instance, if it were desired to test H_0: $\theta \le 0$ versus H_1: $\theta > 0$, any symmetric choice of θ_0 and θ_1 (i.e., $\theta_0 = -\theta_1$) would result in an SPRT of the same form. It may also be possible, in some situations, to actually minimize $r(\pi, \mathbf{d}^{A,B})$ over the choice of θ_0 and θ_1, as well as over A and B.

7.6. Minimax Sequential Procedures

A *minimax sequential procedure*, as one would expect, is a procedure which minimizes $\sup_\theta R(\theta, \mathbf{d})$ among all proper sequential procedures. (As in the nonsequential situation, it may be necessary to consider randomized procedures. We will avoid this issue, however.) The same techniques used in Chapter 5 to find minimax rules, particularly those based on Theorems 17 and 18 of that chapter, are valid and useful in the sequential setting also.

EXAMPLE 16. Assume that $X_1, X_2, \ldots$ is a sequential sample from a $\mathcal{N}(\theta, \sigma^2)$ distribution, σ^2 known, and that it is desired to estimate θ under loss $L(\theta, a, n) = (\theta - a)^2 + cn$. It was shown in Example 5 that the Bayes procedure for π_m, the $\mathcal{N}(0, m)$ prior, is to take a sample of size n_m, where n_m is the integer n which minimizes $\sigma^2 m/(\sigma^2 + nm) + nc$, and then use the Bayes estimator. The Bayes risk of the procedure was shown to be

$$r(\pi_m) = \frac{\sigma^2 m}{\sigma^2 + mn_m} + n_m c. \tag{7.86}$$

Consider now the procedure $\mathbf{d}^*$, which takes a sample of size n^*, where n^* is the integer n which minimizes $\sigma^2/n + nc$, and then estimates θ by $\bar{x}_{n^*}$. Clearly

$$R(\theta, \mathbf{d}^*) = E_\theta[(\theta - \bar{X}_{n^*})^2 + n^* c] = \frac{\sigma^2}{n^*} + n^* c.$$

Now it is easy to see that, for large enough m, n_m cannot be zero. Hence (7.86) implies that

$$\lim_{m \to \infty} r(\pi_m) = \frac{\sigma^2}{\lim_{m \to \infty} n_m} + c \lim_{m \to \infty} n_m \geq \frac{\sigma^2}{n^*} + n^* c = R(\theta, \mathbf{d}^*).$$

It thus follows from the analog of Theorem 18 of Chapter 5 that $\mathbf{d}^*$ is minimax.

In estimation problems for which there is an equalizer (constant risk) decision rule, it frequently happens (as above) that the minimax sequential procedure is simply the optimal fixed sample size minimax decision rule. (This is demonstrated in considerable generality in Kiefer (1957), through invariance arguments.) Other than in such problems, however, finding minimax sequential procedures is quite hard. For example, to find the minimax sequential test of a simple null hypothesis versus a simple alternative, the Bayes SPRT which is an equalizer procedure must be found. In other words, one must find $0 \leq \pi_0 \leq 1$ $(\pi_0 = \pi(\theta_0))$ for which the resulting Bayes SPRT, $\mathbf{d}^{A,B}$, satisfies $R(\theta_0, \mathbf{d}^{A,B}) = R(\theta_1, \mathbf{d}^{A,B})$. This is usually fairly difficult. One of the few situations in which a minimax procedure that is actually sequential (i.e., not fixed sample size) can be explicitly calculated is given in Exercise 57.

7.7. The Evidential Relevance of the Stopping Rule

7.7.1. Introduction

We have seen that the stopping rule plays a large role in calculation of frequentist measures, such as error probabilities and risk. On the other hand, we saw in Subsection 7.4.3 that, once the decision to stop sampling has been made, the Bayes action taken is unaffected by the stopping rule used to reach that point. Furthermore, it can similarly be shown that *any* posterior measure of risk or accuracy is *completely unaffected* by the stopping rule.

The determination of the proper role of the stopping rule, in making final decisions and conclusions, thus becomes an interesting issue. Indeed it will be seen to be a profound and apparently paradoxical issue of extreme practical importance. As an indication of the apparently paradoxical aspects, note that the Bayesian position allows stopping based on data that "looks good," and then completely ignores this reason for stopping when it comes to drawing conclusions. The extreme practical importance of the issue is illustrated by clinical trials. A frequentist is essentially a slave to the original protocol (which includes choice of the stopping rule to be used) decided upon; any change in the protocol, in response to unexpected developments (such as all of the first 20 treatment patients dying), destroys the possibility of giving valid frequentist conclusions. The Bayesian, on the other hand, can stop earlier than expected or continue longer than expected, and still produce valid Bayesian answers.

We begin the study of these issues, in Subsection 7.7.2, by discussing the Stopping Rule Principle (a consequence of the Likelihood Principle), which implies that the stopping rule is indeed irrelevant to final conclusions. The practical implications of this are discussed in Subsection 7.7.3, followed by a discussion, in Subsection 7.7.4, of criticisms that have been levied against analyses which ignore the stopping rule. Subsection 7.7.5 discusses situations where the stopping rule matters, even to a Bayesian.

7.7.2. The Stopping Rule Principle

The Stopping Rule Principle. *The stopping rule,* $\boldsymbol{\tau}$, *should have no effect on the final reported evidence about* θ *obtained from the data.*

Example 15 of Subsection 1.6.4 provides an illustration of this principle. The binomial and negative binomial experiments can be interpreted as sequential experiments involving independent Bernoulli ($\mathcal{B}(1, \theta)$) trials;

the binomial experiment would arise from using the fixed sample size stopping rule which stops after 12 observations, while the negative binomial experiment would arise from using the stopping rule which says "stop when 3 tails (zeros) have been observed." It was argued in Example 15 that, because of the Likelihood Principle, the same conclusions should be reached in either case when the data consists of 9 heads (ones) and 3 tails (zeros). This implies that the reason for stopping at 9 heads and 3 tails should be irrelevant to final conclusions about θ.

To demonstrate directly that the Likelihood Principle implies the Stopping Rule Principle, suppose $\mathbf{X}^n = (X_1, \ldots, X_n)$ has density $f_n(\mathbf{x}^n \mid \theta)$; then the density of $(N, \mathbf{X}^N)$ (recall that N is the stopping time, i.e., the sample size at which stopping happens to occur) can be shown to be

$$f(n, \mathbf{x}^n \mid \theta) = \left[\prod_{j=0}^{n-1} (1 - \tau_j(\mathbf{x}^j)) \right] \tau_n(\mathbf{x}^n) f_n(\mathbf{x}^n \mid \theta). \tag{7.87}$$

(In other words, the identity

$$P_\theta(N = n, \mathbf{X}^n \in A) = \int_A f(n, \mathbf{x}^n \mid \theta) d\mathbf{x}^n \tag{7.88}$$

can be shown to hold.) But the likelihood function for θ is thus $l(\theta) = f(n, \mathbf{x}^n \mid \theta)$ (for the final $(n, \mathbf{x}^n)$, of course), and this is proportional (as a function of θ) to $f_n(\mathbf{x}^n \mid \theta)$ for *any* stopping rule, $\boldsymbol{\tau}$. The Likelihood Principle thus states that all evidence about θ is contained in $f_n(\mathbf{x}^n \mid \theta)$, which does not depend on $\boldsymbol{\tau}$.

The Stopping Rule Principle was first espoused by Barnard (1947, 1949), whose motivation at the time was essentially a reluctance to allow an experimenter's *intentions* to affect conclusions drawn from data. In Example 15 of Subsection 1.6.4, for instance, suppose the experimenter recorded that he obtained 9 heads and 3 tails in independent Bernoulli trials, but died before recording why he decided to stop the experiment at that point. We cannot then analyze the data from a frequentist perspective; we would need to know the stopping rule, i.e., *we would need to know what was going on in the experimenter's head* at the time. (See also Lindley and Phillips (1976).) It is troubling to have to involve the experimenter's personal thoughts in analyzing hard data. (We presume here that there is no doubt that the data was generated by independent Bernoulli trials.)

The fact that the Stopping Rule Principle follows from the Likelihood Principle was demonstrated in Birnbaum (1962) and Barnard, Jenkins, and Winsten (1962). Pratt (1965) argued that the Stopping Rule Principle holds even when no likelihood function exists. (See Berger and Wolpert (1984) for details.) Other excellent discussions of this principle can be found in Anscombe (1963a), Edwards, Lindman, and Savage (1963), Bartholomew (1967), and Basu (1975).

7.7.3. Practical Implications

It is a rare sequential experiment in which all contingencies are anticipated in advance. Far more common is to encounter "surprises," and in a sequential setting these surprises may call for unplanned alterations of the experiments.

EXAMPLE 17. A clinical trial is conducted in which two treatments, A and B, are to be compared. Patients are randomly allocated to one or the other treatment as they enter the trial.

Case 1. Suppose the originally selected stopping rule was to continue the trial until $n = 200$ patients had been treated. Imagine, however, that after 100 patients have been treated, it is observed that 80% of those receiving treatment A were cured, and only 20% of those receiving treatment B were cured. The evidence then seems fairly overwhelming that treatment A is better, and yet a frequentist *cannot* stop the trial and make a conclusion; the originally specified stopping rule must be followed to achieve frequentist validity.

Case 2. Because of the above problem, it is fairly common to plan *interim analyses* in clinical trials (through appropriate choice of the stopping rule). For instance, the stopping rule in this trial could be to (i) stop after $n = 100$ if the difference in cure rates between the two treatments exceeds 30%, and (ii) take all $n = 200$ observations otherwise. This is a well-defined stopping rule for which frequentist measures, such as error probabilities, could be calculated. Imagine, now, that, at the interim analysis, it is observed that the cure rates of the two treatments are very close, but that many treatment A patients have had highly severe and dangerous side effects. One would rather naturally desire to stop the experiment at this point, conclude that the treatments have roughly the same cure rate, and recommend treatment B for use, but this can again not be done from a frequentist perspective; after-the-fact informal stopping rules are simply not allowed in frequentist evaluations.

The complexities of actual clinical trials make pre-experimental complete specification of the stopping rule (i.e., complete anticipation of all possible contingencies) almost impossible. And even fairly simple stopping rules can cause the frequentist analysis to be extremely difficult.

The morass of problems, caused by the need for flexible optional stopping, could be viewed as an unavoidable difficulty in statistics, except that the Stopping Rule Principle says the difficulty *should not exist*: the evidence about θ from the data should *not* depend on the reason for stopping. In either Case 1 or Case 2 of Example 17, one should be able to stop the experiment early if desired, with no ensuing problems in the analysis. Edwards, Lindman, and Savage (1963) thus state that,

"The irrelevance of stopping rules to statistical inference restores a simplicity and freedom to experimental design .... Many experimenters would like to feel free to collect data until they have either conclusively proved their point, conclusively disproved it, or run out of time, money, or patience."

It should be emphasized that the Stopping Rule Principle does *not* say that one can stop in any desired fashion and then perform a frequentist analysis as if, say, a fixed sample size experiment had been done. Almost any kind of optional stopping tends to have a drastic effect on frequentist evaluations (see Subsection 7.7.4). What the Stopping Rule Principle (or more precisely the Likelihood Principle) does say is that one should ignore the stopping rule, and then do an analysis compatible with the Likelihood Principle (e.g., a Bayesian analysis).

The major reason for this subsection was to show that the Likelihood Principle and its consequences are not just "negative" results of the type "it is philosophically in error to use frequentist measures," but actually have the potential of introducing vast simplifications and improvements in practical statistical analyses. Of course simplicity does not by itself justify a statistical procedure, and if someone could demonstrate that these principles are wrong, then they are wrong. But the combination of their very convincing justifications and wonderful practical implications makes them very difficult to resist.

We have been discussing the special issue of making a decision or conclusion given that sampling has stopped. Most of the chapter, however, was devoted to the technically more complex issue of deciding whether or not to take further observations in a sequential setting. The Stopping Rule Principle or Likelihood Principle need not apply when facing such "design" questions, as indicated in Subsection 1.6.4 and in the following example.

Example 18. Imagine that each observation is of the form $X_i = (Z_i, Y_i)$. Here Y_i is 1 or 0, where $P(Y_{i+1}=1 \mid Y_i=1) = 1 - P(Y_{i+1}=0 \mid Y_i=1) = \frac{1}{2}$ and $P(Y_{i+1}=0 \mid Y_i=0)=1$. (Define $Y_0=1$.) When $Y_i=1$, Z_{i+1} will be independent of the previous Z_i and will have a $\mathcal{N}(\theta, 1)$ distribution. When $Y_i=0$, on the other hand, Z_{i+1} will be zero. (This could correspond to a situation in which a piece of equipment is used to obtain the important observations Z_i, and Y_i tells whether the equipment will work the next time ($Y_i=1$) or has irreparably broken ($Y_i=0$).)

Imagine now that $x_1, \ldots, x_n$ have been observed, and that $y_i = 1$ for $i \leq n-1$. The likelihood function for θ is then $\mathcal{N}(\bar{z}_n, 1/n)$ (since all the z_i are valid observations). The Likelihood Principle thus says that all decisions or inferences concerning θ should involve the data only through this likelihood function. It is obvious, however, that knowledge of y_n may be crucial in deciding what to do. If $y_n = 1$, it may be desirable to take another observation. If $y_n = 0$, on the other hand, taking another observation would be a waste of time (the equipment is broken), so an immediate decision should be made.

It is possible to make the Likelihood Principle generally applicable, by the (natural) expedient of letting θ include not only unknown parameters of the model, but also all future possible (random) observables, such as the future X_i in Example 18. This idea is discussed in Berger and Wolpert (1984); the details are less important than the understanding that the Likelihood Principle only applies to decisions (including "design" decisions) or conclusions about θ, so that θ must include all unknowns relevant to the decision.

From a practical perspective in sequential analysis, it is nice to know when the likelihood function for a *model parameter* θ suffices also for deciding whether to stop or continue sampling (i.e., it is nice to know when one can ignore problems like that in Example 18). A broad class of such "nice" situations have been determined in Bahadur (1954), in which conditions are given whereby sufficient statistics for a model parameter θ are all that is needed to determine whether or not to stop sampling. (Since the likelihood function is always a sufficient statistic, it then suffices for making the decision to stop or continue.) The conditions are, in particular, satisfied by independent observations from a distribution in the exponential family.

7.7.4. Criticisms of the Stopping Rule Principle

To many, the Stopping Rule Principle is counter-intuitive, because it seems as if an experimenter could "bias" results by appropriate choice of a stopping rule.

EXAMPLE 19. Assume that $X_1, X_2, \ldots$ is a sequential sample from a $\mathcal{B}(1, \theta)$ distribution, and that it is desired to estimate θ under loss $L(\theta, a, n) = (\theta - a)^2 + nc$. Imagine that a believer in the Likelihood Principle decides to use a Bayesian approach with the noninformative prior $\pi(\theta) = \theta^{-1}(1-\theta)^{-1}$. After observing $x_1, \ldots, x_n$, it follows that he will estimate θ by $\bar{x}_n$, regardless of the stopping rule used.

A classical statistician might disagree with this analysis, arguing that, among other things, a very biased stopping rule could have been used. For example, the stopping rule could have been

$$\boldsymbol{\tau}\text{: stop sampling if } X_1 = 1\text{, and otherwise after observing } X_2.$$

For this stopping rule,

$$E_{\theta,\boldsymbol{\tau}}[\bar{X}_N] = \theta(1) + (1-\theta)[\tfrac{1}{2}\theta + 0(1-\theta)] = \theta + \tfrac{1}{2}\theta(1-\theta).$$

The Bayes decision rule would thus be biased, in a classical sense, by the amount $\frac{1}{2}\theta(1-\theta)$. The point here is that the experimenter might have wanted θ to appear large, and so purposely used a stopping rule, such as $\boldsymbol{\tau}$, for which $\bar{X}_N$ would tend to overestimate θ.

The criticism of the Stopping Rule Principle in Example 19 is, of course, dependent on the view that it is necessarily bad to have an estimator which is biased from a frequentist perspective. Even most frequentists no longer hold this view (there being too many counterexamples in which the unbiased estimator is silly); let us, therefore, concentrate on a more troubling example, which has a very long history (cf. Armitage (1961) and Basu (1975)).

EXAMPLE 20. Suppose $X_1, X_2, \ldots$ are independent $\mathcal{N}(\theta, 1)$ random variables, and that a confidence or credible set for θ is desired. Consider the stopping rule $\boldsymbol{\tau}^k$, defined by $\tau_0 = 0$, and

$$\tau_j^k(\mathbf{x}^j) = \begin{cases} 1 & \text{if } |\bar{x}_j| \geq k/\sqrt{j}, \\ 0 & \text{if } |\bar{x}_j| < k/\sqrt{j}, \end{cases}$$

where $\bar{x}_j = j^{-1} \sum_{i=1}^{j} x_i$. In other words, one stops sampling when the standardized sample mean exceeds k in absolute value; note that the Law of the Iterated Logarithm implies that this is even sure to happen if $\theta = 0$ (for *any* fixed k), so that $\boldsymbol{\tau}^k$ does define a proper stopping rule.

Consider now the, say, noninformative prior Bayesian who is presented with the information that this stopping rule was used, and is given the data. As usual, let n denote the actual sample size obtained. The posterior distribution (likelihood) for θ is then a $\mathcal{N}(\bar{x}_n, n^{-1})$ density, and the 95% HPD credible set for θ is

$$C_n(\bar{x}_n) = \left(\bar{x}_n - \frac{(1.96)}{\sqrt{n}}, \bar{x}_n + \frac{(1.96)}{\sqrt{n}} \right).$$

Note that, in accordance with the Stopping Rule Principle, the Bayesian has (automatically) not involved the stopping rule in this calculation.

The paradoxical feature of this example is that, by choosing $k > 1.96$, the experimenter can ensure that $C_n(\bar{x}_n)$ *does not contain zero*; thus, as a classical confidence procedure, $\{C_n\}$ will have zero coverage probability at $\theta = 0$ (and, by continuity, will have small coverage probability for θ near zero). It thus seems that the experimenter can, through sneaky choice of the stopping rule, "fool" the Bayesian into believing that θ is not zero.

There are many responses to the "paradox" in the above example, but there is no obvious resolution. By this we mean that the frequentist and conditional (Bayesian) ways of thinking are so different that a frequentist and a Bayesian can each look at the above example and feel that it supports his position. Thus Savage said in Savage *et al.* (1962),

> "I learned the stopping rule principle from Professor Barnard, in conversation in the summer of 1952. Frankly, I then thought it a scandal that anyone in the profession could advance an idea so patently wrong, even as today I can scarcely believe that some people resist an idea so patently right."

How would a Bayesian try to explain why he would view $C_n(\bar{x}_n)$ as correct in Example 20? Basically, he would state that, while he too knows the Law of the Iterated Logarithm and hence that $\theta=0$ could have yielded this data, it is simply much more likely that θ actually happens to be near $\bar{x}_n$. This is predicated, of course, on the assumption that use of a noninformative prior is appropriate (i.e., that no value of θ is considered more likely than any other, as a first approximation). If the Bayesian has *any* reason at all to suspect that $\theta=0$ is a real possibility (i.e., has positive prior probability), then the analysis will change drastically.

EXAMPLE 20 (continued). Suppose the Bayesian assigns positive probability π_0 to the point $\theta=0$, and (for instance) gives the $\theta\neq 0$ density $(1-\pi_0)$ times a $\mathcal{N}(0, A)$ density. Then, (4.17) gives, as the posterior probability that $\theta=0$,

$$\left[1+\frac{(1-\pi_0)}{\pi_0}\,\frac{\exp\{\frac{1}{2}(\sqrt{n}\bar{x}_n)^2/[1+(nA)^{-1}]\}}{(1+nA)^{1/2}}\right]^{-1}. \tag{7.89}$$

The situation in which a frequentist would suspect that $\theta=0$, and be glad he was considering the stopping rule of the experimenter, is when n turned out to be large (since, when $\theta=0$ and k is moderate to large, it can be shown that n will tend to be huge; moderate n would occur otherwise). But notice the behavior of (7.89) for large n: it will be the case that $\sqrt{n}|\bar{x}|\cong k$ if n is large (the change from $\sqrt{n-1}|\bar{x}_{n-1}|$ to $\sqrt{n}|\bar{x}_n|$ is likely to then be small, and the former was less than k), so that (7.89) is approximately

$$\left[1+\frac{(1-\pi_0)}{\pi_0}\,\frac{\exp\{\frac{1}{2}k^2\}}{(1+nA)^{1/2}}\right]^{-1}\cong 1.$$

Thus, if the Bayesian suspects that θ could be zero, and n turns out to be very large, he will actually become convinced that θ is zero.

With the above analysis in mind, let us return to the issue of "fooling" the Bayesian who disregards the stopping rule. Note first that the possibility of "fooling" exists only when the experimenter is privy to information that the Bayesian does not have (e.g., that θ is a distinguished point). Of course, any experimenter who withholds information needed by the analyst can be expected to succeed in deception, if he is at all clever. It is interesting that, in Example 20, the Bayesian needs to know only if $\theta=0$ is a distinguished point, a fact which might be pretty hard to hide in practice. (The experimenter might well have the acknowledged goal of disproving the "theory" that $\theta=0$.) On the other hand, experimenters quite routinely fail to report their stopping rule (undoubtedly more out of ignorance than out of an attempt to deceive), and the frequentist analyst can be massacred by this hard-to-detect deception. (If, in Example 20, it is observed that $\sqrt{n}|\bar{x}_n|=3$, say, with no mention made of the stopping rule, it is the rare frequentist who would not assume a fixed sample size experiment, and reject the possibility that

$\theta = 0$; the Bayesian, by contrast, uses something like (7.89) to evaluate $\theta = 0$, and is much less prone to reject.)

This kind of discussion tends to be inconclusive, because the situation being considered is too complicated for clear intuitive resolution. That is why we find the simple examples and arguments in Section 1.6 of such value; they guided us to the Likelihood Principle (and hence to the Stopping Rule Principle), while pointing out that frequentist answers could be silly. When now we encounter another major clash between frequentist answers and the Likelihood Principle, it is not all that hard to decide which to trust.

As a final argument, note that rejection of the Stopping Rule Principle implies rejection of the Likelihood Principle, which implies rejection of either the Conditionality or Sufficiency Principles (see Subsection 1.6.4). It is possible to rephrase this clash in an interesting fashion, namely to observe that the intuitive belief that final conclusions should depend on the stopping rule is inconsistent with the frequentist concept of admissibility.

EXAMPLE 21. Take the example where $X_1, X_2, \ldots$ are independent Bernoulli (θ) random variables, and suppose we want to estimate θ under squared-error loss. Let $\boldsymbol{\tau}^1$ be the binomial stopping rule yielding a fixed sample of size 12, and let $\boldsymbol{\tau}^2$ be the negative binomial stopping rule which stops upon reaching 3 zeros (failures). Suppose we feel that the stopping rule *should* affect our final estimate of θ, so that, for some $(x_1, \ldots, x_{12})$ which contains 3 zeros,

$$\delta^1(x_1, \ldots, x_{12}) \neq \delta^2(x_1, \ldots, x_{12}),$$

where δ^i is the estimator of θ that is selected when $\boldsymbol{\tau}^i$ is known to obtain.

Consider now the "mixed" experiment, consisting of deciding to use $\boldsymbol{\tau}^1$ with probability $\frac{1}{2}$, and $\boldsymbol{\tau}^2$ with probability $\frac{1}{2}$. This is a well-defined sequential experiment, the data now consisting of $(i, x_1, \ldots, x_n)$, where i is 1 or 2 as $\boldsymbol{\tau}^1$ or $\boldsymbol{\tau}^2$ is selected. If one really felt that the choice of $\boldsymbol{\tau}$ should affect the estimate of θ, then, for some $(x_1, \ldots, x_{12})$ which contains 3 zeros,

$$\delta(1, x_1, \ldots, x_{12}) \neq \delta(2, x_1, \ldots, x_{12}), \tag{7.90}$$

where δ is the estimator for the "mixed" experiment. (A follower of the Conditionality Principle would, of course, use $\delta(i, x_1, \ldots, x_{12}) = \delta^i(x_1, \ldots, x_{12})$, but we need not assume this.) But, in the mixed experiment, it can be shown that i is *not* a part of the sufficient statistic for θ. Hence, if (7.90) is satisfied and since the loss is convex, δ is inadmissible. The details are left for the exercises.

More general discussions of the conflict between violation of the Likelihood Principle and admissibility can be found in Berger and Wolpert (1984) and Berger (1984d).

A final issue that can be raised is that of Bayesian robustness. In Section 4.7 we saw that frequentist measures can be useful in Bayesian robustness investigations, especially when posterior robustness (sensitivity analysis) is not feasible. It can thus be argued (cf. Rubin (1984)) that the stopping rule can play a role (through frequentist measures) in Bayesian robustness studies. We have yet to see convincing examples of this. Indeed, the difficulty of frequentist analyses involving the stopping rule makes it unlikely that the frequentist approach to Bayesian robustness can actually be simpler, in practical problems, than the posterior robustness approach (which can ignore the stopping rule in reaching final conclusions).

7.7.5. Informative Stopping Rules

The stopping rules defined in Section 7.3 have been called *noninformative* by Raiffa and Schlaifer (1961) (see also Roberts (1967), Berger and Wolpert (1984), and Barlow and Shor (1984)), as opposed to other types of stopping rules which can be *informative.* The usual definition of a noninformative stopping rule is something to the effect that the posterior distribution of θ, given $\mathbf{x}^n$, should not depend on the stopping rule used to reach $\mathbf{x}^n$. (We refrain from being too precise, for reasons to be discussed shortly.) One important interpretive feature of this definition is that the label "noninformative" does *not* mean that the stopping time N carries no information about θ. Indeed, it is not uncommon for N to be highly related to θ. In Example 20 (continued), for instance, the posterior probability that $\theta=0$ (see (7.89)) will be most affected by the magnitude of n, at least when n is large. What "noninformative" does mean is that the stopping time provides no information about θ *in addition to* that contained in the observed likelihood, $f_n(\mathbf{x}^n \mid \theta)$, and in the prior.

It is easy to create stopping times that are formally informative, by failing to recognize or model all the data. Here is an example (from Berger and Wolpert (1984)).

Example 22. Suppose $X_1, X_2, \ldots$ are independent Bernoulli (θ) random variables, with $\theta=0.49$ or $\theta=0.51$. The observations, however, arrive randomly. Indeed, if $\theta=0.49$, the observations arrive as a Poisson process with mean rate of 1 per second, while if $\theta=0.51$, the observations will arrive as a Poisson process with mean rate of 1 per hour. The "stopping rule" that will be used is to stop the experiment when 1 minute has elapsed. One can here introduce $Y=$ time, and write down the stopping rule in terms of Y and the X_i.

It is intuitively clear that this stopping rule cannot be ignored since, if one ends up with 60 observations, knowing whether the experiment ran for 1 minute or $2\frac{1}{2}$ days is crucial knowledge, telling much more about which value of θ is true than will the 60 observations themselves. Thus the stopping rule could be called informative.

The trouble with this analysis is that Y should actually be considered to be part of the observation. It can be incorporated in the overall probability model, but, to stay in the sequential setting, let us suppose that we can record the actual times of observation arrivals. Thus let Y_i denote the time of arrival (in minutes) of observation X_i, so that we have a true sequential experiment with observations (X_1, Y_1), (X_2, Y_2), ... The stopping rule is then given by

$$\tau_m((x_1, y_1), \ldots, (x_m, y_m)) = \begin{cases} 0 & \text{if } y_m < 1, \\ 1 & \text{if } y_m \geq 1, \end{cases}$$

and is of the form given in Section 7.3 (and is hence noninformative). The information (about θ) contained in the $\{y_i\}$ will now be in the likelihood function (where it belongs).

It is possible to have partly unknown stopping rules that depend on θ. As a simple example, the "stopping rule" could be to run the sequential experiment until a sensitive instrument becomes uncalibrated, and the reason for this happening could be related to θ in some fashion other than just through the measurements, X_i. In such situations, the $\tau_j(\mathbf{x}^j)$ are really $\tau_j(\mathbf{x}^j \mid \theta)$, and the overall density is (instead of (7.87))

$$f(n, \mathbf{x}^n \mid \theta) = \left[\prod_{j=0}^{n-1} (1 - \tau_j(\mathbf{x}^j \mid \theta)) \right] \tau_n(\mathbf{x}^n \mid \theta) f_n(\mathbf{x}^n \mid \theta).$$

Clearly the likelihood function for θ then depends on the stopping rule, so the stopping rule is informative.

One final type of informative stopping rule should be mentioned, namely stopping rules which affect the *prior* distribution of a Bayesian. In Example 20, for instance, one might have no prior knowledge about θ, itself, but the *choice* by the experimenter of $\boldsymbol{\tau}^k$ as a stopping rule might lead one to suspect that the experimenter views $\theta = 0$ as a real possibility; the Bayesian might then decide to use a prior which gives $\theta = 0$ a positive prior probability. This phenomenon is not really due to any information provided by the stopping rule, however, but rather to information about the experimenter's prior that is revealed by his choice of a stopping rule.

7.8. Discussion of Sequential Loss Functions

Loss functions more general than that considered in this chapter are sometimes needed. In particular, it is sometimes important to allow the loss to depend on the observations. (An extreme example of this occurred in early tests of rocket propellant, in which a very bad batch of fuel would cause the testing equipment to blow up. Hence the observation (fuel quality) could have a significant effect on the experimental cost.) It can be checked, however, that most of the methods developed in this chapter, particularly the Bayesian techniques of Section 7.4, adapt, with obvious modifications, to this situation.

A more serious and difficult concern is the problem of observational cost. In Section 7.1 we introduced the general loss $L(\theta, a, n, s)$, where s denotes some prescribed method of sampling. Subsequently, however, only fixed sample size and one-at-a-time sequential sampling were considered. It may well be optimal to instead use a *sequential batch* plan, in which, say, an initial group of n_1 observations is taken, based on which a second group of size n_2 is taken, etc. This will often be cheaper than a straight sequential sampling plan, because it is often more efficient to take a number of observations at the same time than to take them separately. This will especially be true if a fairly difficult analysis is needed at each stage to decide if sampling should be continued (and if the cost of the statistician is not negligible). For these (and other reasons), it is actually quite common in clinical trials to use such sequential batch plans (the analyses after each batch being called "interim analyses.") One-observation-at-a-time sequential analysis can only be said to be nearly optimal when the observations are very expensive (so that minimizing the total number of observations taken is paramount), or when observations present themselves one at a time (say, products coming off a production line) with sufficient time between observations to conduct whatever analysis is needed.

Actually, even for the general loss $L(\theta, a, n, s)$, one could theoretically use the techniques of Section 7.4 to calculate the optimal procedure. In calculating the j-step look ahead Bayes risks, $r_j(\pi^n, n)$, however, one is now forced to consider all possible ways of selecting batches (of observations) of various sizes out of j additional observations. The calculational complexity quickly becomes extreme.

One attractive sequential batch sampling plan (which can often be mathematically handled) is to choose, at each stage, a batch of a fixed size m. (When $m = 1$, this corresponds to the usual sequential sampling.) The batches can then be considered to themselves be observations in a new sequential problem, and the techniques of this chapter can be brought to bear to find the best procedure, $\mathbf{d}_m$, in this new problem. By minimizing, in some sense, over m, the best fixed-batch-size sequential sampling plan can be found. An example follows.

EXAMPLE 23. Assume that a sequential sample $X_1, X_2, \ldots$ of $\mathcal{N}(\theta, 1)$ random variables is available, and that it is desired to test H_0: $\theta = -\frac{1}{2}$ versus H_1: $\theta = \frac{1}{2}$ under "0–1" decision loss. The cost of taking a batch of observations of size m is C_m.

Note that

$$\bar{X}_i^m = \frac{1}{m} \sum_{j=(i-1)m+1}^{im} X_j$$

is a sufficient statistic for θ, based on $X_{(i-1)m+1}, \ldots, X_{im}$. It can also be shown that any stopping rule need only depend on $X_{(i-1)m+1}, \ldots, X_{im}$

through $\bar{X}_i^m$. It follows that if batches of size m are to be taken sequentially, one need only consider $\bar{X}_1^m, \bar{X}_2^m, \ldots$, which are i.i.d. $\mathcal{N}(\theta, 1/m)$. The problem thus reduces to an ordinary sequential problem with observations $\bar{X}_1^m, \bar{X}_2^m, \ldots$, "0–1" decision loss, and observational cost C_m. This can be dealt with as in Section 7.5. Suppose, for example, that a Bayesian approach is taken, and that $\mathbf{d}_m$ is the Bayes SPRT, with Bayes risk $r(\pi, \mathbf{d}_m)$. This Bayes risk can then be minimized over m, to obtain the optimal batch size.

Exercises

In all of the exercises, assume that the utility function is linear, so that the overall loss is the sum of the decision loss and the sampling cost. In problems involving a test between two hypotheses H_0 and H_1, let a_i denote accepting H_i $(i=0, 1)$.

Section 7.2

1. Assume that $X_1, X_2, \ldots$ is a sequential sample from a $\mathcal{N}(\theta, 1)$ density, that it is desired to estimate θ under squared-error decision loss, and that the cost of a fixed sample of size n is $C(n)=n(0.01)$. If θ has a $\mathcal{N}(1, 4)$ prior distribution, find the optimal fixed sample size rule and its Bayes risk.

2. Do the preceding exercise for
 (a) $C(n)=[\log(1+n)](0.01)$.
 (b) $C(n)=n^{1/2}(0.01)$.

3. Consider the situation of Exercise 1, but assume that the decision loss is $L(\theta, a)=|\theta-a|$. Find the optimal fixed sample size rule and its Bayes risk.

4. Assume that $X_1, X_2, \ldots$ is a sequential sample from a $\mathcal{B}(1, \theta)$ density, that it is desired to estimate θ under decision loss $L(\theta, a)=(\theta-a)^2/\theta(1-\theta)$, and that the cost of a fixed sample of size n is $C(n)=nc$. If θ has a $\mathcal{U}(0, 1)$ prior density, approximate the optimal fixed sample size.

5. Assume that $X_1, X_2, \ldots$ is a sequential sample from a $\mathcal{P}(\theta)$ density, that it is desired to estimate θ under decision loss $L(\theta, a)=(\theta-a)^2/\theta$, and that the cost of a fixed sample of size n is $C(n)=nc$. If θ has a $\mathcal{G}(\alpha, \beta)$ prior density, approximate the optimal fixed sample size.

6. Assume that $X_1, X_2, \ldots$ is a sequential sample from a $\mathcal{G}(1, \theta)$ density, that it is desired to estimate θ under decision loss $L(\theta, a)=(\theta-a)^2/\theta^2$, and that the cost of a fixed sample of size n is $C(n)=nc$. If θ has an $\mathcal{IG}(\alpha, \beta)$ prior density, approximate the optimal fixed sample size.

7. Assume that $X_1, X_2, \ldots$ is a sequential sample from a $\mathcal{Ge}((1+\theta)^{-1})$ density, that it is desired to estimate θ under decision loss $L(\theta, a)=(\theta-a)^2/\theta(1+\theta)$, and that the cost of a fixed sample of size n is $C(n)=nc$. If θ has the prior density

$$\pi(\theta \mid \alpha, \beta)=\frac{\Gamma(\alpha+\beta)}{\Gamma(\alpha)\Gamma(\beta)}\theta^{\alpha-1}(\theta+1)^{-(\alpha+\beta)}I_{(0,\infty)}(\theta),$$

approximate the optimal fixed sample size.

8. Assume that $X_1, X_2, \ldots$ is a sequential sample from a $\mathcal{N}(\theta, 1)$ density, and that it is desired to test H_0: $\theta = 0$ versus H_1: $\theta = 1$. The decision loss is $L(\theta_0, a_0) = L(\theta_1, a_1) = 0$, $L(\theta_0, a_1) = 1$, and $L(\theta_1, a_0) = 2$. The cost of a fixed sample of size n is $C(n) = n(0.1)$. If the prior probabilities of θ_0 and θ_1 are $\frac{2}{3}$ and $\frac{1}{3}$, respectively, approximate the optimal fixed sample size.

9. Assume that $X_1, X_2, \ldots$ is a sequential sample from a $\mathcal{N}(\theta, 1)$ density, and that it is desired to test H_0: $\theta \le 0$ versus H_1: $\theta > 0$. The decision loss is zero for a correct decision and $|\theta|$ for an incorrect decision. The cost of a fixed sample of size n is $C(n) = n(0.1)$. If the prior density of θ is $\mathcal{N}(0, 4)$, approximate the optimal fixed sample size.

10. Assume that $X_1, X_2, \ldots$ is a sequential sample from a $\mathcal{N}(\theta, \sigma^2)$ density (σ^2 known), and that it is desired to test H_0: $\theta \le \theta_0$ versus H_1: $\theta \ge \theta_1$ ($\theta_0 < \theta_1$) under "0–1" decision loss. It is possible for θ to be between θ_0 and θ_1, but no decision loss will then be incurred, regardless of the action taken. Assume that θ has a $\mathcal{N}(\mu, \tau^2)$ prior density, and let δ_n^π denote the Bayes rule for a fixed sample of size n. Show (pretending n is a continuous variable) that

$$\frac{d}{dn} r(\pi, \delta_n^\pi) = \frac{-\sigma\tau}{2\pi(\sigma^2 + n\tau^2)n^{1/2}} \exp\left\{-\frac{1}{2\tau^2}\left[\left(\frac{\theta_0+\theta_1}{2} - \mu\right)^2 + \frac{(\theta_1-\theta_0)^2}{4}\right]\right\}$$
$$\times \exp\left\{-\frac{1}{2}\left[\frac{\sigma^2}{n\tau^4}\left(\frac{\theta_0+\theta_1}{2} - \mu\right)^2 + \frac{n}{4\sigma^2}(\theta_1-\theta_0)^2\right]\right\}.$$

Section 7.4

11. In the situation of Exercise 4 of Chapter 1, assume that a computer consultant can be hired to help in the prediction of the team's winning proportion θ. The consultant will report $X \sim \mathcal{B}(1, \theta)$. (This is a rather silly report, but you will appreciate the ease in calculation.) How much money would you expect the consultant's report to be worth to you?

12. Find a Bayes sequential decision procedure for the situation in Exercise (a) 4, (b) 5, (c) 6, (d) 7.

13. Assume that $X_1, X_2, \ldots$ is a sequential sample from a $\mathcal{B}(1, \theta)$ density, and that it is desired to test H_0: $\theta = \frac{1}{4}$ versus H_1: $\theta = \frac{1}{2}$. The decision loss is $L(\theta_0, a_0) = L(\theta_1, a_1) = 0$, $L(\theta_0, a_1) = 10$, and $L(\theta_1, a_0) = 20$, while the cost of each observation is $c_i = 1$.
    (a) Letting π_0 denote the prior probability of θ_0, find $\mathbf{d}^1$, the Bayes 1-truncated procedure.
    (b) If $\pi_0 = \frac{2}{3}$, find $\mathbf{d}^2$.

14. In the situation of Example 7, find the Bayes 3-truncated procedure, $\mathbf{d}^3$, when $c = 0.01$.

15. Assume that $X_1, X_2, \ldots$ is a sequential sample from a $\mathcal{P}(\theta)$ density, that $L(\theta, a, n) = (\theta - a)^2 + n(\frac{1}{12})$, and that θ has a $\mathcal{G}(1, 1)$ prior density. Find the Bayes 3-truncated procedure, $\mathbf{d}^3$.

16. Assume that $X_1, X_2, \ldots$ is a sequential sample from a $\mathcal{B}(1, \theta)$ density, and that a decision between two possible actions, a_0 and a_1, must be made. The decision

loss is $L(\theta, a_0) = \theta$ and $L(\theta, a_1) = 1 - \theta$, while each observation costs $c_i = \frac{1}{12}$. Find the Bayes 2-truncated procedure, $\mathbf{d}^2$, when θ has a $\mathcal{U}(0, 1)$ prior density.

17. Prove Theorem 1.

18. Assume that $X_1, X_2, \ldots$ is a sequential sample from a $\mathcal{B}(1, \theta)$ density, that $L(\theta, a, n) = (\theta - a^2) + \sum_{i=1}^{n} [0.01(1 + 5/i)]$, and that θ has a $\mathcal{U}(0, 1)$ prior density. Use Theorem 3 to show that the Bayes sequential procedure must be truncated at $m = 17$.

19. Prove Theorem 4.

20. Use Theorem 4 to show that the procedure found in Exercise 14 is actually the sequential Bayes procedure, $\mathbf{d}^\pi$.

21. Use Theorem 4 to show that the procedure found in Exercise 16 is actually the sequential Bayes procedure, $\mathbf{d}^\pi$.

22. Find the 1-step look ahead procedure, $\mathbf{d}_{\mathrm{L}}^1$, for the situation of (a) Example 7, (b) Exercise 13 (with $\pi_0 = \frac{2}{3}$), (c) Exercise 15, (d) Exercise 16.

23. Find the 2-step look ahead procedure, $\mathbf{d}_{\mathrm{L}}^2$, for the situation of (a) Example 6, (b) Exercise 15, (c) Exercise 16.

24. For the situation of Exercise 9, describe how the fixed sample size look ahead procedure, $\mathbf{d}_{\mathrm{FS}}$, would be implemented. (Note the relevance of (7.11).) Carry out the procedure for the sequential sample 0.5, 2, 1, 1.5, 0, 2.5, 2, 1, ...

25. Show that in Example 10, Case 2, the fixed sample size look ahead procedure, $\mathbf{d}_{\mathrm{FS}}$, is as indicated.

26. Assume that $X_1, X_2, \ldots$ is a sequential sample from a $\mathcal{P}(\theta)$ density, that $L(\theta, a, n) = (\theta - a)^2 + nc$, and that θ has a $\mathcal{G}(\alpha, \beta)$ prior density. Show that the fixed sample size look ahead procedure, $\mathbf{d}_{\mathrm{FS}}$, is the same as the 1-step look ahead procedure, $\mathbf{d}_{\mathrm{L}}^1$.

27. Assume that $X_1, X_2, \ldots$ is a sequential sample from a $\mathcal{P}(\theta)$ density, that $L(\theta, a, n) = (\theta - a)^2 + \frac{1}{4}n$, and that θ has a $\mathcal{G}(1, 1)$ prior density. Find the Bayes 3-inner truncated procedure, $\mathbf{d}_{\mathrm{I}}^3$.

28. Find the 1-step inner look ahead procedure, $\mathbf{d}_{\mathrm{IL}}^1$, for the situation of (a) Example 6, (b) Example (7), (c) Exercise 15, (d) Exercise 16.

29. Find the 2-step inner look ahead procedure, $\mathbf{d}_{\mathrm{IL}}^2$, for the situation of (a) Example 6, (b) Exercise 15.

30. Prove that the inequalities in (7.24) of Subsection 7.4.8 are valid.

31. Prove that Theorem 5 is true for a general loss, $L(\theta, a, n)$, providing the following condition is satisfied:

$$\lim_{n \to \infty} E^\pi E_\theta^X [L(\theta, \delta_n^\pi(X), n) - \inf_a L(\theta, a, n)] = 0.$$

32. Prove that Theorem 6 is true for a general loss, $L(\theta, a, n)$, providing the condition in Exercise 31 is satisfied.

33. Complete the analysis in Example 13, showing that $h^*(r)$ (defined by (7.43)) is a solution to (7.39) for $r > 1$.

34. Assume that $X_1, X_2, \ldots$ is a sequential sample from a $\mathscr{B}(1, \theta)$ density, that it is desired to test H_0: $\theta = \frac{1}{3}$ versus H_1: $\theta = \frac{2}{3}$ under "0–1" loss, that each observation costs c, and that $0 < \pi_0 < 1$ is the prior probability that $\theta = \frac{1}{3}$. Show that (7.44) is satisfied, and find an asymptotically pointwise optimal procedure.

35. Verify that (7.47) is satisfied and find an asymptotically pointwise optimal procedure for the situation of (a) Example 7, (b) Exercise 26.

36. In the situation of Example 2, verify that (7.47) is satisfied for $\beta = k/2$.

Section 7.5

37. It is desired to test the null hypothesis that a sequential sample has a $\mathscr{U}(0, 2)$ common density versus the alternative hypothesis that the common density is $\mathscr{U}(1, 3)$. The cost of each observation is c, and the decision loss is "0–K_i" loss. Let π_i denote the prior probability of H_i.
    (a) Show that the Bayes sequential procedure is either $\mathbf{d}_1$ or $\mathbf{d}_2$, where $\mathbf{d}_1$ is the procedure which makes an immediate Bayes decision, and $\mathbf{d}_2$ is the procedure which starts sampling and, at stage n, is given by the following:

    if $x_n \le 1$, stop sampling and accept H_0;
    if $x_n \ge 2$, stop sampling and accept H_1;
    if $1 < x_n < 2$, continue sampling.

    (b) Calculate the Bayes risks of $\mathbf{d}_1$ and $\mathbf{d}_2$.

38. It is desired to test the null hypothesis that a sequential sample has a $\mathscr{U}(0, 2)$ common density versus the alternative hypothesis that the common density is $\mathscr{U}(0, 1)$. The cost of each observation is c, and the decision loss is "0–K_i" loss. Let π_i denote the prior probability of H_i.
    (a) Show that the Bayes sequential procedure is one of the procedures $\mathbf{d}_J$ (J a nonnegative integer) defined as follows. The procedure $\mathbf{d}_0$ is simply the immediate Bayes decision. For $J \ge 1$, $\mathbf{d}_J$ starts sampling; stops sampling when stage $n = J$ is reached, deciding H_0 if $x_J \ge 1$ and deciding H_1 otherwise; and at stage $n < J$ is given by the following:

    if $x_n \ge 1$, stop sampling and decide H_0;
    if $x_n < 1$, continue sampling.

    (b) For $\mathbf{d}_J$ (with $J \ge 1$), show that $\alpha_0 = 2^{-J}$, $\alpha_1 = 0$, $E[N \mid H_0] = 2(1 - 2^{-J})$, and $E[N \mid H_1] = J$.
    (c) Find the Bayes sequential procedure if $K_0 = 2$, $K_1 = 1$, $c = \frac{1}{15}$, and $\pi_0 = \frac{1}{3}$.
    (d) Find the Bayes sequential procedure if $K_0 = 2$, $K_1 = 1$, $c = \frac{1}{15}$, and $\pi_0 = \frac{1}{8}$.

39. Assume that $X_1, X_2, \ldots$ is a sequential sample from the density

$$f(x \mid \theta) = e^{-(x-\theta)} I_{(\theta,\infty)}(x),$$

and that it is desired to test H_0: $\theta = \theta_0$ versus H_1: $\theta = \theta_1$ ($\theta_0 < \theta_1$). The cost of each observation is c, and the decision loss is "0-K_i" loss. Let π_i denote the prior probability of H_i.

(a) Show that the Bayes sequential procedure is one of the procedures $\mathbf{d}_J$ (J a nonnegative integer) defined as follows. The procedure $\mathbf{d}_0$ is simply the immediate Bayes decision. For $J \geq 1$, $\mathbf{d}_J$ is the procedure which starts sampling; stops sampling when stage $n = J$ is reached, deciding H_0 if $x_J \leq \theta_1$ and deciding H_1 otherwise; and at stage $n < J$ is given by the following;

$$\begin{aligned} &\text{if } x_n \leq \theta_1, \quad \text{stop sampling and decide } H_0; \\ &\text{if } x_n > \theta_1, \quad \text{continue sampling.} \end{aligned}$$

(b) For $\mathbf{d}_J$ (with $J \geq 1$), show that

$$\beta(\theta) = P_\theta(\text{deciding } H_1) = \begin{cases} 1 & \text{if } \theta \geq \theta_1, \\ e^{J(\theta - \theta_1)} & \text{if } \theta < \theta_1, \end{cases}$$

and that

$$E_\theta N = \begin{cases} J & \text{if } \theta \geq \theta_1, \\ \dfrac{1 - \exp\{J(\theta - \theta_1)\}}{1 - \exp\{(\theta - \theta_1)\}} & \text{if } \theta < \theta_1. \end{cases}$$

(c) if $0 < \pi_0 < 1$, show that, for small enough c, the Bayes sequential procedure is $\mathbf{d}_{J^*}$, where

$$J^* \cong (\theta_1 - \theta_0)^{-1} \log\left\{\frac{(\theta_1 - \theta_0)\pi_0}{\pi_1}\left[\frac{K_1}{c} - \frac{1}{1 - \exp(\theta_0 - \theta_1)}\right]\right\}.$$

40. Assume that $X_1, X_2, \ldots$ is a sequential sample from the $\mathscr{E}(\theta)$ density, and that it is desired to test H_0: $\theta = \frac{1}{2}$ versus H_1: $\theta = 1$.
(a) Using the Wald approximations, find the SPRT for which $\alpha_0 = \alpha_1 = 0.05$.
(b) Determine $\tilde{\beta}(\frac{2}{3})$ and $\tilde{\beta}(\log 2)$ for the SPRT in (a).
(c) Calculate $\tilde{E}_{1/2}N$, $\tilde{E}_1 N$, $\tilde{E}_{\log 2}N$, and $\tilde{E}_{2/3}N$ for the SPRT in (a).

41. Assume that $X_1, X_2, \ldots$ is a sequential sample from the $\mathscr{P}(\theta)$ density, and that it is desired to test H_0: $\theta = 1$ versus H_1: $\theta = 2$.
(a) Using the Wald approximations, find the SPRT for which $\alpha_0 = 0.05$ and $\alpha_1 = 0.1$.
(b) Calculate $\tilde{\beta}(\theta)$ and $\tilde{E}_\theta N$ at $\theta = 1$, $\theta = 2$, and $\theta = 1/\log 2$, for the SPRT in (a).

42. Assume that $X_1, X_2, \ldots$ is a sequential sample from the $\mathscr{G}e(\theta)$ density, and that it is desired to test H_0: $\theta = \frac{1}{3}$ versus H_1: $\theta = \frac{2}{3}$.
(a) Using the Wald approximations, find the SPRT for which $\alpha_0 = \alpha_1 = 0.1$.
(b) Show that $t_\theta = \log([1 - \theta]^{-1} - 1)/\log 2$, and give formulas for $\tilde{\beta}(\theta)$ and $\tilde{E}_\theta N$.

43. Assume that $X_1, X_2, \ldots$ is a sequential sample from the $\mathscr{N}(0, \sigma^2)$ density, and that it is desired to test H_0: $\sigma^2 = 1$ versus H_1: $\sigma^2 = 2$.
(a) Using the Wald approximations, find the SPRT for which $\alpha_0 = \alpha_1 = 0.01$.
(b) Calculate $\tilde{\beta}(\sigma^2)$ and $\tilde{E}_{\sigma^2}N$ at $\sigma^2 = 1$, $\sigma^2 = 2$, and $\sigma^2 = 2 \log 2$, for the SPRT in (a).

44. Consider the situation of Example 15.
    (a) Show that, for the SPRT with $a=-jv$ and $b=kv$, $\beta(\theta)$ and $E_\theta N$ are given *exactly* by

$$\beta(\theta)=\begin{cases}\dfrac{1-[\theta/(1-\theta)]^j}{[(1-\theta)/\theta]^k-[\theta/(1-\theta)]^j} & \text{if } \theta\neq\frac{1}{2},\\[2ex] \dfrac{j}{k+j} & \text{if } \theta=\frac{1}{2},\end{cases}$$

$$E_\theta N=\begin{cases}\dfrac{1}{(2\theta-1)}\,\dfrac{k\{1-[\theta/(1-\theta)]^j\}-j\{[(1-\theta)/\theta]^k-1\}}{[(1-\theta)/\theta]^k-[\theta/(1-\theta)]^j} & \text{if } \theta\neq\frac{1}{2},\\[2ex] jk & \text{if } \theta=\frac{1}{2}.\end{cases}$$

    (b) If $\theta_0=\frac{1}{3}$, each observation costs $c=1$, the decision loss is "0–20" loss, and the prior probability of θ_0 is $\frac{1}{2}$, find the Bayes sequential procedure.

45. It is known that one of two given coins is fair, and that the other coin has probability $\frac{2}{3}$ of coming up heads when flipped. It is desired to determine which of the coins is the fair coin. The coins, tossed simultaneously and independently, produce a sequential sample $(X_1, Y_1), (X_2, Y_2), \ldots$, where X_i and Y_i are each 0 or 1, according as to whether the respective coin is a tail or a head. Determine the form of an SPRT for this problem, and calculate (exactly) the corresponding α_0, α_1, $E_{\theta_0}N$, and $E_{\theta_1}N$.

46. Prove part (ii) of Theorem 11.

47. Prove that (7.73) is true under the condition that $P_{\theta_j}(|Z_i|<\infty)=1$ for $j=0, 1$.

48. For $0<\alpha_0<1$ and $0<\alpha_1<1$, define $A^{**}=\alpha_1$, $B^{**}=1/\alpha_0$, and let α_0^{**} and α_1^{**} denote the true error probabilities of the SPRT with stopping boundaries A^{**} and B^{**}. Using Theorem 12, show that $\alpha_0^{**}<\alpha_0$ and $\alpha_1^{**}<\alpha_1$.

49. Consider the situation of Exercise 39.
    (a) For any $a<0$ and $b=J(\theta_1-\theta_0)$ (J a positive integer), determine the SPRT $\mathbf{d}_{a,b}$ and calculate the Wald approximations $\tilde{\alpha}_0$, $\tilde{\alpha}_1$, $\tilde{E}_{\theta_0}N$, and $\tilde{E}_{\theta_1}N$. (Note that the Z_i^* in (7.61) should be used.)
    (b) Show that as $a\to-\infty$, the Wald approximations to α_i and $E_{\theta_i}N$ $(i=0, 1)$ converge to the correct values.

50. For the situation considered in Section 7.5, namely testing a simple null hypothesis versus a simple alternative hypothesis, show that the asymptotically pointwise optimal procedure defined in Subsection 7.4.10 (you may assume (7.44) is satisfied) is an SPRT. Find the stopping boundaries of this SPRT, and compare with the small c approximations in (7.84).

51. Assume that $X_1, X_2, \ldots$ is a sequential sample from a $\mathcal{N}(\theta, 1)$ density, and that it is desired to test H_0: $\theta=0$ versus H_1: $\theta=1$. The loss in incorrectly deciding H_0 is 30, in incorrectly deciding H_1 is 15, and is 0 otherwise. The cost of each observation is $c=0.001$, and the prior probability that θ equals 0 is $\frac{2}{3}$. Find the Bayes sequential procedure. (You may use the small c approximations to the stopping boundaries, given in (7.84).)

52. In the situation of Exercise 42, assume that the decision loss is "0–20" loss, that each observation costs $c=0.001$, and that the prior probability that θ equals $\frac{1}{3}$ is $\frac{1}{3}$. Find the Bayes sequential procedure. (You may use the small c approximations to the stopping boundaries, given in (7.84).)

Section 7.6

53. In the situation of Exercise 4, show that a minimax sequential procedure is to choose the integer n which minimizes $n^{-1}+nc$, take a sample of size n, and estimate θ by $n^{-1}\sum_{i=1}^{n} x_i$.

54. In the situation of Exercise 5, show that a minimax sequential procedure is to choose the integer n which minimizes $n^{-1}+nc$, take a sample of size n, and estimate θ by $n^{-1}\sum_{i=1}^{n} x_i$.

55. In the situation of Exercise 6, show that a minimax sequential procedure is to choose the integer n which minimizes $(n+1)^{-1}+nc$, take a sample of size n, and estimate θ by $(n+1)^{-1}\sum_{i=1}^{n} x_i$.

56. In the situation of Exercise 7, show that a minimax sequential procedure is to choose the integer n which minimizes $(n+1)^{-1}+nc$, take a sample of size n, and estimate θ by $(n+1)^{-1}\sum_{i=1}^{n} x_i$.

57. In the situation of Example 13, show that the following procedure is minimax: stop sampling for the first n for which $(b_n^*-a_n^*)^2\leq(2c)^{1/3}$, and estimate θ by $(a_n^*+b_n^*)/2$, where $a_n^*=\max_{1\leq i\leq n}\{x_i\}-\frac{1}{2}$ and $b_n^*=\min_{1\leq i\leq n}\{x_i\}+\frac{1}{2}$.

Section 7.7

58. Suppose that $X_1, X_2, \ldots$ is a sequential sample from a $\mathcal{N}(\theta, 1)$ distribution, and that it is desired to test H_0: $\theta=0$ versus H_1: $\theta\neq 0$. The experimenter reports that he used a proper stopping rule, and obtained the data 3, −1, 2, 1.
    (a) What could a frequentist conclude?
    (b) What could a conditionalist conclude?

59. Prove that (7.88) is true if the indicated densities $\{f_n\}$ exist.

60. Let $X_1, X_2, \ldots$ be a sequential sample from the $\mathcal{P}(\theta)$ density. Suppose the stopping rule is to stop sampling at time $n\geq 2$ with probability $[\sum_{i=1}^{n-1} x_i/\sum_{i=1}^{n} x_i]$, $n=2, 3, \ldots$ (define $0/0=1$). Suppose the first five observations are $\{3, 1, 2, 5, 7\}$, and sampling then stops. Find the likelihood function for θ.

61. To see that classical significance is affected by interim analysis, suppose we can observe independent $\mathcal{N}(\theta, 1)$ observations, X_1 and X_2. We are interested in the significance level for testing H_0: $\theta=0$ versus H_1: $\theta\neq 0$.
    (a) For the stopping rule "take both observations," show that an $\alpha=0.05$ level test is to reject if
    $$|X_1+X_2|>(1.96)(\sqrt{2})\cong 2.77.$$
    (b) For the stopping rule "take only the first observation," show that an $\alpha=0.05$ level test is to reject if $|X_1|>1.96$.
    (c) For the procedure "observe X_1 and stop and reject if $|X_1|>1.96$; otherwise observe X_2 and reject if $|X_1+X_2|>2.77$," show that the significance level (the probability of rejecting when $\theta=0$) is 0.0831.

62. In the situation of Example 20, determine the $100(1-\alpha)\%$ HPD credible set for θ when $\theta=0$ has prior probability $\pi_0>0$ and the $\theta\neq 0$ have prior density $(1-\pi_0)$ times a $\mathcal{N}(0, A)$ density. (Note that the set will for sure include the point $\theta=0$.)

63. In the situation of Example 21, verify that δ is inadmissible if it satisfies (7.90).

Section 7.8

64. Assume that $X_1, X_2, \ldots$ is a sequential sample from a $\mathcal{N}(\theta, \sigma^2)$ density (σ^2 known), and that it is desired to estimate θ under squared-error decision loss. It is possible, at any stage, to take observations in batches of any size. A batch of m observations costs $c \log(m+1)$. If the parameter θ has a $\mathcal{N}(\mu, \tau^2)$ prior density, find the optimal sampling procedure.

65. It is desired to test the null hypothesis that a sequential sample has a $\mathcal{U}(0, 2)$ common density versus the alternative hypothesis that the common density is $\mathcal{U}(1, 3)$. The decision loss is 0 for a correct decision and 20 for an incorrect decision. It is possible, at any stage, to take observations in batches. A batch of m observations costs $m^{1/2}$. The prior probability of each hypothesis is $\frac{1}{2}$.

    Let $\mathbf{d}_0$ denote the procedure which makes an immediate Bayes decision. Let $\mathbf{d}_m$ denote the procedure which takes successive samples (batches) of size m, stopping and deciding H_0 (deciding H_1) if any observation in the batch is less than one (greater than two), and taking another batch otherwise.
    (a) Why can attention be restricted to consideration of $\mathbf{d}_0$ and the $\mathbf{d}_m$?
    (b) Show that the Bayes risk of $\mathbf{d}_m$ is
    $$r(\pi, \mathbf{d}_m)=\frac{m^{1/2}}{1-2^{-m}}.$$
    (c) Show that $\mathbf{d}_2$ is the optimal procedure.

CHAPTER 8

Complete and Essentially Complete Classes

We have previously observed that it is unwise to repeatedly use an inadmissible decision rule. (The possible exception is when an inadmissible rule is very simple and easy to use, and is only slightly inadmissible.) It is, therefore, of interest to find, for a given problem, the class of acceptable (usually admissible) decision rules. Such a class is often much easier to work with, say in finding a sequential Bayes, minimax or a Γ-minimax decision rule, than is the class of all decision rules. In this chapter, we discuss several of the most important situations in which simple reduced classes of decision rules have been obtained. Unfortunately, the subject tends to be quite difficult mathematically, and so we will be able to give only a cursory introduction to some of the more profound results.

8.1. Preliminaries

We begin with definitions of the needed concepts.

Definition 1. A class $\mathscr{C}$ of decision rules is said to be *essentially complete* if, for any decision rule δ not in $\mathscr{C}$, there is a decision rule $\delta' \in \mathscr{C}$ which is R-better than or R-equivalent to δ.

Definition 2. A class $\mathscr{C}$ of decision rules is said to be *complete* if, for any decision rule δ not in $\mathscr{C}$, there is a decision rule $\delta' \in \mathscr{C}$ which is R-better than δ.

Definition 3. A class $\mathscr{C}$ of decision rules is said to be *minimal complete* if $\mathscr{C}$ is complete and if no proper subset of $\mathscr{C}$ is complete.

The following lemmas are instructive in the assimilation of the above concepts. The proofs of the first two are left as exercises.

Lemma 1. *A complete class must contain all admissible decision rules.*

Lemma 2. *If an admissible decision rule δ is not in an essentially complete class $\mathscr{C}$, then there must exist a decision rule δ' in $\mathscr{C}$ which is R-equivalent to δ.*

Lemma 3. *If a minimal complete class $\mathscr{C}$ exists, it is exactly the class of admissible decision rules.*

PROOF. Lemma 1 implies that the class of admissible rules is a subset of $\mathscr{C}$. It remains only to show that if $\delta \in \mathscr{C}$, then δ is admissible. This will be established by contradiction. Thus assume that $\delta \in \mathscr{C}$ and that δ is inadmissible. Note first that there then exists a $\delta' \in \mathscr{C}$ which is R-better than δ. This is because there exists a rule δ'' which is R-better than δ (inadmissibility of δ), and also a rule $\delta' \in \mathscr{C}$ which is R-better than δ'' if $\delta'' \notin \mathscr{C}$ (completeness of $\mathscr{C}$). Hence let $\mathscr{C}'$ be the set of all rules in $\mathscr{C}$, except δ. Clearly $\mathscr{C}'$ is a complete class, since if δ was used to improve upon a rule not in $\mathscr{C}$, then δ' could just as well have been used. But $\mathscr{C}'$ is a proper subset of $\mathscr{C}$, contradicting the assumption that $\mathscr{C}$ is minimal complete. Hence δ must be admissible. □

The above definitions and results apply to sequential problems as well as to fixed sample size problems. We will, for the most part, consider only fixed sample size situations, however, since sequential problems tend to be more complex. Indeed only in Subsection 8.6.3 will the sequential situation be explicitly considered.

8.2. Complete and Essentially Complete Classes from Earlier Chapters

8.2.1. Decision Rules Based on a Sufficient Statistic

Theorem 1 in Section 1.7 established that the class of randomized decision rules based on a sufficient statistic is an essentially complete class. This class will usually not be a complete class. The widespread acceptance of the principle of sufficiency indicates, however, that in practice it is quite acceptable to reduce to an essentially complete class, rather than to a complete class. Indeed, an essentially complete class will typically be smaller than a complete class, and hence may be preferable.

8.2.2. Nonrandomized Decision Rules

In certain situations, the class of nonrandomized decision rules is a complete (or essentially complete) class. One such situation is that of convex loss, as was indicated by Theorem 3 in Section 1.8. Another common situation in which this is the case is that of finite action problems in which the densities $f(x|\theta)$ are all continuous. Certain results of this nature can be found in Dvoretsky, Wald, and Wolfowitz (1951) and Balder (1980).

8.2.3. Finite Θ

Several complete class theorems for situations involving finite Θ were given in Chapter 5. The most basic such result was Theorem 8 in Subsection 5.2.4, which can be restated as follows.

Theorem 1. *If Θ is finite and the risk set S is bounded from below and closed from below, then the set $\mathscr{C}$ of all decision rules which have risk points on $\lambda(S)$, the lower boundary of S, is a minimal complete class.*

PROOF. The fact that $\mathscr{C}$ is complete is an immediate consequence of Theorem 8 of Chapter 5. That $\mathscr{C}$ is minimal complete follows from Theorem 9 of Chapter 5 and Lemma 1 of this chapter. □

The following theorem is an immediate consequence of Theorem 1 above and Theorem 11 of Chapter 5.

Theorem 2. *If Θ is finite and the risk set S is bounded from below and closed from below, then the set of Bayes decision rules is a complete class, and the set of admissible Bayes decision rules is a minimal complete class.*

Theorem 2 is typically the more useful of the above complete class theorems. This is because Bayes rules are usually quite easy to characterize in problems which involve finite Θ. To apply the above theorems it is, of course, necessary to verify that S is closed from below and bounded from below. Several general conditions which imply this were given in Subsection 5.2.4.

8.2.4. The Neyman–Pearson Lemma

The famous Neyman–Pearson lemma (Neyman and Pearson (1933)) was, to a large extent, responsible for establishing the school of thought that led to decision theory. The Neyman–Pearson lemma can indeed be considered to have been the first complete class theorem. It is concerned with testing a simple null hypothesis H_0: $\theta=\theta_0$, versus a simple alternative hypothesis, H_1: $\theta=\theta_1$. For convenience, we will assume that the relevant densities, $f(x|\theta_0)$ and $f(x|\theta_1)$, of the observation X are either both continuous or both discrete. The loss is taken to be "0-K_i" loss (i.e., incorrectly deciding a_i costs K_i, while a correct decision costs 0). As usual in testing, a decision rule will be represented by a test function, $\phi(x)$, which denotes the probability of rejecting the null hypothesis when x is observed. (Note that randomized decision rules are thus being considered.) Also, $\alpha_0(\phi)$ and $\alpha_1(\phi)$ will denote the probabilities of type I and type II error respectively, and we define

$$\alpha^*=P_{\theta_0}(f(X\,|\,\theta_1)>0).$$

This problem was discussed in Subsection 5.3.3, which the reader might profitably reread at this point. Indeed the Neyman–Pearson lemma is basically a rigorous statement of the ideas discussed in that subsection.

Theorem 3 (Neyman–Pearson Lemma). *The tests of the form*

$$\phi(x)=\begin{cases}1 & \text{if } f(x|\theta_1)>Kf(x|\theta_0),\\ \gamma(x) & \text{if } f(x|\theta_1)=Kf(x|\theta_0),\\ 0 & \text{if } f(x|\theta_1)<Kf(x|\theta_0),\end{cases} \tag{8.1}$$

where $0\le\gamma(x)\le 1$ *if* $0<K<\infty$ *and* $\gamma(x)=0$ *if* $K=0$, *together with the test*

$$\phi(x)=\begin{cases}1 & \text{if } f(x|\theta_0)=0,\\ 0 & \text{if } f(x|\theta_0)>0,\end{cases} \tag{8.2}$$

(corresponding to $K=\infty$ *above), form a minimal complete class of decision rules. The subclass of such tests with* $\gamma(x)\equiv\gamma$ *(a constant) is an essentially complete class.*

For any $0\le\alpha\le\alpha^*$, *there exists a test* ϕ *of the form* (8.1) *or* (8.2) *with* $\alpha_0(\phi)=\alpha$, *and any such test is a most powerful test of size* α *(i.e., among all tests* ϕ *with* $\alpha_0(\phi)\le\alpha$, *such a test minimizes* $\alpha_1(\phi)$*).*

PROOF. From Lemma 10 of Subsection 5.3.3 and Theorem 2 of this chapter, we know that the admissible Bayes rules form a minimal complete class. As in Subsection 5.3.3, it is easy to check that if π_0, the prior probability of θ_0, satisfies $0<\pi_0<1$, then the Bayes rules are precisely the tests of the form (8.1) with $0<K<\infty$. When $\pi_0=0$, any test which satisfies $\phi(x)=1$ if $f(x|\theta_1)>0$ is a Bayes test, since then $r(\pi,\phi)=\alpha_1(\phi)=0$. Only the test of the form (8.1) with $K=0$ and $\gamma(x)=0$ is admissible, however, since it minimizes $\alpha_0(\phi)$ among all tests with $\alpha_1(\phi)=0$. When $\pi_0=1$, it can similarly be shown that the test in (8.2) is the admissible Bayes test. Thus the tests in (8.1) and (8.2) are precisely the admissible Bayes tests, and hence form a minimal complete class.

To establish the essential completeness of the class of tests of the form (8.1) and (8.2) with $\gamma(x)$ constant, it suffices to show that for any test of the form (8.1), there is an R-equivalent test of the same form with $\gamma(x)$ constant. Thus let ϕ be of the form (8.1), with given K and $\gamma(x)$, and define $A_0=\{x: f(x|\theta_1)<Kf(x|\theta_0)\}$, $A_1=\{x: f(x|\theta_1)>Kf(x|\theta_0)\}$, and $A_2=\{x: f(x|\theta_1)=Kf(x|\theta_0)\}$. Note that

$$\alpha_0(\phi)=E_{\theta_0}[\phi(X)]=P_{\theta_0}(A_1)+\int_{A_2}\gamma(x)dF(x|\theta_0) \tag{8.3}$$

and

$$\begin{aligned}\alpha_1(\phi)=E_{\theta_1}[1-\phi(X)]&=P_{\theta_1}(A_0)+\int_{A_2}(1-\gamma(x))dF(x|\theta_1)\\ &=P_{\theta_1}(A_0\cup A_2)-\int_{A_2}\gamma(x)dF(x|\theta_1)\\ &=P_{\theta_1}(A_0\cup A_2)-K\int_{A_2}\gamma(x)dF(x|\theta_0).\end{aligned} \tag{8.4}$$

Assume now that $\beta=\int_{A_2}dF(x|\theta_0)>0$, since if $\beta=0$ the result is vacuously correct. Consider the new test, ϕ', which is of the form (8.1) with the given K and

$$\gamma'(x)\equiv\beta^{-1}\int_{A_2}\gamma(x)dF(x|\theta_0).$$

It is easy to check, using (8.3) and (8.4), that $\alpha_0(\phi)=\alpha_0(\phi')$ and $\alpha_1(\phi)=\alpha_1(\phi')$. Hence $R(\theta_0,\phi)=K_1\alpha_0(\phi)=R(\theta_0,\phi')$ and $R(\theta_1,\phi)=K_0\alpha_1(\phi)=R(\theta_1,\phi')$, establishing the R-equivalence of ϕ and ϕ'.

The conclusion that the tests in (8.1) and (8.2) are most powerful of their size follows immediately from the fact that they are admissible tests. The existence of most powerful size α tests of the form (8.1) and (8.2), for $0\le\alpha\le\alpha^*$, follows from (8.3) and the fact that the risk set is closed and convex. □

Example 18 in Subsection 5.3.3 demonstrates an application of the Neyman–Pearson lemma.

8.3. One-Sided Testing

From the Neyman–Pearson lemma it is possible to derive complete class theorems for certain testing situations that concern what are called one-sided tests. Such tests involve testing hypotheses of the form H_0: $\theta\in\Theta_0$ versus H_1: $\theta\in\Theta_1$, where Θ_0 and Θ_1 are subsets of the real line and Θ_0 is to the left (or right) of Θ_1. We begin the development by considering "0–1" loss, so that the risk function of a test ϕ is determined by the error probabilities of the test, or equivalently the power function

$$\beta_\phi(\theta) = E_\theta[\phi(X)] = P_\theta(\text{rejecting } H_0).$$

The crucial concept needed is that of a uniformly most powerful test of size α.

Definition 4. A test ϕ of H_0: $\theta \in \Theta_0$ versus H_1: $\theta \in \Theta_1$ is said to have *size* α if

$$\sup_{\theta \in \Theta_0} E_\theta[\phi(X)] = \alpha.$$

A test ϕ_0 is said to be *uniformly most powerful* (UMP) of size α if it is of size α, and if, for any other test ϕ of size at most α,

$$E_\theta[\phi_0(X)] \geq E_\theta[\phi(X)]$$

for all $\theta \in \Theta_1$.

The concept of a uniformly most powerful test is a fairly natural extension to composite hypotheses of the idea of a most powerful test for simple hypotheses. Whereas most powerful tests virtually always exist, however, there is no reason to expect that UMP tests need exist, and indeed they only exist for certain special distributions and hypotheses. The most important class of distributions for which they sometimes exist is the class of distributions with monotone likelihood ratio. In the discussion and applications of monotone likelihood ratio, we will assume that the distributions have densities (either continuous or discrete) on R^1, and that Θ is a subset of R^1.

Definition 5. The distribution of X is said to have *monotone likelihood ratio* if, whenever $\theta_1 < \theta_2$, the likelihood ratio

$$\frac{f(x|\theta_2)}{f(x|\theta_1)}$$

is a nondecreasing function of x on the set for which at least one of the densities is nonzero. (As usual, a nonzero constant divided by zero is defined to be infinity.)

The distribution in the above definition would perhaps more logically be called a distribution with *nondecreasing likelihood ratio.* If faced with a distribution for which the likelihood ratio is nonincreasing, one need only make the change of variables $Y = -X$ or the reparametrization $\eta = -\theta$ to arrive at a distribution with nondecreasing likelihood ratio.

EXAMPLE 1. The most common distributions with monotone likelihood ratio are those from the *one-parameter exponential family*, i.e., those with densities of the form

$$f(x|\theta) = c(\theta)h(x)\exp\{Q(\theta)T(x)\},$$

where θ and x are in R^1 and c, h, Q, and T are real valued functions. Clearly

$$\frac{f(x|\theta_2)}{f(x|\theta_1)}=\frac{c(\theta_2)}{c(\theta_1)}\exp\{[Q(\theta_2)-Q(\theta_1)]T(x)\}.$$

For $\theta_1<\theta_2$, this is nondecreasing in x(on the set where $h(x)$ is nonzero) providing both $Q(\theta)$ and $T(x)$ are nondecreasing or nonincreasing. Note that if one defines $Y=T(X)$ and $\eta=Q(\theta)$, then the distribution of Y given η will always have monotone likelihood ratio.

The one-parameter exponential family includes such standard distributions as the normal (with either mean or variance fixed), the Poisson, the binomial, and the gamma (either parameter fixed). Also, a random sample from a distribution in the one-parameter exponential family will admit a sufficient statistic which itself has a distribution in the one-parameter exponential family. This ensures a wide range of applicability for this class of distributions.

EXAMPLE 2. The $\mathcal{U}(\theta, \theta+1)$ distribution has monotone likelihood ratio. To see this, note that if $\theta_1<\theta_2$, then

$$\frac{f(x|\theta_2)}{f(x|\theta_1)}=\begin{cases}\infty & \text{if } \theta_1+1\le x \text{ and } \theta_2<x<\theta_2+1,\\ 1 & \text{if } \theta_1<x<\theta_1+1 \text{ and } \theta_2<x<\theta_2+1,\\ 0 & \text{if } \theta_1<x<\theta_1+1 \text{ and } x\le\theta_2.\end{cases}$$

This is clearly nondecreasing in x.

EXAMPLE 3. The $\mathscr{C}(\theta, 1)$ distribution does *not* have monotone likelihood ratio, as can be seen by noting that

$$\frac{f(x|\theta_2)}{f(x|\theta_1)}=\frac{1+(x-\theta_1)^2}{1+(x-\theta_2)^2}$$

converges to one as $x\to+\infty$ or as $x\to-\infty$.

The following theorem presents the key result concerning UMP tests.

Theorem 4. *Assume that the distribution of X has monotone likelihood ratio, and that it is desired to test H_0: $\theta\le\theta_0$ versus H_1: $\theta>\theta_0$. Consider tests of the form*

$$\phi(x)=\begin{cases}1 & \text{if } x>x_0,\\ \gamma & \text{if } x=x_0,\\ 0 & \text{if } x<x_0,\end{cases} \tag{8.5}$$

where $-\infty\le x_0\le\infty$ and $0\le\gamma\le 1$. The following facts are true:

(i) *The power function, $\beta_\phi(\theta)=E_\theta[\phi(X)]$, is nondecreasing in θ.*
(ii) *Any such test is* UMP *of its size, providing its size is not zero.*
(iii) *For any $0\le\alpha\le 1$, there exists a test of the form* (8.5) *which is* UMP *of size α.*

PROOF. Let θ_1 and θ_2 be any points such that $\theta_1 < \theta_2$. The Neyman–Pearson lemma states that any test of the form

$$\phi(x) = \begin{cases} 1 & \text{if } f(x|\theta_2) > Kf(x|\theta_1), \\ \gamma(x) & \text{if } f(x|\theta_2) = Kf(x|\theta_1), \\ 0 & \text{if } f(x|\theta_2) < Kf(x|\theta_1), \end{cases} \tag{8.6}$$

for $0 \le K < \infty$, is most powerful of its size for testing H_0: $\theta = \theta_1$ versus H_1: $\theta = \theta_2$. We must first show that, since the distribution of X has monotone likelihood ratio, any test of the form (8.5) can be written as in (8.6), providing its size is nonzero. To see this when $x_0 = -\infty$, simply set $K = 0$ and $\gamma(x) = 1$ in (8.6). To verify this correspondence when $x_0 > -\infty$, note that the assumptions of monotone likelihood ratio and nonzero size imply that $x_0 < \infty$ and that $f(x_0|\theta_1) > 0$. Hence, simply define $K = f(x_0|\theta_2)/f(x_0|\theta_1)$ and

$$\gamma(x) = \begin{cases} 1 & \text{if } x > x_0 \quad \text{and} \quad f(x|\theta_2) = Kf(x|\theta_1), \\ \gamma & \text{if } x = x_0, \\ 0 & \text{if } x < x_0 \quad \text{and} \quad f(x|\theta_2) = Kf(x|\theta_1). \end{cases}$$

From the assumption of monotone likelihood ratio it follows that the test in (8.6), with this choice of K and $\gamma(x)$, is the same as the test in (8.5).

If now $\phi(x)$ is a most powerful size $\alpha (\alpha > 0)$ test of θ_1 versus θ_2, then since $\phi_0(x) \equiv \alpha$ is also a size α test, it follows that

$$E_{\theta_2}[\phi(X)] \ge E_{\theta_2}[\phi_0(X)] = \alpha = E_{\theta_1}[\phi(X)].$$

Of course, when $E_{\theta_1}[\phi(X)] = \alpha = 0$, it is trivially true that $E_{\theta_2}[\phi(X)] \ge E_{\theta_1}[\phi(X)]$. Hence in all cases, a test ϕ of the form (8.5) satisfies $E_{\theta_2}[\phi(X)] \ge E_{\theta_1}[\phi(X)]$, when $\theta_1 < \theta_2$. This establishes the first conclusion of the theorem.

To prove the second part of the theorem, consider the problem of testing H_0: $\theta = \theta_0$ versus H_1: $\theta = \theta_1$, where $\theta_1 > \theta_0$. By the above argument and the Neyman–Pearson lemma, a test ϕ of the form (8.5) is most powerful of its size, α, providing $\alpha > 0$. Since $\alpha = E_{\theta_0}[\phi(X)]$, this is true no matter what θ_1 is. Hence the tests in (8.5) are UMP size α tests of H_0: $\theta = \theta_0$ versus H_1: $\theta > \theta_0$, providing $\alpha > 0$. Finally, since $\beta_\phi(\theta)$ is nondecreasing, $E_\theta[\phi(X)] \le \alpha$ for all $\theta < \theta_0$. Thus ϕ is a size α test of H_0: $\theta \le \theta_0$ versus H_1: $\theta > \theta_0$, and part (ii) of the theorem follows directly.

Part (iii) of the theorem is established, for $\alpha > 0$, by finding a most powerful size α test of H_0: $\theta = \theta_0$ versus H_1: $\theta = \theta_1$ (which exists by the Neyman–Pearson lemma), writing the test in the form (8.5), and applying part (ii). For $\alpha = 0$, the Neyman–Pearson lemma gives that a most powerful test of H_0: $\theta = \theta_0$ versus H_1: $\theta = \theta_1 (\theta_1 > \theta_0)$ is

$$\phi(x) = \begin{cases} 1 & \text{if } f(x|\theta_0) = 0, \\ 0 & \text{if } f(x|\theta_0) > 0. \end{cases} \tag{8.7}$$

Now $\{x: f(x|\theta_0) > 0\} = \{x: f(x|\theta_1)/f(x|\theta_0) < \infty\}$, and this latter set, by the definition of monotone likelihood ratio, is either $\{x: x < x_0\}$ or $\{x: x \le x_0\}$, where $-\infty < x_0 \le \infty$ is some constant. It follows that the test in (8.7) can be written as in (8.5). The proof that this test is uniformly most powerful proceeds exactly along the line of the proof of part (ii). □

EXAMPLE 4. Assume that $X_1, X_2, \ldots, X_n$ is a sample from the $\mathcal{N}(\theta, 1)$ distribution, and that it is desired to test H_0: $\theta \le \theta_0$ versus H_1: $\theta > \theta_0$. Clearly $\bar{X} \sim \mathcal{N}(\theta, 1/n)$ is sufficient for θ, and its distribution has monotone likelihood ratio. Using Theorem 4 and the continuity of the density, it follows that a UMP size α test is to reject H_0 when $\bar{x} > x_0 = \theta_0 + n^{-1/2} z(1-\alpha)$, where $z(1-\alpha)$ is the $(1-\alpha)$-fractile of the $\mathcal{N}(0, 1)$ distribution.

Note that, by symmetry, Theorem 4 also applies to testing H_0: $\theta \ge \theta_0$ versus H_1: $\theta < \theta_0$, with, of course, the form of the tests in (8.5) being changed to

$$\phi(x) = \begin{cases} 1 & \text{if } x < x_0, \\ \gamma & \text{if } x = x_0, \\ 0 & \text{if } x > x_0. \end{cases}$$

The following corollary to Theorem 4 will lead directly to the desired complete class theorem for one-sided testing.

Corollary 1. *Assume that the distribution of X has monotone likelihood ratio. For every test ϕ and every $\theta_0 \in \Theta$, there then exists a test ϕ' of the form* (8.5) *for which*

$$\begin{aligned} E_\theta[\phi'(X)] &\le E_\theta[\phi(X)] \quad \text{for } \theta \le \theta_0, \\ E_\theta[\phi'(X)] &\ge E_\theta[\phi(X)] \quad \text{for } \theta > \theta_0. \end{aligned} \tag{8.8}$$

PROOF. Define $\alpha = E_{\theta_0}[\phi(X)]$, and choose ϕ' to be a size α test of the form (8.3). Since ϕ' is UMP for testing H_0: $\theta \le \theta_0$ versus H_1: $\theta > \theta_0$, it is true that $E_\theta[\phi'(X)] \ge E_\theta[\phi(X)]$ for $\theta > \theta_0$. Since, by symmetry, $1 - \phi'$ is UMP of size $1 - \alpha$ for testing H_0: $\theta \ge \theta_0$ versus H_1: $\theta < \theta_0$, it similarly follows that $E_\theta[1 - \phi'(X)] \ge E_\theta[1 - \phi(X)]$ for $\theta < \theta_0$, completing the proof. □

We finally come to the complete class theorem. It will be given for a fairly general loss function that embodies the spirit of one-sided testing. Assume that $\mathcal{A} = \{a_0, a_1\}$, and that $L(\theta, a_i)(i = 0, 1)$ satisfies

$$\begin{aligned} L(\theta, a_1) - L(\theta, a_0) &\ge 0 \quad \text{if } \theta < \theta_0, \\ L(\theta, a_1) - L(\theta, a_0) &\le 0 \quad \text{if } \theta > \theta_0. \end{aligned} \tag{8.9}$$

Clearly action a_0 is preferred for $\theta < \theta_0$, while a_1 is preferred for $\theta > \theta_0$. Action a_0 thus corresponds to accepting H_0: $\theta < \theta_0$, while action a_1 corresponds to accepting H_1: $\theta > \theta_0$. Because of this correspondence, we will continue to write a decision rule as a test, $\phi(x)$, which should now, however, be interpreted as the probability of taking action a_1 after observing x. Note that, for $\theta = \theta_0$, (8.9) places no restriction on the loss. The complete class theorem follows, and is due to Karlin and Rubin (1956).

Theorem 5. *If the distribution of X has monotone likelihood ratio and the loss function is as in* (8.9), *then*

(i) *the class of tests in* (8.5) *is an essentially complete class; and*
(ii) *any test of the form* (8.5) *is admissible, providing $\{x: f(x|\theta) > 0\}$ is independent of θ and there exist numbers θ_1 and θ_2 in Θ, with $\theta_1 \le \theta_0 \le \theta_2$, such that $L(\theta_1, a_1) - L(\theta_1, a_0) > 0$ and $L(\theta_2, a_1) - L(\theta_2, a_0) < 0$.*

PROOF. Observe first that the risk function of a test ϕ^* can be written as

$$\begin{aligned} R(\theta, \phi^*) &= L(\theta, a_0)(1 - E_\theta[\phi(X)]) + L(\theta, a_1)E_\theta[\phi(X)] \\ &= L(\theta, a_0) + [L(\theta, a_1) - L(\theta, a_0)]E_\theta[\phi(X)]. \end{aligned} \tag{8.10}$$

For any test ϕ, let ϕ' be the test of the form (8.5) which satisfies (8.8). By (8.10),

$$R(\theta, \phi) - R(\theta, \phi') = [L(\theta, a_1) - L(\theta, a_0)](E_\theta[\phi(X)] - E_\theta[\phi'(X)]). \tag{8.11}$$

For $\theta > \theta_0$, (8.8) and (8.9) imply that both terms on the right-hand side of (8.11) are nonpositive; for $\theta < \theta_0$, both terms are nonnegative; and for $\theta = \theta_0$, $E_\theta[\phi(X)] = E_\theta[\phi'(X)]$. Hence $R(\theta, \phi) \ge R(\theta, \phi')$, implying that ϕ' is R-better than or R-equivalent to ϕ. The class of tests of the form (8.5) is thus essentially complete. The proof of part (ii) of the theorem will be left as an exercise. □

For an application of the above complete class theorem, see Example 14 in Subsection 5.3.2.

8.4. Monotone Decision Problems

The key features of the decision problem discussed in the preceding section were that the distribution of X had monotone likelihood ratio and that the actions and corresponding losses had a particular order. In this section, we consider more general problems with these same basic properties, and again obtain simple and useful essentially complete classes of decision rules. The case in which the action space is finite is considered first.

8.4.1. Monotone Multiple Decision Problems

A multiple decision problem is a problem in which only a finite set of actions, $\mathscr{A} = \{a_1, a_2, \ldots, a_k\} (k \ge 2)$ is available. It will be convenient, in such problems, to continue the use of testing notation, so that a randomized decision rule will be represented by

$$\boldsymbol{\phi} = (\phi_1(x), \ldots, \phi_k(x)),$$

where $\phi_i(x)$ is the probability that action a_i is chosen when $X = x$ is observed. Note that $\sum_{i=1}^k \phi_i(x) = 1$ for all x, and that

$$R(\theta, \boldsymbol{\phi}) = \sum_{i=1}^{k} L(\theta, a_i) E_\theta[\phi_i(X)]. \tag{8.12}$$

A monotone multiple decision problem can be loosely thought of as a problem in which the parameter space is a subset of the real line and is divided into ordered intervals, with action a_i corresponding to the decision that θ is in the ith interval. When $k = 2$, this is simply the one-sided testing situation discussed in Section 8.3. A typical example of a monotone problem in which $k = 3$ is a one-sided testing problem in which a third action—indecision—is allowed. Imagine, for example, that θ represents the true difference in performance between two drugs being tested. It is quite standard in such a situation to decide either that drug 1 is better (a_1), that there is no conclusive evidence supporting either drug (a_2), or that drug 2 is better (a_3). These actions can be thought of as corresponding to three regions of the parameter space, say $(-\infty, \theta_1]$, (θ_1, θ_2), and $[\theta_2, \infty)$, although this interpretation is not strictly necessary. What is necessary, for the problem to be monotone, is that the loss must be larger the more "incorrect" the action is. In the above example, for instance, if drug 1 is indeed significantly better than drug 2, then a_2 should incur less loss than a_3, since it is closer to being correct. This notion is made precise in the following definition.

Definition 6. A multiple decision problem in which Θ is a subset of R^1 is said to be *monotone* if, for some ordering of $\mathscr{A}$, say $\mathscr{A} = \{a_1, \ldots, a_k\}$, there exist numbers $\theta_1 \le \theta_2 \le \cdots \le \theta_{k-1}$ (all in Θ) such that the loss function satisfies (for $i = 1, \ldots, k-1$)

$$\begin{aligned} L(\theta, a_i) - L(\theta, a_{i+1}) \le 0 \quad \text{for } \theta < \theta_i, \\ L(\theta, a_i) - L(\theta, a_{i+1}) \ge 0 \quad \text{for } \theta > \theta_i. \end{aligned} \tag{8.13}$$

When $k = 2$, the relationship in (8.13) is exactly that specified by (8.9). This does, therefore, appear to be a natural extension of the one-sided testing situation. Note also, as a direct consequence of the inequalities in (8.13), that action a_i is preferred if $\theta_{i-1} < \theta < \theta_i$.

The intuitively natural decision rules, for a monotone multiple decision problem in which the observation X is in R^1 and has a distribution with monotone likelihood ratio, are those rules which divide $\mathscr{X}$ into k ordered intervals, and choose action a_i if x is in the ith interval. For simplicity, we only consider the situation in which $\mathscr{X} = R^1$.

Definition 7. For a monotone multiple decision problem with $\mathscr{X}=R^1$, a decision rule $\boldsymbol{\phi}=(\phi_1,\ldots,\phi_k)$ is said to be *monotone* if

$$\phi_i(x)=\begin{cases}0 & \text{if } x<x_{i-1},\\ \gamma'_{i-1} & \text{if } x=x_{i-1},\\ 1 & \text{if } x_{i-1}<x<x_i,\\ \gamma_i & \text{if } x=x_i,\\ 0 & \text{if } x>x_i,\end{cases}$$

where $x_0, x_1, \ldots, x_k$ satisfy $-\infty=x_0\le x_1\le x_2\le\cdots\le x_k=+\infty$, and $0\le\gamma_i\le 1$ and $0\le\gamma'_i\le 1$ for $i=1,\ldots,k-1$.

The essentially complete class theorem for monotone multiple decision problems is due to Karlin and Rubin (1956), and can be considered to be a generalization of Theorem 5(i).

Theorem 6. *For a monotone multiple decision problem in which the distribution of X has monotone likelihood ratio and $\mathscr{X}=R^1$, the class of monotone decision rules is essentially complete.*

Proof. The theorem will be established by constructing, for any decision rule $\boldsymbol{\psi}=(\psi_1(x),\ldots,\psi_k(x))$, a monotone decision rule, $\boldsymbol{\phi}$, which is R-better than or R-equivalent to $\boldsymbol{\psi}$. To begin the construction, define

$$\begin{aligned}\psi^k(x)&=0,\\ \psi^j(x)&=\sum_{i=j+1}^{k}\psi_i(x)\quad \text{for } j=0,\ldots,k-1.\end{aligned}\tag{8.14}$$

Clearly $0\le\psi^j(x)\le 1$, so that ψ^j can be considered to be a test of H_0: $\theta\le\theta_j$ versus H_1: $\theta>\theta_j$. Using Corollary 1 in Section 8.3, it follows that for $j=1,\ldots,k-1$ there exists a one-sided test, ϕ^j, of the form

$$\phi^j(x)=\begin{cases}1 & \text{if } x>x_j,\\ \gamma''_j & \text{if } x=x_j,\\ 0 & \text{if } x<x_j,\end{cases}\tag{8.15}$$

such that

$$E_\theta[\phi^j(X)]-E_\theta[\psi^j(X)]\begin{cases}\le 0 & \text{for } \theta<\theta_j,\\ =0 & \text{for } \theta=\theta_j,\\ \ge 0 & \text{for } \theta>\theta_j.\end{cases}\tag{8.16}$$

(Expression (8.16) will also hold for $j=0$ and $j=k$ if we choose $\phi^0(x)\equiv 1$ and $\phi^k(x)\equiv 0$.) Since the ψ^j are nonincreasing in j, it can be shown that the ϕ^j can be chosen to be nonincreasing in j. Noting the form of the ϕ^j, it follows that we can assume that $x_1\le x_2\le\cdots\le x_{k-1}$ and $\gamma''_{j-1}\ge\gamma''_j$ if $x_{j-1}=x_j$. Hence the test $\boldsymbol{\phi}=(\phi_1,\ldots,\phi_k)$, defined by

$$\phi_i(x)=\phi^{i-1}(x)-\phi^i(x),\qquad i=1,\ldots,k,\tag{8.17}$$

is a monotone decision rule. To complete the proof, we show that $R(\theta, \boldsymbol{\psi}) \geq R(\theta, \boldsymbol{\phi})$.

Observe, using (8.12), (8.14), and (8.17), that

$$\begin{aligned} R(\theta, \boldsymbol{\psi}) - R(\theta, \boldsymbol{\phi}) &= \sum_{i=1}^{k} L(\theta, a_i)\{E_\theta[\psi_i(X)] - E_\theta[\phi_i(X)]\} \\ &= \sum_{i=1}^{k} L(\theta, a_i)\{(E_\theta[\phi^i(X)] - E_\theta[\psi^i(X)]) \\ &\quad - (E_\theta[\phi^{i-1}(X)] - E_\theta[\psi^{i-1}(X)])\} \\ &= \sum_{i=1}^{k-1} [L(\theta, a_i) - L(\theta, a_{i+1})](E_\theta[\phi^i(X)] - E_\theta[\psi^i(X)]). \end{aligned}$$

Using (8.13) and (8.16) it is easy to see that in each term of the above sum, the expressions $L(\theta, a_i) - L(\theta, a_{i+1})$ and $E_\theta[\phi^i(X)] - E_\theta[\psi^i(X)]$ are either the same sign or the last expression is zero. Hence $R(\theta, \boldsymbol{\psi}) \geq R(\theta, \boldsymbol{\phi})$ for all $\theta \in \Theta$, completing the proof. □

Theorem 5(ii) showed that a monotone decision rule is virtually always admissible when $k=2$. This need not be the case when $k \geq 3$, however, as is shown in the Exercises. Conditions under which monotone rules are admissible can be found in Karlin and Rubin (1956), Karlin (1956), Karlin (1957a), and Karlin (1957b).

The above theorem is particularly useful, in that the proof provides an explicit method of improving upon a nonmonotone decision rule. The construction of an improvement is quite simple since it depends on the loss function only through the "cut points" $\theta_1, \ldots, \theta_{k-1}$.

EXAMPLE 5. Based on the observation $X \sim \mathcal{N}(\theta, 1)$, it is desired to decide whether $\theta < -1$ (action a_1), $|\theta| \leq 1$ (action a_2), or $\theta > 1$ (action a_3). The loss function is

$$L(\theta, a_1) = \begin{cases} 0 & \text{if } \theta < -1, \\ \theta + 1 & \text{if } |\theta| \leq 1, \\ 2(\theta+1) & \text{if } \theta > 1, \end{cases}$$

$$L(\theta, a_2) = \begin{cases} 0 & \text{if } |\theta| \leq 1, \\ |\theta| - 1 & \text{if } |\theta| > 1, \end{cases}$$

$$L(\theta, a_3) = \begin{cases} 0 & \text{if } \theta > 1, \\ 1 - \theta & \text{if } |\theta| \leq 1, \\ 2(1-\theta) & \text{if } \theta < -1. \end{cases}$$

Consider the decision rule $\boldsymbol{\psi}$, defined by

$$\psi_1(x) = I_{(-3,-1)}(x), \qquad \psi_2(x) = I_A(x), \qquad \psi_3(x) = I_{(1,3)}(x), \tag{8.18}$$

where $A = (-\infty, -3] \cup [-1, 1] \cup [3, \infty)$.

It is easy to check that (8.13) is satisfied for the above loss, with $\theta_1 = -1$ and $\theta_2 = 1$, so that the decision problem is monotone. The decision rule $\boldsymbol{\psi}$ is not monotone, however, and so can be improved upon using the construction in the proof of Theorem 6. We begin by defining, as in (8.14),

$$\psi^0(x) = 1, \qquad \psi^1(x) = I_{(-\infty,-3]\cup[-1,\infty)}(x),$$
$$\psi^2(x) = I_{(1,3)}(x), \qquad \psi^3(x) = 0.$$

The next step is to find one-sided UMP tests, ϕ^j, of the same size as ψ^j, for testing H_0: $\theta \le \theta_j$ versus H_1: $\theta > \theta_j$, $j = 1, 2$. The desired UMP tests are clearly of the form

$$\phi^j(x) = \begin{cases} 1 & \text{if } x > x_j, \\ 0 & \text{if } x < x_j, \end{cases}$$

and to attain equality in size between ϕ^j and ψ^j it must be true that

$$P_{-1}(X > x_1) = P_{-1}(X < -3) + P_{-1}(X > -1) = 0.5227$$

and

$$P_1(X > x_2) = P_1(1 < X < 3) = 0.4773.$$

From a table of normal probabilities it is easy to calculate that $x_1 = -1.057$ and $x_2 = 1.057$.

To complete the construction, one merely calculates the improved rule $\boldsymbol{\phi}$, using (8.17). In this case, $\boldsymbol{\phi}$ is given by

$$\phi_1(x) = I_{(-\infty,-1.057)}(x), \qquad \phi_2(x) = I_{[-1.057,1.057]}(x),$$
$$\phi_3(x) = I_{(1.057,\infty)}(x).$$

Other examples and results concerning monotonization in testing can be found in Brown, Cohen, and Strawderman (1976), van Houwelingen (1976), and Stijnen (1980, 1982).

8.4.2. Monotone Estimation Problems

In this subsection, we consider the situation in which $\mathscr{A}$ is a closed subset of R^1, Θ is an interval in R^1, and the distribution of X has monotone likelihood ratio.

Definition 8. In the above situation, a decision problem is said to be *monotone* if the loss function $L(\theta, a)$ is such that, for each θ,

(a) $L(\theta, a)$ attains its minimum as a function of a at a point $a = q(\theta)$, where q is an increasing function of θ;
(b) $L(\theta, a)$, considered as a function of a, increases as a moves away from $q(\theta)$.

The usual examples of monotone decision problems of the above type are estimation problems. If, for example, it is desired to estimate θ under loss $L(\theta, a) = W(|\theta - a|)$, where W is increasing, it is easy to check that the conditions in Definition 8 are satisfied with $q(\theta) = \theta$.

Definition 9. A *randomized monotone decision rule* $\delta^*(x, \cdot)$ is a randomized rule with the following property: if $x_1 > x_2$ and A_1 and A_2 are open sets in $\mathscr{A}$ with A_1 lying to the left of A_2, then either $\delta^*(x_1, A_1) = 0$ or $\delta^*(x_2, A_2) = 0$.

A *nonrandomized monotone decision rule* $\delta(x)$ is simply a rule which is nondecreasing in x.

The definition of a randomized monotone decision rule can be seen to imply, as a special case, that of a nonrandomized monotone decision rule, and is also consistent with the definition of a monotone rule for the finite action case.

Theorem 7. *For a monotone decision problem, as defined in Definition* 8, *the class of* (*randomized*) *monotone decision rules is an essentially complete class.*

The above theorem was developed in Karlin and Rubin (1956), to which the reader is referred for a detailed proof. It is also possible to *construct* an improvement on a nonmonotone decision rule, in this situation, as shown in Rubin (1951). (See also Brown, Cohen, and Strawderman (1976), van Houwelingen (1977), and van Houwelingen and Verbeek (1980).) Indeed, for any given nonmonotone (randomized) decision rule δ_0^*, an R-better or R-equivalent monotone (randomized) decision rule δ_M^* can be shown (under suitable conditions) to be that monotone rule which satisfies, for each fixed $a \in \mathscr{A}$, the equality

$$\int \delta_M^*(x, (-\infty, a])dF^X(x|q^{-1}(a)) = \int \delta_0^*(x, (-\infty, a])dF^X(x|q^{-1}(a)), \tag{8.19}$$

where q^{-1} is the inverse function of q, defined by $q^{-1}(q(a)) = a$. The expressions in (8.19) are simply the overall probabilities that the respective rules will select an action in $(-\infty, a]$ when $\theta = q^{-1}(a)$ (the point at which $L(\theta, a)$, considered as a function of θ, is minimized).

When X has a continuous density, (8.19) can be considerably simplified. This is because δ_M^* will then be a nonrandomized rule, say $\delta_M(x)$, so that (8.19) will become

$$P_{q^{-1}(a)}(\delta_M(X) \le a) = V(\delta_0^*, a) \equiv \int \delta_0^*(x, (-\infty, a])dF^X(x|q^{-1}(a)). \tag{8.20}$$

Even greater simplification is possible when $\delta_M(x)$ is strictly increasing.

(Since δ_M is monotone, recall that it must at least be nondecreasing.) Indeed, it is then clear that

$$P_{q^{-1}(a)}(\delta_M(X) \leq a) = P_{q^{-1}(a)}(X \leq \delta_M^{-1}(a)) = F(\delta_M^{-1}(a) | q^{-1}(a)),$$

where δ_M^{-1} is the inverse function of δ_M, and F, as usual, is the cumulative distribution function of X. In this situation, (8.20) thus becomes

$$F(\delta_M^{-1}(a) | q^{-1}(a)) = V(\delta_0^*, a). \tag{8.21}$$

Now this relationship must hold for any $a \in \mathscr{A}$, so that, in particular, it must hold for $a = \delta_M(x)$, the action to be taken when x is observed. Plugging this into (8.21), we arrive at the conclusion that $\delta_M(x)$ is the action a for which

$$F(x | q^{-1}(a)) = V(\delta_0^*, a). \tag{8.22}$$

Example 6. Assume that $X \sim \mathcal{N}(\theta, 1)$, and that it is desired to estimate θ under a loss $W(|\theta - a|)$, where W is an increasing function. It is easy to check that this is a monotone decision problem in the sense of Definition 8, and that q (and hence q^{-1}) is the identity function (i.e., $q(\theta) = \theta$).

Consider the decision rule $\delta_0(x) = -cx$, where $c > 0$. This is clearly nonmonotone, since it is decreasing in x. Hence we can find an improved monotone rule via the preceding construction.

Letting F_0 denote the cumulative distribution function of the $\mathcal{N}(0, 1)$ distribution, note first that

$$\begin{aligned} V(\delta_0^*, a) &= P_{q^{-1}(a)}(\delta_0(X) \leq a) = P_a(-cX \leq a) \\ &= P_a(X \geq -a/c) \\ &= 1 - F_0(-a(1 + c^{-1})) \\ &= F_0(a(1 + c^{-1})). \end{aligned}$$

Also,

$$F(x | q^{-1}(a)) = F(x | a) = F_0(x - a),$$

so that (8.22) becomes

$$F_0(x - a) = F_0(a(1 + c^{-1})).$$

It follows that $x - a = a(1 + c^{-1})$, or equivalently that $a = x/(2 + c^{-1})$. The monotone rule which is better than δ_0 is thus $\delta_M(x) = x/(2 + c^{-1})$.

An important and natural application of the above "monotonization" technique is to empirical Bayes estimation, since many natural empirical Bayes estimators are not necessarily monotone. (See, for instance, Example 16 in Subsection 4.5.4.) The solution to (8.22), or more generally (8.19), will rarely be obtainable in closed form for such problems, but numerical calculation is quite feasible. Examples can be found in van Houwelingen (1977), Stijnen (1980, 1982), and van Houwelingen and Stijnen (1983).

8.5. Limits of Bayes Rules

In Theorem 2 of Subsection 8.2.3 it was seen that the Bayes procedures form a complete class. Such a result holds in a number of other circumstances, including certain of the situations discussed in Sections 8.3 and 8.4 and many situations in which Θ is closed and bounded. (See Theorem 12 in Section 8.8 for one such result; others can be found in Brown (1976), Balder, Gilliland, and van Houwelingen (1983), and Diaconis and Stein (1983).)

Unfortunately, it is not in general true that the Bayes rules form a complete class. If, for instance, $X \sim \mathcal{N}(\theta, 1)$ is observed and it is desired to estimate θ under squared-error loss, then $\delta(x)=x$ is an admissible estimator (as will be seen in Subsection 8.9.2). This decision rule is *not* a Bayes rule (though it is a generalized Bayes rule with respect to $\pi(\theta)\equiv 1$); and, since it is admissible, the Bayes rules cannot form a complete class.

It is true, in very great generality, that limits (in various senses) of Bayes rules form a complete class. Such results are, unfortunately, too advanced mathematically to be presented here. The interested reader can consult Wald (1950), Le Cam (1955), Stein (1955a), Farrell (1966, 1968a), Kusama (1966), Portnoy (1972), LeCam (1974), Brown (1976, 1977), Balder (1982, 1983, 1984), and Diaconis and Stein (1983) for a number of such results and related applications.

It is often possible to characterize, in a directly usable fashion, the nature of limits of Bayes rules; when this is the case, knowing that the limits of Bayes rules form a complete class can yield very useful complete class theorems. Sections 8.6 and 8.7 present (without proof) a number of useful complete class theorems that can be derived in this fashion. Indeed, the "limits of Bayes" technique has been by far the most fruitful method of deriving complete class theorems.

It should be noted that other very general characterizations of complete classes exist, characterizations based on functional analytic interpretation of risk functions. Many of the previously mentioned references indeed work within such a general framework. There have also been more purely analytical attempts to characterize complete classes (generalizations of "differentiate and set equal to zero"); see Kozek (1982) (which also has an interesting application to binomial estimation) and Isii and Noda (1983) for development.

One of the most useful complete class results that can often be obtained is that the class of all Bayes and generalized Bayes rules forms a complete class. We will encounter several situations in which such a theorem is valid.

It will, henceforth, be convenient to understand the term "generalized Bayes" to also include "Bayes."

8.6. Other Complete and Essentially Complete Classes of Tests

In this section we briefly discuss, without detailed proofs, several other interesting complete and essentially complete class results concerning testing.

8.6.1. Two-Sided Testing

It is frequently of interest to test hypotheses of the form H_0: $\theta = \theta_0$ versus H_1: $\theta \neq \theta_0$, or H_0: $\theta_1 \leq \theta \leq \theta_2$ versus H_1: $\theta < \theta_1$ or $\theta > \theta_2$. As in Section 8.3, an essentially complete class of tests can be obtained for this situation, providing the distribution of X is of a certain type. The relevant type here is called *Pólya type 3*, and includes the distributions in the one-parameter exponential family.

The essentially complete class theorem can be stated in terms of a general two-action decision problem which corresponds to the above testing situation. Indeed, the needed conditions on the structure of the problem are that $\mathscr{A} = \{a_0, a_1\}$ and that the loss function satisfies

$$\begin{aligned} L(\theta, a_1) - L(\theta, a_0) &\geq 0 \quad \text{if } \theta_1 < \theta < \theta_2, \\ L(\theta, a_1) - L(\theta, a_0) &\leq 0 \quad \text{if } \theta < \theta_1 \text{ or } \theta > \theta_2. \end{aligned} \tag{8.23}$$

It is then true that an essentially complete class of decision rules is the class of *two-sided tests* of the form

$$\phi(x) = \begin{cases} 1 & \text{if } x < x_1 \quad \text{or} \quad x > x_2, \\ \gamma_i & \text{if } x = x_i, \quad i = 1, 2, \\ 0 & \text{if } x_1 < x < x_2. \end{cases}$$

For a development of this theory, and generalizations to other two-action problems, see Karlin (1956).

8.6.2. Higher Dimensional Results

In the situation of the previous subsection, if X has a continuous density, then the class of two-sided tests is simply the class of tests whose acceptance region (for H_0) is an interval. (The boundary points of the acceptance region are immaterial for a continuous density.) The natural generalization of an interval to higher dimensions is a convex set, and indeed complete class results concerning such a generalization have been obtained. We state here only the simplest meaningful version of such a complete class result, one due to Birnbaum (1955).

Assume that $\mathbf{X}$ is a random vector in R^p with a continuous density from the p-dimensional exponential family, i.e.,

$$f(\mathbf{x}|\boldsymbol{\theta}) = c(\boldsymbol{\theta})h(\mathbf{x})\exp\{\boldsymbol{\theta}'\mathbf{x}\},$$

where $\boldsymbol{\theta} \in R^p$. (The multivariate normal distribution with known covariance matrix can be transformed so as to have a density of this form.) It is desired to test H_0: $\boldsymbol{\theta} = \boldsymbol{\theta}_0$ versus H_1: $\boldsymbol{\theta} \neq \boldsymbol{\theta}_0$, under, say, "0-1" loss. As shown in Birnbaum (1955), it is then true, under quite weak conditions, that the tests with convex acceptance regions (for H_0) form a minimal complete class. A rough outline of the proof is as follows.

First, it can be shown that all Bayes rules have convex acceptance regions, from which it follows easily that all limits of Bayes rules have convex acceptance regions. Since the limits of Bayes rules form a complete class, as alluded to in the previous section, it can be concluded that the tests with convex acceptance regions form a complete class. To prove that this complete class is minimal complete, it suffices to show that any test ϕ with a convex acceptance region A is admissible. This last fact is established by showing that for any other test ϕ' with a convex acceptance region A' whose size is no larger than that of ϕ (i.e., $E_{\boldsymbol{\theta}_0}[\phi'(\mathbf{X})] \leq E_{\boldsymbol{\theta}_0}[\phi(\mathbf{X})]$), there exists a sequence of points $\{\boldsymbol{\theta}_i\}$ with $|\boldsymbol{\theta}_j| \to \infty$ such that $E_{\boldsymbol{\theta}_i}[\phi'(\mathbf{X})] < E_{\boldsymbol{\theta}_i}[\phi(\mathbf{X})]$ for large enough i. In terms of risk functions, this means that if $R(\boldsymbol{\theta}_0, \phi') \leq R(\boldsymbol{\theta}_0, \phi)$, then

$$R(\boldsymbol{\theta}_i, \phi') = E_{\boldsymbol{\theta}_i}[1 - \phi'(\mathbf{X})] > E_{\boldsymbol{\theta}_i}[1 - \phi(\mathbf{X})] = R(\boldsymbol{\theta}_i, \phi)$$

for large enough i, proving the admissibility of ϕ. The sequence $\{\boldsymbol{\theta}_i\}$ is chosen as follows. First, it can be shown that there exists a convex set $B \subset A$ such that B is disjoint from A'. Let H be a supporting hyperplane to B, at say $\mathbf{x}_0 \in B$, which separates B and A'. Finally, let l_0 be the half line perpendicular to H at $\mathbf{x}_0$, and going from $\mathbf{x}_0$ to infinity through B. Choose the $\boldsymbol{\theta}_i$ to be points along l_0. The reason this works is that, as $|\boldsymbol{\theta}| \to \infty$ along this half line, the distribution of $\mathbf{X}$ gives much more mass to B (and hence A) than to A'.

There have been a number of significant generalizations of this result, including the establishment of complete class theorems for various multivariate one-sided testing problems. Such generalizations can be found in Matthes and Truax (1967), Farrell (1968a), Eaton (1970), Ghia (1976), and Marden (1982); several of these establish particularly useful complete class theorems of the type "the generalized Bayes rules form a complete class" (although sometimes a few extra rules have to be thrown in). There is also a large literature on complete classes of tests within the class of invariant tests. See Marden (1983) for results and other references.

8.6.3. Sequential Testing

Complete and essentially complete class results are of great potential usefulness in sequential analysis, due to the difficulties in finding optimal stopping rules. If the class of possible stopping rules can be substantially reduced, many otherwise intractable calculations may become feasible.

One quite useful sequential result was obtained by Sobel (1953) for the situation of testing H_0: $\theta \le \theta_0$ versus H_1: $\theta > \theta_0$, based on a sequential sample from the density (on R^1)

$$f(x|\theta) = c(\theta)h(x)\exp\{\theta x\} \tag{8.24}$$

in the one-parameter exponential family. The loss was assumed to be of the form

$$L(\theta, a, n) = L(\theta, a) + C(n),$$

i.e., the sum of a decision loss and a sampling cost. Sobel (1953) assumed that: (i) $C(n) \to \infty$; (ii) the loss specifies a region about θ_0 in which one is indifferent between H_0 and H_1; and (iii) the loss is zero for a correct decision and is nondecreasing in $\theta - \theta_0$ and $\theta_0 - \theta$ for incorrectly deciding H_0 and H_1, respectively. Brown, Cohen, and Strawderman (1979, 1980) show, however, that only Condition (iii) is actually needed for the essentially complete class result.

The basic result for the above situation is that the procedures of the following form constitute an essentially complete class of sequential procedures. Define $s_n = \sum_{i=1}^{n} x_i$, and let $\{a_1, a_2, \ldots\}$ and $\{b_1, b_2, \ldots\}$ be sequences of constants such that $a_i \le b_i$ $(i = 1, 2, \ldots)$. The procedure based on these constants is to

$$\begin{aligned} &\text{decide } H_0 && \text{if } s_n < a_n, \\ &\text{decide } H_1 && \text{if } s_n > b_n, \\ &\text{continue sampling} && \text{if } a_n < s_n < b_n, \end{aligned} \tag{8.25}$$

where randomization between deciding H_0 and continuing sampling is allowed if $s_n = a_n$, and randomization between deciding H_1 and continuing sampling is allowed if $s_n = b_n$.

Note that tests of the form (8.25) are a natural generalization of the sequential probability ratio test discussed in Section 7.5; indeed the SPRT of H_0: $\theta = \theta_0$ versus H_1: $\theta = \theta_1$, for the density (8.24), is of the form (8.25) with

$$a_n = (\theta_1 - \theta_0)^{-1}\left[a - n\log\left\{\frac{c(\theta_1)}{c(\theta_0)}\right\}\right],$$

$$b_n = (\theta_1 - \theta_0)^{-1}\left[b - n\log\left\{\frac{c(\theta_1)}{c(\theta_0)}\right\}\right].$$

For this reason, sequential tests of the form (8.25) are often called *generalized sequential probability ratio tests.* The term *monotone sequential tests* is also common and natural.

The proof of the above essentially complete class result is based on showing that Bayes procedures and limits of Bayes procedures are of the form (8.25), or are at least R-equivalent to procedures of the form (8.25). Since, again, the limits of Bayes procedures form a complete class, the result follows. For details, see Brown, Cohen, and Strawderman (1979).

Several other interesting complete and essentially complete class results for sequential problems have been obtained by Brown, Cohen, and Stawderman (1980). These results again apply to the situation in which the overall loss is the sum of a decision loss and the cost of observation. It is first shown that for a wide variety of sequential problems in which the closure of the null hypothesis is closed and bounded, the Bayes or generalized Bayes procedures form a complete or essentially complete class, the exact result depending on various features of the problem. One minor, but useful, consequence of such a result is that when dealing with a situation in which the Bayes or generalized Bayes procedures are known to be nonrandomized, the nonrandomized sequential procedures will form a complete or essentially complete class.

Another result obtained in Brown, Cohen, and Strawderman (1980) is that the generalized Bayes tests form an essentially complete class for the one-sided testing problem considered by Sobel. Since the generalized Bayes tests are a subset of the class of rules of the form (8.25), this result provides a smaller essentially complete class than that obtained by Sobel. Explicit conditions concerning admissibility and inadmissibility of these rules can be found in Brown, Cohen, and Samuel-Cahn (1983).

Sequential complete class theorems in other settings also exist. In the selection and ranking scenario, for instance, such results can be found in Bechhofer, Kiefer, and Sobel (1968) and Gupta and Miescke (1984).

8.7. Complete and Essentially Complete Classes in Estimation

The complete class theorems that have been developed in estimation settings are mainly of the type "the generalized Bayes rules (or certain extensions thereof) form a complete class." Subsection 8.7.1 discusses one such result, while Subsection 8.7.2 indicates the potential uses of such complete class theorems.

8.7.1. Generalized Bayes Estimators

Suppose $\mathbf{X}=(X_1,\ldots,X_p)^t$ has a density from the p-dimensional exponential family on $\mathscr{X}\subset\mathbb{R}^p$; thus

$$f(\mathbf{x}|\boldsymbol{\theta})=c(\boldsymbol{\theta})h(\mathbf{x})\exp\{\boldsymbol{\theta}^t\mathbf{x}\}, \tag{8.26}$$

where $h(\mathbf{x})>0$ on $\mathscr{X}$. (As usual, we assume that this is a density with respect to Lebesgue measure in the continuous case, and with respect to counting measure in the discrete case.) The *natural parameter* is $\boldsymbol{\theta}$, and the *natural parameter space* is (in the continuous case for convenience)

$$\Theta=\left\{\boldsymbol{\theta}: \int_{\mathscr{X}} h(\mathbf{x})\exp\{\boldsymbol{\theta}^t\mathbf{x}\}d\mathbf{x}<\infty\right\}. \tag{8.27}$$

Suppose that $\boldsymbol{\theta}$ is to be estimated under the quadratic loss $L(\boldsymbol{\theta},\boldsymbol{\delta})=(\boldsymbol{\theta}-\boldsymbol{\delta})^t\mathbf{Q}(\boldsymbol{\theta}-\boldsymbol{\delta})$, and that $\pi(\boldsymbol{\theta})$ is a (generalized) prior distribution. Then the (generalized) Bayes estimator for $\boldsymbol{\theta}$ can be shown (under mild conditions) to be the posterior mean,

$$\begin{aligned}\boldsymbol{\delta}^{\pi}(\mathbf{x})&=\frac{1}{m(\mathbf{x}|\pi)}\int_{\Theta}\boldsymbol{\theta}f(\mathbf{x}|\boldsymbol{\theta})dF^{\pi}(\boldsymbol{\theta})\\&=\nabla\log m(\mathbf{x}|\pi)-\nabla\log h(\mathbf{x});\end{aligned} \tag{8.28}$$

here $m(\mathbf{x}|\pi)=\int f(\mathbf{x}|\boldsymbol{\theta})dF^{\pi}(\boldsymbol{\theta})$ is the marginal, $\nabla g=((\partial/\partial x_1)g,\ldots,(\partial/\partial x_p)g)^t$ is the gradient of g, and the last equality can be verified by differentiating under the integral sign.

Theorem 8. *Suppose* $\mathbf{X}$ *is a continuous random variable having a density as in* (8.26), *and that* Θ *is closed. Then any admissible estimator is a generalized Bayes rule.*

PROOF. In this situation, it is well known that the limits of Bayes rules do form a complete class (see Section 8.5). If $\boldsymbol{\delta}$ is a limit of Bayes rules $\{\boldsymbol{\delta}^{\pi_i}\}$, then it can be shown that a subsequence of $\{\pi_i\}$ converges to a generalized prior π, and that $\boldsymbol{\delta}$ is generalized Bayes with respect to π. See Berger and Srinivasan (1978) for details. □

The first version of Theorem 8 was for $p=1$, and was developed in Sacks (1963). Brown (1971) proved Theorem 8 for the p-variate normal distribution, and it is his proof which forms the basis of the proof in the more general case. Berger and Srinivasan (1978) actually established Theorem 8 in a more general context, making it applicable (with certain limitations) to discrete problems and problems with open Θ. Caridi (1983) was able to show that the theorem holds in a very wide variety of cases involving nonquadratic loss; indeed, the only nontechnical condition needed is that the loss be unbounded in all directions.

For the p-variate normal mean situation, but with the "control" loss $(\boldsymbol{\theta}^t\boldsymbol{\delta}-1)^2$, Zaman (1981) characterizes the limits of Bayes rules. The resulting complete class is not exactly the set of generalized Bayes rules, but the ideas are similar.

For discrete problems, it seems to be the case that the "correct" characterization of complete classes of estimators usually involves not generalized

Bayes rules, but what are called "stepwise Bayes" rules. Development of such characterizations for finite $\mathscr{X}$ (often in the multinomial situation) can be found in Brown (1981, 1984), Meeden and Ghosh (1981), and Ighodaro, Santner, and Brown (1982). These results can often also be applied to nonparametric estimation problems. For the multivariate Poisson situation of Example 21 in Subsection 5.4 (and for the same distribution, but with sum-of-squares error loss), Brown and Farrell (1985a) give a complete class involving stepwise Bayes rules.

8.7.2. Identifying Generalized Bayes Estimators

The primary use of a result, such as Theorem 8, is to identify inadmissible estimators; estimators inadmissible because they are not generalized Bayes.

EXAMPLE 7. Suppose $\mathbf{X} \sim \mathcal{N}_p(\boldsymbol{\theta}, \mathbf{I})$, $p \geq 3$, and that it is desired to estimate $\boldsymbol{\theta}$ under sum-of-squares error loss. The positive part James–Stein estimator (see Subsections 4.8.2 and 5.4.3),

$$\boldsymbol{\delta}^{\text{J-S+}}(\mathbf{x}) = \left(1 - \frac{p-2}{|\mathbf{x}|^2}\right)^+ \mathbf{x}$$

(where $a^+ = \max\{a, 0\}$), cannot be generalized Bayes. To see this, observe that (8.28) becomes (for the normal distribution)

$$\boldsymbol{\delta}^\pi(\mathbf{x}) = \nabla \log\left[(2\pi)^{-p/2} e^{-|\mathbf{x}|^2/2} \int e^{\mathbf{x}'\boldsymbol{\theta}} e^{-|\boldsymbol{\theta}|^2/2} dF^\pi(\boldsymbol{\theta})\right] - \nabla \log\left[e^{-|\mathbf{x}|^2/2}\right] \tag{8.29}$$

$$= \nabla \log \int e^{\mathbf{x}'\boldsymbol{\theta}} e^{-|\boldsymbol{\theta}|^2/2} dF^\pi(\boldsymbol{\theta}).$$

This integral is a Laplace transform (of $\exp\{-\frac{1}{2}|\boldsymbol{\theta}|^2\} dF^\pi(\boldsymbol{\theta})$), which is known to be infinitely differentiable. Hence $\boldsymbol{\delta}^\pi$ must be infinitely differentiable to be generalized Bayes. But $\boldsymbol{\delta}^{\text{J-S+}}$ is not differentiable at $|\mathbf{x}|^2 = (p-2)$, so it cannot be generalized Bayes or admissible. (It is interesting to note, however, that no one has yet succeeded in finding a better estimator.)

A more complete characterization of generalized Bayes rules can often be given. As an example, the following theorem gives the conditions under which an estimator of a p-variate normal mean (quadratic loss) is generalized Bayes. For convenience of application, we state the theorem in the case of arbitrary (but known) covariance matrix $\boldsymbol{\Sigma}$; it can be shown that Theorem 8 still holds for this situation. The theorem was given in Berger and Srinivasan (1978); Strawderman and Cohen (1971) developed a special case.

Theorem 9. *Suppose that* $\mathbf{X} \sim \mathcal{N}_p(\boldsymbol{\theta}, \boldsymbol{\Sigma})$, $\boldsymbol{\Sigma}$ *known, and that it is desired to estimate* $\boldsymbol{\theta}$ *under a quadratic loss. Then an estimator,* $\boldsymbol{\delta}(\mathbf{x})$, *is generalized*

Bayes (and hence potentially admissible) if and only if

(a) *the vector function* $\mathbf{g}(\mathbf{x}) = \mathbf{\Sigma}^{-1}\boldsymbol{\delta}(\mathbf{x})$ *is continuously differentiable, and the* $(p \times p)$ *matrix* $\mathbf{J}_{\mathbf{g}}(\mathbf{x})$, *having* (i, j) *element* $(\partial/\partial x_i) g_j(\mathbf{x})$, *is symmetric;*
(b) $\exp\{r(\mathbf{x})\}$ *is a Laplace transform (i.e.,* $\exp\{r(\mathbf{x})\} = \int \exp\{\boldsymbol{\theta}'\mathbf{\Sigma}^{-1}\mathbf{x}\} dG(\boldsymbol{\theta})$ *for some measure* G*), where* $r(\mathbf{x})$ *is the real valued function which has* $g(\mathbf{x})$ *as a gradient.*

Proof. The analog of (8.28) for this situation is

$$\boldsymbol{\delta}^{\pi}(\mathbf{x}) = \mathbf{\Sigma}\nabla \log m(\mathbf{x}|\pi) + \mathbf{x}$$
$$= \mathbf{\Sigma}\nabla \log \int \exp\{\boldsymbol{\theta}'\mathbf{\Sigma}^{-1}\mathbf{x}\} dG(\boldsymbol{\theta}),$$

where $dG(\boldsymbol{\theta}) = \exp\{-\frac{1}{2}\boldsymbol{\theta}'\mathbf{\Sigma}^{-1}\boldsymbol{\theta}\} dF^{\pi}(\boldsymbol{\theta})$. Thus

$$\mathbf{g}(\mathbf{x}) = \mathbf{\Sigma}^{-1}\boldsymbol{\delta}^{\pi}(\mathbf{x}) = \nabla \log \int \exp\{\boldsymbol{\theta}'\mathbf{\Sigma}^{-1}\mathbf{x}\} dG(\boldsymbol{\theta}). \tag{8.30}$$

If, now, we are given an estimator $\boldsymbol{\delta}$, then $\mathbf{g}(\mathbf{x}) = \mathbf{\Sigma}^{-1}\boldsymbol{\delta}(\mathbf{x})$ must satisfy an equation of the form (8.30) to be generalized Bayes. First, (8.30) says that $\mathbf{g}$ must be the gradient of a real valued function, $r(\mathbf{x})$. Condition (a) is a standard necessary and sufficient condition for this to be the case. Next, (8.30) implies that this real valued function must be the log of a Laplace transform, which is Condition (b). □

Note that $r(\mathbf{x})$ can be found by calculating the line integral of $\mathbf{g}(\mathbf{x})$ along any path from a fixed point $\mathbf{x}_0$ to $\mathbf{x}$. The verification that $\exp\{r(\mathbf{x})\}$ is (or is not) a Laplace transform can usually be carried out using techniques from Widder (1946) and Hirschmann and Widder (1955).

Example 8. As a simple example, consider the estimator $\boldsymbol{\delta}(\mathbf{x}) = \mathbf{A}\mathbf{x}$, where $\mathbf{A}$ is a $(p \times p)$ matrix. Clearly $g(\mathbf{x}) = \mathbf{\Sigma}^{-1}\mathbf{A}\mathbf{x}$ is continuously differentiable, and its matrix of first partial derivatives is $\mathbf{\Sigma}^{-1}\mathbf{A}$. Hence, for Condition (a) of Theorem 9 to be satisfied, it must be true that $\mathbf{A}$ is of the form $\mathbf{\Sigma}\mathbf{B}$, where $\mathbf{B}$ is a symmetric $(p \times p)$ matrix. Condition (b) is also relatively easy to check in this situation. Indeed $r(\mathbf{x})$ can be seen to be $r(\mathbf{x}) = \mathbf{x}'\mathbf{B}\mathbf{x}/2$, and $\exp\{r(\mathbf{x})\}$ is a Laplace transform of a generalized prior if and only if $\mathbf{B}$ is positive semidefinite. Thus, to be potentially admissible, $\boldsymbol{\delta}(\mathbf{x})$ must be of the form $\boldsymbol{\delta}(\mathbf{x}) = \mathbf{\Sigma}\mathbf{B}\mathbf{x}$, with $\mathbf{B}$ positive semidefinite.

8.8. Continuous Risk Functions

For a number of decision-theoretic results, it is important to know when decision rules have continuous risk functions. Theorem 9 in Section 4.8, Theorem 12 of this section, and Theorem 13 in the next section are typical examples of results for which this knowledge is needed. There are a variety of results establishing the continuity of risk functions; some proving that all decision rules have continuous risk functions and some showing that the decision rules with continuous risk functions form a complete class (and hence that any admissible decision rule has a continuous risk function). Unfortunately, these results and their proofs generally involve measure-theoretic notions. We will, therefore, merely state two quite useful theorems of this nature, and give another complete class result as an application.

Theorem 10. *Suppose $\Theta \subset R^m$ and that $L(\theta, a)$ is a bounded function which is continuous in θ for each $a \in \mathscr{A}$. Suppose also that X has a density $f(x|\theta)$ which is continuous in θ for each $x \in \mathscr{X}$. Then all decision rules have continuous risk functions.*

The most restrictive condition in the above theorem is the condition that the loss function be bounded. When Θ is unbounded, many standard losses, such as squared-error loss, will not satisfy this condition. The boundedness condition can be relaxed, as shown in the following theorem, at the expense of requiring more of the density.

Theorem 11. *Suppose that $\mathscr{X}$, Θ, and $\mathscr{A}$ are subsets of R^1, with $\mathscr{A}$ being closed, and that the distribution of X has monotone likelihood ratio. Suppose also that $f(x|\theta)$ is continuous in θ for each $x \in \mathscr{X}$, and that the loss function $L(\theta, a)$ is such that*

(a) *$L(\theta, a)$ is continuous in θ for each $a \in \mathscr{A}$;*
(b) *$L(\theta, a)$ is nonincreasing in a for $a \le \theta$ and is nondecreasing in a for $a \ge \theta$;*
(c) *there exist functions $K_1(\theta_1, \theta_2)$ and $K_2(\theta_1, \theta_2)$ on $\Theta \times \Theta$ which are bounded on all bounded subsets of $\Theta \times \Theta$, and such that*

$$L(\theta_2, a) \le K_1(\theta_1, \theta_2) L(\theta_1, a) + K_2(\theta_1, \theta_2)$$

for all $a \in \mathscr{A}$.

Then the decision rules with continuous, finite valued risk functions form a complete class.

For proofs of the above theorems, and other such results, see Brown (1976). Ferguson (1967) and Berk, Brown, and Cohen (1981) also give similar theorems.

Example 9. Let $X \sim \mathcal{N}(\theta, 1)$ and assume that it is desired to estimate θ under squared-error loss. The only condition in Theorem 11 which is not obviously satisfied is Condition (c). To check this condition, note that

$$\begin{aligned}(\theta_2 - a)^2 &= ([\theta_2 - \theta_1] + [\theta_1 - a])^2 \\ &\le 2(\theta_2 - \theta_1)^2 + 2(\theta_1 - a)^2.\end{aligned}$$

Hence Condition (c) is satisfied with $K_1(\theta_1, \theta_2)=2$ and $K_2(\theta_1, \theta_2)=2(\theta_2-\theta_1)^2$. It can be concluded that the decision rules with continuous risk functions form a complete class.

As an example of a complete class result which requires continuity of the risk functions, we state the following theorem. This theorem can be found, in considerably greater generality and with a proof, in Brown (1976) or Diaconis and Stein (1983).

Theorem 12. *Assume that $\mathscr{A}$ and Θ are closed and bounded subsets of Euclidean space and, as usual, that X has a continuous or a discrete density. Assume also that $L(\theta, a)$ is a continuous function of a for each $\theta \in \Theta$, and that all decision rules have continuous risk functions. Then the Bayes rules form a complete class.*

8.9. Proving Admissibility and Inadmissibility

Most of the complete classes that have so far been discussed are not minimal complete classes. For instance, not all generalized Bayes rules are admissible in the situation described in Subsection 8.7.1; thus the generalized Bayes rules do not form a minimal complete class. This section is a brief introduction to some of the literature on establishing whether or not such potentially admissible estimators really are admissible. Unfortunately, we have space only for presentation of a few of the general techniques that have been developed for investigating admissibility, and must, for the most part, restrict consideration to estimation examples.

8.9.1. Stein's Necessary and Sufficient Condition for Admissibility

One of the most fruitful characterizations of admissible rules is what has come to be called *Stein's necessary and sufficient condition for admissibility* developed in Stein (1955b). Generalizations and alternate formulations of this result can be found in Le Cam (1955), Farrell (1968a, 1968b), and Diaconis and Stein (1983). Unfortunately, the result in its full generality involves mathematics beyond the level of this text. In Farrell (1968a), however, a more concrete version of the general result was developed, one which is more amenable to discussion at our level. This result can be roughly stated as follows.

Result 1. *Under suitable conditions, a decision rule δ is admissible if and only if there exists a sequence $\{\pi_n\}$ of (generalized) prior distributions such that*

(a) *each π_n gives mass only to a closed and bounded set (possibly different for each n), and hence has finite total mass;*
(b) *there is a closed and bounded set $C \subset \Theta$ to which each π_n gives mass one;*
(c) $\lim_{n\to\infty}[r(\pi_n, \delta) - r(\pi_n, \delta^n)] = 0$, *where δ^n is the Bayes decision rule with respect to π_n. (Note that since π_n has finite mass, there is usually no difficulty in talking about Bayes rules and Bayes risks.)*

For precise theorems of the above nature, and proofs, see Farrell (1968a) and Farrell (1968b). In the remainder of the section we explore the meaning and consequences of Result 1 (and other useful general techniques for investigating admissibility).

8.9.2. Proving Admissibility

Being a sufficient condition for admissibility, Result 1 obviously provides a tool for verifying the admissibility of a decision rule. The use of this sufficient condition actually predates Stein (1955b), first appearing in Blyth (1951). The rationale for the sufficient condition is quite elementary, and is given in the proof of the following easier to use version of the sufficient condition. This version can essentially be found in Farrell (1964) and Brown (1971).

Theorem 13. *Consider a decision problem in which Θ is a nondegenerate convex subset of Euclidean space (i.e., Θ has positive Lebesgue measure), and in which the decision rules with continuous risk functions form a complete class. Then an estimator δ_0 (with a continuous risk function) is admissible if there exists a sequence $\{\pi_n\}$ of (generalized) priors such that*

(a) *the Bayes risks $r(\pi_n, \delta_0)$ and $r(\pi_n, \delta^n)$ are finite for all n, where δ^n is the Bayes rule with respect to π_n;*
(b) *for any nondegenerate convex set $C \subset \Theta$, there exists a $K > 0$ and an integer N such that, for $n \geq N$,*

$$\int_C dF^{\pi_n}(\theta) \geq K;$$

(c) $\lim_{n\to\infty}[r(\pi_n, \delta_0) - r(\pi_n, \delta^n)] = 0$.

Proof. Suppose δ_0 is not admissible. Then there exists a decision rule δ' such that $R(\theta, \delta') \leq R(\theta, \delta_0)$, with strict inequality for some θ, say θ_0. Since the rules with continuous risk functions form a complete class, it can be assumed that δ' has a continuous risk function. Since $R(\theta, \delta_0)$ is also continuous, it follows that there exist constants $\varepsilon_1 > 0$ and $\varepsilon_2 > 0$ such that $R(\theta, \delta') < R(\theta, \delta_0) - \varepsilon_1$ for $\theta \in C = \{\theta \in \Theta: |\theta - \theta_0| < \varepsilon_2\}$. Using this, Conditions (a) and (b), and the fact that $r(\pi_n, \delta^n) \leq r(\pi_n, \delta')$, it can be concluded that for $n \geq N$,

$$\begin{aligned} r(\pi_n, \delta_0) - r(\pi_n, \delta^n) &\geq r(\pi_n, \delta_0) - r(\pi_n, \delta') \\ &= E^{\pi_n}[R(\theta, \delta_0) - R(\theta, \delta')] \\ &\geq \int_C [R(\theta, \delta_0) - R(\theta, \delta')] dF^{\pi_n}(\theta) \\ &\geq \varepsilon_1 \int_C dF^{\pi_n}(\theta) \geq \varepsilon_1 K. \end{aligned}$$

This contradicts Condition (c) of the theorem. Hence δ_0 must be admissible. □

EXAMPLE 10 (Blyth (1951)). Suppose that $X \sim \mathcal{N}(\theta, 1)$ and that it is desired to estimate θ under squared-error loss. We seek to prove that the usual estimator, $\delta_0(x) = x$, is admissible.

The conditions of Theorem 13 will clearly be satisfied for this situation (see Example 9), once a suitable sequence $\{\pi_n\}$ is found. A convenient choice for π_n is the unnormalized normal density

$$\pi_n(\theta) = (2\pi)^{-1/2} \exp\left\{\frac{-\theta^2}{2n}\right\}.$$

If C is a nondegenerate convex subset of Θ, then clearly

$$\int_C \pi_n(\theta) d\theta \geq \int_C \pi_1(\theta) d\theta = K > 0,$$

so that Condition (b) of Theorem 13 is satisfied for this choice of the π_n. (Note that standard $N(0, 1/n)$ prior densities would not satisfy this condition.) A straightforward Bayesian calculation, similar to that with normal priors, shows that $r(\pi_n, \delta_0) = \sqrt{n}$ and $r(\pi_n, \delta^n) = \sqrt{n}n/(1+n)$, verifying Condition (a). Finally,

$$\begin{aligned} \lim_{n\to\infty} [r(\pi_n, \delta_0) - r(\pi_n, \delta^n)] &= \lim_{n\to\infty} \left[\sqrt{n}\left(1 - \frac{n}{n+1}\right)\right] \\ &= \lim_{n\to\infty} \left[\frac{\sqrt{n}}{1+n}\right] = 0. \end{aligned}$$

Condition (c) of Theorem 13 is thus satisfied, and it can be concluded that δ_0 is admissible.

The above technique for proving admissibility is very similar to the technique presented in Theorem 18 of Subsection 5.3.2 for proving that a decision rule is minimax. The admissibility technique is more difficult, however, in that the necessity for the π_n to satisfy both Conditions (b) and (c) of Theorem 13 tends to be very restrictive. Indeed, in general, very elaborate (and difficult to work with) choices of the π_n are needed. See

Stein (1959), James and Stein (1960), Farrell (1964), Berger (1976a), Brown (1979), Rukhin (1984b, 1985a, b), and Vardeman and Meeden (1984). Brown (1979) gives a general heuristic discussion concerning the choice of the π_n, and Rukhin (1984b, 1985a, b) shows how the π_n can sometimes be chosen as approximate solutions to a certain functional equation.

There exist many general theorems providing explicit characterizations of admissible generalized Bayes rules. Some are given in Subsection 8.9.4. The following very general theorem is from Brown and Hwang (1982), to which the reader is referred for a proof. (The proof is based on application of Theorem 13.)

Theorem 14. *Suppose* $\mathbf{X}$ *has a density from the exponential family, as described in* (8.26), *and that* Θ (*in* (8.27)) *is equal to* $\mathbb{R}^p$. *It is desired to estimate the mean of* $\mathbf{X}$, *namely* $\boldsymbol{\psi}(\boldsymbol{\theta}) = E_{\boldsymbol{\theta}}(\mathbf{X})$, *under loss* $L(\boldsymbol{\theta}, \boldsymbol{\delta}) = |\boldsymbol{\psi} - \boldsymbol{\delta}|^2$. *Let* $\pi(\boldsymbol{\theta})$ *be a differentiable generalized prior density for which* (*letting* ∇ *denote gradient as usual*)

(i) $\int |\nabla \pi(\boldsymbol{\theta})| f(\mathbf{x}|\boldsymbol{\theta}) d\boldsymbol{\theta} < \infty$ *for all* $\mathbf{x} \in \mathscr{X}$;
(ii) $\int_{|\boldsymbol{\theta}| \geq 2} [\pi(\boldsymbol{\theta}) / (|\boldsymbol{\theta}| \log(|\boldsymbol{\theta}|))^2] d\boldsymbol{\theta} < \infty$; *and*
(iii) $\int [|\nabla \pi(\boldsymbol{\theta})|^2 / \pi(\boldsymbol{\theta})] d\boldsymbol{\theta} < \infty$.

Then the generalized Bayes estimator of $\boldsymbol{\psi}$, *with respect to* π, *is admissible.*

Example 11. Consider $\pi(\boldsymbol{\theta}) \equiv 1$. It can then be shown (the proof is left as an exercise), that the generalized Bayes estimator of $\boldsymbol{\psi}$ is given by $\boldsymbol{\delta}^{\pi}(\mathbf{x}) = \mathbf{x}$. Conditions (i) and (iii) of Theorem 14 are trivially satisfied by this choice of π. And, transforming to polar coordinates, Condition (ii) is satisfied if

$$\int_2^{\infty} r^{p-1} \left[\frac{1}{(r \log r)^2} \right] dr < \infty;$$

this requires $p \leq 2$. The conclusion is that $\boldsymbol{\delta}^0(\mathbf{x}) = \mathbf{x}$ is an admissible estimator if $p \leq 2$.

Example 12. Suppose $\mathbf{X} \sim \mathcal{N}_p(\boldsymbol{\theta}, \mathbf{I})$ (so that $\boldsymbol{\psi}(\boldsymbol{\theta}) = E_{\boldsymbol{\theta}}[\mathbf{X}] = \boldsymbol{\theta}$) and that $\boldsymbol{\delta}(\mathbf{x}) = \mathbf{B}\mathbf{x}$, $\mathbf{B}$ a given $(p \times p)$ positive semidefinite matrix. (In Example 8, it was seen that $\mathbf{B}$ must be positive semidefinite for $\boldsymbol{\delta}$ to be generalized Bayes, and hence potentially admissible.) It is straightforward to see that $\boldsymbol{\delta}$ is, indeed, generalized Bayes with respect to $\pi(\boldsymbol{\theta}) = \exp\{-\frac{1}{2}\boldsymbol{\theta}^t \mathbf{A} \boldsymbol{\theta}\}$, where

$$\mathbf{A} = (\mathbf{B}^{(-1)} - \mathbf{I})^{(-1)},$$

"(−1)" denoting generalized inverse. If any characteristic root of $\mathbf{A}$ is negative, it is easy to show that Condition (ii) of Theorem 14 is violated (the exponential blows up the integral). Thus all roots of $\mathbf{B}$ must be less than or equal to 1. It will be left as an exercise to show that the conditions of Theorem 14 are satisfied if and only if at most two of these roots equal 1. (This result was first established by Cohen (1966).)

Brown and Hwang (1982) established versions of Theorem 14 which are valid for $\Theta \subset \mathbb{R}^p$, and for certain generalizations of quadratic loss. Theorem 14 is also related to a result in Zidek (1970). There were many precursors to Theorem 14, starting with Karlin (1958), who discussed admissibility of linear estimators in one-dimensional exponential families. Subsequent papers, establishing admissibility of classes of linear estimators of the mean vector in exponential families include Cheng (1964), Joshi (1979), Rao (1976), Ghosh and Meeden (1977), Spruill (1981), LaMotte (1982), DasGupta (1984a), Mandelbaum (1984), Srinivasan (1984b), and Brown and Farrell (1985b). (Not all of these are special cases of Brown and Hwang (1982), and some also have inadmissibility results.) There have also been papers dealing with nonlinear estimators in the exponential family; see Subsection 8.8.4, DasGupta and Sinha (1980), and Ralescu and Ralescu (1981).

Finally, it should be mentioned that quite general admissibility theorems are known for many problems outside the exponential family, primarily location and other invariant problems. Some examples can be found in Brown (1966), Brown and Fox (1974a, b), Berger (1976a), and Brown (1979) (which contains an excellent general discussion).

8.9.3. Proving Inadmissibility

The most common technique of establishing inadmissibility of, say, a generalized Bayes estimator, δ^π, is to simply construct an improved estimator. Thus, in Example 46 of Subsection 4.8.2, it was indicated that $\boldsymbol{\delta}^\pi(\mathbf{x}) = \mathbf{x}$ can be beaten by the James–Stein estimator in estimating a p-variate normal mean, $p \geq 3$. Many more such results and references were given in Section 5.4.

The problems in constructing an improved estimator are (i) guessing what form the improved estimator will take, and (ii) carrying out the technical proof of dominance. (For some ideas concerning these issues, see Brown (1979).) When an unbiased estimator of risk exists (see Subsection 5.4.2), it can often be used to carry out the construction process. Many of the references in Section 5.4 provide examples. We give here one particularly simple result that can be obtained in this fashion. The theorem was developed in Berger (1982c), based on a more general result from Berger (1980a) (see also Hoffman (1984)).

Theorem 15. *Consider the one-dimensional exponential family setup, with X having density $f(x|\theta) = c(\theta)h(x)\exp\{\theta x\}$ with respect to Lebesgue measure on a (possibly infinite) interval (a, b). It is desired to estimate the natural parameter, θ, under squared-error loss, and improvement is sought over the estimator*

$$\delta^0(x) = \frac{d}{dx}\log m_0(x) - \frac{d}{dx}\log h(x). \tag{8.31}$$

(Generalized Bayes estimators are of this form; see (8.28).) *Assume that*

(i) $E_\theta\left|\frac{d}{dX}\log h(X)\right|^2<\infty$;

(ii) *either* $\varphi_1(x)=\int_a^x \frac{1}{m_0(y)}\,dy$ *or* $\varphi_2(x)=\int_x^b \frac{1}{m_0(y)}\,dy$ *is a finite function*;

(iii) *for* φ^*, *a finite function from* (ii),

(a) $E_\theta\left|\frac{d}{dX}\log\varphi^*(X)\right|^2<\infty$, *and*

(b) $\lim_{x\to a}\left\{h(x)e^{\theta x}\left(\frac{d}{dx}\log\varphi^*(x)\right)\right\}=\lim_{x\to b}\left\{h(x)e^{\theta x}\left(\frac{d}{dx}\log\varphi^*(x)\right)\right\}=0.$

Then δ^0 *is inadmissible, and a better estimator is given, for* $0<\lambda<1$, *by*

$$\delta(x)=\delta^0(x)+2\lambda\frac{d}{dx}\log\varphi^*(x). \tag{8.32}$$

PROOF. We only outline the proof, leaving the details for an exercise. The first step is to establish the "unbiased estimator of risk difference," between $\delta(x)=\delta^0(x)+\gamma(x)$ and $\delta^0(x)$,

$$R(\theta,\delta)-R(\theta,\delta^0)=E_\theta\left[2\frac{d}{dX}\gamma(X)+2\gamma(X)\frac{d}{dX}\log m_0(X)+\gamma^2(X)\right]. \tag{8.33}$$

This can be verified using integration by parts, as discussed in Subsection 5.4.2. One then only needs to show that (8.32) yields a γ for which the integrand in (8.33) is negative. □

EXAMPLE 13. It is well known that the best unbiased estimator of θ is $\delta^0(x)=-(d/dx)\log h(x)$, which is of the form (8.31) with $m_0(x)\equiv 1$. Thus, if either a or b is finite, Condition (ii) of Theorem 15 is satisfied; (8.32) then yields an improved estimator, subject to the mainly technical Conditions (i) and (iii).

As a specific example, suppose X has the gamma density, on $(a,b)=(0,\infty)$,

$$f(x|\theta)=\frac{(-\theta)^\alpha}{\Gamma(\alpha)}x^{\alpha-1}e^{\theta x},$$

where α is known and $\theta\in\Theta=(-\infty,0)$. (Note that $\theta=-1/\beta$, where β is from our usual parametrization of the gamma density in Appendix 1.) Then $\varphi_1(x)=\int_0^x(1)dy=x$ is finite, and Conditions (i) and (iii) require only $\alpha>2$ (here $h(x)=x^{\alpha-1}$). Thus, for $\alpha>2$, the best unbiased estimator, $\delta^0(x)=-(\alpha-1)/x$, is inadmissible, and a better estimator is given by (for $0<\lambda<1$)

$$\delta(x)=\delta^0(x)+2\lambda\frac{d}{dx}\log x=-\frac{(\alpha-1-2\lambda)}{x}.$$

Although constructive proofs of inadmissibility are by far the most common, there do exist certain indirect methods of establishing inadmissibility of generalized Bayes estimators, based on careful utilization of Stein's Necessary and Sufficient Condition for Admissibility and other general closure theorems. Examples can be found in Zidek (1976), Brown (1980a), and Berger (1980c).

8.9.4. Minimal (or Nearly Minimal) Complete Classes

The ultimate aim of this line of investigation is, of course, the characterization of a minimal complete class. We previously encountered some such characterizations in testing problems; here several such results for estimation problems are reviewed.

I. *Estimating a Multivariate Normal Mean*

If $\mathbf{X} \sim \mathcal{N}_p(\boldsymbol{\theta}, \boldsymbol{\Sigma})$, $\boldsymbol{\Sigma}$ known, and it is desired to estimate $\boldsymbol{\theta}$ under a quadratic loss, Brown (1971) more or less completely characterized the minimal complete class of generalized Bayes estimators. (Recall from Subsection 8.7.1 that the generalized Bayes estimators form a complete class.) The simplest of Brown's characterizations is that for $\boldsymbol{\Sigma} = \mathbf{I}$, and is given in the following theorem.

Theorem 16. *Suppose that $\mathbf{X} \sim \mathcal{N}_p(\boldsymbol{\theta}, \mathbf{I}_p)$ and that it is desired to estimate $\boldsymbol{\theta}$ under sum-of-squares error loss. Then a generalized Bayes estimator of the form*

$$\boldsymbol{\delta}(\mathbf{x}) = (1 - h(|\mathbf{x}|))\mathbf{x}$$

is

(i) *inadmissible if there exist $\varepsilon > 0$ and $K < \infty$ such that, for $|\mathbf{x}| > K$,*

$$h(|\mathbf{x}|) \leq \frac{(p-2-\varepsilon)}{|\mathbf{x}|^2}; \tag{8.34}$$

(ii) *admissible if there exist $K_1 < \infty$ and $K_2 < \infty$ such that $|\mathbf{x}| h(|\mathbf{x}|) \leq K_1$ for all $\mathbf{x}$ and, for $|\mathbf{x}| > k_2$,*

$$h(|\mathbf{x}|) \geq \frac{(p-2)}{|\mathbf{x}|^2}.$$

EXAMPLE 14. Many generalized Bayes estimators have been developed for the situation in Theorem 16; estimators which, for large $|\mathbf{x}|$, behave like

$$\boldsymbol{\delta}^{\pi}(\mathbf{x}) \cong \left(1 - \frac{c}{|\mathbf{x}|^2}\right)\mathbf{x}.$$

(The estimator (4.114) can be shown to be of this type in the symmetric case, with $c = p+1$; other such estimators are given in Strawderman (1971) and many of the references in Section 5.4.) Clearly, if $c < p-2$, such estimators are inadmissible, while, if $c > p-2$, they are admissible.

Brown (1971) presented an interesting alternative characterization of the admissible and inadmissible generalized Bayes rules, for this problem, in terms of the recurrence of an associated diffusion process. (See also Johnstone and Lalley (1984).) Srinivasan (1981) removes some of the technical restrictions that were required in Brown (1971), and gives somewhat simpler arguments.

II. *Estimating Poisson Means*

In Example 21 of Section 5.4, we considered the problem of estimating a vector of Poisson means, $\boldsymbol{\theta} = (\theta_1, \ldots, \theta_p)^t$, under the weighted quadratic loss $L(\boldsymbol{\theta}, \boldsymbol{\delta}) = \sum_{i=1}^{p} \theta_i^{-1}(\delta_i - \theta_i)^2$. The data was $\mathbf{X} = (X_1, \ldots, X_p)^t$, the X_i being independent $\mathcal{P}(\theta_i)$ random variables. Johnstone (1984, 1985) more or less completely characterizes the admissible and inadmissible generalized Bayes rules for this problem. One of the characterizations is an interesting analog of the diffusion characterization in the normal problem; admissibility and inadmissibility of a generalized Bayes estimator is now, however, related to the recurrence of an associated birth and death process. The following theorem gives one of the easier characterizations from Johnstone (1985), in what is essentially the analog of the symmetric case in the normal problem.

Theorem 17. *A generalized Bayes estimator of the form*

$$\boldsymbol{\delta}(\mathbf{x}) = (1 - h(S))\mathbf{x},$$

where $S = \sum_{i=1}^{p} X_i$, *is*

(i) *inadmissible if there exists* $\varepsilon > 0$ *and* $k < \infty$ *such that, for* $S > K$,

$$h(S) \le \frac{(p-1-\varepsilon)}{S}; \tag{8.35}$$

(ii) *admissible if there exists* $K_1 < \infty$ *and* $K_2 < \infty$ *such that* $\sqrt{S}h(S) \le K_1$ *for all* $\mathbf{x}$, *and, for* $S > K_2$,

$$h(S) \ge \frac{(p-1)}{S}. \tag{8.36}$$

EXAMPLE 15. In Corollary 2 of Subsection 5.4.4, the estimators

$$\boldsymbol{\delta}(\mathbf{x}) = \left(1 - \frac{c}{b+s}\right)\mathbf{x}$$

were considered. The subclass with $b = c$ can be shown to be generalized

Bayes. Theorem 17 can then be used to show that, if $b = c < p-1$, then $\boldsymbol{\delta}$ is inadmissible, while if $b = c > p-1$, then $\boldsymbol{\delta}$ is admissible. The details will be left as an exercise. (Note that admissibility of $\boldsymbol{\delta}$ for $b = c = p-1$ can be established using the more refined results in Johnstone (1985).)

III. *Other Examples*

Srinivasan (1982) generalizes the (nearly) minimal complete class characterization for estimating a p-variate normal mean, to other continuous exponential family situations for which $\Theta = \mathbb{R}^p$. A different type of generalization of the normal mean result is given in Berger, Berliner, and Zaman (1982) and Srinivasan (1984); (nearly) minimal complete classes of rules are obtained for estimating the multivariate normal mean under the "control loss" $L(\boldsymbol{\theta}, \boldsymbol{\delta}) = (\boldsymbol{\delta}'\boldsymbol{\theta} - 1)^2$.

A fairly complete characterization of admissible and inadmissible rules for estimating one coordinate of a location vector is given in Berger (1976a, d). A heuristic outline of such a characterization for estimating the entire location vector is given in Brown (1979).

The conditions (8.34) and (8.35) are examples of what Hwang (1982b, c) has called STUBs ("semi-tail upper bounds" on the class of admissible estimators). Hwang develops STUBs for a number of other situations, thus obtaining roughly "half" of the desired minimal complete class characterization.

Exercises

Section 8.1

1. Prove Lemma 1.
2. Prove Lemma 2.
3. Prove that if $\mathscr{C}$ is a complete class and contains no proper essentially complete subclass, then $\mathscr{C}$ is a minimal complete class.
4. Prove that if the class of admissible rules is complete, it is minimal complete.

Section 8.2

5. Give an example, in terms of the risk set S, in which the conclusion of Theorem 1 is violated if S is not
   (a) closed from below,
   (b) bounded from below.
6. Use a modification of Exercise 24 of Chapter 5 (namely, a different choice of P) to show that the conclusion of Theorem 1 can be true even when S is not closed from below.

7. Find the minimal complete class of decision rules for a sample of size $n=2$ in the situation of
   (a) Exercise 45 of Chapter 5,
   (b) Exercise 46 of Chapter 5.

8. If $X \sim \mathscr{C}(\theta, 1)$, show that the test

$$\phi(x)=\begin{cases}1 & \text{if } 1<x<3,\\ 0 & \text{otherwise,}\end{cases}$$

is most powerful of its size for testing H_0: $\theta=0$ versus H_1: $\theta=1$.

9. Let ϕ be a most powerful test of size $0<\alpha<1$ for testing H_0: $\theta=\theta_0$ versus H_1: $\theta=\theta_1$. Show that $E_{\theta_1}[\phi(X)]>\alpha$, providing the distributions of X under θ_0 and θ_1 differ.

Section 8.3

10. Verify that the following distributions have monotone likelihood ratio:
    (a) $\mathscr{B}e(\alpha, \beta)$, β fixed,
    (b) $\mathscr{B}e(\alpha, \beta)$, α fixed (MLR in $-X$),
    (c) $\mathscr{U}(0, \theta)$,
    (d) $f(x|\alpha, \beta)=(2\beta)^{-1}\exp\{-|x-\alpha|/\beta\}$, β fixed,
    (e) $f(x|\theta)=\exp\{-(x-\theta)\}I_{(\theta,\infty)}(x)$.

11. Find a counterexample to part (ii) of Theorem 4 if the size of the test is allowed to be zero.

12. If $X \sim \mathscr{C}(0, \theta)$, show that the sufficient statistic $|X|$ has monotone likelihood ratio.

13. Assume that $X_1, \ldots, X_n$ is a sample from the $\mathscr{U}(\theta, \theta+1)$ distribution, and that it is desired to test H_0: $\theta \le 0$ versus H_1: $\theta>0$. Defining $t_1=\min\{x_i\}$ and $t_2=\max\{x_i\}$, show that

$$\phi(t_1, t_2)=\begin{cases}0 & \text{if } t_1<1-\alpha^{1/n} \text{ and } t_2<1,\\ 1 & \text{otherwise.}\end{cases}$$

is a UMP test of size α.

14. Prove part (ii) of Theorem 5.

Section 8.4

15. Assume that $X \sim \mathscr{N}(\theta, 1)$, that $\Theta=[-1, 1]$, that $\mathscr{A}=\{a_1, a_2, a_3\}$, and that $L(\theta, a_1)=(\theta+1)^2$, $L(\theta, a_2)=\theta^2$, and $L(\theta, a_3)=(\theta-1)^2$. Show that this is a monotone multiple decision problem.

16. Assume that $X \sim \mathscr{B}(5, \theta)$, that $\Theta=[0, 1]$, that $\mathscr{A}=\{a_1, a_2, a_3\}$, and that the loss function satisfies (8.13) with $\theta_1=\frac{1}{3}$ and $\theta_2=\frac{2}{3}$. Find a monotone decision rule which is R-better than or R-equivalent to the decision rule $\boldsymbol{\psi}$, defined by

$$\psi_1(x)=I_{\{0,3\}}(x), \qquad \psi_2(x)=I_{\{1,4\}}(x), \quad \psi_3(x)=I_{\{2,5\}}(x).$$

17. In the situation of Exercise 15, show that the monotone decision rule $\boldsymbol{\phi}$, defined by
$$\phi_1(x) = I_{(-\infty,-0.1)}(x), \qquad \phi_2(x) = I_{[-0.1,0.1]}(x), \qquad \phi_3(x) = I_{(0.1,\infty)}(x)$$
is inadmissible.

18. In Definition 9, show that the definition of a nonrandomized monotone decision rule is consistent with the definition of a randomized monotone decision rule.

19. Assume that $X \sim \mathcal{N}(\theta, 1)$, that $\Theta = \mathcal{A} = (-\infty, \infty)$, and that $L(\theta, a) = (\theta - a)^2$. Someone proposes the estimator $\delta_0(x) = -cx + b$, where $c > 0$. Find an R-better monotone estimator.

20. Assume that $X \sim \mathcal{N}(\theta, 1)$, that $\Theta = \mathcal{A} = (-\infty, \infty)$, and that $L(\theta, a) = (\theta - a)^2$. Someone proposes the estimator
$$\delta_0(x) = \begin{cases} 0 & \text{if } x < -1, \\ x & \text{if } x \geq -1. \end{cases}$$
Show that an R-better monotone estimator, δ_M, can be described as follows. Let F_0 denote the cumulative distribution function of the $\mathcal{N}(0, 1)$ distribution, let F_0^{-1} denote the functional inverse of F_0, and define $\tau = F_0^{-1}(\frac{1}{2} - F_0(-1))$. If $-\infty < x < \tau$, then $\delta_M(x)$ is that action $a(-1 < a < 0)$ which satisfies
$$F_0(x - a) = \tfrac{1}{2} - F_0(-(1 + a));$$
if $\tau \leq x < 0$, then $\delta_M(x) = 0$; and if $x \geq 0$, then $\delta_M(x) = x$.

Section 8.5

21. If $X \sim \mathcal{N}(\theta, 1)$ and θ is to be estimated under squared-error loss, show that $\delta(x) = x$ is *not* a proper prior Bayes rule.

Section 8.6

22. Assume that $X \sim \mathcal{B}e(\theta, 1)$, that $\mathcal{A} = \{a_0, a_1\}$, and that the loss function is of the form (8.23). A test ϕ is proposed for which $E_{\theta_1}[\phi(X)] = 0.5$ and $E_{\theta_2}[\phi(X)] = 0.3$. Find a test that is R-better than or R-equivalent to ϕ.

23. Verify that the $\mathcal{N}_p(\boldsymbol{\theta}, \mathbf{I})$ distribution is in the exponential family, and identify $c(\boldsymbol{\theta})$, $h(\mathbf{x})$, and Θ.

24. Suppose that $\mathbf{X} \sim \mathcal{N}_p(\boldsymbol{\theta}, \mathbf{I})$, $p \geq 2$, and that it is desired to test H_0: $\boldsymbol{\theta} = \mathbf{0}$ versus H_1: $\boldsymbol{\theta} \neq \mathbf{0}$, under "0–1" loss. Which of the following rejection regions yield admissible, and which yield inadmissible tests:
(a) $\{\mathbf{x}: |\mathbf{x}| > K\}$;
(b) $\{\mathbf{x}: |\mathbf{x}| < K\}$;
(c) $\{\mathbf{x}: x_1 > K\}$;
(d) $\{\mathbf{x}$: at least one coordinate is positive$\}$;
(e) $\{\mathbf{x}$: all coordinates are positive$\}$.

Section 8.7

25. Verify equation (8.28). (You may assume that all integrals exist, and that differentiating under the integral sign is valid.)

26. Assume that $\mathbf{X} \sim \mathcal{N}_p(\boldsymbol{\theta}, \boldsymbol{\Sigma})(p \geq 2)$, where $\boldsymbol{\Sigma}$ is known, and that it is desired to estimate $\boldsymbol{\theta}$ under a quadratic loss. Consider the estimator

$$\boldsymbol{\delta}(\mathbf{x}) = (\boldsymbol{\Sigma}^{-1} + (\mathbf{x}^t\mathbf{C}\mathbf{x})^{-1}\mathbf{A})^{-1}\boldsymbol{\Sigma}^{-1}\mathbf{x}, \tag{8.37}$$

where $\mathbf{C}$ is positive definite. Show that unless $\mathbf{A}$ is a constant multiple of $\mathbf{C}$, such an estimator cannot be generalized Bayes, and hence cannot be admissible. (The estimators of the form (8.37) are called *adaptive ridge regression* estimators, and have been proposed for use as estimators of regression coefficients in standard linear regression.)

27. Suppose that $\mathbf{X} \sim \mathcal{N}_p(\boldsymbol{\theta}, \mathbf{I})$, and that it is desired to estimate $\boldsymbol{\theta}$ under a quadratic loss. If $\boldsymbol{\delta}(\mathbf{x}) = \phi(\mathbf{x})\mathbf{x}$ is generalized Bayes, where ϕ is a real valued function, show that ϕ can never be negative.

Section 8.8

28. Verify that the following loss functions satisfy Condition (c) of Theorem 11:
    (a) $L(\theta, a) = |\theta - a|^k$, $k > 0$, $\Theta = \mathscr{A} = R^1$;
    (b) $L(\theta, a) = \exp\{|\theta - a|\}$, $\Theta = \mathscr{A} = R^1$;
    (c) $L(\theta, a) = (\log[a/\theta])^2$, $\Theta = \mathscr{A} = [1, \infty)$.

29. If $L(\theta, a) = \exp\{(\theta - a)^2\}$, where $\Theta = \mathscr{A} = R^1$, show that Condition (c) of Theorem 11 is violated.

30. Assume that $X \sim \mathscr{B}(1, \theta)$ is observed, and that it is desired to estimate $\theta \in [0, 1]$ under squared-error loss. Since the loss is convex, attention can be restricted to nonrandomized decision rules. A nonrandomized rule can be written as a vector (y, z), where $0 \leq y \leq 1$ and $0 \leq z \leq 1$ are the estimates of θ if $x = 0$ and $x = 1$ are observed, respectively.
    (a) Find a Bayes rule with respect to a given prior distribution π, and show that any prior distribution, π', with the same first two moments as π has the same Bayes rule.
    (b) Plot the set of all nonrandomized Bayes rules.
    (c) Show that the set of rules found in part (b) is a complete class.
    (d) Show that the set of rules found in part (b) is a minimal complete class. (*Hint*: Unless π gives mass one to $\theta = 0$ or to $\theta = 1$, show that the Bayes rule is unique. When π gives mass one to $\theta = 0$ or to $\theta = 1$, show directly that any Bayes rule is admissible.)

Section 8.9

31. Assume that $X \sim \mathcal{N}(\theta, 1)$, and that it is desired to estimate θ under loss $L(\theta, a) = |\theta - a|$. Show that $\delta_0(x) = x$ is an admissible estimator.

32. If $\pi(\boldsymbol{\theta}) \equiv 1$ in the situation of Theorem 14, show that the generalized Bayes estimator of $\boldsymbol{\psi}$ is $\boldsymbol{\delta}^\pi(\mathbf{x}) = \mathbf{x}$.

33. In Example 12, prove that the conditions of Theorem 14 are satisfied if and only if at most two of the characteristic roots of $\mathbf{B}$ equal 1.

34. Suppose that $X \sim \mathcal{P}(\psi)$; note that the natural parameter is $\theta = \log \psi$. It is desired to estimate ψ under squared-error loss, and

$$\pi(\theta) = (1+\theta)^r I_{(0,\infty)}(\theta).$$

Find conditions on r such that the generalized Bayes rule with respect to π is admissible.

35. In the situation of Example 13, suppose that $\alpha > 2$ and that δ is generalized Bayes with respect to $\pi(\theta) = 2/[(1+\theta^2)\pi]$. Construct an estimator with smaller risk than δ. (It is interesting that π is a *proper prior*, and yet δ is inadmissible.)

36. In Example 15, verify the claims for $b = c < p-1$ and $b = c > p-1$.

APPENDIX 1

Common Statistical Densities

For convenience, we list here several common statistical densities that are used in examples and exercises throughout the book. The listings are brief, giving only the name of the density, the abbreviation that will be used for the density, the sample space $\mathscr{X}$, the range of the parameter values, the density itself, useful moments of the density, important special cases, and a brief explanation if needed for clarification.

In the listings, det $\mathbf{B}$ will stand for the determinant of the matrix $\mathbf{B}$; the symbol $I_A(z)$ is defined by

$$I_A(z) = \begin{cases} 1 & \text{if } z \in A, \\ 0 & \text{if } z \notin A, \end{cases}$$

and is called the *indicator function* on the set A; and $\Gamma(\alpha)$ is the usual gamma function, defined by

$$\Gamma(\alpha) = \int_0^\infty e^{-x} x^{\alpha-1}\, dx.$$

We first list the continuous densities, followed by the discrete densities.

I. Continuous

1. Univariate Normal ($\mathcal{N}(\mu, \sigma^2)$): $\mathscr{X} = R^1$, $-\infty < \mu < \infty$, $\sigma^2 > 0$, and

$$f(x|\mu, \sigma^2) = \frac{1}{(2\pi)^{1/2}\sigma} e^{-(x-\mu)^2/2\sigma^2}.$$

Mean $= \mu$, Variance $= \sigma^2$.

2. p-Variate Normal ($\mathcal{N}_p(\boldsymbol{\mu}, \boldsymbol{\Sigma})$): $\mathscr{X} = R^p$, $\boldsymbol{\mu} = (\mu_1, \ldots, \mu_p)' \in R^p$, $\boldsymbol{\Sigma}$ is a $(p \times p)$ positive definite matrix, and

$$f(\mathbf{x}|\boldsymbol{\mu}, \boldsymbol{\Sigma}) = \frac{1}{(2\pi)^{p/2}(\det \boldsymbol{\Sigma})^{1/2}} e^{-(\mathbf{x}-\boldsymbol{\mu})'\boldsymbol{\Sigma}^{-1}(\mathbf{x}-\boldsymbol{\mu})/2}.$$

Mean $= \boldsymbol{\mu}$, Covariance matrix $= \boldsymbol{\Sigma}$.

3. Uniform ($\mathcal{U}(\alpha, \beta)$): $\mathscr{X} = (\alpha, \beta)$, $-\infty < \alpha < \infty$, $\alpha < \beta < \infty$, and

$$f(x|\alpha, \beta) = \frac{1}{\beta - \alpha} I_{(\alpha,\beta)}(x).$$

Mean $= \frac{1}{2}(\alpha + \beta)$, Variance $= (\beta - \alpha)^2/12$.

4. Gamma ($\mathcal{G}(\alpha, \beta)$): $\mathscr{X} = (0, \infty)$, $\alpha > 0$, $\beta > 0$, and

$$f(x|\alpha, \beta) = \frac{1}{\Gamma(\alpha)\beta^\alpha} x^{\alpha-1} e^{-x/\beta} I_{(0,\infty)}(x).$$

Mean $= \alpha\beta$, Variance $= \alpha\beta^2$.

Special Cases:

(a) Exponential ($\mathscr{E}(\beta)$): the $\mathcal{G}(1, \beta)$ density.
(b) Chi-square with n degrees of freedom ($\chi^2(n)$): the $\mathcal{G}(n/2, 2)$ density.

5. Beta ($\mathscr{B}e(\alpha, \beta)$): $\mathscr{X} = [0, 1]$, $\alpha > 0$, $\beta > 0$, and

$$f(x|\alpha, \beta) = \frac{\Gamma(\alpha + \beta)}{\Gamma(\alpha)\Gamma(\beta)} x^{\alpha-1}(1-x)^{\beta-1} I_{[0,1]}(x).$$

Mean $= \alpha/(\alpha + \beta)$, Variance $= \alpha\beta/(\alpha + \beta)^2(\alpha + \beta + 1)$.

6. Cauchy ($\mathscr{C}(\alpha, \beta)$): $\mathscr{X} = R^1$, $-\infty < \alpha < \infty$, $\beta > 0$, and

$$f(x|\alpha, \beta) = \frac{\beta}{\pi[\beta^2 + (x - \alpha)^2]}.$$

Mean and Variance do not exist.

7. F distribution with α and β degrees of freedom ($\mathscr{F}(\alpha, \beta)$): $\mathscr{X} = (0, \infty)$, $a > 0$, $\beta > 0$, and

$$f(x|\alpha, \beta) = \frac{\Gamma[(\alpha + \beta)/2]\alpha^{\alpha/2}\beta^{\beta/2}}{\Gamma(\alpha/2)\Gamma(\beta/2)} \cdot \frac{x^{\alpha/2-1}}{(\beta + \alpha x)^{(\alpha+\beta)/2}} I_{(0,\infty)}(x).$$

Mean $= \beta/(\beta - 2)$ if $\beta > 2$, Variance $= 2\beta^2(\alpha + \beta - 2)/\alpha(\beta - 4)(\beta - 2)^2$ if $\beta > 4$.

8. t distribution with α degrees of freedom, location parameter μ, and scale parameter $\sigma^2(\mathcal{T}(\alpha, \mu, \sigma^2))$: $\mathcal{X} = R^1$, $\alpha > 0$, $-\infty < \mu < \infty$, $\sigma^2 > 0$, and

$$f(x|\alpha, \mu, \sigma^2) = \frac{\Gamma[(\alpha+1)/2]}{\sigma(\alpha\pi)^{1/2}\Gamma(\alpha/2)} \left(1 + \frac{(x-\mu)^2}{\alpha\sigma^2}\right)^{-(\alpha+1)/2}$$

Mean $= \mu$, if $\alpha > 1$, Variance $= \alpha\sigma^2/(\alpha - 2)$ if $\alpha > 2$.

Note: $(X - \mu)^2/\sigma^2 \sim \mathcal{F}(1, \alpha)$.

9. p-Variate t distribution with α degrees of freedom, location vector $\boldsymbol{\mu}$, and scale matrix $\boldsymbol{\Sigma}(\mathcal{T}_p(\alpha, \boldsymbol{\mu}, \boldsymbol{\Sigma}))$: $\mathcal{X} = R^p$, $\alpha > 0$, $\boldsymbol{\mu} \in R^p$, $\boldsymbol{\Sigma}$ is a $(p \times p)$ positive definite matrix, and

$$f(\mathbf{x}|\alpha, \boldsymbol{\mu}, \boldsymbol{\Sigma}) = \frac{\Gamma[(\alpha+p)/2]}{(\det \boldsymbol{\Sigma})^{1/2}(\alpha\pi)^{p/2}\Gamma(\alpha/2)} \left[1 + \frac{1}{\alpha}(\mathbf{x}-\boldsymbol{\mu})^t\boldsymbol{\Sigma}^{-1}(\mathbf{x}-\boldsymbol{\mu})\right]^{-(\alpha+p)/2}.$$

Mean $= \boldsymbol{\mu}$ if $\alpha > 1$, Covariance matrix $= \alpha\boldsymbol{\Sigma}/(\alpha - 2)$ if $\alpha > 2$.

Note: $(1/p)(\mathbf{X}-\boldsymbol{\mu})^t\boldsymbol{\Sigma}^{-1}(\mathbf{X}-\boldsymbol{\mu}) \sim \mathcal{F}(p, \alpha)$.

10. Inverse Gamma $(\mathcal{IG}(\alpha, \beta))$: $\mathcal{X} = (0, \infty)$, $\alpha > 0$, $\beta > 0$, and

$$f(x|\alpha, \beta) = \frac{1}{\Gamma(\alpha)\beta^\alpha x^{(\alpha+1)}} e^{-1/x\beta} I_{(0,\infty)}(x).$$

Mean $= 1/\beta(\alpha - 1)$ if $\alpha > 1$, Variance $= 1/\beta^2(\alpha-1)^2(\alpha-2)$ if $\alpha > 2$.
Note: $1/X \sim \mathcal{G}(\alpha, \beta)$.

11. Dirichlet $(\mathcal{D}(\boldsymbol{\alpha}))$: $\mathbf{x} = (x_1, \ldots, x_k)^t$ where $\sum_{i=1}^k x_i = 1$ and $0 \le x_i \le 1$ for all i, $\boldsymbol{\alpha} = (\alpha_1, \ldots, \alpha_k)^t$ where $\alpha_i > 0$ for all i, and (defining $\alpha_0 = \sum_{i=1}^k \alpha_i$),

$$f(\mathbf{x}|\boldsymbol{\alpha}) = \frac{\Gamma(\alpha_0)}{\prod_{i=1}^k \Gamma(\alpha_i)} \prod_{i=1}^k x_i^{(\alpha_i - 1)}.$$

Mean $(X_i) = \alpha_i/\alpha_0$, Variance $(X_i) = (\alpha_0 - \alpha_i)\alpha_i/\alpha_0^2(\alpha_0+1)$, Covariance $(X_i, X_j) = -\alpha_i\alpha_j/\alpha_0^2(\alpha_0+1)$.
Note: This is really only a $(k-1)$-dimensional distribution because of the restriction on the x_i. In taking expectations with respect to this density, therefore, replace x_k by $1 - \sum_{i=1}^{k-1} x_i$ and integrate over $x_1, \ldots, x_{k-1}$.

12. Pareto $(\mathcal{P}a(x_0, \alpha))$: $\mathcal{X} = (x_0, \infty)$, $0 < x_0 < \infty$, $\alpha > 0$, and

$$f(x|x_0, \alpha) = \frac{\alpha}{x_0}\left(\frac{x_0}{x}\right)^{\alpha+1} I_{(x_0,\infty)}(x).$$

Mean $= \alpha x_0/(\alpha - 1)$ if $\alpha > 1$.

II. Discrete

1. Binomial ($\mathcal{B}(n, p)$): $\mathcal{X} = \{0, 1, 2, \ldots, n\}$, $0 \le p \le 1$, $n = 1, 2, \ldots$, and

$$f(x|n, p) = \binom{n}{x} p^x (1-p)^{(n-x)},$$

where

$$\binom{n}{x} = \frac{n!}{(x!)(n-x)!}.$$

Mean $= np$, Variance $= np(1-p)$.

Here, X is the number of successes in n independent trials when p is the probability of a success at each individual trial.

2. Poisson ($\mathcal{P}(\lambda)$): $\mathcal{X} = \{0, 1, 2, \ldots\}$, $\lambda > 0$, and

$$f(x|\lambda) = \frac{e^{-\lambda}\lambda^x}{x!}.$$

Mean $= \lambda$, Variance $= \lambda$.

3. Negative Binomial ($\mathcal{NB}(\alpha, p)$): $\mathcal{X} = \{0, 1, \ldots\}$, $0 < p \le 1$, $\alpha > 0$, and

$$f(x|\alpha, p) = \frac{\Gamma(\alpha + x)}{\Gamma(x+1)\Gamma(\alpha)} p^\alpha (1-p)^x.$$

Mean $= \alpha(1-p)/p$, Variance $= \alpha(1-p)/p^2$.

When α is an integer, X is the number of failures in a sequence of independent trials performed until α successes are observed, where p is the probability of a success at each individual trial.

Special Case: Geometric ($\mathcal{G}e(p)$): the $\mathcal{NB}(1, p)$ density.

4. Multinomial ($\mathcal{M}(n, \mathbf{p})$): $\mathbf{x} = (x_1, x_2, \ldots, x_k)^t$ where $\sum_{i=1}^k x_i = n$ and each x_i is an integer between 0 and n, $\mathbf{p} = (p_1, \ldots, p_k)^t$ where $\sum_{i=1}^k p_i = 1$ and $0 \le p_i \le 1$ for all i, and

$$f(\mathbf{x}|\mathbf{p}) = \frac{n!}{\prod_{i=1}^k (x_i!)} \prod_{i=1}^k p_i^{x_i}.$$

Mean $(X_i) = np_i$, Variance $(X_i) = np_i(1-p_i)$, Covariance $(X_i, X_j) = -np_ip_j$.

If an independent sample of size n is drawn from a population of k types, where p_i is the probability that a single observation is of the ith type, then X_i is the number of individuals of the ith type in the sample.
Note: $k = 2$ gives the $\mathcal{B}(n, p)$ distribution, with $p = p_1 = 1 - p_2$.

APPENDIX 2

Supplement to Chapter 4

I. Definition and Properties of H_m

The function H_m is defined for $v > 0$ by

$$H_m(v) = \begin{cases} \dfrac{v^{m/2}}{\left[\dfrac{m}{2}\right]! \left\{ e^v - \sum_{i=0}^{(m/2-1)} \dfrac{v^i}{i!} \right\}} & \text{if } m \text{ is even,} \\[2ex] \dfrac{v^{m/2}}{\Gamma\left(\dfrac{m}{2}+1\right)\left\{ e^v[2\Phi(\sqrt{2v})-1] - \sum_{i=0}^{(m-3)/2} \dfrac{v^{(i+1/2)}}{\Gamma(i+3/2)} \right\}} & \text{if } m \text{ is odd,} \end{cases}$$

where Φ is the standard normal c.d.f. and the summation in the last expression is defined to be zero when $m = 1$. Note that

$$H_1(v) = \frac{2\sqrt{v/\pi}}{\{e^v[2\Phi(\sqrt{2v})-1]\}}$$

and

$$H_2(v) = \frac{v}{[e^v - 1]}.$$

A good approximation to H_m, for most purposes, is

$$H_m(v) \cong \frac{1-w}{1-w^{2[(m+6)/\pi)]^{1/2}}},$$

where $w = 2v/(m+2)$. (This approximation is defined to be

$$1/\{2[(m+6)/\pi]^{1/2}\} \quad \text{if } w = 1.)$$

Some expressions for H_m in terms of standard special functions are

$$H_m(v) = \frac{2v^{m/2}e^{-v}}{[m\gamma(m/2, v)]},$$

where γ is the incomplete gamma function

$$\gamma\left(\frac{m}{2}, v\right) = \int_0^v e^{-t}t^{(m-2)/2}\,dt;$$

and

$$H_m(v) = \left[e^v\Gamma\left(\frac{m}{2}+1\right)v^{-m/2}F_m(2v)\right]^{-1},$$

where F_m is the c.d.f. of the chi-square distribution with m degrees of freedom.

A useful infinite series expansion of H_m is

$$H_m(v) = \left[1 + \sum_{i=1}^{\infty} \frac{v^i}{(m/2+1)(m/2+2)\cdots(m/2+i)}\right]^{-1},$$

and a recurrence relation is

$$H_{m+2}(v) = \frac{2v}{(m+2)}\left(\frac{1}{H_m(v)} - 1\right)^{-1}.$$

II. Development of (4.121) and (4.122)

There are several possible routes to these formulas. The easiest, technically, is to define $\mathbf{P} = \mathbf{\Sigma B}'(\mathbf{B\Sigma B}')^{-1}\mathbf{B}$, and write

$$\mathbf{X} = \mathbf{PX} + (\mathbf{I} - \mathbf{P})\mathbf{X}, \qquad \boldsymbol{\theta} = \mathbf{P}\boldsymbol{\theta} + (\mathbf{I} - \mathbf{P})\boldsymbol{\theta}.$$

Note that $\mathbf{PX}$ and $(\mathbf{I}-\mathbf{P})\mathbf{X}$ are independent, and that $(\mathbf{I}-\mathbf{P})\boldsymbol{\theta}$ is in the null space of $\mathbf{B}$. Since we are assuming that there is no prior knowledge about the null space of $\mathbf{B}$ (in Θ), it is reasonable to choose the constant noninformative prior for $(\mathbf{I}-\mathbf{P})\boldsymbol{\theta}$. Furthermore, if we assume "independence" of $\mathbf{P}\boldsymbol{\theta}$ and $(\mathbf{I}-\mathbf{P})\boldsymbol{\theta}$ *a priori*, then the independence (conditionally given $\boldsymbol{\theta}$) of $\mathbf{PX}$ and $(\mathbf{I}-\mathbf{P})\mathbf{X}$ means that the posterior distribution of $(\mathbf{I}-\mathbf{P})\boldsymbol{\theta}$ given $\mathbf{X}$ is simply the posterior given $(\mathbf{I}-\mathbf{P})\mathbf{X}$, and the posterior of $\mathbf{P}\boldsymbol{\theta}$ given $\mathbf{X}$ is simply the posterior given $\mathbf{PX}$.

Since $(\mathbf{I}-\mathbf{P})\mathbf{X}$ is (singular) normal with mean $(\mathbf{I}-\mathbf{P})\boldsymbol{\theta}$ and covariance matrix $(\mathbf{I}-\mathbf{P})\mathbf{\Sigma}(\mathbf{I}-\mathbf{P})' = (\mathbf{I}-\mathbf{P})\mathbf{\Sigma}$, the posterior for $(\mathbf{I}-\mathbf{P})\boldsymbol{\theta}$ can be seen to be (singular) normal with mean $(\mathbf{I}-\mathbf{P})\mathbf{x}$ and covariance $(\mathbf{I}-\mathbf{P})\mathbf{\Sigma}$.

To find the posterior mean and covariance matrix for $\mathbf{P}\boldsymbol{\theta}$ given $\mathbf{PX}$, it suffices to first find the same quantities for $\mathbf{B}\boldsymbol{\theta}$ and $\mathbf{BX}$ (and then transform by $\mathbf{H} = \mathbf{\Sigma B}'(\mathbf{B\Sigma B}')^{-1}$). Now $\mathbf{BX} \sim \mathcal{N}_k(\mathbf{B}\boldsymbol{\theta}, \mathbf{B\Sigma B}')$, and $\mathbf{B}\boldsymbol{\theta}$ has prior mean $\mathbf{d}$

and covariance matrix $\mathbf{C}$. Imbuing $\mathbf{B\theta}$ with the robust prior $\pi_k(\mathbf{B\theta})$ (with p replaced by k, $\boldsymbol{\mu}$ replaced by $\mathbf{d}$, $\mathbf{A}$ replaced by $\mathbf{C}$, and $\boldsymbol{\Sigma}$ replaced by $\mathbf{B\Sigma B}'$), Lemma 3 (with $\mathbf{x}$ then replaced by $\mathbf{Bx}$) yields the posterior mean, $\boldsymbol{\mu}^{\pi_k}(\mathbf{Bx})$, and covariance matrix, $\mathbf{V}^{\pi_k}(\mathbf{Bx})$, for $\mathbf{B\theta}$. Thus the posterior mean and covariance matrix for $\mathbf{P\theta} = \mathbf{HB\theta}$ are, respectively,

$$\mathbf{H}\boldsymbol{\mu}^{\pi_k}(\mathbf{Bx}) \quad \text{and} \quad \mathbf{HV}^{\pi_k}(\mathbf{Bx})\mathbf{H}'.$$

Finally, the posterior mean and covariance matrix of $\boldsymbol{\theta}$, with respect to the implied prior π^*, are given by

$$\begin{aligned}\boldsymbol{\mu}^{\pi^*}(\mathbf{x}) &= E^{\pi^*(\boldsymbol{\theta}|\mathbf{x})}[\boldsymbol{\theta}] = E^{\pi^*(\boldsymbol{\theta}|\mathbf{x})}[\mathbf{P\theta}] + E^{\pi^*(\boldsymbol{\theta}|\mathbf{x})}[(\mathbf{I}-\mathbf{P})\boldsymbol{\theta}] \\ &= \mathbf{H}\boldsymbol{\mu}^{\pi_k}(\mathbf{Bx}) + (\mathbf{I}-\mathbf{P})\mathbf{x},\end{aligned}$$

and (using the posterior independence of $\mathbf{P\theta}$ and $(\mathbf{I}-\mathbf{P})\boldsymbol{\theta}$)

$$\begin{aligned}\mathbf{V}^{\pi^*}(\mathbf{x}) &= E^{\pi^*(\boldsymbol{\theta}|\mathbf{x})}[(\boldsymbol{\theta}-\boldsymbol{\mu}^{\pi^*}(\mathbf{x}))(\boldsymbol{\theta}-\boldsymbol{\mu}^{\pi^*}(\mathbf{x}))'] \\ &= E[(\mathbf{P\theta}-\mathbf{H}\boldsymbol{\mu}^{\pi_k})(\mathbf{P\theta}-\mathbf{H}\boldsymbol{\mu}^{\pi_k})' + \{(\mathbf{I}-\mathbf{P})(\boldsymbol{\theta}-\mathbf{x})\} \\ &\quad \times\{(\mathbf{I}-\mathbf{P})(\boldsymbol{\theta}-\mathbf{x})\}'] \\ &= \mathbf{HV}^{\pi_k}(\mathbf{Bx})\mathbf{H}' + (\mathbf{I}-\mathbf{P})\boldsymbol{\Sigma}.\end{aligned}$$

Algebra reduces $\boldsymbol{\mu}^{\pi^*}$ and $\mathbf{V}^{\pi^*}$ to (4.121) and (4.122).

III. Verification of Formula (4.123)

By definition, the posterior distribution of $\boldsymbol{\theta}$ and σ^2, given $\mathbf{x}$ and s^2, is

$$\pi_p^*(\boldsymbol{\theta}, \sigma^2|\mathbf{x}, s^2) = \frac{\pi_p(\boldsymbol{\theta}|\sigma^2)\sigma^{-2}f(\mathbf{x}|\boldsymbol{\theta}, \sigma^2)h(s^2|\sigma^2)}{m(\mathbf{x}, s^2)}, \tag{A2.1}$$

where h is the density of S^2 and $m(\mathbf{x}, s^2)$ is the joint marginal density of $\mathbf{X}$ and S^2. Note that

$$\pi_p(\boldsymbol{\theta}|\sigma^2)f(\mathbf{x}|\boldsymbol{\theta}, \sigma^2) = \pi_p^*(\boldsymbol{\theta}|\mathbf{x}, \sigma^2)m(\mathbf{x}|\sigma^2), \tag{A2.2}$$

where $\pi_p^*(\boldsymbol{\theta}|\mathbf{x}, \sigma^2)$ is the posterior distribution of $\boldsymbol{\theta}$ given $\mathbf{x}$ and σ^2, while $m(\mathbf{x}|\sigma^2)$ is the marginal density of $\mathbf{x}$ given σ^2.

The approximation that will be made consists of (i) replacing $m(\mathbf{x}|\sigma^2)$ by $m(\mathbf{x}|\hat{\sigma}^2)$, where $\hat{\sigma}^2 = s^2/(m+2)$, and (ii) replacing $\pi_p^*(\boldsymbol{\theta}|\mathbf{x}, \sigma^2)$ by a normal distribution (to be denoted $f(\boldsymbol{\theta})$) with mean $\boldsymbol{\mu}^*(\mathbf{x}, s^2)$ and covariance matrix $(\sigma^2/\hat{\sigma}^2)\mathbf{V}^*(\mathbf{x}, s^2)$. These approximations will be discussed after the derivation is complete. Using these approximations in (A2.2) and hence (A2.1), and defining

$$k(\mathbf{x}, s^2) = \frac{m(\mathbf{x}|\hat{\sigma}^2)}{m(\mathbf{x}, s^2)},$$

results in the following approximation to $\pi_p^*(\boldsymbol{\theta}, \sigma^2|\mathbf{x}, s^2)$:

$$\hat{\pi}_p(\boldsymbol{\theta}, \sigma^2|\mathbf{x}, s^2) = k(\mathbf{x}, s^2) f(\boldsymbol{\theta}) \sigma^{-2} h(s^2|\sigma^2).$$

To make decisions and inferences about $\boldsymbol{\theta}$, the marginal posterior distribution of $\boldsymbol{\theta}$ given $\mathbf{x}$ and s^2 is needed. This will be given by

$$\hat{\pi}_p(\boldsymbol{\theta}|\mathbf{x}, s^2) = k(\mathbf{x}, s^2) \int_0^\infty f(\boldsymbol{\theta}) \sigma^{-2} h(s^2|\sigma^2)\, d\sigma^2.$$

For notational convenience in evaluating this integral, let

$$\mathcal{Q}(\boldsymbol{\theta}, \mathbf{x}, s^2) = [\boldsymbol{\theta} - \boldsymbol{\mu}^*(\mathbf{x}, s^2)]' \mathbf{V}^*(\mathbf{x}, s^2)^{-1} [\boldsymbol{\theta} - \boldsymbol{\mu}^*(\mathbf{x}, s^2)],$$

and define $k^*(\mathbf{x}, s^2)$ as $k(\mathbf{x}, s^2)$ multiplied by all multiplicative constants and factors (not involving $\boldsymbol{\theta}$ or σ^2) from the densities f and h. Then

$$\begin{aligned} \hat{\pi}_n(\boldsymbol{\theta}|\mathbf{x}, s^2) &= k^*(\mathbf{x}, s^2) \int_0^\infty \sigma^{-p} \exp\left\{-\frac{\hat{\sigma}^2}{2\sigma^2} \mathcal{Q}(\boldsymbol{\theta}, \mathbf{x}, s^2)\right\} \sigma^{-m} \\ &\quad \times \exp\left\{-\frac{s^2}{2\sigma^2}\right\} \sigma^{-2}\, d\sigma^2 \\ &= k^*(\mathbf{x}, s^2) \int_0^\infty \sigma^{-(m+p+2)} \exp\left\{-\frac{s^2}{2\sigma^2}\left[1 + \frac{\mathcal{Q}(\boldsymbol{\theta}, \mathbf{x}, s^2)}{m+2}\right]\right\} d\sigma^2. \end{aligned}$$

Making the change of variables

$$z = \frac{s^2}{2\sigma^2}\left[1 + \frac{\mathcal{Q}(\boldsymbol{\theta}, \mathbf{x}, s^2)}{m+2}\right]$$

results in the expression

$$\hat{\pi}_p(\boldsymbol{\theta}|\mathbf{x}, s^2) = k^*(\mathbf{x}, s^2) \int_0^\infty \frac{[2^{(m+p)/2}][z^{(m+p-2)/2}] e^{-z}}{s^{(m+p)}[1 + \mathcal{Q}(\boldsymbol{\theta}, \mathbf{x}, s^2)/(m+2)]^{(m+p)/2}}\, dz.$$

Noting that $\int_0^\infty z^{(m+p-2)/2} e^{-z}\, dz = \Gamma([m+p]/2)$, and defining $\tilde{k}(\mathbf{x}, s^2)$ as $k^*(\mathbf{x}, s^2)$ multiplied by all new factors not involving $\boldsymbol{\theta}$ above, we finally have

$$\hat{\pi}_p(\boldsymbol{\theta}|\mathbf{x}, s^2) = \frac{\tilde{k}(\mathbf{x}, s^2)}{[1 + \mathcal{Q}(\boldsymbol{\theta}, \mathbf{x}, s^2)/(m+2)]^{(m+p)/2}},$$

which is the expression in (4.123).

The rationale behind the approximations (i) and (ii) is that in the expression

$$\begin{aligned} \boldsymbol{\Sigma}^{1/2}(\boldsymbol{\Sigma} + \mathbf{A})^{-1} \boldsymbol{\Sigma}^{1/2} &= (\mathbf{I}_p + \boldsymbol{\Sigma}^{-1/2} \mathbf{A} \boldsymbol{\Sigma}^{-1/2})^{-1} \\ &= (\mathbf{I}_p + \boldsymbol{\Sigma}_0^{-1/2} \mathbf{A} \boldsymbol{\Sigma}_0^{-1/2}/\sigma^2)^{-1}, \end{aligned}$$

replacing σ^2 by $\hat{\sigma}^2$ is a reasonable approximation, in that any resultant error is like an error in specifying $\mathbf{A}$. We have seen many indications that the procedures considered are robust against this type of error. The same type of argument holds for expressions like $\boldsymbol{\Sigma}(\boldsymbol{\Sigma} + A)^{-1}$ and $(\mathbf{X} - \boldsymbol{\mu})'(\boldsymbol{\Sigma} +$

$\mathbf{A})^{-1}(\mathbf{X}-\boldsymbol{\mu})$. It can be seen that all terms in $m(\mathbf{x}|\sigma^2)$, $\boldsymbol{\mu}^{\pi_p}(\mathbf{x})$, and $\mathbf{V}^{\pi_p}(\mathbf{x})$ are of this type, except for a multiple, $\boldsymbol{\Sigma}$, of $\mathbf{V}^{\pi_p}(\mathbf{x})$. This multiple cannot be approximated without significantly altering the distribution, so it is left as is, introducing the factor σ^2 in the approximation to the covariance matrix. The replacing of $\pi_p^*(\boldsymbol{\theta}|\mathbf{x}, \sigma^2)$ by a normal distribution with the same approximate mean and covariance matrix is the same type of approximation used to good effect previously.

APPENDIX 3

Technical Arguments from Chapter 7

I. Verification of Formula (7.8)

Letting $K'(n)$ denote the derivative of $K(n)$ (assuming it exists) it is clear that

$$\frac{d}{dn} P_\theta(\bar{X}_n < K(n)) = \frac{d}{dn} \int_{-\infty}^{K(n)} \left(\frac{n}{2\pi\sigma^2}\right)^{1/2} \exp\left\{-\frac{n}{2\sigma^2}(y-\theta)^2\right\} dy$$

$$= \left(\frac{n}{2\pi\sigma^2}\right)^{1/2} \exp\left\{-\frac{n}{2\sigma^2}(K(n)-\theta)^2\right\} K'(n)$$

$$+ \int_{-\infty}^{K(n)} \frac{1}{2\sigma(2\pi n)^{1/2}} \exp\left\{-\frac{n}{2\sigma^2}(y-\theta)^2\right\} dy$$

$$- \int_{-\infty}^{K(n)} \left(\frac{n}{2\pi\sigma^2}\right)^{1/2} \left(\frac{(y-\theta)^2}{2\sigma^2}\right) \exp\left\{-\frac{n}{2\sigma^2}(y-\theta)^2\right\} dy.$$

Integrating by parts shows that

$$\int_{-\infty}^{K(n)} (y-\theta)\left(\frac{-n(y-\theta)}{\sigma^2}\right) \exp\left\{-\frac{n}{2\sigma^2}(y-\theta)^2\right\} dy$$

$$= (y-\theta)\exp\left\{-\frac{n}{2\sigma^2}(y-\theta)^2\right\}\Bigg|_{-\infty}^{K(n)} - \int_{-\infty}^{K(n)} \exp\left\{-\frac{n}{2\sigma^2}(y-\theta)^2\right\} dy$$

$$= (K(n)-\theta)\exp\left\{-\frac{n}{2\sigma^2}(K(n)-\theta)^2\right\} - \int_{-\infty}^{K(n)} \exp\left\{-\frac{n}{2\sigma^2}(y-\theta)^2\right\} dy.$$

Hence

$$\frac{d}{dn}P_\theta(\bar{X}_n < K(n)) = \left(\frac{n}{2\pi\sigma^2}\right)^{1/2}\exp\left\{-\frac{n}{2\sigma^2}(K(n)-\theta)^2\right\}K'(n)$$
$$+\int_{-\infty}^{K(n)}\frac{1}{2\sigma(2\pi n)^{1/2}}\exp\left\{-\frac{n}{2\sigma^2}(y-\theta)^2\right\}dy$$
$$+\frac{1}{2\sigma(2\pi n)^{1/2}}\left[(K(n)-\theta)\exp\left\{-\frac{n}{2\sigma^2}(K(n)-\theta)^2\right\}\right.$$
$$\left.-\int_{-\infty}^{K(n)}\exp\left\{-\frac{n}{2\sigma^2}(y-\theta)^2\right\}dy\right]$$
$$=(2\pi n\sigma^2)^{-1/2}\exp\left\{-\frac{n}{2\sigma^2}(K(n)-\theta)^2\right\}$$
$$\times[nK'(n)+\tfrac{1}{2}(K(n)-\theta)].$$

Assuming that differentiation under the integral sign is valid in (7.7), it follows from the above result that

$$\frac{d}{dn}r(\pi, \delta_n^\pi) = \int_{\Theta_1} L(\theta, a_0)(2\pi n\sigma^2)^{-1/2}\exp\left\{-\frac{n}{2\sigma^2}(K(n)-\theta)^2\right\}$$
$$\times[nK'(n)+\tfrac{1}{2}(K(n)-\theta)]dF^\pi(\theta)$$
$$-\int_{\Theta_0} L(\theta, a_1)(2\pi n\sigma^2)^{-1/2}\exp\left\{-\frac{n}{2\sigma^2}(K(n)-\theta)^2\right\}$$
$$\times[nK'(n)+\tfrac{1}{2}(K(n)-\theta)]dF^\pi(\theta).$$

Formula (7.8) follows from this and (7.6).

II. Verification of Formula (7.10)

It is first necessary to determine $K(n)$. Recall that $\pi^n(\theta|\bar{x}_n)$ is a $\mathcal{N}(\mu_n(\bar{x}_n), \rho_n)$ density (see (7.3)). Note also, from the definitions of the posterior density and the marginal density, that

$$\left(\frac{2\pi\sigma^2}{n}\right)^{-1/2}\exp\left\{-\frac{n}{2\sigma^2}(K(n)-\theta)^2\right\}\pi(\theta) = f(K(n)|\theta)\pi(\theta)$$
$$= m_n(K(n))\pi^n(\theta|K(n)). \tag{A3.1}$$

Using this in (7.6) and eliminating common constants gives

$$\int_{\theta_0}^{\infty} W(|\theta-\theta_0|)\exp\left\{-\frac{1}{2\rho_n}[\theta-\mu_n(K(n))]^2\right\}d\theta$$
$$=\int_{-\infty}^{\theta_0} W(|\theta-\theta_0|)\exp\left\{-\frac{1}{2\rho_n}[\theta-\mu_n(K(n))]^2\right\}d\theta. \tag{A3.2}$$

From the symmetry of the situation, it is clear that equality is achieved in (A3.2) when

$$\mu_n(K(n)) = \theta_0, \tag{A3.3}$$

or (using (7.3)) when

$$K(n) = \left(1 + \frac{\sigma^2}{n\tau^2}\right)\theta_0 - \frac{\sigma^2}{n\tau^2}\mu. \tag{A3.4}$$

Using (A3.1) and (A3.3) in (7.8), and choosing $K^* = \theta_0$, gives

$$\begin{aligned}
\frac{d}{dn} r(\pi, \delta_n^\pi) = \frac{m_n(K(n))}{2n} \Bigg[\int_{-\infty}^{\theta_0} & W(|\theta - \theta_0|)(\theta - \theta_0)\pi^n(\theta|K(n))d\theta \\
& - \int_{\theta_0}^{\infty} W(|\theta - \theta_0|)(\theta - \theta_0)\pi^n(\theta|K(n))d\theta \Bigg] \\
= \frac{m_n(K(n))}{2n} \Bigg[\int_{-\infty}^{\theta_0} & W(|\theta - \theta_0|)(\theta - \theta_0)(2\pi\rho_n)^{-1/2} \\
& \times \exp\left\{-\frac{1}{2\rho_n}(\theta - \theta_0)^2\right\} d\theta \\
& - \int_{\theta_0}^{\infty} W(|\theta - \theta_0|)(\theta - \theta_0)(2\pi\rho_n)^{-1/2} \\
& \times \exp\left\{-\frac{1}{2\rho_n}(\theta - \theta_0)^2\right\} d\theta \Bigg].
\end{aligned}$$

A change of variables ($y = (\theta - \theta_0)\rho_n^{-1/2}$) establishes that

$$\frac{d}{dn} r(\pi, \delta_n^\pi) = -n^{-1} m_n(K(n)) \int_0^\infty W(\rho_n^{1/2} y)\rho_n^{1/2} y (2\pi)^{-1/2} e^{-y^2/2}\, dy. \tag{A3.5}$$

Recalling that

$$m_n(K(n)) = \left[2\pi\left(\tau^2 + \frac{\sigma^2}{n}\right)\right]^{-1/2} \exp\left\{-\frac{1}{2(\tau^2 + \sigma^2/n)}(K(n) - \mu)^2\right\},$$

and noting from (A3.4) that

$$[K(n) - \mu]^2 = \left[\left(1 + \frac{\sigma^2}{n\tau^2}\right)\theta_0 - \frac{\sigma^2}{n\tau^2}\mu - \mu\right]^2 = \left(1 + \frac{\sigma^2}{n\tau^2}\right)^2 (\theta_0 - \mu)^2,$$

a little algebra shows that (A3.5) is equivalent to (7.10).

Bibliography

Aitchison, J., and Dunsmore, I. R., 1975. *Statistical Prediction Analysis.* Cambridge University Press, Cambridge.

Akaike, H., 1978. A new look at the Bayes procedure. *Biometrika* **65**, 53–59.

Alam, K., and Hawkes, J. S., 1979. Minimax property of Stein's estimator. *Commun. Statist.—Theory and Methods* **A8**, 581–590.

Albert, J. H., 1981. Simultaneous estimation of Poisson means. *J. Multivariate Anal.* **11**, 400–417.

Albert, J. H., 1983. A gamma-minimax approach for estimating a binomial probability. Technical Report, Department of Mathematics and Statistics, Bowling Green State University, Bowling Green, Ohio.

Albert, J. H., 1984a. Empirical Bayes estimation of a set of binomial probabilities. *J. Statist. Comput. Simul.* **20**, 129–144.

Albert, J. H., 1984b. Estimation of a normal mean using mixtures of normal priors. Technical Report, Department of Mathematics and Statistics, Bowling Green State University, Bowling Green, Ohio.

Albert, J. H., and Gupta, A. K., 1982. Mixtures of Dirichlet distributions and estimation in contingency tables. *Ann. Statist.* **10**, 1261–1268.

Allais, M., and Hagen, O., 1979. *Expected Utility Hypotheses and the Allais Paradox.* Reidel, Dordrecht.

Alpert, M., and Raiffa, H., 1982. A progress report on the training of probability assessors. In *Judgement Under Uncertainty: Heuristics and Biases,* D. Kahneman, P. Slovic, and A. Tversky (Eds.). Cambridge University Press, Cambridge.

Alvo, M., 1977. Bayesian sequential estimates. *Ann. Statist.* **5**, 955–968.

Amster, S. J., 1963. A modified Bayes stopping rule. *Ann. Math. Statist.* **34**, 1404–1413.

Anscombe, F. J., 1963a. Sequential medical trials. *J. Amer. Statist. Assoc.* **58**, 365–383.

Anscombe, F. J., 1963b. Bayesian inference concerning many parameters with reference to supersaturated designs. *Bull. Int. Statist. Inst.* **40**, 721–733.

Antoniak, C. E., 1974. Mixtures of Dirichlet processes with applications to Bayesian nonparametric problems. *Ann. Statist.* **2**, 1152–1174.

Armitage, P., 1961. Contribution to the discussion of C. A. B. Smith, "Consistency in statistical inference and decision." *J. Roy. Statist. Soc.* (Ser. B) **23**, 1–37.

Arnold, S. F., 1981. *The Theory of Linear Models and Multivariate Analysis.* Wiley, New York.

Arrow, K. J., 1966. *Social Choice and Individual Values* (2nd edn.). Wiley, New York.

Arrow, K. J., Blackwell, D., and Girshick, M. A., 1949. Bayes and minimax solutions of sequential decision problems. *Econometrika* **17**, 213-244.

Atchison, T. A., and Martz, H. F. (Eds.), 1969. *Proceedings of the Symposium on Empirical Bayes Estimation and Computing in Statistics.* Mathematics Series No. 6, Texas Technical University, Lubbock.

Bacharach, M., 1975. Group decisions in the face of differences of opinion. *Manage. Sci.* **22**, 182-191.

Bahadur, R. R., 1954. Sufficiency and statistical decision functions. *Ann. Math. Statist.* **25**, 423-462.

Bahadur, R. R., 1958. A note on the fundamental identity of sequential analysis. *Ann. Math. Statist.* **29**, 534-543.

Balder, E. J., 1980. An extension of the usual model in statistical decision theory with applications to stochastic optimization problems. *J. Multivariate Anal.* **10**, 385-397.

Balder, E. J., 1983. Mathematical foundations of statistical decision theory. Preprint No. 199, Mathematical Institute, University of Utrecht, The Netherlands.

Balder, E. J., 1984. A general approach to lower semicontinuity and lower closure in optimal control theory. *SIAM J. Control and Optimization* **22**, 570-598.

Balder, E. J., Gilliland, D. C., and van Houwelingen, J. C., 1983. On the essential completeness of Bayes empirical Bayes decision rules. *Statistics and Decisions* **1**, 503-509.

Baranchik, A. J., 1970. A family of minimax estimators of the mean of a multivariate normal distribution. *Ann. Math. Statist.* **41**, 642-645.

Barlow, R. E., Mensing, R. W., and Smiriga, N. G., 1984. Combination of expert's opinions based on decision theory. In *Proceedings of the International Conference on Reliability and Quality Control*, A. Basu (Ed.).

Barlow, R. E., and Shor, S. W. W., 1984. Informative stopping rules. Report ORC 84-1, Operations Research Center, University of California, Berkeley.

Barnard, G. A., 1947. A review of "Sequential Analysis" by Abraham Wald. *J. Amer. Statist. Assoc.* **42**, 658-669.

Barnard, G. A., 1949. Statistical inference (with discussion). *J. Roy. Statist. Soc.* (Ser. B) **11**, 115-139.

Barnard, G. A., 1962. Comments on Stein's "A remark on the likelihood principle." *J. Roy. Statist. Soc.* (Ser. A) **125**, 569-573.

Barnard, G. A., 1967. The use of the likelihood function in statistical practice. In *Proc. Fifth Berkeley Symp. Math. Statist. Probab.* **1**, 27-40, University of California Press, Berkeley.

Barnard, G. A., 1980. Pivotal inference and the Bayesian controversy. In *Bayesian Statistics*, J. M. Bernardo, M. H. DeGroot, D. V. Lindley, and A. F. M. Smith (Eds.). University Press, Valencia.

Barnard, G. A., Jenkins, G. M., and Winsten, C. B., 1962. Likelihood inference and time series. *J. Roy. Statist. Soc.* (Ser. A) **125**, 321-372.

Barndorff-Nielsen, O., 1971. *On Conditional Statistical Inference.* Matematisk Institute, Aarhus University, Aarhus.

Barndorff-Nielsen, O., 1978. *Information and Exponential Families in Statistical Theory.* Wiley, New York.

Barnett, V., 1973. *Comparative Statistical Inference* (2nd edn., 1982). Wiley, New York.

Bartholomew, D. J., 1967. Hypothesis testing when the sample size is treated as a random variable. *J. Roy. Statist. Soc.* **29**, 53-82.

Basu, D., 1975. Statistical information and likelihood (with discussion). *Sankhyā* (Ser. A) **37**, 1-71.

Bayes, T., 1783. An essay towards solving a problem in the doctrine of chances. *Phil. Trans. Roy. Soc.* **53**, 370-418.

Bechhofer, R. E., Kiefer, J., and Sobel, M., 1968. *Sequential Identification and Ranking Procedures.* University of Chicago Press, Chicago.

Becker, G. M., DeGroot, M. H., and Marschak, J., 1964. Measuring utility by a single-response sequential method. *Behav. Sci.* **9**, 226-232.

Berger, J., 1975. Minimax estimation of location vectors for a wide class of densities. *Ann. Statist.* **3**, 1318-1328.

Berger, J., 1976a. Admissibility results for generalized Bayes estimators of coordinates of a location vector. *Ann. Statist.* **4**, 334-356.

Berger, J., 1976b. Minimax estimation of a multivariate normal mean under arbitrary quadratic loss. *J. Multivariate Anal.* **6**, 256-264.

Berger, J., 1976c. Admissible minimax estimation of a multivariate normal mean with arbitrary quadratic loss. *Ann. Statist.* **4**, 223-226.

Berger, J., 1976d. Inadmissibility results for generalized Bayes estimators of coordinates of a location vector. *Ann. Statist.* **4**, 302-333.

Berger, J., 1979. Multivariate estimation with nonsymmetric loss functions. In *Optimizing Methods in Statistics*, J. S. Rustagi (Ed.). Academic Press, New York.

Berger, J., 1980a. Improving on inadmissible estimators in continuous exponential families with applications to simultaneous estimation of gamma scale parameters. *Ann. Statist.* **8**, 545-571.

Berger, J., 1980b. A robust generalized Bayes estimator and confidence region for a multivariate normal mean. *Ann. Statist.* **8**, 716-761.

Berger, J., 1980c. A modification of Brown's technique for proving inadmissibility. In *Recent Developments in Statistical Inference and Data Analysis*, K. Matusita (Ed.). North-Holland, Amsterdam.

Berger, J., 1982a. Selecting a minimax estimator of a multivariate normal mean. *Ann. Statist.* **10**, 81-92.

Berger, J., 1982b. Bayesian robustness and the Stein effect. *J. Amer. Statist. Assoc.* **77**, 358-368.

Berger, J., 1982c. Estimation in continuous exponential families: Bayesian estimation subject to risk restrictions and inadmissibility results. In *Statistical Decision Theory and Related Topics III*, S. S. Gupta and J. Berger (Eds.). Academic Press, New York.

Berger, J., 1984a. The robust Bayesian viewpoint (with discussion). In *Robustness of Bayesian Analysis*, J. Kadane (Ed.). North-Holland, Amsterdam.

Berger, J., 1984b. The frequentist viewpoint and conditioning. In *Proceedings of the Berkeley Conference in Honor of Kiefer and Neyman*, L. LeCam and R. Olshen (Eds.). Wadsworth, Belmont.

Berger, J., 1984c. A review of J. Kiefer's work on conditional frequentist statistics. To appear in *The Collected Works of Jack Kiefer*, L. Brown, I. Olkin, J. Sacks, and H. Wynn (Eds.). Springer-Verlag, New York.

Berger, J., 1984d. In defence of the likelihood principle: axiomatics and coherency. To appear in *Bayesian Statistics II*, J. M. Bernardo, M. H. DeGroot, D. Lindley, and A. Smith (Eds.). North-Holland, Amsterdam.

Berger, J., 1984e. Bayesian salesmanship. In *Bayesian Inference and Decision Techniques with Applications: Essays in Honor of B. deFinetti*, P. K. Goel and A. Zellner (Eds.). North-Holland, Amsterdam.

Berger, J., 1986. *Multivariate Estimation: Bayes, Empirical Bayes, and Stein Approaches.* SIAM, Philadelphia.

Berger, J., and Berliner, L. M., 1983. Robust Bayes and empirical Bayes analysis with ε-contaminated priors. Technical Report #83-35, Department of Statistics, Purdue University, West Lafayette.

Berger, J., and Berliner, L. M., 1984. Bayesian input in Stein estimation and a new minimax empirical Bayes estimator. *J. Econometrics* **25**, 87–108.

Berger, J., Berliner, L. M., and Zaman, A., 1982. General admissibility and inadmissibility results for estimation in a control problem. *Ann. Statist.* **10**, 838–856.

Berger, J., Bock, M. E., Brown, L. D., Casella, G., and Gleser, L., 1977. Minimax estimation of a normal mean vector for arbitrary quadratic loss and unknown covariance matrix. *Ann. Statist.* **5**, 763–771.

Berger, J., and Das Gupta, A., 1985. Bayesian inference for a normal mean using robust priors, calculation of t convolutions, and scientific communication. Technical Report, Department of Statistics, Purdue University, West Lafayette.

Berger, J., and Dey, D. K., 1983. Combining coordinates in simultaneous estimation of normal means. *J. Statist. Planning and Inference* **8**, 143–160.

Berger, J., and Dey, D. K., 1985. Truncation of shrinkage estimators of normal means in the nonsymmetric case. In *Multivariate Analysis VI*, P. R. Krishnaiah (Ed.). North-Holland, Amsterdam.

Berger, J., and Haff, L. R., 1983. A class of minimax estimators of a normal mean vector for arbitrary quadratic loss and unknown covariance matrix. *Statistics and Decisions* **1**, 105–129.

Berger, J., and Sellke, T., 1984. Testing of a point null hypothesis: the irreconcilability of significance levels and evidence. Technical Report #84-27, Department of Statistics, Purdue University, West Lafayette.

Berger, J., and Srinivasan, C., 1978. Generalized Bayes estimators in multivariate problems. *Ann. Statist.* **6**, 783–801.

Berger, J., and Wolpert, R., 1983. Estimating the mean function of a Gaussian process and the Stein effect. *J. Multivariate Anal.* **13**, 401–424.

Berger, J., and Wolpert, R., 1984. *The Likelihood Principle.* Institute of Mathematical Statistics Monograph Series, Hayward, California.

Berger, R., 1979. Gamma minimax robustness of Bayes rules. *Commun. Statist.* **8**, 543–560.

Berger, R., 1981. A necessary and sufficient condition for reaching a consensus using DeGroot's method. *J. Amer. Statist. Assoc.* **76**, 415–418.

Berk, R. H., Brown, L. D., and Cohen, A., 1981. Properties of Bayes sequential tests. *Ann. Statist.* **9**, 678–682.

Berkson, J., 1938. Some difficulties of interpretation encountered in the application of the chi-square test. *J. Amer. Statist. Assoc.* **33**, 526–542.

Berliner, L. M., 1983. Improving on inadmissible estimators in the control problem. *Ann. Statist.* **11**, 814–826.

Berliner, L. M., 1984a. Robust Bayesian analysis with applications in reliability. Technical Report, Department of Statistics, Ohio State University, Columbus.

Berliner, L. M., 1984b. A decision-theoretic structure for robust Bayesian analysis with applications to the estimation of a multivariate normal mean. In *Bayesian Statistics II*, J. M. Bernardo, M. H. DeGroot, D. V. Lindley, and A. F. M. Smith (Eds.). North-Holland, Amsterdam.

Bernardo, J. M., 1979a. Expected information as expected utility. *Ann. Statist.* **7**, 686–690.

Bernardo, J. M., 1979b. Reference posterior distributions for Bayesian inference (with discussion). *J. Roy. Statist. Soc.* **41**, 113–147.

Bernardo, J. M., and Bermudez, J. D., 1984. The choice of variables in probabilistic classification. In *Bayesian Statistics II*, J. M. Bernardo, M. H. DeGroot, D. V. Lindley, and A. F. M. Smith (Eds.). North-Holland, Amsterdam.

Berry, D., 1982. Optimal stopping regions with islands and peninsulas. *Ann. Statist.* **10**, 634–636.

Berry, D., and Christensen, R., 1979. Empirical Bayes estimation of a binomial parameter via mixture of Dirichlet processes. *Ann. Statist.* **7**, 558–568.

Berry, D., and Fristedt, B., 1985. *Bandit Problems: Sequential Allocation of Experiments.* Chapman & Hall, London.

Bhattacharya, P. K., 1966. Estimating the mean of a multivariate normal population with general quadratic loss function. *Ann. Math. Statist.* **37**, 1819–1824.

Bibby, J., and Toutenburg, H., 1977. *Prediction and Improved Estimation in Linear Models.* Wiley, New York.

Bickel, P. J., 1981. Minimax estimation of the mean of a normal distribution when the parameter space is restricted. *Ann. Statist.* **9**, 1301–1309.

Bickel, P. J., 1984. Parametric robustness or small biases can be worthwhile. *Ann. Statist.* **12**, 864–879.

Bickel, P. J., and Yahav, J. A., 1967. Asymptotically pointwise optimal procedures in sequential analysis. In *Proc. Fifth Berkeley Symp. Math. Statist. Probab.* **1**, 401–413. University of California Press, Berkeley.

Birnbaum, A., 1955. Characterizations of complete classes of tests of some multiparametric hypotheses, with applications to likelihood ratio tests. *Ann. Math. Statist.* **26**, 21–36.

Birnbaum, A., 1962. On the foundations of statistical inference (with discussion). *J. Amer. Statist. Assoc.* **57**, 269–326.

Birnbaum, A., 1977. The Neyman–Pearson theory as decision theory and as inference theory: with a criticism of the Lindley–Savage argument for Bayesian theory. *Synthese* **36**, 19–49.

Bishop, Y., Fienberg, S., and Holland, P., 1975. *Discrete Multivariate Analysis: Theory and Practice.* MIT Press, Cambridge, Massachusetts.

Blackwell, D., 1951. On the translation parameter problem for discrete variables. *Ann. Math. Statist.* **18**, 105–110.

Blackwell, D., and Girshick, M. A., 1954. *Theory of Games and Statistical Decisions.* Wiley, New York.

Blum, J. R., and Rosenblatt, J., 1967. On partial *a priori* information in statistical inference. *Ann. Math. Statist.* **38**, 1671–1678.

Blyth, C. R., 1951. On minimax statistical decision procedures and their admissibility. *Ann. Math. Statist.* **22**, 22–42.

Bock, M. E., 1975. Minimax estimators of the mean of a multivariate normal distribution. *Ann. Statist.* **3**, 209–218.

Bock, M. E., 1982. Employing vague inequality information in the estimation of normal mean vectors (estimators that shrink to closed convex polyhedra). In *Statistical Decision Theory and Related Topics III*, S. S. Gupta and J. Berger (Eds.). Academic Press, New York.

Bondar, J. V., 1977. On a conditional confidence property. *Ann. Statist.* **5**, 881–891.

Bondar, J. V., and Milnes, P., 1981. Amenability: a survey for statistical applications of Hunt-Stein and related conditions on groups. *Z. Wahrsch. verw. Gebiete* **57**, 103–128.

Bordley, R. F., 1982. A multiplicative formula for aggregating probability assessments. *Manage. Sci.* **28**, 1137–1148.

Borel, E., 1921. La theorie du jeu et les equations integrales a noyau symetrique. *C. R. Acad. Sci. Paris* **173**, 1304–1308.

Box, G. E. P., 1980. Sampling and Bayes inference in scientific modelling and robustness (with discussion). *J. Roy. Statist. Soc.* (Ser. A) **143**, 383–430.

Box, G. E. P., and Tiao, G. C., 1973. *Bayesian Inference in Statistical Analysis.* Addison-Wesley, Reading, Massachusetts.

Brandwein, A., and Strawderman, W., 1980. Minimax estimators of location parameters for spherically symmetric distributions with concave loss. *Ann. Statist.* **8**, 279–284.

Brier, G. W., 1950. Verification of forecasts expressed in terms of probabilities. *Month. Weather Rev.* **78**, 1–3.

Brockett, P. L., Charnes, A., and Paick, K. H., 1984. A method for constructing a unimodal inferential or prior distribution. Working Paper 83/84-2-26, Department of Finance, University of Texas, Austin.

Brown, L. D., 1966. On the admissibility of invariant estimators of one or more location parameters. *Ann. Math. Statist.* **37**, 1087–1136.

Brown, L. D., 1971. Admissible estimators, recurrent diffusions, and insoluble boundary-value problems. *Ann. Math. Statist.* **42**, 855–903.

Brown, L. D., 1973. Estimation with incompletely specified loss functions. *J. Amer. Statist. Assoc.* **70**, 417–427.

Brown, L. D., 1976. *Notes on Statistical Decision Theory.* (Unpublished lecture notes, Ithaca.)

Brown, L. D., 1977. Closure theorems for sequential-design processes. In *Statistical Decision Theory and Related Topics II*, S. S. Gupta and D. S. Moore (Eds.). Academic Press, New York.

Brown, L. D., 1978. A contribution to Kiefer's theory of conditional confidence procedures. *Ann. Statist.* **6**, 59–71.

Brown, L. D., 1979. An heuristic method for determining admissibility of estimators—with applications. *Ann. Statist.* **7**, 960–994.

Brown, L. D., 1980a. A necessary condition for admissibility. *Ann. Statist.* **8**, 540–544.

Brown, L. D., 1980b. Examples of Berger's phenomenon in the estimation of independent normal means. *Ann. Statist.* **8**, 572–585.

Brown, L. D., 1981. A complete class theorem for statistical problems with finite sample spaces. *Ann. Statist.* **9**, 1289–1300.

Brown, L. D., 1984. Admissibility in discrete and continuous invariant nonparametric estimation problems, and in their multinomial analogues. Technical Report, Department of Mathematics, Cornell University, Ithaca, New York.

Brown, L. D., 1985. *Foundations of Exponential Families.* Institute of Mathematical Statistics Monograph Series, Hayward, California.

Brown, L. D., Cohen, A., and Samuel-Cahn, E., 1983. A sharp necessary condition for admissibility of sequential tests—necessary and sufficient conditions for admissibility of SPRT's. *Ann. Statist.* **11**, 640–653.

Brown, L. D., Cohen, A., and Strawderman, W. E., 1976. A complete class theorem for strict monotone likelihood ratio with applications. *Ann. Statist.* **4**, 712–722.

Brown, L. D., Cohen, A., and Strawderman, W. E., 1979. Monotonicity of Bayes sequential tests. *Ann. Statist.* **7**, 1222–1230.

Brown, L. D., Cohen, A., and Strawderman, W. E., 1980. Complete classes for sequential tests of hypotheses. *Ann. Statist.* **8**, 377–398.

Brown, L. D., and Farrell, R. H., 1985a. Complete class theorems for estimation of multivariate Poisson means and related problems. *Ann. Statist.* **13**.

Brown, L. D., and Farrell, R. H., 1985b. All admissible linear estimators of a multivariate Poisson mean. *Ann. Statist.* **13**, 282–294.

Brown, L. D., and Fox, M., 1974a. Admissibility of procedures in two-dimensional location parameter problems. *Ann. Statist.* **2**, 248–266.

Brown, L. D., and Fox, M., 1974b. Admissibility in statistical problems involving a location or scale parameter. *Ann. Statist.* **4**, 807–814.

Brown, L. D., and Hwang, J. T., 1982. A unified admissibility proof. In *Statistical Decision Theory and Related Topics III*, S. S. Gupta and J. Berger (Eds.). Academic Press, New York.

Brown, L. D., and Purves, R., 1973. Measurable selections of extrema. *Ann. Statist.* **1**, 902–912.

Brown, P. J., and Zidek, J. V., 1980. Adaptive multivariate ridge regression. *Ann. Statist.* **8**, 64–74.

Buehler, R. J., 1959. Some validity criteria for statistical inference. *Ann. Math. Statist.* **30**, 845–863.
Buehler, R. J., 1971. Measuring information and uncertainty. In *Foundations of Statistical Inference*, V. P. Godambe and D. A. Sprott (Eds.). Holt, Rinehart, and Winston, Toronto.
Buehler, R. J. 1976. Coherent preferences. *Ann. Statist.* **4**, 1051–1064.
Bunke, O., 1977. Mixed model, empirical Bayes, and Stein estimators. *Math. O.F. Statist. Ser. Statist.* **1**, 55–68.
Caridi, F., 1983. Characterization of limits of Bayes procedures. *J. Multivariate Anal.* **13**, 52–66.
Casella, G., 1980. Minimax ridge regression estimation. *Ann. Statist.* **8**, 1036–1056.
Casella, G., and Hwang, J. T., 1983. Empirical Bayes confidence sets for the mean of a multivariate normal distribution. *J. Amer. Statist. Assoc.* **78**, 688–698.
Casella, G., and Strawderman, W., 1981. Estimating a bounded normal mean. *Ann. Statist.* **9**, 868–876.
Chaloner, K., 1984. Optimal Bayesian experimental design for linear models. *Ann. Statist.* **4**, 283–300.
Chen, C. F., 1979. Bayesian inference for a normal dispersion matrix and its applications to stochastic multiple regression analysis. *J. Roy. Statist. Soc.* (Ser. B) **41**, 235–248.
Chen, S. Y., 1983. Restricted risk Bayes estimation. Ph.D. Thesis, Purdue University, West Lafayette.
Cheng, P., 1964. Minimax estimates of parameters of distributions belonging to the exponential family. *Chinese Math. Acta.* **5**, 277–299.
Cheng, P., 1982. Admissibility of simultaneous estimation of several parameters. *J. Syst. Sci. Math. Sci.* **2**, 176–195.
Chernoff, H., 1959. Sequential design of experiments. *Ann. Math. Statist.* **30**, 755–770.
Chernoff, H., 1968. Optimal stochastic control. *Sankhyā* (Ser. A) **30**, 221–252.
Chernoff, H., 1972. *Sequential Analysis and Optimal Design.* SIAM, Philadelphia.
Chernoff H., and Moses, L. E., 1959. *Elementary Decision Theory.* Wiley, New York.
Chou, J. P., 1984. Multivariate exponential families: an identity and the admissibility of standard estimates. Ph.D. Thesis, Department of Mathematics, University of California, Los Angeles.
Chow, Y. S., Robbins, H., and Siegmund, D., 1971. *Great Expectations: The Theory of Optimal Stopping.* Houghton-Mifflin, Boston.
Chuang, D. T., 1984. Further theory of stable decisions. In *Robustness of Bayesian Analyses*, J. Kadane (Ed.). North-Holland, Amsterdam.
Clevenson, M., and Zidek, J., 1975. Simultaneous estimation of the mean of independent Poisson laws. *J. Amer. Statist. Assoc.* **70**, 698–705.
Cohen, A., 1966. All admissible estimators of the mean vector. *Ann. Math. Statist.* **37**, 458–470.
Cohen, A., and Sackrowitz, H. D., 1984. Decision theory results for vector risks with applications. *Statistics and Decisions Supplement No. 1: Recent Results in Estimation Theory and Related Topics*, 159–176. Oldenbourg, Munich.
Coleman, A., 1982. *Game Theory and Experimental Games: The Study of Strategic Interaction.* Pergamon Press, New York.
Copas, J. B., 1969. Compound decisions and empirical Bayes (with discussion). *J. Roy. Statist. Soc.* (Ser. B) **31**, 397–425.
Cornfield, J., 1969. The Bayesian outlook and its applications. *Biometrics* **25**, 617–657.
Cox, D. R., 1958. Some problems connected with statistical inference. *Ann. Math. Statist.* **29**, 357–372.
Cox, D. R., and Hinkley, D. V., 1974. *Theoretical Statistics.* Chapman & Hall, London.
Cressie, N., 1982. A useful empirical Bayes identity. *Ann. Statist.* **10**, 625–629.

Dalal, S. R., and Hall, W. J., 1983. Approximating priors by mixtures of natural conjugate priors. *J. Roy. Statist. Soc.* (Ser. B) **45**, 278–286.

Das Gupta, A., 1984a. Simultaneous estimation in the multiparameter gamma distribution under weighted quadratic losses. Technical Report #84-6, Department of Statistics, Purdue University, West Lafayette.

Das Gupta, A., 1984b. Simultaneous estimation of arbitrary scale-parameter problems under arbitrary quadratic loss. Technical Report #84-32, Department of Statistics, Purdue University, West Lafayette.

Das Gupta, A., and Sinha, B. K., 1980. On the admissibility of polynomial estimators in the one-parameter exponential family. *Sankhyā* (Ser. B) **42**, 129–142.

Dawid, A. P., 1973. Posterior expectations for large observations. *Biometrika* **60**, 664–666.

Dawid, A. P., 1982. Intersubjective statistical models. In *Exchangeability in Probability and Statistics*, G. Koch and F. Spizzichino (Eds.). North-Holland, Amsterdam.

Dawid, A. P., Stone, M., and Zidek, J. V., 1973. Marginalization paradoxes in Bayesian and structural inference (with discussion). *J. Roy. Statist. Soc.* (Ser. B) **35**, 189–233.

Deely, J. J., and Lindley, D. V., 1981. Bayes empirical Bayes. *J. Amer. Statist. Assoc.* **76**, 833–841.

de Finetti, B., 1937. Foresight: Its logical laws, its subjective sources. Translated and reprinted in *Studies in Subjective Probability*, H. Kyburg and H. Smokler (Eds.), 1964, 93–158. Wiley, New York.

de Finetti, B., 1962. Does it make sense to speak of "Good probability appraisers"? In *The Scientist Speculates*, I. J. Good (Ed.). Basic Books, New York.

de Finetti, B., 1972. *Probability, Induction, and Statistics.* Wiley, New York.

de Finetti, B., 1974, 1975. *Theory of Probability*, Vols. 1 and 2. Wiley, New York.

DeGroot, M. H., 1970. *Optimal Statistical Decisions.* McGraw-Hill, New York.

DeGroot, M. H., 1973. Doing what comes naturally: interpreting a tail area as a posterior probability or as a likelihood ratio. *J. Amer. Statist. Assoc.* **68**, 966–969.

DeGroot, M. H., 1974. Reaching a consensus. *J. Amer. Statist. Assoc.* **69**, 118–121.

Dempster, A. P., 1968. A generalization of Bayesian inference. *J. Roy. Statist. Soc.* (Ser. B) **30**, 205–248.

Dempster, A. P., 1973. The direct use of likelihood for significance testing. In *Proceedings of the Conference on Foundational Questions in Statistical Inference*, O. Barndorff-Nielsen et al. (Eds.). University of Aarhus, Aarhus.

Dempster, A. P., 1975. A subjectivist look at robustness. *Bull. Int. Statist. Inst.* **46**, 349–374.

Dempster, A. P., 1976. Examples relevant to the robustness of applied inference. In *Statistical Decision Theory and Related Topics III*, S. S. Gupta and J. Berger (Eds.). Academic Press, New York.

Dempster, A. P., Laird, N. M., and Rubin, D. B., 1977. Maximum likelihood from incomplete data via the EM algorithm (with discussion). *J. Roy. Statist. Soc.* (Ser. B) **39**, 1–38.

Dempster, A. P., Selwyn, M. R., Patel, C. M., and Roth, A. J., 1984. Statistical and computational aspects of mixed model analysis. *Appl. Statist.* **33**, 203–214.

De Robertis, L., and Hartigan, J. A., 1981. Bayesian inference using intervals of measures. *Ann. Statist.* **9**, 235–244.

de Waal, D., Groenewald, P., van Zyl, D. and Zidek, J., 1985. Multi-Bayesian estimation theory. *Statistics and Decisions* **3**.

Dey, D. K., and Berger, J., 1983. On truncation of shrinkage estimators in simultaneous estimation of normal means. *J. Amer. Statist. Assoc.* **78**, 865–869.

Dey, D. K., and Srinivasan, C., 1985. Estimation of covariance matrix under Stein's loss. *Ann. Statist.* **13**.

Diaconis, P., and Freedman, D., 1980. Finite exchangeable sequences. *Ann. Probab.* **8**, 745–764.

Diaconis, P., and Freedman, D., 1981. On the statistics of vision: the Julesz conjecture. *J. Math. Psych.* **24**, 112–138.

Diaconis, P., and Freedman, D., 1982. Bayes rules for location problems. In *Statistical Decision Theory and Related Topics III*, S. S. Gupta and J. Berger (Eds.). Academic Press, New York.

Diaconis, P., and Freedman, D., 1983. Frequency properties of Bayes rules. In *Scientific Inference, Data Analysis, and Robustness*, G. E. P. Box, T. Leonard, and C. F. Wu (Eds.). Academic Press, New York.

Diaconis, P., and Freedman, D., 1984. Partial exchangeability and sufficiency. Technical Report, Department of Statistics, Stanford University, Stanford.

Diaconis, P., and Stein, C., 1983. *Lectures on Statistical Decision Theory*. Unpublished Lecture Notes, Stanford University, Stanford.

Diaconis, P., and Ylvisaker, D., 1979. Conjugate priors for exponential families. *Ann. Statist.* **7**, 269–281.

Diaconis, P., and Ylvisaker, D., 1984. Quantifying prior opinions. In *Bayesian Statistics II*, J. M. Bernardo, M. H. DeGroot, D. V. Lindley, and A. F. M. Smith (Eds.). North-Holland, Amsterdam.

Diaconis, P., and Zabell, S., 1982. Updating subjective probability. *J. Amer. Statist. Assoc.* **77**, 822–830.

Diamond, G. A., and Forrester, J. S., 1983. Clinical trials and statistical verdicts: probable grounds for appeal. *Ann. Intern. Med.* **98**, 385–394.

Dickey, J. M., 1967. Expansions of t densities and related complete integrals. *Ann. Math. Statist.* **38**, 503–510.

Dickey, J. M., 1968. Three multidimensional integral identities with Bayesian applications. *Ann. Math. Statist.* **39**, 1615–1627.

Dickey, J. M., 1971. The weighted likelihood ratio, linear hypotheses on normal location parameters. *Ann. Math. Statist.* **42**, 204–223.

Dickey, J. M., 1973. Scientific reporting. *J. Roy. Statist. Soc.* (Ser. B) **35**, 285–305.

Dickey, J. M., 1974. Bayesian alternatives to the F-test and least-squares estimate in normal linear model. In *Studies in Bayesian Econometrics and Statistics*, S. E. Fienberg and A. Zellner (Eds.). North-Holland, Amsterdam.

Dickey, J. M., 1976. Approximate posterior distributions. *J. Amer. Statist. Assoc.* **71**, 680–689.

Dickey, J. M., 1977. Is the tail area useful as an approximate Bayes factor? *J. Amer. Statist. Assoc.* **72**, 138–142.

Dickey, J. M., 1980a. Beliefs about beliefs, a theory of stochastic assessments of subjective probabilities. In *Bayesian Statistics*, J. M. Bernardo, M. H. DeGroot, D. V. Lindley, and A. F. M. Smith (Eds.). University Press, Valencia.

Dickey, J. M., 1980b. Approximate coherence for regression models with a new analysis of Fisher's broadback wheatfield example. In *Bayesian Analysis in Econometrics and Statistics*, A. Zellner (Ed.). North-Holland, Amsterdam.

Dickey, J. M., and Chong-Hong, C., 1984. Subjective-probability modeling with elliptical distributions. In *Bayesian Statistics II*, J. M. Bernardo, M. H. DeGroot, D. Lindley, and A. Smith (Eds.). North-Holland, Amsterdam.

Dickey, J. M., and Freeman, P. R., 1975. Population-distributed personal probabilities. *J. Amer. Statist. Assoc.* **70**, 362–364.

DuMouchel, W. M., and Harris, J. E., 1983. Bayes methods for combining the results of cancer studies in humans and other species (with discussion). *J. Amer. Statist. Assoc.* **78**, 293–315.

Dvoretsky, A., Wald, A., and Wolfowitz, J., 1951. Elimination of randomization in certain statistical decision procedures and zero-sum two-person games. *Ann. Math. Statist.* **22**, 1-21.

Eaton, M. L., 1970. A complete class theorem for multidimensional one-sided alternatives. *Ann. Math. Statist.* **41**, 1884-1888.

Eaton, M. L., 1982. A method for evaluating improper prior distributions. In *Statistical Decision Theory and Related Topics III*, S. S. Gupta and J. Berger (Eds.). Academic Press, New York.

Eaton, M. L., 1983. *Multivariate Statistics.* Wiley, New York.

Eberl, W. Jr., and Moeschlin, O., 1982. *Mathematische Statistik.* deGruyter, Berlin.

Edwards, A. W. F., 1972. *Likelihood.* Cambridge University Press, Cambridge.

Edwards, A. W. F., 1974. The history of likelihood. *Int. Statist. Rev.* **42**, 9-15.

Edwards, W., 1968. Conservatism in human information processing. In *Formal Representation of Human Judgement*, B. Kleinmuntz (Ed.). Wiley, New York.

Edwards, W., Lindman, H., and Savage, L. J., 1963. Bayesian statistical inference for psychological research. *Psychol. Rev.* **70**, 193-242.

Efron, B., and Hinkley, D., 1978. Assessing the accuracy of the maximum likelihood estimator: observed versus expected Fisher information. *Biometrika* **65**, 457-482.

Efron, B., and Morris, C., 1971. Limiting the risk of Bayes and empirical Bayes estimators—Part I: the Bayes case. *J. Amer. Statist. Assoc.* **66**, 807-815.

Efron, B., and Morris, C., 1972a. Limiting the risk of Bayes and empirical Bayes estimators—Part II: The empirical Bayes case. *J. Amer. Statist. Assoc.* **67**, 130-139.

Efron, B., and Morris, C., 1972b. Empirical Bayes on vector observations—an extension of Stein's method. *Biometrika* **59**, 335-347.

Efron, B., and Morris, C., 1973. Stein's estimation rule and its competitors—an empirical Bayes approach. *J. Amer. Statist. Assoc.* **68**, 117-130.

Efron, B., and Morris, C., 1976a. Families of minimax estimators of the mean of a multivariate normal distribution. *Ann. Statist.* **4**, 11-21.

Efron, B., and Morris, C., 1976b. Multivariate empirical Bayes and estimation of covariance matrices. *Ann. Statist.* **4**, 22-32.

Ellsburg, D., 1961. Risk, ambiguity, and the Savage axioms. *Quart. J. Econ.* **75**, 644-661.

Ewing, G. M., 1969. *Calculus of Variations with Applications.* Norton, New York.

Faith, R. E., 1978. Minimax Bayes and point estimators of a multivariate normal mean. *J. Multivariate Anal.* **8**, 372-379.

Farrell, R. H., 1964. Estimators of a location parameter in the absolutely continuous case. *Ann. Math. Statist.* **35**, 949-998.

Farrell, R. H., 1966. Weak limits of sequences of Bayes procedures in estimation theory. *Proc. Fifth Berkeley Symp. Math. Statist. Probab.* **1**, 83-111. University of California Press, Berkeley.

Farrell, R. H., 1968a. Towards a theory of generalized Bayes tests. *Ann. Math. Statist.* **38**, 1-22.

Farrell, R. H., 1968b. On a necessary and sufficient condition for admissibility of estimators when strictly convex loss is used. *Ann. Math. Statist.* **38**, 23-28.

Ferguson, T. S., 1967. *Mathematical Statistics: A Decision-Theoretic Approach.* Academic Press, New York.

Ferguson, T. S., 1973. A Bayesian analysis of some nonparametric problems. *Ann. Statist.* **1**, 209-230.

Fine, T., 1973. *Theories of Probability.* Academic Press, New York.

Fishburn, P. C., 1964. *Decision and Value Theory.* Wiley, New York.

Fishburn, P. C., 1965. Analysis of decisions with incomplete knowledge of probabilities. *Oper. Res.* **13**, 217-237.

Fishburn, P. C., 1981. Subjective expected utility: a review of normative theories. *Theory and Decision* **13**, 139–199.

Fishburn, P. C., Murphy, A. H., and Isaacs, H. H., 1968. Sensitivity of decisions to probability estimation errors: a re-examination. *Oper. Res.* **16**, 253–268.

Fisher, R. A., 1920. A mathematical examination of the methods of determining the accuracy of an observation by the mean error, and by the mean square error. *Mon. Notic. Roy. Astron. Soc.* **80**, 757–770.

Fisher, R. A., 1922. On the mathematical foundations of theoretical statistics. *Philos. Trans. Roy. Soc. London* (Ser. A) **222**, 309–368.

Fisher, R. A., 1935. The fiducial argument in statistical inference. *Ann. Eugenics* **6**, 391–398.

Fisher, R. A., 1959. *Statistical Methods and Scientific Inference* (2nd edn.). Oliver and Boyd, Edinburgh.

Fraser, D. A. S., 1968. *The Structure of Inference.* Wiley, New York.

Fraser, D. A. S., 1977. Confidence, posterior probability, and the Buehler example. *Ann. Statist.* **5**, 892–898.

Fraser, D. A. S., 1979. *Inference and Linear Models.* McGraw-Hill, New York.

Fraser, D. A. S., and Mackay, J., 1976. On the equivalence of standard inference procedures. In *Foundations of Probability Theory, Statistical Inference and Statistical Theories of Science, Vol. II*, W. L. Harper and C. A. Hooker (Eds.). Reidel, Boston.

Freedman, D. A., and Purves, R. A., 1969. Bayes methods for bookies. *Ann. Math. Statist.* **40**, 1177–1186.

French, S. G., 1980. Updating of belief in the light of someone else's opinion. *J. Roy. Statist. Soc.* (Ser. A) **143**, 43–48.

French, S. G., 1981. Consensus of opinion. *Eur. J. Oper. Res.* **7**, 332–340.

French, S. G., 1984. Group consensus probability distributions: a critical survey. In *Bayesian Statistics II*, J. M. Bernardo, M. H. DeGroot, D. V. Lindley, and A. F. M. Smith (Eds.). North-Holland, Amsterdam.

Gardenfors, P., and Sahlin, N. E., 1982. Unreliable probabilities, risk taking, and decision making. *Synthese* **53**, 361–386.

Gatsonis, C. A., 1984. Deriving posterior distributions for a location parameter: a decision-theoretic approach. *Ann. Statist.* **12**, 958–970.

Gaver, D. P., 1985. Discrepancy-tolerant hierarchical Poisson event-rate analyses. Technical Report, Naval Postgraduate School, Monterey.

Geertsema, J. C., 1983. Recent views in the foundational controversy in statistics. *South African Statist. J.* **17**, 121–146.

Geisser, S., 1971. The inferential use of predictive distributions. In *Foundations of Statistical Inference*, V. P. Godambe and D. A. Sprott (Eds.). Holt, Rinehart, and Winston, Toronto.

Geisser, S., 1980. A predictivist primer. In *Bayesian Analysis in Econometrics and Statistics*, A. Zellner (Ed.). North-Holland, Amsterdam.

Geisser, S., 1984a. On prior distributions for binary trials. *American Statist.* **38**, 244–251.

Geisser, S., 1984b. On the prediction of observables: a selective update. In *Bayesian Statistics II*, J. M. Bernardo, M. H. DeGroot, D. V. Lindley, and A. F. M. Smith (Eds.). North-Holland, Amsterdam.

Geisser, S., and Eddy, W. F., 1979. A predictive approach to model selection. *J. Amer. Statist. Assoc.* **74**, 153–160.

Genest, C., 1984. A characterization theorem for externally Bayesian groups. *Ann. Statist.* **12**, 1100–1105.

Genest, C., and Schervish, M., 1983. Modelling expert judgements for Bayesian updating. Technical Report No. 304, Department of Statistics, Carnegie-Mellon University, Pittsburgh.

George, E. I., 1985. Shrinkage towards multiple points and subspaces. Technical Report, Graduate School of Business, University of Chicago, Chicago.

Ghia, G. D., 1976. Truncated generalized Bayes tests. Ph.D. Thesis, Department of Statistics, Yale University, New Haven.

Ghosh, B. K., 1970. *Sequential Tests of Statistical Hypotheses.* Addison-Wesley, Reading, Massachusetts.

Ghosh, J. K., Sinha, B. K., and Joshi, S. N., 1982. Expansions for posterior probability and integrated Bayes risk. In *Statistical Decision Theory and Related Topics III*, S. S. Gupta and J. Berger (Eds.). Academic Press, New York.

Ghosh, M., 1983. Estimates of multiple Poisson means: Bayes and empirical Bayes. *Statistics and Decisions* **1**, 183-195.

Ghosh, M., Hwang, J. T., and Tsui, K., 1983. Construction of improved estimators in multiparameter estimation for discrete exponential families (with discussion). *Ann. Statist.* **11**, 351-376.

Ghosh, M., Hwang, J. T., and Tsui, K., 1984. Construction of improved estimators in multiparameter estimation for continuous exponential families. *J. Multivariate Anal.* **14**, 212-220.

Ghosh, M., and Meeden, G., 1977. Admissibility of linear estimators in the one-parameter exponential family. *Ann. Statist.* **5**, 772-778.

Ghosh, M., and Meeden, G., 1984. A new Bayesian analysis of a random effects model. *J. Roy. Statist. Soc.* (Ser. B) **83**.

Gilliland, D. C., Boyer, J. E., Jr., and Tsao, H. J., 1982. Bayes empirical Bayes: finite parameter case. *Ann. Statist.* **10**, 1277-1282.

Giron, F. J., and Rios, S., 1980. Quasi-Bayesian behavior: a more realistic approach to decision making? In *Bayesian Statistics II*, J. M. Bernardo, M. H. DeGroot, D. V. Lindley, and A. F. M. Smith (Eds.). University Press, Valencia.

Gitting, J. C., 1979. Bandit processes and dynamic allocation indices (with discussion). *J. Roy. Statist. Soc.* (Ser. B) **41**, 148-177.

Gleser, L. J., 1979. Minimax estimation of a normal mean vector when the covariance matrix is unknown. *Ann. Statist.* **7**, 838-846.

Goel, P. K., 1983. Information measures and Bayesian hierarchical models. *J. Amer. Statist. Assoc.* **78**, 408-410.

Goel, P. K., and DeGroot, M. H., 1981. Information about hyperparameters in hierarchical models. *J. Amer. Statist. Assoc.* **76**, 140-147.

Goldstein, M., 1980. The linear Bayes regression estimator under weak prior assumptions. *Biometrika* **67**, 621-628.

Goldstein, M., 1981. Revising previsions: a geometric interpretation (with discussion). *J. Roy. Statist. Soc.* (Ser. B) **43**, 105-130.

Goldstein, M., 1984. Separating beliefs. In *Bayesian Inference and Decision Techniques with Applications: Essays in Honor of Bruno de Finetti*, P. K. Goel and A. Zellner (Eds.). North-Holland, Amsterdam.

Good, I. J., 1950. *Probability and the Weighing of Evidence.* Charles Griffin, London.

Good, I. J., 1952. Rational decisions. *J. Roy. Statist. Soc.* (Ser. B) **14**, 107-114.

Good, I. J., 1953. On the population frequencies of species and the estimation of population parameters. *Biometrika* **40**, 237-264.

Good, I. J., 1958. Significance tests in parallel and in series. *J. Amer. Statist. Assoc.* **53**, 799-813.

Good, I. J., 1962. Subjective probability as a measure of a nonmeasurable set. In *Logic, Methodology and Philosophy of Science.* Stanford University Press, Stanford, California.

Good, I. J., 1965. *The Estimation of Probabilities: An Essay on Modern Bayesian Methods.* M.I.T. Press, Cambridge, Massachusetts.

Good, I. J., 1973. The probabilistic explication of evidence, surprise, causality, explanation, and utility. In *Foundations of Statistical Inference*, V. P. Godambe and D. A. Sprott (Eds.). Holt, Rinehart, and Winston, Toronto.

Good, I. J., 1976. The Bayesian influence, or how to sweep subjectivism under the carpet. In *Foundations of Probability Theory, Statistical Inference, and Statistical Theories of Science*, Vol. II, W. L. Harper and C. A. Hooker (Eds.). Reidel, Boston.

Good, I. J., 1980. Some history of the hierarchical Bayesian methodology. In *Bayesian Statistics II*, J. M. Bernardo, M. H. DeGroot, D. V. Lindley, and A. F. M. Smith (Eds.). University Press, Valencia.

Good, I. J., 1983a. *Good Thinking: The Foundations of Probability and Its Applications.* University of Minnesota Press, Minneapolis.

Good, I. J., 1983b. The robustness of a hierarchical model for multinomials and contingency tables. In *Scientific Inference, Data Analysis, and Robustness*, G. E. P. Box, T. Leonard, and C. F. Wu (Eds.). Academic Press, New York.

Good, I. J., 1984. Notes C140, C144, C199, C200, and C201. *J. Statist. Comput. Simul.* **18**, **19**, **20**.

Govindarajulu, Z., 1981. *The Sequential Statistical Analysis of Hypothesis Testing, Point and Interval Estimation, and Decision Theory.* American Sciences Press, Columbus, Ohio.

Gupta, S. S. and Hsiao, P., 1981. On Γ-minimax, minimax, and Bayes procedures for selecting populations close to a control. *Sankhyā* (Ser. B) **43**, 291–318.

Gupta, S. S. and Miescke, K. J., 1984. Sequential selection procedures—a decision-theoretic approach. *Ann. Statist.* **12**, 336–350.

Gupta, S. S., and Panchapakesan, S., 1979. *Multiple Decision Procedures: Theory and Methodology of Selecting and Ranking Populations.* Wiley, New York.

Gutmann, S., 1982a. Stein's paradox is impossible in problems with finite sample space. *Ann. Statist.* **10**, 1017–1020.

Gutmann, S., 1982b. Stein's paradox is impossible in the nonanticipative context. *J. Amer. Statist. Assoc.* **77**, 934–935.

Haff, L. R., 1976. Minimax estimators of the multinormal mean: autoregressive priors. *J. Multivariate Anal.* **6**, 265–280.

Haff, L. R., 1977. Minimax estimators for a multinormal precision matrix. *J. Multivariate Anal.* **7**, 374–385.

Haff, L. R., 1978. The multivariate normal mean with intraclass correlated components: estimation of urban fire alarm probabilities. *J. Amer. Statist. Assoc.* **60**, 806–825.

Haff, L. R., 1980. Empirical Bayes estimation of the multivariate normal covariance matrix. *Ann. Statist.* **8**, 586–597.

Haff, L. R., and Johnson, R. W., 1984. The superharmonic condition for simultaneous estimation of means in exponential families. Technical Report, Department of Mathematics, University of California, San Diego.

Hald, A., 1971. The size of Bayes and minimax tests as function of the sample size and loss ratio. *Skand. Aktuar Tidskr.*, 53–73.

Halmos, P. R., 1950. *Measure Theory.* van Nostrand, New York. Second edition 1974. Springer-Verlag, New York.

Hammersley, J. M., and Handscomb, D. C., 1964. *Monte Carlo Methods.* Wiley, New York.

Hartigan, J., 1964. Invariant prior distributions. *Ann. Math. Statist.* **35**, 836–845.

Hartigan, J. A., 1969. Linear Bayesian methods. *J. Roy. Statist. Soc.* (Ser. B) **31**, 446–454.

Hartigan, J. A., 1983. *Bayes Theory.* Springer-Verlag, New York.

Heath, D., and Sudderth, W., 1978. On finitely additive priors, coherence, and extended admissibility. *Ann. Statist.* **6**, 333–345.

Heyde, C. C., and Johnstone, I. M., 1979. On asymptotic posterior normality for stochastic processes. *J. Roy. Statist. Soc.* (Ser. B) **41**, 184–189.

Hildreth, C., 1963. Bayesian statisticians and remote clients. *Econometrica* **31**, 422–438.

Hill, B., 1965. Inference about variance components in the one-way model. *J. Amer. Statist. Assoc.* **60**, 806–825.

Hill, B., 1974. On coherence, inadmissibility, and inference about many parameters in the theory of least squares. In *Studies in Bayesian Econometrics and Statistics*, S. Fienberg and A. Zellner (Eds.). North-Holland, Amsterdam.

Hill, B., 1975. A simple general approach to inference about the tail of a distribution. *Ann. Statist.* **3**, 1163–1174.

Hill, B., 1977. Exact and approximate Bayesian solutions for inference about variance components and multivariate inadmissibility. In *New Developments in the Application of Bayesian Methods*, A. Aykac and C. Brumat (Eds.). North-Holland, Amsterdam.

Hill, B., 1980a. Robust analysis of the random model and weighted least squares regression. In *Evaluation of Econometric Models*. Academic Press, New York.

Hill, B., 1980b. Invariance and robustness of the posterior distribution of characteristics of a finite population, with reference to contingency tables and the sampling of species. In *Bayesian Analysis in Econometrics and Statistics*, A. Zellner (Ed.). North-Holland, Amsterdam.

Hill, B., 1982. Comment on "Lindley's Paradox," by G. Shafer. *J. Amer. Statist. Assoc.* **77**, 344–347.

Hill, B., and Lane, D., 1984. Conglomerability and countable additivity. In *Bayesian Inference and Decision Techniques with Applications*, P. K. Goel and A. Zellner (Eds.). North-Holland, Amsterdam.

Hinde, J., and Aitkin, M., 1984. Canonical likelihoods: a new likelihood treatment of nuisance parameters. Technical Report, Centre for Applied Statistics, University of Lancaster, Lancaster.

Hinkley, D. V., 1979. Predictive likelihood. *Ann. Statist.* **7**, 718–728.

Hinkley, D. V., 1983. Can frequentist inference be very wrong? A conditional "yes". In *Scientific Inference, Data Analysis, and Robustness*, G. E. P. Box et al. (Eds.). Academic Press, New York.

Hirschmann, I. I., and Widder, D. V., 1955. *The Convolution Transform*. Princeton University Press, Princeton, N.J.

Hoadley, B. 1981. Quality management plan (QMP). *Bell System Tech. J.* **60**, 215–273.

Hodges, J. L. Jr., and Lehmann, E. L., 1952. The use of previous experience in reaching statistical decisions. *Ann. Math. Statist.* **23**, 396–407.

Hodges, J. L. Jr., and Lehmann, E. L., 1954. Testing the approximate validity of statistical hypotheses. *J. Roy. Statist. Soc.* (Ser. B) **16**, 261–268.

Hoeffding, W., 1960. Lower bounds for the expected sample size and the average risk of a sequential procedure. *Ann. Math. Statist.* **31**, 352–368.

Hoffman, K., 1984. Admissibility and inadmissibility of estimators in the one-parameter exponential family. *Math. Oper. Statist.*

Hogarth, R. M., 1975. Cognitive processes and the assessment of subjective probability distributions. *J. Amer. Statist. Assoc.* **70**, 271–294.

Hooper, P. M., 1982. Invariant confidence sets with smallest expected measure. *Ann. Statist.* **10**, 1283–1294.

Hora, R. B., and Buehler, R. J., 1966. Fiducial theory and invariant estimation. *Ann. Math. Statist.* **37**, 643–656.

Howard, R. A., and Matheson, J. E., 1984. Influence diagrams. In *The Principles and Applications of Decision Analysis*, Vol. II, R. A. Howard and J. E. Matheson (Eds.). Strategic Decisions Group, Menlo Park, California.

Huber, P. J., 1972. Robust statistics: a review. *Ann. Math. Statist.* **43**, 1041-1067.

Huber, P. J., 1973. The use of Choquet capacities in statistics. *Bull. Int. Statist. Inst.* **45**, 181-191.

Huber, P. J., 1981. *Robust Statistics*. Wiley, New York.

Hudson, H. M., 1974. Empirical Bayes estimation. Technical Report #58, Department of Statistics, Stanford University, Stanford, California.

Hudson, H. M., 1978. A natural identity for exponential families with applications in multivariate estimation. *Ann. Statist.* **6**, 473-484.

Hudson, H. M., 1985. Adaptive estimators for simultaneous estimation of Poisson means. *Ann. Statist.* **13**, 246-261.

Hui, S., and Berger, J., 1983. Empirical Bayes estimation of rates in longitudinal studies. *J. Amer. Statist. Assoc.* **78**, 753-760.

Huzurbazar, V. S., 1976. *Sufficient Statistics*. Marcel Dekker, New York.

Hwang, J. T., 1982a. Improving upon standard estimators in discrete exponential families with applications to Poisson and negative binomial cases. *Ann. Statist.* **10**, 857-867.

Hwang, J. T., 1982b. Semi-tail upper bounds on the class of admissible estimators in discrete exponential families, with applications to Poisson and negative binomial distributions. *Ann. Statist.* **10**, 1137-1147.

Hwang, J. T., 1982c. Certain bounds on the class of admissible estimators in continuous exponential families. In *Statistical Decision Theory and Related Topics III*, S. S. Gupta and J. Berger (Eds.). Academic Press, New York.

Hwang, J. T., 1985. Universal domination and stochastic domination—decision theory simultaneously under a broad class of loss functions. *Ann. Statist.* **13**, 295-314.

Hylland, A., and Zeckhauser, R., 1981. The impossibility of Bayesian group decision making with separate aggregation of beliefs and values. *Econometrica* **79**, 1321-1336.

Ighodaro, A., Santner, T., and Brown, L., 1982. Admissibility and complete class results for the multinomial estimation problems with entropy and squared-error loss. *J. Multivariate Anal.* **12**, 469-479.

Isii, K., and Noda, K., 1983. A vector space approach to obtaining a condition for the admissibility of statistical decision functions. Technical Report, Institute of Statistical Mathematics, Tokyo.

Jackson, D. A., Donovan, T. M., Zimmer, W. J., and Deely, J. J., 1970. Γ-minimax estimators in the exponential family. *Biometrika* **57**, 439-443.

Jackson, P. H., Novick, M. R., and Thayer, D. T., 1971. Estimating regressions in *m*-groups. *Brit. J. Math. Statist. Psychol.* **24**, 129-153.

James, W., and Stein, C., 1960. Estimation with quadratic loss. In *Proc. Fourth Berkeley Symp. Math. Statist. Probab.* **1**, 361-380. University of California Press, Berkeley.

Jaynes, E. T., 1968. Prior probabilities. *IEEE Transactions on Systems Science and Cybernetics*, **SSC-4**, 227-241.

Jaynes, E. T., 1980. Marginalization and prior probabilities. In *Bayesian Analysis in Econometrics and Statistics*, A. Zellner (Ed.). North-Holland, Amsterdam.

Jaynes, E. T., 1983. *Papers on Probability, Statistics, and Statistical Physics*. A reprint collection. R. D. Rosenkrantz (Ed.). Reidel, Dordrecht.

Jeffrey, R., 1968. Probable knowledge. In *The Problem of Inductive Logic*, I. Lakatos (Ed.). North-Holland, Amsterdam.

Jeffrey, R., 1983. *The Logic of Decision* (2nd edn.). University of Chicago Press, Chicago.

Jeffreys, H., 1957. *Scientific Inference.* Cambridge University Press, London.
Jeffreys, H., 1961. *Theory of Probability* (3rd edn.). Oxford University Press, London.
Jennrich, R. I., and Sampson, P. F., 1976. Newton-Raphson and related algorithms for maximum likelihood variance component estimation. *Technometrics* **18**, 11-18.
Jewell, W. S., 1979. Stochastically-ordered parameters in Bayesian prediction. Report ORC 79-12, Operations Research Center, University of California, Berkeley.
Jewell, W. S., 1983. Enriched multinormal priors revisited. Technical Report. Operations Research Center, University of California, Berkeley.
Jewell, W. S., and Schnieper, 1984. Credibility approximations for Bayesian prediction of second moments. Report ORC 84-3, Operations Research Center, University of California, Berkeley.
Johnson, R. A., 1967. An asymptotic expansion for posterior distributions. *Ann. Math. Statist.* **38**, 1899-1906.
Johnson, R. A., 1970. Asymptotic expansions associated with posterior distributions. *Ann. Math. Statist.* **41**, 851-864.
Johnson, R. W., 1984. Simultaneous estimation of generalized Pearson means. Ph.D. Thesis, Department of Mathematics, University of California, San Diego.
Johnstone, I., 1984. Admissibility, difference equations, and recurrence in estimating a Poisson mean. *Ann. Statist.* **12**, 1173-1198.
Johnstone, I., 1985. Admissible estimation, Dirichlet principles and recurrence of birth-death chains in Z_+^p. *Z. Wahrsch. verw. Gebiete* **69**.
Johnstone, I., and Lalley, S., 1984. On independent statistical decision problems and products of diffusions. *Z. Wahrsch. verw. Gebiete.* **68**, 29-47.
Johnstone, I., and Velleman, P., 1984. Efficient scores, variance decompositions, and Monte Carlo swindles. Technical Report No. 221, Department of Statistics, Stanford University, Stanford, California.
Jones, A. J., 1980. *Game Theory: Mathematical Models of Conflict.* Wiley, New York.
Joshi, V. M., 1969. On a theorem of Karlin regarding admissible estimates for exponential populations. *Ann. Math. Statist.* **40**, 216-223.
Judge, G., and Bock, M. E., 1978. *Implications of Pre-Test and Stein Rule Estimators in Econometrics.* North-Holland, Amsterdam.
Kadane, J. B., 1980. Predictive and structural methods for eliciting prior distributions. In *Bayesian Analysis in Econometrics and Statistics*, A. Zellner (Ed.). North-Holland, Amsterdam.
Kadane, J. B., 1984. *Robustness of Bayesian Analyses.* North-Holland, Amsterdam.
Kadane, J. B., and Chuang, D. T., 1978. Stable decision problems. *Ann. Statist.* **6**, 1095-1110.
Kadane, J. B., Dickey, J. M., Winkler, R. L., Smith, W. S., and Peters, S. C., 1980. Interactive elicitation of opinion for a normal linear model. *J. Amer. Statist. Assoc.* **75**, 845-854.
Kahneman, D., Slovic, P., and Tversky, A., 1982. *Judgement Under Uncertainty: Heuristics and Biases.* Cambridge University Press, New York.
Karlin, S., 1956. Decision theory of Pólya type distributions; case of two actions, I. In *Proc. Third Berkeley Symp. on Statist. Probab.* **1**, 115-129. University of California Press, Berkeley.
Karlin, S., 1957a. Pólya type distributions, II. *Ann. Math. Statist.* **28**, 281-308.
Karlin, S., 1957b. Pólya type distributions, III: admissibility for multi-action problems. *Ann. Math. Statist.* **28**, 839-860.
Karlin, S., 1958. Admissibility for estimation with quadratic loss. *Ann. Math. Statist.* **29**, 406-436.
Karlin, S., and Rubin, H., 1956. The theory of decision procedures for distributions with monotone likelihood ratio. *Ann. Math. Statist.* **27**, 272-299.
Keener, R. W., 1980. Renewal theory and the sequential design of experiments with two states of nature. *Commun. Statist.* **A9(16)**, 1699-1726.

Keener, R. W., 1984. Second-order efficiency in the sequential design of experiments. *Ann. Statist.* **12**, 510-532.

Keeney, R. L., and Raiffa, H., 1976. *Decisions with Multiple Objectives.* Wiley, New York.

Kelley, T. L., 1927. *The Interpretation of Educational Measurements.* World Books, New York.

Kempthorne, P. J., 1983. Minimax-Bayes compromise estimators. In *Business and Economics Statistics Proceedings of the American Statistical Association*, Washington, D.C.

Kempthorne, P. J., 1984. Numerical specification of discrete least favorable prior distributions. Research Report S-98, Department of Statistics, Harvard University, Cambridge, Massachusetts.

Kiefer, J., 1957. Invariance, minimax sequential estimation, and continuous time processes. *Ann. Math. Statist.* **28**, 573-601.

Kiefer, J., 1966. Multivariate optimality results. In *Multivariate Analysis*, P. R. Krishnaiah (Ed.). Academic Press, New York.

Kiefer, J., 1976. Admissibility of conditional confidence procedures. *Ann. Statist.* **4**, 836-865.

Kiefer, J., 1977a. Conditional confidence statements and confidence estimators (theory and methods). *J. Amer. Statist. Assoc.* **72**, 789-827.

Kiefer, J., 1977b. The foundations of statistics—are there any? *Synthese* **36**, 161-176.

Kiefer, J., and Sacks, J., 1963. Asymptotically optimum sequential inference and design. *Ann. Math. Statist.* **34**, 705-750.

Kloeck, T., and van Dijk, H. K., 1978. Bayesian estimates of equation system parameters; an application of integration by Monte Carlo. *Econometrica* **46**, 1-19.

Kozek, A., 1982. Towards a calculus for admissibility. *Ann. Statist.* **10**, 825-837.

Kraft, C. H., Pratt, J. W., and Seidenberg, A., 1959. Intuitive probability on finite sets. *Ann. Math. Statist.* **30**, 408-419.

Kudo, H., 1955. On minimax invariant estimates of the transformation parameter. Natural Science Report 6, 31-73, Ochanomizu University, Japan.

Kudo, H., 1967. On partial prior information and the property of parametric sufficiency. *Proc. Fifth Berkeley Symp. Statist. Probab.* **1**, University of California Press, Berkeley.

Kuo, L., 1983. Bayesian bio-assay design. *Ann. Statist.* **11**, 886-895.

Kuo, L., 1985. Computation of mixtures of Dirichlet processes *SIAM J. Sci. Statist. Comput.* **6**.

Kusama, T., 1966. Remarks on admissibility of decision functions. *Ann. Inst. Statist. Math.* **18**, 141-148.

Laird, N. M., 1978. Nonparametric maximum likelihood estimation of a mixing distribution. *J. Amer. Statist. Assoc.* **73**, 805-811.

Laird, N. M., 1982. The computation of estimates of variance components using the EM algorithm. *J. Statist. Comput. Simul.* **14**, 295-303.

Laird, N. M., 1983. Empirical Bayes estimates using the nonparametric maximum likelihood estimate for the prior. *J. Statist. Comput. Simul.* **15**, 211-220.

La Motte, L. R., 1982. Admissibility in linear estimation. *Ann. Statist.* **10**, 245-255.

Lane, D. A., and Sudderth, W. D., 1983. Coherent and continuous inference. *Ann. Statist.* **11**, 114-120.

Laplace, P. S., 1812. *Theorie Analytique des Probabilites.* Courcier, Paris.

Larson, R. E., and Casti, J. L., 1978 and 1982. *Principles of Dynamic Programming, Parts I and II.* Marcel Dekker, New York.

Lauritzen, S. L., 1982. *Statistical Models As Extremal Families.* University Press, Aalborg.

Lauritzen, S. L., 1984. Extreme point models in statistics. *Scand. J. Statist.* **11**, 65-91.

Leamer, E. E., 1978. *Specification Searches*. Wiley, New York.
Leamer, E. E., 1982. Sets of posterior means with bounded variance prior. *Econometrica* **50**, 725-736.
LeCam, L., 1955. An extension of Wald's theory of statistical decision functions. *Ann. Math. Statist.* **26**, 69-81.
LeCam, L., 1956. On the asymptotic theory of estimation and testing hypotheses. *Proc. Third Berkeley Symp. Math. Statist. Probab.* **1**, University of California Press, Berkeley.
LeCam, L., 1974. Notes on asymptotic methods in statistical decision theory. Centre de Recherches Mathematiques, Universite de Montreal.
LeCam, L., 1977. A note on metastatistics or "an essay toward stating a problem in the doctrine of chances." *Synthese* **36**, 133-160.
Lehmann, E. L., 1947. On families of admissible tests. *Ann. Math. Statist.* **18**, 97-104.
Lehmann, E. L., 1959, 1985. *Testing Statistical Hypotheses*, 1st edn. (1959); 2nd edn. to appear. Wiley, New York.
Lehmann, E. L., 1983. *Theory of Point Estimation*. Wiley, New York.
Lehn, J., and Rummel, F., 1985. Gamma-minimax estimation of a binomial probability under squared-error loss. *Statistics and Decisions* **3**.
Lempers, F. B., 1971. *Posterior Probabilities of Alternative Linear Models*. University of Rotterdam Press, Rotterdam.
Leonard, T., 1976. Some alternative approaches to multiparameter estimation. *Biometrika* **63**, 69-76.
Leonard, T., 1984. Some data-analytic modifications to Bayes-Stein estimation. *Ann. Inst. Statist. Math.* **36**, 11-22.
Lepage, R., 1979. Smoothing Wald's test. Technical Report, Department of Statistics, Michigan State University, East Lansing.
Lin, P. E., and Mousa, A., 1983. Minimax estimation of a multivariate normal mean under a convex loss function. *Australian J. Statist.* **25**, 463-466.
Lindley, D. V., 1957. A statistical paradox. *Biometrika* **44**, 187-192.
Lindley, D. V., 1961. The use of prior probability distributions in statistical inference and decision. In *Proc. Fourth Berkeley Symp. Math. Statist. Probab.* University of California Press, Berkeley.
Lindley, D. V., 1965. *Introduction to Probability and Statistics from a Bayesian Viewpoint* (Parts 1 and 2). Cambridge University Press, Cambridge.
Lindley, D. V., 1971a. *Bayesian Statistics, A Review*. SIAM, Philadelphia.
Lindley, D. V., 1971b. The estimation of many parameters. In *Foundations of Statistical Inference*, V. P. Godambe and D. A. Sprott (Eds.). Holt, Rinehart, and Winston, Toronto.
Lindley, D. V., 1977. A problem in forensic science. *Biometrika* **64**, 207-213.
Lindley, D. V., 1980. Approximate Bayesian methods. In *Bayesian Statistics II*, J. M. Bernardo, M. H. DeGroot, D. V. Lindley, and A. F. M. Smith (Eds.). Valencia Press, Valencia.
Lindley, D. V., 1982a. Scoring rules and the inevitability of probability. *Int. Statist. Rev.* **50**, 1-26.
Lindley, D. V., 1982b. The improvement of probability judgements. *J. Roy. Statist. Soc.* (Ser. A) **145**, 117-126.
Lindley, D. V., 1984. Reconciliation of discrete probability distributions. In *Bayesian Statistics II*, J. Bernardo, M. H. DeGroot, D. V. Lindley, and A. F. M. Smith (Eds.). Valencia Press, Valencia.
Lindley, D. V., and Novick, M. R., 1981. The role of exchangeability in inference. *Ann. Statist.* **9**, 45-58.
Lindley, D. V., and Phillips, L. D., 1976. Inference for a Bernoulli process (a Bayesian view). *Amer. Statist.* **30**, 112-119.

Lindley, D. V., and Singpurwalla, N. D., 1984. Reliability and fault tree analysis using expert opinions. Technical Report, Department of Operations Research, The George Washington University, Washington, DC.

Lindley, D. V., and Smith, A. F. M., 1972. Bayes estimates for the linear model. *J. Roy. Statist. Soc.* (Ser. B) **34**, 1-41.

Lindley, D. V., Tversky, A., and Brown, R. V., 1979. On the reconciliation of probability assessments (with discussion). *J. Roy. Statist. Soc.* (Ser. A) **142**, 146-180.

Lindsay, B. G., 1981. Properties of the maximum likelihood estimator of a mixing distribution. In *Statistical Distributions in Scientific Work*, C. Taillie et al. (Eds.). Vol. 5, 95-109.

Lindsay, B. G., 1983. The geometry of mixture likelihoods, Part II: the exponential family. *Ann. Statist.* **11**, 783-792.

Lord, F. M., 1969. Estimating true-score distributions in psychological testing (an empirical Bayes estimation problem). *Psychometrika* **34**, 259-299.

Lord, F. M., and Cressie, N., 1975. An empirical Bayes procedure for finding an interval estimate. *Sankhyā* (Ser. B) **37**, 1-9.

Lorden, G., 1977. Nearly-optimal sequential tests for finitely many parameter values. *Ann. Statist.* **5**, 1-21.

Lorden, G., 1984. Nearly optimal sequential tests for exponential families. Technical Report, Department of Mathematics, California Institute of Technology, Pasadena.

Louis, T. A., 1982. Finding the observed information matrix when using the EM algorithm. *J. Roy. Statist. Soc.* (Ser. B) **44**, 226-233.

Louis, T. A., 1984. Estimating a population of parameter values using Bayes and empirical Bayes methods. *J. Amer. Statist. Assoc.* **79**, 393-398.

Luce, R. D., and Raiffa, H., 1957. *Games and Decisions*. Wiley, New York.

Lush, J. L., 1937. *Animal Breeding Plans*. Iowa State University Press, Ames.

Magwire, C. A., 1953. Sequential decisions involving the choice of experiments. Ph.D. Thesis, Stanford University, Stanford, California.

Mandelbaum, A., 1984. All admissible linear estimators of the mean of a Gaussian distribution on a Hilbert space. *Ann. Statist.* **10**, 1448-1466.

Marazzi, A., 1985. On constrained minimization of the Bayes risk for the linear model. *Statistics and Decisions* **3**.

Marden, J. I., 1982. Minimal complete classes of tests of hypotheses with multivariate one-sided alternatives. *Ann. Statist.* **10**, 962-970.

Marden, J. I., 1983. Admissibility of invariant tests in the general multivariate analysis of variance problem. *Ann. Statist.* **11**, 1086-1099.

Maritz, J. S., 1970. *Empirical Bayes Methods*. Methuen, London.

Martz, H. F., and Waller, R. A., 1982. *Bayesian Reliability Analysis*. Wiley, New York.

Matthes, T. K., and Truax, D. R., 1967. Tests of composite hypotheses for the multivariate exponential family. *Ann. Math. Statist.* **38**, 681-697.

McConway, K. J., 1981. Marginalization and linear opinion pools. *J. Amer. Statist. Assoc.* **76**, 410-414.

Meeden, G., and Ghosh, M., 1981. Admissibility in finite problems. *Ann. Statist.* **9**, 846-852.

Meeden, G., and Isaacson, D., 1977. Approximate behavior of the posterior distribution for a large observation. *Ann. Statist.* **5**, 899-908.

Meinhold, R. J., and Singpurwalla, N. D., 1983. Bayesian analysis of a commonly used model for describing software failure. *The Statistician* **32**, 168-173.

Menges, G., 1966. On the Bayesification of the minimax principle. *Unternehmensforschung* **10**, 81–91.

Morris, C., 1977. Interval estimation for empirical Bayes generalizations of Stein's estimator. In *Proceedings of the 22nd Conference on the Design of Experiments in Army Research Development and Testing.* (ARO Report 77-2).

Morris, C., 1982. Natural exponential families with quadratic variance functions. *Ann. Statist.* **10**, 65–80.

Morris, C., 1983a. Parametric empirical Bayes inference: Theory and applications. *J. Amer. Statist. Assoc.* **78**, 47–65.

Morris, C., 1983b. Parametric empirical Bayes confidence sets. In *Scientific Inference, Data Analysis, and Robustness*, G. E. P. Box, T. Leonard, and C. F. Wu (Eds.). Academic Press, New York.

Morris, P. A., 1977. Combining expert judgements: a Bayesian approach. *Manage. Sci.* **23**, 679–693.

Morris, P. A., 1983. An axiomatic approach to expert resolution. *Manage. Sci.* **29**, 24–31.

Muirhead, R. J., 1982. *Aspects of Multivariate Statistical Theory.* Wiley, New York.

Nachbin, L., 1965. *The Haar Integral.* van Nostrand, New York.

Nash, J. F. Jr., 1950. The bargaining problem. *Econometrica* **18**, 155–162.

Naylor, J. C., and Smith, A. F. M., 1982. Applications of a method for the efficient computation of posterior distributions. *Appl. Statist.* **31**, 214–225.

Naylor, J. C., and Smith, A. F. M., 1983. A contamination model in clinical chemistry: an illustration of a method for the efficient computation of posterior distributions. *The Statistician* **32**, 82–87.

Nelson, W., 1966. Minimax solution of statistical decision problems by iteration. *Ann. Math. Statist.* **37**, 1643–1657.

Neyman, J., 1957. "Inductive behavior" as a basic concept of philosophy of science. *Rev. Math. Statist. Inst.* **25**, 7–22.

Neyman, J., 1967. *A Selection of Early Statistical Papers of J. Neyman.* University of California Press, Berkeley.

Neyman, J., 1977. Frequentist probability and frequentist statistics. *Synthese* **36**, 97–131.

Neyman, J., and Pearson, E. S., 1933. On the problem of the most efficient tests of statistical hypotheses. *Philos. Trans. Roy. Soc.* (Ser. A) **231**, 289–337.

Novick, M. R., and Hall, W. J., 1965. A Bayesian indifference procedure. *J. Amer. Statist. Assoc.* **60**, 1104–1117.

Novick, M. R., Jackson, P. H., Thayer, D. T., and Cole, N. S., 1972. Estimating multiple regressions in *m*-groups; a cross-validation study. *Brit. J. Math. Statist. Psychol.* **25**, 33–50.

O'Bryan, T. E., 1979. Rates of convergence in a modified empirical Bayes estimation problem involving Poisson distributions. *Commun. Statist.—Theory and Methods* **A8**, 167–174.

O'Hagan, A., 1979. On outlier rejection phenomena in Bayes inference. *J. Roy. Statist. Soc.* (Ser. B) **41**, 358–367.

O'Hagan, A., 1981. A moment of indecision. *Biometrika* **68**, 329–330.

Olkin, I., and Selliah, J. B., 1977. Estimating covariance in a multivariate normal distribution. In *Statistical Decision Theory and Related Topics II*, S. S. Gupta and D. S. Moore (Eds.). Academic Press, New York.

Oman, S. D., 1983. A class of modified Stein estimators with easily computable risk functions. *J. Statist. Planning and Inference* **7**, 359–369.

Peng, J. C. M., 1975. Simultaneous estimation of the parameters of independent Poisson distributions. Technical Report No. 78, Department of Statistics, Stanford University, Stanford, California.

Pilz, J., 1983. *Bayesian Estimation and Experimental Design in Linear Regression Models.* Teubner-Texte, Leipzig.

Pitman, E. J. G., 1939. The estimation of location and scale parameters of a continuous population of any given form. *Biometrika* **30**, 391-421.

Polasek, W., 1982. Local sensitivity analysis and matrix derivatives. In *Operations Research in Progress*, G. Feichtinger et al. (Eds.). Reidel, Dordrecht.

Polasek, W., 1984. Multivariate regression systems: Estimation and sensitivity analysis of two-dimensional data. In *Robustness of Bayesian Analyses*, J. Kadane (Ed.). North-Holland, Amsterdam.

Polasek, W., 1985. Sensitivity analysis for general and hierarchical linear regression models. In *Bayesian Inference and Decision Techniques with Applications*, P. K. Goel and A. Zellner (Eds.). North-Holland, Amsterdam.

Portnoy, S., 1971. Formal Bayes estimation with application to a random effects model. *Ann. Math. Statist.* **42**, 1379-1402.

Portnoy, S., 1972. On fundamental theorems in decision theory. Unpublished Notes, Department of Statistics, Harvard University, Cambridge, Massachusetts.

Pratt, J. W., 1962. Discussion of A. Birnbaum's "On the foundations of statistical inference." *J. Amer. Statist. Assoc.* **57**, 269-326.

Pratt, J. W., 1965. Bayesian interpretation of standard inference statements (with discussion). *J. Roy. Statist. Soc.* (Ser. B) **27**, 169-203.

Pratt, J. W., Raiffa, H., and Schlaifer, R., 1965. *Introduction to Statistical Decision Theory* (preliminary edn.). McGraw-Hill, New York.

Press, S. J., 1978. Qualitative controlled feedback for forming group judgements and making decisions. *J. Amer. Statist. Assoc.* **73**, 526-535.

Press, S. J., 1980. Bayesian computer programs. In *Bayesian Analysis in Econometrics and Statistics*, A. Zellner (Ed.). North-Holland, Amsterdam.

Press, S. J., 1982. *Applied Multivariate Analysis: Using Bayesian and Frequentist Measures of Inference* (2nd edn.). Kreiger, New York.

Press, S. J., 1984. Multivariate analysis (Bayesian). In *Encyclopedia of Statistical Sciences*, N. Johnson and S. Kotz (Eds.). Wiley, New York.

Press, S. J., and Shigemasu, K., 1984. Bayesian MANOVA and MANOCOVA under exchangeability. *Can. J. Statist* **12**.

Proschan, F., and Singpurwalla, N. D., 1979. Accelerated life testing—a pragmatic Bayesian approach. In *Optimization in Statistics*, J. Rustagi (Ed.). Academic Press, New York.

Proschan, F., and Singpurwalla, N. D., 1980. A new approach to inference from accelerated life tests. *IEEE Transactions of Reliability R-29*, 98-102.

Rabena, M., 1983. Sensitivity analysis in Bayesian decision. Publicacione #2, Departamento de Bioestadistica, Universidad de Valencia.

Raiffa, H., 1968. *Decision Analysis: Introductory Lectures on Choices under Uncertainty*. Addison-Wesley, Reading, Massachusetts.

Raiffa, H., and Schlaifer, R., 1961. *Applied Statistical Decision Theory*. Division of Research, Graduate School of Business Administration, Harvard University, Boston.

Ralescu, D., and Ralescu, S., 1981. A class of nonlinear admissible estimators in the one-parameter exponential family. *Ann. Statist.* **9**, 177-183.

Ramsay, J. O., and Novick, M. R., 1980. PLU robust Bayesian decision theory: point estimation. *J. Amer. Statist. Assoc.* **75**, 901-907.

Ramsey, F. P., 1926. Truth and probability. Reprinted in *Studies in Subjective Probability*, H. E. Kyburg and H. E. Smokler (Eds.). Wiley, New York.

Rao, C. R., 1976. Estimation of parameters in a linear model. *Ann. Statist.* **4**, 1023-1037.

Rao, C. R., 1977. Simultaneous estimation of parameters—a compound decision problem. In *Statistical Decision Theory and Related Topics II*, S. S. Gupta and D. S. Moore (Eds.). Academic Press, New York.

Ray, S. N., 1965. Bounds on the maximum sample size of a Bayes sequential procedure. *Ann. Math. Statist.* **36**, 859-878.

Reilly, A., and Sedransk, N., 1984. Bayesian analysis of growth curves. Technical Report, Department of Mathematics and Statistics, State University of New York, Albany.

Reinsel, G. C., 1983. Mean square error properties of empirical Bayes estimators in a multivariate random effects generalized linear model. Technical Report, Department of Statistics, University of Wisconsin, Madison.

Reinsel, G. C., 1984. Estimation and prediction in a multivariate random effects generalized linear model. *J. Amer. Statist. Assoc.* **79**, 406–414.

Robbins, H., 1951. Asymptotically subminimax solutions of compound statistical decision problems. In *Proc. Second Berkeley Symp. Math. Statist. Probab.* **1**. University of California Press, Berkeley.

Robbins, H., 1955. An empirical Bayes approach to statistics. In *Proc. Third Berkeley Symp. Math. Statist. Probab.* **1**, 157–164. University of California Press, Berkeley.

Robbins, H., 1964. The empirical Bayes approach to statistical decision problems. *Ann. Math. Statist.* **35**, 1–20.

Robbins, H., 1983. Some thoughts on empirical Bayes estimation. *Ann. Statist.* **1**, 713–723.

Roberts, H. V., 1965. Probabilistic prediction. *J. Amer. Statist. Assoc.* **60**, 50–62.

Roberts, H. V., 1967. Informative stopping rules and inference about population size. *J. Amer. Statist. Assoc.* **62**, 763–775.

Robinson, G. K., 1979a. Conditional properties of statistical procedures. *Ann. Statist.* **7**, 742–755.

Robinson, G. K., 1979b. Conditional properties of statistical procedures for location and scale parameters. *Ann. Statist.* **7**, 756–771.

Rosenkrantz, R. D., 1977. *Inference, Method, and Decision: Towards a Bayesian Philosophy of Science.* Reidel, Boston.

Rubin, D. B., 1984. Bayesianly justifiable and relevant frequency calculations for the applied statistician. *Ann. Statist.* **12**, 1151–1172.

Rubin, H., 1951. A complete class of decision procedures for distributions with monotone likelihood ratio. Technical Report 9, Department of Statistics, Stanford University, Stanford, California.

Rubin, H., 1971. A decision-theoretic approach to the problem of testing a null hypothesis. In *Statistical Decision Theory and Related Topics*, S. S. Gupta and J. Yackel (Eds.). Academic Press, New York.

Rubin, H., 1974. Decision-theoretic approach to some multivariate problems. In *Multivariate Analysis II*, P. R. Krishnaiah (Ed.). Academic Press, New York.

Rubin, H., 1976. Some fast methods of generating random variables with preassigned distributions I. General acceptance—rejection procedures. Mimeograph Series #467, Department of Statistics, Purdue University, West Lafayette.

Rubin, H., 1977. Robust Bayesian estimation. In *Statistical Decision Theory and Related Topics II*, S. S. Gupta and D. S. Moore (Eds.). Academic Press, New York.

Rubin, H., 1985. A weak system of axioms for "rational" behavior and the non-separability of utility from prior. Technical Report, Purdue University, West Lafayette.

Rubinstein, R. Y., 1981. *Simulation and the Monte Carlo Method.* Wiley, New York.

Rukhin, A. L., 1978. Universal Bayes estimators. *Ann. Statist.* **6**, 345–351.

Rukhin, A. L., 1983. A class of minimax estimators of a normal quantile. *Statist. Probab. Letters* **1**, 217–222.

Rukhin, A. L., 1984a. Universal estimators of a vector parameter. *J. Multivariate Anal.* **14**, 135–154.

Rukhin, A. L., 1984b. Estimation of a ratio of normal parameters. Technical Report #84-25, Department of Statistics, Purdue University, West Lafayette.

Rukhin, A. L., 1985a. Admissibility and minimaxity results in estimation of exponential quantiles. *Ann. Statist.* **13**.

Rukhin, A. L., 1985b. Estimation of a linear function of the normal mean and variance. *Sankhyā*.
Rukhin, A. L., and Strawderman, W. E., 1982. Estimating a quantile of an exponential distribution. *J. Amer. Statist. Assoc.* **77**, 159–162.
Rummel, F., 1983. Iterative berechnung von gamma-minimaxschatzverfahren. Dissertation, Technischen Hochschule Darmstadt, Darmstadt.
Sacks, J., 1963. Generalized Bayes solutions in estimation problems. *Ann. Math. Statist.* **34**, 751–768.
Savage, L. J., 1954. *The Foundations of Statistics*. Wiley, New York.
Savage, L. J., 1961. The subjective basis of statistical practice. Technical Report, Department of Statistics, University of Michigan, Ann Arbor.
Savage, L. J. et al., 1962. *The Foundations of Statistical Inference*. Methuen, London.
Savage, L. J., 1971. Elicitation of personal probabilities and expectations. *J. Amer. Statist. Assoc.* **66**, 783–801.
Schacter, R. D., 1984. Automating probabilistic inference. Technical Report, Department of Engineering—Economic Systems, Stanford University, Stanford, California.
Schervish, M. J., 1983. Combining expert judgements. Technical Report No. 294, Department of Statistics, Carnegie-Mellon University, Pittsburgh.
Schwarz, G., 1962. Asymptotic shapes of Bayes sequential testing regions. *Ann. Math. Statist.* **33**, 224–236.
Schwarz, G., 1968. Asymptotic shapes for sequential testing of truncation parameters. *Ann. Math. Statist.* **39**, 2038–2043.
Sclove, S. L., 1968. Improved estimators for coefficients in linear regression. *J. Amer. Statist. Assoc.* **63**, 596–606.
Sen, P. K., and Saleh, A. K. Md. E., 1985. On some shrinkage estimators of multivariate location. *Ann. Statist.* **13**, 272–281.
Shafer, G., 1976. *A Mathematical Theory of Evidence*. Princeton University Press, Princeton.
Shafer, G., 1981. Constructive probability. *Synthese* **48**, 1–60.
Shafer, G., 1982a. Belief functions and parametric models (with discussion). *J. Roy. Statist. Soc.* (Ser. B) **44**, 322–352.
Shafer, G., 1982b. Constructive decision theory. Technical Report, Department of Mathematics, University of Kansas, Lawrence.
Shafer, G., 1982c. Lindley's paradox. *J. Amer. Statist. Assoc.* **77**, 325–351.
Shapiro, C., and Wardrop, R., 1980. Bayesian sequential estimation for one-parameter exponential families. *J. Amer. Statist. Assoc.* **75**, 984–988.
Shinozaki, N., 1980. Estimation of a multivariate normal mean with a class of quadratic loss functions. *J. Amer. Statist. Assoc.* **75**, 973–976.
Shinozaki, N., 1984. Simultaneous estimation of location parameters under quadratic loss. *Ann. Statist.* **12**, 322–335.
Shubik, M., 1982. *Game Theory and the Social Sciences: Concepts and Solutions*. MIT Press, Cambridge, Massachusetts.
Siegmund, D., 1985. *Sequential Analysis: Tests and Confidence Intervals*. Springer-Verlag, New York.
Simon, G., 1976. Computer simulation swindles with applications to estimates of location and dispersion. *Appl. Statist.* **25**, 266–274.
Smith, A. F. M., 1973a. Bayes estimates in the one-way and two-way models. *Biometrika* **60**, 319–330.
Smith, A. F. M., 1973b. A general Bayesian linear model. *J. Roy. Statist. Soc.* (Ser. B) **35**, 67–75.
Smith, A. F. M., 1978. In discussion of Tanner, J. C., "Long-term forecasting of vehicle ownership and road traffic." *J. Roy. Statist. Soc.* (Ser. A) **141**, 50–51.

Smith, A. F. M., 1983. Bayesian approaches to outliers and robustness. In *Specifying Statistical Models*, J. F. Florens et al. (Eds.). Springer-Verlag, New York.
Smith, A. F. M., 1984. Present position and potential developments: some personal views Bayesian statistics. *J. Roy. Statist. Soc.* (Ser. A) **147**, 245–259.
Smith, A. F. M., and Spiegelhalter, D. J., 1980. Bayes factors and choice criteria for linear models. *J. Roy. Statist. Soc.* (Ser. B) **42**, 213–220.
Smith, C. A. B., 1961. Consistency in statistical inference and decision. *J. Roy. Statist. Soc.* (Ser. B) **23**, 1–25.
Smith, C. A. B., 1965. Personal probability and statistical analysis. *J. Roy. Statist. Soc.* (Ser. A) **128**, 469–499.
Smith G., and Campbell, F., 1980. A critique of some ridge regression methods (with discussion). *J. Amer. Statist. Assoc.* **75**, 74–103.
Sobel, M., 1953. An essentially complete class of decision functions for certain standard sequential problems. *Ann. Math. Statist.* **24**, 319–337.
Solomon, D. L., 1972. Gamma-minimax estimation of a multivariate location parameter. *J. Amer. Statist. Assoc.* **67**, 641–646.
Spiegelhalter, D. J., and Knill-Jones, R., 1984. Statistical and knowledge-based approaches to clinical decision support systems, with an application in gastroenterology. *J. Roy. Statist. Soc.* (Ser. A) **147**, 35–77.
Spruill, M. C., 1982. Admissibility of the natural estimator of the mean of a Gaussian process. *J. Multivariate Anal.* **12**, 568–574.
Srinivasan, C., 1981. Admissible generalized Bayes estimators and exterior boundary value problems. *Sankhyā* (Ser. A) **43**, 1–25.
Srinivasan, C., 1982. A relation between normal and exponential families through admissibility. *Sankhyā* **44**, 423–435.
Srinivasan, C., 1984a. A sharp necessary and sufficient condition for inadmissibility of estimators in a control problem. *Ann. Statist.* **12**, 927–944.
Srinivasan, C., 1984b. On admissible linear estimators in exponential families. Technical Report, University of Kentucky, Lexington.
Steele, J. M., and Zidek, J. V., 1980. Optimal strategies for second guessers. *J. Amer. Statist. Assoc.* **75**, 596–601.
Stein, C., 1946. A note on cumulative sums. *Ann. Math. Statist.* **17**, 498–499.
Stein, C., 1955a. Inadmissibility of the usual estimator for the mean of a multivariate normal distribution. In *Proc. Third Berkeley Symp. Math. Statist. Probab.* **1**, 197–206. University of California Press, Berkeley.
Stein, C., 1955b. A necessary and sufficient condition for admissibility. *Ann. Math. Statist.* **26**, 518–522.
Stein, C., 1959. The admissibility of Pitman's estimator of a single location parameter. *Ann. Math. Statist.* **30**, 970–979.
Stein, C., 1962. A remark on the likelihood principle. *J. Roy. Statist. Soc.* (Ser. A) **125**, 565–568.
Stein, C., 1965. Approximation of improper prior measures by prior probability measures. In *Bernouilli-Bayes-Laplace Festschrift*, 217–240. Springer-Verlag, New York.
Stein, C., 1966. An approach to the recovery of inter-block information in balanced incomplete block designs. In *Festschrift for J. Neyman*, F. N. David (Ed.). Wiley, New York.
Stein, C., 1973. Estimation of the mean of a multivariate distribution. In *Proceedings of the Prague Symposium on Asymptotic Statistics.*
Stein, C., 1975. Estimation of a covariance matrix. Reitz Lecture, IMS-ASA Annual Meeting (and unpublished lecture notes).
Stein, C., 1981. Estimation of the mean of a multivariate normal distribution. *Ann. Statist.* **9**, 1135–1151.

Stein, M. L., 1984. An efficient method of sampling for statistical circuit design. Technical Report, Mathematical Sciences Department, IBM T. J. Watson Research Center, Yorktown Heights.

Stewart, L., 1979. Multiparameter univariate Bayesian analysis. *J. Amer. Statist. Assoc.* **74**, 684-693.

Stewart, L., 1983. Bayesian analysis using Monte Carlo integration—a powerful methodology for handling some difficult problems. *The Statistician* **32**, 195-200.

Stewart, L., and Johnson, J. D., 1972. Determining optimum burn-in and replacement times using Bayesian decision theory. *IEEE Transactions on Reliability R-21*, 170-175.

Stigler, S., 1982. Thomas Bayes' Bayesian inference. *J. Roy. Statist. Soc.* (Ser. A) **145**, 250-258.

Stijnen, Th., 1980. On the asymptotic behavior of monotonized empirical Bayes rules. Doctoral Thesis. University of Utrecht, The Netherlands.

Stijnen, Th., 1982. A monotone empirical Bayes estimator and test for the one-parameter continuous exponential family based on spacings. *Scand. J. Statist.* **9**, 153-158.

Stone, M., 1963. Robustness of nonideal decision procedures. *J. Amer. Statist. Assoc.* **58**, 480-486.

Stone, M., 1970. Necessary and sufficient conditions for convergence in probability to invariant posterior distributions. *Ann. Math. Statist.* **41**, 1349-1353.

Stone, M., 1971. Strong inconsistency from uniform priors—with comments. *J. Amer. Statist. Assoc.* **71**, 114-125.

Stone, M., 1979. Review and analysis of some inconsistencies related to improper priors and finite additivity. In *Proceedings of Sixth International Congress on Logic, Methodology, and Philosophy of Science*, L. J. Cohen, J. Los, H. Pfeiffer, and K. P. Podewski (Eds.). North-Holland, Amsterdam.

Strawderman, W. E., 1971. Proper Bayes minimax estimators of the multivariate normal mean. *Ann. Math. Statist.* **42**, 385-388.

Strawderman, W. E., 1974. Minimax estimation of location parameters for certain spherically symmetric distributions. *J. Multivariate Anal.* **4**, 255-264.

Strawderman, W. E., 1978. Minimax adaptive generalized ridge regression estimators. *J. Amer. Statist. Assoc.* **73**, 623-627.

Strawderman, W. E., and Cohen, A., 1971. Admissibility of estimators of the mean vector of a multivariate normal distribution with quadratic loss. *Ann. Math. Statist.* **42**, 270-296.

Strenio, J. F., Weisberg, H. I., and Bryk, A. S., 1983. Empirical Bayes estimation of individual growth-curve parameters and their relationship to covariates. *Biometrics* **39**, 71-86.

Susarla, V., 1982. Empirical Bayes theory. In *Encyclopedia of Statistical Sciences*, Vol. 2, S. Kotz and N. Johnson (Eds.). Wiley, New York.

Takada, Y., 1979. A family of minimax estimators in some multiple regression problems. *Ann. Statist.* **7**, 1144-1147.

Thisted, R. A., 1982. Decision-theoretic regression diagnostics. In *Statistical Decision Theory and Related Topics III*, S. S. Gupta and J. Berger (Eds.). Academic Press, New York.

Thomas, L. C., 1984. *Games, Theory and Applications.* Wiley, New York.

Tiao, G. C., and Tan, W. Y., 1965. Bayesian analysis of random-effects models in the analysis of variance. I. Posterior distribution of variance components. *Biometrika* **52**, 37-53.

Tiao, G. C., and Tan, W. Y., 1966. Bayesian analysis of random-effects models in the analysis of variance. II. Effects of autocorrelated errors. *Biometrika* **53**, 477-495.

Tiao, G. C., and Zellner, A., 1964a. On the Bayesian estimation of multivariate regression. *J. Roy. Statist. Soc.* (Ser. B) **26**, 277-285.

Tiao, G. C., and Zellner, A., 1964b. Bayes theorem and the use of prior knowledge in regression analysis. *Biometrika* **51**, 219-230.

Tierney, L., and Kadane, J. B., 1984. Accurate approximations for posterior moments and marginals. Technical Report #326, Department of Statistics, Carnegie-Mellon University, Pittsburgh.

Tjur, T., 1978. Statistical inference under the likelihood principle. Preprint 1, Institute of Mathematical Statistics, University of Copenhagen, Copenhagen.

Tsui, K. W., 1979. Multiparameter estimation of discrete exponential distributions. *Can. J. Statist.* **7**, 193-200.

Tsui, K. W., and Press, S. J., 1982. Simultaneous estimation of several Poisson parameters under k-normalized squared-error loss. *Ann. Statist.* **10**, 93-100.

van Der Merwe, A. J., Groenewald, P. C. N., Nell, D. G., and van Der Merwe, C. A., 1981. Confidence intervals for a multivariate normal mean in the case of empirical Bayes estimation using Pearson curves and normal approximations. Technical Report 70, Department of Mathematical Statistics, University of the Orange Free State, Bloemfontein.

van Dijk, H. K., and Kloeck, T., 1980. Further experience in Bayesian analysis using Monte Carlo integration. *J. Econometrics* **14**, 307-328.

van Dijk, H. K., and Kloeck, T., 1984. Experiments with some alternatives for simple importance sampling in Monte Carlo integration. In *Bayesian Statistics II*, J. M. Bernardo, M. H. DeGroot, D. V. Lindley, and A. F. M. Smith (Eds.). North-Holland, Amsterdam.

van Houwelingen, J. C., 1976. Monotone empirical Bayes tests for the continuous one-parameter exponential family. *Ann. Statist.* **4**, 981-989.

van Houwelingen, J. C., 1977. Monotonizing empirical Bayes estimators for a class of discrete distributions with monotone likelihood ratio. *Statistica Neerlandica* **31**, 95-104.

van Houwelingen, J. C., and Stijnen, Th., 1983. Monotone empirical Bayes estimators for the continuous one-parameter exponential family. *Statistica Neerlandica* **37**, 29-43.

van Houwelingen, J. C., and Verbeek, A., 1980. On the construction of monotone symmetric decision rules for distributions with monotone likelihood ratio. Preprint 149, Department of Mathematics, University of Utrecht, The Netherlands.

Vardeman, S., and Meeden, G., 1984. Admissible estimators for the total of a stratified population that employ prior information. *Ann. Statist.* **12**, 675-684.

Vardi, Y., 1979. Asymptotic optimality of certain sequential estimators. *Ann. Statist.* **7**, 1034-1039.

Villegas, C., 1977. On the representation of ignorance. *J. Amer. Statist. Assoc.* **72**, 651-654.

Villegas, C., 1981. Inner statistical inference, II. *Ann. Statist.* **9**, 768-776.

Villegas, C., 1984. On a group structural approach to Bayesian inference. Technical Report 84-11, Department of Mathematics, Simon Fraser University, Burnaby.

Vinod, H. D., and Ullah, A., 1981. *Recent Advances In Regression Methods.* Marcel Dekker, New York.

Von Neumann, J., 1928. Zur theorie der gesellschaftspiele. *Math. Annalen* **100**, 295-320.

Von Neumann, J., and Morgenstern, O., 1944. *Theory of Games and Economic Behavior* (3rd edn. 1953). Princeton University Press, Princeton, NJ.

Wagner, C., 1982. Allocation, Lehrer models, and the consensus of probabilities. *Theory and Decision* **14**, 207-220.

Wald, A., 1947. *Sequential Analysis.* Wiley, New York.

Wald, A., 1950. *Statistical Decision Functions.* Wiley, New York.

Wald, A., and Wolfowitz, J., 1948. Optimum character of the sequential probability ratio test. *Ann. Math. Statist.* **19**, 326-339.
Walker, A. M., 1969. On the asymptotic behavior of posterior distributions. *J. Roy. Statist. Soc.* (Ser. B) **31**, 80-88.
Wallace, D. L., 1959. Conditional confidence level properties. *Ann. Math. Statist.* **30**, 864-876.
Weerahandi, S., and Zidek, J. V., 1981. Multi-Bayesian statistical decision theory. *J. Roy. Statist. Soc.* (Ser. A) **144**, 85-93.
Weerahandi, S., and Zidek, J. V., 1983. Elements of multi-Bayesian decision theory. *Ann. Statist.* **11**, 1032-1046.
Weiss, L., 1961. *Statistical Decision Theory.* McGraw-Hill, New York.
Whitney, A. W., 1918. The theory of experience rating. *Proceedings of the Casualty Actuarial Society* **4**, 274-292.
Whittle, P., 1982. *Optimisation over Time: Dynamic Programming and Stochastic Control*, Vol. 1. Wiley, New York.
Widder, D. V., 1946. *The Laplace Transform.* Princeton University Press, Princeton, NJ.
Williams, J. D., 1954. *The Compleat Strategyst.* McGraw-Hill, New York.
Winkler, R. L., 1967a. The quantification of judgement: some methodological suggestions. *J. Amer. Statist. Assoc.* **62**, 1105-1120.
Winkler, R. L., 1967b. The assessment of prior distributions in Bayesian analysis. *J. Amer. Statist. Assoc.* **62**, 776-800.
Winkler, R. L., 1968. The consensus of subjective probability distributions. *Manage. Sci.* **15**, B61-B75.
Winkler, R. L., 1972. *Introduction to Bayesian Inference and Decision.* Holt, Rinehart, and Winston, New York.
Winkler, R. L., 1980. Prior information, predictive distributions, and Bayesian model-building. In *Bayesian Analysis in Econometrics and Statistics*, A. Zellner (Ed.). North-Holland, Amsterdam.
Winkler, R. L., 1981. Combining probability distributions from dependent information sources. *Manage. Sci.* **27**, 479-488.
Wolfenson, M. and Fine, T. L., 1982. Bayes-like decision making with upper and lower probabilities. *J. Amer. Statist. Assoc.* **77**, 80-88.
Woodroofe, M., 1976. Frequentist properties of Bayesian sequential tests. *Biometrika* **63**, 101-110.
Woodroofe, M., 1978. Large deviations of likelihood ratio statistics with applications to sequential testing. *Ann. Statist.* **6**, 72-84.
Woodroofe, M., 1980. On the Bayes risk incurred by using asymptotic shapes. *Commun. Statist.—Theory and Methods* **9**, 1727-1748.
Woodroofe, M., 1981. A.P.O. rules are asymptotically nondeficient for estimation with squared-error loss. *Z. Wahrsch. verw. Gebiete* **58**, 331-341.
Woodroofe, M., 1982. Empirical Bayes estimation of the mean of normal distribution with convex loss. In *Statistical Decision Theory and Related Topics III*, S. S. Gupta and J. Berger (Eds.). Academic Press, New York.
Zacks, S., 1971. *The Theory of Statistical Inference.* Wiley, New York.
Zaman, A., 1981. A complete class theorem for the control problem, and further results on admissibility and inadmissibility. *Ann. Statist.* **9**, 812-821.
Zellner, A., 1971. *An Introduction to Bayesian Inference in Econometrics.* Wiley, New York.
Zellner, A., 1977. Maximal data information prior distributions. In *New Methods in the Applications of Bayesian Methods*, A. Aykac and C. Brumat (Eds.). North-Holland, Amsterdam.
Zellner, A., 1984a. Posterior odds ratios for regression hypotheses: general considerations and some specific results. In *Basic Issues in Econometrics.* University of Chicago Press, Chicago.

Zellner, A., 1984b. *Basic Issues in Econometrics.* University of Chicago Press, Chicago.

Zellner, A., and Highfield, R. A., 1983. Calculation of maximum entropy distributions and approximation of marginal posterior distributions. Technical Report, Graduate School of Business, University of Chicago, Chicago.

Zellner, A., and Siow, A., 1980. Posterior odds ratios for selected regression hypotheses. In *Bayesian Statistics*, J. M. Bernardo, M. H. DeGroot, D. V. Lindley, and A. F. M. Smith (Eds.). University Press, Valencia.

Zeytinoglu, M., and Mintz, M., 1984. Optimal fixed size confidence procedures for a restricted parameter space. *Ann. Statist.* **12**, 945–957.

Zheng, Z., 1982. A class of generalized Bayes minimax estimators. In *Statistical Decision Theory and Related Topics III*, S. S. Gupta and J. Berger (Eds.). Academic Press, New York.

Zidek, J., 1969. A representation of Bayes invariant procedures in terms of Haar measure. *Ann. Inst. Statist. Math.* **21**, 291–308.

Zidek, J., 1970. Sufficient conditions for the admissibility under squared-error loss of formal Bayes estimators. *Ann. Math. Statist.* **41**, 446–456.

Zidek, J., 1971. Inadmissibility of a class of estimators of a normal quantile. *Ann. Math. Statist.* **42**, 1444–1447.

Zidek, J., 1976. A necessary condition for the admissibility under convex loss of equivariant estimators. Technical Report No. 113, Stanford University, Stanford, California.

Zidek, J., 1978. Deriving unbiased risk estimators of multinormal mean and regression coefficient estimators using zonal polynomials. *Ann. Statist.* **6**, 769–782.

Zidek, J., 1984. Multi-Bayesianity. Technical Report No. 05, Department of Statistics, University of British Columbia, Vancouver.

Notation and Abbreviations

The symbols listed here are those used throughout the book. Symbols specific to a given chapter are, for the most part, not listed. This glossary is for reference only, in that virtually all symbols are also explained when first used in the book. A page number after an entry indicates where in the book a more extensive explanation can be found.

Mathematical Notation

R^1, R^2, R^p: the line, the plane, p-dimensional Euclidean space.

$\mathbf{y}$, $\mathbf{A}$: bold-faced letters are column vectors or matrices. The (i, j)th element of a matrix $\mathbf{A}$ will be denoted A_{ij}.

$\mathbf{y}^t$, $\mathbf{A}^t$: the "t" denotes transpose.

$\mathbf{A}^{-1}$, det $\mathbf{A}$, tr $\mathbf{A}$: the inverse, determinant, and trace of a matrix $\mathbf{A}$.

$\mathbf{I}_p$: the $(p \times p)$ identity matrix.

$\mathbf{1}, \mathbf{0}$: the vectors of all ones and zeros, respectively.

$|\mathbf{y}|$: the Euclidean norm of $\mathbf{y} \in R^p$ (absolute value if $y \in R^1$). Thus $|\mathbf{y}| = (\sum_{i=1}^{p} y_i^2)^{1/2}$.

$(a, b), [a, b], (a, b], [a, b)$: the open interval from a to b, the closed interval, and half-open intervals.

$\{a, b\}$: the set consisting of the elements a and b.

B^c, $\bar{B}$: the complement and closure of the set B.

$I_B(y)$: the indicator function on the set B (equals one if $y \in B$ and zero otherwise).

$g'(y), g''(y)$: the first and second derivatives of the function g.

$\log y$: the logarithm of y to the base e.

min, max, inf, sup, lim: minimum, maximum, infimum, supremum, and limit.

$\cap$, $\cup$: intersection, union.

$\cong$: approximately equal to.

$\overline{\mathbf{xy}}$: the line segment (in R^p) joining the points $\mathbf{x}$ and $\mathbf{y}$.

$\mathbf{x}_n \to \mathbf{x}_0$: the sequence of points $\{\mathbf{x}_n\}$ converges to $\mathbf{x}_0$, i.e.,

$$\lim_{n\to\infty} |\mathbf{x}_n - \mathbf{x}_0| = 0.$$

$\#$: the number of.

$\Gamma(\alpha)$: the gamma function.

$\binom{n}{x}$: defined as $n!/[x!\,(n-x)!]$.

Probabilistic Notation

X, $\mathscr{X}$, x: the random variable X, the sample space, a realization of X (p. 3).

θ, Θ: a parameter (or "state of nature"), the parameter space (p. 3).

$f(x|\theta)$, $F(x|\theta)$: the density of X given θ, the cumulative distribution function of X given θ (p. 4).

$\int h(x)\,dF(x|\theta)$: see page 4.

$P(A)$ or $P(X \in A)$: the probability of the set A (p. 3).

$E[h(X)]$: the expectation of $h(x)$ (p. 4).

Subscript on P or E (such as E_θ): a parameter value under which the probability or expectation is to be taken (p. 4).

Superscript on P, E, or F (such as E^X): clarifies the relevant random variable or distribution (p. 4).

$z(\alpha)$: the α-fractile of the distribution of the random variable X (pp. 79, 162).

$\langle a \rangle$, $\langle \delta \rangle(x, A)$: "$\langle\ \rangle$" is a notational device used to denote the probability distribution which selects the enclosed quantity with probability one. Thus $\langle \delta \rangle(x, A) = I_A(\delta(x))$.

$\mathscr{E}n$: entropy (p. 91).

Statistical Notation

H_0, H_1: null hypothesis, alternative hypothesis.

$\phi(x)$: a test function; i.e., $\phi(x)$ is the probability of rejecting the null hypothesis (zero or one for a nonrandomized text) when x is observed.

$\beta_\phi(\theta)$ or $\beta(\theta)$: the power function of a test; i.e., the probability of rejecting the null hypothesis when θ is the parameter value.

$\alpha_0(\phi)$ or α_0, $\alpha_1(\phi)$ or α_1: the probabilities of Type I and Type II errors, respectively, for testing H_0: $\theta=\theta_0$ versus $H_1:\theta=\theta_1$. Thus $\alpha_0=\beta(\theta_0)$ and $\alpha_1=1-\beta(\theta_1)$.

$\bar{x}$: the sample mean; i.e., $\bar{x}=1/n\sum_{i=1}^n x_i$.

$l(\theta)$: the likelihood function (i.e., $l(\theta)=f(x|\theta)$) (p. 27).

$\pi(\theta)$: the prior density for θ (pp. 3, 4).

$m(x|\pi)$ or $m(x)$: the marginal density of X (p. 95).

$\pi(\theta|x)$: the posterior distribution of θ given x (p. 126).

$\mu^\pi(x)$, $V^\pi(x)$: posterior mean and variance (p. 136).

C or $C(x)$: a confidence or credible set for θ (pp. 140, 414).

$I(\theta)$: expected Fisher information (pp. 88, 224).

$\hat{I}(x)$: observed (or conditional) Fisher information (p. 224).

Decision Theoretic Notation

a, $\mathscr{A}$: action, action space (p. 3).

$L(\theta, a)$: the loss function (pp. 3, 59).

$\mathscr{R}$: reward space (p. 47).

$U(r)$ or $U(\theta, a)$: utility function (pp. 47, 58).

$\rho(\pi^*, a)$: the Bayesian expected loss of a with respect to π^* (p. 8), or posterior expected loss when π^* is the posterior distribution (p. 159).

a^{π^*}: a Bayes action (p. 16).

δ, δ^*: nonrandomized decision rule (p. 9), randomized decision rule (p. 12).

$L(\theta, \delta^*)$: the loss function for a randomized decision rule (p. 13).

$R(\theta, \delta)$ and $R(\theta, \delta^*)$: the risk function (expected loss) of a decision rule (pp. 9 and 14).

$\mathscr{D}$: the class of nonrandomized decision rules with $R(\theta, \delta)<\infty$ for all θ (p. 11).

$\mathscr{D}^*$: the class of randomized decision rules with $R(\theta, \delta^*)<\infty$ for all θ (p. 14).

$r(\pi, \delta)$ and $r(\pi, \delta^*)$: the Bayes risk of a decision rule (p. 11).

δ^π: a Bayes decision rule (p. 17).

$r(\pi)$: the Bayes risk of π (i.e., $r(\pi)=r(\pi, \delta^\pi)$) (p. 17).

$r_\Gamma(\delta)$: Γ-Bayes risk of δ (p. 213).

Author Index

Subject Index